THE CHEMICAL CENTURY

Molecular Manipulation and
Its Impact on the 20th Century

THE CHEMICAL CENTURY

Molecular Manipulation and Its Impact on the 20th Century

Richard J. Sundberg, PhD

AAP APPLE ACADEMIC PRESS

Apple Academic Press Inc.
3333 Mistwell Crescent
Oakville, ON L6L 0A2 Canada

Apple Academic Press Inc.
9 Spinnaker Way
Waretown, NJ 08758 USA

© 2017 by Apple Academic Press, Inc.

First issued in paperback 2021

Exclusive worldwide distribution by CRC Press, a member of Taylor & Francis Group
No claim to original U.S. Government works

ISBN-13: 978-1-77463-618-3 (pbk)
ISBN-13: 978-1-77188-366-5 (hbk)

Library and Archives Canada Cataloguing in Publication

Sundberg, Richard J., 1938-, author
The chemical century : molecular manipulation and its impact on the 20th century / Richard J. Sundberg, PhD.

Includes bibliographical references and index.
Issued in print and electronic formats.
ISBN 978-1-77188-366-5 (hardcover).--ISBN 978-1-77188-367-2 (pdf)
1. Chemistry--History--20th century. I. Title.
QD15.S85 2016   540.9'04   C2016-904239-1   C2016-904240-5

Library of Congress Cataloging-in-Publication Data

Names: Sundberg, Richard J., 1938-
Title: The chemical century : molecular manipulation and its impact on the 20th century / Richard J. Sundberg, PhD.
Description: Toronto : Apple Academic Press, 2017. | Includes bibliographical references and index.
Identifiers: LCCN 2016027320 (print) | LCCN 2016027475 (ebook) | ISBN 9781771883665 (hardcover : alk. paper) | ISBN 9781315366265 (ebook)
Subjects: LCSH: Chemistry--History--20th century.
Classification: LCC QD15 .S88 2016 (print) | LCC QD15 (ebook) | DDC 540.9/04--dc23
LC record available at https://lccn.loc.gov/2016027320

Apple Academic Press also publishes its books in a variety of electronic formats. Some content that appears in print may not be available in electronic format. For information about Apple Academic Press products, visit our website at **www.appleacademicpress.com** and the CRC Press website at **www.crcpress.com**

ABOUT THE AUTHOR

Richard J. Sundberg, PhD

Richard J. Sundberg, PhD, is a Professor of Chemistry at the University of Virginia and has been teaching for 50 years. He was a Fulbright Scholar (France) in 1978–79, and he was honored as a Distinguished Alumni by the University of Minnesota in 2001. He has published about 100 research papers and five monographs in the area of heterocyclic organic compounds and has directed 25 PhD theses. His past books include *Advanced Organic Chemistry* (all editions) (co-authored with Francis A. Carey), Springer; *Indoles, Best Synthetic Methods*, Academic Press, 1996; and *The Chemistry of Indoles*, Academic Press, 1970.

CONTENTS

LIST OF ABBREVIATIONS

AA	aldosterone antagonists
ABS	alkylbenzene sulfonates
ABS	Applied Biosystems
ACAC	acetyl coenzyme-A carboxylase
ACE	angiotensin converting enzyme
ACE-I	angiotensin-converting enzyme inhibitors
Ach	acetyl choline
AD	Alzheimer's disease
ADA	American Dental Association
ADHD	attention deficit hyperactivity disorder
ADI	acceptable daily intake
AIOC	Anglo-Iranian Oil Company
ARB	angiotensin receptor blockers
BACs	bacterial artificial chromosomes
CCB	calcium channel blockers
CDC	Center for Disease Control
CDER	Center for Drug Evaluation and Research
CFCs	chlorofluorocarbons
CMR	Committee on Medical Research
DAPG	Deutsche-Amerikanische Petroleum Gesselschaft
DES	diethylstibestrol
DHEA	dehydroepiandrosterone
DMAC	dimethylacetamide
DMF	dimethylformamide
DNA	deoxyribonucleic acid
DOD	Department of Defense
DOE	Department of Energy
DRI	dietary reference intake
EPS	expanded polystyrene
FDA	Food and Drug Administration
GAD	generalized anxiety disorder
GE	General Electric Company
GFFATM	Global Fund to Fight AIDS, Tuberculosis and Malaria
HDPE	high-density polyethylene
HFCS	high-fructose corn syrup
HGP	Human Genome Project

HGS	Human Genome Sciences
HHS	Health and Human Services
IARC	International Agency for Research on Cancer
ICH	International Conference on Harmonization
ICI	Imperial Chemical Industries
IGF	Interessengemeinshchaft Farbenindustrie
IND	Investigational New Drug Application
IOL	intraocular lenses
IU	international units
IVF	*in vitro* fertilization
LAS	linear alkylbenzene sulfonates
LDL	low-density lipoprotein
LDPE	low-density polyethylene
LH	luteinizing hormone
LSD	lysergic acid diethylamide
MAO	monoamine oxidase
MAOI	monoamine oxidase inhibitors
MCP	methylcyclopropene
MCP	monocalcium phosphate
MIRV	multiple independent re-entry vehicles
MLB	Major League Baseball
MRC	Medical Research Council
MTBE	methyl t-butyl ether
MW	molecular weight
NAD	nicotinamide-adenine-dinucleotide
NASA	National Aeronautics and Space Administration
NDA	new drug application
NIH	National Institutes of Health
NO	nitric oxide
NSAIDS	nonsteroidal anti-inflammatory drugs
NTBC	2-[2-nitro-4-trifluoromethyl-benzoyl]-1,3-cyclohexanedione
OMB	Office of Management and Budget
OSRD	Office of Research and Development
PAS	para-aminosalicylic acid
PBB	polybrominated biphenyls
PCB	polychlorinated biphenyls
PE	polyethylene
PEG	polyethylene glycol
PETN	pentaerythritol tetranitrate
PHS	public health service
PIs	protease inhibitors
POPs	persistent organic pollutants

PP	polypropylene
PPAR	peroxisome proliferation activated receptors
PS	polystyrene
PVC	polyvinyl chloride
RAAS	renin–angiotensin–aldosterone system
RDA	Recommended Dietary Allowances
RNA	ribonucleic acid
RTI	reverse transcriptase inhibitors
SPF	sunscreen protection factor
SSRI	selective serotonin reuptake inhibitors
STR	short tandem repeats
TB	tuberculosis
TEL	tetraethyllead
TFA	trans fatty acids
TIGR	The Institute for Genomic Research
TNT	trinitrotoluene
UCSF	University of California, San Francisco
UNFCCC	United Nations Framework Convention on Climate Change
UOP	Universal Oil Products
USA	United Space Alliance
UV	ultraviolet
VGSC	voltage-gated sodium channels
WARF	Wisconsin Alumni Research Foundation
WGS	whole genome shotgun
WHO	World Health Organization
YAC	yeast artificial chromosome

PREFACE

This material was developed for the course "The Chemical Century" that I have taught at the University of Virginia since the spring of 2010. The original impetus came from the Parent's Fund, which provided support to develop seminar courses that might help students appreciate the career opportunities in various fields. The material evolved to an overview of some of the major ways in which chemistry has had an impact over the course of the twentieth century. It is organized under the broad headings: (1) Power from Molecules; (2) Making Things We Use; (3) Chemistry and Food; (4) Molecules for the Treatment of Illness; and (5) Molecular Biology and Its Applications.

The investment in chemical education and research in the second half of the nineteenth century gave rise to chemical technology that became the foundation of a major worldwide industry. It also generated a network of educational and research institutions dedicated to expanding and applying chemical knowledge. Along with advances in transportation, electronics and communication, food production, and medical science, chemical technology became an integral part of our current experience. I have made an effort to identify some of the major advances resulting from chemistry.

The material is very much an overview. It emphasizes the sources and applications of particular types of compounds and the relationship between major structural features and the applications. Considerable information on chemical structure is given in all the sections. The intent is not that this material be memorized, but rather that it provide some acquaintance with the chemical substances that might be known by name, but without a sense of structure. In areas such as herbicides, insecticides, and the drug categories, representative structures are given, and, at least in outline, the relationship between the structure and activity. I have also included information on some of the broader economic, environmental, medical and sociological consequences of chemical technology. Because of the nature and scope of the material, I have relied on reviews, especially those that have provided an historical perspective. These sources are cited as bibliographies for the various sections at the end of each chapter. Some more specific results or examples are cited by endnotes. Brief biographical sketches or points of background are given as footnotes. There are "Topics" sections that provide more extended biographical information or specific cases related to the subject under discussion.

I have summarized information about the background of particular discoveries and innovations. Names, such as Alfred Nobel; Vladimir Ipatieff and Vladimir

Haensel; F. Sherwood Rowland and Mario Molina; Hermann Staudinger and Wallace Carothers; Fritz Haber and Carl Bosch; Alexander Fleming, Howard Florey and Ernst Chain; Dorothy Crowfoot Hodgkins; Russell Marker and Carl Djerassi; and Frederick Banting, come from the distant past for current students, but their discoveries still have powerful impact. Names, such as James Watson and Francis Crick; Marshall Nirenberg and H. Gobind Khorana; Kerry Mullis and Michael Smith; Francis Collins and J. Craig Venter; may strike a response, but the chemical context of their contributions may not be appreciated.

For many students some of the salient examples of chemical missteps are also unknown. Most have heard of DDT and can relate the story in outline, but cases such as the sulfanilamide elixir, thalidomide, or impure tryptophan supplements are unfamiliar. I hope the inclusion of this historical material, both heroic and regrettable, will help students appreciate the role of chemical science in their current experience. It has been my sense that even students with fairly sophisticated chemistry backgrounds are prone to make some of the broad cultural assumptions about "chemicals" without considering fundamental chemical information. In teaching the course, I have incorporated exercises that ask students to critically evaluate information from a variety of sources in a chemical context.

The student reaction to the course has been very positive, with a common comment being that it has provided a connection between the more detailed aspects of the chemistry curriculum and everyday experience. A fairly wide range of students has been interested in the course. In addition to chemists and chemical engineers, students with interests in biomedical and environmental sciences, food and nutrition, and science policy have enrolled. This material may also be useful to nonstudents interested in an acquaintance with chemical discovery, innovation, and application. Over the century it became clear that all chemical inputs into organisms and ecosystems have effects in addition to the immediately intended ones. Hydrocarbon fuels brought increasing atmospheric carbon dioxide levels. Chlorofluorocarbons opened the ozone hole. Herbicide and insecticides promoted resistant weeds and bugs. Antibiotics selected for resistant strains. Drugs have side effects. Access to genetic information raised questions about its proper use. The response to these issues has required input from the economic and political spheres.

I believe the material can help inform the interested public and the economic and political structures that must address these issues.

ACKNOWLEDGEMENTS

As mentioned in the Preface, this book is the outcome of an effort funded by the Parent's Fund at the University of Virginia, and I would like to thank them for their support. My first PhD student, Peter A. Bukowick, read most of the chapters and provided comments from his perspective of experience in chemical management. Beth Blanton-Kent, Physical Sciences Librarian, provided valuable assistance in tracking difficult references and sources. Mr. Jarrad Reiner and Ms. Cynthia Knight gave invaluable help in dealing with technical aspects of the manuscript preparation. Finally, I would like to thank the students who took the course. Their questions and papers often initiated fruitful paths of inquiry and their positive response provided encouragement to prepare the material for publication.

—**Richard J. Sundberg**
May, 2016

INTRODUCTION

Around the middle of the nineteenth century, humankind began to master a new skill. Chemists deduced the molecular structure of many substances and discovered ways to manipulate and modify these structures. By understanding molecular structure and chemical reactivity, it became possible to make new compounds and materials. These activities accelerated rapidly as the twentieth century progressed. Many new substances were made from starting materials such as coal, petroleum, and cellulose. The word *synthetic* came into use to encompass these new materials. Chemists also learned how to relate properties to chemical structure and were able to tailor materials to have specific desirable properties. The list of such substances is long, and the purpose of this book is to discuss some representative examples and their uses.

At the start of the twentieth century most energy came from coal or wood. The petroleum age had begun a few decades earlier, but oil was not yet a major factor in industry or transportation. There was a rather short list of materials that could be used to make the items used in daily living. These included wood, glass, and ceramics, metals such as iron, steel, brass, and bronze, natural fibers such as cotton, linen, silk, and wool and a few more exotic materials such as ivory. By the middle of the twentieth century, dozens of different synthetic polymers were in use. Similarly, the array of consumer goods that filled the aisles of stores became innumerable. At the beginning of the century many people around the world were undernourished. By the end of the century, obesity was a bigger problem in much of the world. In 1900, there were only a handful of effective medicines and no one knew how they worked. There were no antibiotics and no effective treatments for diseases such as diabetes, tuberculosis, or pneumonia. By the end of the century about 5500 drugs had been approved by the US FDA.

All the properties of any specific material are determined by the forces and motions that are inherent in its particular arrangement of atoms. That is, the properties of any substances are determined by its *molecular structure*. Chemists can't change these properties. Chemists can, however, change the structure of molecules, thereby obtaining modified properties. It was this ability that created the myriad of new materials. For example, a change in gasoline composition can improve engine performance. Brittle polymers can be given physical strength by incorporation of rubber-like molecules. Sweet taste can be separated from high caloric intake by making low-calorie sweeteners. A particularly important example is the ability to optimize the effectiveness of drugs by modifying their molecular structure. We will focus attention on several useful kinds of molecules and see how they are made or modified and how their properties are related to their structure.

1. STRUCTURAL FORMULAS—THE LANGUAGE OF CHEMISTRY

Chemistry is a very precise and highly organized branch of science. The structures of molecules can be determined with high precision. For example, the molecule benzene C_6H_6 is known to be a perfect hexagon with the carbon (C) atoms separated from one another by 1.40×10^{-8} cm and from the hydrogen by 1.10×10^{-8} cm. The six carbon atoms and six hydrogen atoms are held together by 30 electrons that are represented as lines (bonds). Carbon is such a common element that structural pictures often omit it. For example, benzene can be drawn without the C atoms being explicitly shown. Sometimes hydrogen, too, is omitted, the understanding being that there are enough hydrogens to account for any missing bonds, up to the limit of four.

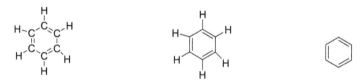

Benzene showing all carbons, hydrogens and bonds

Benzene with the carbons not shown.

Benzene with only the carbon framework shown.

In this book, we will deal with the chemical structures of molecules such as vitamins, hormones, and several types of drugs. At first glance, their structures may look quite complex, but they convey a great deal of information in a simple way. Each letter or, in the case of carbon each line intersection, represents an atom. For the most part, we will see only a few atoms from the periodic table, especially hydrogen (H), carbon (C), nitrogen (N), and oxygen (O). A few of the other elements such as chlorine (Cl), phosphorus (P), silicon (Si), and sulfur (S) will appear occasionally. We will also see some structures containing ions that have charges associated with them, especially metals such as sodium Na (Na^+), calcium (Ca^{2+}), iron (Fe^{2+} or Fe^{3+}), and copper (Cu^{1+} or Cu^{2+}), as well as anions, such as the halides, as well as sulfate and phosphate (both of which, incidentally, are perfectly tetrahedral in shape).

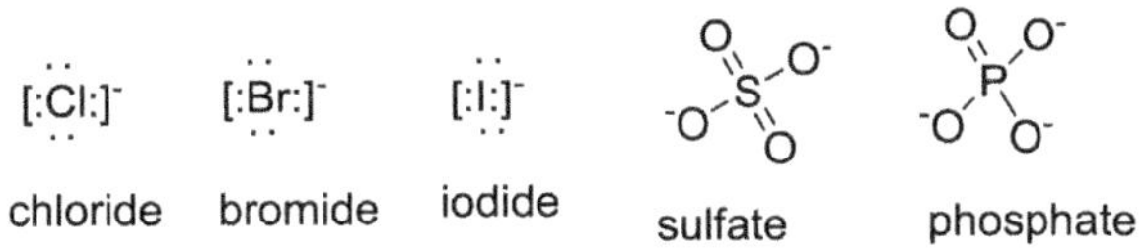

[:Cl:]⁻	[:Br:]⁻	[:I:]⁻	sulfate	phosphate
chloride	bromide	iodide		

Pure chemical substances can be characterized by a molecular structure, an unambiguous description of the bonds in the molecules. There can double, triple, and very rarely, quadruple bond, represented by multiple lines. The elements are represented by their chemical symbols from the periodic table. The structures below show some relatively simple and familiar molecules.

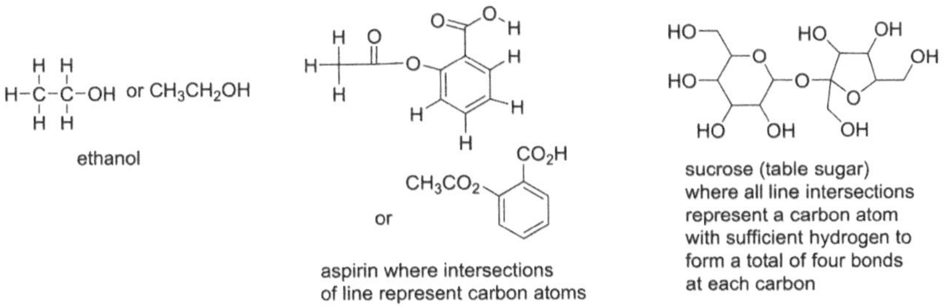

ethanol

aspirin where intersections
of line represent carbon atoms

sucrose (table sugar)
where all line intersections
represent a carbon atom
with sufficient hydrogen to
form a total of four bonds
at each carbon

Chemists have been very thorough about recording the substances they find and make. This has been done mainly by two organizations. *Beilstein's Handbuch*, which began publishing in 1880 in Germany and the *Chemical Abstracts Service*, an affiliate of the American Chemical Society, which began publication in 1907. Both archives have been converted to digital databases from which information on essentially all chemicals reported in the scientific literature is available to subscribers. It is estimated that about 110,000 chemical substances had been described at the beginning of the twentieth century. By, the end of the century, that number was about 28,500,000. Systematic rules for naming these compounds have been established. Despite the precise system of structural nomenclature, many chemical compounds have several names. One reason for this is that the systematic chemical names are often long and complicated and it is easy to make mistakes in their assignment. As a result, shortened names are often used. For example, all of the amino acids found in proteins (there are 20) have names that are used widely by chemists, biomedical scientists and nutritionists. They also have three- and one-letter abbreviations, which are shown in Scheme 1.

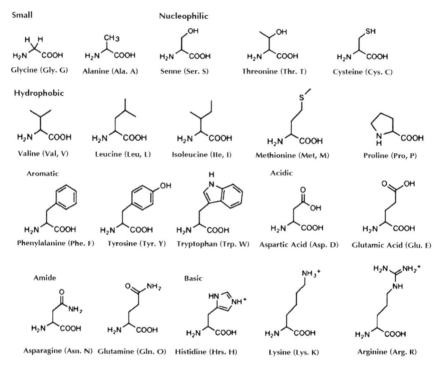

SCHEME 1. Structures, names, and abbreviations of the amino acids.

Consumer and industrial chemical products are often formulations of a number of substances and are identified by brand names. Drugs have both generic and proprietary (brand) names. The former are assigned as part of the drug approval process and the latter are selected by the seller. This multiplicity of names can lead to confusion for a person who is unfamiliar with them. Internet search systems can usually find the official name and chemical structure from one of the other names for substances that are used in commerce or medicine.

Some molecules are too large to depict as structural formulas. This is particularly true of biological molecules such as proteins and nucleic acids. In this case, several types of condensed formulas are used. For proteins, the sequence of amino acids is given using either a three- or one-letter abbreviation for each of the amino acids. This works well for most polypeptides, because they are linear molecules formed by amide bonds between the successive amino acids. Figure 1 shows the amino acid sequence for the important molecule insulin, which contains two polypeptide chains connected by several disulfide bonds.

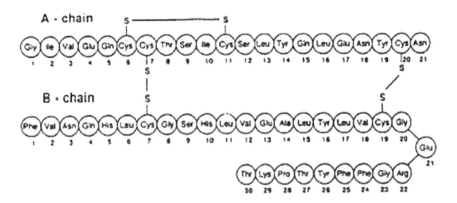

FIGURE 1   Amino acid sequence of insulin.

These linear designations are not helpful in visualizing the three-dimensional shape of the molecules, however. To do so, symbolic structural representations that depict the peptide chain bent to correspond to the shape of the molecule are used. Proteins often incorporate regions of organized structure called α-helix or β-sheet. These are depicted as spirals and ribbons, respectively. Figure 2 gives the structure of a small enzyme, *ribonuclease A*, in this format.

FIGURE 2   Representation of the enzyme ribonuclease.

Similar issues arise with the genetic material deoxyribonucleic acid (DNA). The structures are represented by the sequence of letters that identify the sequence of the nucleic acids. For DNA, these are A, C, G, and T. For RNA they are A, C, G, and U. Again, the sequence conveys no three-dimensional imagery and when that is needed, the sequence is depicted as strands, as shown in Figure 3.

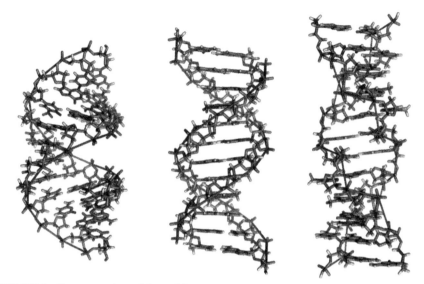

FIGURE 3   Representation of three different forms of DNA. Left-to-right: A; B; Z.

2.   THE DEVELOPMENT OF THE CHEMICAL INDUSTRY

The ability to make new materials with useful properties led to the development
of several large worldwide industries. These include petroleum refining, production of rubber and polymers, as well as manufacture of consumer goods such
as soaps and detergents. Chemistry also had a profound impact on the production and distribution of food by providing fertilizers and pesticides. The ability
to isolate, identify, and synthesize new molecules also led to development of
the pharmaceutical industry and the availability of a wide variety of drugs to
combat illness.

The modern chemical industry had its origin in Britain and continental
Europe. Processes for a few fundamental compounds, such as acids and bases,
were developed by the eighteenth century. During the second half of the nineteenth century an industry aimed primarily at the production of dyes developed.
The first discovery by W. H. Perkin, while a student at the Royal College of
Chemistry in 1856 was accidental, but Perkin followed up and commercialized
a new dye called *mauve.* By the late 1800s, the German industry had accomplished the synthesis of naturally occurring dyes such as alizarin and indigo. The
industry also made synthetic dyes from chemicals derived from coal. Late in the
nineteenth century, the German dye companies, especially Bayer and Hoechst,
began to introduce pharmaceutical products. One, aspirin, would remain on the

market throughout the century, eventually providing evidence of several beneficial effects. Another, heroin, mistakenly thought to be a nonaddictive form of morphine, was introduced as a cough suppressant. It was soon withdrawn, but later became one of the most notorious of the illicit narcotics. So, in the space of just a few years, there were examples of both the benefits and dangers of molecular manipulation.

In 1855, a group of entrepreneurs and investors was encouraged by a favorable report from a Yale chemistry professor about the potential applications of "rock oil" from western Pennsylvania. The group financed the drilling of an oil well by Edwin Drake. This was the beginning of the petroleum and petrochemical industries that came to dominate the world economy during the twentieth century.

Alfred Nobel, the son of a Swedish designer and mechanic, learned how to manipulate nitroglycerin and other powerful explosives with relative safety in the late 1860s. The resulting products were used to dig the Suez and Panama canals and to construct the networks for road and rail transportation. In the United States, the DuPont Company, founded in 1803 by an aristocratic family that had fled the French Revolution, branched out from its explosives business to become a supplier of paints and coatings and, in the 1930s, the birthplace of nylon.

In 1898, a prominent British scientist, William Crooke predicted that Europe would run out of food as the result of exhaustion of its main source of nitrogen fertilizer, which was then imported from the deserts of Chile. He looked to chemists to solve the problem, and it was in 1914 by Fritz Haber in association with Carl Bosch of the German chemical company Badische Anilin und Sodafabrik (now BASF). The Haber–Bosch process for reaction of nitrogen from air with hydrogen from water and coal or natural gas, became a model for production of chemicals on a massive scale. The resulting ammonia is not only used as a fertilizer but also as the starting material for the explosives used in military weapons. From its beginnings in dyes, petroleum, explosives, and ammonia, the chemical industry grew to encompass hundreds of companies, many of them worldwide in scope.

The ability to make molecules with useful properties brought many changes to the way people live. An entirely new class of materials, collectively known as *synthetic polymers* became available. For many applications they could replace, and often improve upon, traditional materials such as glass, metal, stone, and wood in making the items of material culture. Chemistry also changed many aspects of our personal lives—how we style our hair, what we wear, and how we clean our houses and clothes. Our purpose is to follow some of these developments—how they occurred, and their consequences, both anticipated and unanticipated. We want to develop a big picture relating the crucial discoveries to the impact on our everyday experience.

3. SOURCES OF CHEMICALS

The air, land, and sea around us all have chemical compositions. The fuel we burn, the food we eat, the clothes we wear, and all of the numerous things we use in our daily life have molecular structures. Stated simply, everything is molecular. By carrying out chemical reactions or mixing substances together in specific ways, chemists can produce materials with new properties. There are, however, only a few basic sources that serve as the starting point for the amazing variety of materials. Nitrogen from air and hydrogen from natural gas are converted to fertilizer and explosives. Minerals, substance found on the earth's surface or in the sea are another source. Examples include salt (alkali and chlorine), sand (glass, silicon, and silicones), and sulfur (sulfuric acid). One fundamental source of carbon compounds is fossil fuel deposits, specifically coal, petroleum, and natural gas. The other primary source is plants and animals, ultimately fueled by the sun. Plants provide huge quantities of carbohydrates such as cellulose, starch and sugars, as well as vegetable oils. Animals provide primarily fats and proteins and in some cases fibers, such as wool and silk. As we trace the story of chemistry in the twentieth century, we will see how these fundamental starting materials came to be used in increasingly sophisticated ways to feed and clothe people, improve the convenience of daily life, cure some diseases, wage war, and profoundly change the relationship between humans and the environment.

4. THE MOLECULES OF LIFE

As the twentieth century progressed, the substances and processes that constitute living cells and organisms came to be understood in molecular terms. Work in the late nineteenth and early twentieth century began to establish the structures of carbohydrates, fats, proteins, and nucleic acids. Biochemists began to unravel the reactions that occur in living cells and to purify and characterize the enzymes that catalyze them. The 1950s, 1953 to be precise, opened a new facet of molecular science. The recognition by James D. Watson and Francis Crick of the hydrogen-bonded, base-paired helical structure of DNA began an era of molecular interpretation of the entire scope of life processes. The era of *molecular biology* began. The techniques of molecular biology led to sequencing, amplification, cloning, and mutation of genetic material. These techniques provided entirely new insight into biological processes and also provided a new means to modify structure based on biological processes. The completion of the first draft of the human genome in 2003 achieved one landmark goal.

Another landmark achievement occurred in 1953 when Frederick Sanger determined the amino acid sequence of insulin (see Figure 1), setting the stage for what would eventually be massive amounts of data on the amino acid sequence of thousands of proteins. The three-dimensional structure of proteins determined by X-ray crystallography soon followed as illustrated by the work of John Kendrew (myoglobin, 1960), David Blow (lysozyme, 1967), and Max Perutz (hemoglobin, 1968). In 1976, the number of recorded protein structures was about 15. In the final quarter of the century, an explosion of information on the organization and function of biological systems occurred. The mechanisms by which proteins (enzymes) catalyze reactions, DNA encodes genetic information, and messenger molecules and their receptors control cellular function were unraveled. At the end of the century, there were 13,600 protein structures in the protein database. By 2010, the number was 70,000.

The expanded knowledge of biological structure and function permitted many advances in treatment of disease. One example is the isolation (1921), amino acid sequence (1953), synthesis (1963), and three-dimensional structure (1969) of insulin. Insulin production shifted from isolation from animals to genetically modified bacteria in 1986 and by 2005, mutated forms that could control the rate of release were available. The era of synthetic antibiotics began in the first decade of the twentieth century when Paul Erhlich demonstrated the concept by treatment of syphilis. The concept received great impetus from the synthesis of the sulfa drugs by Gerhard Dogmak in the 1930s. The successful production of penicillin (1938–1942) provided dramatic evidence that some naturally occurring molecules could cure disease. Hundreds of synthetically modified penicillins and other natural, modified, and synthetic antibiotics were introduced in the succeeding 60 years.

Insulin and penicillin were followed by the isolation and synthesis of many other kinds of therapeutics. These drugs have improved our health, lengthened our lifespan, changed the number of children we have, and increased their chance of survival to adulthood. Chemistry also enabled increased food production and improved storability. Chemistry helped to unravel the story of nutrition and to improve most peoples' diets. An immediate result was a burst in the world's population, followed in much of the world by a declining birthrate and overall aging of the population. The main cause of mortality shifted from infectious diseases to those associated with lifestyle and aging.

5. ADAPTING TO THE CONSEQUENCES

With the development of synthetic materials, a new issue arose. The natural world contains a myriad of molecules ranging from the essential vitamins in foods to deadly toxins and venoms found in some plants and animals. Living systems, including viruses, microbes, plants, animals, and humans can respond and adapt to encounters with new substances, and therein lies an important question. What are the consequences to ourselves and the world around us of the many new molecules? A key aspect of the introduction of new substances and materials is the temporal relationship to ecological and evolutionary mechanisms. Natural chemicals have coexisted with humans and other evolving species, providing signals that generate evolutionary responses. The availability of synthetic materials introduced unprecedented numbers and amounts of new molecules that can be beyond the capacity of a successful evolutionary response by some organisms. Whether in an individual cell or an ecosystem, the compounds impose a new influence on the system. That influence has often been the desired beneficial one, but not infrequently there are other consequences that were unexpected, and sometimes harmful. It is clear, for example, that wide use of antibiotics has promoted the survival of resistant strains of microorganisms. Similarly, use of herbicides and insecticides generated resistant weeds and bugs.

Homo sapiens, with one of the longest reproductive cycles, must rely on intellect and ingenuity to respond to environmental changes that can occur in less than a generation. The consequences for individuals, society, and the environment are sometimes evident, but that is not always the case. The introduction of chlorinated insecticides resulted in rapid population reduction in several species that were observed within the matter of few years. On other hand, the depletion of polar ozone levels by chlorofluorocarbons was not evident and was initially predicted on the basis of kinetic modeling, but dramatically confirmed when physical measurements became available. Individuals can react adversely to drugs and in some cases the effects may be sufficient to require withdrawal from the market. We have learned that we need to be vigilant in looking for unanticipated effects.

The most profound change that occurred over the past century has been an increase of some 40% in the CO_2 content of the atmosphere. At the start of the century, the CO_2 level was about 280 ppm. Since 1960, the CO_2 content has been precisely measured, and it has increased from 315 to 400 ppm. Humans can learn and apply the Laws of Nature, but not change them. We can measure the amount of energy that arrives from the sun. The Laws of Thermodynamic tell us how much energy is stored in a particular material and how much work we can obtain from it. Human ingenuity can find ways to exploit this energy. But the Law of Conservation of Matter tells us that combustion of one ton of hydrocarbon fuel produces about three tons of carbon dioxide, no matter how inconvenient that may be. Human intelligence and logic can detect, and to some extent predict, the consequences. It remains to be seen how humans will respond and adapt.

PART I
Power from Molecules

CHAPTER 1

EXPLOSIVES AND PROPELLANTS: POWER TO BREACH MOUNTAINS, WAGE WAR, AND VISIT THE MOON

ABSTRACT

Beginning with the formulation of nitroglycerin (dynamite, 1866) and nitro-cellulose (Cordite, Ballistite, Poudre B, 1880s) chemical synthesis became a key part of production of explosives and propellants. The new materials provided a major tool for construction and mining into the 20th century. The demands of WWI and WWII led to other explosives such as trinitrotoluene (TNT), the nitramines and various mixed explosives. After WWII both military and space programs required powerful fuels for missiles and rocket engines. These included solid fuels incorporating polymers, aluminum and ammonium perchlorate as well as liquid fuels based on hydrocarbons, hydrazines or liquid hydrogen. By the end of the century, the explosives used by terrorists had become a major focus.

Molecules contain energy that can be released as force and heat. A particularly dramatic form of release of chemical energy is an *explosion*. Explosives seem a good place to start discussing the role of chemical technology in the 20th century. One key discovery, the invention of dynamite by Alfred Nobel, took place in 1866. Dynamite and other explosives played a major role in economic development of the 20th century, especially in mining and construction. Nobel's network of companies became one of the first international industries based on chemical technology. Explosives played a large role in the conduct of the two World Wars and the many others that dominated the political history of the century. In 1969, propellants helped send the first humans to the moon. The Nobel Prizes, first awarded in 1902, and financed by Alfred Nobel's profits from explosives, provide one means of recognizing major discoveries in chemistry and other areas of science during the century.

1.1 STRUCTURES AND PROPERTIES OF EXPLOSIVES AND PROPELLANTS

We will be considering *chemical explosives* and *propellants*, that is, substances from which the energy is released by chemical reactions. This can be contrasted, for example, to an explosion that might occur in an over-heated steam boiler that ruptures as the result of physical force generated by heat and pressure, not from a chemical transformation. It also excludes explosions due to nuclear reactions. The energy released in an explosive chemical reaction generates heat and pressure. Explosives can be characterized by the *detonation velocity*, the rate of movement of the detonation. When this is greater than the speed of sound (340 m/s), a *shock wave* is created that rapidly propagates the explosion through the material. For most of the common explosives, the detonation velocity is in the range of 7000–9000 m/s. Explosives are classified as primary or secondary explosives. *Primary explosives* are very heat or shock sensitive and can be used to detonate other powerful but more stable materials, called *secondary explosives*. Secondary explosives do not easily detonate by ignition (burning). Rather, they require a shock wave for detonation. *Propellants* burn rapidly but do not detonate and the heat and gas released in a confined space are converted into kinetic energy, as with a bullet, artillery shell, or rocket. This is called a *deflagration*. The fuel and oxidant are present in the composition, but the rate of burning is insufficient to create a shock wave. Propellants for bullets and munitions are designed to be exhausted as the projectile leaves the barrel. Among the factors controlling the rate of burn of a propellant are particle size and shape. Some specific structures and names for explosives and propellants are given in Scheme 1.1. They can be classified into three groups, nitrate esters, polynitroaromatics, and nitramines. Two salts, ammonium nitrate and ammonium perchlorate are also important components of explosives and propellants.

So what makes an explosive explosive? Why is "nitro" or "nitrate" associated with most of the common explosives, including nitroglycerine (in dynamite), trinitrotoluene (TNT), nitramines (RDX, HMX), and ammonium nitrate? Explosives need both fuel and oxidizing power. The nitro and nitrate groups provide the latter. Explosives must have high potential energy in thermodynamic terms, positive heats of formation relative to the reaction products. The high potential energy of nitro compounds is related to the very high stability of molecular nitrogen (N_2), which is formed in explosions. For any given explosive reaction, the heat generated can be calculated by the thermodynamic relationship:

Heat of reaction = heat of formation (products) – heat of formation (reactants)

Thus, two critical requirements for an explosive are the ability to produce a relatively large number of gaseous molecules and release a large amount of heat. The balanced explosion equations can be used to calculate the amount of gas that will be produced by the explosion. The heat released by explosion can be calculated by comparing the thermodynamic stability of the products to that of the reactants. The heats of formation of the common explosives are given in Scheme 1.1.

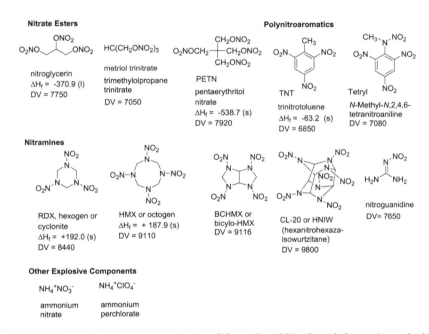

SCHEME 1.1 Structures, names, heats of formation (kJ/mol) and detonation velocities (m/s) of some explosives.

Predictions about the products of an explosion can be made by application of principles of stoichiometry in what are known as the Kistiakowsky–Wilson rules.[a] The rules are based on the *principle of conservation of matter*, which states that atoms are neither created nor destroyed by chemical reactions.

[a]George Kistiakowsky was a physical chemist. He was born in Russian-controlled Ukraine in 1900. He fought against the Bolsheviks during the Russian revolution and in 1920 fled Russia. He received a Ph. D. in Berlin in 1925. He came to the United States in 1926 and joined the faculty of Harvard in 1928. During WWII, he was chair of the committee on Explosives of the National Defense Research Committee. He was also associated with the Manhattan Project and led the development of the explosive device that was used to trigger the first atomic bombs. He served as Science Adviser to Pres. Dwight Eisenhower in 1959 and was director of the Office of Science and Technology until 1961. Kistiakowski was involved in early strategic planning for use and then control of nuclear weapons. Later in his career, he became active in efforts at disarmament and banning of nuclear weapons. The Wilson of the Kistiakowsky–Wilson rules is E. Bright Wilson, a Harvard colleague of Kistiakowsky. Although known primarily as a molecular spectroscopist, Wilson served as the Director of the Underwater Explosives Research Laboratory during WWII. The work of the laboratory was directed toward antisubmarine explosive devices.

Explosions are sufficiently energetic to rupture all of the bonds, generating atoms that recombine to form stable molecules. For CHNO explosives, the main products are H_2O, CO_2 or CO, N_2, and possibly C or H_2, depending on how much oxygen is present. One step in the process of estimating detonation energy is to determine if there is sufficient oxygen for complete conversion to H_2O and CO_2. The general formula for an explosive reaction is

$$C_nH_xN_yO_z \longrightarrow n\ CO_2 + x/_2\ H_2O + y/_2\ N_2 + \mu\ O_2$$

where μ is the number of excess or deficient oxygen atoms for formation of CO_2 and H_2O. This can be determined by balancing the equation or by applying the following formula:

$$\Omega = \frac{[z - 2n - (x/_2)] \times 1600}{MW}$$

which gives Ω the oxygen deficiency as a percentage of MW, the molecular weight of the explosive. The maximum energy is available if the oxygen deficiency is zero, since all C is then converted to CO_2 and all H to H_2O. This goal can be achieved by mixing explosives. For example TNT has a rather high oxygen deficiency that can be improved by mixing it with NH_4NO_3, which has an oxygen excess. The key elements in determining explosive power are the formation of gas, which provides the mechanical force of expansion, and the release of energy as heat.

Clearly there is another facet besides composition and potential energy content that is crucial. That is the matter of *sensitivity*! An explosive must be stable until it is intentionally detonated. Generally a flame, spark, friction, pressure, shock, or an initiating detonation is used to trigger the explosion. When this occurs and a shockwave is generated, the entire amount of explosive is rapidly detonated. Once initiated, the detonation is self-propagating, so the energy required for initiation determines sensitivity. The initiation involves generation of a "hot spot." If the hot spot generates more heat than can be dissipated to the surroundings, the temperature and rate of reaction increases. If the reaction becomes sufficiently fast, it changes from a deflagration to a detonation and the shock wave progressing through the secondary explosive propagates the detonation. To some extent, the sensitivity of the explosive is related to its Arrhenius activation energy for thermal decomposition. The ease of initiation depends on the strength of the *weakest bond* in the explosive, which can be called the *trigger bond*. The "hot spots" are often associated with discontinuities (defects) in the crystal structure of the explosive. As a consequence, crystal structure and physical form can affect sensitivity. A crystal structure that places more strain on a trigger bond will increase sensitivity.

Detonators are used to set off the explosive charge. Detonators must meet even more stringent conditions than explosives. They must have sufficient power to trigger explosion of the main charge but also be stable enough for safe handling. The current version of detonators can be traced back to Alexander J. Forsyth, a Scottish clergyman who developed a detonation system for sporting rifles in 1807. He used a mixture (such as $KClO_3$, charcoal, and sulfur) that was sensitive to the shock of a metallic hammer and able to set off a gun-powder charge. Alfred Nobel adapted the basic idea to produce a blasting cap for dynamite in 1867, consisting of a mixture of $Hg(OCN)_2$ (mercury fulminate) and gun powder. Various improvements have been made over the years with current detonators being such substances as lead azide, lead styphnate, or PETN. Figure 1.1 shows some the designs of common initiation devices or "blasting caps."

$Pb(N_3)_2$

lead azide

lead styphnate

$C(CH_2ONO_2)_4$

PETN

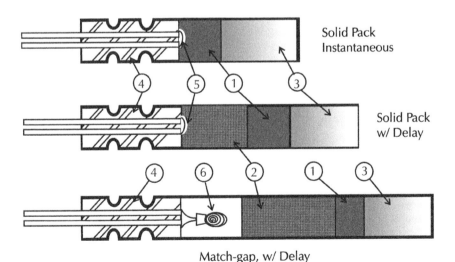

Match-gap, w/ Delay

FIGURE 1.1 Examples of Blasting Caps and Detonators. (top) Blasting cap using a fuse; (b) percussion primer; (c) hot wire detonator. (1) = Lead azide; (2) = pyrotechnic; (3) secondary explosive; (4) insulating plug; (5) = bridgewire; (6) electrical igniter. Adapted from Cooper, P. W.; Kurowski, S. R. *Introduction to Technology of Explosives*; Wiley-VCH: New York, 1996; p 108.

In the use of explosives for specific purposes such as mining, construction, demolition, or military operations, the placement and shaping of the charge is a critical factor. For example, in military use, detonation after penetration of armor may be the desired performance. The optimization of performance involves matching the timing and direction of the explosive force with the requirement of specific applications.

The elements of an explosion are summed up in a rhyme attributed to a "second inspector" in a British munitions plant in 1963.[1]

Initiators fire the chain
Acceleration boards the train
Fierce and fast reactions zip
Ingredients, self-sufficient, whip
The pace beyond *combustion*
Past the point of *deflagration*
Atoms fly with mounting pressure
Explosion then becomes the measure
Ah! but for some that's not enough
For they are made of rougher stuff
And still the pace goes up and up
Until it reaches ceiling, top
The pace by now extremely hot
No more acceleration can be got
Energy loosed in *shocking wave*
Atoms agitated so behave
With truly violent reputation
The label then is *detonation*

1.2 PRODUCTION OF EXPLOSIVES AND PROPELLANTS

1.2.1 BEFORE DYNAMITE

Black powder or gunpowder probably originated in China in the ninth century. It was used, among other purposes, to ward off evil spirits by making firecrackers. The recipe for gunpowder seems to have been introduced to England by Roger Bacon (1214–1292). He probably found the information in Arabic literature.[b]

---

[b]Bacon, extremely well educated for his time, was a Franciscan friar who had studied at Oxford and Paris. He believed that mathematics and experiment could reveal truth and was at times in conflict with his order. He thought technological advances could be achieved, not by magic, but by means of manipulation of natural forces learned by experiment. In one document, he encoded a recipe for gunpowder and later described what we would call a firecracker. It is likely that other Medieval scholars, such as Albertus Magnus (St. Albert the Great), had also learned the composition of gunpowder from Arabic writings.

The optimum formula of gunpowder is about 75% KNO_3, 10% sulfur, and 15% charcoal, as a finely powdered mixture. Gunpowder is set off by fire or a very hot rod. It provides a relatively slow deflagration that can be used to propel a projectile such as a bullet or cannonball from an enclosed tube. Manufacture of gunpowder requires the three components. In England, the original source of KNO_3 was bacterial decomposition of nitrogen-containing organic matter such as animal waste. It can be obtained from nitrate-rich soil by mixing with wood ash (potash, K_2CO_3), which allows the soluble KNO_3, also know as "saltpeter" or "nitre," to be leached out. The KNO_3 is then obtained by evaporation. It must be pure and dry to function well in gunpowder. Local manufacture was largely replaced by imports from India later in the 1600s. Early in the 19th century, large $NaNO_3$ deposits were discovered in Chile, and the subsequent discovery of KCl deposits in Germany were the basis for manufacture of potassium nitrate in the second half of the 19th century.

The process for making black powder initially involved grinding the three components together in a mortar but this was dangerous because friction from the grinding could ignite the material. In the 1400s, heavy stone wheels were developed for grinding the mixture into a powder. The process was called "corning." The pressure of the grinding melts the sulfur and it acts as a binder for the KNO_3 and the charcoal. The optimum properties for various uses were obtained by controlling the quality of the charcoal and the particle size of the gunpowder.

Bofors Industries, founded in Sweden in 1646 became the major producer of black powder in Europe. Use of black powder in mining began in the 1600s, but it can be used only for relatively soft rock. In France, Antoine-Laurent Lavoisier was responsible for the manufacture of gunpowder and conducted his experiments on combustion at the Arsenal of Paris.[c] During the American Revolution, gunpowder was imported from France and the Netherlands and was also produced by small local mills. Pierre Samuel du Pont de Nemous, a contemporary of Lavoisier, arrived with his family in America in 1800. Pierre S. du Pont had been arrested during the French Revolution and decided to relocate to America. Pierre Samuel's son, Eleuthere Irene, had worked with Lavoisier at the Paris Arsenal and knew the methods for production of gun powder. E. I. du Pont de Nemours and associates established a powder mill on Brandywine Creek, near Wilmington, DE, which became the DuPont company. The first powder was

---

[c]Antoine-Laurent Lavoisier was born into a wealthy Parisian family in 1743. He studied science with prominent scholars of the time. His early work included geological surveys and maps. Lavoisier's fundamental chemical studies included the recognition of hydrogen as an element and the role of oxygen in combustion and respirations. He showed that air was mainly oxygen and nitrogen and that diamond was a form of carbon. He formulated the Law of Conservation of Mass. His results were published in *Traite elementaires de chimie*, which was the first text book of chemistry to correctly identify many elements. Through his father-in-law, he was a member of the Ferme Generale, an exploitive tax collection enterprise, and was falsely charged with treason during the French Revolution, although he had made attempts to reform the system. He was guillotined along with his father-in-law. Later, the conviction was reversed and his accusers were eventually executed as well.

produced in 1804. Two sons of E. I. du Pont, Alfred Victor (from 1837 to 1850) and Henry (from 1850 to 1889), managed the company for most of the 19th century.

Sulfur deposits are found worldwide, especially in volcanic areas and the main European source was Sicily. It was isolated from sulfur-containing rock by fire that melted part of the sulfur, which was collected by allowing it to run down-hill. At the end of the 19th century, Herman Frasch[d] patented a method for recovery of sulfur from underground deposits in Texas and Louisiana, as shown in Figure 1.2. A series of three concentric pipes was used. Hot water was pumped down the outer pipe, melting the sulfur. Hot air was pumped down the center pipe, forcing the molten sulfur up the middle tube. When black powder was replaced by nitrates for explosives in the second half 19th century, a new need arose for the sulfur. It was oxidized to sulfuric acid, which is required to make both nitrate esters such as nitroglycerin and nitroaromatics such as TNT.

After the Civil War, a cartel called the "Powder Trust" was formed by the major gunpowder producers in the eastern United States. The du Ponts also purchased several small producers and achieved a dominant position in gunpowder manufacture. Later, when dynamite arrived on the scene, the du Ponts purchased California Powder, one of Alfred Nobel's US licensees. The company made a variant of dynamite called "White Hercules" that contained nitrated sugar as well as nitroglycerine and KNO_3. DuPont also began the manufacture of "smokeless powder," a form of nitrated cellulose also called "guncotton" (see Section 1.2.2). By 1900, nearly all of the production of gunpowder and explosives in the United States was under the control of the du Pont family in a company called the Eastern Dynamite Corporation. In 1907, the company was charged under the Sherman Anti-Trust Act and was eventually split into three companies, DuPont, Atlas, and Hercules. In the period after 1910, the DuPont Company began to diversify its business. We will hear more about this in Chapter 3.

---

[d]Herman Frasch was born on Christmas Day, 1851 in Oberrot, Schwabia, in Germany. In 1868, at the age of 16, he emigrated to America. He worked as a trainee and clerk in a Philadelphia drugstore and learned chemistry from Johannes Maisch, who was also a Professor at the Philadelphia College of Pharmacy. By 1874, Frasch had formed his own firm, the Philadelphia Technical Laboratory. He obtained his first patent in 1875, for a method of obtaining tin from iron mining tailings by reduction with carbon and founded a company to commercialize the technology. Deposits of sulfur were discovered in Louisiana in 1865. They were several hundred feet thick, but covered by alluvial quicksand. Several companies had attempted to mine the sulfur, but without success. In some cases, workers had been killed by release of hydrogen sulfide. Frasch decided to attempt to mine the sulfur by melting it. He drilled a 10 in. pipe and then inserted 6 in. and 3 in. pipes. Pressurized water at 335° was pumped down the outer pipe. Many were skeptical of the plan, but it succeeded. In his address accepting the Perkin Medal in 1912, Frasch said in reference to the skeptics: "This severe criticism while not agreeable, did not carry very much weight with me. I felt that I had given the subject more thought than my critics, and I went about my work as best I could, thoroughly convinced that he who laughs last, laughs best." The liquid sulfur was quite heavy and corrosive and required special pumps and valves. By 1903, the wells were producing 35,000 tons of sulfur and exports began in 1904. Frasch also made major contributions to the refining of petroleum, which is discussed in Chapter 2.

Sources: Hass, H. B. *ChemTech* **1976**, *8*, 88–105; Botsch, W. *Chem. Unserer Zeit* **2001**, *35*, 324–331.

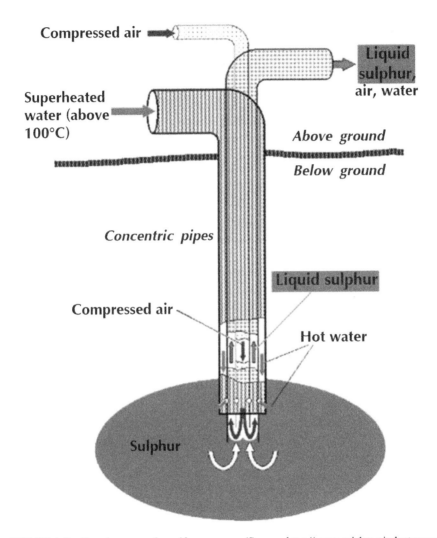

FIGURE 1.2 Frasch process for sulfur recovery (Source: http://www.mishraq.industry.gov. iq/eregions.htm)

By the early 20th century technological advances provided new sources for the starting materials for making gunpowder. In 1909, Fritz Haber developed the high temperature and pressure method for nitrogen reaction with hydrogen and the process was commercialized at BASF for manufacture of fertilizer (see Section 8.1.3). Besides providing fertilizer for agriculture, the ammonia could be oxidized to nitric acid, providing an alternative source of nitrate salts.

1.2.2 NITRATE ESTERS

1.2.2.1 NITROGLYCERIN AND DYNAMITE

Nitroglycerin, the explosive component of dynamite, was synthesized for the first time in 1846. The process involves mixing glycerol, nitric acid, and sulfuric acid. When the reaction is complete the material is poured into water, from which the nitroglycerin separates. It is further washed with alkaline water to remove acids. A distinguishing feature of nitroglycerin relative to black powder is that the substance is both *fuel and oxidant* and the explosion can be an *intramolecular* process. The detonation of nitroglycerin is much faster and more powerful than that of gun powder. Alfred Nobel formed a company, which included his father Immanuel, to manufacture nitroglycerin for use as an explosive. Production began in 1863 and soon reached quantities in the tons. However, nitroglycerin is both difficult to reliably detonate and prone to unintended explosive decomposition. Its practical use depended on solving these problems. The first successful approach to detonation was to use a small amount of gunpowder. In 1867, Alfred Nobel patented a blasting cap that consisted of gun powder and mercury fulminate in a copper tube attached to a thin tube containing nitroglycerin. The cap was ignited by a fuse or, later, an electric spark.

Nitroglycerin in liquid form is treacherous to handle or transport and Alfred's younger brother Emil was killed in an explosion at the factory near Stockholm in 1864. The company's German factory was destroyed by an explosion in 1866 and several explosions occurred in ships and warehouses containing nitroglycerin. Nobel recognized he would need to find the cause of these explosions and prevent them. Nobel found that mixing nitroglycerin with a highly adsorbent powder called kieselguhr[e] allowed it to be handled much more safely. Nobel gave the material the name Dynamite. He soon established facilities for manufacture in Germany, Norway, and Finland. By 1871 he had established a factory near Glasgow in Scotland and another in the United States. Soon thereafter, factories were established in France, Spain, Italy, and Portugal.

There are many variations of dynamite. It can contain KNO_3 or NH_4NO_3. Some mixtures contained paraffin or wood pulp. Dynamite had a profound influence on mining and civil engineering. Projects such as the Suez Canal and Panama Canal were made possible by its use. Countless accidents also occurred. Beginning in the 1920s, dynamite was reformulated to include ethylene glycol dinitrate. This reduced the freezing point of the nitrate esters below that encountered in mines and other cold environments and considerably improved the safety of dynamite.

---

[e]Kieselguhr, also known as diatomaceous earth, is a highly absorbent form of fossilized silicon dioxide formed from algal deposits. It is mined in many places around the world.

1.2.2.2 NITROCELLULOSE—GUNCOTTON AND SMOKELESS POWDER

Nitrocellulose, also called *guncotton* or *pyroxyline*, is made by mixing cellulose with HNO_3 and H_2SO_4. It is a partial nitrate ester of the polymeric carbohydrate. Like nitroglycerin, it is a dangerous material and impurities can lead to spontaneous explosions. The first efforts to use it as a military explosive were made by an Austrian artillery officer, Wilhelm von Lenk, who recognized that the stability of the material could be improved by extensive washing. Frederick Abel,[f] Chief Chemist of the British War Office, in 1865 patented a process for shredding and washing the material that made it much safer to handle. In the 1870s, there was further development in France at the Depot Central des Poudres et Saltpetres where Paul Vielle and Emile Serrau developed Poudre B (Powder B). Their work was guided by thermochemical studies on explosives carried out by Marcellin Berthelot during the Franco-Prussian war. From these studies, Vielle recognized that the physical form of the guncotton would influence its burning properties and made a denser, gelatinized form. The French Army adopted Poudre B in 1886.

Alfred Nobel developed a blasting gelatin called *Ballistite* in 1888 that is a mixture of nitrocellulose and nitroglycerin. By using weaker acid, cellulose is less completely nitrated than for guncotton. Nobel found that a 1:1 mixture of nitroglycerin and this nitrocellulose containing some camphor formed a gel that was excellent for blasting rock and was resistant to moisture. The story is told that Nobel's invention of Ballistite was triggered by a laboratory accident in which he cut his finger. He used collodion, then a common laboratory sealant, to coat the finger, but during the night the idea occurred to mix collodion (which is a solution of nitrocellulose in 3:1 ether–alcohol), with nitroglycerin. Ballistite was soon used to construct the 9-mile St. Gotthard tunnel in the Alps. Ballistite and similar materials also became important military explosives (see below).

At nearly the same time, Abel and James Dewar (also the inventor of the vacuum bottle) patented a mixture of 58% nitroglycerin, 37% guncotton, and 5% petroleum jelly. This material eventually came to be called *Cordite*. It was adopted as a military explosive and production began in 1891. The nitrocellulose was somewhat more highly nitrated than that used by Nobel in Ballistite.

---

[f]Frederick Abel was born in London in 1827. At the age of 17, Abel began studies at the Royal Polytechnic Institute but transferred in 1845 to the newly founded Royal College of Chemistry. In 1851, he succeeded Michael Faraday as Lecturer in Chemistry at the Royal Military Academy, Woolwich. He eventually became chief chemist at the Woolwich Arsenal and served in that capacity from 1854–1888 and led the effort to adapt the growing chemical technology to the needs of the military. He studied guncotton from 1868–1875, eventually resulting in the development of the Cordite used in rifles and artillery munitions through WWI. From 1887–1901, he was involved in the development of the Imperial Institute, a museum and laboratory devoted to the study and commercialization of materials from the British colonies.

Source: Mauskopf, S. H. *Bull. History Chem.* **1999**, *24*, 1–15.

It was mixed using acetone as a solvent and extruded as a cord, thus the name. The similarity of Ballistite and Cordite, and the circumstance that there had been some communication between Nobel and Abel, led to patent infringement litigation by Nobel. Nobel lost and this experience solidified a long-standing animosity toward the patent and legal system. Ballistite and Cordite and related materials quickly replaced gunpowder for use in small arms and other military applications. The outbreak of WWI required rapid increase in production of explosives. During WWI, a shortage of acetone needed to make Cordite developed. Most of it came from wood by making and then pyrrolyzing calcium acetate. Chaim Weizmann discovered a microorganism, *Clostridium acetobutylium*, that could ferment starch to acetone.[g] Winston Churchill, then First Lord of the Admiralty, and others encouraged the development of this method of acetone production, which supplemented the supply from wood. The production of acetone by fermentation was one of the first examples of use of a biochemical reaction for production of a chemical on an industrial scale.

The expansion of mining, construction of railways, roads, canals, and tunnels in the last quarter of the 19th century created a huge demand for Dynamite, Ballistite and related products, and made Alfred Nobel rich. Nobel's will, written in 1895, and the prizes that he endowed in chemistry, physics, physiology, or medicine, literature and peace, reflect his life's dedication to science in the service of mankind, his strong personal interest in literature and poetry, and his belief that peace could be achieved. Nobel's will named the institutions to administer the Prizes in Chemistry, Physics, Physiology or Medicine, Literature and Peace, but not any of the mechanics. He died in 1896 and, after some legal wrangling, the prizes were initiated in 1901. (The prize in economics was added later, endowed by the Swedish Central Bank.)

The impact of explosives in the late 1880s is summarized in the following lines from Rudyard Kipling:

Do you wish to make the mountains bare their head
And lay their new-cut forests at your feet?
Do you want to turn a river in its bed,
Or plant a barren wilderness with wheat?
Shall we pipe aloft and bring your water down

---

[g]Chaim Weizman was born in what is now Belarus in 1874. He studied chemistry in Darmstadt, Germany and Fribourg, Switzerland, and received a Ph. D. in 1899. He moved to the University of Manchester in the United Kingdom in 1904 and became a British citizen in 1910. He was a pioneer in using microorganisms to carry out chemical reactions. He served as the Director of the British Admiralty laboratories during WWI. Weizman was an ardent Zionist and was acquainted with both Arthur Balfour and Winston Churchill. In 1917, while Foreign Minister, Balfour issued a statement of support for a Jewish state in Palestine, then a British Protectorate. Weizman was instrumental in the foundation of the Hebrew University in Jerusalem and Technion, the Israeli Institute of Technology. The Weizman Institute, a scientific research institute, was named in his honor in 1949. Weizman served as president of Israel from its founding in 1948 until his death in 1952.

From the never-failing cisterns of the snows,
To work the mills and tramways in your town
And irrigate your orchard as it flows?
    It is easy! Give us dynamite and drills!
    Watch iron-shouldered rocks lie down and quake
As the thirsty desert-level floods and fills,
And the valley we have damned becomes a lake.[2]

1.2.2.3 PENTAERYTHRITOL TETRANITRATE

Pentaerythritol tetranitrate (PETN) is chemically closely related to nitroglycerin and was first prepared in 1891. The preparation of PETN from pentaerythritol
is similar to that of nitroglycerin. It became commercially available in the 1930s
and is used in detonating cords and blasting caps. The pure material is very
sensitive to friction or impact. PETN is used as a mixture with TNT to make
"Pentolite." One of the explosives used by terrorists is "Semtex," which contains PETN and RDX in a styrene–butadiene copolymer. It was manufactured
in Czechoslovakia during the Viet Nam war. Excess material at the end of the
war was sold to Libya, and from there found its way to various terrorist groups.

TOPIC 1.1 ALFRED NOBEL

Alfred Nobel was born in Stockholm in 1833. His father, Immanuel, was a
self-taught designer, builder, and inventor. As a result of a series of mishaps,
Immanuel was forced into bankruptcy that same year. In 1837, Immanuel left
his family in Stockholm to manufacture a naval mine he had invented in St.
Petersburg, Russia. He was successful in this endeavor and developed a work-
shop that produced a number of other materials. While Immanuel was in St.
Petersburg, the Nobel family lived in relative poverty. In 1842, Immanuel moved
his family to St. Petersburg. The family was quite prosperous at this time, and
Alfred Nobel received education from several tutors, including chemists.

   In 1856, the Russian defeat in the Crimean War brought an end to the Nobel
family's prosperity. Immanuel, his wife Andriette, Alfred and the youngest son
Emil returned to Stockholm. The older sons, Robert and Ludvig reorganized
the family finances in Russia and eventually developed a productive oil field in
Baku (see Topic 2.2). After he and his father returned to Stockholm, Alfred continued to experiment with nitroglycerin and made some progress with mixtures
of nitroglycerin and gunpowder. This preliminary success led to major financing
from a Paris bank interested in large construction projects. Nevertheless, manufacture of nitroglycerin was treacherous, and in 1864, the plant in Stockholm

exploded, killing Nobel's younger brother Emil and several others. A factory was rebuilt further removed from Stockholm. A factory was also built near Hamburg, Germany, but this factory also exploded in 1866 and again in 1870.

Alfred Nobel traveled extensively pursuing his business interests but never married. He spoke German, French, English, and Russian, as well as his native Swedish. He moved in the highest circles of business and finance but was also well known among the writers and intellectuals in France. He had homes in Paris, Italy, and Sweden but at the time of his death he was not a legal resident of any country. He was a supporter of international peace efforts, although in the press he was depicted as a war profiteer. In fact, dynamite was mainly used for civilian purposes, although Ballistite was primarily a military explosive. Several factors loom large in the story of dynamite and related explosives. One is Nobel's persistence. He pursued nitroglycerin and its application relentlessly. He surely recognized the physical danger in this endeavor, but even his brother Emil's death did not deter him. He was also an astute businessman intent not only on securing the economic rewards of dynamite, but also that his efforts would be vindicated and recognized. One of his biographers has speculated that this persistence arose at least in part from his childhood poverty and the insecurity it brought. Several circumstances helped set the stage for Nobel's discoveries. One was his father Immanuel's experience with explosives as a designer and manufacture of mines. Nobel's science tutors in St. Petersburg knew of nitroglycerin and discussed it with both Immanuel and Alfred. Nobel studied in France in 1849, and he shared lab space with Ascanio Sobrero, the Italian who was the first to synthesize nitroglycerin in 1846. The combination of hard work, keen intelligence, and propitious circumstances contributed to Nobel's ability to achieve his goal of putting nitroglycerin to productive use.

Nobel was in failing health in the 1890s and lived mainly in Paris and Italy. Scandal had erupted around his French company and one of his close associates Paul F. Barbe. Nobel suffered a considerable financial set-back, but he remained a very wealthy man. The press again attacked Nobel and he relocated to San Remo in Italy. He wrote his will in his own hand in Swedish and signed it on November 27, 1895, evidently without legal advice. He suffered a stroke at his home in San Remo and died on December 10, 1896. His will named his young personal assistant Ragnar Sohlman and a Swedish industrialist Rudolf Lilljeqvist as executors. Over the next several years, they attended to implementing the terms of the will. The will was contested by some of Robert Nobel's family in Sweden, but supported by his nephew Emanuel, then the head of the Nobel's Russian oil company. The will was eventually upheld by the Swedish courts and the government cooperated in the establishing of the Nobel Foundation. The Swedish Academy (Literature), Swedish Academy of Sciences (Chemistry, Physics), Karolinska Institute (Medicine or Physiology), and the Norwegian Parliament (Peace) took responsibility for the award of the prizes.

Sources: Schueck, H.; Sohlman, R.; Oesterling, A.; Liljestrand, G.; Westgren, A.; Siegbahn, M.; Schou, A.; Stahle, N. K. *Nobel, the Man and His Prizes*; Elsevier: Amsterdam, 1962; pp 17–72; Fant, K. *Alfred Nobel*; translated by Ruuth, M. Arcade Publishing: New York, 1993.

1.2.3 POLYNITROAROMATICS

The first polynitroaromatic used for artillery shells was picric acid or its ammonium salt. The British adopted picric acid, called Lyddite, in the 1890s and used it in the Boer War in South Africa. Picric acid, however, has some notable disadvantages. It is corrosive to shells and the salts formed can act as detonators. The relatively high melting point of picric acid also makes it difficult to handle as a melt.

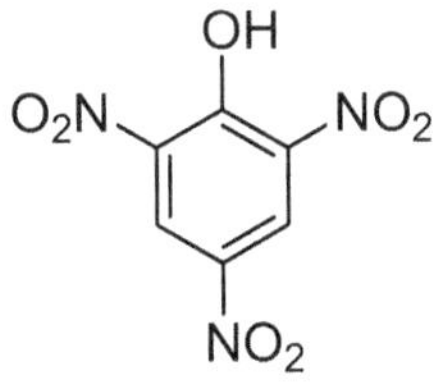

picric acid

Germany adopted TNT for artillery shells in 1902. The German chemical industry was able to produce TNT from toluene obtained from coal tar. Although a less powerful explosive than picric acid, it was much easier to manufacture and use in shells. TNT is easy to cast as a melt and is very stable in the cast form. Its main drawback is that it is difficult to detonate and requires a "booster." At the beginning of WWI, the Germans had a great advantage in artillery as a result of both the size of their guns and the amount and quality of ammunition available. As this became apparent, the British switched to TNT. TNT production was increased from 20 t/week in 1914 to 1000 t/week by 1916. Both the French and British also imported large amounts of TNT made by DuPont. TNT mixed with NH_4NO_3, known as *Amatol*, was also put into production. The optimized method for production of TNT uses step-wise nitration in a circulating pump system. The nitrating acid mixture and toluene are introduced into the pump and part of the reaction product is drawn off and separated by gravity into the organic and acid layers. The acid can be recycled after increasing its strength and the partially nitrated toluene is sent through successive reactors. The final product is washed with water, sodium sulfite solution, and again by water. It is dried by hot air and finally processed from the molten state. Tetryl, which is *N*-methyl-*N*,2,4,6-tetranitroaniline, and a close chemical relative of picric acid, was also used in artillery shells during WWI. Tetryl is too sensitive to use alone as an explosive, but can be used in a combination with TNT that is called Tetrylol.

TOPIC 1.2 THE ROLE OF ARTILLERY MUNITIONS IN WWI

Chemistry played a key role during WWI in terms of artillery ammunition. The military strategy of both the allied English and French and the Germans assumed a quick war and initial planning was for a war of less than 6 months. When, instead, the war bogged down into trench warfare and artillery bombardment, both the quality and quantity of artillery ammunition became important factors. The critical element for Germany was a source of nitric acid to produce ammunition for both small arms and artillery. The main source of nitrates at the time was imports from Chile, and the British blockage cut off these supplies. As we will see in Section 8.1.3, the development of new catalytic processes for making both ammonia and nitric acid were developed at the Badische Anilin und Sodafabrik in Germany. This gave the Germans their needed supply of nitric acid to make nitrocellulose, nitroglycerin, and TNT.

The British also faced severe problems. Many of the chemical starting materials had been imported from Germany. At the beginning of the war, the British used picric acid in their artillery shells, which was more difficult to produce and handle than TNT. Picric acid was made by nitration of phenol obtained from coal distillation. The picric acid was also manufactured from coal tar benzene, via chlorobenzene and 2,4-dintrochlorobenzene, a relatively cumbersome process. The British soon shifted to TNT for artillery ammunition. It required toluene as a starting material, which was obtained both from coal distillation and from crude petroleum imported from Indonesia. TNT production was increased by 50-fold between 1914 and 1916 by construction of new plants and introducing improved processes. The British eventually adopted Amatol, a mixture of TNT and ammonium nitrate for their artillery ammunition.

Sources: For a contemporary account of the production of explosives see, *J. Soc. Chem. Ind.* **1919,** *38,* 366R–369R; see also MacLeod, R. *Ann. Sci.* **1993,** *50,* 455–481.

1.2.4 NITRAMINES

Another class of explosives is nitramines. These explosives contain nitro groups attached at nitrogen atoms in ring structures. These include hexogen (also called cyclonite or RDX) and octogen (HMX) as well as more recently developed examples, such as BCHMX and HNIW (see Scheme 1.1 for structures). Both hexogen and octogen are ultimately derived from formaldehyde and ammonia. In the manufacturing process, the starting material is hexamethylene tetramine, which is a combined form of the starting materials. The ring system undergoes opening that becomes irreversible after the nitrogen is nitrated. One synthetic method is called the Bachmann Process.[h] It produces both hexogen and octogen, using acetic anhydride, ammonium nitrate, and a small amount of nitric acid. This RDX–HMX mixture can be used in combination with TNT. Pure HMX is used in special applications.

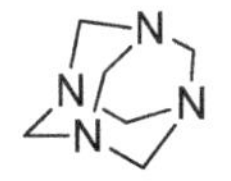

hexamethylene
tetramine

HNIW is conceptually related to hexogen and octogen. In this case, the carbon source is glyoxal, which reacts with benzyl amine. The benzyl groups are removed by reduction, followed by nitration.

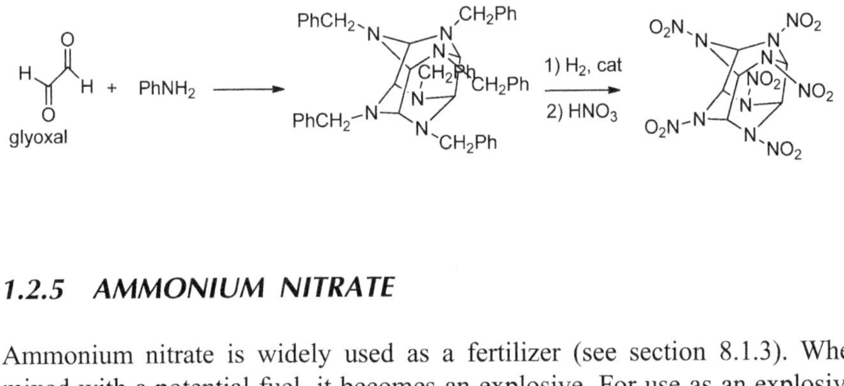

1.2.5   AMMONIUM NITRATE

Ammonium nitrate is widely used as a fertilizer (see section 8.1.3). When mixed with a potential fuel, it becomes an explosive. For use as an explosive, the NH_4NO_3 is produced in the form of spherical particles called *prills*. A concentrated solution in water is sprayed down a tower. As the droplets fall through

---

[h]Werner E. Bachmann was a Professor of Chemistry at the University of Michigan. He received his undergraduate training there in chemical engineering but undertook graduate work in Chemistry and received his Ph. D. in 1926. He remained as a member of the faculty and became the Moses Gomberg Distinguished Professor in 1947. His major research effort was in steroids and polycyclic compounds, but during WWII he was a participant in the NDRC program in chemistry. Working with his Ph. D. student, John C. Sheehan, he developed an improved process for manufacture of RDX and HMX. RDX and HMX were used in bombs and torpedo shells. Bachmann received a Presidential Certificate of Merit (United States) and the King's Medal (United Kingdom) in recognition of this work.

Source: Wilds, A. L. *J. Org. Chem.* 1954, *19*, 129–130.

the tower into a rising stream of air they dry and cool into the prills. At the base of the tower they land on rubber pads and are transported by conveyors and elevators for packaging and shipment. The material can also be coated with a wax and mixed with clay to prevent clumping. Beginning in the 1950s, a mixture of NH_4NO_3 with fuel oil, known as *ANFO*, became widely used in mining in place of dynamite, being cheaper than other explosives. Ammonium nitrate–fuel mixtures can be produced as slurries or emulsions. The oxidizer is present as a concentrated water solution dispersed in the fuel phase. The mixtures are typically 60–80% by weight NH_4NO_3, 7–15% water, and 4–6% oil or wax. One of the technological advances that enabled the use of AFNO was the ability to drill larger bore holes using tungsten carbide bits that became available in the 1950s. In the 1970s, further improvements were made by including powdered aluminum and methylammonium nitrate in the explosive mixture. It was NH_4NO_3, coated with a wax that exploded in the Texas City Harbor in 1947. More than 2000 t loaded on a freighter was detonated by a fire. Nearly 600 people were killed in the explosions that followed when the fire spread to nearby waterfront facilities. Timothy McVeigh used NH_4NO_3 mixed with nitromethane and diesel fuel to blow up the federal building in Oklahoma City, OK in 1994. The mixture was placed in 55 gallon barrels and detonated by a delayed timing fuse.

1.2.6 NITROGUANIDINE AND NITROUREA

Among the problems associated with Cordite (see Section 1.2) was excessive flash and erosion of gun barrels. Paul Vieille and others found that nitroguanidine reduced the problem. Further work occurred in the 1920s at the Woolwich Arsenal in the United Kingdom. A process was developed using calcium cyanamide as the starting material. The crystallization process that was eventually developed involved spraying super-heated solutions into drying towers, similar to the process for ammonium nitrate. Nitroguanidine mixed with NH_4NO_3 and paraffin was used in mortar shells in WWII.

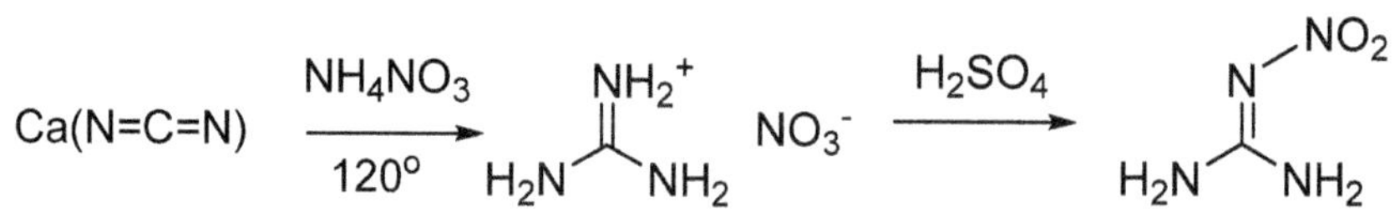

Other related materials that have been considered as explosive oxidants are ammonium dinitramide, NH_4^+ $[N(NO_2)_2]^-$ and hydrazinium trinitromethide, $NH_2NH_3^+$ $[C(NO_2)_3]^-$. The salt of urea and nitric acid, known as both urea nitrate and uronium nitrate, was used in the 1993 bombing of the World Trade Center in New York. It can be used with aluminum, as well as hydrocarbon-based fuels. It is not very stable and this has limited its use in commercial explosives.

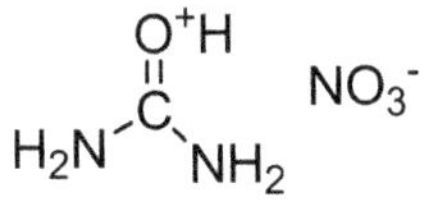

$$\underset{H_2N}{\overset{O^+H}{\underset{\parallel}{C}}}\underset{NH_2}{} \qquad NO_3^-$$

uronium nitrate

1.2.7 CHLORATE AND PERCHLORATE SALTS

Second in importance to nitrates and nitro compounds as explosive oxidants are chlorate and perchlorate salts. As with the nitrogen compounds, they contain a high oxidation state (Cl +5 and Cl +7, respectively) and provide the potential energy in reactions with fuels. They are also easily detonated. Sodium or potassium chlorate can be produced by oxidation of chloride or chlorine. Potassium chlorate ($KClO_3$) was a component of the original detonating caps for guns. During WWI, it was used mixed with hydrocarbon wax in devices such as grenades. Because it can be made from easily available materials, it has been used as an explosive by terrorists in home-made bombs (see Section 1.6). It has also been used as an oxygen source, such as in space vehicles, but it is very sensitive to purity and has been the source of a number of explosive accidents. Ammonium perchlorate is also a powerful oxidizer. It can be used either with hydrocarbon-based fuels or with aluminum. It is used extensively in composite rocket propellants (see Section 1.3.2).

1.2.8 MIXED EXPLOSIVES

There are names associated with various mixtures of explosives. Among the more common are the following:

Amatol: TNT and NH_4NO_3
Pentolite: PETN and TNT
Semtex: RDX and PETN
Tetrylol: TNT and Tetryl
Torpex: TNT, RDX, and powdered aluminum
Tritonal: TNT with powdered aluminum.

1.2.9   PLASTIC EXPLOSIVES

Plastic explosives are mixture of secondary explosives with flexible binding material that also serves as fuel. The materials have the advantage of being able to be molded in place for particular applications. The plastic explosives also tend to be highly stable to handling and resistant to accidental detonation. They are used both in military and civilian applications. They use either nitrate esters or nitramines as the main explosive. The most common current plastic explosive, referred to as C-4, contains about 90% RDX, 5–6% plasticizer, 2% polyisobutylene binder, and 1.5% fuel oil. Another widely used plastic explosive, Semtex, is manufactured by the Czech company Explosia. It contains a mixture of PETN (50–75%), RDX (5–40%), a styrene–butadience polymer (9%), and a plasticizer (8–9%). The plasticizers for both C-4 and Semtex are diesters such as dioctyl adipate or dioctyl phthalate. As we will learn in Section 1.6, Semtex attained notoriety as an explosive of terrorists.

1.3   MILITARY APPLICATIONS OF EXPLOSIVES AND PROPELLANTS

Both WWI and WWII were fought mainly with conventional weapons ranging from small arms to bombs that were powered by the explosive described in Section 1.2. The desired effect of military weapons is achieved by the destructive force of the blast and generation of shrapnel. The ammunition can be designed to maximize these effects. Many new military applications for explosives and propellants were developed in the second half of the 20th century. These ranged from tactical rockets to intercontinental ballistic missiles armed with nuclear warheads. For rockets and missiles, high energy compounds are combined with fuels to provide a powerful but controllable thrust. It is crucial that propellants not explode, so they must not readily undergo the *deflagration to detonation transition*. Military uses of rockets and missiles often present particularly challenging environments. Air craft carriers for example, involve storage of ammunition and fuel, arriving and departing air craft, as well as the potential of hostile fire. The goal is *insensitive munitions* that are unlikely to explode under the conditions of operation.

1.3.1   SMALL ARMS, ARTILLERY, AND BOMBS

Nitrocellulose remained the main propellant in small arms throughout the 20th century. Various refinements were introduced. One called Improved Military Rifle Powder, used nitrocellulose 97.4%, K_2SO_4 (2%), and diphenylamine

(0.6%) as a stabilizer. Another, called Ball Powder contained 89% nitrocellulose, 9% nitroglycerin, and 0.9% diphenylamine. It was produced by extruding an emulsion of the propellants into water under conditions where it precipitated as spheres. The terms *single base, double base*, and *triple base* are used to indicate how many high energy compounds are contained in the ammunition. For example, the double base composition of 60% RDX and 40% TNT is widely used in artillery ammunition and bombs.

1.3.2 ROCKETS AND MISSILES

Rockets are believed to have been used for military purposes as early as the 13th century in China, India, and the Arab world. They were fueled with gunpowder, but had no guidance systems. The British used rockets in the War of 1812 and this was the origin of the phrase "the rockets' red glare" in the National Anthem. Both the North and South used rockets in the civil war, but without major effect. The energy used to propel military rockets is provided by chemical reactions. As with explosives, the reactions involve a fuel and an oxidizer and the transformation generates both the thermal energy of the chemical reaction and the physical force of the formation and expansion of gaseous reaction products. The ultimate requirement of a rocket fuel is a powerful and reliable thrust. The thrust depends primarily on the velocity of the exiting gases. The thrust is increased by high fuel density and formation of low molecular weight gaseous products, because this maximizes the volume of gas that is produced.

The primary impetus for the development of rocket systems after WWII was for military purposes and each branch of the armed services developed systems that were appropriate for its needs. Systems for the placement of satellites for intelligence purposes were also developed. In the 1960s, rocket engines were used for launching vehicles for space exploration. When the international tensions decreased in the 1980s, resulting in limitation on large nuclear weapons delivery systems, some of the rocket engines were adapted to civilian purposes such as launching communication, global position, and weather satellites.

There are three basic designs for rockets and missiles, depending on the physical state of the fuel and oxidant. If both are liquids they can be metered into the engine to provide the desired level of thrust. These are called *liquid fuel rockets*. Solid propellants contain both the fuel and oxidant. The oxidant can be imbedded in a matrix of the fuel or prepared as an intimate mixture. Solid fuel rockets usually cannot be controlled subsequent to ignition. Hybrid rockets consist of a solid phase and a liquid phase. The degree of thrust can be controlled by metering in the liquid phase. Figure 1.3 shows these basic designs.

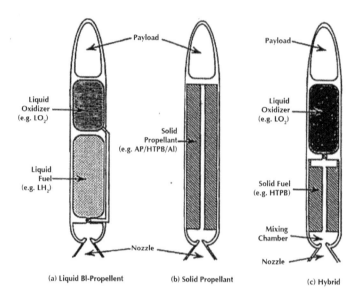

(a) Liquid BI-Propellent   (b) Solid Propellant   (c) Hybrid

FIGURE 1.3   Rocket Fuel types from Kuo, K. K. *Proceedings of the 24th International Pyrotechnics Seminar*, 1–41 (1998). AP = ammonium perchlorate; HTPB = hydroxyl terminated polybutadiene; Al = aluminum; LO_2 = liquid oxygen; LH_2 = liquid hydrogen.

1.3.2.1   LIQUID FUELS

The first modern liquid fuel missile was the German A-4 rocket (also known as the V-2) used in WWII. It used 75% ethanol/25% water as fuel and liquid oxygen as the oxidant. Later, Russian and American missiles used kerosene and liquid O_2 for the launch stage. Hydrocarbon fuel and liquid O_2 were used in the Atlas, Delta II, Titan I and the first stage of the Saturn V rocket motors.

Other systems use liquid H_2 as fuel. The Saturn/Apollo rockets used kerosene and liquid O_2 in the first stage and liquid H_2 and O_2 in the second and third stages. Both liquid H_2 and O_2 must be stored under pressure at low temperature and are called *cryogenic fuels*. The boiling point of liquid H_2 is −253°C and liquid O_2 −183°C. As a result, these fuels require special storage and transfer facilities and are most appropriate for fixed installations, such as those used in launching satellites and space vehicles. The advantage of liquid H_2 is that it provides the maximum thrust per unit weight. The Space Shuttles used liquid H_2 and O_2 as the primary fuel. Liquid H_2 is also used in Ariane 5, Delta-IV, and Centaur rocket engines. Methane (bp −182.5 °C) is a potential alternative liquid fuel. Like hydrogen, it requires low temperature and pressurized storage and transfer. So far, is has not been used in any operational rocket systems.

Hydrazine derivatives are also used as fuels in rockets and missiles. The oxidants used are nitric acid or N_2O_4. These materials ignite on contact and are called *hypergolic fuels*. The Space Shuttles used NH_2NH_2/N_2O_4 for the thrusters. In 1964, Shell introduced an iridium catalyst that permitted use of NH_2NH_2 without an oxidant. Derivatives of hydrazine, in particular CH_3NNH_2 and $(CH_3)_2NNH_2$ can be used in place of NH_2NH_2. The main engines of the Russian Proton Rockets use *unsym*-dimethylhydrazine and N_2O_4. The Titan-II rocket used a 50:50 mixture of H_2NNH_2 and $(CH_3)_2NNH_2$. The hydrazine fuel mixtures are corrosive and very toxic.

The reaction equations of the major fuel combinations are given below. All the reactions generate gas and are exothermic. The order of total expansive power is $H_2/O_2 > C_nH_{2n+2} + O_2 > NH_2NH_2/N_2O_4 > NH_2NH_2$.

$$2\,H_2 + O_2 \longrightarrow 2\,H_2O$$

$$C_nH_{2n+2} + \frac{2n + 1}{2}\,O_2 \longrightarrow n\,CO_2 + n{+}1\,H_2O$$

$$2\,NH_2NH_2 + N_2O_4 \longrightarrow 3\,N_2 + 4\,H_2O$$

$$2\,NH_2NH_2 \xrightarrow{\text{cat}} 2\,NH_3 + N_2 + H_2$$

A critical characteristic of the rocket fuel systems is the *specific impulse* (I_{sp}). This is a measure of the impulse (force) per unit weight of fuel and is in units of seconds. The specific impulse is proportional to $(T/mw)^{1/2}$, where T is temperature of the chamber and mw is the mean molecular with of the expelled gases. Thus, specific impulse is favored by low molecular weight products. Values for some of the fuel systems are given below.

Fuel	I_{sp} (s)
$H_2 + O_2$	455
RP-1 (kerosene) + O_2	358
$NH_2NH_2 + N_2O_4$	305

1.3.2.2 SOLID ROCKET FUELS

The origins of solid rocket fuels can be traced to the work of Alfred Nobel, Paul Vieille, and Frederick Abel on nitrocellulose in the late 1800s (see Section 1.2.2.2). Work began anew after WWII at locations such as the Guggenheim

Aeronautical Laboratory (later The Jet Propulsion Laboratory) at Cal Tech and naval propellant research laboratories at China Lake, CA and Indian Head, MD. The research at Cal Tech resulted in the formation of the Aerojet Company by John W. Parsons, Theodore von Karman, Edward Forman, Frank Molina, and Martin Summerfield. The initial propellant, conceived by Parsons, was asphalt mixed with $KClO_4$. The idea evidently came to Parsons (or, by some accounts, to his wife) on seeing the pouring of an asphalt roof. There has been considerable development and evolution since then. Several kinds of materials are used. One component is a high-energy oxidant, usually NH_4ClO_4 or NH_4NO_3. Alternatively, compounds such as nitrocellulose, nitroglycerin, or one of the nitramines can be used. These substances are combined with an organic polymer that is used to provide both solid form and serve as fuel. Aluminum metal is also included as a fuel. The proper function of a solid fuel rocket depends on a reliable *deflagration.* This requires that the fuel–oxidant mixture be in a physical state that allows smooth burning. Cracks and imperfections cause failures. The shape of the container, to some extent, can control the thrust, but once ignited, a solid fuel rocket cannot be shut down.

Several kinds of polymer matrices are used. (We will discuss polymers in detail in Chapter 5.) Early polymers included polystyrene, polyesters, polyvinyl chloride, and polysulfides. In the 1960s, a copolymer of butadiene, acrylic acid, and acrylonitrile was introduced. At present, the most common are hydroxyl-terminated polybutadiene, polyethylene glycol, or polypropylene glycol. These polymers have terminal hydroxyl groups that can be cross-linked with isocyanates. An advantage of this cross-linking system is that it does not generate any small molecule byproducts that might lead to bubbles or other imperfections in the solid. The development of various polymeric binders was enabled by the advances that were taking place in the polymer field, and were applied to the specific purposes of solid rocket fuels.

1.3.2.3   HYBRID ROCKET FUELS

In hybrid rockets, either the fuel or the oxidant is a liquid and the other is a solid. The rate of burn can be controlled by metering the liquid component. The solid component, most often the fuel, is generally inert and can be handled fairly simply as compared to a cryogenic liquid fuel. Solid fuels can provide relatively high density, which is an advantage in overall rocket design. At the present time, however, the technology of hybrid rocket systems is not so well developed as either the solid fuel or liquid fuel version and they are not widely employed.

1.3.3 EXAMPLES OF MILITARY ROCKET AND MISSILE SYSTEMS

The Germans initiated use of missiles during WWII. Exploratory study had begun in the mid-1930s under Wernher von Braun, but serious military efforts began during the war. The German A-4 rockets were fueled by aqueous ethanol (25:75) and liquid O_2. The first successful test, after a number of failures, was in October 1942. The missiles were launched from mobile systems. The design incorporated an inertial guidance system and radio-control. Nevertheless, the accuracy was quite low, so that targeting was to cities, not specific targets. The first military launches were in September 1944. At this point, they had acquired the name V-2, *vergeltstungswaffe 2, vengeance weapon,* and German propaganda justified them in terms of a response to the extensive allied bombing that was taking place by then. London and Antwerp were the major targets and more than 3000 V-2 missiles were fired. British intelligence successfully created disinformation to the effect that the rockets were overshooting and after the Germans recalibrated the targeting most of the missiles fell short of London.

A group of German rocket scientists including Wernher von Braun formed the nucleus of the post WW-II US program for military and space rockets. The group was first housed at Ft. Bliss, TX, USA and was initially top secret under the code name "Operation Paperclip." Tests were conducted at the White Sands Proving Grounds in New Mexico, initially using V-2 rocket parts to retro-engineer the rocket. The Russian were known to have captured other German rocket experts and were presumed to be pursuing rocket development as well. During 1946, modified V-2 rockets were tested at White Sands and achieved altitudes of 100–200 km. By 1948, altitudes of nearly 400 km were reached with speeds of nearly 9000 km/h. The headquarters for rocket development, which was under control of the Army, was relocated to Redstone Arsenal near Huntsville, AL, USA.

1.3.3.1 SOLID FUELED ROCKETS FOR MOBILE AND TACTICAL WEAPONS

Solid fuels are favored for most short-range missiles. An advantage of solid fuel rockets is that they can be transported with relatively safety and launched quickly without the need for sophisticated equipment. Some of the solid fuels are called "double-base" and contain nitroglycerin and nitrocellulose. They are thus descendants of Nobel's Ballistite and Abel's Cordite. Examples of US weapons based on solid fuel rockets were Pershing, Patriot, and Sidewinder missiles. The Polaris, Poseidon, and Trident missiles used on submarines also used solid fuel rockets. Pershing missiles were deployed between 1960 and 1991. They were operated by Army Artillery units and were launched from tracked vehicles.

The original version of the missile used an inertial guidance system and could be (and was) equipped with nuclear warheads. They were deployed, mainly in Germany through the 1980s. The Pershings were taken out of service starting in 1988 as the result of an agreement with the Soviet Union to reduce intermediate range nuclear weapons.

The Patriot missile was originally developed and deployed as a mobile anti-air craft weapon. It had a solid-fuel rocket engine and a radar guidance system that allowed it to "lock-on" targets. Various upgrades made the missile capable of intercepting incoming missiles. Patriots were used during the Persian Gulf War in 1991, but its effectiveness was controversial. The missile was launched against incoming Iraqi SCUD missiles in Iraq, Saudi Arabia, and Israel. In one instance, a SCUD targeted at an occupied barracks in Dharan, Saudi Arabia, killed 28 soldiers. The site was defended by a Patriot which detected the incoming missile and fired. However a mis-calibration in the timing system in the software caused a miss. Patriots were again used in the war with Iraq in 2003 and had good success against short-range tactical missiles.

Sidewinders are short-range air-to-air heat-seeking weapons used on fighter planes and gunship helicopters. More than 100,000 have been built, of which roughly 1% have been used in combat. They are used by many different countries. The missile system is relatively simple and has been in production since 1953. It is scheduled to remain in service through 2055, which if it occurs, will be beyond its centennial anniversary. The Sidewinder concept was originally developed at the Naval Weapons Station in China Lake CA. The Air Force adopted Sidewinders to replace its air-to-air missile in the 1960s. Sidewinders were used with excellent success rates by the Israeli Air Force against Syrian fighters in the 1982 Arab–Israeli war. The current version of the Sidewinder was introduced in 2003. The missile can be targeted simply by the gunner looking at the target. The guidance system remains a heat-seeking lock-on device that can adjust the missile trajectory to movement of the target.

Polaris, Posiedon, and Trident Missiles were (and are) deployed on the US submarine fleet. The Polaris was a solid-fuel rocket designed for underwater launch. The missile is ejected by high-pressure steam and the rocket ignites after leaving the submarine. The first version deployed in 1961 carried three nuclear warheads and had a range of 2500 nautical miles. The submarine-based missiles were thought of as "second-strike" deterrents, that is, as a response to a first-strike that might disable ground-based missiles. The Polaris was replaced by the Poseidon. Like Polaris, Poseidon was a two-stage solid fuel rocket. It had the capability of carrying up to 14 multiple independently targeted (MIRV) nuclear warheads. Continued improvements led to Trident-I and eventually to the current version Trident-II. The Trident-II is ejected by an explosive charge that flash-vaporizes water. There are three solid-fuel stages that ignite after the missile emerges from the water. There are currently 14 submarines equipped with Trident-II missiles, and they are scheduled to remain in operation until 2027.

1.3.3.2   SOLID-FUELED INTERCONTINENTAL BALLISTIC MISSILES

The United States developed two major families of solid-fueled ICBMs, the Minutemen and Peacekeepers. The first was Minuteman-I, on which work began in 1957. The prime contractor was Boeing and the program was under the Air Force Strategic Air Command. The missiles were located in hardened underground silos dispersed around the country to minimize vulnerability to enemy (Russian) attack. The first two versions, Minutemen I and II carried single nuclear warheads. By 1967, 1000 Minutemen missiles were deployed. These were eventually replaced by Minutemen-III, which carried three nuclear warheads. The accuracy of the guidance systems was improved during the course of development. The Minutemen-III missiles have three solid-fuel rocket stages. They also have a liquid-fuel propulsion system associated with the payload containing the guidance system and warheads. As of 2010, 450 Minutemen-III remained in service. They have been modified, including replacing the solid fuel to extend the usable lifetime. The Minutemen were to be replaced by Peacekeeper missiles that could carry 10 nuclear warheads. They were designed for super-hardened silos, in response to increased power of Russian missiles and warheads. Eventually, 50 Peacekeeper missiles were deployed, but as a result of the collapse of the Soviet Union and arms reduction negotiations with the Russian Federation, they were decommissioned as of 2005. The rockets are being used for the launch of intelligence and communication satellites as Minotaur IV.

1.3.3.3   LIQUID-FUELED ROCKETS AND INTERCONTINENTAL BALLISTIC MISSILES

The Redstone rockets were the first family of rockets to be developed by Wernher von Braun and his team. They were designed as surface-to-surface weapons and were developed under the auspices of the US Army. The engines were manufactured by the Rocketdyne Division of North American Aviation. They were direct descendants of the German V-2 rockets and used 75% ethanol and liquid O_2 as fuel. They were deployed by US forces in Germany beginning in 1958 and had a range of up to about 200 mi. They had an inertial guidance system and were targeted prior to launch and could be armed with either conventional or nuclear warheads. The Redstones were deactivated and replaced in 1964.

The name Jupiter was applied to modified Redstone rockets that were augmented with solid-fuel boosters. The Jupiter name was first applied to a medium range missile whose development was initiated jointly by the Army and Navy, but transferred to the Air Force in 1956. They were fueled by RP-1 (kerosene) and liquid O_2. They had originally been designed for use on ships and were not particularly well-suited to Air Force needs. They were deployed in Europe from 1961 to 1963, armed with nuclear warheads. A version called Jupiter-C was used to launch the first several capsules containing monkeys to test the ability of primates to survive launch and weightlessness (see Section 1.4.1).

The first Thor rockets were developed for the Air Force to serve as intermediate range (1500 miles) ballistic missiles. The main engine was fueled by kerosene (RP-1) and liquid O_2. Thor rockets were used to launch Explorer 6, which was the first satellite to successfully transmit pictures of the moon to earth. In the 1960s, Thor rockets, augmented by engines using hypergolic fuel, were used to launch navigational and intelligence satellites. The Delta rockets were modifications of the Thors and there were several versions. In the 3-stage version, stage 1 was fueled by RP-1 and liquid O_2. The second stage was hypergolic *unsym*-dimethylhydrazine and nitric acid. A third stage was a solid fuel rocket known as Altair. The Delta-II used nine solid fuel rocket boosters, six of which ignited at launch. They were used extensively for launch of GPS satellites. The Thor–Deltas were used from 1969 through 1978 and National Aeronautics and Space Administration (NASA) made 84 launches, of which 77 were classified as completely successful.

Titan rocket engines were used both as military ICBMs and in space exploration. They were used in the manned Gemeni flights and for interplanetary exploration of Mars, Jupiter, Saturn, Uranus, and Neptune. The Titan-I became operational in 1962 and was fueled by RP-1 (kerosene) and liquid O_2. The next version, Titan-II used hydrazine, *unsym*-dimethylhydrazine and N_2O_4, a hypergolic combination. These were in place in ICBM silos from the 1960s through the 1980s. The Titan-III was a modification with solid fuel booster rockets. The Titan-IV version was used mainly to launch surveillance satellites. The Titans were phased out in 2006. The toxicity and somewhat dangerous nature of the hydrazine fuel mixture was a problem throughout their time of use.

1.4   ROCKET ENGINES FOR SPACE EXPLORATION

1.4.1   THE SPACE RACE

The Naval Ordnance Bureau commissioned a report in 1945 which concluded that a multistage hydrogen/oxygen-fueled rocket would be able to achieve orbit.

Space travel began to catch the public's imagination. In 1952–1953, *Colliers* magazine published a number of articles illustrating travel to the moon. In 1955–1957, Disney produced programs on travel to the Moon and Mars. In 1954, Wernher von Braun urged the United States to launch a small satellite, recognizing the Russians might soon do so, as had been indicated by intelligence. At the time, space exploration was under the control of the Navy and based on a rocket called the Vanguard that was behind schedule. The Army had a rocket, Jupiter-C, that was capable of a satellite launch, but was not intended for that purpose.

On October 4, 1957 the Soviet Union successfully launched Sputnik and also a second satellite carrying a dog on November 3 of that year. A hurried US attempt with the Vanguard on December 6 failed. President Dwight D. Eisenhower turned to the von Braun team. On January 31, 1958, a Jupiter-C rocket launched the first American satellite, Explorer 1. Geiger counters on board discovered the Van Allen radiation belt. But by mid-year 1958, the Russians had launched another Sputnik with a 3000 lb payload. Press coverage and public fear generated by the Russian success was extreme, because it indicated the Russian could launch ICBMs carrying nuclear weapons. President Eisenhower and Congress launched efforts to "catch-up" with the Russians. Working with Senator Lyndon Johnson, plans were made to form a civilian agency to manage space exploration. In July, 1958, Congress authorized the formation of NASA, and it came into being in October, 1958. Late in 1959, responsibility for manned space exploration was transferred to NASA, while military projects were assigned to the Air Force. In August, Eisenhower authorized NASA to undertake manned flight based on the Atlas ICBM. The project was called Mercury. Limits on the thrust available defined the size and shape of the Mercury vehicle. Tests with the unoccupied capsule failed in the Fall of 1959 and the Summer of 1960, but the capsule survived.

John F. Kennedy became president on January 29, 1961, having narrowly defeated Richard M. Nixon in the November election. During the campaign Kennedy had invoked a "missile gap" and argued that the United States was trailing in the "space race." Two weeks after the election, a test of a Redstone missile with the Mercury capsule failed spectacularly. In February 1961, Kennedy named James E. Webb as the NASA administrator. Kennedy instructed Vice-President Lyndon B. Johnson to report on the status of the space program. Johnson's report in April 1961 recommended undertaking manned flight to the Moon. Webb and Robert M. McNamara, the Secretary of Defense, took on the responsibility for guiding the space program.

On April 12, 1961, the Soviet Union launched Yuri Gagarin into orbit for 89 min and recovered him, making him the first human to orbit the earth. A test of the Atlas–Mercury combination on April 25 failed, but on May 5, 1961, Alan Shepard was launched on a 15-min ride by a Redstone missile. President

Kennedy was present for the launch. On July 21, Virgil (Gus) Grissom flew the second suborbital flight, but had to be rescued when the re-entry vehicle sank in the Atlantic. Early in August, Russian G. S. Titov completed 17 orbits of the earth. These developments spurred Kennedy to set in motion an effort to land a man on the moon within the decade. He told Congress "no single space project...will be so difficult or expensive to accomplish." Politics made Houston, TX, USA the center of manned space activity. Dr. George E. Mueller became Associate Administrator for Manned Space Flight. Responsibility for development of the necessary rocket engines was assigned to the Marshall Space Flight Center, under the direction of Wernher von Braun. One of the key tenets of von Braun's management of both the Marshall operation and of contractors was intimate knowledge by the managers of the technical aspects of the systems for which they were responsible.

The Atlas was used in the first manned space flights of the Mercury program. The original Atlas rocket engines used hydrocarbon (RP-1, kerosene) and liquid O_2. There were three engines, two of which were jettisoned during launch. The original tanks were of thin stainless steel and maintained shape only if filled or pressurized. Beginning in 1960, an upper Agena stage using a hypergolic fuel was added. The fuel mixture was *unsym*-dimethylhydrazine and fuming nitric acid. This version was used for many orbital and space launches, such as the Mariner (Mars) program. On September 13, 1961, the Atlas–Mercury combination put an unoccupied capsule into orbit. On November 29, Enos the Chimp was put in orbit and recovered and finally on February 20, 1962, John Glenn became the first American to orbit the earth, launched by an Atlas rocket. A total of six manned Mercury missions were successfully completed, the last on May 15–16, 1963.

On November 22, 1963, while in Texas to dedicate the Aerospace Health Center, John F. Kennedy was assassinated. Lyndon Johnson became the President.

1.4.2 MISSIONS TO THE MOON

Various plans to get to the moon were considered. Wernher von Braun originally favored a huge rocket that could land on and return from the moon. Another possibility was a cluster of rockets. This approach had the advantage of using existing rocket engines that were well-tested. The Air Force was working on a large new rocket engine called F-1 in conjunction with Rocketdyne. The plan that eventually emerged used a smaller vehicle able to descend to and return from the moon from a spacecraft orbiting the moon. This was called *lunar orbit rendezvous*. The advantage of this plan was that it minimized the overall payload that was required but the risk was the absolute requirement for successful lift off from the moon by the lunar landing vehicle. The manned moon project was named Apollo and various rocket configurations were tested. The configuration eventually chosen for the main launch vehicle was called Saturn-V, and consisted of five F-1 engines in the first stage, four or five F-1 engines for the second stage fueled by liquid hydrocarbon (RP-1) and a J-2 liquid H_2-fueled engine for the third stage.

As development proceeded, there were many problems, especially combustion instability, and high-temperature corrosion. These problems were approached by systematic analysis and engineering of solutions. The extreme conditions in the thrust chambers required very strong metal alloys. Because of the hardness of the metal, special tools and conditions were needed for machining and protective glazing. The final components were subjected to extensive testing such as radiographic, ultrasonic, and infrared examination for defects and uniformity of glazing. As eventually developed, each F-1 engine used 24,611 gallons/min of liquid O_2 (LOX) and 15,471 gallons/min of RP-1, a ratio of 2.27:1. The temperature in the engines reaches 5970°C. The mechanical power for pumps and generators was provided by a second gas generating chamber fueled by RP-1 and LOX. The contract for assembly of the stage 1 of Saturn V, consisting of five F-1 engines, was awarded to Boeing and they were manufactured at the Michoud Assembly Facility in Louisiana, a converted ordnance facility used in WWII and the Korean war. The engines could be shipped by truck or barge to the Cape Canaveral launching site. Eventually C-133 cargo planes were also able to deliver individual F-1 engines. Figure 1.4 shows a diagram of the first stage of Saturn V.

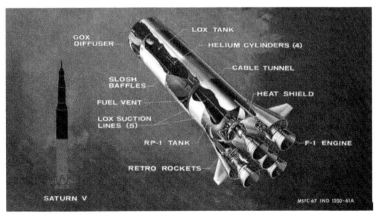

FIGURE 1.4 Cutaway diagram of first stage of the Saturn V incorporating five F-1 engines. (From history/nasa.gov.)

Many of the concepts for travel to the moon were tested by Gemini missions during 1965 and 1966. They involved a two-man space craft that had the capacity for manipulation is space. The Russian spacecraft of the time did not have that capacity. The Gemini missions also demonstrated the ability to perform docking of vehicles and the ability to perform "space walks." On January 27, 1967, the first Apollo capsule caught fire during flight rehearsal, killing the three astronauts on board, Roger Chafee, Virgil (Gus) Grissom, and Edward H. White. The cause was found to be poor design and faulty wiring. On April 23, 1967, the first Russian Soyuz I space-craft flew; the cosmonaut, Vladimir Komarov, was killed on reentry.

On December 21, 1968, Apollo VIII, lifted off with a 3-man crew, orbited the moon 10 times, and returned safely to earth. Two more orbiting mission occurred on March 3 and May 18, 1969. On July 16, 1969, Apollo XI lifted off and the lunar vehicle landed on the moon on July 20, 1969. Five more missions occurred, one of which, Apollo XIII, encountered great difficulty. One of the J-1 engines shut down prematurely. Despite this, the mission to the moon continued. While orbiting the moon, an oxygen tank in the service module exploded. Only by careful conservation of oxygen and manual control of reentry, was the vehicle able to return safely to Earth. Congressional support of NASA was high at the beginning of the Apollo program and the budget grew 9-fold between 1960 and 1965. The first cut, of 1.5%, occurred in 1966. By 1968, the cuts reached 20%. The public, distracted by Viet Nam, was also losing interest. With opposition to the Viet Nam war generating chaos, Lyndon Johnson decided not to seek reelection, and in November, 1968, Richard Nixon was elected to succeed him. Unlike Johnson, Nixon had little personal interest in space and resented its association with John F. Kennedy. Nixon decided to end the manned lunar program. The last Apollo mission, Apollo XVII occurred on December 7–19, 1972.

1.4.3   THE SPACE STATION AND SPACE SHUTTLES

The Nixon administration replaced much of the upper echelons at NASA. James Fletcher served as NASA administrator from 1971 to 1977. He was committed to the concept of manned flight and the space shuttle and space station. However, in order to win financial support, he argued that the space program would be economically justifiable. The number of flights for commercial and military purposes was projected to be much higher than was feasible. The space shuttle was promoted as a joint Air Force–NASA project with both military and civilian missions. The Office of Management and Budget (OMB) imposed cost/benefit analyses that forced cost-cutting wherever possible.

The first US space station, Skylab, was launched into orbit on May 14, 1973, by a two-stage Saturn-V rocket. The orbit was about 435 km above the earth but the craft was damaged during the launch. On May 25, 1973, a repair mission was launched with a Saturn IB rocket. An Apollo module docked with Skylab for 28 days and effected repairs. Two other extended missions of 59 and 84 days were completed in 1973 and 1974. Skylab eventually descended from orbit in July 1979, with fragments of debris falling in Australia.

Throughout, the Nixon administration budgetary pressures increased. After Nixon's resignation and Gerald Ford's brief presidency, Jimmy Carter became president in 1976. Carter had no particular interest in space, and his Vice-President, Walter Mondale, had long been skeptical of NASA. Budget pressures were exacerbated by the rampant inflation generated by the 1974 oil crisis (see Topic 2.4). Eventually, the projected space shuttle fleet was reduced from seven to four. Furthermore, there were major technical problems. On March 24, 1979, the space shuttle *Columbia* was flown from California to Florida attached to a Boeing 727. In the process, many heat shield tiles fell off.

Cooperation with the Soviet Union in space had begun with an agreement between President Nixon and Chairman Leonid Brezhnev in 1972. Both nations, however, maintained independent efforts, including surveillance, intelligence, and military applications. The Russian Salyut was an orbiting single module space station, of which nine were launched between 1971 and 1982. The Salyut stations supported several long-term missions and the first in-space crew change-overs. They were launched as empty modules that then docked with Soyuz space craft. Several of the Salyut launches encountered problems, including loss of the crew from Salyut X when their Soyuz vehicle leaked on re-entry. The early space stations had relatively short durations but Salyut 7 remained in orbit from 1982 to 1991. The Russian Salyut program was followed by construction of the multimodule Mir Station between 1986 and 1996. It eventually consisted of seven modules and was occupied for about 12.5 years, with the longest individual stay being 437 days. It was taken from orbit in March, 2001, with fragments falling in the South Pacific. The Salyut vehicles and Mir modules were launched by huge Proton rockets. They had originally been conceived as super ICBMs and had the capacity to lift 5–6 t of payload. They have continued in use in the launching of components of the International Space Station and are used to launch commercial satellites. The fuel is a hypergolic mixture of *unsym-*dimethylhydrazine and dinitrogen tetroxide.

The Space Shuttle program was part of the *Space Transportation System* within NASA. The first test flights occurred in 1981 and the shuttle was phased out in 2011, 30 years later. On April 12, 1981, *Columbia* lifted off being the first real space ship. A few tiles were noted to be missing, but it landed safely in California on April 14, 1981. The assembly at launch for the shuttle consisted of the main fuel tank, two solid rocket boosters and the shuttle itself, called the *Orbiter Vehicle.* The main engines were fueled by liquid H_2 and liquid O_2. The solid fuel boosters provide slightly more than 70% of the initial thrust for lift-off. The solid fuel boosters contained 70% ammonium perchlorate, 16% aluminum, 12% polybutadiene–acrylic acid, and 2% epoxy binder. The casings are jettisoned, but recovered for reuse. The orbiter maneuvered using engines fueled by monomethylhydrazine and N_2O_4. Eventually, six shuttles were built. The first, the *Enterprise* was used only for testing. The *Columbia, Challenger, Discovery,* and *Atlantis* were then built. The *Endeavor* was built to replace the *Challenger*, after it was lost in 1986. There were a total of 134 flights, of which two were lost. The shuttles were used to service the international space station, launch satellites and interplanetary probes and for various scientific experiments. They were able to carry a payload of 15 t and were used for both government and commercial projects.

The space shuttle and space station had been justified on the basis of economic development and an exaggerated flight schedules. The commitment of the Reagan administration was not primarily to manned space exploration, but to military applications. In 1983, OMB under David Stockman, rejected NASA's request for a fifth shuttle and instead increased funding for spare parts for the existing craft. The result was the cannibalization of the existing shuttles for each subsequent flight. The schedule of shuttle flights was heavy and the work crews were under great pressure. There was a persistent problem with O-ring seals on the solid booster rockets. Damage had been noted on 10 of the 23 flights through the end of 1985.

Challenger flight 51-L was originally scheduled to launch, on Sunday, January 27, 1986, which happened to coincide with Super-Bowl Sunday, but was postponed because of unfavorable weather forecasts. The weather got colder and the seas were rough. Nevertheless, on January 28, the countdown began. Ice formed on the shuttle. There was concern about the O-ring seals in cold weather, and Thiokol engineers in Utah recommended that the flight not be launched if the temperature was below 53°F. The temperature was below 30°F. During the launch, O-rings failed, flames from the solid fuel booster rockets damaged the shuttle and debris pierced the main H_2 and O_2 tanks. The *Challenger* exploded about 75 s after launch. The cabin separated from the main vehicle and plunged into the sea at >200 mph killing all seven crew members, including Christa McAuliffe, an elementary school teacher selected for the flight.

After flights were resumed, the schedule was reduced to about 10–12 per year and a new orbiter, the *Endeavor* was built. Intelligence and commercial projects were largely shifted to expendable rockets. It was in this period that one of the most embarrassing, but ultimately most significant, undertakings was launched. The Hubble Space telescope was first orbited in 1990, but it was only after it was in orbit that it was found to have been incorrectly shaped. In 1993, the telescope was repaired and began furnishing incomparable information about deep space.

In July 1992, Presidents George H. W. Bush and Boris Yeltsin signed a new agreement to cooperate in space and this was followed in 1993 by an agreement to build the International Space Station. The first components were launched with a Russian Proton rocket in 1998. The first crew entered the station in November 2000 and it has been continuously occupied since then. The station travels about 17,000 mph and completes 15.7 orbits per day. Currently, it is operated by an international consortium consisting of NASA and the Russian, European, Japanese, and Canadian space agencies. The current projected lifetime of the Space Station is until 2020.

Daniel Goldin was appointed NASA administrator in April, 1992 and was continued in this post by the Clinton administration when it took over in January 1993. Goldin's goals of "faster, better, cheaper" involved privatization, decentralization, and down-sizing with more reliance on contractors. NASA's work force dropped from 30,000 in 1993 to under 18,000 in 2002. The contracting structure had also been changed from multiple contractors (more than 80) to a single prime contractor United Space Alliance (USA), a joint venture of Boeing and Lockheed Martin. The contract gave USA responsibility for all aspects of the shuttle program including safety. As a result of the downsizing of NASA and the outsourcing of much of the effort, NASA had reduced internal expertise, which made contractor communication vitally important. As the century drew to a close, attention was focused on the International Space Station, which would require an expanded schedule of shuttle flights. When George W. Bush became president in January 2001, Sean O'Keefe replaced Goldin as administrator. O'Keefe set a goal of Feb. 2004 for the completion of the US core component of the Space Station.

A second Space Shuttle disaster occurred in February, 2003, when Columbia broke up on reentry. The safety review prior to Space Shuttle launches required the identification of any problems that could interfere with a successful mission. Among the problems that had been noted was foam insulation shedding and heat shield damage, which had been documented in 65 of the 79 launches for which usable photographs were available. After each instance the tile was repaired, but the underlying cause of the problem was not corrected. Mission ST-112, immediately preceding the *Columbia* flight had experienced particularly serious tile damage. On January 16, the day after the launch, of *Colombia*, the Marshall Center recognized that a large piece of foam had hit the space shuttle during launch. The Columbia began its descent at 8:15 on the morning of February 1, 2003, and began its entry into the earth's atmosphere about 30 minutes later over the Pacific Ocean. At 8:55, the first electronic indication of malfunction was noted when the left wing hydraulic sensor dropped off-scale. Ground observers in California and Nevada saw flashes of light over the next few minutes. Other sensors recorded unusual readings and by 9:00 all communication was lost. The space ship completely disintegrated over Texas and debris was eventually collected and analyzed. The investigation concluded that the damage to heat shield tiles had been sufficient to allow the wing to burn through during reentry.

The final three space shuttle flight took place in 2011. Flights to the International Space Station now depend on Russian vehicles. After about 2015, Ariane-5 engines will be the main rockets used for transport to the International Space Station. They will be launched from Arianespace's facility in French Guiana. The Ariane is fueled by a liquid H_2/O_2 engine and solid fuel boosters that use NH_4ClO_4 (68%), Al (18%), and polybutadiene (14%). Upgraded Soyuz engines, using kerosene and liquid O_2 fuel, will also be launched from this site in the future.

1.5   OTHER USES OF EXPLOSIVES AND PROPELLANTS

1.5.1   PYROTECHNICS AND FIREWORKS

Pyrotechnics for flares and fireworks must emit light. This is achieved by including metal that can be thermally excited. The most basic compositions contain aluminum, magnesium, and titanium metal, which gives off white light. The fuel is usually some form of carbon, such as charcoal. The oxidants are NH_4NO_3, $NaNO_3$, chlorates, or perchlorates. Colors are obtained by inclusion of metal chlorides. Copper gives blue, strontium gives red, and barium is used for green. The starburst effect in fireworks is achieved with individual particles that are ignited by delay fuses. Pyrotechnic flares are also used as defensive devices in air craft subject to ground-to-air missile fire. The flares are released and the heat and light they generate can divert the targeting apparatus of the missiles, which is usually based on IR sensors.

1.5.2   PROPELLANTS FOR AUTOMOBILE AIR BAGS

Automobile air bags and related safety restraints are constructed by using a material that rapidly produces a large volume of gas. Desirable properties are low toxicity, insensitivity to shock, friction or heat, and safe disposal methods. The propellant used is NaN_3, mixed with a nitrate salt and either CuO or Fe_2O_3. Besides N_2 the decomposition produces Na_2O which reacts with water to give NaOH.

$$2\ NaN_3 + CuO \longrightarrow Na_2O + Cu + 3\ N_2$$

$$10\ NaN_3 + 2\ KNO_3 \longrightarrow 5\ Na_2O + K_2O + 16\ N_2$$

Sodium azide is quite toxic, with a lethal dose of about 10 mg/kg. It causes severe vasodilation and hypotension. By the mid 1990s, production reached about 5 million kg/year, with 50–100 g in each driver air bag and around 250 g in side air bags. Relatively, little is known about the disposal of NaN_3 in the process of automobile salvage.[3] In view of these disadvantages of NaN_3, other compounds are being explored, for example guanidinium tetrazolate with $Cu_2(OH)_2(NO_3)_2$ as the oxidant.

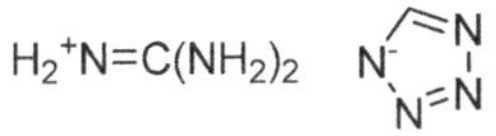

$$H_2{}^+N=C(NH_2)_2$$

guanidinium tetrazolate

Pilot ejection seats in military aircraft are another example of controlled use of propellants.

1.5.3 EXPLOSIVES IN METAL FABRICATION

Explosives for binding metals together were developed at DuPont in the 1960s. The first large application was making bimetallic coins by binding copper–nickel surfaces to a copper core. Explosives are also used in certain metal fabrication processes to weld together sheets and tubes. The method is used to make high-performance tubing used in petroleum refining and reforming. The explosion effects bonding in two ways. The extreme pressure removes any loose oxides that coat the metal surfaces and the extreme pressure creates a thin but strong bond between the two metals. The explosive force also generates a pattern of small waves on the surfaces that enhances the bonding.

1.6 EXPLOSIVES AND TERRORISM

The final quarter of the 20th century witnessed expansion of another use of explosives, that is, for terrorism. Bombs and grenades had been used throughout the 20th century in politically motivated attacks against targeted individuals and groups, but the final quarter of the century saw numerous attacks on random locations and groups. This motivated the development of means of deterring access and detecting explosive materials used by terrorists. Most of the materials used for terrorism are the same as used for civilian and military use, but limited technical resources usually restrict terrorists to relatively crude devices.

1.6.1   EXPLOSIVES USED IN TERRORISM

A number of explosives have been used by terrorists for making bombs. The combination of ammonium nitrate and hydrocarbon fuel is potentially explosive. It was the explosive used by Timothy McVeigh in the bombing of the Murrah Federal Building in Oklahoma City in 1995. Another of the explosives used by terrorists is Semtex, which contains RDX and PETN in a styrene–butadiene copolymer. It was manufactured in Czechoslovakia during the Viet Nam war. Excess material at the end of the war was sold to Libya, and from there it was distributed to various terrorist groups. Semtex was used by Libyan operatives to blow up PanAm Flight 103 over Lockerbie Scotland in 1988 and was also used by terrorist of the Provisional Irish Republican Army.[4] More recently, PETN has been used as a liquid explosive in bombs intended to blow up airliners. In the cases of the "shoe bomber" (2001) and the "underwear bomber" (2009), the terrorists failed to detonate the explosives. Potassium chlorate, which is commercially available or can be made easily from bleach, has been used extensively by terrorists. Another type of explosive material that can be made fairly easily is a mixture of acetone peroxides, mainly the trimeric material, triacetone peroxide.[5] The acetone peroxides are extremely sensitive and powerful explosives and can be used to detonate nitrate explosives.

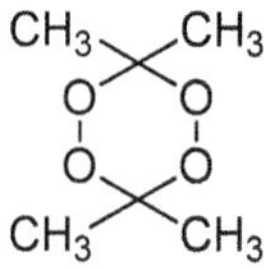

diacetone peroxide 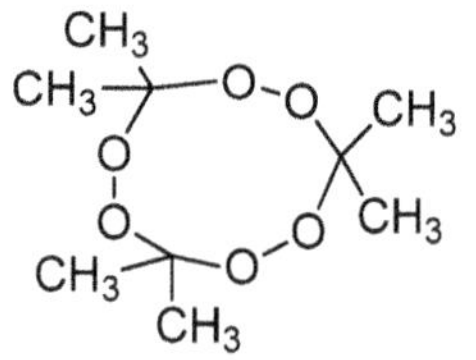   triacetone peroxide

1.6.2   DETECTION OF EXPLOSIVES

The spread of terrorist use of explosives has necessitated the development of detection systems.[6] For basic screening purposes sensitivity, speed, reliability, and cost are major considerations. *Ion mobility spectrometry* is used in the common airport luggage screening systems. A swipe of the luggage can absorb traces of explosives. The instruments are calibrated to recognize ions generated by common explosive and emit a signal if a characteristic ion is detected. Chemical detection can be based either on identifying vapor from the explosive or residual particulate material. The former is limited by the low vapor pressure of most explosives, the use of plastic-bonded explosives (which further reduces the vapor emission), and the enclosure of the explosives in sealed containers. Fluorescence or fluorescence-quenching are sensitive methods for detecting some explosives, such as the polynitroaromatic compounds. Nitrate esters and nitramines can be detected based on chemical treatments that generate nitrite ion, which can be detected colorimetrically.

Trained dogs are the most reliable method of searching for explosives at the present time.[7] This requires training of dogs to recognize odors associated with various explosive types. Currently, the DOD requires that dogs detect explosive samples with 95% accuracy and <5% false positives. Police dog certification generally requires >90% accuracy. Typically it is not the explosive itself, but rather more volatile materials that are recognized. These may be quite complex mixtures, including products of decomposition of the explosive or residues of preparation. Dogs are also trained not to respond to common interfering odors, such as those from personal care products, snacks, vitamins, or over-the-counter medicines. Training typically requires 6–8 weeks and is done using food and praise rewards. Most dogs are handled by a single trainer. Courts usually require expert testimony corroborating the reliability of the results under the specific conditions.

Various other detection methods are under exploration. One is biosensors, usually based on antibodies to specific explosives. An antibody response can be linked to a subsequent enzymatic assay that produces color, fluorescence, or luminescence. Antibody-based detection is usually highly specific for a particular explosive. Another approach to biosensors is to use nitro reductase enzymes. Because nitro groups are so prevalent in explosives, this method can detect a broader range of materials. Other methods of detection are based on selective adsorption of explosives to surfaces where they then can be detected by various spectroscopic methods. Explosive detection also comes into play in detecting environmental contamination from manufacture and use, and in the detection of land mines.

KEYWORDS

- **chemical explosives**
- **nitrate esters**
- **nitroaromatics**
- **nitramines**
- **solid-fuel engines**
- **liquid-fuel engines**
- **military missiles**
- **engines for space exploration**
- **terrorism**

BIBLIOGRAPHY

Cooper, P. W.; Kurowski, S. R. *Introduction to Technology of Explosives*; Wiley-VCH: New York, 1996; Dolan, J. E.; Langer, S. L. Eds.; *Explosives in the Service of Man*; Special Publication 203, Royal Society of Chemistry: Cambridge UK, 1997; Zukas, J. A.; Walters, W. P. *Explosive Effects and Applications.* Springer: New York, 1998; Brown, G. I. *The Big Bang.* Sutton Publishing: UK, 1998; Meyer, R.; Koehler, J.; Homburg, A. *Explosives*, 5th ed. Wiley-VCH: Weinheim, 2002; Akhavan, J. *The Chemistry of Explosive*, 2nd ed. Royal Chemical Society: Cambridge, UK, 2004; Buchanan, B. J., Ed. *Gunpowder, Explosive and the States*, Ashgate Publishing: Aldershot, UK, 2006; Hopler, R. B. *Forensic Investigation of Explosives.* 2nd ed.; Beveridge, A. Ed.; CRC Press: Boca Raton, FL, 2012; pp 1–18.

BIBLIOGRAPHY BY SECTIONS

Section 1.3: Forbes, F. S.; Van Splinter, P. A. *Encyclopedia of Physical Science and Technology*, 3rd ed. Academic Press, 2000; pp 741–777; Jain, J. R. *J. Sci. Ind. Res.* **2002**, *61*, 899–911; Edwards, T. *J. Propul. Power* **2003**, *19*, 1089–1107; Caveny, L. H.; Geisler, R. L.; Ellis, R. A.; Moore, T. L. *J. Propul. Power* **2003**, *19*, 1038–1065; Davenas, A. *J. Propul. Power* **2003**, *19*, 1108–1128; Altman, D.; Holzman, A. *Prog. Astronaut. Aeronaut.* **2007**, *218*, 1–36; Shusser, M. *Wiley Encyclopedia of Composites*, Vol. 1; Nicolais, L., Borzeccheillo, A., Eds.; Wiley: New York, 2012; pp 527–544.

Section 1.4.2: Young, A. *The Saturn V F-1 Engine, Powering Apollo into History.* Springer-Praxis: Chichester, UK, 2009.

Section 1.4.3: Trento, J. J. *Prescription for Disaster.* Crown Publishing: New York, 1987; Tompkin, P. K. *Apollo, Challenger, Columbia: The Decline of the Space Program.* Roxbury Publishing: Los Angeles, CA, 2005; Mahler, J. G. *Organizational Learning at NASA: The Challenger and Columbia Accidents*; Georgetown University Press: Washington, 2009.

Section 1.5.1: Lancaster, R. *Chem. Rev. (Deddington, UK)* **1992**, *2*, 2–6.

Section 1.5.3: Antalffy, L. P.; Young, G. A. *Pressure Vessels and Piping: Manufacturing and Performance*; Raj, B., Choudhary, B. K., Albert, S. K., Eds.; Narosa Publishing House: New Delhi, 2009; pp 1–10.

Section 1.6: National Research Council. *Making the Nation Safer*; National Academy Press: Washington, DC, 2002; pp 107–134; Moore, D. S. *Rev. Sci. Instr.* **2004**, *75*, 2499–2512; Singh, S. *J. Hazard. Mater.* **2007**, *144*, 15–28.

REFERENCES

1. Jeacocke, J. *Explosives in the Service of Man*, Vol. 203; Dolan, J. E.; Langer, S. S., Eds.; The Royal Society of Chemistry Special Publication: Cambridge, UK, 1997; p 23.
2. From Kipling, R. *The Secret of Machines*, 1911.
3. Chang, S.; Lamm, S. H. *Int. J. Toxicol.* **2003**, *22*, 3175–3186; Betterton, E. A. *Crit. Rev. Environ. Sci. Technol.* **2003**, *33*, 423–458.
4. Feraday, A. W. *Advances in Analysis and Detection of Explosives, Sept. 1992, Jerusalem Israel*; Kluwer Academic: Boston, MA, 1993; pp 67–72.

5. Espinosa-Fuentes, E. A.; Pena-Quevado, A. J.; Pacheco-Loondono, L. C.; Infante-Castillo, R.; Hernandez-Rivera, S. P. *Explosive Materials*; Jansen, T. J., Ed.; 2011; pp 259–212.
6. Caygill, J. S.; Davis, F.; Higson, S. P. J. *Talanta* **2012,** *88,* 14–29.
7. Furton, K. G.; Myers, L. J. *Talanta* **2001,** *54,* 487–500.

CHAPTER 2

HYDROCARBONS AS FUELS AND PETROCHEMICALS: SHAPING THE PAST, DOMINATING THE PRESENT, COMPLICATING THE FUTURE

ABSTRACT

Drilling for and refining of petroleum provided hydrocarbon fuels that became the dominant energy source for transportation and industry in the 20th century. The introduction and improvement of internal combustion engines required new refining procedures based on chemical change, including thermal and catalytic cracking, platforming, and alkylation. Diesel and jet engines also needed specific fuel compositions. Petroleum economics has been characterized by intermittent periods of perceived shortages and excess supply, further complicated by political factors. This, along with concern about the climatic effects of fossil fuel use, has prompted investigation of alternative sources of liquid fuels, including coal, natural gas, and biomaterials. The refining processes also provide alkenes, such as ethylene, propylene, and the butenes, that are used to make petrochemicals. After WWII, evidence began to accumulate that fossil fuel use was increasing the atmospheric carbon dioxide concentration and almost certainly contributing to global warming. This added yet another complication to the economic and political factors of petroleum production and use.

No type of chemical substances had greater impact on the economics and politics of the 20th century than hydrocarbons found as petroleum deposits and used as fuel and a source of petrochemicals. The combination of hydrocarbon fuels and the internal combustion engine, and later, jet and turbine engines, revolutionized transportation and commerce. The many polymers made from petrochemicals added an entirely new class of materials for making useful products. Access to petroleum resources was a fundamental issue in global politics. As the century came to a close, another facet of hydrocarbon use, carbon dioxide generation and accumulation, and its potential for causing climate change became a topic of urgency.

Hydrocarbon deposits were known and used in some parts of the ancient world. Asphalt and pitch were used as sealants in construction and for boats in the Middle-East by Babylonians and Phoenicians. The Byzantine Empire's "Greek Fire" was a heated mixture of petroleum wax and rosin, somewhat like modern Napalm[a] in its properties. In parts of China, natural gas was collected via bamboo pipes and used for illumination. Distillation of oil for lamps from Caspian Sea petroleum was known as early as the ninth century. This technology was unknown in Europe until reported by explorers such as Marco Polo. Distillation of petroleum in Romania began as early as the 13th century. In the 16th century, wells were dug by hand in the area of Baku in Azerbaijan. An oil field in Alsace (France) was worked in the 1700s. The hydrocarbon source in most cases was seepage from rocks, and hence the name petroleum, that is, "rock oil."

The exploration of the New World led to discovery of petroleum deposits in the Americas. Early explorers found pitch deposits in Cuba and Trinidad and used them for sealing boats. Spanish explorers found the natives in Mexico and South America using petroleum for medicinal purposes. French and English explorers encountered oil seepages in the Appalachians in the 1700s and reported that the Native Americans used them in the treatment of various ailments and injuries. Petroleum was also frequently observed in conjunction with salt wells in the region, where it was considered an undesired byproduct. One company, the American Medicinal Oil Company of Burkesville, KY, however, bottled and sold medicinal oil from a salt well. It was Samuel M. Kier who noted that the Pennsylvania rock oil seepage and the oil from salt wells appeared to have the same properties. He was the first person to purify and sell the rock oil for medicinal purposes. Kier also began to refine the material to produce fuel for lamps.

There were other sources of liquid hydrocarbon fuels in the middle of the 19th century. In Scotland, James Young had begun distillation of shale to produce "coal oil" and it was being used in place of whale oil or tallow as a fuel in lamps. In Canada, Abraham Gesner[b] distilled hydrocarbons from coal and called the material *kerosene*. In the 1850s, he formed the North American Kerosene and Gas Light Company in New York, which was later renamed the New York Kerosene Company. Technical advances improved the quality of the material. These included distillation and treatment with sulfuric acid and caustic soda. By the late 1850s, the company had the capacity to produce 650,000 gallons of

---

[a]Napalm is a gelled form of hydrocarbon fuel developed as a military weapon during WWII. In its current form, it is made from gasoline, benzene, and polystyrene and is used in flame-throwers. It has the property of adhering to and igniting the target.

[b]Abraham Gesner was a native of Nova Scotia who was primarily interested in geology. He received an M.D. in London and supported his studies in geology with the income from his practice. He published various geological studies of the Maritime Provinces of Canada. In 1846, while studying a shale mineral, he found that heating produced various hydrocarbons, including the fraction he called "kerosene."

kerosene annually. It also produced lubricants from the higher boiling fractions. The methodology was rapidly expanded and other companies were formed in locations such as Pittsburgh and Cincinnati. Coal oil production peaked at around 250,000 barrels/year (b/y) in 1860, with at least 55–60 plants in operation. By 1861, many coal–oil refiners began to shift to petroleum as their source of kerosene. The basic distillation and purification processes that had been developed were readily adapted to petroleum.

The story of the petroleum industry in the United States begins in western Pennsylvania and nearby parts of Ohio, West Virginia, and Kentucky. In this region, certain streams had a coating of oil. Its only use until the 1850s was as patent medicines sold under names such as "Seneca Oil" or "American Medicinal Oil." A Titusville, PA physician turned lumberman, Frank B. Brewer, leased the rights to one of the local oil springs. Brewer brought a sample of the oil to his alma mater, Dartmouth, where he showed it to his uncle, a medical professor, and to O. P. Hubbard, the chemistry professor. A few weeks later another Dartmouth alumnus, George H. Bissell, a lawyer based in New York, saw and was intrigued by the material. Bissell was well acquainted with the coal–oil business. Bissell interested a group of investors from New Haven CT in the possibility of developing the Pennsylvania location. They sent a sample to Prof. Benjamin Silliman at Yale in 1855. Silliman reported that the rock oil had potential for use in illumination, lubrication and as a sealing wax. On the basis of this favorable information, the Pennsylvania Rock Oil Company was formed with funds from the New Haven investors, led by banker James M. Townsend. Townsend hired Edwin L. Drake to operate the site. On inspecting the site, Drake evidently conceived the idea of drilling rather than collecting the seepage. Drake had no previous experience at drilling, although drilling for salt was well established. Drake assembled a steam boiler to power the drilling and constructed the wooden derrick. The first salt driller he hired never arrived on the scene, but the next spring he hired an experienced salt driller W. A. Smith. Smith supervised the drilling and insertion of iron pipe. On Sunday, 28 August, 1859, Smith checked the well and found that oil had risen into the pipe, which at the time was at a depth of about 70 feet. The initial flow is thought to have been about 10 gallons a day. In 1861, the first deeper wells that provided flowing oil came into operation, some with production in the range of several thousand barrels/day (b/d). Such wells usually had fairly short production periods, but could be extended by pumping. The total production swelled to 2 million barrels in 1861 and prices dropped from $2/b to $0.10/b. Nevertheless, frenzied development continued through the mid-1860s.

There were contemporaneous developments in Europe. An early refinery was in operation in the Alsatian field by 1857. Refining stills were also used in Russia (Baku) and Romania by the early 1860s. The fraction boiling up to about 120°C was called *naphtha* and was not very useful at the time. This is now the fraction used for gasoline. The major product until the early 20th century was kerosene for lamps and stoves. The kerosene boiling range was about 150–200°C. The next higher boiling fraction, called "gas oil," corresponds to the current diesel and fuel oil fractions. The name "gas oil" comes from the use of this fraction for production of "town gas" for lighting lamps in certain cities. The higher boiling fractions were used as lubricating oils and the solid paraffin wax for making candles and as a sealant.

During the 20th century hydrocarbon fuels became the primary source of energy for transportation, manufacturing, agriculture, electricity generation and indoor heating/cooling. The advantages of petroleum hydrocarbons include comparatively ease of extraction and transportation, which made them relatively cheap, and compatibility with the crucial energy-consuming systems. As a result, an extensive infrastructure for extracting, refining and distributing hydrocarbon fuels developed, including liquid and gas pipelines, rail and motor transport, and ocean-going tankers. This infrastructure has continued to expand in most of its aspects throughout the 20th century. The expanded use of petroleum was closely related to the invention of the internal combustion engine, automobiles and airplanes, and the resulting expansion of manufacturing and transportation. Table 2.1 shows vehicle registrations in the United States and worldwide during the 20th century.

TABLE 2.1   Non-military Motor Vehicles in Service (Thousand).

Year	United States[a]	World[b]
1900	8	
1910	77	
1920	8132	
1930	23,035	
1940	27,466	
1950	40,399	
1960	61,671	126,955
1970	89,244	246,368
1980	121,601	411,113
1990	133,701	582,982
2000	133,621	751,830

[a]US Department of Transportation.

[b]Ward's motor vehicle facts and figures.

The combustion of wood, coal and hydrocarbons generates carbon dioxide (CO_2). The expansion of petroleum use led to a very significant increase in atmospheric levels of the combustion product CO_2, which has been measured very accurately since 1960. This increase is shown in Figure 2.1.

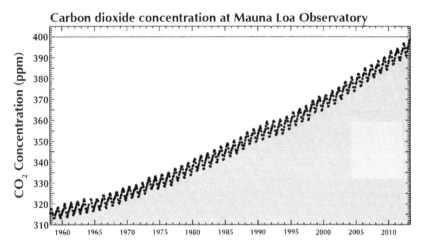

FIGURE 2.1   Monthly average carbon dioxide concentrations at Mauna Loa Observatory.

In the last quarter of the 20th century, the environmental, economic, and political consequences of hydrocarbon use and carbon dioxide emissions became evident. Political and economic systems are struggling to deal with the consequences of the combustion of enormous amounts of hydrocarbons.

2.1   PETROLEUM REFINING AND PROCESSING

Hydrocarbons contain only carbon–hydrogen and carbon–carbon bonds. Hydrocarbon molecules have characteristic shapes and sizes that are determined by their *molecular structure*. Hydrocarbons can be *straight-chain*, consisting of an unbranched string of carbons. They can also have *branched chains* or they can be *cyclic*. Hydrocarbons are also classified as *saturated*, *unsaturated* or *aromatic*. Saturated hydrocarbons contain only single bonds. Unsaturated hydrocarbons have one or more carbon–carbon double or triple bonds and can react in the presence of a catalyst to add hydrogen, thereby becoming saturated hydrocarbons. *Aromatic* hydrocarbons contain at least one six-membered ring with three double bonds. Aromatic hydrocarbons are also capable of adding hydrogen, but they are quite unreactive and require vigorous reaction conditions. Some examples of these structures are shown in Scheme 2.1.

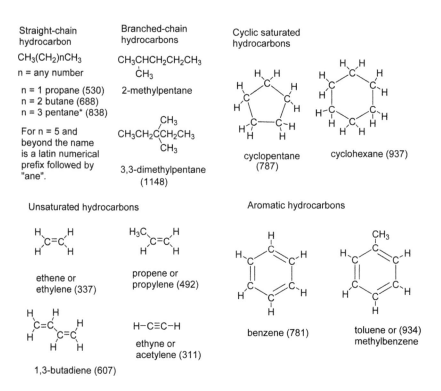

SCHEME 2.1   Structures and heats of combustion (kcal/mol) of some hydrocarbons.

Petroleum hydrocarbons contain chemical potential energy that is ultimately derived from the sun. They were formed by transformation of once-living deposits over eons of time and are called *fossil fuels*, connoting both their ancient origin and nonrenewability. Both liquid (petroleum) and gaseous hydrocarbons (natural gas) were formed and trapped in certain kinds of geological formations. The potential energy is released from hydrocarbons by combustion. Complete combustion yields only carbon dioxide and water. Because of the differing ratios of carbon and hydrogen in the hydrocarbons, slightly different ratios of CO_2 and water are obtained as illustrated by the equation below. The amount of heat evolved is also a function of the structure and is a physical constant for each compound and is called the *heat of combustion*. The heat of combustion indicates the amount of thermal energy produced by the oxidation. The values are given in units of kcal/mol for each of the structures in Scheme 2.1.

$$C_nH_m + (n + m/4)\, O_2 \longrightarrow n\, CO_2 + m/2\, H_2O + \text{heat of combustion}$$

The amount of heat and carbon dioxide produced depends on the composition of the fuel. The equations that follow show the results for graphite (pure carbon—a rough approximation for *coal*), methane (*natural gas*) and octane (approximating *gasoline*). We can see that the energy produced per ton of fuel is natural gas > gasoline > coal, while the amount of CO_2 produced is in the opposite order. Combustion of coal produces almost four tons of CO_2 for each ton of coal burned, but even "clean-burning" methane produces 2.75 times its weight in CO_2 by consuming 4 times its weight in oxygen. For liquid hydrocarbons, the ratio is about 3:1.

$$CH_4 + 2\ O_2 \longrightarrow CO_2 + 2\ H_2O$$
$$MW\ 16 \quad MW\ 32 \qquad MW\ 44 \quad MW\ 18$$

Energy and amount of carbon dioxide produced per ton from fossil carbon fuels.

	Btu/ton (thousands)	Tons of CO_2/ Ton of Fuel
$C + O_2 \longrightarrow CO_2$	28.2*	3.7
$CH_4 + 2\ O_2 \longrightarrow CO_2 + 2H_2O$	47.8	2.75
$C_8H_{20} + 13\ O_2 \longrightarrow 8\ CO_2 + 10\ H_2O$	41.2	3.1

* Depending on the type of caol, the range is 25-30 thousand Btu per ton.

2.1.1   EARLY DISTILLATION AND PURIFICATION METHODS

Crude petroleum is a complex mixture of hydrocarbons and other compounds and must be refined to obtain useful fuels. The hydrocarbons containing up to four carbon atoms are gases. The smallest hydrocarbon, CH_4, methane is the main component of natural gas. The C_3 and C_4 hydrocarbons, propane and butane, are liquids under moderate pressure and are called *liquefied petroleum gas*. Hydrocarbons with five or more carbon atoms are liquids, while still larger molecules are oils, greases, and waxes. The most basic step in petroleum refining is distillation. Fractions are separated on the basis of boiling point, which depends on molecular size and shape. Distillation separates the petroleum mixture into different fractions, but does not modify the structure of the hydrocarbons. A residue (which is used for making asphalt) remains after all the volatile materials have distilled.

Several technological advances in the refining process occurred late in the 19th century. One was the use of a series of stills held at increasing temperature so that various fractions could be collected. These had been used earlier in Europe but were introduced in the United States at the Atlantic Refinery Company in Pennsylvania and also by Royal Dutch-Shell in California. They were soon replaced by vertical condensing columns with temperature control that permitted collection of successively higher boiling fractions. These stills contained "dephlegmators" which were water-cooled plates that assisted condensation of the distillate and reduced the amount of liquid droplets entrained in the vapor. These vertical columns allowed for continuous distillation as opposed to batch distillation that had been used previously. Figure 2.2 shows a conceptual representation of the distillation process.

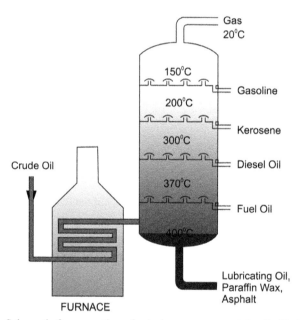

FIGURE 2.2 Schematic for separation of petroleum components by distillation.

Each of the crude distillation fractions could be further purified. The gasoline and kerosene fractions were treated with H_2SO_4, which resulted in separation of reactive, for example, unsaturated, components. The upper layer was then washed with caustic soda and finally allowed to settle before redistillation, packaging, and shipping. Kerosene was sold in several grades. There were a variety of specifications, depending on the sales location, pertaining to properties such as flash point and specific gravity. The kerosene was shipped to wholesalers and then to retailers who sold it to households. Exports of kerosene were a major outlet for refined petroleum. Along with cotton and foodstuffs (e.g., flour), it was among the top US exports. More than half of the kerosene being produced was exported. Several other products were sold, lubricating oils being second in importance to kerosene. Highly refined products such as "mineral oil," "petrolatum" and *Vaseline* were sold for cosmetic and medicinal purposes, the latter brand manufactured by Chesebrough Manufacturing. Waxes were used as sealants, especially in home canning, and for making candles.

Crude oil production and refining during the 19th and early 20th century was plagued by repeated periods of gross excess and low prices as new fields were discovered. In part this was due to the law covering mineral rights which, in the United States, assigned them to the surface owner. Because oil is fluid, drilling can easily drain oil from adjacent properties. This circumstance forced each owner to begin drilling immediately and to recover as much oil as possible. The result was periodic production gluts, low prices and excess investment in storage and transportation facilities. By the 1920s, oil industry geologists recognized that much higher levels of recovery could be achieved if fields were developed cooperatively and the production divided on the basis of surface acreage, a concept called *proration*. H. L. Doherty, who founded Cities Service, proposed in 1924 that fields be developed on a rational basis using proration to allocate the production and royalties. Several fields were prorated voluntarily, usually those controlled by larger companies, and in Texas proration was established under the auspices of the Texas Railroad Commission, which had the responsibility for oil field regulation. The discovery of the huge East Texas fields in 1930 resulted in the greatest glut ever. By 1931, the field was producing 1 million barrels daily in the midst of the Great Depression. Proration schemes collapsed and the price dropped to $0.10/b. The depression also affected gasoline sales. By 1933, vehicle registration had dropped by 2.5 million and the price of gasoline was about $0.12/gal. The large number of service stations built during the 1920s resulted in price-cutting and "gas wars." The 1933 National Industrial Recovery Act established federal regulation of oil production, but it was declared unconstitutional by the Supreme Court in 1935. In 1934, Secretary of the Interior Harold Ickes proposed that the industry be regulated as a public utility, but that proposal, too, failed. Finally in 1935, the major oil-producing states entered a compact designed to decrease physical waste and to promote rational development.

TOPIC 2.1 THE ORIGINS OF "BIG OIL"—EXXONMOBIL, CHEVRON, CONOCO-PHILLIPS, SHELL, BP, AND TOTAL

The discovery of oil in Western Pennsylvania in 1859 started the long history of the worldwide petroleum industry, often called "Big Oil." The first of the large oil companies was the Standard Oil Trust, organized by John D. Rockefeller and his associates in 1882. Rockefeller controlled most of the refining capacity in Cleveland, OH and expanded the trust until it controlled more than 80% of US refining by 1890. The Trust also controlled much of the pipeline capacity to the refineries in the Eastern half of the country. It had major refineries in Bayonne, NJ, Whiting, IN, Baton Rouge, LA, and Kansas City, MO. In 1900, it expanded into California and built refineries near San Francisco and Los Angeles.

In the late 1800s, the Standard Oil Trust became the target of the press and progressive politicians bent on "trust-busting." Hearings in several states established that the Trust controlled 90 separate companies and 80–90% of the refined petroleum business.[c] In 1906, the federal government filed suit against the Trust and its officers, alleging violation of the Sherman Anti-Trust Act of 1890. The hearings lasted for several years but in March 1914, the Supreme Court upheld lower court findings of violation of the act and ordered that the Trust be disbanded. It was broken up into several companies, the largest being SO(NJ), SO(NY), SO(IN), SO(OH), SO(CA), Vacuum Oil, and Atlantic Refining (ARCO). These companies and their successors still constitute some of the largest oil companies in the world. The current ExxonMobil includes the old SO(NJ), SO(NY), and Vacuum Oil. The former SO(CA) now operates as Chevron and also includes several companies that were not part of the original Trust, including Gulf, Texaco, and Union Oil. Other companies from the Trust including SO(IN), SO(OH), and ARCO are currently part of BP (see below). The Trust had also founded international companies, including The Anglo-American Oil Company (1888) in Britain and Deutsche-Amerikanische Petroleum Gesselschaft (DAPG, 1890) in Germany, as well as companies in Holland and Italy.

Major discoveries of petroleum were made in the early 1900s in Kansas and Oklahoma. These fields gave rise to numerous companies, the largest survivor being Conoco–Phillips. Conoco was purchased by DuPont in 1981 but sold as a public company in 1998. Conoco and Phillips merged in 2002. The refining/marketing (Phillips 66) and production/exploration (Conoco–Phillips) components were separated in 2011.

Shell Oil was formed in 1906 by combination of Royal Dutch Shell and a British shipping and trading company. The Dutch company had discovered oil in Sumatra, where "earth wax" was used to make torches. The British company was a shipping and trading company that provided kerosene to Asia, mainly from the Baku region in Russia. The driving force behind Shell was Henri Deterding who served as its managing director from 1906 to 1936. Shell entered the United

---

[c]This record was compiled by Ida M. Tarbell and originally published in *McClures Magazine*, then as a two-part book first published in 1904. The book was reissued in 1925 and again in 1950. The latter edition, I. M. Tarbell, *The History of the Standard Oil Company*, Peter Smith, New York, NY, 1950, is the source of most of the information on the Standard Oil Trust through 1900. An extensive history of the company has been published in three parts, R. W. Hidy and H. E. Hidy, *Pioneering in Big Business 1882–1911*; G.S. Gibbs and E. H. Knowlton, *The Resurgent Years 1911–1927*; H. M. Larson, E. H. Knowlton and C. S. Popple, *New Horizons*.

States in 1912, starting on the West Coast with a refinery near San Francisco. It also built refineries in Oklahoma, near New Orleans and near St. Louis, MO. By 1930, Shell was selling gasoline nationwide and was behind only SO(NJ) and SO(IN) in US sales.

BP (formerly British Petroleum) is a global oil company whose origins were in Iran. William Knox D'Arcy obtained a concession to explore for oil from the Shah in 1901, and the Anglo-Persian Oil Company was formed in 1909 to exploit the discoveries. The name was changed to the Anglo-Iranian Oil Company (AIOC) in 1935. The company was nationalized by the Iranian Prime Minister Mohammed Mossadegh in 1951. In 1953, Mossadegh was ousted in a coup engineered by the United States and United Kingdom, which returned the Shah to power. At that point, an agreement was reached to share the profits of the Iranian oil fields 50:50 with the government. The fields were operated by a consortium, that included the former AIOC, renamed British Petroleum, with other participants including Shell and Compagnie Francaises des Petroles. BP now includes the former US companies SOHIO, American Oil (AMOCO) and Atlantic Richfield (ARCO), all of which were parts of the Standard Oil Trust. It has production in both Alaska and the Gulf of Mexico.

Total was called Compagnie Frances des Petroles when founded in 1925. It has incorporated several other French and European companies includ-ing Petofina (Belgium) and Elf Aquitaine (France) and has production in the Middle-East and Africa.

The huge petroleum reserves found in the Arabian Peninsula in the 1960s were developed by the large petroleum companies by agreements with the local governments. The fields in Saudi Arabia were developed by a consortium of major companies that operated under the name ARAMCO. The company was purchased by the Saudi Government in 1974 and since that time the petroleum companies have transported and refined the petroleum purchased from the Saudi government (see Topic 2.4).

TOPIC 2.2 THE NOBEL FAMILY AND RUSSIAN OIL

When Immanuel Nobel returned to Sweden in 1859, his son Ludvig remained in Russia and eventually reorganized the bankrupt businesses (see Topic 1.1). He made both civilian products and arms. In the 1870s, Ludvig sent his older brother Robert to Baku in Azerbaijan to investigate a source of wood for an expanding rifle factory. Baku was already the site of a petroleum industry of sorts. Oil sands and oil pits existed and oil was shipped to Russia and Persia for use as fuel. The oil was produced under contracts to the government that could be revoked at any time, and production methods were crude and inefficient. Robert Nobel purchased a small refinery in 1873. He moved to

Baku and began to learn the refining process and, by using his knowledge of Chemistry, soon was able to produce the highest quality product available in Baku. In 1876, Ludvig and his son Emanuel came to Baku. Ludvig immediately began to analyze and plan for rationalization and expansion of the refining business. He built a pipeline to the refinery area and found it was paid for in a year by reduced transportation costs. He designed and had constructed in Sweden, small tankers to transport oil across the Caspian Sea. By 1885, he had about a dozen tankers in operation on the Caspian and Baltic. Soon, tankers were in operation throughout the world, with the first trans-Atlantic tanker being commissioned by Heinrich Riedemann of DAPG (see Topic 2.1) in 1884 and in operation by 1886.

In 1879, a company known as Branobel was formed. Ludvig had 54%, an associate, Peter Bilderling 31%, Alfred 3.3% and Robert 3.3%, with others holding small shares. Robert soon became disenchanted with the operation and retired to Sweden. Alfred was concerned about rapid expansion and financing, but was not directly involved in management of the company. Ludvig remained in control and was optimistic about the future, even as both over-production and American competition began to complicate the situation. The Baku crude had relatively little of the naphtha fraction, producing mainly kerosene and fuel oil. The latter was shipped north via the Volga and used for heating, as well as in locomotives and ship boilers. The Nobels used sulfuric acid and caustic soda treatment for kerosene purification, and manufactured both the acid and caustic. Their efficiency allowed them to become the most profitable refiners in the Caucasus region.

In 1883, the Branobel company found itself over-extended and in competition with Standard Oil, among others. Alfred Nobel made a major loan to the company and joined the board. The management of the oil company became increasingly stressful and Ludvig's health broke in 1887. He suffered from heart disease and tuberculosis. He died in 1888 at the age of 57. A few months earlier, management of the oil company was turned over to his son Emanuel, then 29. The Rothschild family of France also invested in the Russian industry, forming a company known as BNITO. Standard Oil was eager to expand its American monopoly into Europe. In 1895, Emanuel considered an agreement with Standard Oil and the Rothschild's BNITO. The proposal essentially divided the world oil market, 75% to Standard Oil, the rest to be shared by BNITO and the Russian producers, including the Nobels, but no agreement was reached. The pace of industrialization in Russia increased in the 1890s and the Nobel company prospered. In 1900, it accounted for about 10% of world production, behind only Standard Oil.

But trouble was brewing. The government of Czar Nicholas II was often incompetent, particularly so in the Caucasus. Intolerance of ethnic minorities was promoted throughout Russia and the Czar's government seized the property of the Armenian Church. Soon, the Armenians were in revolt. Oversupply of oil, economic depression and shortages of food occurred in 1901. Baku became a center for distribution of antigovernment literature, including that of Lenin. A series of strikes throughout the country in 1904 was brutally suppressed. Ethnic violence between Christian Armenians and Muslim Tatars erupted in Baku. Refineries and wells in the Baku region were burned by the raging mobs. The Nobel installations survived with only minor damage, but the industry as a whole had been badly set back. Strong repression remained in place throughout Russian from 1905 to 1910 until some semblance of order was restored.

With the mobilization for WWI, production and transportation came under the direction of the government. Russian defeats led to the abdication of the Czar Nicholas II in March 1917 and chaos in St. Petersburg. In November, the Bolsheviks under Lenin took power. In August 1918, the petroleum industry was nationalized. All the of the Nobel family managed to escape from Russia. Immediately after the revolution, many believed that the Communist regime would collapse and that the confiscated properties would be restored. Standard Oil(NJ) negotiated for half the rights to all Nobel properties in Russia and paid $6.6 million to the Nobels. Similarly, Henri Deterding of Royal Dutch-Shell bought rights to other confiscated properties. These investments eventually proved worthless. With the aid of western equipment and loans, the Soviet-controlled oil fields eventually returned to productivity, and by 1927, the Soviets were profitably exporting oil to Western Europe.

The exiled Nobels first lived in Paris, but when hopes for return to Russia faded, they returned to Sweden. Emanuel became a Swedish citizen in 1923. Emanuel was one of the administrators of Alfred's estate and supported his intentions to leave the bulk of his estate for the establishment of the Nobel Prizes, while the descendants of Robert Nobel initially sought to challenge the will. The last remaining Nobel assets in the oil business, in Poland and Finland, were liquidated in 1937 and 1959, respectively.

Source: Tolf, R. W. *The Russian Rockefellers*. Hoover Institution Press: Stanford, CA, 1976.

2.1.2 THERMAL CRACKING

At the beginning of the 20th century, the technology of petroleum refining was based on distillation and purification procedures that did not change the chemical composition of the product. The arrival of automobiles dramatically changed the demand pattern for refined petroleum and presented the need to increase the yield of the gasoline fraction. At the same time, kerosene began to be replaced for both heat and light by electricity. This shift in demand required modification of the refining process to produce more of the lower boiling gasoline (naphtha) fraction. The process involves splitting molecules to reduce the average molecular weight (MW) and is called *cracking*. The first approach was *thermal cracking*, a form of pyrolysis. A relatively crude form of "cracking" was practiced in the late 1800s. When the distillate reached a certain temperature, the still residue was held at that temperature for some time, then increased. Evidently during this hold period, some pyrolysis occurred, generating addition amounts of the kerosene fraction. The overall yield was poor and much carbon was deposited as coke.

The first thermal cracking process to be commercialized was introduced by SO(IN) in 1913. William M. Burton,[d] a Ph.D. chemist trained at Johns Hopkins University, joined SO(IN) in 1890 and worked on the desulfurization of "sour" crude with Herman Frasch. In 1896, Burton became manager of the Whiting, IN refinery. He was very unusual at the time in having both scientific training and practical refinery experience. Beginning about 1906, Burton established a group of three chemists (Robert E. Humphreys, F. M. Rogers, and O. E. Bransky) all trained at Johns Hopkins, to study thermal cracking under pressure. The result was "Burton stills," which were constructed of riveted steel and heated by fire. Burton's innovation was the use of pressure. The chemical principle is that reaction is faster at the higher temperature made possible by the pressure. The temperature control was crude, consisting of slowing or speeding the addition of fuel to the fires. The feedstock was the gas oil fraction boiling >250°C. The stills operated at 370–400°C and pressures of 90 psi. The stills held about 8000 gallons of gas oil and the batch process ran for about 24 h. An important feature of the apparatus was a tube that sloped downward and led back to the still. This tube condensed the heavier distillate and returned it to the still. There was extensive deposition of carbon (coke) that needed to be removed after each run by manual cleaning. The cleaning was facilitated by a grid near the bottom of the still that collected the coke deposits. Later, fractionating columns were added to separate the cracked product. The cracked gasoline had some characteristics that were unfavorable. It tended to form gums and to become acidic on storage as the result of oxidation and polymerization of unsaturated compounds in the material. Its odor was different from straight-run gasoline. These problems were solved by treating the cracked gasoline with sulfuric acid, followed by alkali,

---

[d]William M. Burton was born in 1865 and received a B. S. degree at Western Reserve in 1886. He received a Ph. D. from Johns Hopkins University in 1889, where his mentor was Ira Remsen. He worked for a time with Herman Frasch at a Standard Oil laboratory in Cleveland but then became the research chemist at the Whiting Indiana refinery, which later became Standard Oil (Indiana). The thermal cracking process was patented in January 1913. Contemporary accounts of Burton's work can be found in his acceptance address of the Gibbs medal in 1918 (Burton, W. M. *J. Ind. Eng. Chem.* 1918, *10*, 484–486) and in reviews of the process: Egloff, G. *Ind. Eng. Chem.* 1923, *15*, 580–583; Boyd, T. A. *Ind. Eng. Chem.* 1924, *16*, 1004–1006; Wilson, R. E. *Ind. Eng. Chem.* 1928, *20*, 1099–1101; Keith, P. C.; Montgomery, W. B. *Ind. Eng. Chem.* 1934, *26*, 190–194. Burton served as president of SO(IN) from 1918 until retiring in 1927.

as was done for certain grades of kerosene. Figure 2.3 shows a diagram of the Burton still.

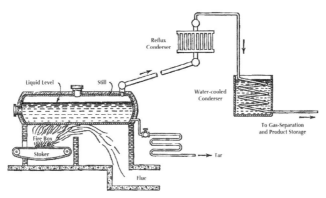

FIGURE 2.3   Design of Burton still (From Beaton, K. *Enterprise in Oil*; Appleton-Century-Croft: New York, 1957; p. 238).

A significant improvement on the Burton still was made by Edgar M. Clark, the refinery superintendent at the Woods River, IL, USA refinery of SO(IN). Unlike Burton, Clark had no technical training but had come to understand the cracking process through his observations while operating the stills. He conceived of using a bank of pipes as the cracking zone. This design had several advantages, among them increased operating pressure and reduced coking. The tubes could be cleaned by an auger and this reduced the down time between runs. The yield of gasoline from gas oil was 30–40%. The run time was increased to 3–4 days. A diagram of the still is shown in Figure 2.4.

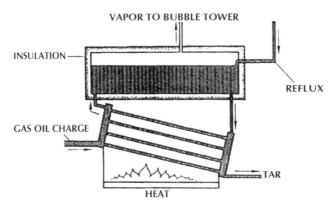

FIGURE 2.4   Burton–Clark still. (From Peters, A. W.; Flank, W. H.; Davis, B. H. *Am. Chem. Soc. Symp. Ser.* **2009**, *1000*, 103–187. © American Chemical Society).

Standard Oil (IN) decided to license the cracking process, first to other companies of the old Standard Oil Trust, but later to others as well. The agreements protected the SO(IN) sales area in the upper Midwest from competition. From the first introduction of the Burton stills in 1913, the yields of the cracking process approximately doubled, and by 1923, 1.5 billion gallons of cracked gasoline was produced annually. The Burton process was dominant in the industry until about 1925.

Standard (NJ) began research on thermal cracking in 1919. During WWI, SO(NJ) had trouble obtaining crude suitable for use in Burton crackers. Edgar Clark was hired away from SO(IN) (see above). On Clark's recommendation, Frank A. Howard, an engineer and patent attorney, was chosen to organize an independent research organization. Howard and Clark established consulting arrangements with Ira Remsen of Johns Hopkins, Robert Millikan of Caltech, and Warren K. Lewis of the new chemical engineering department at MIT. The SO(NJ) organization emphasized research on use of heavy crude that was unsuitable for the Burton process. A design called the "Tube and Tank" process was developed and put in operation by a licensee, Beacon Oil, in 1921. Several units were also installed at the Bayway, NJ, USA refinery, where Clark was manager. Figure 2.5 shows a design of the tank and tube process.

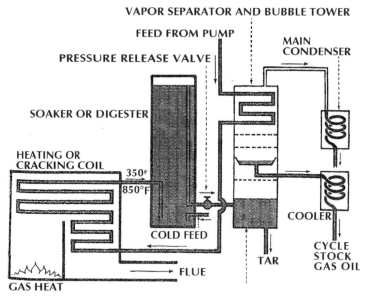

FIGURE 2.5   Tank and tube process from Williams, H. F.; Andreano, R. L.; Daum, A. R.; Klose, G. C. *The American Petroleum Industry, The Age of Energy, 1899–1959*; Northwestern University Press, 1959; p 385.

Several other modifications of thermal cracking were introduced. The most economically significant of the modified processes was the Dubbs process, which used heated tubes for the pyrolysis reaction and minimized the amount of petroleum in the high temperature zone. A crucial aspect of the process was the so-called clean recirculation, which removed the heaviest crude before it was recycled to the cracking tubes. This design cut down on the generation of coke and allowed a longer run cycle, up to 10 days. The origin of the Dubbs process was complicated and controversial. In 1910, J. Ogden Armour was making about $15 million annually from the meat packing business and decided to diversify into other industries. He purchased an asphalt company located in Independence, KS, USA, which owned the patents of Jesse Dubbs, a California oil refiner. Among Dubbs' patents was one filed in 1909 describing the continuous processing of crude oil by heat and pressure. The purpose of the original Dubbs application was to break emulsions characteristic of California crude and it did not involve thermal cracking. One of the Armour attorneys, Frank L. Belknap, saw the opportunity to develop the patent to cover the cracking process as being operated by SO(IN). Between 1909 and 1915, when the patent was finally issued, he added claims that more closely approached the technology of the Burton stills. Armour, Dubbs, and several associates formed a company, Universal Oil Products (UOP) to exploit the potential of the patent. At that point they had not operated any cracking facility. Dubbs' son, C. P. Dubbs (The C. P. is for Carbon Petroleum) and Gustav Egloff began development work to make a continuous cracking process operational and obtained a series of patents. One, issued in 1919, covered the concept of "clean recirculation." They eventually succeeded in operating a process at 820°C and 135 psi. The Dubbs process was a case of technology developed to use a patent, the reverse of the normal relationship. The Dubbs process is outlined in Figure 2.6.

The technology was licensed first to Royal Dutch-Shell. The first test run at the Shell Wood River, IL refinery in December 1921 exploded killing two and injuring several others, including C. P. Dubbs. After modifications, six units were installed in 1922. Improved units were installed at Shell's Dominguez, CA location in 1927. The improved units could operate for over two weeks before needing to be shut down for cleaning. The Dubbs process could also accept heavier feeds, even crude oil, further improving the economics of operation. The Dubbs process quickly became the dominant technology.

Lawsuits had been filed as early as 1916 claiming infringements of the various thermal cracking methods, but they remained unsettled. The Patent Office had granted many seemingly conflicting patents. In 1923, SO(NJ), SO(IN), and Texaco agreed to cross-licensing of their cracking patents. In 1924, the Justice Department challenged this agreement on the basis of restraint of trade, but after court decisions in 1924, and an appeal in 1927, the Supreme Court finally rejected this charge. A suit by UOP claiming infringement of the Dubbs patent remained outstanding. In 1931, Shell, SO(CA), SO(IN), SO(NJ), and Texaco bought UOP from its stockholders for $25 million. In 1944, the companies that owned UOP transferred ownership to the Petroleum Research Fund, with the profits to be dedicated to research in petroleum chemistry and engineering. UOP continued to operate as a process development firm. In 1959, the company was sold to the public for $72.5 million. The proceeds provided permanent financing for the Petroleum Research Fund, which continues to support research.

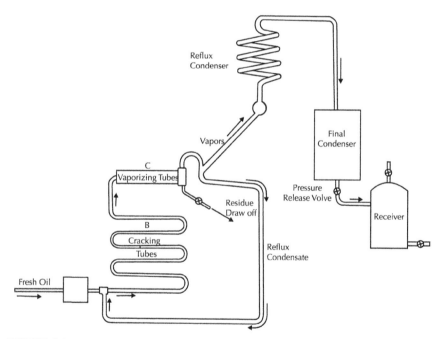

FIGURE 2.6   Dubbs continuous circulation cracker (From A. W. Peters, W. H. Flank and B. H. Davis, *Am. Chem. Symp. Ser.* **2009,** *1000,* 103 – 187. © American Chemical Society. Reprinted with permission.).

2.1.3   CATALYTIC CRACKING

The next major advance in petroleum refining was the introduction of catalysts that could lower the temperature and increase the selectivity for more desirable components. One factor driving this innovation was the need for better gasoline performance as the compression ratios in automobile increased. Another factor was the growing need for high octane aviation fuel.

2.1.3.1   HOUDRY PROCESS

The first successful catalytic process was spearheaded by Eugene Houdry. He was born in Paris in 1892 and trained as an engineer. He went to work at his father's steel company, but when WWI began he joined a tank brigade and fought in the first tank battle of the war. After the war he rejoined the company, but soon became very interested in auto racing, which was being used to test automotive innovations. He was also keenly aware of the weak position of France in petroleum resources. He began to investigate production of gasoline from lignite in association with E. A. Prudhomme, a French chemist working in Italy. This effort was not successful, but in 1924, they also began to investigate use of catalysts to obtain gasoline from heavy oil. Many materials were examined and in 1927, they found that a mixture of silica and alumina clays promoted formation of gasoline. The catalyst was quickly deactivated by carbon deposits, but could be regenerated if burned in air. The research continued until 1929, but Houdry was unable to convince the French government or industry to invest.

In 1930, Houdry came to the United States and was able to interest Vacuum Oil, a former Standard Trust company, in the process and a pilot plant was set up in Paulsboro NJ. The depression and merger of Vacuum with Socony (the old SO-NY) diverted the company's interest and Houdry sought additional support from Sun Oil, which was owned and managed by the Pew family. The pilot plant was moved to Marcus Hook PA and the Houdry Process Corporation was owned jointly by Houdry, Sun and Socony-Vacuum. Production was begun by Socony-Vacuum in 1936 (2000 bpd) and in 1937 by Sun (12,000 bpd). A key feature of the process was that the amount of branched-chain hydrocarbon increased, leading to an increased octane number of 91. Sun sold the product as a premium gasoline. The crackers operated at about 480°C and 30 psi pressure. The reactors were subjected to fast cycles of operation and catalyst regeneration, because the catalyst was rapidly deactivated by deposition of carbon. The catalyst was regenerated by hot air that oxidized the deposited coke. The length of a run between catalyst reactivation in the Houdry unit depended on the feed stock, with heavier oil requiring cycles on the scale of minutes. Automated control was necessary. Continuous operation required at least three units that could be phased between the cracking, purging, and reactivation stages. By 1939, US capacity was >200,000 bpd and the Houdry process had a cost advantage of about $ 0.05/barrel compared with thermal cracking. It also had solid patent protection, because the use of a catalyst was an innovation. During WWII, total capacity reached 300,000 bpd and the Houdry units were used almost entirely for aviation fuel. Synthetic silica–alumina catalysts were also introduced, replacing natural clays. These modified Houdry processes were competitive up until the mid-1950s, at which point they were supplanted by fluid bed catalytic crackers. A design for a Houdry Process unit is shown in Figure 2.7.

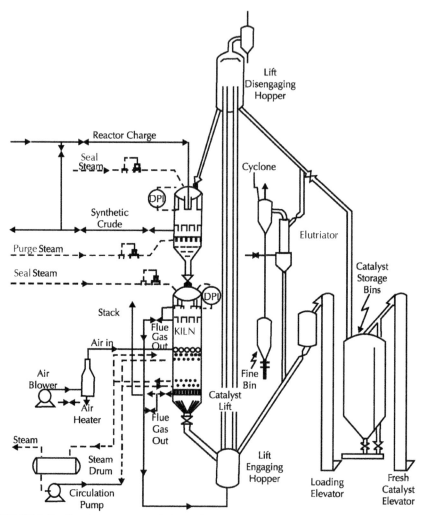

FIGURE 2.7 Houdry process catalytic cracker. (From Peters, A. W.; Flank, W. H.; Davis, B. H. *Am. Chem. Soc. Symp. Ser.* **2009**, *1000*, 103–187. © American Chemical Society).

Although the original discovery of the catalysts had been entirely empirical, further study began to elucidate the nature of the catalysis. Work at UOP and elsewhere led to the recognition of acidity as a key function of the catalysts. The acidic catalysts promote rearrangement of the hydrocarbon chains, increasing the amount of branched chain and aromatic compounds. The reaction is driven by the greater *thermodynamic stability* of these compounds, compared to straight chain hydrocarbons. The key chemical intermediates, carbocations, were also becoming recognized through the work of Frank Whitmore[e] at Pennsylvania State University.

2.1.3.2 FLUID BED CRACKING

Another major improvement in catalytic cracking was made late in the 1930s when *fluid bed reactors* were introduced. In this method, finely powdered catalyst circulates through the hydrocarbon stream and catalyst life is extended. The major instigation for the research came from SO(NJ). SO(NJ) had begun research on hydrogenation of hydrocarbons (hydrocracking, see Section 2.1.3.4) and acquired access to the considerable information that the German combine IGF had developed on the hydrogenation of coal (see Section 2.2.1). In 1930, 17 oil firms joined a consortium called the Hydro Patents Company. However, the depression and new discoveries of crude lessened the interest in coal hydrogenation as a source of gasoline. When the Houdry process came on stream, SO(NJ) turned its attention to catalytic cracking, with the intent of finding a process not covered by the Houdry patents. In 1938, several major firms including SO(IN), Texaco, Shell, BP, IGF, UOP, and William Kellogg Company agreed to collaborate on research in catalytic cracking through a consortium called Catalytic Research Association. Together, at that time they had a research staff of nearly 1000. At the SO(NJ) labs in Baton Rouge, LA, attention turned to the idea of using a powdered catalyst at the suggestion of consultants W. K. Lewis and G. R. Gilliland of the MIT Chemical Engineering Department. Lewis and Gilliland had demonstrated the feasibility of the concept and patented it. A successful pilot plant was put in operation in 1940. A major modification made was for the recovery of catalyst. The powdered catalyst was recovered from the hydrocarbon vapor by a centrifugal cyclone separator. After regeneration, the catalyst was returned to reacting system. The hot catalyst also served as a heat transfer medium. A 12,000 bpd unit began operating in Baton Rogue in May 1942. Other units were soon constructed in New Jersey and Texas. The concept of the Fluid Bed Catalytic Cracker is shown in Figure 2.8.

---

[e]Frank Whitmore was born in Massachusetts in 1887 and received his Bachelors (1911) and Doctorate (1914) in chemistry from Harvard. He was on the faculty of Northwestern University from 1920 to 1929. He then moved to Pennsylvania State University where he was the Dean of the School of Chemistry and Physics and led the development of the sciences at Penn State. Whitmore developed the fundamental concepts of carbocations as reaction intermediates. He received the Willard Gibbs Medal in 1945.

Source: Marvel, C. S. *Bibliographic Memoirs*; National Academy of Science: Washington, DC, 1954; pp 289–311.

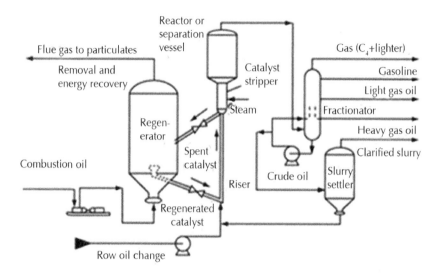

FIGURE 2.8   Conceptual diagram of fluid bed cracker from OSHA Instruction TED 1.15, 1996. (From OSHA Instruction TED 1.15, 1996.)

WWII put enormous demands on the production of fuel and high octane aviation fuel was particularly critical. Large pipelines were constructed from the Gulf Coast area to the Chicago and New Jersey refining complexes. Both to conserve gasoline and rubber, rationing of gasoline for civilian use was implemented in 1942. Crude oil production increased from 3.7 bpd in 1940 to 4.8 bpd at the end of the war. The development of catalytic cracking was of major importance to the success of the war-time effort, because it not only provided gasoline of higher octane rating but also provided the alkene byproducts necessary to make aviation fuel by alkylation (see Section 2.1.5). Government subsidies were available for construction of the units. Most of the companies that had participated in the catalytic research consortium built fluid bed crackers, as did several other refiners. By 1955, the fluid bed crackers were the dominant technology, accounting for about 2.3 million bpd of capacity.

2.1.3.3   CATALYSTS FOR CRACKING

The original catalysts used in the Houdry Process were natural clays that had been treated with sulfuric acid. Since the 1960s, the most prominent catalysts for cracking have been *zeolites*. Zeolites are crystalline aluminosilicates containing tetrahedral $(AlO_2)^-$ and SiO_2 units. Zeolites are porous materials with channels and cavities having dimensions in the molecular (nanometer) range. They have many applications. The largest commercial applications are in detergents, selective adsorbents and catalysts. They can function as ion exchange materials and this is their purpose in detergents (see Section 6.2.1). They also can act as "molecular sieves" by selectively adsorbing molecules based on size and shape. They function as catalysts by combining shape-selectivity with chemical reactivity, in particular acidity. Some zeolites occur naturally as minerals. A large number of zeolites have been prepared by synthesis. Union Carbide developed synthetic zeolites at its Linde division. This work began in the late 1940s and the initial purpose was for separation of N_2 and O_2 from air. The synthetic zeolites are prepared from aluminum salts or freshly prepared $Al(OH)_3$ and a silicon source such as silicates, silicic acid, silica gel or silicate ethers. The aluminum units are anionic but the silicon units are neutral. The ratio of Si to Al units must be >1 to avoid adjacent anionic units. The ratio of Si to Al is important, with the optimum ratio being about 3. Under specified ratios and conditions, these materials form amorphous units that can crystallize into specific structures. Specific shapes can be obtained by including template structures, such as alkylammonium ions, that are later removed by pyrolysis or chemical reactions. The final forms of the catalysts are heated to drive off excess water or other undesired material. Inclusion of phosphoric acid leads to aluminophosphates. Zeolites can also be made in the presence of fluoride ion at acidic pH, which results in new structural types. Various companies including Union Carbide, American Cyanamid, Nalco, Filtrol and Davison, a division of W. R. Grace, developed catalysts, often in collaboration with one of the major oil companies. By the 1960s, zeolites became the dominant catalyst in fluid bed crackers.

There are a wide variety of potential structures for zeolites. The Si centers are neutral, but the anionic Al centers must be balanced by cations. If the cations are protons, the zeolites are acidic. The Si:Al ratio determines several properties of the zeolites. At ratios of 1–1.5, they are relatively hydrophilic. The surface becomes increasingly hydrophobic as the ratio is raised. Acid strength increases with the Si:Al ratio, but the number of acidic sites decreases. The acidic sites are created by replacing a metal cation, usually Na^+, with NH_4^+, then driving off ammonia.

$$Na^+Z^- + NH_4^+ \longrightarrow NH_4^+ Z^- \xrightarrow{\text{heat}} H^+ Z^- + NH_3$$

Zeolites can contain both protic (Brønsted) and Lewis acid sites. The former involve a proton associated with an oxygen bridging between an alumina and silicon. The latter exist where ligand vacancies are present at aluminum or another metal doped into the system.

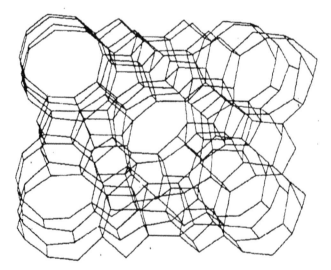

The various types of zeolites can assume regular structures based on 4-, 5-, 6- and 8-membered rings. These rings are incorporated into chains, sheets and channels that are part of 3-dimensional arrays having still larger rings. These 3-dimensional arrays provide the channels that characterize the structure and properties of the individual zeolites. Synthetic zeolites can be prepared with very high Si:Al ratios. One such zeolite is ZSM-5, shown in Figure 2.9. The selectivity of ZSM-5 is such that cracking of straight chain parafins occurs in preference to branched isomers. The catalyst functions by generating alkenes and carbocations, which then combine. One of the properties of the ZSM-5 zeolite is that the shape of its cavities resists formation of the precursors of carbon deposits (coking) and it, therefore, has a long catalyst life.[1]

FIGURE 2.9   The structural framework of ZSM-5. (From Olson, D. H.; Kokotallo, G. T.; Lawton, S. L.; Meier, W. M. *J. Phys. Chem.* **1981**, *85*, 2238–2243. © American Chemical Society).

The catalytic selectivity of the zeolites depends on matching the shape of cavity with either reactant, product, or *transition state*. If only one of two molecules can reach the active catalytic site, only it can react. If only one of several products can be formed at the site, it will be formed preferentially. If the catalytic site favors a particular transition state, the corresponding product will be formed. The zeolite must permit continuous entry, reaction, and exit at the active site to function as a catalyst.

2.1.3.4 HYDROCRACKING

Hydrocracking had its origin in IGF's process for hydrogenation of coal to hydrocarbons (see Section 2.2.1.1). The Humble Oil subsidiary of SO(NJ) operated a plant beginning in 1937. However, hydrocracking did not become economically important at that time. The catalysts used for hydrocracking have acidic components for cracking, but also include hydrogenation catalysts. The hydrogenation catalysts include either a noble metal (Pt, Pd) or a transition metal sulfide (Mo, W, Co, Ni). The process typically operates at 300–450°C and at 80–250 bar of H_2 pressure. These conditions result in the saturation of intermediates and alkenes formed in the cracking process. Hydrocracking is used for heavy and high-sulfur crudes that do not perform well under the usual cracking conditions. Polycyclic aromatics and nitrogen compounds are converted to saturated hydrocarbons, generating NH_3 and H_2S. As of the early 2000s, worldwide capacity was about 3.8 million bpd.

Hydrocracking

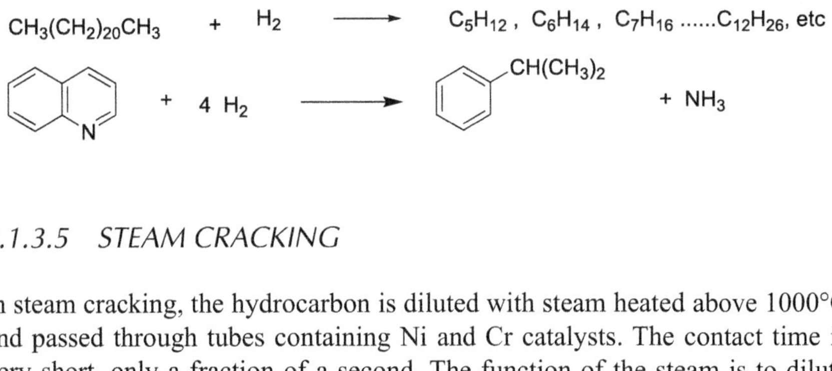

$CH_3(CH_2)_{20}CH_3$ + H_2 $\longrightarrow$ C_5H_{12}, C_6H_{14}, C_7H_{16} ......$C_{12}H_{26}$, etc

+ $4\ H_2$ $\longrightarrow$ $\overset{CH(CH_3)_2}{}$ + NH_3

2.1.3.5 STEAM CRACKING

In steam cracking, the hydrocarbon is diluted with steam heated above 1000°C and passed through tubes containing Ni and Cr catalysts. The contact time is very short, only a fraction of a second. The function of the steam is to dilute the hydrocarbon, which promotes alkene formation. These conditions produce a large proportion of alkenes, particularly ethylene, propylene, and butadiene from ethane, propane, and butane. When higher MW feed stocks are used, benzene and other aromatic hydrocarbons such as toluene are major products. Steam-cracking for the production of ethylene and propylene grew out of research sponsored by Union Carbide at the Mellon Institute in the 1920s. The first commercial unit began operation in South Charleston WV in 1925. We will say more about steam cracking in the section on petrochemicals.

2.1.4 CATALYTIC REFORMING

Straight-run distillate in the gasoline range has an octane number of only about 50, which was too low for automobiles after 1920. While thermal cracking results in some improvement, it alone is not sufficient. More significant improvement was achieved by the Houdry and fluid bed catalytic cracking methods. The improvement results from formation of branched, cyclic and aromatic compounds, which have high octane numbers. Even more extensive structural change occurs in *reforming*. Reforming catalyzes reactions that reorganize the carbon structure and increase the proportion of branched chain, cyclic, and aromatic components. Reforming improves engine performance, as measured by the octane rating. An early version of catalytic reforming called *hydroforming* was introduced in the late 1930s and involved the use of molybdena-alumina or chromia-alumina catalysts. The product from this treatment contained large amounts of aromatic compounds. During WWII, hydroforming was a major source of toluene for production of TNT (see Section 1.2.3).

A major advance in catalytic reforming was made at UOP in the early 1940s. The company was not directly involved in refining but rather undertook fundamental research and process development in petroleum chemistry. Vladimir Ipatieff,[f] a distinguished Russian chemist who emigrated at the age of 64, joined UOP and was also a professor at Northwestern University. One of his students,

---

[f]Vladimir Ipatieff was born in Moscow in 1867. After studying at several military schools, he was admitted to artillery school in 1886 and was commissioned an officer in 1887. He then became an instructor in math and artillery at a military school outside Moscow and established a small chemical laboratory in his home. He also studied organic chemistry in St. Petersburg, preparing a dissertation in the lab of A. E. Favorsky. In 1895, Ipatieff received a 1-year fellowship to study in the lab of Adolf von Baeyer in Munich. In 1897, he returned to St. Petersburg and became Professor of Chemistry and Explosives at the Artillery Academy. He was a member of the Explosives and Artillery Committee of the Russian army.

Ipatieff began study of reactions at high temperature (500–600°C), which was well outside the range considered feasible for organic compounds. He discovered reactions such as the dehydration of alcohols and noted catalytic activity of materials such as iron and alumina. By 1914, he had received considerable recognition and had the military rank of lieutenant general. During WWI, he chaired the committee on preparation of explosives. After the revolution of 1917, he was asked to organize the transition of the chemical industry to peacetime and from 1921 to 1926 served as chair of the committee for Scientific and Technical Administration. In 1927, he was awarded a Lenin Prize by the USSR. In 1927, he was authorized to spend 3.5 months at the Bayer nitrogen works in Berlin as a consultant. By 1929, perhaps because of his time in Germany, Ipatieff began to hear rumors of his impending arrest. In 1930, he and his wife fled Russia, leaving their belongings behind.

In 1931, at age 64 and speaking little English, Ipatieff began work at UOP as Director of Chemical Research and also taught a course at Northwestern. His research at UOP included development of phosphoric acid on kieselguhr as a catalyst for dimerization of isobutene and, with Herman Pines, on acid-catalyzed alkylation of isobutane by alkenes. He initiated the laboratory in high-pressure chemistry at Northwestern, which was supported by UOP and was subsequently named in his honor. In the mid-1930s, the Russian Ambassador tried to persuade him to return to Russia, but he became a US citizen in 1937. His Russian citizenship was revoked and he was expelled from the Russian Academy of Sciences.

Source: Haensel, V. *Am. Chem. Soc. Symp. Ser.* **1983**, *222*, 141–152.

Vladimir Haensel,[g] worked at UOP and joined the company full-time upon completion of his Ph. D. in 1939. Haensel found that at high temperature and H_2 pressure a reforming catalyst based on platinum did not require frequent regeneration. This process became known as *platforming*. The catalyst includes a small amount of platinum (~0.1%) and fluoride or chloride (<1.0%) on alumina. This catalytic system produces large amounts of branched chain saturated hydrocarbons. The reforming process also generated substantial amounts of aromatic compounds. The catalysts are referred to as *dual function catalysts* and promote two reactions. One is hydrogenation and dehydrogenation. These reactions are catalyzed by the platinum. The second function is to promote rearrangement. The Lewis acid character of the alumina is essential to this reaction. The reactions are *thermodynamically controlled*, that is the formation of the branched chain and aromatic compounds occurs because they are *more stable* than their precursors under the operating conditions. The practical effect is to both increase the yield of the desired gasoline fraction and to improve its performance (octane) characteristics. Several kinds of structural changes occur during reforming. One is *hydrocracking* whereby larger molecules are converted to smaller saturated ones. Another is *rearrangement* (isomerization), which increases the proportion of branched chain compounds. *Dehydrocyclization* results in the formation of cyclic compounds. *Aromatization* gives benzene and its alkyl derivatives. The latter reactions produce hydrogen as a byproduct. By careful study of the relationship between temperature, hydrogen pressure, and catalysts composition, Haensel and his associates were able to optimize the product mix.

The UOP platforming technology was introduced in 1949 and was rapidly adopted by both small independents and the large oil companies. Further improvement was made in the 1960s, by the introduction of rhenium–platinum and iridium–platinum *bimetallic catalysts*. Many specific catalysts that have the

[g]Vladimir Haensel was born in Freiburg, Germany in 1914 just before the outbreak of WWI. His father was a professor of economics in Moscow from 1903 to 1928. The family left Russia in 1928, eventually coming to the Chicago area, where the senior Haensel taught economics at Northwestern. Vladimir Haensel received a B. S. in engineering from Northwestern in 1935 and an M. S. in Chemical Engineering from MIT in 1937. He then joined UOP. From 1939 to 1946, while employed by UOP, he worked as an assistant to Vladimir Ipatieff at the High Pressure Laboratory at Northwestern.

In 1947, Haensel began to work on catalysts for upgrading gasoline performance. The hydrogenation–dehydrogenation properties of platinum and other noble metals were known, and Haensel believed a combination with the alumina type catalysts used in the Houdry and fluid bed processes might be effective. Because of the high cost of platinum, both low catalyst loading and long catalyst life were essential. Haensel succeeded in making catalysts with low overall platinum content but high surface concentration. He also found that catalyst activity was promoted by halides. The resulting catalysts substantially improved octane rating. The platforming process also produced net hydrogen and removed sulfur impurities as H_2S. Haensel later contributed to the development of the catalytic converter to reduce automobile exhaust emissions.

Haensel served as Director of Process Research (1960–1969), Director of Research (1969–1972), and Vice-President for Science and Technology (1972–1979) at UOP. In 1979, Haensel became a professor of chemical engineering at the University of Massachusetts, Amherst, while continuing as a consultant with UOP. Haensel was awarded both the Perkin Medal of the American Chemical Society and the Charles Stark Draper Prize of the National Academy of Engineering. He was also awarded the National Medal of Science in 1973.

ability to upgrade gasoline to meet performance and regulatory requirements have been developed and commercialized. Several factors increased the demands on reforming processes in the 1980s. One was the impending removal of tetraethyllead (TEL) as an antiknock component (see Section 2.3.1). The amount of aromatic compounds permissible in gasoline was also being restricted because benzene and some of the other aromatics are carcinogenic.

2.1.5 ALKYLATION

Both thermal and catalytic cracking produce gaseous C_3 and C_4 hydrocarbons as byproducts. These compounds are too low boiling to include in gasoline but they can be combined to give larger molecules. This process is called *alkylation* and results from the combination of an alkene and alkane. The main products of the alkylation processes are highly branched C_8 hydrocarbons. Three fundamental chemical reactions are involved: (1) protonation of the alkene to form a carbocation; (2) reaction of the carbocation with alkene to form a "dimeric carbocation;" and (3) hydride transfer from isobutane to the dimeric cation, producing alkane and a new *t*-butyl cation. In the case of a 2-butene, the main product is 2,2,4-trimethylpentane. The reaction path is shown in Scheme 2.2. There are three critical chemical aspects of this process. (1) The alkylation produces alkanes in the gasoline range from byproducts of cracking. (2) Isobutane has a crucial structural feature, a tertiary hydrogen, that makes the hydride transfer step feasible. (3) The highly branched product has a high octane rating.[h]

The first commercial alkylation plant was put in operation in 1938 by the Humble Oil subsidiary of SO(NJ). The process grew rapidly during WWII because it produced high octane fuel suitable for aviation. By 1946, the capacity was 170,000 bpd. Later growth in demand came from increased compression ratios in automobile engines, and by 1970 the capacity was 800,000 bpd, and represented about 6% of total gasoline used in the United States. The importance of alkylation product, also called *alkylate*, increased with the phase-out of TEL as an antiknock additive and the reduction of the aromatic content to meet environmental regulations. As of 2000, about 11–13% of gasoline was alkylate and the production was about 1 million bpd in the United States.

---

[h]The alkylation process was based on the research of Herman Pines. Pines was born in Poland in 1902. He was educated at the Ecole Superiure de Chimie Industrielle in Lyon, France and received an undergraduate degree in chemical engineering in 1927. He emigrated to the United States and joined UOP in 1930. He also undertook graduate work in Chemistry, earning a Ph. D. from Northwestern in 1935. Between 1941 and 1952, he concurrently held a faculty position at Northwestern, and in 1953 he became director of the Ipatieff Laboratory at Northwestern. He retired in 1970 but continued his research at Northwestern until 1995.

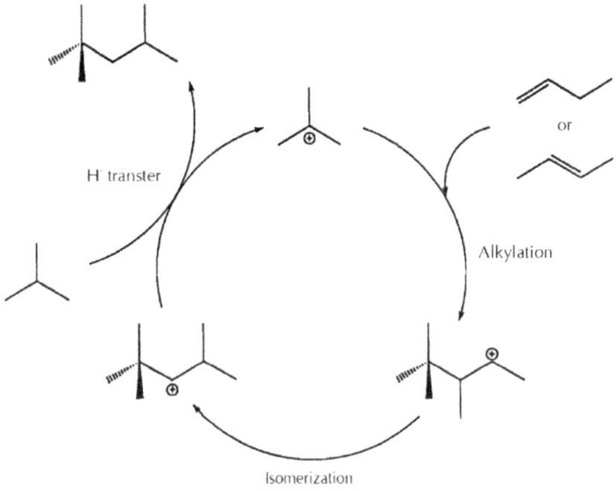

SCHEME 2.2   Reactions in formation of 2,2,4-trimethylpentane by alkylation.

Alkylation is also used to convert benzene to alkyl benzenes, especially eth-ylbenzene and isopropylbenzene (cumene). The former is the precursor of sty-rene and the latter can be converted to phenol and acetone. These reactions are examples of the *Friedel–Crafts reaction* and use strongly acidic catalysts.[i] The

[i]Although the work of Friedel and Crafts was not directly involved with the petroleum industry, the ground-work they laid is fundamental to the alkylation and reforming reactions that are currently used. Friedel and Crafts discovered the catalytic activity of $AlCl_3$, the quintessential *Lewis acid*, though the term was not then in use. The present day alkylation and reformation catalysts utilize the carbocation chemistry that $AlCl_3$ evokes, and for a time $AlCl_3$ itself was used in the petroleum industry.

Charles Friedel was born in Strasbourg France in 1832. Friedel was a student at the University in Strasbourg for a time when Louis Pastuer was on the faculty. Friedel received degrees from the Sorbonne in Mathematics in 1854 and Science in 1855. He studied Chemistry in the laboratory of Charles Adolphe Wurtz, a leading chemist of the time. In that laboratory, he encountered James Mason Crafts, an American trained as a mining engineer, who came to Wurtz's lab in 1861. Friedel received his Doctorate in Chemistry and Mineralogy in 1869 and by that time had a considerable scientific reputation. In 1871, he was appointed to a position in mineralogy at the Ecole Normale Superieure and became professor of mineralogy at the Sorbonne in 1876. He succeeded Wurtz as the Professor of Chemistry in 1884. Among his major accomplishment was the preparation and characteriza-tion of many ketones by heating calcium carboxylates. He also characterized fundamental silicon compounds, including the chlorides, the tetraalkylsilanes and the first examples of what we now call silicones.

James Mason Crafts was born in Boston, MA, USA in 1839. He received a Bachelors degree at the Lawrence Scientific School at Harvard in 1858 and then studied in Frieburg, Heidelberg, and Paris. While he was in Paris, he studied in Charles Wurtz's lab and collaborated with Charles Friedel. Crafts returned to the United States in 1866 and was the first Chemistry Professor at Cornell University. In 1871, he became the professor of analyti-cal and organic chemistry at MIT. In 1874, he returned to Paris, expecting to stay only a short time, but in fact remained there until 1892. It was during this time that he and Friedel studied the chemistry of $AlCl_3$. Crafts described the discovery as resulting from a study of the reaction of halides with aluminum metal. They soon recognized that $AlCl_3$ was being formed and that it was the active catalyst. They proceeded to demonstrate the alkylation and acylation of benzene and other aromatics in the presence of $AlCl_3$. At the time, of course, the concept of "Lewis acids" was unknown nor were the "carbocations" recognized. However, Friedel and Crafts did recognize the wide potential of their new reaction. Their French and British patents claimed the upgrading of petroleum as a result of treatment by $AlCl_3$. Crafts returned to MIT in 1892 and served as president of the Institute from 1898 to 1900. He then returned to research and made high precision measurements of temperature and vapor pressure and evaluated the empirical relationships between them.

Sources: Crafts, J. M. *J. Chem. Soc.* **1900,** *77*, 993–1019; Cross, C. R. *Biographic Memoirs, National Academy of Sciences*, Vol IX, 1919, pp 159–177; Willemart, A. *J. Chem. Ed.* **1949,** *26*, 3–9.

original catalyst for ethylbenzene was $AlCl_3$ or alternatively Al_2O_3–BF_3. Cumene was made from benzene and propene using a phosphoric acid–BF_3 catalyst on a solid support. Solid zeolite catalysts are now used for both of these reactions.

2.1.6   SPECIAL REQUIREMENTS FOR AVIATION FUEL

The demand for aviation fuels began in WWI and increased rapidly during WWII. During WWII, the government subsidized the construction of several new catalytic cracking units to meet the demand. Production increased from about 54 million gallons/year in 1934 to about 25 million gallons *per day* by 1945. Aviation fuels impose some specific demands. For the piston engines of WWII, high power and a high octane rating were required. Aviation fuel began at an octane rating of 87 in the 1930s but the specifications increased to 100 in 1934 and 115 by the 1940s. Isooctane prepared by dimerization of isobutene or combination of isobutene and isobutane was an important component. Isopropyl benzene (cumene), which was also used to upgrade aviation fuel, was made by Friedel–Crafts alkylation of benzene with propene. The availability of high octane fuel gave British pilots an advantage in the battle for control of the air early in WWII.

With the advent of jet engines, the fuel changed from gasoline to kerosene. Jet aviation fuel has a maximum in the C_{10}–C_{12} range and a total range of C_8–C_{18}. Among the requirements for jet fuel are a low freezing point for high altitude operation (Jet A-1 has a freeze point of $-47°C$). Jet fuel contains a number of additives. One is an icing-inhibitor to prevent fuel line freeze-up. The compound used is diethyleneglycol monomethyl ether, known as FSII in aviation fuel terminology, at 0.15 vol%. Dispersants, antioxidants, metal chelators, and static charge dissipaters are also used. An example of the latter is Stadis-450, which contains salts of dinonylnaphthalene sulfonic acid.

The loss of TWA Flight 800 on July 17, 1996 is believed to have been caused by an explosion in the nearly empty central fuel tank. The plane exploded a few minutes after departing from New York for Paris. It was determined that the vapor mixture in the tank was potentially explosive, exacerbated by a relatively high temperature caused by warming of the tank by output of air conditioners while the plane was on the ground for several hours. About 95% of the wreckage was recovered and the plane reconstructed during the investigation into the cause of the explosion. The source of the ignition was never positively identified. One possibility is wiring associated with the fuel tank monitors. Inspection of planes of similar vintage showed various types of wiring damage. Another possibility was heat from fuel transfer pumps.

2.2 ALTERNATIVE SOURCES OF LIQUID FUELS

Petroleum is a finite resource and its location and accessibility around the world is uneven. These circumstances have, from time to time, spurred interest in alternative sources of liquid fuels that would be compatible with the existing equipment and infrastructure for hydrocarbons. The main potential sources are coal, natural gas, and biomass.

2.2.1 LIQUID HYDROCARBONS FROM COAL

Coal is a complex mixture of carbon-containing compounds, including hydrocarbons and derivative structures. The compounds are mainly polycyclic and of relatively high MW and low hydrogen content. Coal also contains oxygen, nitrogen, and sulfur. Lower boiling aromatic compounds such as benzene and alkylbenzenes can be obtained by pyrolysis of coal. The distillate also contains nitrogen compounds such as pyridine and quinoline. The yield of product is low. The hydrogen content of coal is 3–6%, whereas in the octane fraction of gasoline it is 15.8%. Thus conversion of coal to saturated hydrocarbons requires the addition of hydrogen.

2.2.1.1 HYDROGENATION OF COAL

Friedrich K. R. Bergius, working in Germany between 1910 and 1925, developed methods for high-pressure hydrogenation of coal to liquid hydrocarbons. The process was commercialized in Germany on a modest scale after WWI. The initial version used a slurry of coal dust in heavy oil that was hydrogenated to produce a hydrocarbon mixture. The gasoline fraction was isolated by distillation and the heavy oil fraction was recycled. This process was also used in Great Britain between the wars, but was not commercially developed in the United States because of the availability of petroleum. This process was operated extensively in Germany during WWII. About 60% of German hydrocarbon fuel was produced by coal hydrogenation, with 30% coming from petroleum.

2.2.1.2 GASIFICATION OF COAL

The technology for gasification of coal goes back to the late 1700s. It was initially used to produce "town gas," which contains mainly CH_4, H_2, and CO, for use in street lighting.

$$C_mH_nN_xO_yS_z \longrightarrow CH_4 + CO + H_2 + H_2O + NH_3 + H_2S$$

Town gas eventually became a source of home illumination and heating in urban areas. The underlying process is partial oxidation of the coal to supply the energy for thermal decomposition (pyrolysis) of the rest of the coal. Sulfur and nitrogen containing compounds are also formed, but are removed and used to make byproducts. Much of the early development is attributed to William Murdoch, who worked in the United Kingdom. He used the process for external lighting at several locations in England. By the 1820s, there were systems for distribution of gas for lighting and heating in locations such as London, Baltimore, MD, USA, and Hanover in Germany. By the 1850s, many towns and cities had gas works and distribution systems. The technology for gas generation and distribution remained in place in the United Kingdom into the 1960s.

Modern gasification plants use *Lurgi gasifiers* or similar installations. These were developed in the 1930s. The process uses coal or lignite with steam and air, with the ratios being controlled to support only partial combustion. The energy of the partial combustion is used to maintain the reaction temperature that is necessary. Lurgi gasifiers have been used in a number of locations around the world, particularly in South Africa. A schematic diagram is given in Figure 2.10.

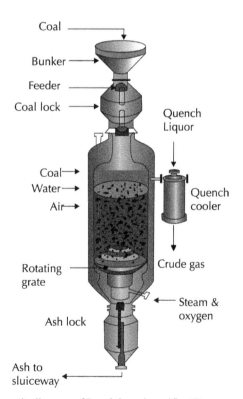

FIGURE 2.10   Schematic diagram of Lurgi dry ash gasifier (From www.netl.gov.).

The same kind of pyrolysis reaction can be applied to other carbon-rich material including biomass, municipal waste or heavy hydrocarbon residues. Gasifiers can be used in closed cycle systems for electricity generation and provides some efficiency improvement compared to combustion of coal for steam generation. The gasifiers can also be used to produce *synthesis gas*, a mixture of CO and H_2 that can be converted to various hydrocarbons and derivatives (see below).

2.2.1.3 HYDROCARBON FORMATION BY THE FISCHER–TROPSCH PROCESS

The current route to hydrocarbons from coal involves the use of carbon monoxide and hydrogen (synthesis gas) produced by coal gasification. The underlying chemistry was developed in Germany in the 1930s and is known as the *Fischer–Tropsch* (F–T) process. Franz Fischer and Hans Tropsch worked at the Kaiser Wilhelm (now Max Planck) Institute for Coal Research at Mulheim.[j] Either cobalt or iron-based catalysts can be used and the first plant based on the process was started up by Ruhrchemie in 1935. Eventually, at the direction of the Nazi government, a number of plants were established for conversion of coal to hydrocarbon fuel by the F–T process. During WWII, about 10% of German fuel was produced by the F–T process. Production reached 4.4 million barrels by 1944, but dropped rapidly as a result of bombing and the collapse of the German war effort. The cost of hydrocarbon production was about twice that of fuel from petroleum at the time. After the war, there was little interest in the F–T

---

[j]Franz Fischer was born in 1877. He began his university studies in 1897 at the Technical University in Munich and received a Ph. D. in 1899 from Giessen for studies on lead storage batteries. He pursued further studies in Paris and then with Wilhelm Ostwald in Leipzig, which included work in electrochemistry. He completed his Habilitation in Freiburg (1901–1904). He returned to Emil Fischer's institute in 1904. Emil Fischer was instrumental in Franz Fischer's appointment as director of the Kaiser Wilhelm Institute for Coal Research, which was opened in 1914. The Institute opened only a few days before the outbreak of WWI, and this resulted in research being initiated to provide liquid fuel from coal. Fischer noted a 1913 patent from BASF that reported formation of hydrocarbons from carbon monoxide and hydrogen over various metal and metal oxide catalysts. Fischer and Hans Tropsch began to study this reaction at various CO/H_2 ratios and at different temperatures and pressures. They initially produced alcohols, aldehydes, carboxylic acids, and ketones, but little hydrocarbon. By 1925–1926, using cobalt catalysts and lower pressure, they were successfully producing hydrocarbons. Pilot pants were constructed in 1932 and 1934.

Source: Koch, H. *Brennst.—Chem.* **1949**, *30*, 3–13; Pilcher, H. *Chem. Ber.* **1967**, *100*, cxxvii–clviii.

   Hans Tropsch was born in 1889 in Bohemia, then part of Austria–Hungary but now in the Czech Republic. He studied in German language universities in Prague. He worked at several locations in the Ruhr from 1916 to 1920 before coming to the Kaiser Wilhelm Institute for Coal Research in 1920. He worked with Franz Fischer and Otto Roelen on the development of the Fischer–Tropsch process. In 1928, he became a professor at the Institute for Coal Research in Prague. In 1931, he came to the United States and was associated with UOP and the related Armour Research Institute. In 1935, because of ill health, he returned to Germany and died shortly thereafter.

Source: Anonymous. *Monit. Pet. Roumain* **1935**, *36*, 1581.

process until the oil price shock of 1974–1975, when attention was again turned to alternative sources of hydrocarbon fuels.

In the first step of the F–T process, coal is converted to a mixture of carbon monoxide (CO) and hydrogen (H_2) that is known as *synthesis gas* (syngas for short). The process requires considerable energy input, which is obtained by burning part of the coal. The chemical sequence consists of pyrolysis, steam reforming and the water-gas shift reaction. Altogether, about half of the carbon in the coal ends up being converted to CO_2 and about half to CO. The synthesis gas is then converted to hydrocarbons by the F–T reaction. The F–T step itself is exothermic so an important factor in design is heat transfer from the reactor. The reaction occurs in the presence of iron or cobalt catalysts and produces a mixture of mainly straight-chain hydrocarbons. Because of the dominance of straight-chain alkanes, the F–T product has a low octane rating and must be subjected to reforming or blended with other hydrocarbons to provide gasoline. On the other hand, the product has excellent qualities as a diesel fuel. Worldwide coal reserves are believed to exceed petroleum reserves by 5–25 times. However, the cost of conversion of coal to hydrocarbons for fuel is currently too high to compete with petroleum. For areas rich in coal, the F–T process may compete economically with petroleum refining as petroleum costs rise.

Coal Gasification: coal + H_2O $\longrightarrow$ CO + H_2

Water-Gas Shift: CO + H_2O $\longrightarrow$ CO_2 + H_2

Fischer-Tropsch n CO + (2n + 2) H_2 $\longrightarrow$ C_nH_{2n+2} + n H_2O

The F–T reaction occurs on the surface of the catalyst by adsorption and reduction of CO. The chain grows by successive steps of addition and reduction of CO. As a result of this mechanism, the product is predominantly straight chain. The product composition can be controlled to some extent by the reaction conditions and catalyst. The low temperature (200–240°C) version uses either Co or Fe catalysts and produces high MW products. The high temperature version over Fe catalysts produces mainly hydrocarbons in the gasoline and diesel fuel range, along with terminal alkenes.

The F–T process was developed extensively by SASOL in South Africa after WWII. The SASOL plants can produce straight chain olefins, alcohols, and carbonyl compounds. SASOL currently operates the F–T process for making diesel fuel, olefins, and waxes. However, most of the plants have been converted to use natural gas, rather than coal, as the source of syngas.

2.2.2 LIQUID HYDROCARBONS FROM NATURAL GAS

Methane from natural gas can be converted to liquid hydrocarbons by the Fischer–Tropsch process. Steam reforming can be used to convert methane to carbon monoxide and hydrogen (synthesis gas). This process is endothermic. Methane can also be converted to synthesis gas by partial oxidation, an exothermic process. The latter reaction was commercialized by ICI in the 1960s.

$$CH_4 + H_2O \longrightarrow C{\equiv}O + 3\,H_2 \qquad 2\,CH_4 + O_2 \longrightarrow 2\,C{\equiv}O + 4\,H_2$$

$$\Delta H = +206\ kcal/mol \qquad\qquad \Delta H = -38\ kcal/mol$$

steam reforming   partial oxidation

The partial oxidation can also be carried out in the presence of CO_2 or H_2O. These reactions are also exothermic so that the heat supplied can be used to operate the reactors. These processes are called *autothermal reforming.*

$$2\,CH_4 + O_2 + CO_2 \longrightarrow 3\,H_2 + 3\,CO + H_2O$$

$$4\,CH_4 + O_2 + 2\,H_2O \longrightarrow 10\,H_2 + 4\,CO$$

It is estimated that natural gas reserves equivalent to 800 billion barrels of oil exist. Methane to hydrocarbon projects have been under consideration since WWII, but have seldom been economically promising. A plant operated at Brownsville, TX, USA from 1951 to 1957. Synthesis gas was produced by partial oxidation of methane, and then converted to liquid hydrocarbon by a F–T process. The plant had a capacity of 350,000 t/year of hydrocarbon fuel. However, at the time it was not economically competitive. Currently, Shell Oil (Bintulu, Malaysia) and Sasol/Chevron (Ras-Laffan, Qatar) operate methane to liquid hydrocarbon plants in locations where large quantities of natural gas are available. The Shell process includes hydrocracking of high MW products back to gasoline/diesel fractions. The Qatar plant uses the Oryx Cobalt-catalyzed F–T process. Figure 2.11 shows the schematic design of the Sasol/Chevron process.

Methanol represents another potential intermediate in the conversion of synthesis gas to hydrocarbon fuel. Methanol can be readily produced from synthesis gas and can be converted to gasoline over zeolite catalysts. However, this process is not currently economically competitive. Formation of methanol directly from methane would be even more desirable, but at present there is no commercially feasible technology for this transformation.

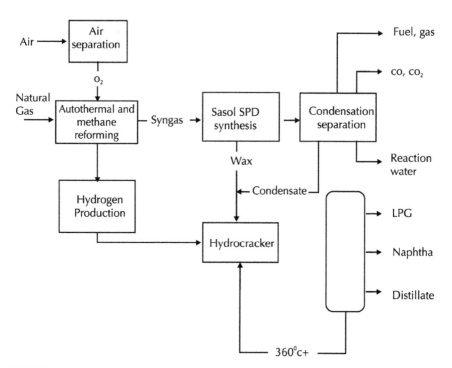

FIGURE 2.11 Schematic diagram for Sasol/Chevron gas-to-liquid Fischer–Tropsch process, (From Leckel, D. *Energy Fuels* **2009**, *23*, 2342–2358. © American Chemical Society).

2.2.3 LIQUID FUEL FROM BIOMASS

Dwindling and more expensive petroleum reserves and the mounting evidence that CO_2 from combustion of fossil fuels contributes to global warming have turned attention to use of biomass as a source of liquid fuel. Because biomass is generated from CO_2 by plants through photosynthesis, its use as a fuel can, in principle, maintain an equilibrium level of CO_2 in the atmosphere. Plant biomass is primarily carbohydrates, such as starch and cellulose that have the composition $C_nH_{2n}O_n$. Biofuels can also be generated from fats and oils obtained from animals or plants which have the composition $C_{n+3}H_{2n+6}O_6$ with $n = 36$–60. Fats and oils have substantially less oxygen content than carbohydrates and therefore have a higher potential energy content. However, they are much less plentiful than carbohydrates.

2.2.3.1   ETHANOL

Currently, ethanol is produced from carbohydrate sources, mainly corn starch and sugar cane. The processes use starch or sugar in a relatively pure form that can be fermented to ethanol. The production of ethanol from sugar cane is more efficient than from corn, both because of its higher ratio of usable carbohydrate and lower energy inputs. Ethanol is used as both a performance additive (oxygenate) and energy source when mixed with hydrocarbons from petroleum. As shown in Figure 2.12, ethanol production in the United States increased about 10-fold between 1980 and 2000. The 1990 Clean Air Act Amendments mandated use of 2–2.7% oxygen content in fuel to reduce air pollution in urban areas. This requires 7.75% by weight of ethanol. The phase out of methyl *t*-butyl ether (MTBE)[k] as an alternate oxygenate in the 1990s further increased the demand for ethanol. The US Energy Policy Act of 2005 mandated that by 2012, at least 7.5 billion gallons of ethanol be blended with gasoline. There are limits, however, to how much corn can be produced in terms of the amount of land and water needed. It is currently a matter of controversy as to whether or not production of ethanol from corn grain provides a net energy gain.

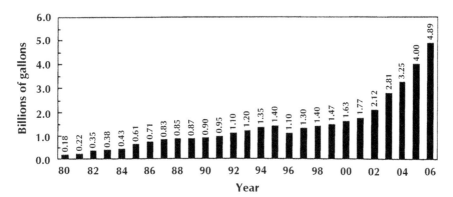

FIGURE 2.12   Ethanol production is the United States, 1980–2006 (From Bai, F. W.; Anderson, W. A.; Moo-Young, M. *Biotechnol. Adv.* **2008**, *26*, 89–105 by permission of Elsevier.)

---

[k]MTBE was used after the phase-out of TEL, but it was soon found to cause ground-water contamination and was banned in many states.

The technology of corn processing had its beginnings in the corn starch industry of the mid-19th century. Production of pure refined glucose (dextrose) began in the 1860s and corn syrup was introduced in the 1920s (see Section 9.2.1). The fermentation process itself goes back to prehistory and is the same as making beer, wine, and other alcoholic beverages. The production of ethanol from corn grain currently involves three steps: (1) cooking to swell and partially hydrolyze the starch, making it accessible to enzymes (gelatinization); (2) digestion by α-amylase enzymes to glucose and other small carbohydrates (liquefaction); and (3) fermentation of the glucose by *Saccharomyces cerevisiae* to produce ethanol. The optimum temperatures for the latter two steps are 65°C and 35°C and so the reactions are normally carried out separately as batch processes. The fermentation step is mildly exothermic. The theoretical yield of ethanol is 51% by weight and 90–93% of this value can be achieved. Carbon dioxide is also produced. Small amounts of glycerol, acetic acid, and lactic acid are the main byproducts.

$$C_6H_{12}O_6 \longrightarrow 2\ C_2H_5OH + 2\ CO_2$$

$$140\ g \qquad\qquad 92\ g \qquad\quad 88\ g$$

The ethanol is produced as a dilute aqueous solution (10–12%) and energy must be used to separate it by distillation. The distillate is an azeotrope containing 4.6% water and for use as a fuel it must be dehydrated. This is accomplished either by a second distillation with benzene, which removes water as a benzene azeotrope, or by drying with molecular sieves. The yeast (*S. cerevisiae*) that is used to ferment glucose must grow as the ethanol is formed. This means that biomass is not only consumed, but also produced in the form of yeast cells. This can be used as an animal feed. Depending on the process used, corn oil, and gluten may also be obtained as byproducts as summarized in Figure 2.13.

Fermentation of sucrose from sugar cane is used extensively in Brazil. Brazil's production of ethanol is about 20% of its petroleum production (2008), and Brazil is a major exporter of ethanol. As of 2000, it was estimated that the cost of ethanol from sugar cane as practiced in Brazil was about $0.20/l as compared to $0.25/l from corn in the United States. Neither the United States nor Brazilian industry currently use the fibrous cellulosic biomass (cobs, stalks, bagasse) produced by growing corn or sugar cane.

Biotechnology is being applied to the production of ethanol from starch. For example, enzymes that can hydrolyze solid starch without the need for the endothermic cooking (liquefaction) step are being developed.[2] Dramatically improved efficiency of ethanol production will require the conversion of *cellulosic biomass*, such as wood and the fibrous portion of corn and sugar cane. Cellulosic material may also be available from plants such as switchgrass that can be grown

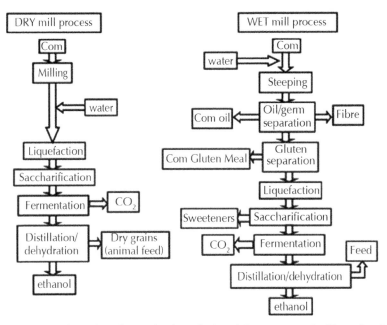

FIGURE 2.13 Flow sheet for production of ethanol from corn grain (From Ramirez de la Piscina, P.; Homs, N. *Chem. Soc. Rev.* **2008**, *37*, 2459–2467 by permission of the Royal Society for Chemistry.)

on land not suitable for corn or sugar cane production. Fast-growing trees might also provide a source of biomass. The carbohydrates present in these materials are largely in the form of *cellulose*. This is a structurally different polymer of glucose and is considerably harder to process. In addition to glucose polymer (35–50%) about 20–30% is branched material containing other carbohydrates. These structures are not amenable to hydrolysis by the enzymes that can hydrolyze starch. Cellulosic biomass also contains noncarbohydrate material, such as lignin, which includes phenolic alcohols, such as *p*-coumaryl, coniferyl, and sinapyl alcohols. There are bacterial and fungal enzymes that can accomplish the hydrolysis of cellulose, but at the present time they are too expensive to use on a commercial scale. Figure 2.14 summarizes conversion of cellulosic biomass to ethanol.

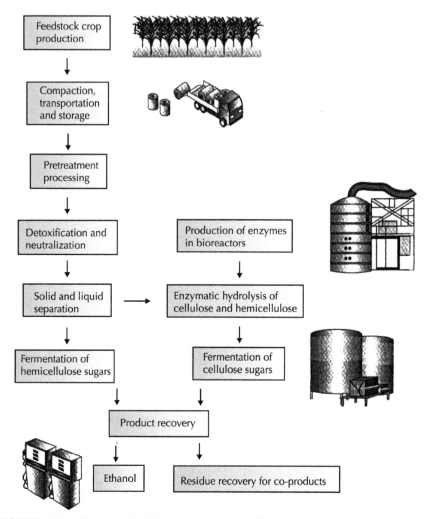

FIGURE 2.14 Steps required for conversion of cellulosic biomass to ethanol. (From Sticklen, M. B. *Nat. Rev. Genet.* **2008,** *9*, 433–443 by permission of Nature Publishing Group.)

A 2006 study compared "energy return" for ethanol produced from corn with that produced from cellulosic biomass. Five of the six studies of corn ethanol calculated positive ratio ranging from 1.29 to 1.62, indicating 29–62% gain in energy content over energy input. Three of the four studies on cellulosic ethanol yield gave values ranging from 4.40 to 6.61, but one study showed a fractional return (0.69), indicating more energy is consumed than produced. Assuming that a "consensus" can be derived from these results, they indicate a significant (roughly 40–50%) gain in usable energy via corn fermentation and potentially much larger (400–600%) for cellulosic ethanol.[3] It should be noted that on the basis of crude oil energy content, the energy return of gasoline is fractional, because of the energy needed to extract, transport, refine, and distribute it. It is the availability of existing petroleum deposits that give petroleum its economic advantage.

It is estimated that the United States could produce as much as 30% of its transportation-fuel needs from cellulosic ethanol, if the process were economically competitive. Current efforts are devoted to improving the economics of production of ethanol from cellulosic material. The process requires pretreatment, usually with acid, to partially hydrolyze the polymer and the use of special enzymes that are able to break down the cellulose. These enzymes may be developed on the basis of genetic engineering. The ideal system would involve a continuous catalytic process that could effect these transformations with minimal generation of other material. Another approach is to convert the biomass to liquid hydrocarbon, rather than ethanol. Carbohydrates such as cellulose are also a potential source of synthesis gas for conversion to hydrocarbons by the Fischer–Tropsch process (see Section 2.2.1.3). As with coal, this requires partial combustion of the material to provide the energy for gasification. The same chemical reactions are involved as in coal gasification, but because biomass has a higher hydrogen content than coal, the process differs in details. The biomass is first converted to a material of suitable moisture content and uniform particle size. It can then be subjected to pyrolysis, which generates a gaseous mixture that can be subjected to steam-reforming. Finally, the composition of the gas can be modified by the water-gas shift reaction. Some potential for improving the efficiency of biomass gasification exists in the use of *supercritical water.* A supercritical fluid is formed above a specific temperature and pressure. For water, these are >221 bar and $>375°C$. The supercritical water has excellent mixing properties.

2.2.3.2 BIODIESEL

Biodiesel fuel can be obtained from materials such as used frying fat, animal tallow, and soybean oil. These substances are all esters of glycerol and can be transesterified to methyl esters, which have higher volatility and can be blended with other types of diesel fuel. The plant-derived material is unsaturated and is subjected to hydrogenation.

$$CH_2O_2C(CH_2)nCH_3$$
$$HC-O_2C(CH_2)nCH_3 \quad \xrightarrow[\text{heat}]{CH_3OH} \quad 3\ CH_3O_2C(CH_2)nCH_3 \ + HOCH_2CHCH_2OH$$
$$CH_2O_2C(CH_2)nCH_3 \qquad\qquad\qquad\qquad\qquad\qquad\qquad\qquad | \\ \qquad\qquad\qquad\qquad\qquad\qquad\qquad\qquad\qquad\qquad\qquad\qquad OH$$

One of the fundamental problems with biofuels is that they compete with food production for agricultural resources. The total amount of fuel required for transportation is currently 4–6 times the amount of biomass produced for food. Thus, biofuel has the potential to crowd-out food production.

2.2.4 PROSPECTS FOR ALTERNATIVE ENERGY SOURCES

Several alternative energy sources have been investigated during the final quarter of the 20th century. Unfortunately, to date they have provided only modest increments to total energy use.

2.2.4.1 SOLAR ELECTRICITY GENERATION

Solar electricity generation is technologically feasible and used on a small scale. The current limitations are the area needed for collection on a large scale, the necessity for storage in areas where sun is intermittent and the relatively high cost of the current devices.

2.2.4.2 FUEL CELLS

There are several designs for fuel cells and they are more efficient than internal combustion engines. Small scale devices are used in several applications. Fuel cells based on hydrocarbons have been developed, but they produce CO_2 as a byproduct. The hydrocarbons are converted to hydrogen by reforming, which can be done in a separate process or incorporated into the fuel cell design. The oxidation of the hydrogen provides the electrical energy produced by the cell.

2.2.4.3 WIND AND WAVES

Both wind and ocean waves are forms of energy derived from the effect of solar energy on the weather. Wind power is feasible as an incremental power source, but requires investment in generation, storage, and transmission infrastructure. Similarly, structures for harnessing the power of waves have been developed as incremental power sources.

2.2.4.4 NUCLEAR FISSION

Nuclear fission, the basis of the current nuclear generation of electricity, is highly efficient and has relatively small direct environmental impact. The disadvantage is that nuclear waste is generated and needs to be stored safely. This is a technically solvable problem, but public fear of radioactive waste and nuclear accidents has retarded nuclear electric power generation in the United States and most of Europe, France being a significant exception. The Chernobyl incident illustrated the potential dangers when coupled with poor design and management. The 2011 Japanese earthquake and tsunami demonstrated that extreme precautions against unexpected events are necessary at nuclear power plants.

2.2.4.5 NUCLEAR FUSION

Nuclear fusion has the potential to generate vast amounts of energy with the production of only minimal amounts of radioactive waste. However, it requires temperatures in excess of 1 million °C and so far has been accomplished only in short experiments.

2.3 PETROCHEMICALS

A second major industry arose based on petroleum, that being petrochemicals. A wide variety of compounds can be obtained by further processing of refined petroleum products. The chemical industry developed first in Germany and Britain (see Chapter 3) but was based on coal. In the 1930s, the United States took the lead in petrochemicals as reactors were designed to take advantage of the availability of small alkenes formed as byproducts of the cracking methods. Among the pioneering companies were Union Carbide, Standard Oil (NJ), Dow Chemical and Shell. The first compounds to be produced in quantity were isopropyl alcohol and acetone, which were made from propene by hydration and subsequent dehydrogenation. Isopropyl alcohol was used as an automobile antifreeze in the 1920s. Acetone was in demand for the preparation of Cordite explosive (see Section 1.3.1) and also as an antiknock additive for aviation fuel.

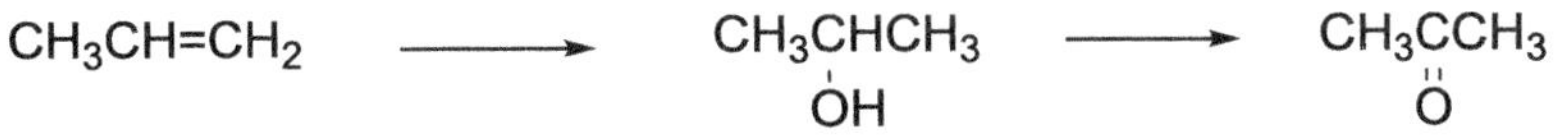

$$CH_3CH=CH_2 \longrightarrow \underset{\underset{OH}{|}}{CH_3CHCH_3} \longrightarrow \underset{\underset{O}{||}}{CH_3CCH_3}$$

Union Carbide saw the opportunity to develop ethylene as a source of chemicals as a result of research on steam-cracking it sponsored at the Mellon Institute in Pittsburgh. The original plant near Charleston WV produced ethylene from ethane and converted it to ethylene oxide and ethylene glycol. The ethylene glycol was sold as Prestone antifreeze. Its much higher boiling point was a great advantage over lower boiling alcohols, such as isopropanol. Several solvents were also produced from ethylene, including dichloroethane, diethylene glycol, and several related methyl and ethyl ethers.

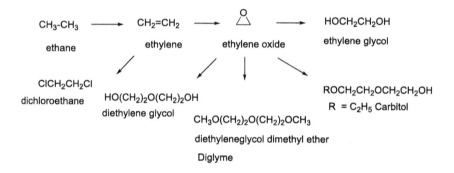

Shell initiated a petrochemical research effort near its Martinez, CA, USA refinery in 1928. The first facility made ammonia using natural gas (see Section 8.1.3). The company also had technology for thermal cracking of hydrocarbons. Petrochemicals products included acetone and methyl ethyl ketone, which were made from propene and butene, respectively, by acidic hydration and oxidation. A high-octane aviation fuel component (isooctane) was made by acid-catalyzed dimerization of isobutene, followed by hydrogenation. Methodology for synthetic glycerin from propene via allyl chloride was also developed, but not commercialized until after WWII.

$CH_3CH=CH_2$ ⟶ CH_3CCH_3 $CH_3CH=CHCH_2$ ⟶ $CH_3CCH_2CH_3$

propene $\overset{\text{O}}{}$ butene $\overset{\text{O}}{}$

            acetone   methyl ethyl ketone

$(CH_3)_2C=CH_2$ ⟶ $(CH_3)_3CHCH_2CH(CH_3)_2$ $CH_3CH=CH_2$ ⟶ $CH_2=CHCH_2Cl$

isobutene   isooctane   propene   allyl chloride

               (2,2,4-trimethylpentane)

Dow's original technology base was halogen chemistry, including the Dow process for making phenol from chlorobenzene which was commercialized in the early 1920s. Another important product was 1,2-dibromoethane, which came into use along with TEL as an antiknock additive (see Section 2.4.1). The halogen source was seawater and Dow decided to build a plant on the Gulf Coast at Freeport, TX, USA in 1940. The initial products included ethylene glycol, 1,2-dichloroethane, 1,1-dichloroethylene, as well as styrene from ethylene and benzene.

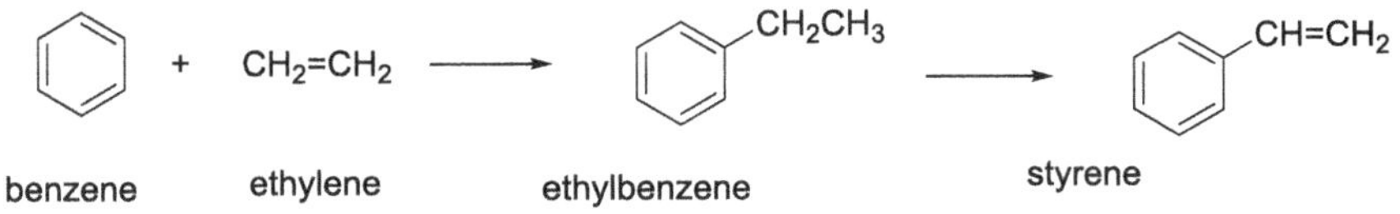

benzene   ethylene   ethylbenzene   styrene

Standard Oil (NJ) had a technology sharing agreement with IGF that provided access to high pressure hydrogenation technology. SO(NJ) developed *hydrocracking* for the upgrading of heavy crudes and also discovered methods for alkylation under highly acidic condition that provided saturated hydrocarbons by combination of an alkene and alkane (see Section 2.1.5). Standard Oil (IN) developed a process, called *hydroforming*, that produced a high ratio of aromatic compounds. Among the compounds produced by this technology was high purity toluene, needed during WWII for the production of TNT (see Section 1.2.3).

The growth of the polymer industry in the 1940s created an expanded need for alkenes as monomers. Thermal cracking gives the smaller alkenes in the order ethylene > propylene > butenes, and operating conditions have relatively little effect on the ratio. The introduction of catalytic crackers increased the ability to control the product mix and produced relatively larger amounts of propene and butenes. Both butadiene and styrene became crucial starting materials for synthetic rubber during WWII (see Section 5.3.2.3). 1,3-Butadiene became available by catalytic dehydrogenation of butane. In addition to Dow's route to styrene by catalytic dehydrogenation of ethylbenzene, Koppers, Union Carbide, and Monsanto also constructed plants for styrene. At the end of the war, these government-financed petrochemical plants were sold at advantageous prices, often to the firms that had constructed and operated them. The vast new capacity for petrochemicals provided an impetus for finding civilian uses. The greatest demand came from the manufacture of polymers (see Chapter 5).

Catalytic reforming also generated significant amounts of aromatic compounds (see Section 2.1.4). In the 1950s as the production of benzene from coal began to decrease, improved processes for isolation of benzene from petroleum were developed. Dow and UOP developed a process that involved preferential extraction into a glycol–water mixture. Later, sulfolane was used as the extraction solvent. With the introduction of polyesters derived from terephthalic acid, *p*-xylene (1,4-dimethylbenzene) became a valuable intermediate and was also isolated by selective extraction.

Many improved processes for key petrochemicals were introduced in the 1950s and early 1960s. SOHIO developed a process for conversion of propene to acrylonitrile. Oxychlorination of ethylene provided a new route to vinyl chloride. These compounds are the monomers for polyacrylonitrile and polyvinyl chloride (PVC) (see Chapter 5).

$$CH_2=CHCl \qquad\qquad CH_2=CHCN$$

vinyl chloride   acrylonitrile

Immediately after WWII, the United States enjoyed the advantage of its intact and greatly expanded petrochemical complex. In contrast, most of the plants in continental Europe had been damaged or destroyed. Furthermore, the existing technology for chemical production in Britain and Europe was based on coal, not petroleum. There were not large supplies of petroleum in Western Europe. The IGF combine had been disbanded to the successor companies, Bayer, BASF, and Hoechst. Europe was also in the process of creating the European Community and the Common Market. European markets had previously been smaller national markets and there was limited experience with international companies. While American companies took first advantage of these opportunities, by the 1960s the European firms had been reorganized and the raw material shifted to petroleum from the Middle-East. Sharing of technology provided for rapid spread of improved processes and the European petrochemical industry rapidly became competitive and international in scope.

A second major change occurred in the 1970s. At the beginning of the decade, crude oil was about $3/barrel and had actually dropped in terms of constant dollars. In 1973, the Arab countries imposed an oil embargo, largely in response to long-festering discontent at their share of the revenues, but triggered by the Arab–Israeli war of that year (see Topic 2.4). Within a few months crude reached $17/barrel, a >500% increase. Petrochemical companies raised their prices accordingly and profits accelerated, but soon inflation and a severe recession occurred. With major new revenues available, the oil-rich countries began to build their own petrochemical complexes, adding to worldwide capacity.

Petrochemical processes have made many new materials available. A wide variety of industries used these chemicals to produce new, better or cheaper materials for the market place. Among the industries, we will consider later are polymers, consumer products (detergents, etc.), and agricultural chemicals. All of these became users of large amounts of petrochemicals.

2.4   ENVIRONMENTAL CONSEQUENCES OF HYDROCARBON FUEL USAGE

2.4.1   ANTIKNOCK ADDITIVES FOR GASOLINE

A major issue with automobile engine performance arose in the 1920s. The problem of knocking presented the first case of an environmental effect associated with hydrocarbon fuels. In 1923, TEL was introduced as an antiknock additive for gasoline. TEL is a very toxic compound and within a year a number of cases of insanity and death had occurred among manufacturing workers exposed to TEL. They were widely reported at the time. Several physicians and public health workers, among them Alice Hamilton,[1] argued that the toxicity of lead should preclude its widespread use. The industry argued that there were no known alternatives. Government hearings ensued to investigate the safety of TEL use. Influenced by the commercial interests supporting its use, including General Motors, Standard Oil (NJ), and DuPont, the Public Health Service declined to ban TEL. TEL was used until the mid-1980s, and millions of tons of lead became widely distributed in the environment.

$$C_2H_5-\overset{\displaystyle C_2H_5}{\underset{\displaystyle C_2H_5}{Pb}}-C_2H_5$$

tetraethyllead

Concern about TEL use continued, however, and a geochemist, Clair Patterson, was able to demonstrate that there had been a 200-fold increase in the amount of lead in the human body accompanying the onset of the technological

---

[1] Alice Hamilton was born in New York in 1869 and grew up in Fort Wayne, IN, USA. She received an M. D. from the University of Michigan in 1893. She studied bacteriology and pathology in Germany from 1895 to 1897 and then became a professor of pathology at the Woman's Medical School of Northwestern University. When it closed, she moved to the Memorial Institute for Infectious Diseases in Chicago. She was a resident of Hull House and an associate of its founder Jane Addams. She became interested in occupational medicine and the hazards associated with certain jobs. She was one of the first American physicians to do research in this area. She served on the Illinois Occupational Disease Commission and was instrumental in the passage of early legislation for workers compensation in Illinois. In 1919, she was appointed an assistant professor in the new Department of Industrial Medicine at Harvard, being the first women appointed to the Harvard Medical School Faculty. Later, she was a Professor in the School of Public Health. She was a vocal critic of TEL during the 1925 controversy. She died in 1970 at the age of 101.

era.[m] Hearings led by Sen. Edmund Muskie (D. Maine) in 1966 eventually focused public attention on the issue and resulted in TEL being phased out between 1976 and 1986. Since the banning of TEL, lead levels found in the blood of children have dropped sharply. There was another incentive for removal of lead from gasoline. The Clean Air Act of 1970 required reduction of the amount of carbon monoxide, nitrogen oxides, and hydrocarbons in automobile exhaust. Meeting these requirements required the introduction of catalytic converters. The proper operation of the catalytic converters is incompatible with leaded gasoline, so the industry, too, had reason to remove the TEL.

TEL was first replaced by MTBE. Soon after MTBE was introduced, it was found as a ground water contaminant resulting from storage tank leakage and spillage. In contrast to hydrocarbons, MTBE has appreciable water solubility and spreads readily from leakage and spillage sites. A controversy followed that resulted in MTBE being banned in many states and introduction of federal regulations on storage tank leakage. At the current time ethanol is the major oxygenated compound used to improve engine performance.

[m]Clair Patterson grew up in Iowa and received a B. S. degree from Grinnell College and an M. S. from the University of Iowa. He then received a Ph. D. from the University of Chicago. In his graduate work on the age of the earth, he found that lead contamination was so ubiquitous that extensive cleaning of all material used to collect, process, and analyze samples was imperative for reliable results. He also worked at Oak Ridge National Laboratory as part of the Manhattan Project, where he worked on analysis of ^{235}U enrichment. He took a position as a geochemist at Caltech and continued research aimed at determining the age of the earth by measuring ratios of lead and uranium isotopes in meteorites. In the course of these studies he recognized that much of the lead in the environment was from man-made sources. He examined lead concentrations in ice cores from the Greenland snow-pack and found a 230-fold increase over the past 3000 years. He also found that lead levels were much higher in surface ocean waters than at great depths. Patterson also concluded that measurement of lead levels in blood alone was misleading, because lead is sequestered in the bones. By studying pre-iron age mummies, he demonstrated that human lead levels were 500–1000 times higher than in the preindustrial age. The publication of this information prompted a counter-attack by the petroleum, automobile, and lead industries. The US Public Health Service and American Petroleum Institute did not renew his research grants. Trustees at Caltech, influenced by the automobile industry, attempted to have him fired. These industries funded research aimed at showing that the high level of lead Patterson found were "normal." Robert Kehoe was the leading spokesman for this view. Patterson challenged the assertion that the "normal" levels were "natural." The National Academy of Sciences (NAS) issued reports in 1972 and 1980 that largely supported the industry view, although the 1980 report included a dissenting section drafted by Patterson. It was not until 1993 that a third NAS report finally acknowledged the extent and some of the consequences of lead contamination.

Patterson's studies also led to the recognition of the hazards associated with other lead-containing materials such as solder in cans for vegetables, lead batteries, and lead-based paints. A strong connection between childhood lead exposure and neurological development was also established. In the 1940s, a Boston pediatrician, Randolph Byers, found evidence that lead intoxication of children was correlated with disruptive behavior and learning deficits. In 1974, a study of around 2500 school children in the Boston area showed a connection between the level of lead in children's teeth and behavior and school performance. The industry attacked these results, impugning the integrity of the research and threatening legal action. Charges of research misconduct were instigated by industry sources, but proven to be unfounded. The results were becoming overwhelming, however, and gradually action was taken to remove lead-based paints (1977), lead solder from canned food (1996), and tetraethyllead from gasoline (1986).

Sources: Flegal, A. R. *Environ. Res., A* **1998**, *78*, 65–70; Nriagu, J. O. *Environ. Res., A* **1998**, *78*, 71–78; Needleman, H. L. *Environ. Res., A* **1998**, *78*, 79–85; Rosen, J. F.; Pounds, J. G. *Environ. Res., A* **1998**, *78*, 140–151; Davidson, C. I., Ed. *Clean Hands: Clair Patterson's Crusade against Environmental Lead Contamination*, Nova Science Publishers, Commack, NY, 1999.

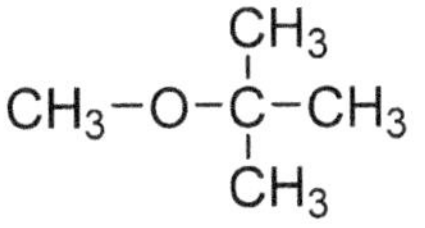

$$
\begin{array}{c}
\quad\quad\ \ CH_3 \\
\quad\quad\ \ | \\
CH_3-O-C-CH_3 \\
\quad\quad\ \ | \\
\quad\quad\ \ CH_3
\end{array}
$$

methyl *t*-butyl ether (MTBE)

2.4.2   AIR QUALITY

The most evident effect of hydrocarbon fuel use on the atmosphere began to become apparent soon after WWII. As early as 1905, Henry A. Des Voeux had coined the term "smog" in a paper entitled "Fog and Smoke," referring to a combination of smoke (at that time from coal) and fog that sometimes enveloped London. By the 1950s, smog was a familiar word in America, first as the basis for jokes about the Los Angeles basin, but not long thereafter it was recognized as a threat to health. Hydrocarbon fuel exhaust contains a variety of pollutants. These include particulate matter, volatile organic compounds, carcinogenic aromatics (e.g., benzene, polycyclic aromatic hydrocarbons), carbon monoxide, and nitrogen dioxide. There are secondary and even tertiary consequences. For example, NO_2 in automobile exhaust reacts with O_2 to form ozone. Ozone, a powerful irritant, damages forests, leading to increased tree death and fuel for wildfires.

Targets for exhaust emissions have gradually moved downward. There are several reasons for this. Targets are influenced by technological feasibility and improved technology has permitted lower limits. Longer range studies have detected adverse effects at lower levels, increasing the pressure for reduction. Public knowledge and interest in controlling emissions has provided an impetus for policy makers to work for lower targets. Public concerns about automobile exhaust resulted in the passage of the Clean Air Act of 1970. By that time, smog was clearly visible from the air over most urban areas and "smog alerts" were common on both the East and West Coasts. The law set exhaust emission targets that forced the automobile industry to introduce catalytic converters and, as a consequence, switch to unleaded gasoline.

2.4.3   CARBON DIOXIDE AND GLOBAL WARMING

The largest and least controllable product of hydrocarbon combustion is carbon dioxide. The amount of carbon dioxide produced is truly mind-boggling. According to US DOE statistics, total CO_2 emissions in the United States for

1990, 1995, and 2000, were, respectively, 5.0, 5.3, and 5.9 billion metric tons. Nearly all of this (2000) came from petroleum (42%), coal (36%), and natural gas (20%). The main sources are electricity generation (38%), transportation (32%), manufacturing (14%), and commercial and residential buildings (9%).

The annual solar flux entering the earth's atmosphere is 385×10^{22} J. About 0.08% of this is harvested by photosynthesis and turned into organic material. Over the eons deposits of this organic material have built up in the form of coal, natural gas, petroleum, and shale oil. These are called *fossil fuels*. During the 20th century, humans expanded the use of these deposits for heating and cooling, generation of electricity, manufacturing, and transportation. Total annual human energy use is currently about 500×10^{18} J, which is roughly 15% of photosynthesis and 0.013% of the total solar energy flux. But with the bonanza that fossil fuels provide, come some problems. The first is that the rate of generation of fossil fuels in the current time frame is negligible. There are finite limits on the amount that is available. A second major problem is the chemical consequences of fossil fuel use, primarily the generation of excess CO_2, relative to previous levels. With this excess generation of CO_2 comes the prospect of global warming.

The amount of CO_2 released per unit of energy produced depends on fuel type. The ratio between natural gas, petroleum, and coal is about 1:1.5:2 (see Section 2.1). The extensive use of coal for electricity generation in the US results in about 40% of CO_2 emission coming from electricity generation. As of 2004, the United States accounted for about 22% of total worldwide CO_2 emission.[4] Since the beginning of the industrial age around 1850, the total amount of CO_2 released is estimated to be about 270×10^9 t. Very accurate measurement of CO_2 levels is possible and the net increase in the atmosphere in that period is about 176×10^9 t. Thus, about 2/3 of the CO_2 has been added to the atmosphere and another 1/3 absorbed by oceans and other components of the ecosphere. As shown in Figure 2.1, the level of CO_2 in the atmosphere is continuing to increase. Figure 2.15 shows carbon dioxide levels for the past 400,000 years.

Carbon dioxide is a "greenhouse gas," which means that it absorbs radiation and converts it to thermal energy, increasing the fraction of energy from the sun that is retained on the earth and its atmosphere. The change in the energy balance leads to net global warming. Carbon dioxide is not the only greenhouse gas. In fact, water is the main greenhouse gas, but it is maintained in a fluctuating range by atmospheric circulation. On the other hand, CO_2 is increasing as a result of use of fossil fuels. Methane and nitrous oxide, N_2O, are also greenhouse gases, although less important than CO_2.

In its 4.5 billion years of existence, planet earth has seen a wide range of climatic conditions including being completely covered with ice and totally ice free. The current polar ice caps date from about 34 million (South) and 10 million (North) years ago. The energy received from the sun is very slowly increasing. There is some irregularity because of variation in sun-spot activity. In addition there is a periodic variation in the amount of radiation adsorbed because the axis of the earth wobbles slightly. These variations have a period of about 25,000 years and every third period results in warming. We are currently in such a period. The glacial periods dominate otherwise, as most recently about 20,000 years ago, when Europe and North America were covered by ice down to about 35° latitude. At that time northern Europe was coved by ice 3000 m thick and the sea level was 120 m lower as the result of the accumulated ice. Ice coverage has a very large effect on temperature, because of its high reflectance. Cloud coverage also leads to cooling by reflecting the sun's energy. Fairly reliable temperature records have been reconstructed for nearly 1 million years and show cyclic variations, with current time being in a relatively warm period that began some 20,000 years ago. Figure 2.15 shows the temperature record for the past 800,000 years.

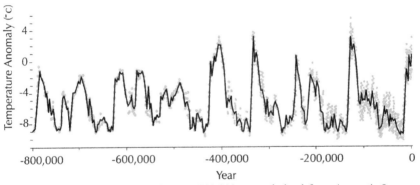

FIGURE 2.15 Temperature for the past 800,000 years derived from Antarctic Ice cores. (From Earthobservatory.NASA.gov.)

During the last quarter of the 20th century, the extent of CO_2 emission and build-up and its role in climate change became a major area of investigation. That CO_2 is emitted and that its level is increasing in the atmosphere is both logical and measureable. The issue is the effect on temperature and climate of this build-up. This is a very difficult problem, but as understanding of influences on climate and how they feed back on one another has developed, it has become possible to model the climate change effects of CO_2 build-up. The current consensus of experts in the field, as represented by the Fourth Assessment Report (2007) of the Intergovernmental Panel on Climate Change, is that "most of the observed increase in global average temperature since the mid-20th century is very likely due to the observed increase in anthropogenic greenhouse gas concentration." It further states that the temperature record "suggests it is likely that the rate and duration of the warming of the 20th century is faster and longer, respectively, than any other time in the last 1000 years."

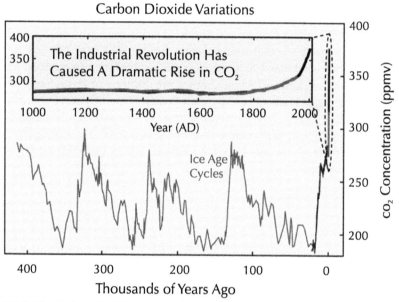

FIGURE 2.16   Carbon dioxide concentration over the Past 400,000 (From nasawavelength. org.)

In 1994, the United Nations Framework Convention on Climate Change (UNFCCC) took effect, in which most countries in the world committed to take steps toward reducing greenhouse gas emissions. The commitments were not specific goals for reduction but, rather to adopt policies to restrict increases and mitigate the potential consequences of greenhouse gas emissions. Some countries have made some progress, but in most, emissions have continued to increase. US emissions increased by nearly 20% in the period 1990–2006. The United Kingdom and a few of the western European countries have recorded declines attributable to reduced use of coal for electricity generation. Most of the former Soviet bloc countries show substantial decreases in the period, partly from decreased economic activity, but also from improved efficiency of energy use. European countries with more rapidly expanding economies such as Ireland, Spain, and Greece show substantial increases. Rapidly developing countries such as India and China are expected to rapidly increase CO_2 emissions as their economies grow. They have been resistant to limitations when their current levels of energy use (and standard of living) are below those of the United States and Europe. In 1997, the UNFCCC adopted a new more ambitious set of goals. In the United States, the protocols of the UNFCCC were rejected by President George W. Bush in 2001 on the basis that they would cause increased energy costs.

TOPIC 2.3 THE HISTORY OF GLOBAL WARMING

The issue of global warming first started to attract interest in the early 1800s, when geologists began to recognize that there had been times when much of the earth had been covered by glaciers and others when tropical vegetation had covered most of the earth. They began to speculate on what the cause of these vast changes in climate might be. The concept of "green house gases," that is, the reabsorption of infrared radiation from the earth's surface, was recognized in 1827. The first person to attempt to determine if CO_2 from fossil fuel might lead to global warming was the Swedish chemist Svante Arrhenius in the late 1800s. His calculations were unable to account for the many inter-related effects. With a biological colleague, however, he was able to show that the amount of CO_2 being produced by coal consumption was significant with respect to the exist-ing CO_2 levels. Arrhenius suggestion was not much more than speculation, and did not have much effect. Several objections were raised. A prominent Swedish physicist, Knut Angstrom, concluded that additional CO_2 would have little ef-fect. The spectra of gases available at the time were not of very high resolu-tion and indicated that the water and CO_2 absorptions overlapped. That being the case, since water was present in much higher concentration, additional CO_2 would have no effect. It was also believed that because the oceans would absorb much more CO_2 than the atmosphere, the increase in the atmosphere would be minimal. Both of these contentions have since been shown to be incorrect. Beyond these specific objections, the mindset of the time was that human activ-ity was so small in relation to other influences as to be negligible. Various other potential influences on temperature were recognized including sun spot cycles, regular variations in the earth's orbit (Milanovich cycles), volcanic eruptions, and shifts in ocean currents. But climatologists, in general, assumed that climate was a constant, at least on the human time scale, and had little interest in study-ing the potential for global warming.

Several specific findings removed the earlier objections to the significance of atmospheric CO_2 as a radiation absorber. One was recognition that, in fact, an increase in CO_2 levels near the earth's surface would give rise to a thicker layer of CO_2, thus increasing the net absorption of energy. Another was improved resolution of the spectra of H_2O and CO_2, which showed they were not overlap-ping broad bands, but rather sharp lines so that CO_2 would significantly increase total absorption. Study of the absorption of CO_2 by the oceans indicated that the ocean surface would rapidly be saturated, and that only on a very long time scale would the oceans act as a CO_2 sink. Furthermore, increasing ocean temperature would decrease the amount of CO_2 absorbed. The first person to propose that an increase in CO_2 was responsible for observable temperature increases was Guy Stewart Callendar, in a paper to the Royal Meteorological Association in 1938. He presented data suggesting a 10% increase in atmospheric CO_2 since 1890 and proposed that it was responsible for increasing temperature. His data on both points were sketchy and raised little concern. The popular press at the time suggested global warming might lead to increased food production.

However, it remained impossible to carry out calculation on how much of an effect man-made CO_2 might have. Only as computers became available were calculations possible. Physicists Lewis D. Kaplan and Gilbert N. Plass used the improved spectroscopic data to show that energy readsorption by CO_2 was significant. Hans Suess and Roger Revelle[n] of the Scripps Institute of Oceanography, used ^{14}C isotopic ratios to measure the amount of CO_2 derived from fossil-fuel in the atmosphere and to measure the rate of exchange of atmospheric and oceanic CO_2. Revelle addressed the problem of the effect of other constituents of the ocean on CO_2 absorption and found that they decreased the value by a factor of about 10. It was also shown that there was little diffusion of CO_2 beyond the surface levels. Revelle had found evidence for the small vertical diffusion previously when studying radioactive fallout from nuclear tests in the Pacific.

Revelle became interested in obtaining accurate continuing measurement of CO_2 in the atmosphere. Charles Keeling joined Revelle's group at Scripps for this purpose. A series of measurements were made, at first including Antarctica, but later confined to Mauna Loa Observatory in Hawaii. The first results became available in 1958 and subsequent measurements showed unequivocally that CO_2 levels were rising at a rate that would lead to doubling in about 30–35 years (see Figure 2.1). The measurements at Mauna Loa have been maintained continuously since 1958, albeit with frequent difficulties in obtaining the necessary

---

[n]Roger Revelle was born in Seattle in 1909 and grew up in Pasadena, CA, USA. He received his undergraduate degree in geology at Pomona College and then did graduate work in geology at UC Berkeley. He received his doctorate at the Scripps Institute of Oceanography in La Jolla, CA and did most of his scientific work there and eventually became the Director of the Institute. He was instrumental in founding the UC campus at San Diego, CA, USA, and one of its colleges is named for him. He became director of the Center for Population Studies Harvard in 1964, but returned to UCSD in the late 1970s.

Beginning in the 1930s, Revelle conducted studies in collaboration with the US Navy and became a reserve Navy officer. He was called to active duty in 1941 and was engaged in research on SONAR. After the war, he was involved in measurement of radioactive waste generated by nuclear explosion tests in the Pacific. He was influential in the development of the Navy Research Program in Oceanography. One of his areas of study was the buffering effect of ocean water on the absorption of CO_2 and he found that only about half of the CO_2 generated by fossil fuels was absorbed by the ocean, rather than the 98% that had been accepted. This led to Revelle's work on the carbon cycle with Hans Suess and Charles Keating. He received a National Medal of Science in 1990.

Source: Munk, W. H. *Proc. Natl. Acad. Sci., USA* **1997**, *94*, 8275–8279.

funds. In 1981, the Reagan administration began efforts to obscure the potential for global warming and moved to sharply reduce funding for related research. Only congressional hearings led by Rep. Albert Gore (D. TN) resulted in public awareness and continuation of funding.

Another fruitful approach was measurement of the level of CO_2 in ice cores in Greenland and Antartica. These were able to measure temperature and CO_2 fluctuations over 400,000 years including the last four glacial cycles. The range found was between 180 ppm and 280 ppm, as compared with the current level of 380 ppm. In other words, CO_2 had *never before* reached current levels. The data also showed that CO_2 levels slightly lagged temperature effects. Thus, CO_2 levels were not the direct cause of the cycles, but rather an agent of the effect. The cause is the Milanovitch cycles in the earth's orbit, which modify the amount of energy the earth receives. Warming is a result of the change in the tilt of the Earth's axis and leads to release of CO_2 from the oceans and tundra, amplifying the effect. Growth or shrinkage of the polar ice caps modifies the reflectance of solar energy. When radiation decreases, leading to more reflective frozen surface and greater absorption of CO_2 by the ocean, the CO_2 levels decrease. The modern anthropomorphic generation of CO_2 is a new phenomenon that is *superimposed* on the astronomical cycle.

With access to reliable data, various models of the effect of CO_2 level on global climate became feasible. Powerful computers and increasing understanding has permitted modeling of the effects that can follow increasing CO_2 levels and the resulting global warming. As of 2005, these models have successfully predicted measured ocean warming. These models can also predict effects such as sea level rise, precipitation pattern changes, spread of ranges for species, including some that are vectors of plant and animal diseases. At this point in time, societal impacts and consequences seem unpredictable.

Source for Topic 2.3: Weart, S. R. *The Discovery of Global Warming*. Harvard University Press: Cambridge, MA, 2008.

2.5 ECONOMICAL, POLITICAL, AND ETHICAL ASPECTS OF FOSSIL FUEL USAGE

2.5.1 DOMESTIC PRODUCTION AND CONSUMPTION IN THE UNITED STATES

The availability of hydrocarbon fuels and the invention of the machines they power triggered enormous growth in industry and transportation. In particular, automobiles brought expanded manufacturing activity, increased mobility and growth of the many services associated with transportation, including tourism.

The development of refining, cracking, reforming, and petrochemicals put the United States in a unique economic position by the end of the 1930s. Western Europe had little petroleum and at that time its chemical industry was based on coal. Japan also had no domestic petroleum and its small chemical industry, too, was based on coal. The United States also had special strength in process design, based on the principles of chemical engineering that made processes continuous and efficient. These circumstances permitted the US petroleum and petrochemical industry to meet the challenges of WWII, including production of aviation fuel and synthetic rubber. At the end of WWII, the petroleum industry faced several large issues. The extensive capacity that had been built during the war needed to be converted to peacetime use. That occurred relatively smoothly. Rationing and price restrictions were removed quickly. The government-owned refining capacity that had been built in conjunction with existing refineries was sold and converted to domestic production. The pipelines that had been built were sold to private companies for conversion to natural gas transmission lines.

But, as it had periodically, concern about the ratio between domestic production and consumption arose. A report of the US Tariff Commission in 1946 frames the issue: "If the recent rate of decline in the ratio of new discoveries to current production should continue, the time would soon come when production would exceed discoveries and known reserves would begin to fall off. Many geologists and oil experts believe the United States will relatively soon be running short of crude oil and will have to rely increasingly on imports. Many others hold that the decline in discoveries during the last few years has been due largely to wartime conditions and that now that the war is over the rate of discovery will rise again. Ability of the United States to supply its petroleum requirements in the future will depend, of course, somewhat on the magnitude of consumption. If national income remains high, consumption within a few years may become almost as large as during the war, and possibly even larger."[5]

The last sentence of that analysis proved to be correct. The economic interests of both the automobile and domestic oil industries converged in high consumption. Embodying the commitment to automotive transportation, the construction of the Interstate Highway System, later to be named for President Dwight D. Eisenhower, began in 1956. The discovery of the vast oil reserves in the Middle-East opened a new source of very cheap petroleum. With this access to cheap foreign oil, another economic force entered the picture. Domestic production, especially by the many small independent companies, could not compete in price with imports. In March, 1959 President Eisenhower instituted limits on oil imports to protect domestic price levels. The regulations restricted imports to about 10% of domestic production.

Crude production in the United States rose steadily after WWII, from about 2 billion b/d in 1945 to 3.5 b/d in 1959. Exports remained rather steady at about 100 ± 20 million b/y. Imports began to rise steadily from 43 million barrels in

1945 to 350 million by 1959. The pace of new discovery slowed, however, and the ratio of proven reserves to annual consumption dropped from 15:1 in 1940 to 10:1 in 1959. At the same time, the cost of discovery and extraction increased substantially. Domestic production remained somewhat below capacity, largely as the result of regulatory efforts aimed at avoiding over-supply of crude and maintaining domestic oil prices high enough to keep marginal wells profitable. In the 1960s, imported oil remained substantially cheaper than domestically produced oil.

Marion K. Hubbert, a Shell geologist, developed methods for estimating the rate of decline of oil production and in 1956 predicted US total reserves would peak in about 1970 at 2 trillion barrels. In fact, US oil production peaked in 1972, but much higher than Hubbert had predicted at just under 10 million bpd and is now (2005) around 5 million bpd, while current consumption is about 20 million bpd. There was a significant increase between 1980 and 1985, corresponding to the development of the Alaska fields, but production has decreased steadily since then, as shown in Figure 2.17. Since 2000, the development of new recovery technology, in particular hydraulic fracturing ("fracking"), has increased production from shale oil deposits, especially in Pennsylvania, Texas, and North Dakota.

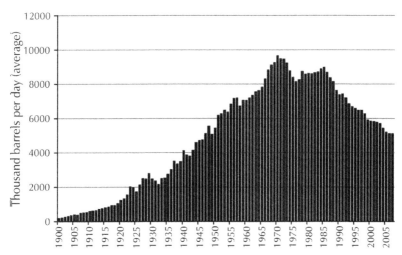

FIGURE 2.17 Average daily petroleum production in the United States, 1900–2007. (From Gleick, P. H.; Palanimappan, M. *Proc. Natl. Acad. Sci. U.S.A.* **2010,** *107*, 11155–11162).

The United States was a net exporter of oil during the first half of the century. Exports peaked during WWII. From that point on imports, began to increase and reached 20% by the 1970s. With decreasing domestic production and increased consumption, imports continued to rise and reached half of consumption in1995. Figure 2.18 shows domestic production and imports for 1950–2012.

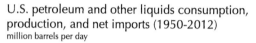

U.S. petroleum and other liquids consumption, production, and net imports (1950-2012)
million barrels per day

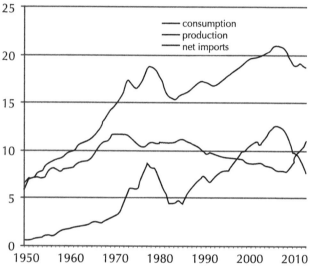

FIGURE 2.18   Domestic production, net imports and consumption, 1950–2010. (From US Energy Information Administration *Monthly Energy Review,* April, 2013).

The availability of "cheap" fuel before 1975 encouraged careless use, automobile efficiency was poor and public transportation neglected. The result is that a century and a half after the 1859 discovery in Pennsylvania, the American economy is utterly dependent on imported petroleum. There have been various efforts to reduce total energy use and reliance on imported petroleum. Most significant are the "Corporate Average Fleet Economy" standards introduced under the leadership of Pres. Gerald Ford in 1975. The standard mandated an improvement of more than double (13–27.5 mpg) for domestically produced automobiles. With transportation accounting for about 70% of petroleum usage, this represents an enormous savings. A 2002 report of the National Academy of Sciences estimates the reduction amounts to about 14%, relative to what would otherwise have been used.[6] In 2007, these standards were revised with a target of 39.0 set by 2016. They were recently (2011) revised to 54.5 mpg for 2025. It was also announced that fuel efficiency for heavy duty trucks and buses would be instituted with the goal of 15–20% increase in efficiency by 2018. Fuel efficiency has improved by about 25% between 1950 and 2000, but has been more than offset by the increased average distance per vehicle by 30%, and by a three-fold increase in the number of vehicles.

Other factors, including price increases and regulation, have led to efforts to reduce industrial and household use of energy. Energy efficient appliances and improved building insulation have contributed to improved efficiency of energy utilization. A consequence of the improved efficiency in energy use is the "decoupling" of hydrocarbon fuel consumption from economic growth as measured by GDP. Figure 2.19 shows that between 1973 and 2001, total fuel consumption dropped by 50% per constant (1996) dollar of GDP. There is a noticeable break in the slope of the curve in Figure 2.19 at about 1985, where energy prices stabilized. This might have two interpretations: (1) incentives for further improvements diminished as oil prices stabilized or (2) that "easy" improvements had been achieved and further improvements became more difficult to achieve.

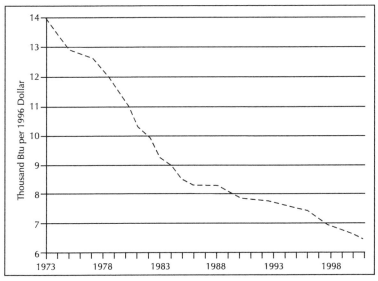

FIGURE 2.19 Consumption of hydrocarbon fuel per constant (1996) dollar of GDP in United States. (From Glover, C.; Behrens, C. E. *Congressional Research Service Report: "Energy: Useful Facts and Figures*, 2003).

Several variables will govern future petroleum consumption in the United States. One is economic and population growth. Another is fuel economy, especially for transportation. A third is total vehicle miles, which is related to both land use patterns and transportation alternatives. At $72/barrel, the daily cost of imported petroleum was nearly $900,000,000 per day in 2006. Most predictions point to higher petroleum costs in the future. Similar factors will effect worldwide growth and, in particular, rapid economic growth in countries such as Brazil, China, and India is expected to add to worldwide demand.

2.5.2 WORLDWIDE PETROLEUM PRODUCTION AND CONSUMPTION

Currently, Saudi Arabia possesses about 20% of conventional petroleum reserves, although information on production is restricted and there is some doubt that the Saudi's currently possess sufficient capacity to make up for short-falls from other producing nations. The estimated maximum capacity for the immediate future is 12.5 million bpd. In contrast, the US EIA in 2004 predicted production of 18.2 million bpd in 2020, based on the assumption that output would continue to increase in response to demand. Current world consumption is about 25 billion bpy. The United States currently uses 25% of global oil. Table 2.2 gives the production, exports, and imports for some of the major petroleum exporting and consuming countries. The United States is the third largest producer of oil, but is by far the largest consumer and as a result is a major net importer of oil at about 12.2 million bpd (2006). Other major net importers are Japan (5.1 million bpd), China (3.4 million bpd), Germany (2.3 million bpd), and France (1.9 million bpd).

TABLE 2.2 Total Oil Production Net Exports, Consumption, and Net Imports in Millions of Barrels/Day.

Production 10^6 bpd		Exports 10^6 bpd		Consumption 10^6 bpd		Imports 10^6 bpd	
Saudi Arabia	10.7	Saudi Arabia	8.7	USA	20.6	USA	12.2
Russia	9.7	Russia	6.7	China	7.3	Japan	5.1
USA	8.4	Norway	2.5	Japan	5.2	China	3.4
Iran	4.1	Iran	2.5	Russia	3.1	Germany	2.5
Mexico	3.7	UAE	2.5	Germany	2.6	S. Korea	2.1
China	3.8	Venezuela	2.2	India	2.2	France	1.9
Canada	3.2	Kuwait	2.2	Canada	2.2	India	1.7
UAE	2.9	Nigeria	2.2	Brazil	2.1	Italy	1.6
Venezuela	2.8	Algeria	1.9	S. Korea	2.1	Spain	1.6
Norway	2.8	Mexico	1.7	Saudi Arabia	2.1	Taiwan	0.9

One mechanism for reducing CO_2 emissions is by more efficient energy use. Figure 2.20 shows an index ratio of CO_2 emission per unit of GDP in various parts of the world and indicates that increased energy efficiency can be compatible with economic growth. The sharp decline in China and the more moderate decline in non-OECD Europe are the result economic expansion and more efficient energy use. Of course, the total emission curve continues to rise because both population and energy use per individual are increasing, especially in the developing world.

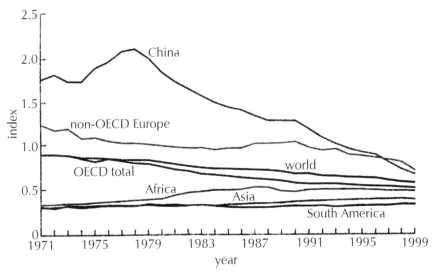

FIGURE 2.20 Trends in fossil fuel emission relative to GDP. The index is the ratio of CO_2 emission relative to GDP. (From International Energy Agency, 2001).

TOPIC 2.4 OPEC CONTROL OF PETROLEUM PRODUCTION AND PRICING

In most of the world, unlike the United States, mineral rights belong to the government, not the surface owner. The exploration and development of international oil fields was for the most part conducted by the large oil companies, Amoco, Chevron, Exxon, Gulf, Mobil, Texaco, British Petroleum, Compagnie Francaise Petroles and Shell, or their respective predecessor and successor companies. Agreements to explore and develop oil fields outside the United States were between oil companies and the local governments. These agreements were highly favorable to the oil companies, but local resentment of the agreements eventually developed. Foreign oil resources were sometimes nationalized, as in Bolivia (1937) and Mexico (1938). Beginning in the late 1940s producing countries began to negotiate more favorable arrangements. This occurred first in Venezuela, where a 1948 agreement specified that the government would receive revenues equal to the profits of the companies. The same arrangement was made with Saudi Arabia in 1950.

Disputes over petroleum rights in Iran began soon after WWII. The British Anglo-Iranian Oil Company (later to become British Petroleum and then BP) was the only significant producer. In 1951, a nationalist government took power and the parliament passed a law nationalizing the company. Disputes followed for several years, during which time Iranian production fell drastically. At the time, crude production was expanding rapidly in the Middle-East and the decreased production had little effect on the world oil supply. Eventually, the nationalist government was overthrown with British and US involvement and the pro-Western government of the Shah Mohammad Reza Pahlavi was installed. An agreement was reached for management of the oil fields by a consortium of companies for 25 years (until 1979) with provision for 5-year extensions to 1999. The consortium included British Petroleum (40%), Shell (14%), Chevron, Exxon, Mobil, Gulf and Texaco (8% each), and Compagnie Francaise Petroles (6%).

OPEC was formed in 1960, with the leadership of Venezuela and Saudi Arabia. At the time, Venezuela was the leading oil-producing state and had just increased its share of revenues from 50:50 to 70:30. In 1962, Indonesia and Libya joined OPEC. The OPEC countries were learning more of the workings of the worldwide petroleum industry and coming to appreciate their place in it. At that time, a general sense had developed that consumption was likely to outstrip production, making access to crude essential.

Discoveries in the Middle-East continued in the 1960s. Other sources of oil were developed in the 1960s, such as in Libya and Nigeria. Some exports from Russia also appeared. Prices generally remained under $2.00/b and usage continued to expand rapidly. This price contrasts with the much higher costs for extraction from off-shore and Arctic fields being developed in the United States and in the North Sea in the 1980s. US domestic production had begun to decline while consumption continued to increase. Western Europe and Japan were dependent on imports of oil from the Middle-East. Oil-producing states began nationalization, with Algeria (1971), Iraq (1972) and Libya (1973) nationalizing all or parts of foreign-owned oil companies. Saudi Arabia negotiated for 20% ownership of Aramco, and eventually completed the purchase of the entire company in 1974. Compensation was negotiated for the nationalized properties, but generally was well below current market value.

By 1973, the United States was producing to its full capacity and importing a third of its oil needs. In October 1973, Egypt and Syria attacked Israel in what is known as the Yom Kippur war. OPEC raised the posted price of oil by 70% from about $3/b to $5.20/b. An embargo against the United States and the Netherlands was imposed. Spot market prices rose from $8–9/b to as high as

$17/b. Consumer countries took steps to decrease use, such as imposing speed limits and limiting weekend sales of gasoline. Annual inflation reached 10% and later in the decade, nearly 20%. In the United States, work on the Alaska pipeline was speeded up. In general, the embargo *per se* had little direct effect, because available supplies were redirected and the embargo ended in 1974. At the end of 1974, Saudi Arabia completed its purchase of Aramco. Control over production was now firmly in the hands of the OPEC nations.

In early 1979, political unrest overtook Iran and the Shah fled the country. Ayatollah Khomeini and Shiite clerics took power. Later in the year, the Iranians took the occupants of the US Embassy hostage. Iranian production was temporarily halted and spot oil prices rose. By the end of the year, they reached $40/b. In October 1980, war broke out between Iran and Iraq, reducing production in both countries. The resulting disruptions in supply led to purchases and increase in inventory against the likelihood of further price increases. The official OPEC price was increased to $28/b in 1979 and $30/b in 1980. When the crisis subsided in 1981, the official price was $34/b. While supply channels were severely disrupted, there were no major shortages during the crisis, but prices had again doubled.

After the 1979 price increase, world consumption of oil dropped sharply. There were several reasons for this. Conservation measures, particularly mpg increases for automobiles, reduced demand. A good deal of electricity generation was shifted to coal and natural gas, and a number of nuclear power plants came on stream. Additional non-OPEC oil sources, including the North Sea, Alaska, and Mexican fields added production. By 1985, the OPEC share of oil production dropped to 15 Mb/d, about half its maximum production. Further price reductions occurred reaching $12/b in 1986. In the late 1980s, OPEC finally arrived at an agreement to limit production with a price of around $18/b. After that, production began to increase again, reaching 25.0 Mb/d by 1990. Prices remained in the $15–25 range through most of the 1990s, correlating with a decade of relative economic stability in the United States.

In August 1990, Iraq invaded Kuwait, the purpose being, among others, to annex Kuwait's oil. By February 1991, a US-led coalition had driven Iraq from Kuwait and imposed sanctions on the Iraq. War with Iraq began again in 2003 when a US-led coalition invaded Iraq to remove Saddam Hussein from power. The war went quickly enough, but political instability persisted. Since 2000, there has been another series of increase in crude oil prices, up to and even above $100/b. Figure 2.21 shows the range of oil prices from 1970 to 2006 in constant 1990 dollars. The drastic increases of 1973–1974, 1979–1980, and after 2003 are evident.

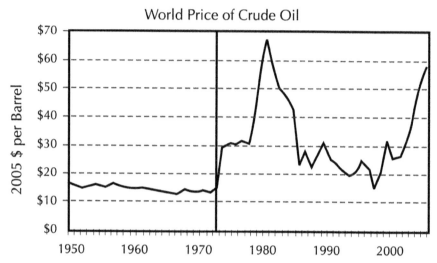

FIGURE 2.21   Crude oil prices in constant 2005 dollars from 1950 to 2000. (From US Energy Information, Administration, 2006.)

2.5.3   ETHICAL ASPECTS OF FOSSIL FUEL CONSUMPTION

What if there is no practical solution to the problem of replacing fossil fuels by renewable energy sources and reducing CO_2 emissions? Some years ago, a biologist, Garrett Hardin, considered the issue of humanity's response to a critical problem for which there is no technical solution. He defined a "technical solution" as one that required no fundamental change in existing systems of values and ethics.[7] His particular emphasis was on population growth and he posited that because the planet and its resources are finite, population growth is ultimately limited. He argued that current economic theory, traceable to Adam Smith's "invisible hand" which asserts that "individual gain" ultimately proves to be in the "public interest," breaks down when resources are limited. Hardin described the situation as the *tragedy of the commons* in terms of the effect of an incremental cow, belonging to a single individual, on a herd of cattle sharing a piece of land (the commons) that was at maximum capacity. The effect of the new cow is beneficial to its owner, but detrimental to all the other users of the commons. Each other herdsman will presumably also make the decision to add another cow, resulting in overgrazing, decreasing production (and in the long run) perhaps desertification and collapse.

The same principles apply to control of pollution or environmental degradation if some members of the community reduce contamination, while others do not. There is relatively little direct benefit to non-polluters. They must bear the cost, while the benefit accrues to the entire group. In this particular instance, the public in most parts of the world has come to the conclusion that environmental degradation is undesirable and have instituted incentives and penalties that work to reduce pollution. Hardin described this as a "social arrangement" instituted by "coercion by mutual agreement." He argued that this is more effective than efforts to modify individual behavior on the basis of appeals to morality, conscience, responsibility or guilt, because these approaches are most likely to benefit the selfish, who would be least likely to respond.

At the present time the use of fossil fuels and its consequence of CO_2 emission and global warming is probably the most urgent example of the *tragedy of the commons*. The atmosphere and its climatic consequence are clearly a "commons." The inputs of fossil fuel production and consumption are widely conceived as "property" and "rights" to be used to the individual's benefit. "Property" is a particular issue for entities such as corporations that produce fuels and that are charged with maximizing profit. They are generally immune to calls of conscience or responsibility and are restrained only by public reaction and regulation. The maximization of economic growth and activity has become an almost universal goal. The consequence is increased energy utilization. This poses several questions. (1) How much fossil fuel is there and how long will it last? (2) How much of an increase in CO_2 can the earth survive? Or, in terms of the commons, at what point will the ecology of the earth collapse under the pressure of over-utilization of fossil fuels? (3) How rapidly can fossil fuels be replaced by renewable energy sources? (4) What "coercion by mutual agreement" can be put in place? No one knows the answers.

KEYWORDS

- petroleum
- distillation
- thermal cracking
- catalytic cracking
- platforming
- alkylation
- coal gasification
- Fischer–Tropsch process
- petrochemicals

- **air quality**
- **carbon dioxide**
- **global warming**
- **petroleum production and pricing**

BIBLIOGRAPHY

Smil, V. *Energy at the Crossroads—Global Perspectives and Uncertainties.* MIT Press: Cambridge, MA, 2003.

Goodstein, D. *Out of Gas.* W. W. Norton: New York, 2004.

Kunstler, J. H. *The Long Emergency: Surviving the Converging Catastrophes of the Twenty-First Century.* Atlantic Monthly Press: New York, 2005.

Tamminen, T. *Lives per Gallon—The True Cost of Our Oil Addiction,* Island Press: Washington, DC, 2006.

Maass, P. *Crude World: The Violent Twilight of Oil.* Alfred Knopf: New York, 2009.

Yergin, D. *The Prize: The Epic Quest for Oil, Money and Power.* Free Trade Press: New York, 2009.

Yergin, D. *The Quest: Energy, Security and the Remaking of the Modern World*, Penguin Books: New York, 2012.

BIBLIOGRAPHY BY SECTION

Section 2.1.2: Peters, A. W.; Flank, W. H.; Davis, B. H. *Am. Chem. Soc. Symp. Ser.* **2009,** *1000,* 103–187.

Section 2.1.3: Fletcher, R. P. *Am. Chem. Soc. Symp. Ser.* **2009,** *1000,* 189–249.

Section 2.1.3.3: Glaeser, R.; Weitkamp, J. *Basic Principles in Applied Catalysis,* Vol. 75; Baerns, M., Ed.; Springer, 2004, pp 161–212.

Section 2.1.3.4: Sullivan, R. F.; Scott, J. W. *Am. Chem. Symp. Ser.* **1983,** *222,* 293–313; Gruia, A. *Practical Advances in Petroleum Processing*; Hsu, C. S.; Robinson, P. R., Ed.; Springer: New York, NY, 2006; pp 219–255.

Section 2.1.3.5: Rostrup-Nielsen, J. *Stud. Surf. Sci. Catal.* **2004,** *147,* 121–126.

Section 2.1.4: Haensel, V. *Am. Chem. Soc. Symp. Ser.* **1983,** *222,* 141–152; Turaga, V. T.; Ramanathan, R. *J. Sci. Ind. Res.* **2003,** *62,* 963–978.

Section 2.1.5: Hommeltoft, S. I. *Appl. Catal. A* **2001,** *221,* 421–428; Olah, G. A.; Molnar, A. *Hydrocarbon Chemistry*, 2nd ed., Wiley-Interscience, Chap. 2, 2003; Albright, L. F. *Ind. Eng. Chem. Res.* **2003,** *42,* 4283–4289; Feller A.; Zuazo, I.; Guzman A.; Barth, J. O.; Lercher, J. A. *J. Catal.* **2003,** *316,* 313–323; Traa, Y.; Weitkamp, J. *Handbook of Heterogeneous Catalysis,* 2nd ed., Vol. 6, 2008, pp 2830–2854.

Section 2.2.1: Shulz, H. *Appl. Catal. A* **1999, 186,** 3–12; Dry, M. E. *Catal. Today* **2002,** *71,* 227–241; Stranges, A. N. *Fischer–Tropsch Synthesis, Catalysis, and Catalysts*; Davis, B.

H.; Occelli, M. L., Eds.; Elsevier: Amsterdam, 2007, pp 1–26; Dancuart, L. P.; Steynberg, A. P. *Fischer–Tropsch Synthesis, Catalysis, and Catalysts*; Davis, B. H.; Occelli, M. L., Eds.; Elsevier: Amsterdam, 2007, pp 379–399; Dry, M. E. *Handbook of Heterogeneous Catalysis*, Vol. 6, Wiley-VCH: Weinheim, 2008, pp 2965–2994; Leckel, D. *Energy Fuels* **2009**, *23*, 2342–2358.

Section 2.2.2: Navarro, R. M.; Pena, M. A.; Fierro, J. L. G. *Chem. Rev.* **2007**, *107*, 3952–3991; Leckel, D. *Energy Fuels* **2009**, *23*, 2342–2358; Sousa-Aguilar, E. F.; Noronha, F. B.; Faro, A. *Catal. Sci. Technol.* **2011**, *1*, 698–713.

Section 2.2.3.1: Dien, B. S.; Bothast R. J.; Nichols, N. N. Cotta, M. A. *Int. Sugar J.* **2002**, *104*, 204–211; Demirbas, A. *Energy Sources* **2004**, *26*, 715–730; Nichols, N. N.; Dien, B. S.; Bothast, R. J.; Cotta, M. A. *Chem. Ind.* **2006**, *112*, 59–78; Edye, L. A.; Doherty, W. O. S.; Blinco, J. A.; Bullock, G. E. *Int. Sugar J.* **2006**, *106*, 19–27; Gray, K. A.; Zhao, L.; Emptage, M. *Curr. Opin. Chem. Biol.* **2006**, *10*, 141–146; Wyman, C. *Trends Biotechnol.* **2007**, *25*, 153–157; Albertazzi, S.; Basile, F.; Trifiro, F. *Renewable Resources Renewable Energy* **2007**, 197–213; Lee, S. *Handbook of Alternative Fuel Technologies*; CRC Press, Boca Raton, FL, 2007; pp 323–341; Sticklen, B. *Nat. Rev. Genet.* **2008**, *9*, 433–443; Vertes, A. A.; Inui, M.; Yukawa, H. *J. Microbiol. Biotechnol.* **2008**, *15*, 16–30.

Section 2.2.3.2: Pahl, G. *Biodiesel: Growing a New Economy*. Chelsea Green Publishing Company: White River Junction, VT, 2008.

Section 2.2.4.5: Nuttall, W. J. *The Nuclear Renaissance*. IOP: Bristol, UK, 2005.

Section 2.3: Spitz, P. H. *Petrochemicals—The Rise of an Industry*. John Wiley & Sons: New York, NY, 1988.

Section 2.4.1: Kovarik, W. *Int. J. Occup. Environ. Health* **2005**, *11*, 384–397; Needleman, H. L. *Environ. Res., A* **2000**, *84*, 20–35.

Section 2.5: Lloyd, B. *Energy Policy* **2007**, *35*, 5806–5818.

REFERENCES

1. Degnan, T. F.; Chitnis, G. K.; Schipper, P. H. *Microporous Mesoporous Mater.* **2000**, *35–36*, 245–252.

2. Shetty, J. K.; Lantero, O. J.; Dunn-Coleman, N. *Int. Sugar J.* **2005**, *107*, 605–610.

3. Hammerschlag, R. *Environ. Sci. Technol.* **2006**, *40*, 1744–1750.

4. U S Department of Energy, Energy Information Administration. Brochure DOE/EIA-X012, 2008.

5. U. S. Tariff Commission. *Petroleum, War Changes in Industry*, Series No. 5, 1946, p 16.

6. National Academy of Sciences. *Effectiveness and Impact of Corporate Average Fuel Efficiency*, 2002.

7. Hardin, G. *Science* **1968**, *162*(3859), 1243–1248.

PART II
Making Things We Use

CHAPTER 3

THE CHEMICAL INDUSTRY: AN OVERVIEW

ABSTRACT

The ability to effect chemical transformations led to the development of a worldwide chemical industry. Important early large scale methods included the Solvay Process for sodium carbonate and catalytic oxidation of sulfur to sulfuric acid. By the final quarter of the 19th century, manufacture of dyes based on coal tar intermediates became the first specialized chemical industry. It was dominated by German and Swiss companies. Early in the 20th century, availability of electricity permitted production of chlorine, calcium carbide and cyanamide on an industrial scale. The Haber–Bosch process for ammonia became operational in 1915. The development of petroleum refining in the United States provided new starting materials. Synthetic polymers including bakelite, alkyd resins, nylon, and polystyrene joined materials such as cellophane and rayon, which were made from cellulose. Expanding prosperity after WW-II led to rapid growth of chemical production in Europe, North America, and Japan. In the final quarter of the century, many chemical companies became international in scope. Production of petrochemicals expanded to other parts of the world, including the Middle-East and Asia. Over-capacity led to competition and consolidation. Environmental consequences of use of certain chemicals became apparent and led to regulation and efforts to minimize the adverse effects of chemical use. Research in the expanding pharmaceutical industry led to the discovery of many new medicines.

In this section, we will look broadly at the development of the chemical industry during the 20th century. Most of the emphasis is on the major American companies but some attention is given to the European and Japanese industries. Most of the largest US chemical companies had continuous, or at least related, names from their founding before or near the beginning of the 20th century until about 1975. After that time, however, mergers, acquisitions and divestitures led to rapid changes and many new names. At about the same time European and Japanese companies appeared in the United States and the chemical industry became international in scope. Some of the important groups of products and processes will be mentioned briefly, but more detailed information will be provided later in the sections devoted to specific kinds of materials, such as polymers, agricultural chemicals, food products and drugs.

3.1   PRIOR TO WWI

Materials made by chemical transformations go back before recorded history. Bronze production began between 3000 and 2000 BC. Glass and potash (K_2CO_3) were being produced in Egypt by 2000 BC. The Egyptians and Romans used various ointments, fragrances and pigments for make-up. Black powder appeared in China sometime around the ninth century (see Section 1.2.1). Chemical transformations were part of the alchemists' repertoire in medieval Europe, although shrouded in mystery and secrecy. In the 1700s, scientific approaches began to develop. Joseph Priestley[a] (England) and Carl Wilhelm Scheele (Sweden) recognized oxygen as a component of the atmosphere and Antoine Laurent Lavoisier explained its role in combustion in the 1770s. Lavoisier published *Traite Elementaires de Chimie* in 1789, in which he identified 33 materials as nontransformable elements. He also recognized the principle of conservation of mass in chemical transformations (see Section 1.2.1 for biographical data on Lavoisier).

A few fundamental chemical processes were done for commerce in the 18th century. Sulfuric acid (H_2SO_4) was made by heating sulfur with KNO_3 in lead-lined containers, a process developed in England by John Roebuck about 1750. Soap was made by reaction of animal fats with the carbonate minerals natron and trona (hydrated forms of Na_2CO_3). Potassium nitrate (KNO_3, saltpeter) was collected from stables and purified for use in the manufacture of black powder (see Section 1.1). Sodium carbonate (Na_2CO_3, also called soda) was needed by the glass and soap industries. In 1775, the French Academy of Science offered a prize for the production of pure soda. Nicolas LeBlanc, a physician, found that it could be produced from salt by heating with chalk ($CaCO_3$), coal, and H_2SO_4. A plant was established and the process was patented in 1791. The plant was confiscated during the French Revolution. LeBlanc spent the rest of his life unsuccessfully seeking restitution and in 1806, he shot himself.

---

[a]Joseph Priestley was a philosopher, educator, theologian, and natural scientist. He was one of the discoverers of oxygen, which he made by decomposing mercuric oxide and tested its ability to sustain life. However, he explained its properties in terms of the "phlogiston theory" and therefore did not fully recognize its significance. Priestly founded the Unitarian Church and served as a clergyman in several English cities. His outspoken philosophy and support of the American and French Revolutions eventually resulted in his leaving England for the United States in the 1790s. The American Chemical Society annually awards the Priestly Medal named in his honor.

LeBlanc Process

$$NaCl + CaCO_3 + 2\ C + H_2SO_4 \longrightarrow Na_2CO_3 + 2\ HCl + CaS = 2\ CO_2$$

Bleaching Powder Process

$$HCl + MnO_2 \longrightarrow Cl_2$$
$$2\ Cl_2 + 2\ CaO \longrightarrow Ca(OCl)_2 + CaCl_2$$

At the beginning of the 19th century, the developing textile industry in Britain created demand for bleaching materials. Bleaching powder, calcium hypochlorite [(Ca(OCl)$_2$], was made by passing chlorine through lime. The chlorine was obtained by oxidation of HCl with manganese dioxide.

Britain was the leading producer of chemicals at the middle of the 19th century. It made the largest amounts of basic chemicals such as soda (Na$_2$CO$_3$), caustic soda (NaOH), and sulfuric acid (H$_2$SO$_4$). The LeBlanc process for soda was well established by the 1830s, but generated both gaseous HCl and solid calcium sulfide as byproducts. The latter was dumped and the acid was released to the air or rivers. The manufacturing sites and surrounding areas suffered from the fumes produced by the process. In 1863, the Alkali Act was passed by the British Parliament requiring the recovery of the acid. The recovered HCl could be used to manufacture bleaching powder. In the 1880s, a method of recycling the CaS was developed.

In 1861, a new method for making soda was patented by the Belgian brothers Albert and Ernest Solvay. The process used tall towers to convert ammonia, salt, and CO$_2$ into NaHCO$_3$, which was then converted to Na$_2$CO$_3$. The ammonia was recovered using lime.

$$NaCl + NH_3 + CO_2 + H_2O \longrightarrow NaHCO_3 + NH_4Cl$$
$$2\ NH_4Cl + CaO \longrightarrow 2\ NH_3\ (recycle) + CaCl_2 + H_2O$$
$$2\ NaHCO_3 \xrightarrow{heat} Na_2CO_3 + CO_2 + H_2O$$

The Solvay process was much cleaner and more efficient than the LeBlanc process and was adopted rapidly in continental Europe. It was introduced into Britain by Ludwig Mond and John Brunner, who founded a company to use the method.[b] The new competition led the companies using the LeBlanc process to

---

[b]Ludwig Mond was born in Germany in 1839. He studied Chemistry, but did not complete a degree. He worked in chemical factories in Holland and England between 1862 and 1867. He founded a company that developed a method to recover the sulfur produced by the Le Blanc soda process. In 1873, Mond and John Brunner founded a company that introduced the Solvay process to England. Later, Mond discovered the metal carbonyls and

join together as United Alkali. These two companies became the foundation of the modern chemical industry in Britain. The Solvay process was operated in the United States by Allied Chemical, Michigan Alkali (later Wyandotte) and by Pittsburg Plate Glass (later PPG).

During the second half of the 19th century, chemistry was put on a sound foundation with recognition of more pure elements, the formulation of the periodic table by Mendeleev, and the determination of the composition and structures of many compounds. In 1865, Friedrich August Kekule proposed a correct structure for benzene and put aromatic chemistry on a solid foundation. Germany had established the most comprehensive system for training chemists, first at Giessen under Justus von Liebig, then in Gottingen, Heidelberg, and Bonn. A Royal School of Chemistry was established in London in 1845 under the influence of Queen Victoria's husband Prince Albert, a German. August Wilhelm Hofmann, a student of Justus Liebig, was the professor. Hofmann studied the composition of coal tar and identified several of the aromatic compounds it contained, most significantly aniline.

The first area affected by chemical synthesis was the dye industry. At the beginning of the 19th century, dyes were obtained from natural sources, some of them quite exotic. Indigo was isolated from a plant from India and carmine red came from an insect, *Coccus cacti*. A red dye from the madder plant was produced in France.

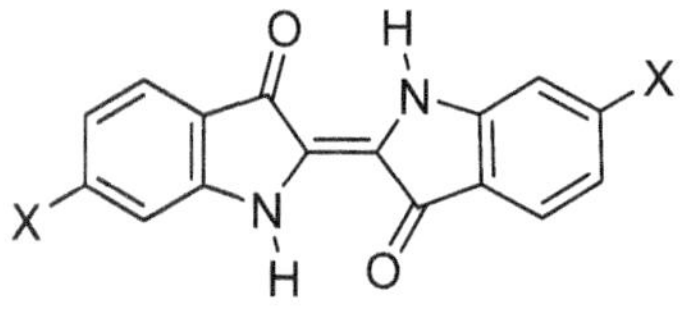

X = H = indigo
X = Br = Royal Purple

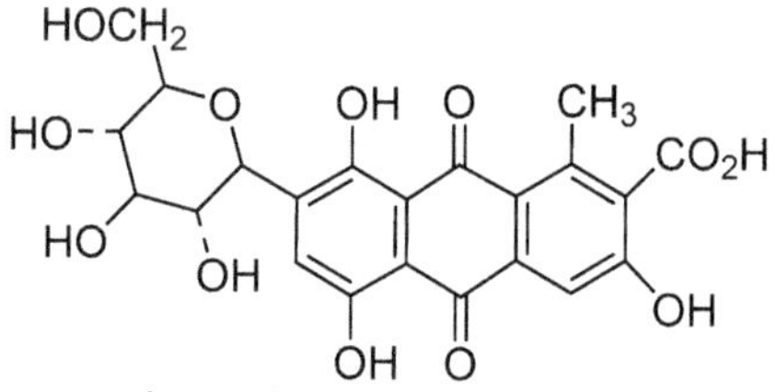

carmine red

Several of A. W. Hofmann's students in London contributed to the synthetic dye industry. William H. Perkin, at age 18, accidentally made the purple dye mauve when he oxidized aniline with potassium dichromate. It was patented in 1856. Other dyes including magenta I (fuchsin), malachite green, and aniline

exploited nickel carbonyl for the manufacture of pure nickel metal. His son Alfred Mond was instrumental in the formation of ICI after WWI.

John Brunner was born in 1842. He did not have a university education but worked in shipping. He eventually became a manager of an alkali works and it was there that he met Ludwig Mond. Brunner was chairman of the Brunner–Mond company from 1891 to 1918. The company became part of ICI in 1926. Brunner was also active in politics and was a Member of Parliament from 1887 to 1910.

yellow were made and eventually led to the formation of the British Alizarine Company. Britain had a plentiful supply of the raw materials needed, which were obtained from coal tar produced as a byproduct of manufacture of illuminating gas and coke for steel-making. Other related dyes were soon discovered, especially in France.

magenta I (fuchsin)   malachite green   aniline yellow

Germany captured the lead in dye-making, the "high-tech" industry of the time, in the last quarter of the 19th century. One reason appears to have been the difference in the education systems of the two countries. German universities initiated strong programs of laboratory experimentation, especially in organic chemistry. As the chemical industry developed in Germany, collaboration with leading universities was common. Friedrich Bayer began producing dyes at Elberfeld in 1863, founding a company which became *Farben Fabrik vormal Friedrich Bayer*, now simply *Bayer.* A second company was set up in Hoechst, near Frankfurt and took the name of its location. A firm manufacturing both dyes and soda at Mannheim and Ludwigshafen became *Badische Anilin und Soda Fabrik*, now BASF. Three Swiss companies, CIBA (Chemische Industrie Basel), Geigy, and Sandoz, also began as dye manufacturers.

By 1890, BASF had an integrated plant that made all the chemicals needed to produce a variety of dyes. Bayer and Hoechst quickly followed suite. These firms also invested in coal production to provide raw materials. The German companies established research laboratories and had chemically trained people in key management positions. A close association between the chemical industry and educational system developed and new understanding was quickly transferred to industrial practice. The Germans developed marketing and technical assistance networks. The larger companies also formed cooperative arrangements to manage competition in both internal and external markets. By the end of the 19th century, the German firms dominated the markets both for dye intermediates and the final products. Just before the outbreak of WWI, they produced 90% of the world dye supply. Until WWI, Germany continued to invest heavily in technical education and research in Chemistry. The Kaiser Wilhelm Institutes for Chemistry, Physical Chemistry, and Coal Research (later to become the Max Planck Institutes) were established shortly before WWI.

The German and Swiss dye companies became the precursors of drug companies. Pasteur's recognition of microorganisms as the cause of infectious diseases led to rational approaches to therapy. Joseph Lister (England) introduced phenol as an antiseptic in 1867. Robert Koch (Germany) successfully stained anthrax bacteria in 1872 and the tuberculosis bacilli in 1882. Both Koch and Paul Ehrlich studied staining of biological specimens with dyes and Ehrlich conceived the idea of killing bacteria with dyes (see Section 11.1). By the late 1800s, German research labs were making new pharmaceuticals, most notably analgesics and antipyretics. The antipyretics phenacetin (1888) and aspirin (1898) were made at Bayer. Hoechst marketed the antifever drug antipyrine and an antidiphtheria serum. The antisyphilis drug Salversan, discovered by Paul Ehrlich and marketed by Hoechst, became the first synthetic material used to treat an infectious disease (see Section 11.1).

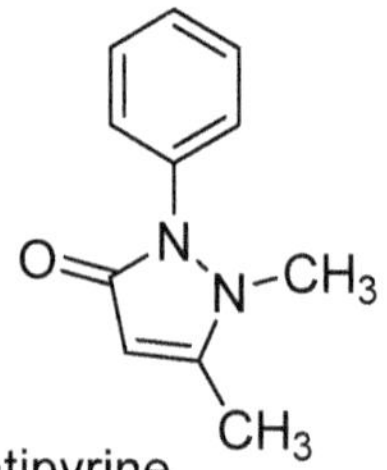

antipyrine 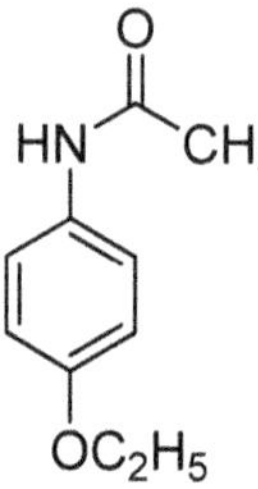   phenacetin 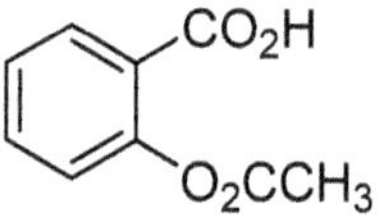   aspirin

In the United States, the processing of pine for turpentine and rosin had begun by the end of the 18th century. Dry distillation of wood also became the source for methanol (wood alcohol). Later, manufacture of paper from wood shavings by boiling with alkali was introduced. Eleuthere Irenee du Pont de Nemours started manufacture of black powder near Wilmington, DE in 1802 using saltpeter from mines in Kentucky (see Section 1.2.1). In 1839, Charles Goodyear in the United States and Thomas Hancock in England developed the process for vulcanization of rubber by heating with sulfur. Production grew from 1000 t in 1850 to 40,000 t by 1900 (see Section 5.3.1).

The first consumer products based on chemistry began to appear in the 1800s. In the United States, William Colgate started making soap in 1806 in New York City. William Procter and James Gamble started soap manufacture in Cincinnati in 1837. The soap was made by the traditional process of alkaline hydrolysis of fat. Improved soaps were developed in Europe. In England, William Lever began to make a soap, called *Sunlight*, using vegetable oil rather than animal fat. His company later moved into the manufacture of margarine (see Section 9.3). Fritz Henkel in Germany developed *Persil*, a soap containing a bleach (sodium perborate). Companies bearing these names remain engaged in the industry to this day.

Advances were made in the manufacture of explosives during the late 19th century. Alfred Nobel made dynamite by stabilizing nitroglycerin by adsorbing it on kieselguhr. Smokeless powder was made by nitration of cellulose and it was found that by pressing it to remove excess acid, stability was improved (see Section 1.2). In 1890, Herman Frasch, a German immigrant, developed a method for melting and extracting sulfur deposits in Louisiana (see Section 1.2.1). At about the same time BASF developed a catalytic (contact) process for oxidation of sulfur to make sulfuric acid.

Near the end of the 19th century electric power became available on a scale able to support industrial processes. In 1886, electrolysis was applied to the production of aluminum by Paul Heroult in France and by Charles Hall in the United States. Heroult's process was commercialized in Switzerland and Scotland, while Hall's company became ALCOA. Electric furnaces were also used to produce calcium carbide used to generate acetylene for street lamps and early automobile headlamps. This process was discovered by Thomas L. Wilson, a Canadian electrical engineer, when he experimented with electrical furnaces to make aluminum and calcium. A plant was set up at Eden, NC on land owned by an investor, James T. Morehead, to take advantage of water for generation of electricity. In addition to use in lamps, CaC_2 was for a time used to produce cyanamide by reaction with nitrogen (see Section 8.1.2). Wilson and Morehead's company eventually became part of Union Carbide, which was formed in 1917 from several smaller companies engaged in electrolytic processes for making products such as calcium carbide, metals, and industrial gases. Electrolytic manufacture of caustic soda (NaOH) and chlorine began in the late 1800s. In the United States, Herbert Dow (Midland, MI), Thomas Mathieson (Saltville, VA) and Elon Hooker (Niagara Falls, NY) produced chlorine. Improvement and scale-up of the contact process for sulfuric acid occurred at General Chemical (later a part of Allied Chemical). At the turn of the century several of the major chemical companies that would dominate the 20th century were in place including American Cyanamid, Dow, DuPont, and Monsanto. The United States chemical industry grew rapidly with population growth in the first quarter of the 20th century.

Up to WWI, Germany had the most advanced chemical industry. Its universities and technical institutes provided strong training and research in chemistry, and there were close associations with the chemical companies. The industry was based on coal. Basic chemicals such as sulfuric acid and sodium carbonate were manufactured in large quantities. Germany dominated the dye industry worldwide. The three largest companies producing dyes were

BASF, Bayer, and Hoechst. All were engaged in research and were beginning to diversify into other areas, especially pharmaceuticals. The same was true of the Swiss companies CIBA, Geigy, and Sandoz. The chemical industry in Britain at the beginning of the 20th century, by comparison, was in relative decline. The basic processes were out-modeled and protected by the colonial system. The British relied on imports from Germany for production of dyes. The strongest industries were soaps (Lever Brothers) and viscose rayon fiber (Courtaulds). Several pharmaceutical firms such as Beecham, May & Baker, and Burroughs–Wellcome were established, but had not yet developed research programs. In France, only a relatively few institutions taught industrial chemistry or engaged in research, the exceptions being The Ecole de Physique et de Chimie in Paris and schools of industrial chemistry in Lyon and Nancy. Most manufacturing facilities were small and scattered. A silk industry existed and there was interest in artificial fiber. The manufacture of rayon began in France in the 1880s (see Section 5.3.2). Both the French and Swiss used hydroelectric power to produce aluminum and calcium carbide. In Norway, hydroelectric power was used to produce cyanamide and nitric acid. Dynamite production was widespread, usually in conjunction with the Nobel Company (see Section 1.2).

The chemical knowledge on which the industry is based grew rapidly in the first quarter of the 20th century. In the United States, Gilbert N. Lewis[c] and Linus Pauling[d] provided a conceptual basis for understanding much of organic and inorganic chemistry. These concepts were rapidly incorporated into chemical education and provided a strong foundation for research progress. The new discipline of chemical engineering provided insight into the means of production of chemicals from petroleum (see Section 2.3). The innovations that were introduced benefitted from patent protection and the retained income

---

[c]Gilbert N. Lewis was born in Massachusetts in 1875, but spent much of his boyhood in Lincoln, NE, USA. He received B. S. (1896) and Ph. D. (1899) degrees from Harvard University, and then studied with Walther Nernst and Wilhelm Ostwald in Germany. He held a junior faculty position at Harvard from 1901 to 1904 and then moved to MIT (1905–1912). He moved to UC Berkeley in 1912 and spent the rest of his career there. He formulated thermodynamic relationships in ways that were useful to understanding chemical equilibria. Lewis developed fundamental concepts of bonding including the concept of covalent bonds, the octet rule and the electron donor–acceptor character of acid–base interactions.

[d]Linus Pauling was born in Portland, OR, USA in 1901. He received a B. S. degree in Chemical Engineering from Oregon State in 1922 and a Ph. D. in Physical Chemistry from CalTech in 1925. He was on the faculty of CalTech until 1964. He was affiliated with UC (Santa Barbara, CA, USA) and UC (San Diego, CA, USA) until 1969 and then was located at Stanford. Pauling formulated several fundamental concepts of chemical structure including hybridization, resonance and electronegativity. These concepts were the basis of his book *The Nature of the Chemical Bond*, first published in 1939. Pauling used X-ray crystallography in his early work, including studies on the structure of hemoglobin. He originated the concept of the α-helix in protein structure and recognized that sickle cell anemia was a "molecular disease". He received the 1954 Nobel Prize in Chemistry, which cited his work on chemical bonds. After WWII, Pauling became a vociferous opponent of nuclear weapons and as a result was accused of being a communist during the "McCarthy Era." He published a book, *No More War* in 1958. He received the 1962 Nobel Peace Prize.

was reinvested. This provided the early innovators with a competitive advantage both in technical information and access to capital and protected their positions of leadership. The companies were often managed by the people responsible for the new technology.

3.2   FROM WWI TO WWII

WWI greatly altered the balance of strength in the chemical industry. There was a major reorganization of the chemical industry in Britain. The British Nobel Company, which had concentrated its efforts on nitroglycerin for dynamite, was not well-positioned to meet the demands for new military explosives, particularly TNT (see Topic 1.2). The deficiencies that the war revealed led to increased funding of chemical research through the Department of Scientific and Industrial Research. Imperial Chemical Industries (ICI) was formed in 1926 from British Nobel, British Dyestuffs, United Alkali, and Brunner–Mond. ICI had effective monopolies in Britain and its colonies. The ICI components remained specialized in explosives, dyes, basic chemicals, and ammonia, respectively. In England and France, production that had been concentrated on the war effort needed to be redirected. In particular, there was excess capacity in cellulose nitrate, cellulose acetate, and chlorine.

At the end of WWI, Germany lost her near monopoly on dyes and chemical intermediates. The Germans lost patent protection on their chemical technology in Britain and France. German-owned production facilities were confiscated in those countries and in the United States. The victorious allies also imposed high tariffs to encourage development of their national industries. WWI had revealed the importance of nitrogen compounds for both explosives and fertilizers. Before the war, Chilean mines were the principal source. The Haber–Bosch process for ammonia, put in operation at BASF in 1915, will be discussed in detail in Section 8.1.3. It provided a means of producing ammonia from the nitrogen in air and hydrogen generated from coal and water. It was the first industrial process to operate at very high temperature and pressure. After the war, ammonia production in Britain was undertaken by the Brunner–Mond Company, using the Haber–Bosch process. France, too, began ammonia production.

In the United States, the initial manufacture of ammonia was by a relatively less efficient process, the high-temperature reaction of nitrogen with CaC_2 to produce cyanamide. However, production from air soon became dominant. DuPont began making nitrates and ammonia in cooperation with a French company L'Air Liquide at Belle WV in the 1920s. The hydrogen was made from coke by the water-gas shift reaction and methanol was produced as a byproduct (see Section 2.2.1.3). Research into new uses of the byproduct methanol resulted in the development of the Zerone brand of antifreeze for automobiles. Union Carbide was also a major producer of ammonia at that time. Shell already produced ammonia in Europe with coal as the hydrogen source. Shell's plant near San Francisco was the first to produce ammonia using methane as the hydrogen source (1920s). The fertilizer was initially made as the sulfate salt, using recycled sulfuric acid from the nearby refining operation.

3.2.1 AMERICAN COMPANIES

Several US firms, for example, DuPont, had profited greatly during WWI and were in a position to invest in plants and research. The end of the war coincided with the rapid expansion of automobile use and the demands it created. The American chemical industry became oriented toward the automobile and other consumer products. Even during the Depression, the chemical industry grew steadily as the result of growing demands from the automobile industry. Tetraethyllead (Ethyl Corporation) and ethylene dibromide (Dow) became major products with the introduction of leaded gasoline (see Section 2.4.1). Automobiles also increased demand for lacquers and paints (DuPont), as well as antifreeze (DuPont, Union Carbide). Other demand arose from the consumer shift to semisynthetic rayon and acetate fibers (see Section 5.2.2). Chemical engineering emerged as a specific discipline in the 1920s and enabled systematic process development and scale up. It became possible to produce key chemical intermediates from petroleum. The United States took the lead in this aspect of the chemical industry.

Several large, broad-based chemical firms emerged in the period between the wars such as Allied Chemical, American Cyanamid, Dow, DuPont, Hercules, Mathieson, Monsanto, and Union Carbide. These chemical firms began to create marketing and technical service departments. One of the first major products to be developed for the automobile industry was the fast-drying lacquer Duco, developed by DuPont. Duco was nitrocellulose dissolved in a solvent such as butanol and contained a plasticizer such as camphor. General Electric developed the *alkyd resins* that were used for automobile and appliance paints. These were esters of polyols (primarily glycerol) and phthalic acid, containing some polyunsaturated carboxylic acids. Formation of ester bonds establishes a three-dimensional network that is hardened by oxidative coupling of the unsaturated groups.

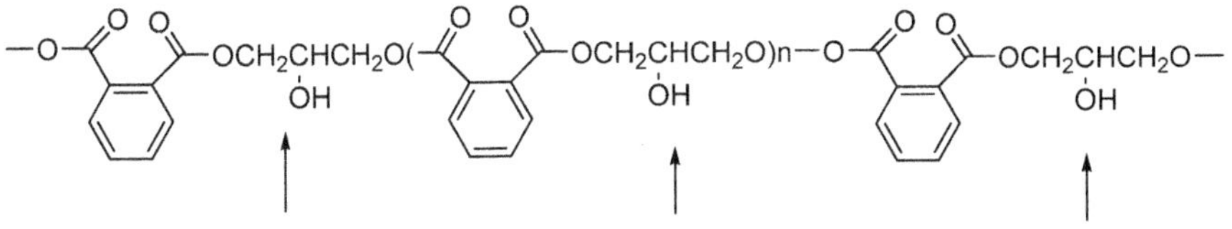

alkyl resin from phthalic acid and glycerol showing points for branching and cross-linking

Rubber manufacturing also received a major impetus from the automobile. The properties of rubber were improved by introduction of accelerators for vulcanization (2-mercaptothiazole, diphenylguanidine, thiuram disulfide) and by the introduction of antioxidants (see Section 5.2.1).

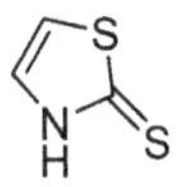

2-mercaptothiazole 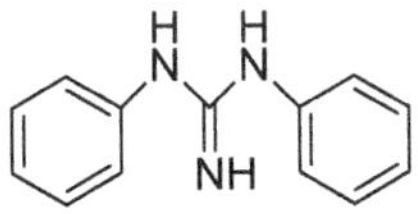   diphenylguanidine 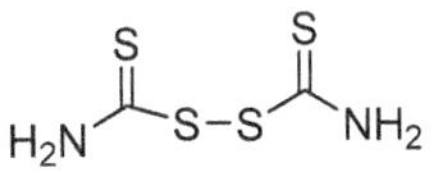   thiuram disulfide

In the United States, cartels were prohibited by antitrust laws, but the growing markets and developing technology led to growth and mergers of chemical companies. By the 1930s, the largest company in terms of sales was Allied Chemical, which produced basic chemicals such as ammonia, sodium carbonate, and inorganic acids. DuPont was the second largest company and had diversified from explosives into fibers (rayon), films (cellophane), and coatings (Duco). Dow, Union Carbide, Standard Oil (NJ), and Shell pioneered in petrochemicals including ethylene, butadiene, styrene, and vinyl chloride. The rubber companies, such as B. F. Goodrich, Goodyear, and Firestone, also began to be involved in chemical research and production. The chemical industry was the leader in establishing research laboratories, accounting for slightly over 25% of those formed between 1900 and 1950.[1]

3.2.2 INTERNATIONAL COMPANIES

The repercussions of WWI brought many changes in the worldwide chemical industry. Over-capacity and competition led to drops in prices of basic chemicals soon after the war ended. Price-setting cartels were established in areas such as dyes, potash and phosphorus. In Germany, eight companies including Bayer, BASF, and Hoechst became affiliated as *Interresengemeinschaft Farbenindustrie Aktien-Gessellschaft* (IGF) in 1925, under the leadership of Carl Duisberg of Bayer and Carl Bosch of BASF. IGF's management moved to eliminate duplication of effort among various firms.

Research innovation at ICI was limited up to WWII and the company was not involved in direct consumer products. ICI had technology and market sharing arrangements with both DuPont and IGF, primarily aimed at maintaining its preferred position in the Commonwealth. Independent firms such as Unilver and Courtauld's focused on consumer products (fats-food-soap and fibers, respectively). In Britain, the absence of any local source of petroleum and tariff/tax laws delayed development of a petrochemical industry. The manufacture of organic chemicals remained based on ethanol and coal.

The French chemical company, Rhone-Poulenc was formed in 1928 by merger of Poulenc Freres and Usine du Rhone, but much of the chemical industry in France remained small and scattered. France also had state-owned chemical producers in ammonia, potash, and petroleum. The French were less involved in international markets than the Germans, British, and Americans. The French excelled in areas such as perfumes and hair dyes, the latter introduced by Eugene Schueller through a firm that became L'Oreal. In Italy, Montecatini became the dominant chemical firm.

IGF joined with the Swiss and French dye-makers and ICI to form a dye cartel in 1932 with the market shares allocated 65:17:8.5:8.5, respectively. IGF also had connections throughout Europe and the United States. For example, IGF had agreements with ICI, Shell, and SO(NJ) in the area of synthesis of hydrocarbons from coal. A German chemist, Friedrich Bergius, first demonstrated the production of gasoline from coal by hydrogenation at high temperature in 1915. This led to formation of a consortium of IGF, Brunner–Mond, and Standard Oil (NJ) to produce gasoline from coal. IGF pursued production of gasoline from coal, even though it was not then economically competitive with petroleum. The rise to power of the Nazis led to large government subsidies and expansion, based on the desire for internal sufficiency (see Section 2.2.1.1). An alternative process was developed by Franz Fischer and Hans Tropsch, working at the Max Planck Institute for Coal Research at Mulhheim, Germany in the 1930s. This involved gasification of coal to synthesis gas and catalytic formation of hydrocarbon (see Section 2.2.1.3). Similarly, synthetic rubber production was encouraged by the German government. While used extensively during WWII, these technologies were not applicable after the war.

The chemical industry in Japan arose within the multicompany combinations called *zaibatsu* that included Mitsubishi, Mitsui, and Sumitomo. Mitsubishi entered the coal chemicals and dye business in the 1930s. Mitsui entered the dye and fabric business during WWI and included Toyo Rayon. Sumitomo's chemical origins were in mining and fertilizers and it began producing ammonia in 1928. These groups typically included several companies with related chemical components. For example, Mitsubishi had not only chemical and petrochemical companies, but also companies in gases, plastics, and rayon. Two other largely chemical firms, Showa Denko and Ube Industries were parts of other groups.

Showa Denko began as a manufacturer of iodine and chlorides and was primarily an electrochemical company and was part of a *zaibatsu* known as the Mori group, which also included a fertilizer company. Ube Industries was formed from mining, cement, iron, and nitrogen producers.

3.2.3   PETROCHEMICALS AND CHEMICAL ENGINEERING

The thermal cracking of petroleum made available large amounts of small unsaturated hydrocarbons, especially ethylene, propylene, and the butenes. As discussed in Section 2.3, these could be converted to alcohols, ketones, and other compounds that found application as solvents and intermediates for the production of other chemicals. Both petroleum (SO(NJ), Shell) and chemical (Union Carbide) companies were involved in production of alkenes by cracking. The United States developed a wide lead in chemical engineering. The concept of chemical engineering as a series of unit operations and processes had taken hold, beginning at the Massachusetts Institute of Technology (MIT) chemical engineering department founded in 1920. The impetus for the department came from Arthur D. Little[e] who had considerable practical experience as a consultant to chemical and paper manufacturers. The chair of the new department, Warren K. Lewis Jr.,[f] further developed the concepts of chemical engineering in the early text *Principles of Chemical Engineering* by Walker, Lewis, and McAdams. The discipline grew and spread rapidly in the United States, especially in the land grant institutions. The chemical engineering approach provided the necessary background for construction and operation of large scale petrochemical and chemical plants. In contrast, neither the German nor British industries had developed the same base for scale-up. Many of the German processes, especially

---

[e]Arthur D. Little was the leader of one of the major consulting firms in the chemical industry. Little studied chemistry at MIT in the early 1880s and then worked for a short time in the paper industry. Under his guidance, the first sulfite-based paper process was introduced at a plant in Rhode Island. Little left the paper industry in 1886 and returned to Boston and formed with Roger B. Griffin a company that became a leading consultant firm to the paper industry. Among the technologies he helped to develop was conversion of cellulose to cellulose acetate fiber (see Section 5.3.2.3). In 1905, after several unsuccessful ventures in chemical manufacturing, Little returned to the consulting business and founded Arthur D. Little Company. The company expanded into other areas of chemistry. Little was a leading advocate of systematic chemical processing, which he called "unit operations" and was instrumental in the foundation of the chemical engineering department at MIT, which took the lead in development of the discipline of chemical engineering. The company remains a major consulting firm with expertise in a broad range of fields, such as energy development, environmental issues, and project management.

[f]Warren K. Lewis, Jr., was born in 1882. He majored in Chemical Engineering then located in the Chemistry Department at MIT. He did graduate work in Physical Chemistry and received a D. Sc. from the University of Breslau in Germany in 1908. His first area of research was the process of distillation. In 1920, he became the head of the newly established Chemical Engineering Department at MIT. Lewis served as a consultant to SO(NJ) and was instrumental in the development of the fluid bed process for catalytic cracking. During WWII, Lewis was involved with the Manhattan project. He was the first chemical engineer to receive the Priestly Medal of the American Chemical Society (1947).

in dyes and pharmaceuticals, were batch processes where scale-up issues were not so critical. The German education system, while otherwise very strong, had not integrated the engineering disciplines into the leading universities, but instead most engineering training was in the Technische Hochschules, which did not provide for doctoral training and research. In Britain, where the universities tended to emphasize the liberal arts, science, and especially engineering were slow to develop.

3.2.4   SYNTHETIC POLYMERS

A major technological advance in the interwar period was the development of synthetic polymers. Up to that time, most "plastics" were modified natural substances such as rubber, cellulose-based coatings and fibers, and protein-derived adhesives. The first synthetic polymer, the thermosetting resin Bakelite, a phenol–formaldehyde condensation product, appeared in 1910. Even the concept of polymers as covalently linked macromolecules remained controversial among chemists until the 1930s. Production of synthetic rubber began in the 1930s in Germany and the Soviet Union, spurred in both cases by limited access to natural rubber. Copolymers of 1,4-butadiene and styrene (Buna-S) or acrylonitrile (Buna-N) found some applications. Polyisobutene and its copolymers with 1,3-butadiene or isoprene were developed by Standard Oil(NJ) in collaboration with IGF. It came to be used for tire inner-tubes. Polyethylene was first made in 1933 at ICI as the result of explorations in high pressure chemistry. The resumption of the study in 1935 led to the recognition of means of catalysis and led to controllable production.

At DuPont, Wallace Carrothers approached the study of polymers from a fundamental perspective. The first product was a polymer of 2-chlorobutadiene, which was commercialized as Neoprene for specialty application requiring resistance to oils. Carrothers also succeeded in preparing polyesters and a bit later polyamides. The latter had remarkable properties of strength and flexibility and became important during WWII. Working to side-step the DuPont patents on polyamides, IGF successfully made nylon-6 from caprolactam shortly before the outbreak of WWII (see Section 5.4.4.1). The Rohm and Haas Company in Germany made polyacrylate esters (acrylics) as shatter-proof glass substitutes and commercialized poly(methyl methacrylate) as Plexiglas. ICI and DuPont soon entered this market using the trade name Lucite. Dow created a market for polystyrene as "Styron." Polyvinyl chloride became a useful material in the form of copolymers with vinyl acetate. These and other polymers saw their first large scale applications during WWII.

3.3   WWII TO 1975

WWII again drastically altered the chemical landscape. The US industry based on petrochemicals expanded capacity rapidly during the war. Explosives, high performance aviation fuel, polyethylene, and nylon were put into military applications. The emulsion polymerization process for production of styrene–butadiene synthetic rubber was developed (see Section 5.3.2.3). Fluid bed cracking and catalytic reforming provided both improved gasoline and the alkenes used in the alkylation process (see Section 2.1.5). The development of nuclear weapons engaged DuPont, Union Carbide, Hooker, and Harshaw and proceeded by the preparation and isotopic fractionation of UF_6 by gaseous diffusion. After the war, nuclear technology was applied to generation of electricity. At the end of WWII, the United States was in a position of unchallenged technical superiority in the petrochemical industry and had a rapidly expanding domestic market. Several factors contributed to the rapid ascendancy of the American chemical industry immediately after WWII. One was the technology and infrastructure that had been developed in war-related programs such as synthetic rubber, aviation fuel, and penicillin. The availability of petroleum and natural gas in the United States made them the preferred feedstock. In contrast, coal remained the principle chemical source in Europe until after WWII. The chemical infrastructure on the continent had largely been destroyed during the war. In 1950, Fortune Magazine in a cover story proclaimed the 20th to be the Chemical Century.[g] The article chronicled the importance of the industry's contribution to GNP, concluding that it had surpassed the automobile industry in total output. The article pointed to the investment in plant construction and research during the decade of 1939–1949 as a major factor in industry growth. The article identified products such as plastics, fibers, detergents, fertilizers, agricultural chemicals, synthetic rubber, and pharmaceuticals as major areas of growth.

The chemical industry in Germany was profoundly changed by WWII. During the war production had turned to explosives, synthetic rubber, and synthetic fuel based on coal. Much of the capacity was destroyed or had become obsolete and major plants were lost to the Russian sector. Production of synthetic rubber and hydrogenation of coal were banned for a time by the victorious Allied governments. IGF was split into Bayer, BASF, and Hoechst in 1950. Germany's share of worldwide chemical production dropped from 22% in 1938 to about 8% by 1950, at which point rapid reconstruction began. German conversion to petrochemicals began in mid-1950s. BASF and Bayer joined forces with international petroleum companies, such as Shell, BP, and SO(NJ) to develop the needed technology. Transition from coal-based to petroleum starting materials was nearly complete by the mid-1960s. The British became involved

---

[g]Pesscatelo, M. *Fortune* **1950**, *LXI*(3), 69–76, 114–122.

in petrochemistry through the Anglo-Iranian Oil Company, later to become British Petroleum. France and Italy also converted to petrochemicals. Chemical processing technology spread rapidly as prosperity returned after WWII. There was extensive licensing and technology sharing among established chemical companies. Also, many independent engineering firms arose that had the capacity to design, construct, and oversee the start-up of new plants.

The German companies entered the polymer field. Hoechst was an early licensee of the Ziegler process for polyethylene (see Section 5.4.2.3) and also polyesters from ICI (see Section 5.4.5). Bayer was an early participant in the polyurethane area (see Section 5.4.6). The German companies also formed joint ventures in the United States, for example Hoechst joined forces with Celanese and Hercules, BASF with Dow and Wyandotte, and Bayer with Monsanto. Many plants and subsidiaries were established outside of Germany and these companies participated in the expanding worldwide market. By the 1970s, Bayer, BASF, and Hoechst again appeared in lists of the top chemical producers in the world.

Among the major technological developments during this period were low-pressure processes for preparation of polyethylene and polypropylene based on the Ziegler coordination catalysts and Natta's discovery of stereospecific polymerization (see Section 5.4.2.3). Although initial development occurred at Montecatini in Italy, Phillips Petroleum introduced a commercial process in 1956 and was eventually granted the US patent rights, but the technology was widely adapted by other companies. Other types of polymers including epoxy resins, polyurethanes, polyesters, and polyacrylonitrile found many applications. The major oil companies also attempted to enter consumer goods areas such as fabrics, plastics and agricultural chemicals, but with some exceptions, did not succeed in the long run.

Development in pharmaceuticals also was rapid. Besides penicillin, antibiotics such as chloramphenicol and the tetracyclines were discovered. Successful treatment of tuberculosis by drugs such as streptomycin, isoniazid, and *p*-aminosalicylic acid became possible. Drugs for treatment of mental illness, depression, and anxiety, such as chlorpromazine, imipramine, and the benzodiazepines were introduced. Antidiabetic drugs such as tolbutamide and phenylbiguanide became available as did diuretics such as the thiazides. (See Chapter 11 for further discussion.)

New agricultural chemicals also appeared. During the first-half of the century, minerals such as sulfur, $CuSO_4$–CaO (boullie bordelaise), and lead arsenate were used as fungicides and insectides on fruit trees and grape vines. Pyrethrins and rotenone were natural products used to control insects in gardens and livestock. The first synthetic fungicides were ditihiocarbamates introduced by DuPont in the late 1930s. In 1939, Paul Mueller, working at Geigy in

Switzerland, recognized the insecticidal activity of DDT (1,1-dichlorodiphenyl-2,2,2-trichloroethane). It came into widespread use after WWII (see Section 8.3.1). 2,4-D (2,4-dichlorophenoxyacetic acid) became the first synthetic herbicide (see Section 8.2.1). DuPont developed monuron and diuron as herbicides. Other herbicides such as atrazine also were introduced.

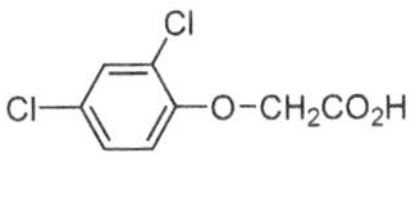

2,4-D:
2,4-dichlorophenoxyacetic acid

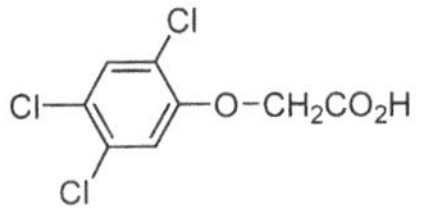

2,4,5-T:
2,4,5--trichlorophenoxyacetic acid

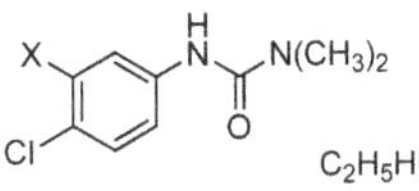

X = H, monuron
X = Cl, diuron

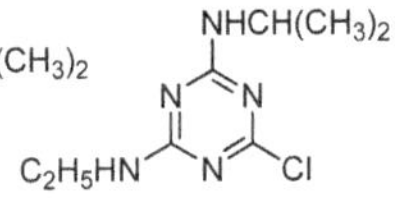

atrazine

The organophosphorus insecticides parathion and malathion were introduced, based in part on British, German, and Russian research on nerve gases. The carbamate insectides, such as carbaryl, were commercialized.

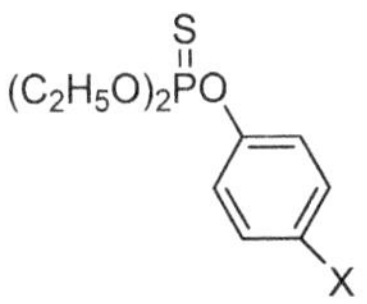

X = H, malathion
X = NO₂ parathion

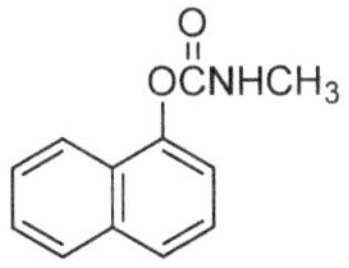

carbaryl

The period 1945–1975 represented the steepest rise in both the improved standard of living and economic benefits of chemical research and process development. Producer's income rose and consumer's costs declined. Fundamental conceptual improvements were made in fuels, materials, medicines, and agricultural chemicals. The chemical industry in the United States grew at twice the rate of the GNP and profits were above those of manufacturers in general. The results were phenomenal in terms of effect on daily life.

The British combine ICI was not as profitable in the immediate post-war years as its American and German competitors. It had developed important new products, especially low-density polyethylene and polyester, but these were quite widely licensed, because ICI did not have strong capacity for product development and marketing. ICI established strength in the pharmaceutical area in the 1960s and 1970s. The pharmaceutical division was spun-off as Zeneca in 1991. In 1997, ICI made a major move toward consumer products when it acquired Unilever.

Japan had little chemical industry at the end of WWII but rapidly developed a petrochemical industry based on imports of petroleum. The conversion to petrochemical feedstocks was engineered by the Japanese government through its Ministry of International Trade and Industry (MITI) and the Ministry of Finance. The rapid development of its domestic economy and exports, especially of automobiles and electronics, created high demand and growth through the 1960s. MITI policies encouraged expansion to larger plants and by 1970 considerable overcapacity had developed in petrochemicals. The oil price increases of 1973 and 1979 were particularly disruptive in Japan. MITI began to work to try to reduce the overcapacity that had become evident.

The public image of chemistry was also beginning to change by the 1970s. From "Better Things for Better Living through Chemistry,"[h] the connection with the Viet Nam war (napalm, defoliants, dioxins), and toxic wastes (Love Canal, 1976; Times Beach 1982) became public concerns.[i] These resulted in passage of several laws designed to limit contamination of air and water. These laws had considerable impact on the operation of the chemical industry. The Clean Air Act, initially passed in 1963 and updated in 1970, 1977, and 1990, authorized regulation of atmospheric contaminants such as ozone, oxides of nitrogen and sulfur, and volatile organic compounds. It is administered on the state level through a system of permits aimed at reducing pollution. The original law also began limiting pollution from automobile exhausts (see Section 2.3.2.1).

The Clean Water Act was originally enacted in 1972 and updated in 1977 and 1987. It regulates discharges from industrial sites, water treatment facilities, and storm water systems. Like the Clean Air Act, it is enforced through permits that limit potential discharge and carries penalties for negligence or purposeful violation. The Comprehensive Environmental Response, Compensation and Liability Act of 1980, often called the "Superfund Law" made provision for clean-up of contaminated sites. It was strengthened by the Superfund Amendments and

---

[h]"Better Things for Better Living through Chemistry" was the slogan of DuPont from 1935 to 1982, when the "through Chemistry" was dropped.

[i]Love Canal was a section of Niagara Falls, NY, USA. It had acquired its name from an effort in the early 1900s by William T. Love to dig a canal connecting two branches of the Niagara River for electricity generation. The project was never completed and in the 1920s, the site was converted into a municipal dump. Among the users was Hooker Chemical Company, a producer of halogenated hydrocarbon products. In 1953, the site was sold to the City of Niagara Falls as a site for a school and residential development. In the 1970s, it was discovered that the site contained many deteriorating barrels filled with chemical waste. The site was evacuated and eventually became one of the first major clean-up sites under the Super-Fund program.

Times Beach was a community near St. Louis, MO, USA. It was in a low-lying area prone to flooding and was originally developed as a site for summer homes. Gradually, a small residential community developed. In 1972, the community hired a waste-handling company to spray dusty roads with oil. The company also had a contract for waste disposal at a Missouri chemical company that had produced chlorinated products, including the defoliant "Agent Orange" used during the Viet Nam war. The company had mixed the chemical waste with oil used in Times Beach and also at local stables, where a number of deaths of horses occurred. In 1982, flooding revealed extensive contamination by dioxin and other polychlorinated compounds. The EPA purchased and evacuated the site, which is currently a state park.

Reauthorization Act of 1986. These laws provide for identification and remediation of hazardous material areas, such as dumping areas or sites of manufacturing activity that left pollution. The law holds organizations and individuals responsible for pollution liable for costs and is supported by a tax on most classes of chemicals. Initially, the industry reacted defensively to these developments, forming organizations such as the American Industrial Health Council to oppose legislation regulating the environmental effects and toxicity of chemical products. Later, another industry group, the American Chemical Council developed the *Responsible Care Program* aimed at minimizing the environmental and safety hazards associated with the industry.[2] The program was updated in 2002, in light of the terrorist attacks of September 11, 2001. It is based on inspection of chemical plants by independent auditors to evaluate safety, environmental, and security measures. There are also international efforts to improve environmental health and safety in the chemical industry. These include the Global Environmental Management Initiative and the International Standards Organization.

3.4   1975–2000

The growth in production of most categories of chemicals was rapid until the mid-1970s, at which time the consequences of increased petroleum prices were felt (see Topic 2.4). Production of many product categories leveled off until about 1982, when consumer products such as paints and soaps resumed growth. Production of heavy chemicals and plastics leveled off as well.[3] Much of the growth in the period after WWII was due to introduction of synthetic replacements for traditional materials. Synthetic fibers and fabrics replaced cotton, wool, and silk. Plastics had become dominant in the packaging industry, replacing paper in many applications. Polyvinyl chloride replaced iron and copper pipe. Acrylic coatings revolutionized the paint industry. Various polymers became parts of appliances, automobiles, and airplanes. Many more refinements were to come, but generally at higher cost/benefit ratios. No new nylon or penicillin miracles occurred. The unexpected consequences of the advances, particularly environmental effects, were becoming apparent and the cost of petroleum raw materials increased dramatically. There were many advances to come having a strong chemical component in the pharmaceutical and biotechnology industries. Spectacular advances were made in computer and information technology. But there were few major innovations in the chemical industry itself.

After 1975, the chemical industry became international and highly competitive. International trade restrictions were reduced by the formation of the European Economic Community and the General Agreement on Trade and Tariffs. American companies, such as Dow, Monsanto, and Hercules, established both production and marketing overseas. In the United States, government antitrust actions required DuPont to license some of it products and to divest its General Motors holdings. DuPont was also required to terminate long-standing technology sharing agreements with ICI. Conglomerates with substantial chemical activities, such as Olin, W. R. Grace, and FMC were formed. European companies made investments and acquisitions in the United States. ICI had plants in many of the former colonial territories such as India and the Dominions (Australia, Canada). The large German companies also invested in the United States. BASF purchased Wyandotte Chemicals in 1969. Hoechst worked with Hercules on polyolefins and purchased Celanese in 1987. Bayer and Monsanto formed Mobay. Bayer purchased Miles Laboratories in 1979 and Sterling Winthrop in 1991, thereby establishing a position in the US pharmaceutical industry. Other European companies established free-standing companies in the United States such as Rohm & Haas and GAF (General Aniline and Film). The Swiss pharmaceutical companies Hoffmann-LaRoche, Ciba-Geigy, and Sandoz all established American affiliates and laboratories. Both American and international oil companies, such as British Petroleum and Shell, were involved in the worldwide petrochemical industry.

The final quarter of the 20th century was characterized by much merger and reorganization activity as the chemical industry faced slower growth, higher costs, and smaller profit margins. Economic analysts argue that the chemical enterprises lost "organizational capacity," that is, the specialized know-how that permits success in a competitive environment. In part, this reflected the increasingly widely available technology base. The innovative concepts of chemistry and chemical engineering that had been applied in the first three quarters of the century were now available around the world. The mature industry of the 1980s saw fewer opportunities for expansion through research and innovation and focused on efficiency and cost reduction. Research expenditures were reduced at many companies. At DuPont, for example, research expenditures dropped from 7.1% of sales in 1970 to 3.6% in 1980.

International competition led to overcapacity in many areas. There were several contributing factors. One being that plant efficiency tends to increase with scale. Also the technology for expansion was readily available, often by licensing from engineering firms. In fundamental areas such as petrochemicals (Middle-East), fertilizers (North Africa and Eastern Europe), and fibers (China, Southeast Asia and Korea), new capacity was built in expanding economies. The oil price increases imposed by OPEC in 1973 and 1979 exacerbated the situation. The largest chemical companies divested less profitable units to newly formed companies. This was particularly true in plastics and fibers. In the United States, the era of corporate raiders and leveraged buyouts took hold, which often led to divestment of major parts of the business of the large chemical companies. One of the consequences was decreased investment in research.

The increased oil prices and overcapacity of the 1970s and 1980s affected the German companies as well. As profitability of commodity chemicals continued to decline into the 1990s, the companies began to move toward higher value products, including pharmaceuticals. Bayer and Hoechst essentially left industrial chemicals to focus on agricultural chemicals and pharmaceuticals, the latter merging with Rhone-Poulenc to form Aventis. Only BASF remained as a producer of basic chemicals, concentrating on highly efficient integrated production. Germany has remained a net exporter of chemicals with exports exceeding imports by about 30%. The Swiss companies Sandoz and Ciba-Geigy combined, eventually splitting into two companies, Novartis in pharmaceuticals and Syngenta in agriculture.

The oil price increases of 1973 and 1979 were particularly difficult for Japan, which depended entirely on imports. Import duties on refined petroleum for cracking exacerbated the problem. In the 1980s, MITI proposed a 36% cut in petrochemical capacity. Reorganization of the Japanese petrochemical industry has remained a difficult problem. Several forces kept marginal facilities in production. Company management was generally geared to expansion and had no strong incentive for restructuring. The banks were reluctant to force closures and write-offs. As a result, the relative profitability of the Japanese firms has remained low and they have relied mainly on the developing economies of SE Asia, which offered the new market provided by the burgeoning electronics industry.[4] The Japanese chemical industry developed some incremental improvements but, unlike in automobiles and electronics, never became fully competitive on a worldwide scale. Among the reasons were the dependence on imported raw materials and the continuing innovation of the American and European chemical industries, which made it difficult for the Japanese companies to fully catch up. Nevertheless, four Japanese companies, Mitsubishi, Mitsui, Sumitomo, and Toray, were among the top 20 companies in worldwide sales in 2005.

Several of the large chemical companies made forays into the pharmaceutical industry in search of higher returns, but most eventually left the field. Dow acquired Marion and Merrell and also initiated joint ventures with Japanese pharmaceutical companies, but these interests were eventually sold to Hoechst in 1995. Monsanto acquired Searle in 1985, but it was spun-off as Solutia in 1996. Solutia subsequently merged with Upjohn and Pharmacia, but in 2005, all were absorbed into Pfizer. In 1991, DuPont formed a joint pharmaceutical venture with Merck, which became DuPont Pharmaceutical in 1992. The business was sold to Bristol Myers Squibb in 2001.

The 1980s saw widespread restructuring of manufacturing industries, including chemicals, in the United States. Several factors came into play. (1) High interest rates put pressure on corporations to achieve higher returns. (2) Methods for takeovers and leveraged buyouts were devised that became highly profitable both for the "raiders" and the financial institutions backing them. (3) Corporate management was forced to focus on short term results. Industries that had large cash flows, but limited opportunity for profitable new investment, were particularly susceptible to takeovers and related activities. The commodity chemical industry was in this situation. The overall results tended to be higher corporate debt loads, decreased investment in new plants and research, slowing of wage gains for workers and employees, except at the very highest level of management. As the decade proceeded, both corporate management and labor saw loss of their relative benefits and pressed for state antitakeover laws. In contrast, at the federal level there was reduced interest in antitrust or financial regulation. By the second half of the 80s the "restructuring fad" had taken on a life of its own, propelled by the financial apparatus established to promote it and the huge profits that were being made. Executive compensation soared in relation to other wages and salaries.

During the 1980–1982 recession, the value of chemical shipments declined 7%, as compared with a 2.5% decline for the economy as a whole. During the decade, the net income relative to both sales and assets were in the range of 5–9%, with the lowest values in the middle of the decade. As a result several broad trends emerged. (1) The largest companies reduced their activity in commodity chemicals, while diversifying into more specialized, higher value products. For example, during the decade of the 1980s, the percent of revenues from basic chemicals at Dow declined from 60% to 48%. (2) A number of plant reductions occurred in the 1980s. In the United States, these were driven by market forces, and in Europe the privatization of government-owned firms also contributed. In Japan, MITI directed capacity reductions. Individual companies tended to narrow the scope of their operations, concentrating on the most profitable and promising. (3) Several new firms, often highly leveraged, were formed to acquire and operate commodity chemical plants. Examples included: (a) Cain Chemicals, which acquired plants from DuPont and ICI Americas. (b) Aristech was spun-off from USX. It was acquired first by Mitsubishi and is now owned by Sunoco. (c) Vista acquired chemical plants from the Conoco part of DuPont. (d) Huntsman Chemical, which acquired petrochemical and polymer plants from Shell and Hoechst, became the largest polystyrene producer. Subsequently, Huntsman has expanded into many other kinds of polymers, coatings, and adhesives. (e) Sterling Chemical acquired many of Monsanto's petrochemical facilities. (f) Quantum, which was formerly the National Distillers and Chemical Corporation, produces polyethylene, industrial ethanol, methanol, and vinyl acetate. It is also a supplier of liquefied propane gas. (g) Ineos was formed by a private equity group that acquired petrochemical and polymer facilities from firms such as BP, ICI, BASF, and Solvay. These firms tended to have high debt to asset ratios. For example, while the industry as a whole saw its debt:asset ratio increase from 0.5 to 0.65 over the decade, the ratio of a sample of newly formed and restructured firms was 0.95.

In the 1990s, buyouts and related financial maneuvers continued. The number of firms decreased. Several of the large chemical companies from the first-half of the century disappeared. Both Allied Chemical and Union Carbide were disassembled. Monsanto dramatically repositioned itself and shifted its emphasis to agricultural chemicals based on synergy with biotechnology. In 1996, its chemical components were spun-off to Sterling Chemical. DuPont and Dow also made major investments and acquisition in agricultural chemistry (see Section 21.2).

3.5 THE INTERNATIONAL CHEMICAL INDUSTRY AT THE END OF THE CENTURY

Table 3.1 gives the 2005 ranking of the world's 20 largest chemical companies in terms of sales.

TABLE 3.1 Worldwide Chemical Sales in 2005.

Company	Sales US$ billions	Country of origin
Dow	46.3	USA
BASF	43.7	Germany
Shell	35.0	Netherlands/UK
ExxonMobil	31.2	USA
Total	27.8	France
DuPont	25.3	USA
Sinopec	21.1	China
Bayer	20.7	Germany
SABIC	18.9	Saudi Arabia
Formosa Plastic	18.7	Taiwan
Lyondell	18.6	Netherlands
Mitsubishi	17.9	Japan
Degussa	14.6	Germany
Mitsui	13.4	Japan
Huntsman	13.0	USA
Ineos	12.4	UK
AkzoNobel	11.8	Netherlands/Sweden
Sumitomo	11.5	Japan
Air Liquide	11.4	France
Toray	11.3	Japan

(Data from Global Top 50, *Chemical and Engineering News*, July 24, 2006).

Some of the long term trends in the chemical industry in relation to the world as a whole can be seen in Tables 3.2 and 3.3. Table 3.2 shows the value of chemical products worldwide from 1913 to 2000. By 1917, the United States accounted for about a third of total production. This reached a high point of around 40% in the early 1950s. Germany accounted for roughly 20% between WWI and WWII, but has accounted for 6–8% since then. Britain's share has steadily declined from 10% in the early part of the century to 3% by 2000. The United States and Western Europe accounted for 55% of production in 2000, as compared with 66% in 1951, indicating the growth in the rest of the world since then.

TABLE 3.2 Production of Chemicals in US $ (Billions) and Share by Country.

Year	USA	%	UK	%	Japan	%	Germany	%	France	%	World total
1913	3.4	34	1.1	11	0.15	2	2.4	24	0.85	9	10.0
1927	9.45	42	2.3	10	0.55	2	3.6	16	1.5	7	22.5
1935	6.8	32	1.95	9	1.3	6	3.7	18	1.6	8	21.0
1938	8.0	30	2.3	9	1.5	6	5.9	22	1.5	6	26.9
1951	71.8	43	14.7	9	6.5	4	9.7	6	5.9	4	166
1970	49.2	29	7.6	4	15.3	9	13.6	8	7.2	4	171
1980	168.3	23	31.8	4	79.2	11	59.3	8	38.6	5	719
1990	309.1	24	44.7	4	162.8	13	100.5	8	66.3	5	1248
2000	460.0	28	50.7	3	218.4	13	100.0	6	73.0	4	1669

(From Murmann, J. P. *Chemical Industries After 1850*, *Oxford Encyclopedia of Economic History* 2002, pp 398–406. By permission of Oxford University Press, USA.)

The same trends appear in chemical exports, as summarized in Table 3.3. At the beginning of the century, the United Kingdom and Western Europe accounted for 80% of exports, with Germany having the largest share. The United States accounted for only 14%. By 1950, the European share had dropped to 60%, while the US share had risen to 35%. The rest of the world, including Japan, accounted for only a little over 1% in 1950. By 2000, the rest of the world accounted for nearly 26% of exports. The United States, by contrast had dropped to 14%. There had been a significant shift to Europe, reflecting the internationalization of production and trade.

TABLE 3.3   Shares of Chemical Exports by Country of Origin 1899–2000.

Country	1899	1913	1929	1937	1950	1959	1990	2000
UK	19.6	20.0	17.5	16.0	17.9	15.0	8.4	6.6
France	13.1	13.1	13.5	9.9	10.1	8.6	9.1	7.8
Germany	35.0	40.2	30.9	31.6	10.4	20.2	17.7	12.1
Western Europe	13.1	13.1	15.3	19.4	20.5	21.1	31.7	32.0
US	14.2	11.2	18.1	16.9	34.6	27.4	13.2	14.1
Canada	0.4	0.9	2.5	2.9	5.2	4.4	1.8	1.6
Japan	0.4	1.0	1.8	3.0	0.8	3.1	5.4	6.1
Other	4.2	0.3	0.4	0.3	0.5	0.2	12.8	19.8
Total US$(billions)	0.26	0.59	1.04	0.98	2.17	5.48	309.2	566.0

(From Murmann, J. P. *Chemical Industries After 1850*, *Oxford Encyclopedia of Economic History* 2002, pp 398–406. By permission of Oxford University Press, USA.)

The data in Tables 3.2 and 3.3 are in percent of totals and this to some extent obscures the very large total growth in both production and exports. The 1913 worldwide production total was estimated at $10 billion US and exports at $590 million, suggesting about 96% of production was used internally. By 1937/1938 total US production had risen to about $27 billion with exports quoted as just under one billion, with still only about 4% of production being exported. For 1950/1951, US production is $71.8 billion and exports are shown as $750 million, indicating exports had dropped to around 1% of production. The 1990 data put US exports at $40 billion compared to production of $309 billion, almost 13%. The shares of individual countries and parts of the world vary somewhat but the trend is the same. For Germany, the 1990 numbers indicate about $100 billion of production of which $17.7 billion was exported. For Japan the comparable 1990 figures are $16.6 billion in exports out of $163 billion of production. For 2000, the percent of exports for the United States, Germany, and Japan were, respectively, 17%, 65%, and 16%. These data, along with the rapidly increasing share of "other" in the exports indicates the great expansion of international trade in the latter part of the century.

3.6   ORIGINS OF THE PHARMACEUTICAL INDUSTRY

We will discuss the pharmaceutical industry in more detail in Chapter 11. However, it is useful to introduce some of the basic story and names of the firms at this point. In contrast to its beginning in Germany in the dye industry, most of the American pharmaceutical companies evolved from producers and wholesalers of pharmaceutical formulations, not from chemical firms. Many of the companies dealt both with prescription (called "ethical" in the first-half of the century) and proprietary (over-the-counter) drugs. As synthetic drugs began to arrive from Germany in the early 1900s, several of the pharmaceutical companies began to initiate research activities. As with dyes, WWI interrupted supplies from Germany and forced the American companies to learn how to make the medicines.

Between the two world wars companies such as Merck, Pfizer, Eli Lilly, Upjohn, Parke-Davis, Squibb, and G. D. Searle were becoming the modern pharmaceutical industry as scientific advances opened the way for rational therapies. Knowledge of vitamin deficiency diseases developed first from empirical observation and then from systematic investigation of the essential substances that became known as vitamins. Biochemical research also began to identify the hormones, the chemical messengers, including adrenalin (1901), thyroxine (1915), insulin (1921), and the steroids (1930s).

At the end of WWI, a few synthetic compounds were in use, for example, antiseptics (phenol), analgesics (aspirin, phenacetin), and anesthetics (ether, procaine, novocaine). Synthetic pharmaceuticals expanded rapidly after WWI. Gerhard Domagk, a physician working at Bayer in Elberfeld, found Prontosil to be active against *Streptococcus*. Soon it was shown that the active substance was sulfanilamide, which was free of patent coverage. Other *sulfa drugs* were soon on the market (see Chapter 12).

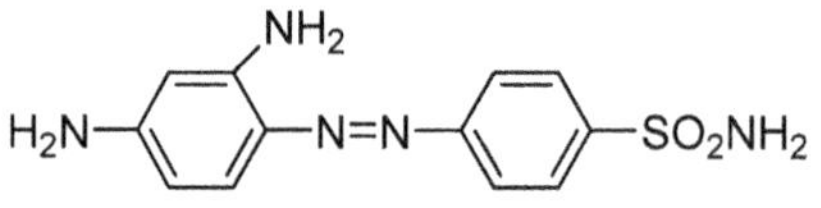

prontosil 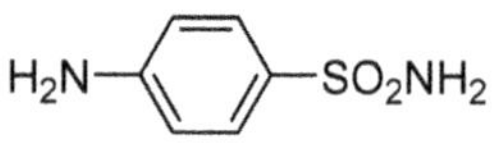   sulfanilamide

First reported by Alexander Fleming in 1928, the mold metabolite penicillin was isolated and shown to be a powerful antibiotic by Howard Florey and Ernst Chain at Oxford University in the late 1930s. It became available as a result of war-time collaboration between the US government and the pharmaceutical industry (see Section 12.1).

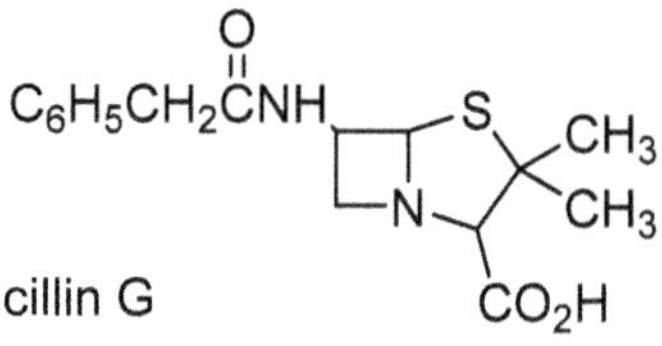

Penicillin G

These discoveries set the stage for the rapid expansion of the pharmaceutical industry in the second half of the century. Synthetic chemistry soon produced antihistamines, tranquilizers, diuretics, and pain-relievers. Following the penicillin path, fermentation and chemical modification of natural products led to many new antibiotics. At the same time, the fields of biochemistry, enzymology, microbiology, and pharmacology were finding ways to rationally identify potential drugs. The large pharmaceutical companies began to invest heavily in research aimed at discovery of new drugs. The United States gained worldwide leadership in the pharmaceutical industry after WWII, particularly as biochemistry became an essential part of drug development and discovery. The United States benefitted from the large government investment in training and research in health-related disciplines. The US pharmaceutical industry also had the advantage of having the largest domestic market.

During the 1990s, Bayer and Hoechst increased their presence in pharmaceuticals, while scaling back in commodity chemicals and polymers. Hoechst acquired Dow's pharmaceutical company Merrell-Dow and Bayer acquired the Winthrop-Sterling Company from Eastman Kodak. Two Swiss companies, Ciba-Geigy and Sandoz merged in 1995 to form Novartis. The French companies RousellUclaf and Rhone-Poulenc merged and then combined with Hoechst to form Aventis. The former Searle (by then a Monsanto spin-off Solutia), Upjohn and Pharmacia were merged and later acquired by Pfizer. ICI's pharmaceutical ventures were merged with the Swedish Company Astra to form Astra-Zeneca. Japanese pharmaceutical companies, which were particularly strong in antibiotics, focused mainly on their domestic market and used licensing agreement to sell products in the North America and Europe. The rank of companies in sales as of 2011 is given in Table 3.4.

TABLE 3.4 Pharmaceutical Company Sales 2011.

Company	2011 Sales $ billions	Country
Pfizer	65.3	US
Novartis	58.6	Switzerland
Sanofi	47.9	France
Merck	42.1	US
GlaxoSmithKline	39.3	UK–US
Roche	39.1	Switzerland
Astra-Zeneca	32.0	UK–Sweden
Johnson & Johnson	24.8	US
Eli Lilly	21.5	US

TABLE 3.4   *(Continued)*

Company	2011 Sales $ billions	Country
Abbott Laboratories	22.5	US
Bayer	18.9	Germany
Teva	17.4	Israel
BristolMyerSquibb	17.1	US
Takeda	17.0	Japan
NovoNordirsk	12.3	Denmark

KEYWORDS

- Solvay process
- dyes
- chlorine
- Haber–Bosch process
- alkyd resins
- polymers
- chemical engineering
- pharmaceuticals
- agricultural pesticides
- Clean Water Act
- Superfund Law
- globalization

BIBLIOGRAPHY

Aftalion, F. *A History of the International Chemical Industry*; 2nd ed. transl. by Benfrey, O. T.; University of Pennsylvania Press: Philadelphia, PA, 2001; Arora, A.; Landau, R.; Rosenberg, N. *Chemicals and Long-term Economic Growth*, Wiley: New York, 1998; Murmann, J. P. *Oxford Encyclopedia of Economic History*; Vol. 1, 2002, pp 398–406; Chandler, A. D. *Shaping the Industrial Century: The Remarkable Story of the Modern Chemical and Pharmaceutical Industries*; Harvard University Press: Cambridge, MA, 2005; Galambos, L.; Hikino, T.; Zamagi, V. *The Global Chemical Industry in the Age of the Petrochemical Revolution*; Cambridge University Press: Cambridge, UK, 2007.

BIBLIOGRAPHY BY SECTIONS

Section 3.3: Carpenter, B. P.; Watson, D. E.; Carpenter, B. C. In *Albright's Chemical Engineering Handbook*; Albright, L. F. Ed.; CRC Press: Boca Raton, FL, 2009; pp 1485–1500.

Section 3.4: Lane, S. J. In *The Deal Decade*; Blair, M. M. Ed.; The Brookings Institution: Washington, DC, 1993; Smith, J. K. *Bus. Econ. Hist.* **1994,** *23*, 152–161.

REFERENCES

1. Arora, A.; Rothenberg, N. *Chemicals and Long-Term Economic Growth*; In Arora, A.; Landau, R.; Rothenberg, N., Eds.; Wiley: New York, 1998, p 81.
2. Yosie, T. F. *Environ. Sci. Technol.* **2003,** *37*, 401A–406A.
3. Arora, A.; Rothenberg, N. In *Chemicals and Long-term Economic Growth*; Arora, A.; Landau, R.; Rothenberg, N., Eds.; Wiley: New York, 1998, p 88.
4. Hikino, T.; Harada, T.; Tokahisa, Y.; Yoshida, J. In *Chemicals and Long-term Economic Growth*; Arora, A.; Landau, R.; Rosenberg, N., Eds.; Wiley: New York, 1998; pp 103–135.

CHAPTER 4

HALOGENATED HYDROCARBONS: PERSISTENCE, TOXICITY, AND PROBLEMS

ABSTRACT

The availability of chlorine by electrolysis made the production of chlorinated hydrocarbons possible on an industrial scale. These included chlorinated methanes, ethanes, and ethylenes. These compounds became widely used as solvents and chemical intermediates. Their advantages included relatively high chemical stability and low flammability. Extended use, however, revealed health hazards and resulted in restriction on exposure and use. The discovery of 1,1,1-trichloro-2,2-*bis*-(4-chlorophenyl)ethane (DDT) in 1939 opened a new era in insect control and contributed to the elimination of malaria in many parts of the world. However, DDT and several related chlorinated insecticides were found to be persistent in the environment and to be detrimental to some species of birds, amphibians and fish. Most uses have been banned by international agreement. Chlorofluoro derivatives of methane and ethane, known as chlorofluorocarbons were found to have excellent properties as refrigerants and aerosol propellants. These, too, turned out to be persistent for decades in the atmosphere and stratosphere and to lead to polar ozone depletion. They were phased out under the Montreal Protocols adopted in 1987. They were replaced by hydrochlorofluorocarbons, which have much shorter half-lives, but contribute to atmospheric warming. They are scheduled to be replaced by 2020.

We will discuss three classes of halogenated hydrocarbons in this chapter: (1) volatile solvents and intermediates; (2) low volatility polyhalogenated hydrocarbons; and (3) halofluorocarbons. Representatives of each of these groups were introduced as commercial products by the middle of the century. But each had significant adverse environmental effects. As a result, they were among the first groups of synthetic chemicals to attract wide-spread public attention as pollutants.

4.1 VOLATILE CHLORINATED SOLVENTS AND INTERMEDIATES

The ability to produce chlorine economically by electrolysis (see Topic 4.1) led to methods for chlorination of hydrocarbons. The chlorinated compounds found several large-scale applications. One of the most widely disseminated uses of halogenated solvents is in dry-cleaning. Industrial degreasing and cleaning of electronic components was also an important use. Other uses included "blowing" of polymeric foams and in pressurized aerosol cans. Among the most widely used solvents was dichloromethane (CH_2Cl_2, also known as methylene chloride) which for many years was the main solvent for decaffeination of coffee (see Section 9.4). Trichloromethane ($CHCl_3$, chloroform), and tetrachloromethane (CCl_4, carbon tetrachloride) were also used as solvents. Chloroform was used as an anesthetic in the latter part of the 19th century (see Section 13.8). Carbon tetrachloride was the major solvent in the dry-cleaning industry from 1930 until 1970, when its chronic toxicity was recognized. Trichloroethylene $Cl_2C{=}CHCl$ (TCE), and tetrachloroethylene $Cl_2C{=}CCl_2$ (also known as perchloroethylene, PCE, or "PERC") replaced CCl_4 and the latter remains a major dry-cleaning solvent. Halogenated hydrocarbons also became important intermediates for the preparation of polymers. In particular, chloroethene $CH_2{=}CHCl$ (vinyl chloride) and 1,1-dichloroethene $CH_2{=}CCl_2$ (vinylidene chloride) are monomers for preparation of polyvinyl chloride (PVC) and saran, respectively (see Section 5.4).

TOPIC 4.1 ELECTROLYTIC PRODUCTION OF CHLORINE

The development of an electrolytic process for production of chlorine was one of the major steps forward in the development of the chemical industry (see Section 3.1). In 2000, over 40 million tons were produced worldwide and used in a wide variety of chemical processes. Over the years since its introduction, many improvements have been made to reduce the energy requirements and to minimize deleterious waste products. In the original version there was a flowing mercury cathode. The sodium metal that was produced formed an amalgam with the mercury. The amalgam was separately hydrolyzed producing hydrogen and sodium hydroxide. The process required an electrochemical potential of about 4.6 V. The reactions were as follows:

anode: $2\,Cl^- - 2e^- \longrightarrow Cl_2$

cathode: $2\,Na^+ + 2e^- \xrightarrow{\ Hg\ } 2Na/Hg \xrightarrow{\ H_2O\ } H_2 + 2\,NaOH + Hg$

The anodes were initially graphite, but in the 1950s, anodes made from titanium coated with ruthenium oxide were introduced, reducing the potential

required to about 3.3 V. The next major improvement was use of asbestos dia-
phrams to separate the anode and cathode cells. In these cells, hydrogen is produced directly at the cathode. The NaOH that is produced contains a substantial
amount of NaCl and must be purified. This problem was solved by the introduction of perfluorinated membranes (Nafion) in the 1970s. These membranes have
the necessary chemical stability and contain sulfonate and carboxylate groups
that restrict the migration of anionic chloride through the membrane. These cells
operate at about 3.1 V.

anode: $2\ Cl^- - 2e^- \longrightarrow Cl_2$

cathode: $2\ Na^+ + 2\ H_2O + 2e^- \longrightarrow 2\ NaOH + H_2$

The most recent version uses O_2 as the oxidant and is called the *oxygen-
depolarized cathode* process. It can be applied either to NaCl or HCl solutions.
The latter is advantageous because HCl is frequently a byproduct of Cl_2 use, and
this provides a means of recovery of chlorine. This process does not produce H_2.
The cathodes contain metal such as Pt or Pd and the process operates at slightly
over 1 V.

anode: $4\ Cl^- - 4e^- \longrightarrow 2\ Cl_2$

cathode $O_2 + 4\ H^+ + 4\ e^- \longrightarrow 2\ H_2O$

Sources: Brooks, W. N. *Chem. Br.* **1986**, *22*, 1095–1098; Fauvarque, J. *Pure Appl. Chem.*
1996, *68*, 1713–1720; Kreysa, G.; Juttner, K. *Chem. Eng.* **2007**, *114*, 50–55.

4.1.1 DRY-CLEANING

Dry-cleaning was introduced in the mid-1800s when it was recognized that nonpolar hydrocarbons could remove greases and oils from fabrics. The first material to be used was turpentine, but it was replaced by benzene, which we now
know to be a carcinogen. When petroleum refining made naphtha available, it
became the solvent of choice, but it represented an extreme fire hazard. In the
1920s, it was replaced by hydrocarbon mixtures with higher flash-points. The
halogenated solvents were introduced in the 1930s. The initial attraction of the
chlorinated solvents in the dry-cleaning industry was the reduced fire hazard.
Writing in 1936, D. H. Killeffer emphasized this advantage.[1] He also recognized two potential problems, corrosion of equipment and possible toxic effects

associated with the halogenated solvents, but the extent of the latter problem was not realized at the time. The availability and relative safety of halogenated and higher boiling hydrocarbon solvents led to the establishment of many small dry-cleaning shops in commercial and mixed-use residential areas.

In dry-cleaning with either hydrocarbons or halocarbon solvents, the material is tumbled with the solvent. The solvent dissolves grease and loosens adhering solids so they can be washed away. The material is usually rinsed with fresh solvent. The advantage of nonpolar solvents is that, unlike water, they do not swell most fabrics and so the items retain their shape. The solvents are recycled and purified by filtration and distillation. The units are designed to minimize loss of the solvent, originally for economic reasons, but now to minimize the environmental contamination. The early machines involved manual transfer of solvent-saturated fabrics from washer to dryer. By 2000, most of these had been replaced by systems that did not expose workers to solvent vapors.

Carbon tetrachloride was banned as a cleaning solvent in 1970 as the result of accumulating evidence of chronic liver and kidney toxicity.[2] At the present time, the main chlorinated solvent used for dry-cleaning is PCE, $CCl_2=CCl_2$. PCE production ranged from 100 to 150 million lb/y from 1965 to 1980, but then began to decline. In 2000, about 70 million pounds were produced. The decline was the result of more efficient machines and better recovery, not a shift to other solvents. About 80–90% is used in the dry-cleaning industry and 85–90% of dry-cleaning establishments use PCE. As of 2003, there was no conclusive evidence that PCE is a human carcinogen, although most of the studies done have limits of either methodology or size.[3] For TCE there is evidence of increased risk for brain, kidney, and liver cancer, as well as non-Hodgkins lymphoma, but the evidence is confounded by factors such as tobacco and alcohol use.[4] There is some evidence of reproductive difficulties and respiratory disease in women exposed to chlorinated solvents.[5] PCE is classified as a potential carcinogen by NIOSH. In 1989, the Occupational Safety and Health Administration (OSHA) attempted to establish a 170 mg/m^3 permissible level for PCE but the US District Court of Appeals in Washington DC ruled against it and the level remains at 680 mg/m^3. The levels permitted in Germany and Japan are 340 and 170 mg/m^3, respectively. Few dry-cleaning shops meet the current standard. The EPA proposed new standards as of May 2007 to take effect in 2010.[6] Sites of dry-cleaning establishments are frequently contaminated, and the Comprehensive Environmental Response, Compensation and Liability Act (Superfund Law, see Section 3.3) assigns responsibility for clean-up to operators and landlords, representing a substantial potential liability.

There are several potential alternatives to halogenated solvents for dry-cleaning. One is called "professional wet cleaning" which uses water and relies on computer-controlled washing machines to achieve the gentle washing required by certain fabrics and garments. Another possibility is use of supercritical carbon dioxide. Carbon dioxide, however, is a weak solvent for most oils and greases and new detergents will be required to improve its effectiveness. Pressurized equipment is also required.

4.1.2 INDUSTRIAL CLEANING SOLVENTS

Many industrial processes involve cleaning of metals and other materials so they can be welded, plated, or painted. In addition to TCE and PCE, a mainstay is Cl_3CCH_3, trichloroethane. Cleaning of metal parts can be done by placing the part in the solvent vapor, where condensation occurs, washing off oil and other debris, and returning it to the heating still. Since 1997, the emissions, but not use *per se*, have been strictly regulated. In addition, the economic incentive to control costs has favored other types of cleaning. There are several alternative methods. One is ultrasonic cleaning, with or without water as a solvent. Biological cleaning, using solutions of microorganisms, is also feasible. Carbon dioxide is also beginning to attract attention in industrial cleaning. It can be used in several ways. Pressurized solid CO_2 can be used for abrasive cleaning. A gentler process uses "CO_2 snow" guns. The CO_2 particles evidently loosen dirt, then evaporate and, except for the green-house effect, the emissions are innocuous. Supercritical CO_2 is also a possibility, although it requires pressurized equipment.

4.1.3 HIGH-PRECISION ELECTRONIC CLEANING

Chemical cleaning is an important operation in the manufacture of microelectronic devices. This industry is now largely centered in Asia, with production in China > Japan > Taiwan > Korea. The United States currently accounts for 15% of world production and Europe less than 10%. Several polymers are used in the structural components, including fiberglass impregnated with epoxy resin and paper-phenolics. High-performance components use Teflon and polyimides. The final products consist of copper circuits and connectors deposited or etched on a series of layers. The manufacturing process can involve chemical etching, although plasma etching is becoming increasingly common. At the present time, halogenated solvents are not used as much as glycol monoethers and dimethyl-formamde (DMF). As in both dry-cleaning and industrial cleaning, supercritical CO_2 seems to have considerable potential for use in microelectronic cleaning.[7] The microelectronic industry has had limited attention in terms of regulation and environmental impact. Relatively little is known about the consequences of the manufacturing processes. In 1997, the EPA and industry began the "Design for the Environment Project" to minimize the environmental impact. The most stringent regulations are in Europe and Asian manufacturers design products to meet the relevant import requirements.

4.1.4   VINYL CHLORIDE AND VINYLIDENE DICHLORIDE

Vinyl chloride ($CH_2=CHCl$) and vinylidene dichloride ($CH_2=CCl_2$) are extensively used in the preparation of PVC and saran, respectively (see Section 5.4.3). In the early 1970s, evidence that workers exposed to vinyl chloride had elevated risk to angiosarcoma of the liver began to appear, including three cases of this exceedingly rare tumor at a B. F. Goodrich plant in Louisville, KY, USA. At that point, the permissible industrial exposure level was decreased from 500 ppm to 1 ppm. Earlier, workers exposed to very high doses were also noted to have acute symptoms. The increased risk for angiosarcoma from heavy exposure to vinyl chloride is about 45-fold. The latency period is around 20 years. As of 1999, about 200 deaths worldwide had been associated with angiosarcoma resulting from vinyl chloride exposure. Because most parts of the world have imposed strict exposure limits, it is expected that the number of cases will decrease. There is also some evidence of increased brain cancer risk, although this remains controversial. As late as 1999, chemical manufacturers were arguing against further restriction of vinyl chloride levels in the workplace. The EPA has been criticized for excessive reliance on industry data and input in establishing risk assessments.[8] The evidence as to whether vinylidene dichloride is carcinogenic is unclear. Vinyl chloride has been found in groundwater near landfills and industrial waste sites. It may arise in these situations, at least in part, from microbiological reduction of the cleaning solvents PCE and TCE.

4.1.5   HEALTH CONSEQUENCES OF EXPOSURE TO VOLATILE CHLORINATED HYDROCARBONS

The general public is seldom exposed to high concentrations of volatile chlorinated hydrocarbons, but they have been widely used in chemical and industrial processes and in the dry-cleaning industry. Many studies have examined the relationship between workplace exposure and health outcomes. Vinyl chloride is classified as a human carcinogen. The chronic kidney and liver toxicity of carbon tetrachloride became evident from their relatively high incidence among workers in the dry cleaning industry. For most of the other volatile chlorinated compounds the evidence is conflicting. Data on exposure is often inexact and workers are frequently exposed to variable combinations of other solvents. The most frequent areas of observed effects include kidney damage, respiratory disease, reproductive effects and for chlorinated ethylenes, cancer of the brain, bladder, kidney, and liver.

4.2 DDT AND OTHER PERSISTENT HALOGENATED COMPOUNDS

4.2.1 DDT

DDT, 1,1,1-trichloro-2,2-*bis*-(4-chlorophenyl)ethane, was the first widely used insecticide. Its insecticidal properties were discovered in 1939 by Paul Mueller, working at Geigy in Switzerland. Its benefits were immediately apparent. At the end of WWII, it was used to control insect-borne typhus in Europe. Mueller received the 1948 Nobel Prize in medicine and physiology for his discovery. DDT was used after WWII for control of malaria-transmitting mosquitoes. Malaria was eradicated in the United States and in Europe and greatly reduced elsewhere in the world. It is estimated that a billion cases of malaria were prevented and about 50 million deaths averted.[9] Annual production in the United States was in the range of 100–200 million pounds in the 1960s. By 1970, some 800 million pounds of DDT were being produced annually worldwide. The dichloro analog is known as DDD. 1,1-Dichloro-2,2'-*bis*-(4-chlorophenyl)ethene, DDE, is the main degradation and metabolic product of DDT and has similar physical and biological properties. Methoxychlor is another insecticide with closely related properties.

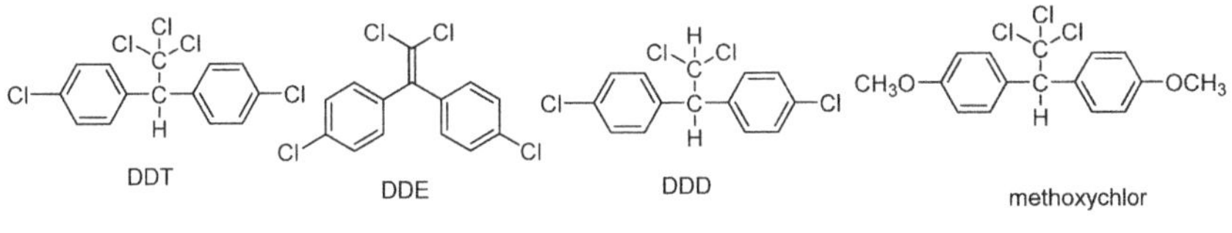

DDT   DDE   DDD   methoxychlor

DDT falls in the class of *polychlorinated aromatic hydrocarbons*. It is extremely stable chemically and insoluble in water. It is classified as *persistent*, meaning it has a long lifetime in the environment. Within a decade of its introduction, unexpected consequences of DDT use began to appear. The eggs of peregrine falcons were noted to be easily broken. Brown pelican populations in southern California plummeted. A major spill of dicofol, a DDT analog, in Lake Apopka, FL, led to genital abnormalities in alligators. DDT application for gnat control at Clear Lake, CA, resulted in the death of grebes. Among the birds, most adversely affected by DDT was the American bald eagle. Beyond persistence, DDT and DDE are subject to *bioaccumulation*, which is the tendency of some organisms to concentrate the material from their food and the environment. These properties soon resulted in ubiquitous presence of DDT and DDE in the food supply, including human breast milk.

The Department of Agriculture and some state and local entities engaged in wide-scale application of DDT. Programs to control the gypsy moth, the beetle vector of Dutch elm disease, and the fire ant were begun. Frequent local cases of bird and fish kills were noted, but little concerted opposition developed immediately. By 1958, several papers documenting significant DDT toxicity to birds had appeared. A study of the grebe kill at Clear Lake, CA, USA, documented astonishing bioaccumulation through the food chain. These developments were among the cases cited in *The Silent Spring*, published by Rachel Carson in 1962.[a] The public interest generated by *Silent Spring* and the contemporaneous discovery of the toxic effects of Thalidomide (see Topic 11.2) led to public hearings. The President's Science Advisory Committee prepared a report, known as the Wiesner report,[b] which recommended changes in regulatory procedures.

The evidence of environmental effects attributed to DDT led to more study and improvement in instrumental methods for its detection showed it to be wide spread. Direct demonstration was made that exposure to DDE led to thinning of eggshells. Curiously, many bird species are not nearly so affected as eagles, falcons, and many fish-eating birds. Domestic chickens, for example, are virtually

---

[a] Rachel Carson received a Masters degree in marine zoology at Johns Hopkins University in 1932. She worked at the US Fish and Wildlife Service from 1936 to 1952. During this time, she published many articles and several best-selling books, including *The Sea Around Us*, which was a New York Times best-seller and National Book Award winner in 1951. Beyond her professional knowledge, Carson learned from personal experiences of friends and others about the effects of aerial spraying of DDT and other chlorinated insecticides in the late 1950s. As she researched an article and then a book on the subject, she became increasingly alarmed at the implications of insecticide use. Despite serious medical problems, Carson completed *Silent Spring* in 1962. It had immediate impact, both *pro* and *con.* Louis McClean, the general counsel of Velsicol Corporation, the manufacturer of the insecticides chlordane and heptachlor, insinuated that *Silent Spring* was part of a communist plot to weaken American agriculture. Later, writing in *Bioscience*, McLean characterized DDT opponents as "identified by the numerous variant views [they] hold about regular foods, chlorination and fluoridation of water, vaccination, public health programs, food additives, medicine, science, and the business community." (*Bioscience* 1967, *17*, 613–617). A review of *Silent Spring* in *Chemical and Engineering News,* cited various authorities on the safety of pesticides, and concluded: "The responsible scientist should read this book to understand the ignorance of those writing on the subject and the education task which lies ahead." (*Chemical and Engineering News,* Oct. 1 1962, pp 61–63). The book also received many positive reviews. In a CBS Reports program on April 3, 1963, Carson appeared with Secretary of Agriculture, Orville Freeman, Surgeon General Luther Terry and others. Immediately thereafter, Senator Abraham Ribicoff announced that hearings would be held. The resulting legislation required consultation with the Department of Interior and State Governments prior to implementation of spraying and began the move to restrict use of persistent insecticides. A year later, in April 1964, Rachel Carson died at the age of 56.

Sources: Lear, L. *Rachel Carson, Witness for Nature*. Holt and Company: New York, 1997; Quaratiello, A. R. *Rachel Carson, A Biography*. Greenwood Press: New York, 2004.

[b] Jerome Wiesner served as chair of President John F. Kennedy's science advisory committee from 1961 to 1964. He was trained as an electrical engineer (B. S. 1937; Ph. D. 1940, University of Michigan). He was a member of the faculty at MIT from 1946 to 1961. He returned to MIT in 1964 and served as dean (1964–66), provost (1966–71), and president (1971–1980).

unaffected. The direct biochemical mechanism is evidently on the glandular system involved in egg-shell formation. Extensive hearings were held in United States, beginning in 1971 and 1972. The hearings were contentious. The "yes or no" culture of the law clashed with scientific conclusions based on correlation and probability. Shortly thereafter, the EPA administrator William Ruckelshaus[c] ruled to terminate DDT use in the United States. Eventually, DDT, along with several other *persistent organic pollutants* (POPs), was banned by international convention, although exceptions remain for use in disease-vector insect control, where no alternative is available. In many areas of the world, malaria-bearing mosquitoes have become resistant to DDT. The decision to restrict DDT use was not without controversy. Agricultural interests and pesticide suppliers were generally resistant to restrictions. Agricultural entomologists were particularly aggrieved,[10] perhaps feeling that the success of their chemical approach to pest control was under attack. Even, Norman E. Borlaug, the 1970 Nobel Peace Prize winner for his work on new strains of grains, maintained that "fear-provoking irresponsible environmentalists were mounting a propaganda campaign against agricultural chemicals."[11] Proponents of continued use of DDT argued that there is no evidence of acute or chronic human toxicity and that current practice adds minimal amounts of DDT to the environment. They cite substantial and rising rates of death from malaria, particularly in Africa, and absence of acceptable alternative approaches. These arguments persist in anti-environmentalist rhetoric.[12]

DDT is one of a number of compounds that are grouped together as POPs. Structurally, they are all heavily halogenated compounds. They share a number of properties, being long-lived in the environment and subject to bioaccumulation in the food chain. The compounds tend to be transported by wind currents from warmer to colder areas because they are more extensively volatilized in warmer regions. As a result they are present in Arctic regions where they have never been used. Bioaccumulation is observed in the Arctic food chain. For example, caribou feed on lichen and concentrate POPs by a factor of 10. Wolves that feed on caribou, in turn concentrate the POPs by another factor of 6. Certain types of fish have especially high concentrations of POPs.

There have been many studies designed to measure the DDT distribution. One such series of studies involves the level in human breast milk. Figure 4.1 shows data from a number of such studies in various parts of the world. The mean values are clearly decreasing, most sharply in the United States and Canada.

---

[c]William D. Ruckelshaus, appointed in 1970 by Richard M. Nixon, was the first administrator of the Environmental Protection Agency. He also served as interim director of the EPA in 1983 in the aftermath of a crisis arising over the Superfund project. Ruckelshaus also served as acting Director of the FBI. He is best known for his actions during the Watergate scandal. Then serving as Deputy Attorney General, he and the Attorney General, Elliott Richardson, resigned their posts when ordered by President Nixon to fire the Watergate Special Prosecutor, Archibald Cox. This signaled the final unraveling of the Nixon presidency.

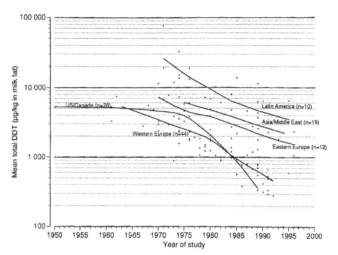

FIGURE 4.1   Mean total DDT concentration (g/kg) in human milk fat from various parts of the world. (From Smith, D. *Int. J. Epidemiol.* **1999,** *28,* 179–188. By permission of Oxford University Press, USA.)

From the point of 60 years of hindsight, the speed and scale on which DDT was released into the environment is astonishing. As noted by Woodwell, Craig, and Johnson in a *Science* article in 1971, mankind was fortunate that toxicity to humans, as far as is known, is quite low.[13] The warnings from toxicity to wildlife were recognized and corrective actions taken fairly quickly.

4.2.2   OTHER POLYHALOGENATED COMPOUNDS

A number of other persistent halogenated compounds have been introduced into the environment. Several, like DDT, are insecticides, but most are more toxic and they have also been banned. Examples are dieldrin, aldrin, endrin, heptachlor, mirex, and toxaphene (a derivative of the terpene camphene having 6–9 chlorines). Like DDT, they have adverse environmental effects.

An extreme case of contamination occurred in 1975 in Hopewell, VA involving the manufacture of kepone, at the time used in ant and roach traps and, in Europe, for manufacture of an insecticide for control of potato beetles. Inadequate safety procedures led to extensive exposure of workers. Although aspects of the problem had been reported to OSHA and EPA, the "dots weren't connected" until a primary care physician Dr. Yi-nan Chou examining a plant worker suspected poisoning. His diagnosis was confirmed by blood analysis at the CDC. This resulted in inspection of the plant and determination of the extent of the contamination. An investigation of past and current workers found 55% exhibited signs of poisoning, with 80% of these exhibiting tremors. Spillage, windblown dust, dumping, and waste disposal through the sewage system led to contamination of the neighborhood, the James River and portions of the Chesapeake Bay. It is estimated that up to 200,000 lb of kepone was released from the site.[14]

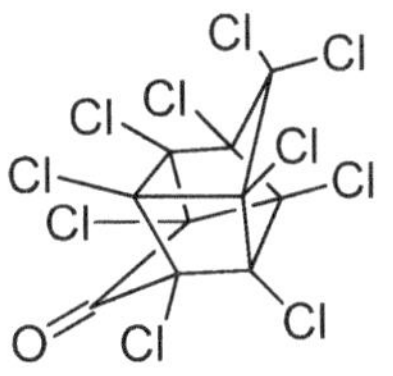

kepone

There have been several other sources of persistent halogenated compounds in the environment. Polychlorinated biphenyls (PCBs) were widely used as hydraulic fluids and heat transfer medium in electric transformers and similar applications. They entered the environment by spillage and careless disposal. Polybrominated diphenyl ethers were used as fire retardants. 2,3,7,8-tetrachlorodibenzo-1,4-dioxin, a contaminant in the defoliant Agent Orange used in the VietNam war, was widely dispersed in Southeast Asia (see Section 8.2.1). The chlorinated dibenzodioxins and dibenzofurans are also produced unintentionally by incineration of waste, paper and pulp-making where chlorine bleach is used, and in production of certain metals.

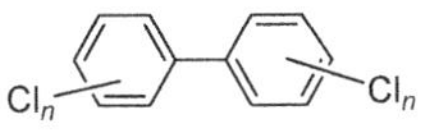

polychlorinated biphenyl 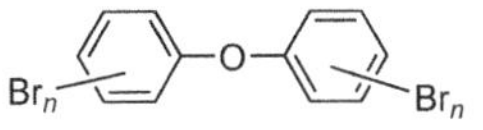   polybrominated diphenyl ether 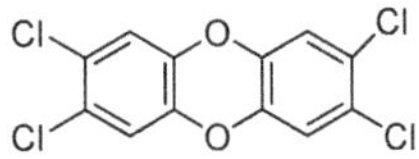   2,3,7,8-tetrachlorodibenzo-1,4-dioxin

There have been several instances of accidental high-dosage exposure to these compounds. Contamination of cooking oil with PCBs has occurred in Japan and Taiwan.[15] Children of pregnant women exposed in these cases had

abnormal pigmentation and abnormal development of teeth and nails. Exposure
to chlordane has occurred as a result of termite treatments.[16] These human exposures, along with many animal studies, suggest that some polyhalogenated compounds have toxicity in the nervous system and that low exposure, especially
during fetal development, may have neurological consequences.

Another chlorinated compound that was widely used is hexachlorophene,
an active topical bacteriacide. At one time, it was commonly used as an antibacterial rinse for newborn infants. However, an accidental mixing with talcum
powder, which led to a severe infant rash, including 36 fatalities, revealed that
the compound had considerable toxicity.[17] Currently, FDA regulations make this
a prescription item.

hexachlorophene

In 2001, many nations signed the Stockholm convention intended to restrict
use and reduce contamination by twelve specific polychlorinated organic pollutants. These were aldrin and dieldrin, chlordane, DDT, endrin, mirex, heptachlor,
and hexachlorobenzene. In addition to these insecticides several other types of
compounds including PCBs, toxaphene, and chlorinated benzodioxins and ben-
zofurans were banned. The Stockholm convention provides for elimination and
disposal. The PCBs may be used through 2025. Certain exemptions are available for uses where no other compound is satisfactory and Australia, Botswana,
and China still use chlordane or mirex for termite control. India and China produce DDT, but China committed to eliminate production by 2014. China and
Brazil use DDT in the production of dicofol.

Nine more substances came under the terms of the agreement in 2010. These
include: α-, β-, and γ- isomers of hexachlorocyclohexane (the γ-isomer is also
known as lindane), hexabromobiphenyl, tetrabromo-, pentabromo-, hexabro-
mo-, and heptabromodiphenyl ether, pentachlorobenzene. Also included were
perfluorooctane sulfonic acid and its sulfonyl fluoride derivative. The latter two
compounds have a different spectrum of usages from the other POPs, being used
in electronic equipment, photoimaging, and textile coatings. The US Senate has
never ratified the Stockholm convention, one of 12 of the original signatories
that have failed to do so. Except for the perfluorooctanesulfonic acids, the compounds are not currently in production or use in the United States and efforts
have been made to clean up highly contaminated sites.

4.2.3 BIOLOGICAL AND ENVIRONMENTAL EFFECTS OF POLYHALOGENATED COMPOUNDS

Studies relating DDT and DDE exposure to adverse health effects in humans are difficult. Essentially universal background exposure means there is no "control group." Retrospective epidemiological studies are sometimes contradictory and are often limited by lack of precise knowledge of the time and extent of exposure. There have been several studies suggesting an association of fetal death with high levels of chlorinated pesticides, but in all cases the level of evidence is considered "limited."[18] The US EPA and IARC have classified PCBs as probable human carcinogens. Laboratory studies have demonstrated that both DDT and DDE are bound at estrogen receptors. They are thus classified as *endocrine disruptors.* The development of sexual characteristics is under the control of estrogen receptors that regulate transcription of genes. It is noteworthy that not all species determine sex at fertilization, as is the case with mammals and birds. Many species of fish and amphibians can undergo sex conversion under the influence of hormonal agents. Several cases of extreme exposure to chlorinated pesticides as a result of accidental spills have led to reproductive abnormalities in fish and amphibians. Figure 4.2 summarizes the possible mechanism of action.

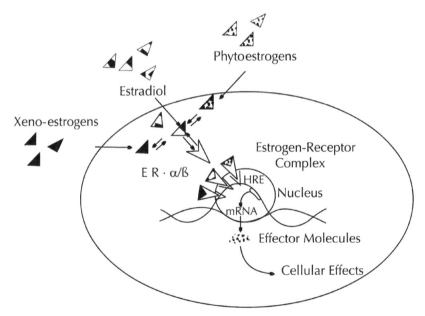

FIGURE 4.2   Conceptual representation of competition from exogenous (xeno-) and plant-derived (phyto)estrogens for estrogen receptors. (From Rossell, M.; Rinehart, K.; Imthurm, B.; Keller, P. J.; Dubey, R. K. *Hum. Reprod. Update* **2000**, *6*, 333–350. By permission of Oxford University Press, USA.)

There is evidence of a number of abnormalities in marine mammals, such as dolphins, seals, and whales, which are associated with polychlorinated hydrocarbons. These include depression of the immune system, reproductive impairment, and abnormalities in bone development and structure. Efforts have been made to construct both temporal and location trends of marine concentration of polychlorinated hydrocarbons. The highest concentrations of DDT in marine mammals have been found in the Mediterranean and off the western US coast. The highest concentrations of PCB are in the Mediterranean and Baltic seas and off the eastern US coast. Concentrations have slowly declined in the Mediterranean (and probably elsewhere) since the major sources of DDT and PCBs were banned, although concentrations remain significant.[19]

TOPIC 4.2 ACCIDENTAL EXPOSURES TO HALOGENATED COMPOUNDS

A. EXPOSURE OF FARMERS AND FARM ANIMALS TO POLYBROMINATED BIPHENYLS

In 1973, an accidental packaging mix-up resulted in polybrominated biphenyls (PBB), a fire retardant, being substituted for magnesium oxide in premixed cattle feed. The two products were produced at the same chemical company and sold under the brand names FireMaster and NutriMaster, respectively. The company ran out of color-coded shipping bags and the products were packaged in plain brown bags with the name stenciled on the bag. Some of the FireMaster (PBB) was erroneously shipped to a feed processing plant, where it was mixed in a number of batches of feed intended for dairy cattle. By the time the contamination was discovered, more than 500 farms in Michigan had been contaminated and some 30,000 cattle, 4500 pigs, 1500 sheep, and 1.5 million chickens were quarantined and destroyed. Large quantities of cheese, butter, dried milk, and eggs were also contaminated.[20]

The animals exposed directly to the contaminated feed revealed acute symptoms including reduced appetites, low milk production, abnormal hoof growth, loss of hair, and abscesses. The incident generated a number of controversies. Farmers were dissatisfied with the level of compensation and lawsuits were filed. There was no scientific basis for establishing a safe level of exposure and limits based on legal definitions were first set at 1.0 ppm, then 0.3 ppm, and eventually, by act of the Michigan legislature, 20 ppb. Farmers with animals below the 0.3 ppm level were not eligible for compensation and sued but lost. The compound was found to be widely distributed, for example in breast milk in Michigan, but not in nearby states. About 4000 people living on or near the farms have been the subject of long-term study by the Michigan Department of Community Health. The results to date suggest a possible increase in spontaneous abortions in women born to exposed mothers.

B. EXPOSURE TO POLYCHLORINATED DIOXIN AT SEVESO, ITALY

An industrial accident in 1976 near Seveso, Italy led to release of considerable amounts of polychlorinated dioxin. The level of exposure varied depending on proximity to the plant. Analysis of blood samples taken at the time of the accident ranged as high as 86 ppm. The health status of those exposed has been followed since the accident. The only immediate effect observed was a skin condition known as chloracne, which was as high as 50% in the area with the highest exposure. Most of the cases were in children. There was also evidence of peripheral neuropathy. After 20 years or more, there was no excess in total mortality, but there was evidence of increase in certain kinds of cancer, including rectal (men only), lung, leukemia, and related neoplasms. Women showed elevated risk of diabetes and respiratory infection mortality. The latter may be due to the known immunotoxicity of dioxin. All the absolute number of cases is small, imposing statistical limits.[21]

4.3   CHLOROFLUOROCARBONS AND RELATED COMPOUNDS

The chlorofluorocarbons (CFCs) are small fully halogenated hydrocarbons that are gases or very volatile liquids with no water solubility. The compounds are designated by a system of letters and numbers. The letter designations are CFC for *chlorofluorocarbons*, HFC for *hydrofluorocarbons*, and HCFC for *hydrochlorofluorocarbons*. The digit to the far right in the code is the number of fluorine atoms. The next digit to the left is the number of hydrogens plus one, the third digit is the number of carbons minus one, but it is not used if there is only one carbon. When the compound can exist as isomers, the code is followed by a letter, with "a" being the most symmetrical isomer and proceeding to higher letters as the isomers are less symmetrical. Bromine compounds are called *halons*. For the halons, the code is simply the number of C, F, Cl, and Br atoms in that order. Chart 4.1 gives some examples of the compounds and codes.

Beginning in the 1930s, these materials became widely used as coolants in refrigerators, freezers, and air-conditioners. They also came to be used in industrial applications such as cleaning electronic components and in "blowing" of foams. They were also used as propellants in aerosols, including medicinal sprays. Their combination of high chemical stability, ease of vaporization, low water solubility, low flammability, and low toxicity make them ideal materials for these applications. These valuable characteristics are mainly due to the high fluorine content of the compounds and are unlikely to be obtained with non-fluorinated compounds. Unfortunately, the very same characteristics lead to very long lifetimes in the atmosphere, and in the 1970s, it was recognized that these compounds could cause ozone depletion in the stratosphere.

CCl_2F_2	CCl_2FCCl_2F	CCl_3CClF_2	CCl_2FCClF_2	CCl_3CF_3
CFC-12	**CFC-112**	**CFC-112a**	**CFC-113**	**CFC-113a**
CHF_3	CH_2F_2	CH_2FCHF_2	CH_3CF_3	CH_2FCH_2F
HFC-23	**HFC-32**	**HFC-143**	**HFC-143a**	**HFC-152**
$CHCl_2F$	$CHClF_2$	$CHCl_2CClF_2$	$CHClFCCl_2F$	CHF_2CCl_3
HCFC-21	**HCFC-22**	**HCFC-122**	**HCFC-122a**	**HCFC-122b**
$CBrClF_2$	$CBrF_3$			
Halon-1211	**Halon-1301**			

CHART 4.1   Examples of codes for chlorofluorocarbons and hydrochlorofluorocarbons.

4.3.1   PREDICTION AND CONFIRMATION OF THE OZONE HOLE

The first inkling that CFCs, and CCl_3F in particular, might be an environmental issue was a report in the journal *Nature* in 1973 that CFCs were present in the atmosphere over the entire Atlantic Ocean from England to Antarctica.[22] Even more startling were the estimates that virtually none of the 1.3 million tons of the material that had been manufactured to date had disappeared, that is, it was accumulating in the atmosphere. At the time, no deleterious properties of the CFC were recognized and the *Nature* article states "the presence of these compounds constitutes no conceivable hazard." The CFCs are stable to the conditions in the lower atmosphere (below 25 km). Below 25 km they are also stable to light from the sun, because the ozone layer absorbs ultraviolet (UV) radiation below 230 nm. Furthermore, because they are insoluble in water, they are not washed out of the atmosphere by rain. They are also chemically inert to oxygen and other oxidants because they contain no C–H bonds.

Ozone (O_3) is produced in the stratosphere 20–30 km above the earth's surface by solar UV radiation. Ozone is both formed and destroyed by UV light in the 280–320 nm range and this establishes an equilibrium concentration of O_3 that can range up to 10 ppm. The maximum concentration is at about 25 km.

The combination reaction (step 2) takes place only if energy can be transferred to a third molecule and that energy is eventually released as heat, leading to a temperature increase at altitudes of 15–50 km. The O_3 that is formed plays the important role of filtering out high-energy solar UV radiation of 230–320 nm, which would otherwise be detrimental to many living systems on the earth's surface.

$$O_2 + UV \longrightarrow 2O$$

$$O + O_2 + M \longrightarrow O_3 + M^*$$

$$M^* \longrightarrow M + heat$$

$$O_3 + UV \longrightarrow O_2 + O$$

In 1973, F. Sherwood Rowland and Mario J. Molina, a post-doctoral student, began to study the atmospheric fate of the CFCs.[d] Their approach was to study model systems to predict the behavior of the CFCs in the atmosphere and troposphere. Rowland and Molina soon discovered that chlorine atoms can initiate a chain reaction that destroys ozone molecules. In contrast to their stability near the earth's surface, CFCs are reactive to solar UV radiation in the stratosphere. At this level, photolysis breaks the carbon–chlorine bonds in CFCs and releas-

---

[d]The 1995 Nobel Prize in Chemistry was shared by Paul Crutzen, F. Sherwood Rowland, and Mario J. Molina for their work on the depletion of ozone by man-made chemicals.

Paul J. Crutzen was born in Amsterdam in 1933. He was educated in Holland, originally as an engineer. He married and relocated to Sweden in 1958. He began work as a computer programmer in the Meteorological Institute of the University of Stockholm in 1959, without any background in the field. He worked mainly on programing models for meteorological systems. By 1963, he qualified for an M. S. degree by taking courses at the Institute and in 1965 completed a Ph. D. thesis on modeling of tropical cyclones. He then began his independent work on ozone in the stratosphere and the effect of nitrogen oxides on ozone levels. He was director of the US National Center for Atmospheric Research in Boulder, CO, USA from 1977 to 1980 and then moved to the Max Planck Institute for Chemistry in Mainz, Germany.

F. Sherwood Rowland was born in 1927 in Delaware, OH, USA. He received a B. A. from Ohio Wesleyan in 1948 and a Ph. D. from the University of Chicago in 1952. He was on the faculty of Princeton (1952–1956) and the University of Kansas (1956–1964) before moving to the newly opened Irvine campus of the University of California in 1964. Most of his early research centered on the chemistry of radioactive molecules and radioactive tracer-labeling. He became interested in atmospheric chemistry in 1970 and began work on CFCs in 1973, after learning of the worldwide distribution of CFCs. His first co-worker in this area was Mario J. Molina. Molina was born in Mexico City in 1943. He studied chemical engineering at the Universidad Nacional Autonoma Mexico (UNAM). After studying in Europe, he returned to UNAM as an assistant professor, but with the continuing goal of graduate study in physical chemistry. He enrolled in UC (Berkeley, CA, USA) in 1968 and received a Ph. D. in 1973. He studied with George C. Pimentel in the area of laser spectroscopy. As a new post-doctoral student in F. S. Rowland's laboratory, he chose to study the atmospheric chemistry of the CFCs, a topic new to both Rowland and Molina. Within a short time, they recognized the environmental consequences of CFC release and began to alert both the scientific community and the public to the problem. Subsequently, Molina has worked on atmospheric chemistry at UC (Irvine, CA, USA), JPL, MIT, and, most recently, UC (San Diego, CA, USA).

es chlorine atoms. Because the chain process regenerates chlorine atoms, any single photolytic event can destroy many ozone molecules. The mechanism is summarized in Figure 4.3. Bromine atoms are even more efficient than chlorine in destroying ozone. This information brought the fire suppressant compounds called "halons" into the group of ozone-depleting halocarbons.

$$Cl\text{-}CX_3 \xrightarrow{\text{UV}} Cl^\bullet + {}^\bullet CX_3$$

chain sequence for destruction of ozone

$$Cl^\bullet + O_3 \longrightarrow ClO^\bullet + O_2$$

$$ClO^\bullet + O \longrightarrow Cl^\bullet + O_2$$

$$O + O_3 \longrightarrow 2\,O_2$$

FIGURE 4.3   Steps in the catalytic cycle for destruction of ozone by chlorine atoms.

Rowland and Molina calculated that these reactions had the potential to re-duce the ozone level in the stratosphere and lead to increased UV exposure in the lower atmosphere. The publication of this information in 1974 and 1975 led to vigorous attacks by manufacturers of CFC aerosols and refrigerants. In 1976, the National Academy of Sciences issued a report confirming the existence of the ozone-depletion process, but which demurred on recommending regulation. While there was some continuing effort toward regulation, this ended with the election of Ronald Reagan in 1980. In the mid-1980s, Rowland, Molina, and oth-ers discovered another process that contributes to ozone depletion. This involves a surface reaction that can take place on ice crystals formed in stratospheric clouds during polar winters. The predictions of their models were confirmed in 1985 when the results of 18 years of continuous ozone measurements obtained by the British Antarctic Survey were published in *Nature*.[23] The data showed a 35% decrease in ozone levels from the late 1950s. Indeed, these reductions, which were soon confirmed by other measurements, exceeded the predictions of the original models and were consistent with the occurrence of surface reac-tions in polar stratospheric clouds. Data from the Antarctic showed depletions above 95%, ranging up to 99% of the Antarctic ozone in September and October. Analysis of data from the northern hemisphere also showed a smaller, but signifi-cant decrease relative to the 1960s. Data from a station in Switzerland beginning in 1931 showed a drop after 1970. Similar comparisons between 1965–1975 and 1976–1986 at Bismark ND and Caribou ME in the United States also showed decreases. Data from 18 stations distributed around the world confirmed that the decrease in ozone level was worldwide. Measurements also were made of ClO concentrations in 1986 and 1987 and gave values and distribution consistent with the expectations of the various models.

Ozone levels have been measured over the Antarctic since the mid-1950s. The observed level for the Antarctic Spring (August–October) began to decrease in the late 1970s. Starting in 1979, it became possible to measure the O_3 level over the entire South Polar region through the Nimbus-7 weather satellite. The maximum O_3 depletion to date was measured in 2006, although 2008 > 2007. Figures 4.4 and 4.5 depict the trend and show the 2008 ozone hole map.

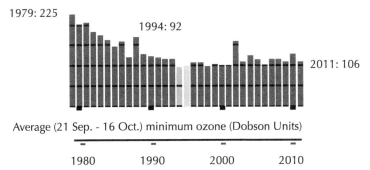

FIGURE 4.4 Minimum Antarctic ozone concentrations for 1979–2011. (From http://www. theozonehole.com/ozoneholehistory.htm.)

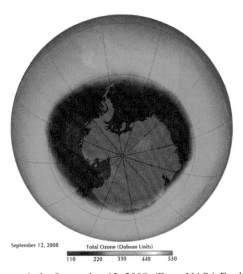

FIGURE 4.5 The ozone hole, September 12, 2008. (From NASA Earth Observatory.)

4.3.2 REPLACEMENTS FOR CFCS

The ozone depletion data were readily visualized as the "ozone hole" and led relatively rapidly to international agreements to ban the manufacture of low molecular weight fully halogenated CFCs. These agreements were formalized by the Montreal Protocols of 1987, which have subsequently been updated. The ban prompted CFC manufacturers to move toward replacement of CFCs by HCFCs and HFCs. The principal advantage of HFCs over HCFC is that they contain no chlorine and have no ozone depletion potential. The HCFCs and HFCs are susceptible to atmospheric oxidation and therefore have reduced potential for reaching the stratosphere. The oxidation products, while noxious, are water soluble. The most widely used of the HFCs is HFC-134a, 1,1,1,2-tetrafluoroethane. The toxicity of the HFCs is very low and current permissible exposure limits are 1000 ppm. These compounds have much shorter atmospheric lifetimes than CFCs and do not reach the stratosphere in nearly as high concentration as the CFCs. However, HFCs have their own problems. In particular, they are "greenhouse gases." On a per molecule basis, they are much stronger absorbers than CO_2, although their concentrations are much lower. These data are summarized in Table 4.1.

TABLE 4.1 Lifetime and Global Warming Potential of CFCs and HFCs.

Compound	Atmospheric lifetime (years)	Ozone depletion potential	Global warming potential
CFC-11	45	1	4750
CFC-12	100	1	10,890
CFC-13	640	1	14,420
CFC-114	300	1	10,040
CFC-115	1700	0.44	7370
HCFC-21	2	0	200
HCFC-22	12	0.04	1700
HCFC-123	1.4	0.014	120
HFC-11		1	4000
HFC-12		1	2400
HFC-23	260	0	12,000
HFC-32	5	0	550
HFC-125	29	0	3400
HFC-134	96	0	1100

TABLE 4.1 *(Continued)*

Compound	Atmospheric lifetime (years)	Ozone depletion potential	Global warming potential
HFC-134a	138	0	1300
HFC-143a	52	0	4300
HFC-152a	1.4	0	120
HFC-227ea	33		3500
HFC-236ea	10		1200

(Data from Tsai, W.-T. *Chemosphere* **2005**, *61*, 1543; US EPA, www.epa.gov/ozone).

In addition to the CFCs and HCFCs, bromomethane is also a significant contributor to ozone depletion. It is extensively used in treatment of soil to eliminate fungi and nematodes. It was included for phase-out with the CFCs in the 1987 Montreal protocols. Its annual use in the United States decreased from about 18 million kg to about 2 million kg by 2001. Use has continued under "critical use exemptions" based on the absence of alternatives. Bromomethane is also used extensively to fumigate warehouses and containers used in international shipments. There is evidence that workers can be exposed to excessive levels in the course of loading and unloading of treated containers.[24]

Figure 4.6 shows a projection of the impact of the CFCs, HCFC, bromomethane, halons, and some other man-made chemicals on the ozone-depleting potential of chlorine and bromine atoms in the stratosphere. The projection indicates that impact of the CFCs will be significant for more than a century after the elimination of their major sources.

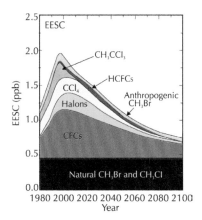

FIGURE 4.6   Past and future abundances of ozone-depleting chlorine compounds. (From Daniel, J. S.; Velders, G. J. M.; Morgenstern, O.; Toohey, D. W.; Wallington, T. J.; Wuebbles, D. J. *Scientific Assessment of Ozone Depletion, WMO-UNEP 2010*, Chap 5.)

The HCFCs are scheduled to be phased out by 2020 and it is not entirely clear what the replacements may be. One possibility is HFEs, which are ethers of hydrofluorocarbon groups. Unsaturated ethers such as trifluoromethyl trifluorovinyl ether, E216, have some promising characteristics, including short atmospheric lifetimes.

$$CF_3-O-CF=CF_2$$

E216

KEYWORDS

- **halogenated hydrocarbons**
- **chlorine**
- **electrolytic process**
- **chloroform**
- **dry-cleaning**
- **persistent organic pesticides**
- **chlorofluorocarbons**
- **ozone destruction**

BIBLIOGRAPHY BY SECTION

Section 4.1: Bolt H. M.; Borlak J. T. 1999. In *Toxicology*; Marquart, H., Ed.; Academic Press, pp 645–657.
Section 4.1.1: Plaa, G. L. *Annu. Rev. Pharmacol. Toxicol.* **2000**, *40*, 43–65; Sinsheimer, P.; Gottlieb R.; Farrar. C. *Environ. Sci. Technol.* **2002**, *36*, 1649–1655; Gold, L. S.; De Roos, A. J.; Waters, M.; Stewart, P. *J. Occup. Environ. Health* **2008**, *5*, 807–839.
Section 4.1.2: Beckman, E. J. *Environ. Sci. Technol.* **2002**, *36*, 347A–353A; Heaton, C.; Northeim, C.; Helminger, A. *Chem. Eng.* **2004**, *111*, 42–48.
Section 4.1.3: LaDou, J. *Int. J. Hyg. Environ. Health* **2006**, *209*, 211–219.
Section 4.1.4: Kielhorn, J.; Melber, C.; Wahnachaffe, U.; Aitio, A.; Mangelsdorf, I. *Environ. Health Perspect.* **2000**, *108*, 579–588; Roberts, S. M.; Jordan, K. E.; Warren, D. A.; Britt, J. K.; James, R. C. *Regul. Toxicol. Pharmacol.* **2002**, *35*, 44–55; Bolt, H. M. *Crit. Rev. Toxicol.* **2005**, *35*, 307–323.
Section 4.1.5: Roberts, S. M.; Jordan, K. E.; Warren, D. A.; Britt, J. K.; Jones, R. C. *Regul. Toxicol. Pharmacol.* **2002**, *35*, 44–55; Bolt, H. M. *Crit. Rev. Toxicol.* **2005**, *35*, 307–323.

Section 4.2.1: Metcalf, R. *J. Agric. Food Chem.* **1973,** *21,* 511–519 ; Dunlap, T. R. *DDT, Scientists, Citizens, and Public Policy,* Princeton University Press: Princeton, NJ, 1981.; Smith, A. G. *Chlorinated Hydrocarbon Insecticides, Handbook of Pesticide Toxicology,* Hays, W. J.; Laws, E. R. Ed.; Academic Press: San Diego, 1991, pp 731–915; Peakall, D. B. *Environ. Rev.* **1993,** *1,* 13–20; Landrigan, P. J.; Sonawane, B.; Mattison, D.; McCally, M.; Garg, A. *Environ. Health Perspect.* **2002,** *1,* A313–A315; Sharma, V. P. *Curr. Sci.* **2003,** *85,* 1532–1537; Wong, M. H.; Leung, A. O. W.; Chan, J. K. Y.; Chow, M. P. K. *Chemosphere* **2005,** *60,* 740–752.

Section 4.2.2: Zitko, V. In *Handbook of Environmental Chemistry;* Fieldler, H. Ed.; Springer Verlag, Part O, 2003; pp 47–90; Lucena, R. A.; Allam, M. F.; Jiminez, S. S.; Villarejo, M. L. J. *Curr. Drug Saf.* **2007,** *2,* 163–172.

Section 4.2.3: Fry, M. *Environ. Health Perspect.* **1995,** *103,* 165–171; Kelce, W. R.; Stone, C. R.; Laws, S. C.; Gray, L. E.; Kempainen, J. A.; Wilson, E. M. *Nature* **1995,** *375,* 581–585; Miyamoto, J.; Klein, W. *Pure Appl. Chem.* **1998,** *70,* 1829–1845; Beard, J. *Sci. Total Environ.* **2006,** *355,* 78–89; Heffron, J. *Biochemist* **1999,** *21,* 28–31; Rosselli, M.; Reinhart, K.; Imthurn, B.; Keller, P. J.; Dubey, R. K. *Hum. Reprod. Update,* **2000,** *6,* 332–350; Attarajan, A.; Maharaj, P. *Br. Med. J.* **2000,** *321,* 1403–1404; Blus, L. J. *Handbook of Ecotoxicology,* 2nd ed., Lewis Publishers: Boca Raton, FL, 2003; Mariussen, E.; Fonnun, F. *Crit. Rev. Toxicol.* **2006,** *36,* 253–289.

Section 4.3.1: Rowland, F. S.; Molina, M. J. *Chem. Eng. News* **1994,** Aug., 8–18; Rowland, F. S. *Angew. Chem. Int. Ed.* **1996,** *35,* 1786–1798; Rowland, F. S. *Philos. Trans. R. Soc.* **2006,** *361,* 769–790.

Section 4.3.2: Sidebottom, H.; Franklin, J. *Pure Appl. Chem.* **1996,** *68,* 1757–1769; McCulloch, A. *J. Fluorine Chem.* **1999,** *100,* 163–173; Wuebbles, D. J.; Jain, A.; Kotamarthi, R.; Naik, V.; Patten, K. O. *Recent Adv. Stratos. Process.* **1999,** 113–143; Good, D. A.; Francisco, J. S. *Chem. Rev.* **2003,** *103,* 4999–5023; Tsai, W.-T. *Chemosphere* **2005,** *61,* 1539–1547.

REFERENCES

1. Killeffer, D. H. *Ind. Eng. Chem.* **1936,** *28,* 640–643.
2. *Federal Register* **1970,** *35,* 4001–4007.
3. Mundt K. A.; Birk T.; Burch M. T. *Int. Arch. Occupation. Environ. Health* **2003,** *74,* 473–491; Hellweg S.; Demou E.; Scheringer M.; McKone T. E.; Hungerbuhler, K. *Environ. Sci. Technol.* **2005,** *39,* 7741–7748.
4. Wartenburg D.; Reyner D.; Scott C. S. *Environ. Health Perspect.* **2000,** *108*(Suppl. 2), 161–175.
5. Ruder A. M. *Ann. N. Y. Acad. Sci.* **2006,** *1076,* 207–227.
6. National Air Emission Standards for Hazardous Air Pollutants: Halogenated Solvent Cleaning. *Fed. Reg.* **2007,** *72,* 25138–25159.
7. King J. W.; Williams L. L.; *Curr. Opin. Solid State Mater. Sci.* **2003,** *7,* 413–424; Weibel G. L.; Ober C. K. *Microelectron. Eng.* **2003,** *65,* 145–152.
8. Sass J. B.; Castleman B.; Wallinga D. *Environ. Health Perspect.* **2005,** *113,* 809–812.
9. Knipling E. F.; *J. Econ. Entomol.* **1953,** *46,* 1.

10. See, for example, Edwards J. G. *21st Century Science & Technology Magazine, Summer*. 1992.
11. Gillette R. *Science* **1971,** *174*, 1108–1110.
12. See, for example, Bate R. *The American*, 2007, Nov. 5 Issue.
13. Woodwell G. M.; Craig P. P.; Johnson H. A. *Science* **1971,** *174*, 1101–1107.
14. Dawson G. W.; Weimer W. C.; Shupe S. J. *AICHE Symp. Ser.* **1979,** *75*, 366–374; Reich M. R.; Spong J. K. *Int. J. Health Serv.* **1983,** *13*, 227–246.
15. Kuratsune M.; Yoshimara T.; Matsuzaka J.; Yamaguchi A. *Environ. Health Perspect.* **1972,** *1*, 119–128; Hsu S. T.; Ma C. I.; Hsu S. K. H.; S. S. Wu; Hsu N. H. M.; Yeh C. C.; Wu S. B. *Environ. Health Perspect.* **1985,** *59*, 5–10; Rogan W. J.; Gladen B C.; Hung K. I. *Science* **1988,** *241*, 334–336.
16. Kilburn K. H.; Thornton J. C. *Environ. Health Perspect.* **1995,** *103*, 690–694; Kilburn K. H. *Southern Med. J.* **1997,** *90*, 299–304.
17. Martin-Bouyer G.; Lebreton R.; Toga M.; Stolley P. D.; Lockhart J. *Lancet* **1982,** *1*(8263), 91–95.
18. Wigle D. T.; Arbuckle T. E.; Walker M.; Wade M. G.; Liu S.; Krewski D. *J. Toxicol. Environ. Health B* **2007,** *10*, 3–39.
19. Aguilar A.; Borrrell A.; Reijnders P. J. H. *Marine Environ. Res.* **2002,** *53*, 425–452; Aguilar A.; Borrellm A. *Marine Environ. Res.* **2005,** *59*, 391–404.
20. Reich M. R. *Am. J. Public Health* **1983,** *73*, 302–313.
21. Pesatori A. C.; Consonni D.; Bachetti S.; Zocchetti C.; Bonzini M.; Baccarelli A.; Bertazzi P. A. *Ind. Health* **2003,** *41*, 127–138.
22. Lovelock J. E.; Maggs R. J.; Wade R. J. *Nature* **1973,** *241*, 194–196.
23. Farman J. C.; Gardiner B. G.; Shanklin J. D. *Nature* **1985,** *315*, 207–210.
24. Baur X.; Yu F.; Poschadel B.; Velldman W.; Vos T. K.-D. *Int. Mar. Health* **2006,** *57*, 46–55.

CHAPTER 5

POLYMERS: MAKING NEW MATERIALS

ABSTRACT

Polymers are high molecular weight materials that came to have a wide variety of uses during the twentieth century. Polymers can be made both by modification of natural materials and by synthesis. Modifications of cellulose provide fibers and films such as rayon and cellophane. Synthetic polymers began with phenol-formaldehyde resin and urea-formaldehyde resin. Polymers based on hydrocarbons include polyethylene, polypropylene and polystyrene. Other types of polymers include polyamides, polyesters, polyurethanes and vinyl polymers. The properties of polymers can be controlled by the method of preparation and subsequent treatment. They can assume many shapes and forms including films, sheets, fibers, foams and coatings. Solid objects can be made by molding.

Polymers are high molecular weight compounds that are the basis for many of the new materials introduced during the twentieth century. Some are modified natural substances, but most now in use are synthetic. They are prepared from small molecules called *monomers* that are usually petrochemicals. One classification of polymers is based on molecular structure. For example, polyolefins are polymers derived from alkenes (olefins). Vinyl and acrylic polymers are formed from alkenes that also have substituents such as halogen, cyano, or ester. Polyamides, polyesters, and polyurethanes are long chain molecules containing amide, ester, or urethane (also called carbamate) bonds. Polymers also can be divided by method of polymerization. For example, polyolefins and vinyl polymers are made by *addition polymerization*, in which the monomers are combined into large molecules. Polyamides and polyesters result from *condensation polymerization*, in which a small molecule, usually water, is formed and removed. Terms, such as *emulsion polymerization* or *coordination polymerization*, describe specific methods for polymerization. Polymers can also be *copolymers* in which two or more monomers are incorporated into the polymer. The two monomers can appear in alternating or random order. Synthetic rubber (SBR), for example, is a random copolymer of 1,3-butadiene (B) and styrene (S). *Block copolymers* can be created in which there are segments of each monomer connected to one another. A *graft copolymer* is formed when a new polymer is attached to (grafted on) an existing polymer.

There are natural polymeric materials. The most abundant are cellulose and starch, which are polymers of the carbohydrate glucose. The former is the main constituent of wood and fibers such as cotton and linen. Chitin, the hard covering in arthropods and insects, is a polymer of 2-acetamidoglucose. It is a byproduct of the shell-fishing industry. The fibers silk and wool are proteins. Proteins are polymers of α-amino acids. Rubber is a natural elastomer and is a polymer of the hydrocarbon isoprene (2-methyl-1,3-butadiene). These natural materials can be modified by chemical processing to provide substances such as cellophane or rayon from cellulose and vulcanized (hardened) rubber. Proteins, deoxyribonucleic acid (DNA), and ribonucleic acid (RNA) are *biopolymers* and have many crucial biological functions. The structures of these polymers are shown in Scheme 5.1.

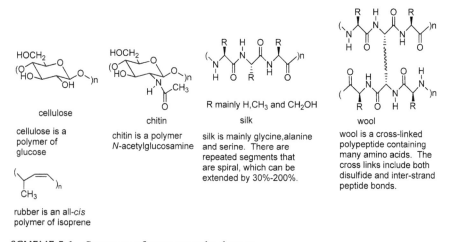

SCHEME 5.1   Structures of some natural polymers.

The development and use of polymers during the twentieth century is very much a story of chemistry. At the beginning of the century only a few derivatives of natural polymers were available. Two natural materials subjected to chemical modification, rubber and celluloid, made their appearance during the nineteenth century, but there were no synthetic polymers. That was soon to change and with profound effects. Polymers and plastics are now ubiquitous materials. Polymers can have many forms and uses: fibers and textiles; films and sheets; foams for insulation; containers such as cups and milk bottles; tires; adhesives; paints and other coatings; molded and shaped objects such as pipes, automobile bumpers, or chairs; composites in which other materials are imbedded. Individual polymers can appear in any or all of these forms. Chemists learned how to make these substances and control their properties.

The word *plastic* is a popular equivalent for polymers. The word is derived from Latin and means the ability to be shaped by molding. This is a characteristic of many polymers, so phrases like "the plastics industry" identify the broad scope of producing and shaping polymers. Over the twentieth century, polymers replaced traditional materials in many applications. Natural fibers such as cotton, silk, wool, and linen were joined by nylon, polyesters, and acrylics. Wood and metals have been supplemented or supplanted by structural polymers and composites. Polymers have two potential advantages: improved properties for specific purposes and, often, lower cost. In some cases, they replace natural materials that are available only in very limited amounts, such as ivory. Plastics are more easily worked and shaped than traditional materials. The techniques for shaping include *compression molding*, *blow molding*, and *injection molding* (see Section 5.4). The first involves pressing a fluid form of the polymer in a mold. In blow molding, a film of softened polymer is inflated to the shape of the mold and cooled. In injection molding, a powdered form of plastic is heated immediately before injection into the mold, in which it cools into the desired shape. *Extrusion* is the major means of producing fibers and films. For fibers, a liquid form of polymer is forced through a small orifice, (spinneret) and is rapidly solidified by a chemical change, cooling or solvent evaporation. Films are produced by forcing the fluid polymer through narrow slits.

Several of the words that we will encounter in this discussion have taken on cultural connotations beyond their chemical origin. For example, *celluloid* is often used in connection with the movie industry, because until about 1920 it was the material used to make movie film. "*Vinyl*" used in the context of music refers to polyvinyl chloride (PVC), which from the1930s to the 1980s was the material used to make records and albums. "*Polyester*" is a common description of fabrics and clothing made from synthetic fibers containing the ester functional group.

5.1   THE NATURE OF POLYMERS

At the beginning of the twentieth century there was no solid understanding of the chemical nature of polymers and at that time the word had a different meaning.[a] Substances that we now recognize as polymers were believed to be colloidal assemblies of normal small molecules. Evidence had accumulated that substances that showed "colloidal" properties, such as starch, rubber, and proteins, also exhibited high apparent molecular weights. It was believed that "secondary valences," some ill-defined attractive capacity beyond covalent bonds, might be the force bringing small molecules into the colloidal state. However, this theory

---

[a]The original meaning of the word polymer was a multiple of a structural formula. While "isomer" meant compounds of the same composition but different structures, "polymer" referred to multiples of a given composition. Thus, ethylene C_2H_4 and butene C_4H_8 would have been classified as polymers of CH_2.

had not given rise to reliable ways of creating polymers from small molecules. The correct formulation of polymers was put forward by Hermann Staudinger, beginning in the 1920s.[b] Staudinger proposed that substances such as rubber, starch and proteins were in fact huge molecules linked by ordinary covalent bonds. He called them *macromolecules*. During much of the 1920s, controversy enveloped the field. Late in the 1920s, the "micellar theory" gained acceptance. It suggested that polymers were indeed long-chain molecules, but not as long as Staudinger believed. Aggregates of relatively small polymers were believed to be held together by special forces, not covalent bonds. Gradually, more detailed structural information became available, particularly by X-ray crystallography, although at the time, it could not decisively measure the molecular weight of polymers. Slowly, understanding and clarity evolved. Staudinger recognized that polymers exhibit a distribution of high molecular weights. He and others showed that physical properties, such as strength, were determined not only by the chemical structure of the polymer, but also by the molecular weight distribution and the particular three-dimensional shape of the polymer molecule.

The ability to make useful polymers is the result of chemistry that provided the means to convert small molecules into large ones and the ability to modify and adapt properties to specific uses. We will follow this story, discussing how polymers are made, transformed into useful materials, and some of the consequences that this technology has brought. In general, the economic result has been the availability of a wider variety of materials to a greater segment of the population. Because they are derived from petroleum and consume energy in their production, their manufacture and use contributes to petroleum consumption, but the amounts are miniscule in comparison with the use of petroleum as fuel in transportation and industry. More significantly, many polymeric materials are intended for a short and single usage, so waste disposal or recycling are important issues. Plastics often contain other materials, such as plasticizers and flame retardants, whose presence may not be obvious to casual users. These

---

[b]Staudinger began to investigate the nature of rubber about 1912. At the time, the first efforts to prepare synthetic rubber were underway. In 1917, in a paper given to a meeting of the Swiss Society for Chemical Industry, he proposed that rubber was a high MW molecule and suggested the correct structure. He further developed this idea and expanded it to the polymers of formaldehyde and styrene in a paper published in 1920 in *Berichte*. In 1922, the first solid experimental support for this idea was obtained when he successfully hydrogenated rubber and found that removal of the double bonds did not dramatically change its properties. This result implied that its properties could not be associated with "secondary valences" resulting from double bonds. Staudinger and his students went on to demonstrate that many polymeric materials, such as polystyrene, poly(vinyl acetate) and polyacrylic acid could be successfully subjected to normal chemical reactions without changing their polymeric character or high molecular weight. He also demonstrated a relationship between viscosity and molecular weight that was consistent with his macromolecular theory. At the end of WWI, Staudinger had written an article opposing gas warfare (see Topic 8.1). After the rise to power of the Nazis, Staudinger's WWI writings became the basis of attempts to remove him from his professorship. The Gestapo interrogated Staudinger on his views and maintained surveillance of his home. His brother, Hans, a politically active economist, was arrested and charged with treason, and eventually fled to America. Staudinger was not permitted to travel to international meetings after 1937. Nevertheless, he continued to work on macromolecules at Freiburg throughout the war and until he retired in 1951. He was awarded the Nobel Prize in Chemistry in 1953.

materials have become a source of concern for their potential physiological and ecological effects.

Polymers, as "plastics," became a ubiquitous part of the material world during the twentieth century. Early in the century, advertisements began to characterize plastics as unique, innovative, man-made, and even "perfectible" products of chemical ingenuity. Beginning with Bakelite, numerous other polymers—polystyrene (PS), synthetic rubber, nylon, polyesters, acrylics, polyethylene (PE), and polypropylene (PP)—as well as numerous copolymers have taken a large place among the things we use on a daily basis. While the component materials were made mainly by large chemical companies such as Celanese, Dow, DuPont, Tennessee Eastman, and Union Carbide, the final products were made by a many companies and were of a wide variability in quality. During WWII, polymer production was largely devoted to the war effort. During and after WWII, production rose rapidly from about 100 t in 1939 to 200 t in 1942, 400 t in 1945, and 1,200 t by 1951. This period coincided with the recognition of the value of polymers for their specific properties. Polymers became a key component of the post-war economy. Scheme 5.2 shows the structures of some major categories of polymers.

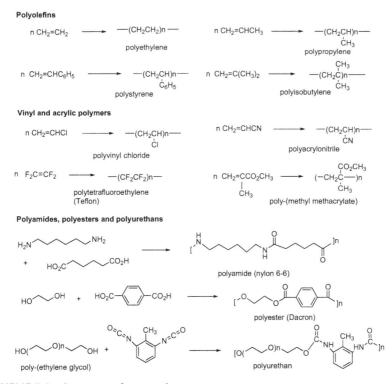

SCHEME 5.2   Structures of some polymers.

The process of polymerization involves forming chemical bonds between the starting small molecules (monomers) to make the high molecular weight polymers. In *addition polymerization*, all of the monomer is converted to polymer. In *condensation polymerization,* small molecules—often water—are removed (eliminated) during the bond formation. Among the polymers in Scheme 5.2, all but the polyesters and polyamides are the result of addition polymerization. Another important distinction is between *chain* and *step-growth* polymerization. Most alkenes and vinyl monomers are polymerized by *chain mechanisms*. In this method, a relatively few initiating events start the growth of chains. The individual chains continue to grow until the chain is terminated. In this type of polymerization, the polymer achieves the desired molecular weight in the presence of unpolymerized monomer. Growing chains do not usually combine. Molecular weight is determined by the rate of *chain initiation and termination.* An important mechanism is *radical chain polymerization.* In this method, a *radical initiator* starts a repeating sequence of reactions that adds monomer units, one-by-one. This is called *propagation.* The average MW of the polymer continually increases and can be controlled by the identity and concentration of initiators and chain terminating agents, which, respectively, start and stop chain growth.

In = initiating radical

propagation step
repeats many times

Step-growth polymerization usually occurs by condensation polymerization. The monomers combine to give first small *oligomers* and the average molecular weight continues to increase as the smaller fragments combine. This type of polymerization requires especially careful control of stoichiometry, because an excess of either monomer will limit the extent of polymerization. In the formation of polyamides and polyesters, heating the monomer units forms bonds by eliminating water. These reactions do not necessarily occur sequentially. Several oligomers can join with one another.

$X\text{-}(CH_2)n\text{-}X$ + $Y\text{-}(CH_2)m\text{-}Y$ $\longrightarrow$ $X\text{-}(CH_2)n\text{-}Z\text{-}(CH_2)m\text{-}Z\text{-}(CH_2)n\text{-}Z\text{-}(CH_2)m\text{-}Y$

x $X\text{-}(CH_2)n\text{-}Z\text{-}(CH_2)m\text{-}Z\text{-}(CH_2)n\text{-}Z\text{-}(CH_2)m\text{-}Y$ $\longrightarrow$ $[\{X\text{-}(CH_2)n\text{-}Z\text{-}(CH_2)m\text{-}Z\text{-}(CH_2)n\text{-}Z\text{-}(CH_2)m\text{-}Y]x$

y $[\{X\text{-}(CH_2)n\text{-}Z\text{-}(CH_2)m\text{-}Z\text{-}(CH_2)n\text{-}Z\text{-}(CH_2)m\text{-}Y]x$ $\longrightarrow$ $\{[\{X\text{-}(CH_2)n\text{-}Z\text{-}(CH_2)m\text{-}Z\text{-}(CH_2)n\text{-}Z\text{-}(CH_2)m\text{-}Y]x\}z$

Polyurethanes are copolymers in which a diol component adds to the diisocyanate portion, extending the chain. Because of the nature of the bond formation, the diol and diisocyanate components alternate along the chain. As with polyamides and polyesters, this may occur between partially polymerized segments. This is an example of an addition polymerization that occurs by a step-growth mechanism.

poly-(ethylene glycol)   polyurethan

As we discuss individual types of polymers, we will learn more details about these processes.

5.2 CHEMICALLY MODIFIED NATURAL POLYMERS

5.2.1 VULCANIZED RUBBER

The story of vulcanized rubber is somewhat less of a chemical story than that of the synthetic polymers. Nature has done the fundamental work in the preparation of natural rubber by polymerizing *isopentenyl pyrophosphate*. The key chemical discovery, vulcanization by sulfur, was made on an entirely empirical basis. Considerable improvement has come from later chemical input, but the basic process is that discovered by Charles Goodyear in 1839. Natural rubber comes from latex produced by tropical trees, principally *Hevea brasiliensis*. The tree's natural habitat is South America and it was encountered by Europeans in 1524. Jean Marie de la Condamine had been sent to Ecuador, to make measurement to determine if the planet was flattened at the poles. Condamine proceeded to explore regions of the Amazon River basin and found rubber being used to make footwear, flexible bottles, and rubber balls for games. The French called the material "caoutchouc," a name used by one of the local tribes. The English name rubber came into being in 1770, when Joseph Priestly noted that it could be used as an eraser by "rubbing" away the marks of a lead pencil. During the nineteenth century the term "India rubber" was widely used in the English-speaking world. In Spanish, rubber is called *goma* or *caucho*, while in German it is *gummi*. Although the tree is native to Brazil, it is now cultivated in Ceylon, Malaysia, and Indonesia. It grows only in humid tropical areas. The latex is an all cis polymer of 2-methyl-1,3-butadiene, which is called *isoprene*. In the plant, the polymer is formed from isopentenyl pyrophosphate. The all *trans*-polymer is called *gutta percha*.[c] It is isolated from a South Asian tree, *Palaquium gutta*.

---

[c]The words *cis* and *trans* refer to the configuration of the double bond. In the *cis* isomer, the major chain extends from the same side of the double bond. In the *trans* isomer, the chain extends from opposite sides of the double bond.

At one time, gutta percha was used as insulation for telegraph wires and cables, including the first trans-Atlantic cable. Until 1900, it was used for the core of golf balls, but it was replaced by rubber. It was also used for many years as a dental filling.

isoprene   isopentenyl pyrophosphate   rubber latex   gutta percha

Rubber falls into the category of an *elastomer*. That is, it can be stretched or deformed but tends to regain its original shape. The property of elasticity is the tendency to regain the original molecular structure after being distorted. K. H. Meyer and others explained this in terms of entropy.[1] The retractive force results from the tendency of rubber to achieve a disordered state. This interpretation was subsequently put on a quantitative basis.

Natural rubber latex can be dissolved in hydrocarbon solvents such as turpentine or kerosene. These solutions can be used for water-proofing fabrics. Rubber-treated cloth and rain gear were developed in Britain in the 1800s, most notably by Charles Macintosh who introduced a rubber layer between two layers of fabric. Thomas Hancock in England found that physical manipulation of rubber made it more soluble and workable and invented a machine called a *masticator* that rotated the rubber around a spiked cylinder. The technology for rubber production expanded in 1839 when Charles Goodyear found that its hardness could be improved by heating with sulfur, a process called *vulcanization.* Goodyear had become obsessed with finding a way of hardening the sticky material. He tried many powders, thinking they might dry the material. Finally, probably partly by accident, he discovered that the combination of sulfur and heat gave a tough yet flexible material. The same observation was made by Thomas Hancock. Each obtained patents in France, Great Britain, and the United States nearly simultaneously in the spring of 1844. The vulcanization process forms sulfide, disulfide, and polysulfide crosslinks. Vulcanization requires the presence of unsaturated bonds in the polymer and is applicable to natural rubber and synthetic materials that contains unsaturation. For many years, the compounding of rubber remained an art with various manufacturers having different recipes for their products. The quality of crude rubber also varied, making the final properties of the product quite variable. Conveyer belts made of rubber were used widely by the second half of the nineteenth century. Rubber usage exploded in the late nineteenth century with the development of pneumatic rubber tires, first for bicycles and then for automobiles. The Dunlop

Rubber Company in Ireland and Michelin in France were the first to introduce detachable tires.

In 1904, it was found that addition of carbon black improved the durability of rubber. Other improvements in rubber production were made later on. Thiocarbanilide was found to act as an *accelerator* and to speed up vulcanization. Various other nitrogen–sulfur compounds such as tetramethyl thiuram disulfide and 4,4'-dithiodimorpholine also came to be used as accelerators. Natural rubber is susceptible to atmospheric oxygen and especially to ozone. Antioxidants are added to increase resistance to oxidative degradation. Thus, a need for "rubber chemicals" developed along with the demand for tires and other forms of rubber.

$$C_6H_5HN-\overset{\overset{\text{S}}{\|}}{C}-NHC_6H_5$$

thiocarbanilide

$$(CH_3)_2N-\overset{\overset{\text{S}}{\|}}{C}-S-S-\overset{\overset{\text{S}}{\|}}{C}-N(CH_3)_2$$

tetramethyl thiuram disulfide

4,4'-dithiodimorpholine

To meet the growing demand for rubber, cultivation was begun in European colonial regions including Ceylon, Malaysia (British), and Indonesia (Dutch). Yields were improved by selection and grafting of superior plant specimens. The rubber latex was usually processed into sheets of "smoked" rubber or coagulated into "crumbs." Both forms were packaged and shipped in bales. In addition to the plantations, the Dutch encouraged many small growers whose product was collected by traders. Although often of poorer quality, this material added significantly to the total output. There was considerable overcapacity during the depression of the 1930s and an international scheme to control production was put in place but was only marginally effective. As indications of war grew in 1939, the United States and Great Britain traded large amounts of surplus cotton (United States) for rubber (United Kingdom) that had been held off the market in an effort to support prices. In 1940, the Rubber Reserve Corporation was created with US government financing to purchase a rubber reserve. These transactions built up a reserve of about 400,000 t. Rubber became a critical strategic issue when Japan took control of most of the natural rubber producing area at the beginning of WWII. Rationing of rubber began almost immediately in the United States. Nearly half a million tons of used rubber was collected after a government appeal and during the course of the war about a million tons was collected and reused. However, it became clear at the beginning of the war that a synthetic substitute would be required to replace natural rubber. The development of synthetic rubber is discussed in Section 5.3.2.4. Rubber use expanded with economic development after WWII and natural rubber was supplemented by synthetic after the war.[2]

5.2.2 FIBERS AND FILMS BASED ON CELLULOSE

The production of cloth from natural fibers such as cotton, flax, silk, and wool was highly developed by the beginning of the eighteenth century. This had not occurred by application of science, but by experience and transfer of techniques from artisan to artisan. The natural fibers are polymers. Cotton and linen are forms of cellulose, a polymer of the carbohydrate glucose. Wool and silk are primarily proteins, polyamides derived from amino acids (see Scheme 5.1 for structures).

5.2.2.1 CELLULOID

Beginning in the nineteenth century, improvements in natural cellulose fibers began to be made based on chemical change. John Mercer (1857) and Horace Lowe (1890) patented procedures for treatment of cotton with caustic soda that improved both its strength and affinity for dyes. The first patents on artificial fibers made from cellulose nitrate, the ingredient in guncotton, appeared in the 1850s. The first material that would fall into the category of a plastic was developed by Henry Parkes and patented in 1861 in England. It was called "Parkesine" and was a mixture of cellulose nitrate and camphor. The camphor functioned as a *plasticizer* (softener) for the cellulose nitrate. The material was lustrous, somewhat resembling ivory. A similar product was developed in the United States, by John W. and Isaiah Hyatt in Albany, NY, USA in 1869. In response to an offer of $10,000 for a replacement for ivory in billiard balls, the Hyatt brothers experimented with nitrated cellulose and camphor and came up with "celluloid." Note that these are the same compounds, except for the nitro-glycerin, used by Frederick Abel and Alfred Nobel at about the same time to make Cordite and Ballistite (see Section 1.2.2.2). The cellulose was less highly nitrated than for guncotton, but nevertheless the production process was haz-ardous and the product was very flammable. Celluloid was used in such items as brush and knife handles, combs, and piano keys (in place of ivory). It was cast from a doughy mass into sheets of variable thickness. Coloring and irregu-lar patterns could be introduced and made to resemble marble or tortoise shell.

Celluloid was also a key material for photographic film. By the first half of the nineteenth century, glass plate coated with silver halide and gelatin or collodion (also a form of cellulose nitrate) had come into use. In 1889, George Eastman introduced celluloid film and a portable camera designed for its use. Eastman also acquired Leo Baekeland's method for making photographic paper. Celluloid became the preferred material for movie film in the early1900s. It was also used extensively for X-ray film. The material is very flammable and was eventually replaced by cellulose acetate.

The first commercial application of cellulose to synthetic fibers was by Louis Marie Hilaire Berniguad in France in the 1880s. His patents introduced the technology of forming fibers by forcing fluid through small holes (spinnerets). This proved to be a critical and enduring technology for most synthetic fibers. The nitrocellulose was dangerous, however, because of its violent flammability. A method for denitration was developed in the late 1800s. This led to the production in France of a bright, shiny material called rayon. By 1900, further improvements had been made and plants were in production in Europe and the United States. However, the flammability of the nitrocellulose intermediate was an inherent disadvantage and these materials were replaced in the early 1900s by other forms of modified cellulose.

5.2.2.2 *VISCOSE RAYON*

In the 1890s, three English chemists, Charles F. Cross, Clayton Beadle, and Edward J. Bevan, produced "viscose" by treating cellulose with strong alkali followed by carbon disulfide. Extensive conversion of hydroxy to xanthate groups occurs. The mixture is then "aged," during which time the reaction partially reverses and viscosity increases. (Thus, the term viscose.) There is also some hydrolysis of the polymer to a lower average MW. At the correct state of viscosity, the solution is forced through spinnerets into acid, which completes the removal of the sulfur groups and reconstitutes the glucose polymer as a fiber. The process was commercialized by the British textile firm Courtalds, which had a subsidiary, American Viscose Corporation, that began fiber production at Marcus Hook, PA, USA in 1910. The details of processing can be used to adjust the properties of viscose rayon. The standard (generic) name *rayon* was adopted by the Federal Trade Commission (FTC) in 1926 to apply to material produced through the viscose process. The FTC added "viscose" to the official designation in 1951.

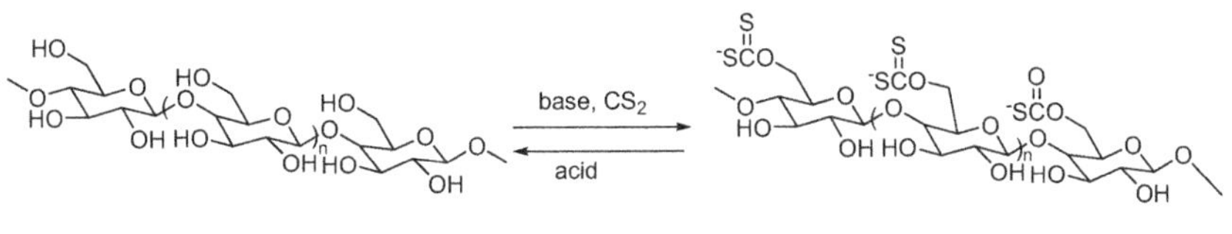

After WWI, DuPont decided to invest in cellulose fibers, hoping to find a synthetic silk substitute. DuPont by this time was the world's largest producer of cellulose nitrate in its explosives business. DuPont acquired "viscose" technology through a joint venture with a French firm and began production of rayon in

1920. Though having many unfavorable qualities, rayon was highly profitable in the 1920s as its introduction coincided with changing fashion trends. As a moderately priced substitute for silk, it provided sheer stockings as an alternative to cotton. During WWII, rayon was used in many applications, including tire cord, because of the shortage of natural fibers. After the war, it faced stiff competition from nylon, polyesters, and acrylics. However, its use in blends with other fabrics has maintained a place in items such as draperies and bedspreads.

5.2.2.3   CELLULOSE ACETATE

Cellulose acetate is made by treating cellulose with acetic anhydride, which converts most of the hydroxy groups to acetate esters. The material with >90% acetylation is called cellulose triacetate. The first useful form of cellulose acetate called "cellulose diacetate" was developed by George Miles in the United States in 1903. It is a mixture with an average acetylation level of about 55%. In Germany, similar applications were due to the research of A. Eichengruen working at Bayer.[3] Eichengruen also developed the first techniques for shaping of objects from cellulose acetate by injection molding. The partial acetylation makes the material soluble in solvents such as acetone. The highly viscous solution that results can be converted into fibers, films, or molded objects. The most extensive commercialization was by two Swiss chemists, Camille and Henri Dreyfus, who founded the Celanese Company in England. One of the uses found for cellulose acetate in WWI was as a lacquer for the light airplanes and zeppelins of the time. It was also used as a non-shattering glass substitute in automobiles, trucks, and planes. During WWI, plants were opened in France and England. A plant was opened in the United States at Cumberland, MD, USA in 1918.

After the war there was extensive overcapacity and the Dreyfus brothers developed cellulose acetate into a fiber called artificial silk. The technology, which involves evaporation of the solvent, is called *dry spinning.* Research optimized the process for both production and spinning of cellulose acetate and eventually resulted in fibers stronger than natural silk. The process also controlled the uniformity of fiber size and cross-section shape, both of which are critical for acceptable performance in fabric. Celanese also developed methods for dyeing the fabric. Eventually, much of the textile use of cellulose acetate was replaced by entirely synthetic fibers such as nylon and polyesters. Cellulose acetate is currently used extensively in cigarette filters. It was also used as one of the first successful membranes for kidney dialysis machines. Cellulose acetate is also used in coating of pills, both to protect the contents from deterioration and as a means of controlled release.

5.2.2.4 CELLOPHANE

The basic process for creating the cellulose-based film *cellophane* was developed by Jacques Brandenberger while working at a textile company in France. A company, La Cellophane Societe, dedicated to production of cellophane was begun in Paris in 1913. DuPont entered the cellophane film business in 1923 with the French company that had formed the Rayon joint venture. DuPont devoted considerable research effort to understanding and improving the process and in the course of this effort developed moisture-proof cellophane. The first steps in production of cellophane are the same as for viscose rayon, although the details of alkali concentration and time differ. In the casting of the film, the viscose is passed through a thin slit into an H_2SO_4–Na_2SO_4 solution that regenerates the cellulose. The film is purified by washings to remove CS_2 and H_2S and is then bleached with NaOCl and H_2O_2. The film is treated with softeners such as PEG that are absorbed by the water contained in the film. The film can also be dyed at this stage. The water content is then reduced by drying. A moisture-proof layer can be added. It consists of a thin film of paraffin wax held in place by a lacquer. This layer is applied from a solution in organic solvents. By incorporating nitrocellulose, a paraffin wax, plasticizer, and blending agent, the cellophane became superior to other coatings, such as wax paper. A later approach to moisture-proofing cellophane involved coating with an impervious polymer, such as PE or polyvinylidene chloride copolymer.

5.3 SYNTHETIC POLYMERS

5.3.1 THERMOSETTING RESINS

5.3.1.1 PHENOL–FORMALDEHYDE RESINS

The first major synthetic polymer was developed by Leo H. A. Baekeland, a Belgium inventor, who came to the United States in1889. He invented a photographic printing process that was sold to Eastman Kodak, bringing Baekeland considerable wealth. Baekeland went on to investigate phenol–formaldehyde resins for several applications.[d] Although phenol–formaldehyde resins had been investigated before, Baekeland found ways to control the condensation process.

---

[d]Leo H. A. Baekeland was born in Belgium in 1863. He received both a bachelors and Ph. D. in chemistry by age 21. He took up a faculty position at the University of Ghent in 1889, but emigrated to the US to pursue the development of paper for photographic film. He succeeded in 1893. In 1899, George Eastman purchased his company and process. Baekeland used the income to establish a private research laboratory and produced a number of other inventions. The most important of these was Bakelite, which was patented in 1909. For additional information on Baekeland see Strom, E. T.; Rasmussen, S. C. *ACS Symp. Ser.* **2011**, *1080*, 1–92.

The initial resin could be heated under pressure to produce a hard material that could not be melted and remolded, so objects were produced by casting. Such materials are called *thermosetting*.

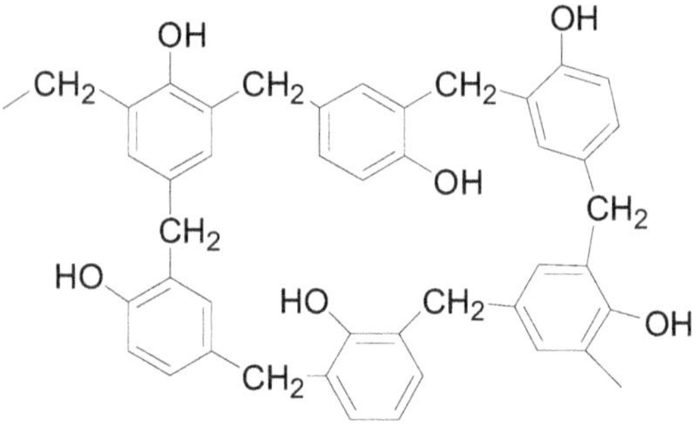

Bakelite is a cross-linked condensation polymer formed
by -CH₂-units linking ortho and para to the phenol group

Bakelite is a dark material and can not be obtained in a range of colors. The material is quite stable and is an excellent insulator. It can be molded to high precision. It can also be used with fillers such as wood and asbestos. Beginning in 1910 a company, General Bakelite, began production and marketing. The early uses included insulators for electrical wiring. Bakelite also found application in the automobile industry. Baekland filed patent infringement suits in 1917 that were decided in his favor in 1921. This led to merger with two of the other major producers to form the Bakelite Corporation in 1922. Formica, a laminated sheet bound with phenol–formaldehyde, was one of the products of the competing firms. Soon, mass production of radios began and Bakelite was a major constituent of both the insides and the cases. Bakelite thus was associated with two of the dominant innovations of the first quarter of the century, automobiles and radios. The heyday of Bakelite also coincided with the era of electrification and introduction of domestic appliances. Bakelite was advertised as the "Material with a Thousand Uses." A book, *The Story of Bakelite* was published in 1924 as part of a publicity campaign to encourage consumers to identify with the brand name. The Bakelite Corporation became part of Union Carbide in 1939.

5.3.1.2   UREA–FORMALDEHYDE RESINS

Urea–formaldehyde resins are hard thermosetting materials, but they are clear so they can be colored. The resins were first produced in the United Kingdom in the 1920s. The resin is formed in three stages. First, urea and formaldehyde are reacted at a roughly 1:2 ratio under alkaline conditions. This results in the introduction of methylol substituents on the urea. Treatment by acid then leads to condensation forming mainly $-CH_2-$ bridges between ureas, but with some ether, $(-CH_2-O-CH_2-)$ links. Finally, additional urea is added to lower the urea–formaldehyde ratio to 1.2 or less and the resin is subjected to vacuum concentration. The resin can be produced as a powder by complete drying. The final processing of the resin involves addition of a curing reagent, such as NH_4Cl or $(NH_4)_2SO_4$, pressing and heating. This treatment further hardens the resin by forming additional covalent bonds. At this point the polymer network provides a hard thermoset material. The properties of specific materials are determined by variables such as the formaldehyde–urea ratio, the temperature and pH, all of which affect the degree of polymerization and cross-linking.

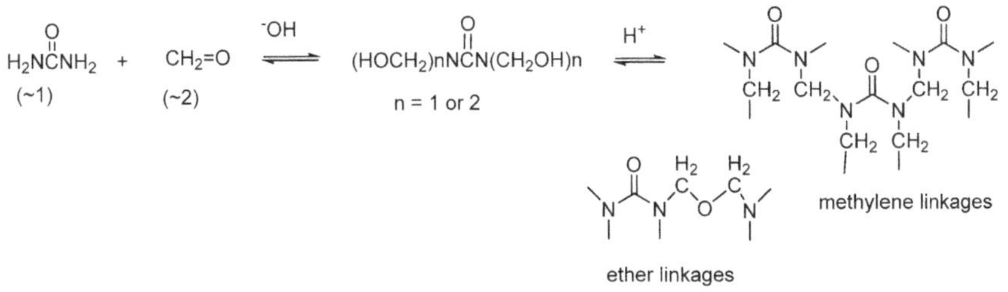

One of the main uses of urea–formaldehyde resins is in the bonding of wood in particle board, fiber board, laminates, and plywood. Two important characteristics of the bonded product are hydrolytic instability and a tendency to release formaldehyde. Both properties are due to the reversibility of the bond-forming reactions. In the presence of moisture, hydrolysis of the polymer can occur, leading to physical deterioration. Formaldehyde can be released by slow hydrolysis. Minimization of formaldehyde emission is achieved by lowering the formaldehyde:urea ratio in the final product. Product resistance to hydrolysis can be improved by addition of melamine, so there are various grades of resins. The melamine increases the cross-linking and also may provide some stability through its buffering effect on acids. Inexpensive colored table-ware is sometimes made of melamine–formaldehyde polymer.

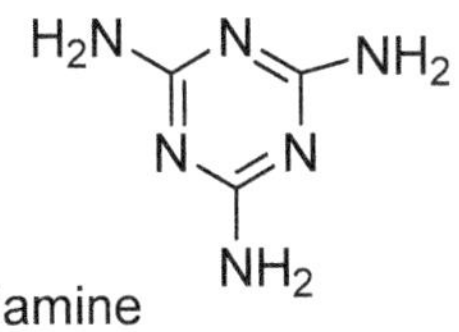

melamine

Urea–formaldehyde and formaldehyde–urea–melamine resins are also used to improve the wet-strength of paper in products such as paper towels, plates, and bags. The melamine resins are used in high-quality papers such as currency and map paper. Paints and coatings can be made from formaldehyde–urea–melamine resins. The formulations include a salt of a volatile amine and a sulfonic acid. As the amine evaporates, the acid is liberated and causes the cross-linking and hardening of the coating.

5.3.1.3   EPOXY RESINS

The epoxy resins contain the epoxide functional group and the polymerization process involves opening the epoxide ring. The most widely used monomer is derived from *bis*-phenol-A and epichlorohydrin. The novolac resins are derived from phenol–formaldehyde oligomers and epichlorohydrin.

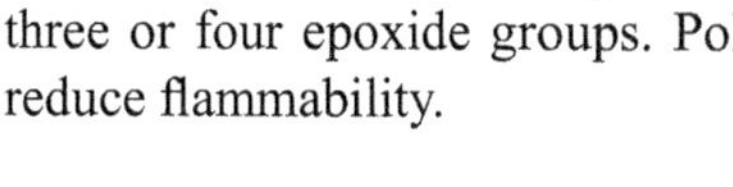

diglycidyl ether of *bis*-phenol-A

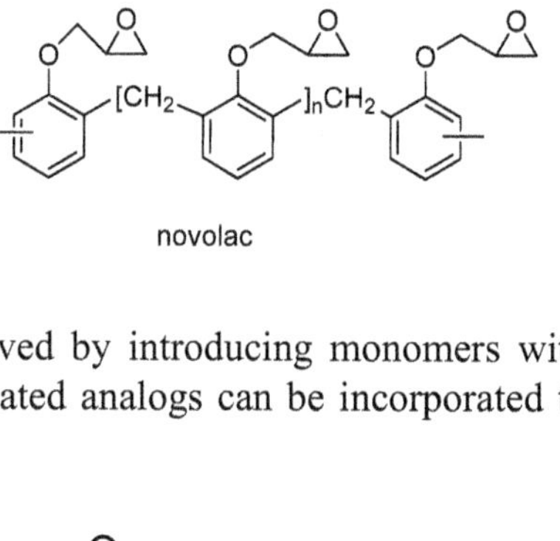

novolac

Additional cross-linking can be achieved by introducing monomers with three or four epoxide groups. Polybrominated analogs can be incorporated to reduce flammability.

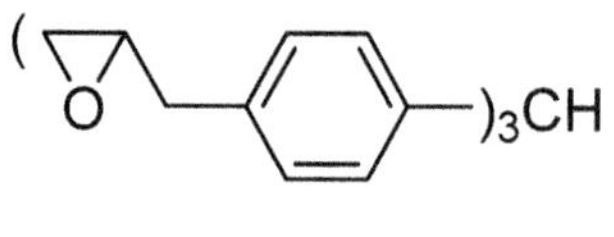

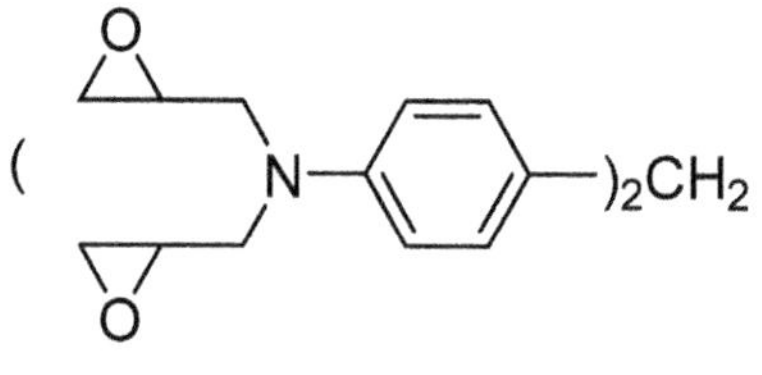

The epoxides are generally supplied as oligomers that are subject of further polymerization, which can be carried out in two ways. A base such as a primary amine can begin the process by opening an epoxide ring. The alkoxide that is formed is a sufficiently strong base to open other epoxide groups, thus creating a cross-linked polymer. Alternatively, a bifunctional amine such as diethylene-triamine can be used. This creates links between epoxide molecules when the amine group reacts with two separate epoxy oligomers. The process is called "hardening" or "curing."

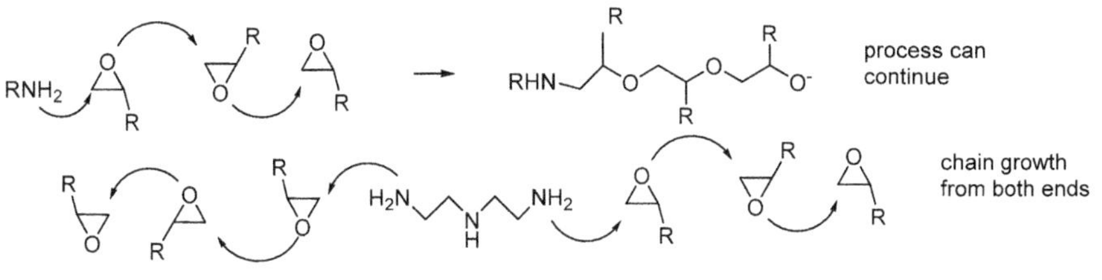

Epoxy resins are used to make adhesives for floor and ceramic tiles, grout and caulk, and for metal–metal and concrete–concrete binding. Epoxy coatings are also used on automobile and truck parts and for coating iron rebar used in concrete. The epoxide polymerization reaction does not generate any byproducts so no shrinkage or bubble formation occurs. High performance applications are found in industrial and marine coatings.

5.3.2   POLYOLEFINS

The polyolefins are formed by addition polymerization of alkenes. The resulting products are high molecular weight saturated hydrocarbons. These materials are quite stable chemically and depending on molecular weight and treatment history, they can have considerable strength. They are *thermoplastic*, which means that they can be softened without decomposition and extruded or molded in the melted form. They are combustible in the sense that the combustion, when initiated, is strongly exothermic. The most widely used polyolefins are PE, PP, PS, and polyisobutylene.

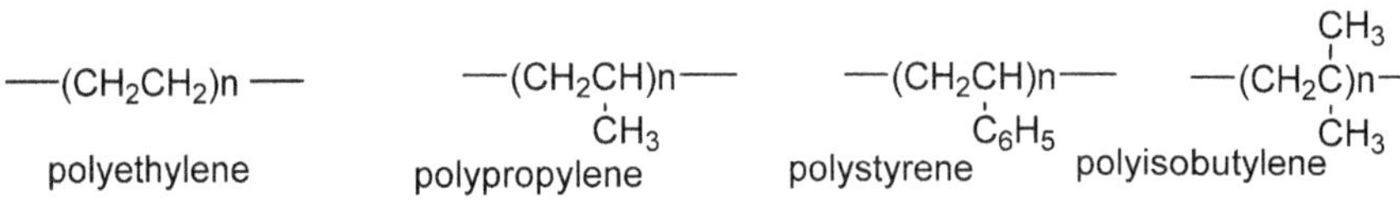

| —(CH$_2$CH$_2$)n— | —(CH$_2$CH)n— | —(CH$_2$CH)n— | —(CH$_2$C)n— |
| polyethylene | $\overset{\vert}{C}H_3$ polypropylene | $\overset{\vert}{C}_6H_5$ polystyrene | $\overset{CH_3}{\underset{CH_3}{\vert}}$ polyisobutylene |

5.3.2.1   POLYSTYRENE

PS (styrene is the common name for 1-phenylethene) was the first olefin to be commercially polymerized. Although its application was not as dramatic as that of nylon, which was introduced at about the same time, it eventually became a major product. Styrene polymerizes easily when exposed to initiators such as peroxides. PS was one of the first substances investigated by Hermann Staudinger in his study of the nature of polymers. Commercial PS production began at the BASF component of IGF in Germany in 1931 and at Dow in the United States in 1938. One of the key chemists involved at BASF was Herman Mark.[e] The BASF process was carried out in towers having internal tubes for thermal control of the exothermic polymerization. At Dow, the polymerization was originally done as a batch process in large cans, and the solid material was ground to a powder. Dow was the primary United States producer of styrene monomer from ethylbenzene and developed PS to expand the market for the styrene. During WWII, when the supply of natural rubber was cut off, an intense program for production of rubber from styrene–butadiene copolymer was initiated (see Section 5.3.2.4). After the war, Dow adopted aspects of the BASF process and produced molten polymer that could be extruded and pelletized. Figure 5.1 shows a conceptual diagram of the process.

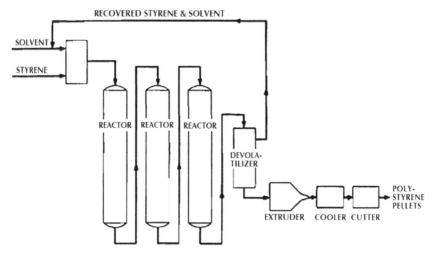

FIGURE 5.1   Continuous Process for Styrene Polymerization (From p. 375, Watson, W. D.; Wallace, T. C. *ACS Symp. Ser.* **1985,** *285*, 363–382. ©1985 American Chemical Society.)

---

[e]Herman F. Mark was born in Vienna in 1895. He served in the Austro-Hungarian army in WWI as a highly-decorated combat officer. After the war, he returned to Vienna and received a doctorate in 1921. Mark began research at the Kaiser Wilhelm Institute in Berlin–Dahlem, using X-ray crystallography to show that cellulose, silk and wool were long chain polymeric molecules. From 1927 to 1932, Mark worked at IGF in Ludwigshafen. His work there included a method to prepare styrene from ethylbenzene, opening a practical route to polystyrene and Buna-S synthetic rubber. Mark was of Jewish ancestry and left Germany in 1932, returning to Vienna as a Professor of Chemistry. He was one of the first academic chemists to systematize teaching in polymer chemistry. When Austria was annexed by Germany in 1938, he and his family fled Austria through Switzerland and France. He first came to Canada, where he worked for the International Pulp and Paper Company. In 1940, he moved to the Brooklyn Polytechnic Institute, becoming a professor in 1942. He founded the Institute of Polymer Research there, serving as its director until 1964. He was awarded the Perkin Medal in 1980 and a National Medal of Science in 1979.

Foam made of PS is a familiar material. The foam is produced by introduction of a volatile compound, usually pentane. The foaming is done by heating, sometimes over several stages, such that the pentane forms bubbles, expanding the polymer which resolidifies on cooling. The material is called *expanded polystyrene* (EPS). The familiar light-weight cups and trays used for disposable food serving and packaging are made of EPS.[4] An alternative form of foamed PS is known as *extruded polystyrene*. It is melted with a foaming agent under pressure and expands as it leaves the extruder. Both materials are excellent insulators and are very buoyant (<5% the density of solid PS). PS foam is used extensively in house construction for foundation, wall and ceiling insulation. In this application, it is treated to increase flame resistance. These materials are easy to break and tougher PS, called *high impact polystyrene* can be produced by incorporating a rubbery material. The first commercial version was made by Dow and incorporated poly-1,3-butadiene. The material is produced in an apparatus that permits stirring and shearing that disperses small particle of the rubber in the polymer.[5] The currently preferred material is a graft copolymer of polybutadiene with styrene and acrylonitrile called ABS (acrylonitrile–butadiene–styrene). The polymer is produced by copolymerizing acrylonitrile and styrene in the presence of polybutadiene under conditions of agitation and shearing. This produces rubber-like particles of the correct size that are covalently bound to the copolymer. Rubber particles can be detected by electron micrography, as shown in Figure 5.2. These particles have a degree

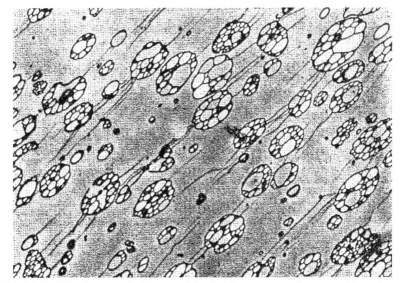

FIGURE 5.2 Distribution of rubber particles in high impact polystyrene (From Amos, J. L. *Polym. Eng. Sci.* **1974**, *14*, 1–11 by permission of Wiley Publishing. © 1974, Society of Plastics Engineers, Inc.)

of elasticity and are believed to absorb deformation forces and reduce the tendency to break. The rubber particles also initiate and terminate small fractures, called "crazes" that relieve distortion without resulting in cracks. This material can be used to produce thermally cast items such as luggage, refrigerator liners, and even boat hulls.

5.3.2.2   POLYETHYLENE AND POLYPROPYLENE

In contrast to styrene, ethylene and propylene are difficult to polymerize by the radical chain mechanism. This is because the phenyl group in the styrene monomer stabilizes the radical intermediate in polymerization. Polymerization of ethylene was first observed at ICI under serendipitous circumstances in 1933 when ethylene was exposed to peroxides at high pressure. While the discovery of polymerization was unintended, the capacity to do high-pressure chemistry had developed as a long-standing research focus at ICI.[6] The initial observation was not immediately followed-up, because it was difficult to reproduce and explosions occurred under some conditions. The role of peroxide catalysts was not appreciated at the time. Beginning in 1935, detailed study of the polymerization was undertaken, led by Michael Perrin. This research led to the recognition of the radical chain nature of the reaction and achieved reproducibility. The polymer formed in this way is not entirely linear because the growing radical chain "bites back" on itself leaving short substituent groups. This material has relatively low density and is called *low-density polyethylene* (LDPE). As operated commercially, the polymerization process uses pressures as high as 50,000 psi and temperatures up to 300°. PE, being a saturated hydrocarbon, is an excellent insulator and is impervious to water. The first use was for coating of underwater cables and during WWII the polymer was used extensively in radar equipment. After the war, the high pressure method was licensed by ICI to Union Carbide and DuPont in the United States. Many kinds of consumer items such as bottles, containers, and bags were rapidly adopted. Tupperware, for example, is made from PE. Specific variations are used in toys and in food and medical packaging.

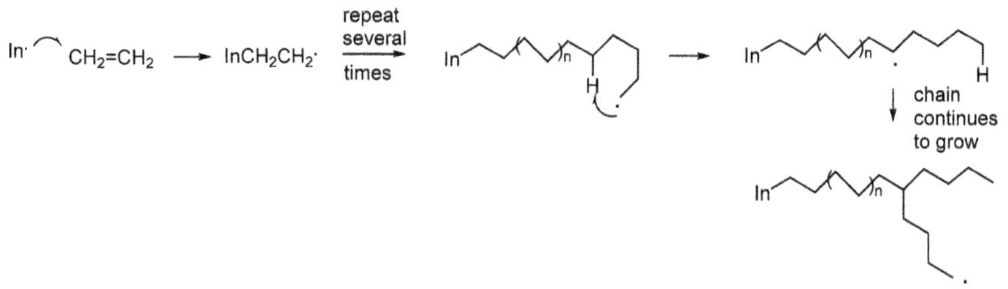

The high pressure–high temperature process remains the basis for production of LDPE and accounts for about half of all types of PE. Worldwide production was about 15 million tons in 2000. About two-thirds of the material is used in films for packaging, including heavy duty pallet covers and agricultural films such as for green-houses. About 15% is used for solid molded materials.

5.3.2.3 COORDINATION POLYMERIZATION OF ETHYLENE AND PROPYLENE

The next major advance in alkene polymerization involved catalysis by organometallic compounds and was made by Karl Ziegler and his group working at the Max Planck Institute at Muhlheim Germany shortly after WWII. Ziegler's group found that a catalyst made from trialkylaluminum and a Lewis acid, such as titanium tetrachloride, resulted in a highly linear polymer. There had been several earlier reports of polymer being formed from ethylene in the presence of organometallics or Lewis acids, but for various reasons the research had not been pursued. Ziegler patented the PE technology, but was not particularly interested in commercial development. He licensed the patent quite broadly, but left it to the individual companies to develop the technology. The crucial feature of the Ziegler catalyst is that polymerization takes place at a metal center. Typically, the metal has two large ligands, such as cyclopentadienide anions (Cp⁻) and two "active" sites. The trialkyl aluminum introduces an alkyl group at one of the active sites. Methylalumoxane activates a second site by removing an anionic ligand that can be replaced by the alkene. Chain growth then continues by a series of *migratory insertion steps* in which the polymer grows out from the metal. The chain length can be controlled by H_2, which serves as a chain transfer agent permitting initiation of a new polymer chain.

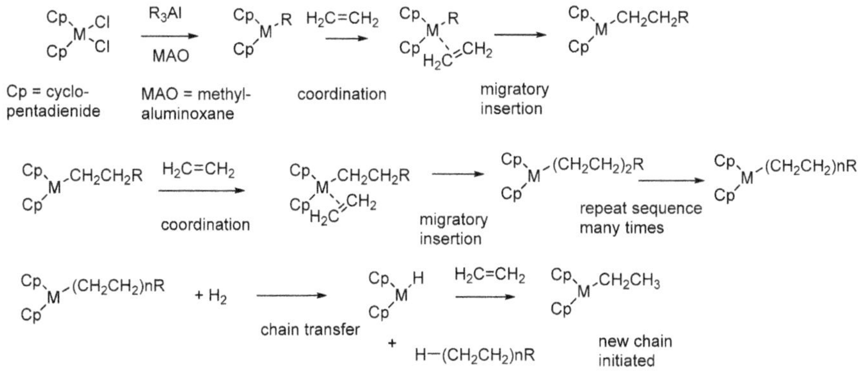

The material produced by metal-catalyzed processes is referred to as *high density polyethylene* (HDPE) and has density >0.940 g/cm³. The high-pressure radical product called LDPE, ranges from 0.90–0.93 g/cm³. Optimization of the physical properties of the linear polymers can be achieved by incorporating small amounts of longer chain alkene, for example, 1-butene. PE modified by incorporation of other alkenes is called *linear low density polyethylene.* PE used for water and natural gas distribution pipes must have long life with resistance to stress-cracking. This property is associated with concentration of short chain branches in the high MW portion of the polymer MW distribution. This outcome can be achieved in tandem reaction systems as shown in Figure 5.3. In the first reactor, the desired branching monomer (e.g., 1-butene) is included so that branched copolymer is formed. Additional polymerization in the second reactor then adds unbranched material and the MW can be controlled by including H_2 as a chain transfer reagent. The catalyst is of the Ziegler–Natta type on silica.[7]

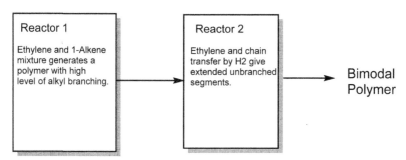

FIGURE 5.3 Schematic of reactors for production of linear low density polyethylene containing highly branched and unchained segments.

The discovery of stereoregular polymers, in particular PP, occurred in Italy in 1954 by the group of Guilio Natta at Milan Polytechnic Institute. Ziegler had shared information about the ethylene polymerization with Natta, but Ziegler did not believe propylene could be polymerized under these conditions. To the contrary, Natta's group observed polymerization of propylene in March 1954 and filed a US patent in June 1955. Working with Ziegler catalyst, Natta's group found that polymerization of propylene led to a polymer that was very crystalline. Further study by X-ray diffraction showed that the polymer was *isotactic*, that is, the configuration at all the methyl substituted sites was the same. Other versions of the catalyst could give *syndiotactic* polymer in which the configuration alternated. *Atactic polymer* has random configuration. These structures are shown in Fig. 5.4.

The stereo-regular polymers have preferred conformations of the polymer chain that results in high crystallinity and increased strength. As with polyethylene, incorporation of small amounts of longer chain alkenes could be used to optimize properties for specific applications. Natta's patent was assigned to the Montecatini Company, which had supported his research. Because of the proximity in time, and the overlapping claims of the patents involved, a lengthy litigation ensued, which is summarized below. Ziegler and Natta shared the 1964 Nobel Prize in Chemistry.[f]

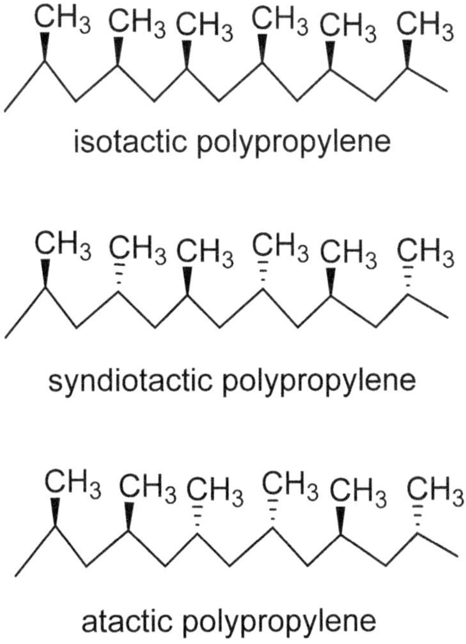

isotactic polypropylene

syndiotactic polypropylene

atactic polypropylene

FIGURE 5.4   Comparison of isotactic, syndiotactic and atactic polypropylene.

[f]Karl Ziegler was born in 1898 and entered the University of Marburg, but his studies were interrupted by service in WWI. He completed his Ph. D. in 1920. He became a professor at the University of Heidelberg in 1926. Among the areas of his research was the synthesis of large ring compounds. He also studied organolithium compounds. In 1943, he became Director of the Max Planck Institute for Coal Research at Muhlheim. His research on the polymerization of ethylene led to the discovery of the catalytic effect of alkylaluminums in conjunction with Ti(IV) salts.

Guilio Natta was born in 1903 and studied chemical engineering at the University of Milan. He became a professor at Pavia in 1933 and Rome in 1935. In 1938, he moved to the Polytechnic Institute of Milan where he was director of the Industrial Chemistry Institute. He conducted research on polyethylene using X-ray crystallography. He extended the use of coordination polymerization to propylene and discovered that it led to stereospecific polymerization.

Related discoveries with other catalysts had been made at Phillips Petroleum (CrO_3–silica–alumina) and Standard Oil(IN) (molybdenum–cobalt–alumina). These procedures produced mixtures from which fractions containing crystalline PP, identified by density, melting point, and infrared spectra, could be isolated.[8] Slightly later, in 1954, chemists at DuPont produced crystalline PP using $TiCl_4$ and Grignard reagents. Phillips Petroleum had filed a composition of matter patent on PP formed over a chromium catalyst on January 27, 1953, predating Ziegler's and Natta's patents. Phillips began producing HDPE using its catalyst in 1957. In 1972, the US Patent office granted a composition of matter patent to Montecatini, based on Natta's work. Eventually, in 1983, as the result of extensive litigation, a Federal Court ruled that priority belonged to Phillips Petroleum. In the meantime several other firms began producing PP. Hercules licensed the Montecatini process and began production in 1957. It later formed a joint venture with Montedison, the Montecatini successor, which in turn became Montell, a subsidiary of Shell.

Various improvements in the technology occurred and many chemical and petrochemical companies began production of HDPE and PP and these products, too, became very competitive economically. Fluid-bed reactors for PE and PP came into use beginning in the 1970s. One of the major technological problems was the morphology of the polymer granule. In the early years, this was difficult to control and led to difficulties in plant operation and product purification and processing. Catalysts supported on $MgCl_2$ were introduced for ethylene polymerization. Among other features, the $MgCl_2$-supported catalysts reduced the amount of catalyst residue in the polymer. Metallocene catalysts led to further improvement in the specificity of polymerization. The inclusion of more complex ligand structures at the catalytically active metal can lead to improved selectivity.[9]

PP can be processed as a fiber that finds use in heavy duty carpet and automobile interiors. The filaments are generally produced by melt-spinning at relatively high temperature (240–300°C). Pigments for coloring can be introduced in the same process. The fibers are drawn (stretched) to specified dimensions. Sheet coatings were also developed. Examples of these are Tyvek and Typar produced by DuPont. These are PE and PP products, respectively. The former is widely used as a moisture proof film for insulation in construction. The use of PE and PP has paralleled economic development around the world. The same can be said about the location of manufacturing and processing plants.

5.3.2.4 SYNTHETIC RUBBERS

Various reports of the production of synthetic rubber by polymerization of isoprene and 1,3-butadiene appeared in the late 1800s, but because the nature of polymers was not understood at the time, interpretation was not based on correct concepts. One such polymerization method was to expose the dienes to sodium metal. When WWI stretched on, Germany ran out of natural rubber and began producing synthetic rubber from 2,3-dimethyl-1,3-butadiene. It was called "methyl rubber." The process was inefficient and the product was of poor quality. The next synthetic material with rubber-like properties to be produced commercially was polyisobutylene. It was discovered first at IGF in 1933 and developed in cooperation with SO(NJ). The catalyst used was BF_3.[10] The product, however, was not particularly elastic and was not used extensively. The Standard Oil group also investigated copolymers of isobutene with 1,3-butadiene or isoprene. These materials could be vulcanized by methods similar to those for natural rubber and became of use during WWII for tire inner-tubes.

Chemists at IGF developed styrene and acrylonitrile copolymers of 1,3-butadiene between WWI and WWII. These were known as Buna-S and Buna-N, respectively. The former could be used for tires and the latter for oil-resistant applications, but in the 1930s, they were not economically competitive with natural rubber. When Hitler came to power, he ordered full development of the synthetic rubbers and they were the main source of rubber in Germany during WWII. The process started with acetylene made either from coal or natural gas. The acetylene was hydrated to acetaldehyde, dimerized to aldol, reduced and dehydrated. Smaller amounts of butadiene were also made via 2-butyne-1,4-diol.

Russia also relied on synthetic rubber between the world wars, because of lack of currency to purchase natural rubber. The Russians used sodium polymerization of butadiene, which was obtained from grain alcohol via acetaldehyde.

The Second World War resulted in a crash program in the United States for production of synthetic rubber. War began in Europe in 1939 and concern about rubber supplies began to grow in the US government. In August 1940, a committee established to assess the situation recommended construction of 100,000 t/year capacity for synthetic rubber. A smaller program, 40,000 t/year was approved in September, 1940. Bids were received from the four major rubber companies, but they were rejected as too costly. Estimates at that time were that available sources of rubber would last for 2–3 years, even if supplies from Southeast Asia were cut off. The rubber companies disagreed with these estimates and decisions and began to press for action. The Office of Production Management also favored a larger plan. By May 1941, construction of four plants with total capacity of 40,000 t/year to be operated by the rubber companies had been agreed upon. The target material was a copolymer of styrene and 1,3-butadiene called *SBR rubber.* One of the issues that remained unsolved was the source of 1,3-butadiene.

Immediately after Pearl Harbor, agreements on patent and technology-sharing were reached and cooperation among the rubber companies began. It soon became apparent that the needed capacity was 800,000 t/year within a time frame of 2 years. Plans were in place by April 1942, about half of the capacity was operated by the four major tire companies, Firestone, Goodyear, Goodrich, and US Rubber and the other half by smaller companies. Congress generated controversy and confusion in July 1942 by passing a law establishing a competing rubber program based on grain alcohol. President Roosevelt vetoed the bill and in response appointed a committee consisting of Bernard Baruch, James B. Conant[g] and Karl T. Compton to review the situation. They reported in September 1942, endorsing the petrochemical-based program, but recommending the appointment of a "rubber director" to coordinate the effort. William M. Jeffers, the president of Union Pacific Railroad, took on the task. Eventually, both petrochemicals and grain alcohol were used to produce butadiene. The alcohol route was put into place more rapidly than the petroleum route, but was more expensive. Later, it became possible to make butadiene by dehydrogenation of butane. The styrene was produced by Friedel–Crafts reaction of benzene and ethylene to produce ethylbenzene that is dehydrogenated to styrene.

[g]James Bryant Conant was born in 1893 in Dorchester, MA, USA near Boston. He entered Harvard in 1910, already having developed an interest in chemistry through his high school teacher and a home laboratory. By 1916, he had a Ph. D. with studies in both physical and organic chemistry. During WWI, he was in the US Chemical Warfare Service. He returned to Harvard as an assistant professor in 1919. His research continued to include elements of physical and organic chemistry. He contributed fundamental studies that provided the foundations of what came to be called physical organic chemistry. He became president of Harvard in 1933. Conant was an important participant in the preparation prior to the outbreak of WWII. He served on the National Defense Research Committee, chaired by Vannever Bush. Conant was responsible for the chemical component, which included the development of nuclear weapons. After the war, Conant returned to Harvard and served as president until 1953. He was on the National Science Board during the formation of the National Science Foundation. After leaving the Harvard presidency, he served as High Commissioner in the US zone of Germany and then as Ambassador to the Federal Republic of Germany. After leaving that post, Conant was primarily engaged in issues of education in the public schools. In 1959, he published a book *The American High School Today*, based on visits he and a small staff made to high schools throughout the country. In 1961, he published *Slums and Suburbs*, in which he warned of the consequences of the poor quality of education in the inner cities. He also wrote *The Education of American Teachers*, in which he criticized the curriculum and teaching in education schools.

The SBR rubber is produced by *emulsion polymerization*. The process requires fatty acid emulsifiers, redox radical catalysts and thiol modifiers. The polymerization is terminated ("short-stopped") at about 63% conversion. The process was originally run at about 50°C, which was called "hot polymerization," but a later modification using redox catalysts lowered the temperature to 5°C, which was called "cold polymerization." The unreacted butadiene and styrene are stripped off and antioxidants are added. The polymer is then coagulated by brine and sulfuric acid. Water is pressed out, followed by drying. This provides "crumbs" that are baled, similarly to natural rubber. The production of some rubber also includes carbon black and mineral oil extenders. The process is summarized in Figure 5.5.

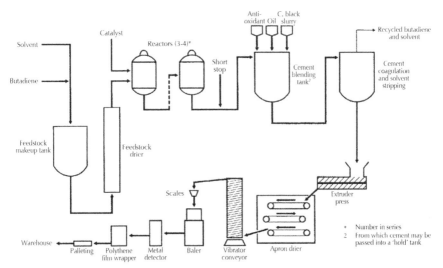

FIGURE 5.5 Schematic Design for SBR Emulsion Polymerization (From Barlow, C.; Jayasuriya, S.; Tuan, C. S. *The World Rubber Industry*; Routledge, London, ©1994; p 103. Used with permission from Taylor & Francis Books UK.)

The first synthetic rubber released to the US public during the war was of poor quality and earned a dubious reputation. By the end of the war, the quality had been greatly improved. As the civilian economy took off after the war, demand for synthetic rubber increased. The Korean war and communist activity in South East Asia renewed concerns about the availability of natural rubber and its price also spiked during this period. Synthetic rubber achieved a competitive place in the market. Plans to dispose of the government-owned plants had begun soon after WWII, but were not finalized until the early 1950s. Most of the plants were sold, often to the existing operators. Just before the end of government ownership in 1954, about 1.2 million tons of rubber were consumed in the United States, nearly equally divided between natural and synthetic. The rate of growth of synthetic rubber use was higher than for natural rubber so that 10 years later, 75% of rubber was synthetic. Exports also grew rapidly under private ownership and reached 300,000 t in 1964. Synthetic rubber plants were built in many places in the world. Between 1965 and 1980, worldwide synthetic rubber production grew to 12.5 million tons, about two-thirds of which was made in the United States, Europe, and Japan. Nearly three-fourth of this usage was for tires and other uses associated with cars and trucks.

5.3.3 HALOGENATED POLYMERS

5.3.3.1 POLYVINYL CHLORIDE

Commercial manufacture of PVC began in the United States at the B. F. Goodrich Company and Union Carbide in the 1930s. The earliest use of polyvinyl chloride was to replace rubber in insulation for electrical wiring. It was also used to make moisture-resistant sheets, such as shower curtains. PVC was also used to make artificial leather for automobiles. A harder version is used to make water and sewage pipes. When mixed with a filler, it produces durable floor tiles. PVC is often used as a co-polymer with an acrylate or methacrylate ester. These copolymers improve the processing of PVC as a melt.

The monomer vinyl chloride was first made by way of acetylene from coal but now is made from ethylene. Originally this was done via chlorination and thermal dehydrochlorination but now is done now by a one step oxychlorination process that avoids the byproduct HCl.

vinyl chloride from ethylene by chlorination

$$H_2C=CH_2 \ + \ Cl_2 \longrightarrow ClCH_2CH_2Cl \xrightarrow{\ > 500^{\circ}C\ } H_2C=CHCl \ + \ HCl$$

vinyl chloride from ethylene by oxychlorination

$$H_2C=CH_2 \ + \ HCl \ + \ 1/2 \ O_2 \longrightarrow H_2C=CHCl \ + \ H_2O$$

Most PVC is made by *suspension polymerization* and the mechanism of polymerization is the radical chain process. In suspension polymerization, polymer particles form during polymerization and are isolated by centrifugation and drying. The morphology of the particles depends on the details of the polymerization process. As with the polyolefins, PVC production and use have closely followed worldwide economic development. The industry was relatively mature in Europe, Japan, and the United States by the 1990s, but has continued to expand in China, India, and other parts of Asia.

5.3.3.2   POLYVINYLIDENE DICHLORIDE

Polyvinylidene dichloride (PVDC) is extensively used for packaging film under the name Saran, originally a trademark of the Dow Chemical Company. One of its outstanding properties is low permeability to both oxygen and water and resistance to oil. PVDC can be used as either an outer or inner layer in multi-layer films to take advantage of its barrier properties. The pure polymer is not sufficiently thermally stable for processing as a melt so it is used in copolymers, mainly with vinyl chloride, but also with acrylates, acrylonitrile, and styrene. PVDC is usually polymerized under aqueous emulsion or suspension conditions. A high degree of crystallinity is usually desired and this can be achieved by controlled heating and cooling as the polymer is processed. Because of its high chlorine content, PVDC increases the ignition resistance of copolymers.

5.3.3.3   POLYTETRAFLUOROETHYLENE (TEFLON)

Polymerization of tetrafluoroethylene was first observed by Roy Plunkett, a chemical engineer working at DuPont in 1938. The brand name *Teflon* was used by DuPont for the generic poly(tetrafluoroethylene) (PTFE). The unique properties of the polymer, including chemical and thermal stability and a very low coefficient of friction were evident, but learning to process the material was challenging. It is very difficult to bond to other materials. Eventually, ways were found to manufacture products that take advantage of its properties. Development of Teflon accelerated during WWII and led to commercial application after the war. By the 1960s, many applications such as Teflon-coated cookware had been introduced. Teflon is used in tubes and seals where high chemical inertness is essential. The "Scotch-Guard" fabrics introduced by the 3M Company are coated with PTFE that reduces adherence of stains and soil and permits easy cleaning.

Teflon is also used as a biomedical material, such as for vascular grafts and joint replacements. The biomedical applications generally involve *expanded PTFE*. This material was developed by Wilbert Gore and is known by the brand name *Gore-Tex*. Gore was a research chemist who worked on Teflon at DuPont. However, he left and with his brother Robert, started a new company to develop the potential of PTFE. He found that extrusion and stretching of PTFE could produce a strong porous form that was considerably less dense than the normal polymer but had excellent strength and permitted the diffusion of small molecules. The material was introduced as vascular tubing by Dr. Ben Eiseman, a friend of Gore's who became intrigued with its possible medical applications. Some of its properties are ideal, for example, the easy diffusion of small molecules and lack of adhesion of most biomolecules. However, it is somewhat less flexible than are natural arteries and does not have an ideal response to the pulses that characterize arterial flow.

Recently, an environmental issue has arisen in connection with use of PTFE. Perfluorooctanoic acid (PFOA) is used as a surfactant in binding Teflon to various surfaces, such as nonstick cookware. Sensitive detection methods have established that PFOA, like DDT, is "everywhere," including human blood. Evidently, like the chlorofluorocarbons (CFCs), it is indestructible. It also appears that PFOA may be carcinogenic. As a result, PTFE manufacturers have agreed to eliminate use of PFOA by 2015.

5.3.3.4 NEOPRENE RUBBER

In 1927, DuPont decided to establish a research laboratory devoted to longer range research. The plan was championed by Charles Stine and he was the first director. After consulting leading chemists to identify areas of high interest, Stine set about to find researchers in the areas of catalysts, colloids, fundamental physical data, organic synthesis, and polymerization. On the recommendation of Roger Adams and J. B. Conant, Wallace H. Carothers, then an instructor at Harvard, was brought to DuPont in 1928 as part of the staffing of the new laboratories. At the time Carothers had not worked on polymers, but in the course of considering an area that would be compatible with DuPont's long-range interest, he decided to make his major effort in polymer chemistry. As his first project, Carothers studied the nature of polymers providing definitive evidence that they were long covalently bonded "macromolecules," the concept championed by Hermann Staudinger.

Elmer K. Bolton had studied vinylacetylene and divinylacetylene earlier in the 1920s in an effort to make a synthetic rubber, but without success. Bolton suggested that Carothers resume this work. One of Carothers's coworkers began by purifying divinylacetylene but it was noted that a lower boiling fraction had formed a rubbery material. Analysis showed it to be $(C_4H_5Cl)_n$, likely a polymer of 2-chloro-1,3-butadiene. This was confirmed by further study and the material was reported in 1931 and was called *neoprene*. DuPont developed a commercial process to manufacture neoprene in the 1930s. The process involved three steps, the dimerization of acetylene to vinylacetylene, the addition of HCl to vinylacetylene and the polymerization. The polymerization eventually was done by emulsion polymerization.

$$H\text{-}C\equiv C\text{-}H \xrightarrow{\text{cat.}} HC\equiv C\text{-}CH=CH_2 \longrightarrow \underset{\underset{Cl}{|}}{H_2C=C\text{-}CH=CH_2} \longrightarrow \underset{\underset{Cl}{|}}{(-CH_2\text{-}C=CHCH_2-)_n}$$

Neoprene rubber had superior chemical resistance and durability and found uses in such materials as telephone wire insulation and automobile hoses. DuPont supplied the material to fabricators and also provided technical assistance. By 1939, the product was profitable and usage soared during WWII. Some of the lessons learned in its production were also applicable to the manufacture of synthetic rubber.

5.3.4   POLYAMIDES

5.3.4.1   NYLON

Elmer K. Bolton succeeded Charles Stine as director of the DuPont Experimental Station in 1930. Bolton was intent on finding practical applications from the Experimental Station's research, particularly under the trying economic conditions of the Great Depression. Carothers' group had made some progress by the discovery of neoprene. Carothers decided to attempt to make large molecules by applying fundamental structural concepts. His idea was to react difunctional molecules, that is, those with reactive groups at both ends of the chain. He chose first to study polyesters, which he believed could be made by reaction of difunctional alcohols with diacids. This is an example of *condensation polymerization*, a concept that had not been incorporated into Staudinger's ideas, all of which involved *addition polymerization*s. The critical technique was the use of a high-vacuum molecular still, which removed the water formed by esterification and drove the reaction toward formation of high molecular weight material. In the course of this work, the first polyesters were made. It was also found that the polyesters could be drawn into fibers having increased strength. These materials, however, had relatively low melting point and were soluble in many organic solvents and were not further developed at the time.

$$HO(CH_2)nOH + HO_2C(CH_2)nCO_2H \longrightarrow [-O(CH_2)nO\overset{O}{\overset{||}{C}}(CH_2)n\overset{O}{\overset{||}{C}}-)x + 2x\ H_2O$$

Attention was then turned to polyamides. Carothers and Donald Coffman succeeded in polymerization of 9-aminononanoic acid to what would now be called nylon-9. Soon thereafter, W. R. Petersen prepared a polyamide derived from pentane-1,5-diamine and 1,10-decanedioic acid (nylon 5-10) and confirmed the outstanding properties of polyamides. Many other amine and acid compositions were examined. In contrast to the discovery of neoprene, which had been at least partly serendipitous, the preparation of the polyamides was directly the result of Carothers fundamental understanding of the polymerization process.

$$H_2N(CH_2)_8CO_2H \longrightarrow [-HN(CH_2)_8\overset{\overset{O}{\|}}{C}-]n$$

Nylon-9

$$H_2N(CH_2)_5NH_2 + HO_2C(CH_2)_8CO_2H \longrightarrow [-HN(CH_2)_5NH\overset{\overset{O}{\|}}{C}(CH)_8\overset{\overset{O}{\|}}{C}-]n$$

Nylon5-10

Elmer Bolton made the decision to commercialize the polyamide from hexane-1,6-diamine and 1,6-hexanedioic acid (nylon 6-6). Although this was not the easiest of the nylons to process, Bolton believed that the starting materials would be the most available and economical. The major work was done by the Ammonia (intermediates), Chemical (polymerization and fiber spinning), and Rayon Departments (conversion of filaments to fabric) of DuPont. Fiber production involved forcing the molten polyamide through spinnerets. The development and promotion of nylon was first targeted at a single market, replacement of silk stockings. Cost analysis indicated that nylon could be produced at about half the cost of natural silk. Charles Stine made the public announcement of nylon in October 1938 at the New York Herald Tribune Annual Forum on Current Problems. Immediate press interest resulted in the concept of a material "stronger than steel made from coal, air and water" entering the public consciousness. Nylon was publicly displayed in early 1939 at both the Golden Gate International Exposition in San Francisco and at the New York World's Fair. Initially, nylon was only one of several synthetic materials displayed by DuPont. The exhibits focused mainly on the company's chemical and scientific progress. The displays hardly satisfied public curiosity that had been raised by the enthusiastic press reports about nylon. Later on, a model wearing nylon stockings (and other synthetics) was added to the display. By the end of the exhibitions, the "Lady of Chemistry" or "Nylon Girl" had become the highlight of the exhibit. Nylon 6-6 was introduced commercially in 1940, and nylon stockings were quickly accepted by the public, but production was soon turned exclusively to war needs, such as the manufacture of parachutes and tents.

After the war, demand for nylon exploded and the product was enormously profitable. Uses expanded to many kinds of fabrics, carpets, and industrial applications. Although the DuPont nylon patents covered many amino acid, diamine, and diacid combinations, they did not cover the conversion of caprolactam to nylon-6. In a 1930 paper, Carothers and G. J. Berchet had reported the inability to polymerize this lactam. This process was successfully developed at IGF in 1938, and in 1939, DuPont signed a cross-licensing agreement with IGF. After WWII competition quickly arose. Under pressure from the government, DuPont licensed its technology to Chemstrand Corporation in 1951, ending its virtual monopoly on nylon in the United States.

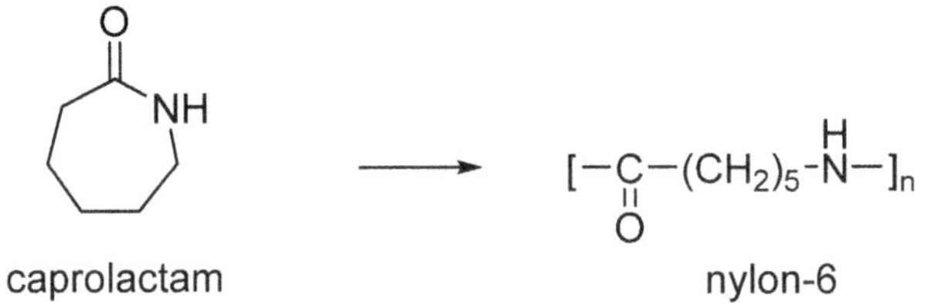

$$[-\underset{\underset{O}{\|}}{C}-(CH_2)_5-\underset{H}{N}-]_n$$

caprolactam   nylon-6

Nylon, like its contemporary penicillin, seemed to be an almost miraculous and nearly perfect product. Its improved qualities, especially the first product "nylons" (stockings), were universally welcomed. Neither its manufacture nor use had any discernable negative consequences. It was a striking example of *"Better Things for Better Living through Chemistry,"* DuPont's motto at the time.

One of the principals that guided Carothers in preparation of polyesters and polyamides was the recognition that substances that could form five-or six-membered ring were unlikely to form long-chain polymers. The natural amino acids both illustrated and contradicted this principal. They readily form six-membered dimers (diketopiperazines) *in vitro*, but form long linear polymers (proteins) in living systems. In 1935, at an international symposium on polymerization, Carothers speculated that the machinery of protein synthesis held the amino acids in such a way that they could only polymerize in a linear fashion.[11] This speculation presaged modern understanding of protein biosynthesis.

The production of nylon 6-6 has evolved during the 70 years since its original introduction. The immediate precursor is the salt formed by adipic acid and 1,6-hexanediamine. The use of the salt ensures precise 1:1 stoichiometry, which is critical to control of the molecular weight of the polymer. Both compounds were originally made from adipic acid obtained by oxidation of cyclohexane. This was done by catalytic procedures going through cyclohexanol and cyclohexanone as intermediates. The adipic acid was converted by high-temperature reaction with ammonia to adiponitrile, which was reduced to the diamine. Later, 1,3-butadiene became the starting point for the diamine. This was originally done via 1,4-dichlorobutane but can also be done by direct addition of HCN to 1,4-butadiene. A competitive process, developed by Monsanto, uses electrolytic dimerization of acrylonitrile.

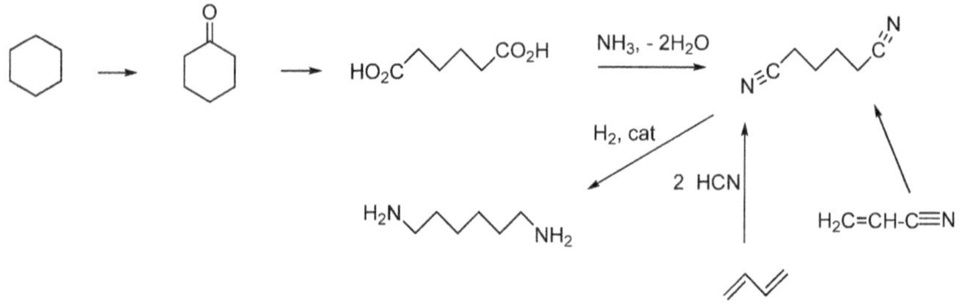

Caprolactam continues to be the key intermediate for preparation of nylon-6. It was originally made by Beckman rearrangement of cyclohexanone oxime. For a time, the oxime was made from nitrocyclohexane. The Beckman rearrangement step generates ammonium sulfate as a byproduct that was sold as a fertilizer. Alternatively, reaction of cyclohexanone with air, ammonia, and hydrogen peroxide to form the oxime, followed by solid phase catalysis can provide caprolactam with water as the only byproduct.[12] In another procedure, caprolactam was made by Baeyer–Villiger oxidation of cyclohexanone to caprolactone, which was then converted to caprolactam by ammonia. Another process goes through phenol, which is made from benzene by hydroxylation using nitrous oxide. The nitrous oxide is produced as a byproduct of the oxidation of cyclohexanone to adipic acid by nitric acid.[13]

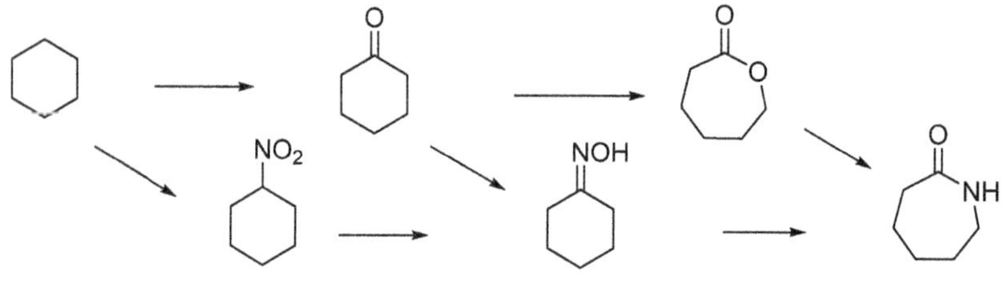

The most recent route was developed by the DSM Company in the Netherlands. It proceeds through two metal-catalyzed carbonylations, reductive amination and cyclization.[14]

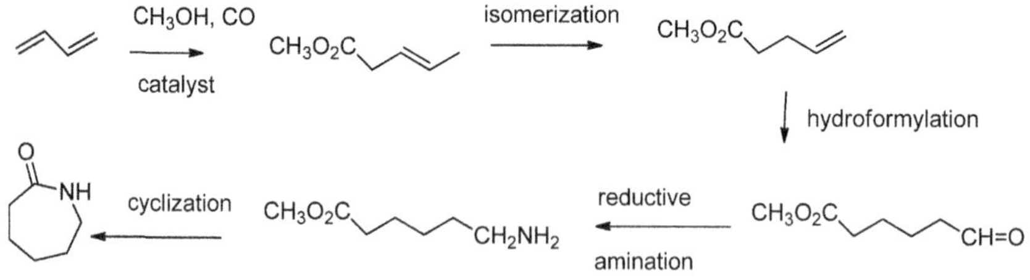

TOPIC 5.1 WALLACE CAROTHERS

Wallace Carothers grew up in Des Moines, IA, and graduated from Tarkio College in Missouri in 1920. He then enrolled at the University of Illinois, working with Carl S. Marvel and earned a Ph. D. in 1924. He served as an instructor at Illinois from 1924–1926. At the time G. N. Lewis' concept of the electron pair bond had been introduced, but Pauling's ideas based on quantum theory had not yet been published (see Section 3.1). Carothers was interested in interpreting reactions in terms of structural and mechanistic reasoning, an approach that was rare in organic chemistry at the time. He joined the faculty of Harvard in 1926. He published single papers from Illinois and Harvard. He suffered from bouts of depression and was probably manic-depressive. He also sometimes used alcohol to excess. In 1928, Carothers moved to DuPont, recruited by Charles Stine to the new fundamental research program. Carothers undertook basic research to determine the nature of polymeric molecules. By 1929, Carothers had about a dozen young Ph. D. chemists from the leading departments in the United States in his laboratory. He seldom worked in the laboratory, but had developed a fundamental understanding of the relationship between difunctional molecules and polymers that guided the research. In 1931, he published a review paper in *Chemical Reviews*, which summarized his results and elaborated the fundamental theory of polymerization. He recognized both the *addition* and *condensation* modes of polymerization. This research led to the synthesis of chloroprene rubber, the first high molecular weight polyesters, and eventually to nylon.

Carothers' personal life, however, became increasingly chaotic. He was involved in a potentially embarrassing affair with a married woman who had connections to the du Pont family. His parents, impoverished by the depression, came to live with him, but in 1934 they returned to Iowa. Later that year, Carothers sought psychiatric help in Baltimore and stayed there for a number of months. In February 1936, Carothers married Helen Sweetman, an assistant in the DuPont patent office. Carothers appears to have been deeply depressed at the time. In June, he was admitted to the psychiatric ward of the Pennsylvania Hospital in Philadelphia. Later that summer, Carothers joined Roger Adams and vacationed in Germany and Austria. At the end of the trip, Carothers remained in Europe for a time, his whereabouts unknown to anyone, including his new wife. When he returned he resumed psychiatric care, and began spending evenings at the psychiatric ward in Philadelphia. He had become fearful of strangers and concerned about his physical health. In late 1936, Carothers' sister, Isobel died at age 36, the result of a strep infection. Soon thereafter, Carothers's psychiatrist alerted one of Carothers DuPont associates, that he felt he could do nothing further to help Carothers and was convinced he would commit suicide. On April 30, 1937, three days after his 41st birthday, and a few days after he learned his wife was pregnant, Wallace Carothers died of self-administered cyanide in a Philadelphia hotel room.

Source: Hermes, M. E. *Enough for One Lifetime*; American Chemical Society: Washington, 1996.

5.3.4.2   ARAMIDES

The *aramides* are polyamides derived from aromatic diamines and diacids. The aromatic rings introduce structural rigidity that can be exploited for applications that require greater strength than can be achieved with nylons. The first aramide introduced was Nomex, which is the polymer of 1,3-diaminobenzene and 1,3-benzenedicarboxylic acid.

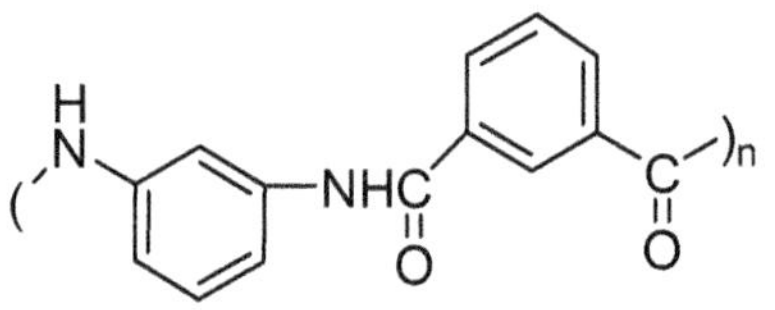

Nomex

Nomex has several applications. It can be produced as a paper by making short fibers. This paper is used to make high-strength fire-resistant honeycomb material used for interior parts of aircraft and high-speed trains. In the manufacture of the honeycomb material, several layers of Nomex paper are printed with a precise adhesive pattern, stacked, pressed, and heated. The stack is then expanded to give the honeycomb core. Finally, it is dipped with a thermosetting resin and cured. It can then be cut to the desired dimension.[15]

Kevlar is a polyamide derived from 1,4-benzenedicarboxylic acid and 1,4-benzenediamine. The structure of the polymer molecules is rod-like because the linear and hydrogen-bonded molecules have little flexibility. The polymer can be dissolved in concentrated sulfuric acid, in which it displays "liquid crystal" behavior. Large aggregates of molecules are organized, yet fluid. The spinning of fibers is done from sulfuric acid solution into a cold water quench, as shown in Figure 5.6. The fibers can be further strengthened by stretching and heat treatment, which gives rise to various grades of the polymer.

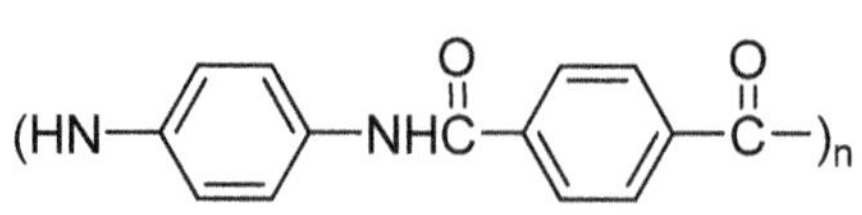

Kevlar is an extremely strong and heat-resistant polymer and is used in applications such as brake and clutch linings and for bullet-proof vests. Other uses include high strength ropes and cables, such as those used in off-shore drilling rigs. It is also used as cord in high performance tires for aircraft.[16] Kevlar composites are used for high pressure gas storage tanks. Sports equipment such as skis, tennis rackets, and kayaks are made from Kevlar composites containing carbon fibers.

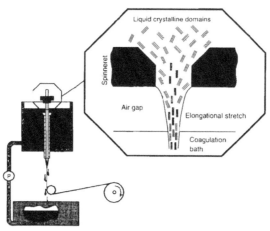

FIGURE 5.6 Process for alignment of aramide polymers, (From p 33, Rebouillat, S. *High Performance Fibres*; Knovel Textile Library, 2001; pp 23–61. By permission from Elsevier.)

An aramide developed in Japan is known as *Technora* and includes a 1:1 mixture of 1,4-diaminobenzene and 3,4′-diaminodiphenyl ether as the diamine. This leads to a more soluble polymer that can be wet spun and then drawn.

Technora

Aramides can also be prepared as foams with densities ranging from 7 to 300 mg/cm³. These foams are used in insulation in high-performance applications such as space craft, aircraft, and submarines. The have excellent thermal stability (up to 300°C) and are flame resistant. They are currently quite expensive, but have potential application for other high performance requirements, such as in piping and ducts.

5.3.5 POLYESTERS

The great commercial success of nylon encouraged research on other synthetic fibers. One goal was to achieve wool-like properties. The main effort occurred in Pioneering Research Laboratory of the Fibers Department led by W. Hale Charch,[h] the chemist who had developed moisture-proof cellophane. Charch felt that progress would be found from two areas of investigation: (1) the techniques for dyeing of hydrophobic polymers and (2) the physical basis for fiber resiliency. The former problem also applied to nylon, because its success had led to demand for dyeable versions. Some work had been done on polyesters by Carothers' group in 1934, but was not pursued at the time because the aliphatic polyesters they made had low melting points compared to the polyamides. The general opinion that polyesters would be too low melting to be successful as fibers persisted at DuPont. Serious work on polyesters was not resumed until near the end of WWII. By then, indications that ICI had developed a polyester began to reach DuPont. The key difference between the early DuPont polyesters and the ICI polymer was the use of the aromatic diacid, terephthalic acid, which gave a much higher melting polymer. ICI had acquired rights to the patent from the original inventor Calico Printers. DuPont, in 1946, upon Charch's advice, purchased US rights to the condensation polymer of ethylene glycol and terephthalic acid, which was being marketed by ICI under the name Terylene. DuPont began to vigorously develop the polyester under its own trade name Dacron. ICI licensed its products to other companies such as Hoechst, Rhone-Poulenc, and Toyo Rayon. Hoechst in turn formed a joint venture with Celanese.

Of the two starting materials for poly(ethylene glycol terephalate) (PET) the terephthalic acid was the more troublesome. Ethylene glycol was already produced in large quantities as automobile anti-freeze and could be made directly from ethylene by catalyzed oxidation. On the other hand, the most logical precursor for terephthalic acid, *para*-xylene was difficult to obtain in pure form. The material was obtained by the combination of catalytic reforming and selective adsorption by zeolites, which takes advantage of its more symmetrical structure in comparison with the *ortho*- and *meta*-isomers. The oxidation, originally done at DuPont with nitric acid, was also improved by use of catalytic processes developed at Shell and Amoco. In the early processes, it was necessary

---

[h]W(illiam) Hale Charch received a chemistry degree from Miami University of Ohio in 1920 and a Ph. D. from Ohio State in 1923. After two years at General Motors, Dr. Charch joined the DuPont laboratory in Buffalo NY. Within two years, he developed a commercially successful process for moisture-proof cellophane that greatly expanded the application of cellophane in packaging. He became associate director of the Rayon Chemical Division and pressed for expansion into fibers beyond cellulose-based material, but until the arrival of nylon, he was unable to obtain much support from the company administration. Charch was instrumental in the development of both Orlon and Dacron. Charch based the search for a wool-like material on fundamental research on the physical basis of the resiliency that characterizes wool. He was also involved in the development of spandex (Lycra).

to purify the terephthalic acid via its dimethyl ester. This step was eliminated when Amoco improved the oxidation to provide the product in sufficient purity for direct polymerization.

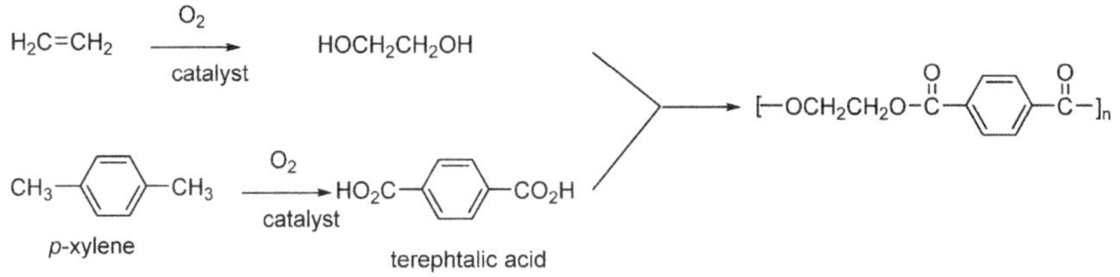

Further research found that by controlling physical manipulations, such as the extent of drawing, it was possible to modify resiliency. Like nylon, the polymer presented dyeing problems. Nevertheless, commercial development was pursued and the first production facility was opened in Kinston, NC in 1953. Product acceptance grew and by 1960, 57 million lbs were sold. For a time, DuPont's Dacron earnings exceeded those from nylon. Innovations in usage and blends came both from textile manufacturers and technical support services at DuPont. One result was cotton blends that turned out to be highly favorable for processing into textiles. The strength of the polymer permitted much faster weaving of fabric than pure cotton. Polyester blends provided the basis for wrinkle-resistant "wash and wear" fabrics. The polymer when heated undergoes "heat-setting" and thereafter tends to maintain its original shape. The success of these synthetic fibers eventually led to the displacement of the cellulose-based rayon and acetate materials. DuPont stopped the manufacture of these products in the late 1950s. Besides textiles, major uses of PET include tire cord, film (Mylar), packaging, and bottles. Polyester reinforced with glass fibers became *fiberglass*.

5.3.6 POLYURETHANES

Polyurethanes are formed by addition polymerization between polyols and diisocyanates. The earliest examples from simple diols and diisocyanates were made at the Bayer component of IGF immediately before WWII. Rapid development occurred after WWII. Various polyols and polyisocyanates were examined and toluene diisocyanate, a mixture of the 2,4- and 2,6-isomers, became the mainstay. Various polyols, particularly oligomers of ethylene oxide and propylene oxides are used. Other structures are also formed, including ureas and isocyanurate rings. The precise composition is controlled by polymerization conditions and leads to a range of physical properties and applications. For example, an increased concentration of isocyanurate rings decreases flammability.

2,4-toluene diisocyanate   2,6-toluene diisocyanate   4,4'-diphenylmethane diisocyanate   urea link   isocyanurate ring

Polyurethanes are extremely versatile and are used as engineering plastics, foams, slabs, elastomers, and coatings. They are used in automobiles, building and appliance insulation, carpets and padding, and in composites, such as fiberglass. The rigid foams incorporate more highly functionalized polyols, such as pentaerythritol and sorbitol to increase cross-linking. Consumption has grown from about 300 thousand tons in 1960 to 8 million in 2000, and represents about 5% of total polymer production worldwide.

Among the successful polymers developed at DuPont in the 1950s and 1960s, was Lycra (also called spandex and elastane), a polyurethane elastomer. Spandex is a segmented polyurethane that gives a very elastic fabric. The Spandex-type fibers achieve this property from the combination of relatively rigid segments of polyurethane with more flexible polyether chains. This type of fiber has found use in undergarments and sports-wear. The diol is usually a polymeric diol, as for example poly(tetramethylene ether) which can be obtained from tetrahydrofuran. The isocyanates used include toluene diisocyanate, naphthalene-1,5-diisocyanate and 4,4'-diphenylmethane isocyanate. If the diisocyanate is used in slight excess so that the initial polymer terminates in unreacted isocyanate groups, further chain elongation can be achieved by a reaction with a diamine. The polymerization is usually catalyzed by mixtures containing amines and tin compounds.

Spandex fibers can be made by either dry or wet spinning. The required solvents are highly polar and high-boiling materials such as N,N-dimethylformamide (DMF) and N,N-dimethylacetamide (DMAC). In dry spinning, these solvents are evaporated as the polymer solution is passed through spinnerets. Wet spinning is done into water which precipitates the polyurethane. In polyurethane foams the interstitial cells are isolated from one another and moisture can not diffuse into the spaces. The foams were originally blown with CFCs, which have ideal properties in terms of low diffusion, high insulation value, and low flammability. However, these were phased out as the result of the Montreal protocols (see Section 4.3). Currently both HCFCs and volatile hydrocarbons such as pentane are used. Some carbon dioxide is generated when isocyanate groups react with water, and this reaction can also serve as a source of gas. Polyurethane foams are used for insulation in many applications such as refrigerated railroad cars, trucks, and ships. Also under development are sealed evacuated panels that can improve insulation properties significantly and reduce the thickness of foam needed.

Polyurethanes are also used to make rigid foams. The main isocyanate in this application is diphenylmethane diisocyanate, which is a mixture of the 2,4′- and 4,4′-isomers. As with polyurethane elastomers, they are linked by oligomeric polyethers usually prepared from epoxides such as ethylene oxide or propylene oxide. The terminal functional groups are determined by the materials used to prepare these polyethers. In rigid foams, aromatic polyester polyols, such as the diester of terephthalic acid and ethylene glycol, are used because they have reduced flammability. Rigid polyurethane foams also contain more isocyanurate rings formed by the trimerization of isocyanate groups. These are formed when the diisocyanate is used in slight excess so that the polymeric molecules have a terminal isocyanate group that can react to form rings. Cross-links can also be formed by urea groups formed when an isocyanate group reacts with water. In some cases, insulating foam can be blown in place during construction. Polyurethane adhesives are made by mixing a pre-polymer having terminal isocyanate groups with polyols. The "curing" process then results in larger and cross-linked polymers. This process can also be used to make composites by including glass or natural fibers.

5.3.7   SUBSTITUTED VINYL AND ACRYLIC POLYMERS

5.3.7.1   POLY(METHYL METHACRYLATE)

Several types of polymers used during WWII became important commercial products after the war. These included poly-(methyl methacrylate), which was marketed as by DuPont as Lucite and by Rohm and Haas as Plexiglas. Poly(methyl methacrylate) is a strong clear material that resembles glass, but is resistant to breakage or shatter. During WWII, air plane cockpits windows were made from the polymer. The material is also used to produce glass-like trays, vases, etc.

5.3.7.2   POLYACRYLONITRILE

DuPont developed a fabric based on poly(acrylonitrile) called Orlon. In the early 1950s, the Rayon Department was split into the Fibers and the Films Departments and W. Hale Church led research in the Fibers Department. His goal was a resilient fiber that would compete with natural wool. Acrylonitrile, the major component of Orlon, was easy enough to polymerize, but it presented several problems. The polymer was very stable to outdoor exposure, but was difficult to dye. Several attempts to include another monomer to improve dyeability led to limited success and were abandoned. The polymer could not be spun from a melt because it decomposed on heating. It was found that solvents such as DMF and DMAC could be used for forming fibers, but they are toxic, requiring special care in process design. Polyacrylonitrile also had another problem in that is was highly flammable. Despite these problems, work toward production continued, first with a pilot plant, then 6.5 and 30 million lb/year plants. When the product was introduced as a stable but nondyeable fabric for uses such as tents and awnings, it failed. Eventually, a 6% copolymer with methyl acrylate was found to be dyeable. This fiber found acceptance in knitwear such as sweaters that became fashionable in the 1950s. A new processing technology, turbo-processing, was also instrumental in the success. By 1956, DuPont increased its Orlon capacity by construction of a new 40 million 1b/year plant at Waynesboro VA. The Chemstrand Corporation, a joint-venture of Monsanto and American Viscose, introduced a similar polyacrylonitrile fiber under the name Acrilan.

5.3.7.3   POLY(VINYL ACETATE) AND POLY(VINYL ALCOHOL)

Vinyl acetate monomer is prepared from ethylene and acetic acid by a Pd-catalyzed process. Vinyl acetate is not a very stable compound. Because it is an enol ester, it is very sensitive to hydrolysis. The polymerization is usually done by a radical chain mechanism in either emulsion or solution. A "back-biting" mechanism similar to that shown for ethylene in Section 5.3.2.2 can also occur by chain transfer to the acetoxy group. When this occurs, a hydrolyzable link is introduced in the polymer chain.

$$(CH_2CH-)_nCH_2CH\cdot \quad\quad\quad\quad (CH_2CH)_nCH_2CH_2$$
$$H_3CCO_2 \quad\quad\quad O_2CCH_3 \quad\longrightarrow\quad O_2CCH_3\ O_2CCH_3$$

$$+\quad H-CH_2CO_2(CHCH_2)_n \quad\quad\quad +\quad \cdot CH_2CO_2(CHCH_2)_n$$

$$\cdot CH_2CO_2(CHCH_2)_n \quad\longrightarrow\quad (CH_2C)_nCH_2CO_2(CHCH_2)_n$$
$$O_2CCH_3 \quad\quad\quad\quad H_3CCO_2 \quad\quad\quad O_2CCH_3$$

One of the main uses for poly(vinyl acetate) is for hydrolytic conversion to the water soluble polymer poly(vinyl alcohol), which is a surfactant and emulsifier. Vinyl alcohol is not a stable chemical entity, so it cannot be polymerized directly. Copolymerization of vinyl acetate with acrylic acid introduces anionic substitutents and the copolymer finds application in paper coating. Hydrophobic groups can be introduced by copolymerization with a long-chain α-olefin or lauryl vinyl ether. These materials give surfactants with high activity and increased viscosity. Copolymers with ethylene, vinyl chloride and methyl methacrylate are also manufactured. Poly(vinyl acetate) is also used as an adhesive. Poly(vinyl acetate) is also used in manufacture of controlled-release tablets and capsules, often in a mixture with poly(*N*-vinylpyrrolidone), which is known as *povidone.* The matrix or coating of the tablet or capsule slowly disintegrates as the polymer dissolves in aqueous media, allowing the active ingredient to diffuse out.[17]

5.3.7.4 POLYVINYL BUTYRAL

Polyvinyl butyral (PVB) is made from polyvinyl alcohol. Reaction with butyr-aldehyde introduces a cyclic acetal (1,3-dioxane ring). PVB is a strong clear adhesive and its major use is as the adhesive between layers in safety glass. The material can contain residual hydroxyl groups that can be used for cross-linking. It is also used as an adhesive with other materials including metals and ceramics. Presumably, bonds are formed with the surfaces by opening of the acetal ring.[18]

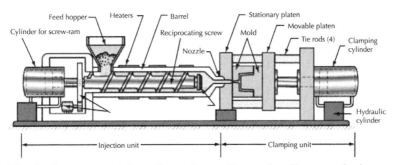

polyvinylbutyral

5.4 FABRICATION OF POLYMERS

We have already discussed formation of fibers and sheets of polymers by com-pression through spinnerets and slits (see Section 5.2.2). Similar principals are used for making continuous shapes such as pipes, as illustrated for PVC in Figure 5.7. The extrusion process is optimized through control of such variables as feed rate, temperature differential, and screw design.[19]

FIGURE 5.7 Screw extruder for molten polymer. (Source: http://www.mechscience.com/injection-moldinginjection-molding-machineinjection-molding-processinjection-molding-on-plastics/)

There are also processes that can make specific shapes for various items. One such process is *extrusion blow molding* as illustrated in Figure 5.8. In this process, the polymer is melted and then subjected to gas that expands the polymer to fit the mold. On cooling the shaped object solidifies.

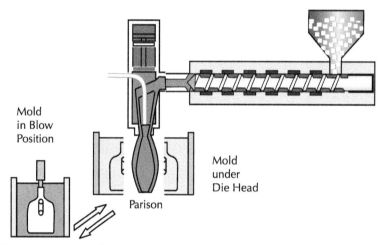

FIGURE 5.8 Blow molding apparatus. (From © Chevron Phillips Chemical Company LP. Image reproduced with permission from Chevron Phillips Chemical Company LP.)

Injection molding is similar but in this case the physical force of a die establishes the shape as shown in Fig. 5.9.

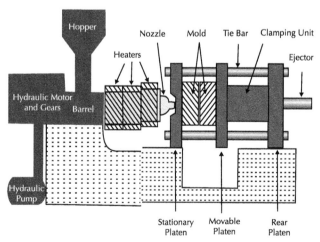

FIGURE 5.9 Injection molding apparatus. (Source: http://hicksplastics.com/Injection_ Molding_Hicks_Plastics_MI.html.)

Foams are produced by introducing gas into a fluid polymer, leaving bubbles that are trapped when the polymer solidifies. Figure 5.10 indicates the structure of a typical foam.

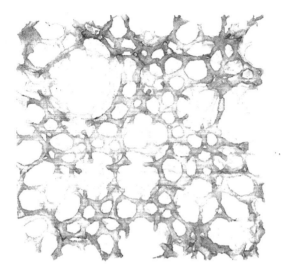

FIGURE 5.10   Electron microscopic image of syndiotactic polystyrene generated in toluene super critical CO_2. (Reprinted with permission from Fang, J.; Kiran, E. *Macromolecules* **2008,** *41*, 7525–7535. © American Chemical Society.)

5.5 FLAMMABILITY OF POLYMERS

Both natural and synthetic fibers are flammable to some degree, depending on the composition, and physical structure of the fiber. The combustion involves pyrolysis of the fiber to produce gaseous fuel and the rate of burning depends on the ease of access of oxygen to the fiber and the rate of generation of combustible gases. Several examples of polymer flammability attracted public attention in the 1970s. In particular, a number of cases of infant death and injury resulted from sleepware catching fire. Polyurethane foams used in airliners were also implicated in fires associated with airplane crashes.

The goal of flame retardants is to minimize the propagation of fire. For example, for electrical equipment, shorts should not lead to fires. Flame retardant properties are also crucial for building insulation. Flame retardants can function in several ways: (1) release of gases that dilute oxygen; (2) inhibition of the combustion process; (3) providing a protective coating to the polymer; or (4) promoting alternative decomposition to inhibit volatile fuel formation. One mechanism of flame retardation is trapping of chain-carrying radicals in the combustion process, especially H˙ and HO˙. Many flame retardants are bromine compounds and their effect can be enhanced by additives, especially antimony trioxide. Quite high bromine levels (~10% by weight) are necessary to meet typical requirements. This requires use of polyhalogenated compounds. Compounds such as decabromodiphenyl ether and *bis*-1,2-(pentabromophenyl)-ethane are used.

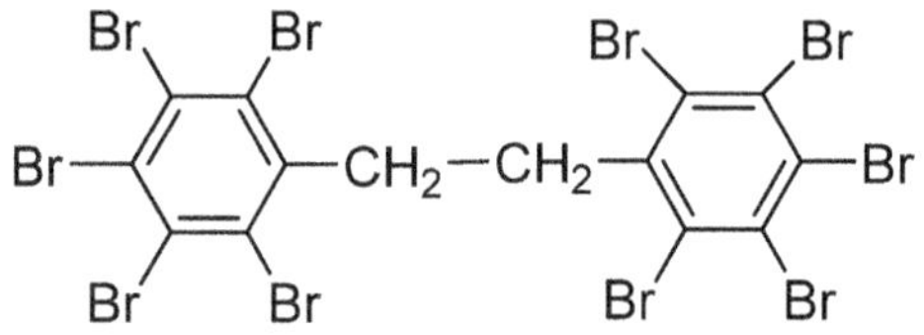

One of the first flame retardants used, *tris*-(2,3-dibromopropyl) phosphate, turned out to be a carcinogen. Currently used compounds tend to be bound covalently on the fabric. Among the compounds used are *tetrakis*-hydroxymethyl) phosphonium salts, *N*-hydroxymethyl dimethylphosphonopropanoamide and vinyl phosphonate oligomers.

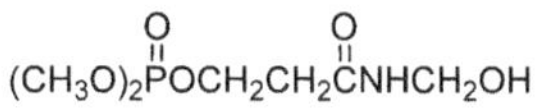

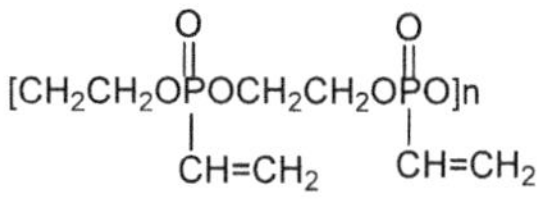

5.6   RECYCLING OF POLYMERS

The recycling of polymers is an important aspect of sustainability. About 50% of plastic is used in disposable packaging and containers. As a result, a huge waste stream is generated. Total annual usage is about 105 kg/person in the United States, 100 kg/person in Europe and 90 kg/person in Japan. Hydrocarbon polymers such as PE, PP, and PS are not easily biodegraded. There are several methods for recycling or recovering value from waste plastic. Thermoplastic materials, in particular PE, PP, and PET are to some extent reusable. If they can be recovered in sufficiently pure form they can be blended with fresh polymer for reuse. Generally speaking, it is difficult to achieve the physical properties of new polymers by use of recycled material. Thermosetting polymer can generally not be reused except as filler in fresh thermosetting resins.

   Another approach is mechanical reuse of polymers. Although waste generated industrially can have uniform composition, waste generated by households is mixed material and requires sorting for any recycling use other than combustion. Technology for sorting is a limiting factor in recycling of polymers recovered after household use. Some automated sorting can be done on the basis of color identification. Other sorting can be done on the basis of density (sink/float separation). Density can also be the basis for separation using cyclones after the polymer has been mechanically processed to provide a suitably uniform size and shape. The hydrocarbon polymers PE and PP can be separated from PVC and PET, but the latter two cannot be separated on the basis of density. Figure 5.11 shows the recycling codes and potential uses of recycled material.

THE PLASTIC IDENTIFICATION CODE

Symbol	Type of Plastic	Properties	Common Uses	Recycled into:
1 PET	PET Polyethylene Terephthalate	Clear, tough, solvent resistant, barrier to gas and moisture, softens at 80°	Soft drink and water bottles, salad domes, biscuit trays, salad dressing and containers	Pillow and sleeping bag filling, clothing, soft drink bottles, carpeting, building insulation
2 HDPE	HDPE High Density Polyethylene	Hard to semi-flexible, resistant to chemicals and moisture, waxy surface, opaque, softens at 75°C, easily coloured, processed and formed	Shopping bags, freezer bags, milk bottles, ice cream containers, juice bottles, shampoo, chemical and detergent bottles, buckets, rigid agricultural pipe, crates	Recycling bins, compost bins, buckets, detergent containers, posts, fencing, pipes, plastic timber
3 PVC	PVC Unplasticised Polyvinyl Chloride PVC-U	Strong, tough, can be clear, can be solvent welded, softens at 80°C	Cosmetic containers, electrical conduit, plumbing pipes and fittings, blister packs, wall cladding, roof sheeting, bottles	Flooring, film and sheets, cables, speed bumps, packaging, binders, mud flaps and mats, new gumboots and shoes
	Plasticised Polyvinyl Chloride PVC-P	Flexible, clear, elastic, can be solvent welded	Garden hose, shoe soles, cable sheathing, blood bags and tubing	
4 LDPE	LDPE Low density Polyethylene	Soft, flexible, waxy surface, translucent, softens at 70°C, scratches easily	Cling wrap, garbage bags, squeeze bottles, irrigation tubing, mulch film, refuse bags	Bin liners, pallet sheets
5 PP	PP Polypropylene	Hard but still flexible, waxy surface, softens at 140°C, translucent, withstands solvents, versatile	Bottles and ice cream tubs, potato chip bags, straws, microwave dishes, kettles, garden furniture, lunch boxes, packaging tape	Pegs, bins, pipes, pallet sheets, oil funnels, car battery cases, trays
PS 6 PS-E	PS Polystyrene PS-E Expanded polystyrene	Clear, glassy, rigid, opaque, semi-tough, softens at 95°C. Affected by fat, acids and solvents, but resistant to alkalis, salt solutions. Low water absorption, when not pigmented is clear, is odour and taste free. Special types of PS are available for special applications.	CD cases, plastic cutlery, imitation glassware, low cost brittle toys, video cases\ Foamed polystyrene cups, takeaway clamshells, foamed meat trays, protective packaging and building and food insulation	Coat hangers, coasters, white ware components, stationery trays and accessories, picture frames, seed trays, building products
7 OTHER PACKAGING	OTHER PACKAGING In packaging, it could be multi-layer materials e.g. PE+PP.	Includes all resins and multi-materials (e.g. laminates). Properties dependent on plastic or combination of plastics.	Automotive and appliance components, computers, electronics, cooler bottles, packaging	Plastic timber, sleepers – looks like wood, used for beach walkways, benches etc.

FIGURE 5.11   Plastic Identification Codes and Recycling.

If recovered polymer cannot be recycled the energy content of polymers can be reclaimed by burning, but this has two undesirable consequences. The combustion product is mainly CO_2 and contributes to the load of greenhouse gases. Also, the polymers contain many additional components such as plasticizers, colorants, adhesives, and fire-retardants and it is necessary to ensure capture of any noxious combustion products. Another possibility is to use polymers as feed stock in gasification process, similar to those used for coal or biomass (see Section 2.3).

5.6.1   POLYOLEFINS

Hydrocarbon polymers such as PE and PP lend themselves to recycling. They are amenable to thermal or steam "cracking" as practiced in petroleum refining. In the early 1990s the DuPont company launched a major effort to use recycled PE in TYVEK. In particular, US EPA regulations pertaining to the postal service spurred inclusion of recycled material for packets and boxes. TYVEK is an advantageous material for express mail packaging because it provides a high level of strength with low weight. The effort to use recycled material faced several challenges: (1) the technical and production challenges of introducing recycled material into a process that used pure PE; (2) locating a reliable source of high-quality segregated PE to recycle; (3) purifying the recycled PE, which contains contaminants such as paper labels, adhesives and grit from collection, and transport; (4) the added production requirements increased the cost of recycled material above the new product. These problems were eventually solved for one segment of the market that for envelopes used for express mail and package services. Three other major segments of the market have been slower to accept recycled materials. These are protective clothing, house wrap, and medical packaging.[20]

5.6.2   POLYESTERS

Polyesters such as PET can be hydrolyzed to the monomers. PET recovered from food/drink containers can be recycled as materials such as fiber for carpets and sheets for heavy-duty packaging.

5.6.3   POLYVINYL CHLORIDE

Current PVC production is estimated at 3.5 million tons/year. About 80% of this is for long-life usage, such as for pipes and construction materials, and does not constitute an immediate recycling problem. PVC presents a particular problem to efforts to recover energy by combustion in that large amounts of HCl are formed and must be removed from the effluent gas. In the mid-1990s, Greenpeace launched a campaign in the UK targeting PVC packaging that included picketing of specific retail chains. This activity set in motion a study sponsored by several of the retailers that concluded that under best practice conditions PVC posed no specific environmental hazard. Subsequently, a group called the PVC Coordinating Group was founded with the support of the two major PVC manufacturers to develop a code of practice that would ensure safe practices were followed. As part of the same undertaking the industry agreed to a study that would identify the conditions for use of PVC under the requirements of "full sustainability." This study identified five basic requirements for "full sustainability" and the challenges that would have to be overcome to meet them. These challenges are (1) carbon neutral manufacturing, including the energy input required; (2) "closed-loop" management of used material including recycling; (3) elimination of net life-cycle emission of persistent or hazardous material; (4) control of additives to eliminate introduction of persistent or hazardous material; (5) improve public awareness to meet these goals. Currently, these goals are formidable but the industry has committed itself to progress in such areas as energy usage reduction, use of renewable energy sources, recycling, and elimination of potential toxic additives such as plasticizers and stabilizers.

KEYWORDS

- polymers
- plastic
- copolymers
- rubber
- addition polymerization
- coordination polymerization
- condensation polymerization
- polyamides
- polyesters
- polyurethanes
- vinyl polymers

BIBLIOGRAPHY

Meikel, J. L. *American Plastic*; Rutgers University Press: New Brunswick, NJ, 1995.
Furukawa, V. *Inventing Polymer Science;* University of Pennsylvania Press: Philadelphia, 1998.
Strom, E. T.; Rasmussen, S. C. *ACS Symp. Ser.* **2011**, *1050*, 1–92.

BIBLIOGRAPHY BY SECTIONS

Section 5.2.1: Barrow, C.; Jayasuriya, S.; Tan, C. S. *The World Rubber Industry*; Routledge, London, 1994; Baker, C. S. L. *Prog. Rubber Plast. Technol.* **1997**, *13*, 203–229; Morawetz, H. *Rubber Chem. Technol.* **2000**, *73*, 405–426; Loadman, J. *Tears of the Tree;* Oxford University Press: Oxford, UK, 2005.
Section 5.2.2.2: Summers, T. A.; Collier, B. J.; Collier, J. R.; Haynes, J. L. *Manmade Fibers: Their Origin and Development*; Elsevier Applied Science: New York, 1993; pp 72–90.
Section 5.2.2.3: Urquhart, A. R. *J. Textile Inst.* **1951**, *42*, 385–394; Treece, L. C.; Johnson, G. I. *Chem. Ind. (Boca Raton, FL)* **1993**, *49*, 241–256.
Section 5.2.2.4: Rosser, C. M. *Colloid Chem.* **1950**, *7*, 641–649; Haskell, V. *Encycl. Polym. Sci. Technol.* **1964**, *3*, 60–79.
Section 5.3.1.2: Dunky, M. *Int. J. Adhes. Adhes.* **1998**, *18*, 95–107; Diem, H.; Matthias, G. *Industrial Polymers Handbook*; Wiley-VCH: New York, 2001; pp 1053–1097; Pizzi, A. *Handbook of Adhesive Technology*; Marcel Dekker: New York, 2003; pp 635–652; Pizzi, A. *Handbook of Adhesive Technology*; Marcel Dekker: New York, 2003; pp 653–680.
Section 5.3.1.3: Bilyear, B.; Brostow, W.; Menard, K. P. *J. Mat. Educ.* **1999**, *21*, 281–286; Muskope, J. W.; McCollister, S. B. *Industrial Polymers Handbook*; Vol. 2, Wiley-VCH, 2001; pp 1099–1127; Goulding, T. M. *Handbook of Adhesive Technology*; 2nd ed., Marcel Dekker: New York, 2003; pp 823–838.
Section 5.3.2.1: Martin, M. F.; Viola, J. P.; Wuensch, J. R. In *Modern Styrenic Polymers*; Scheir, J.; Priddy, D., Eds.; Wiley: Chichester, UK, 2003; pp 247–280; Bohning, J. J. *Chem. Heritage* **2004**, *22*, 8–9, 40–45.
Section 5.3.2.3: McMillan, F. M. *ACS Symp. Ser.* **1985**, *285*, 333–361; Zucchini, U. *Makromol. Chem.* **1993**, *66*, 25–41; Corradini, P. *Macromol. Symp.* **1995**, *89*, 1–11; Okuda, J.; Muelhaupt, R. *Materials Sci. Technol.* **1999**, *20*; In *Synthesis of Polymers*; Dieter Schlueter, A., Ed.; Wiley-VCH, pp 123–162; Galli, P.; Veccellio, G. *Organometallic Catalysts and Olefin Polymerization, Catalysts for a New Millenium*, Springer: Berlin, 2001; pp 169–195; Demirors, M. *100 + Years of Plastics*; Strom, E. T.; Rasmussen, S. C., Eds.; *ACS Symp. Ser.* **2011**, *1080*, 115–145.
Section 5.2.3.4: Herbert, V.; Bisio, A. *Synthetic Rubber: A Project that Had to Succeed*; Greenwood Press: Westport, CT, 1985.
Section 5.3.3.1: Saeki, Y.; Emura, T. *Prog. Polym. Sci.* **2002**, *27*, 2055–2131.
Section 5.3.3.2: Wessling, R. A.; Gibbs, D. S.; Obi, B. E.; Beyer, D. E.; Delassus, P. J.; Howell, B. A. *Kirk–Othmer Encyclopedia of Chemical Technology*; 5th ed., 2007, Vol. 25; pp 691–745.
Section 5.3.3.3: Chandler-Temple, A. R.; Grondahl, L.; Wentrup-Byrne, E. *Chem. Aust.* **2008**, *75*, 3–6; Baquey, C.; Durrieu, M.-C.; Guidoin, R. G. *Fluorine and Health*; Tressaud, A.; Haufe, G. Eds.; Elsevier, 2008; pp 379–406; Yao, J. S. T.; Eskandari, M. K. *Surgery* **2012**, *151*, 126–128.

Section 5.3.3.4: Furukawa, Y. *Inventing Polymer Science*; University of Pennsylvania Press; Philadelphia, PA, 1998.

Section 5.3.4.1: Miller, S. A. *Chem. Process Eng.* **1960**, *50*, 63–72.

Section 5.3.4.2: Rebouillat, S. *High Performance Fibres*; CRC Press: Boca Raton, FL, 2000; pp 23–61.

Section 5.3.5: Hounshell, D. A. *Milestones in 150 Years of the Chemical Industry*; Morris, P. T. J.; Campbell, W. A.; Roberts, H. L., Eds.; Special Publication 96, The Royal Society of Chemistry, 1991.

Section 5.3.6: Frisch, K. C. *60 years of Polyurethanes*; Kresta, J. E.; Eldred, E. W., Eds.; Technomic Publishing Company: Lancaster, PA, 1998; pp 1–20; Clausius, R. *60 years of Polyurethanes;* Kresta, J. E.; Eldred, E. W., Eds.; Technomic Publishing Company: Lancaster, PA; pp 22–34; Eaves, D. *Handbook of Polymer Foams*; Eaves, D., Ed.; Rapra Technology Limited: Shrewsbury, UK, 2004; pp. 55–84; Ulrich, H. *Kirk–Othmer Encyclopedia Chem. Technol.* **2007**, *25*, 454–485.

Section 5.3.7.4: Blomstrom, T. P. *Coatings Technology Handbook*; 3rd ed., Vol. 60, 2006, pp 1–11.

Section 5.5: Price, D.; Horrocks , A. R.; Tunc, M. *Chem. Br.* **1987**, *23*, 235–240.

Section 5.6: Leadbetter, J. *Prog. Polym. Sci.* **2002**, *27*, 2197–2226; Hopewell, J.; Dvorak, R.; Kosior, E. *Philosoph. Trans. Roy. Soc., Sect. B* **2009**, *364*, 2115–2126.

REFERENCES

1. Meyer, K. H.; v. Susich, G.; Valko, E. *Kolloid-Z.* **1932,** *59*, 208–216; Busse, W. F. *J. Phys. Chem.* **1932,** *36*, 2862–2879.

2. Barlow, C.; Jayasuriya, S.; Tan, C. S. *The World Rubber Industry*; Routledge: London, 1994; pp 8–9.

3. Vaupel, E. *Angew. Chem., Int. Ed.* **2005**, *44*, 3344–3355.

4. Klodt, R.-D.; Gougon, B. *Modern Styrenic Polymers: Polystyrene and Styrenic Copolymers;* Scheirs, J.; Priddy, D. B. Eds., 2003; pp 165–201.

5. Amos, J. L. *Polym. Eng. Sci.* **1974**, *14*, 1–11; Greeley, T. R. *ASTM Spec. Publ.,* **1997,** *1320*, 224–239; Barnetson, A. *Handbook of Polymer Foams;* Rapra Technology, 2004; pp 37–54.

6. Gibson, R. O. *Chem. Ind.* **1980**, 635–641.

7. Scheirs, J.; Boehm, L. L.; Boot, J. C.; Levers, P. S. *Trends Polym. Sci.* **1996**, *4*, 408–415.

8. Hogan, J. P.; Banks, R. L. *History of Polyolefins*; Seymour, R. B.; Cheng, T., Eds.; D. Reidel Publishing: Dordrecht, Netherlands, 1986; pp 103–115.

9. Ewen, J. A. *Sci. Am.* **1997,** *May*, 86–91.

10. Thomas, R. M.; Sparks, W. J.; Frolich, P. K.; Otto, M.; Mueller-Cunradi, M. *J. Am. Chem. Soc.* **1940**, *62*, 276.

11. Carothers, W. H. *Trans. Faraday Soc.* **1936**, *32*, 39–49.

12. Hoelerrich, W. E. *Catal. Today* **2000**, *62*, 115–130.

13. Uriarte, A. K.; Rodkin, M. A.; Gross, M. J.; Kharitonov, A. S.; Panov, G. I. *Stud. Sci. Catal.* **1997**, *110*, 857–864.

14. Kiewiet, B.; Koop, K.; van den Berg, H.; van der Ham, L.; Asselbergs, K. *NPT Processtechnol.* **2003**, *10*, 16–18; Sielcken, O.; Haassen, N. F.; Guit, R. P. M.; Tinge, J. T. European Patent, 1251122A1, 2002.

15. Pinzelli, R.; Blomert, D. In *Proc. Intern. Conf. Compos. Sci. Tech.;* Adali, S.; Morozov, E. V.; Verijenko, V. E., Eds.; National Technical Information Service: Springfield, VA, 1998; pp 75–82; Danver, D. *Int. SAMPE Symp. Exhib.* **1997,** *42* (Book 2), 1531–1542.
16. Tanner, D.; Fitzgerald, J. A.; Phillips, B. R. *Prog. Rubber Plast. Technol.* **1989,** *5,* 229–251.
17. Narayan, B. H.; Hall, K. *Pharm. Technol.* **2003,** 34–37.
18. Blomstrom, T. P. *Coatings Technology Handbook,* 3rd ed., 2006; Vol. 60, pp 1–11.
19. Jenkins, S. R.; Powers, J. R.; Hyun, K. S.; Naumovitz, J. A. *J. Plast. Film Sheeting* **1990,** *6,* 90–105.
20. Sharfman, M.; Ellington, R. T.; Meo, M. *J. Ind. Ecol.* **2001,** *5,* 127–145.

CHAPTER 6

HOUSEHOLD AND PERSONAL CARE PRODUCTS: CLEANING UP AND LOOKING GOOD

ABSTRACT

Most household and personal care products contain one or more surfactants. Surfactants are characterized by ionic or polar "head groups" and hydrophobic "tails." They serve to stabilize high surface area systems such as emulsions and foams by reducing surface tension. They can be classified as anionic, cationic, neutral, amphoteric, or zwitterionic based on charge. The classical example of surfactants is soap prepared by alkaline hydrolysis (saponification) of animal or plant fats and oils (oleochemicals). Beginning early in the 20th century, synthetic surfactants, often called detergents, were introduced. These can incorporate a variety of polar or charged head groups and the hydrophobic portions can be obtained from oleochemicals or petrochemicals. Formulation of specific products depends on the anticipated use. Laundry and dishwashing detergents usually function at elevated temperature and may contain materials for stain removal, including enzymes. Interaction with Mg^{2+} and Ca^{2+} ions present in "hard" water is detrimental and "builders" are used to complex these ions. Personal care products may use milder detergents and often include moisturizers (humectants) and oils intended to restore the water and fatty materials removed by the detergents. Shampoos are used with "conditioners" that provide a surface coating to hair to achieve smoothness and sheen. Toothpastes include abrasives, humectants, flavors, and colors. Personal care products may also include material with active biological functions, such as exfoliation, hair growth control, or antibacterial activity. Preservatives may also be present. While many of the materials used in household and personal care products originate in the chemical industry, the final products are usually distributed by companies specializing in consumer products. Advertising plays a key role in this segment of the economy.

The practices and products used for cleaning our clothes, dishes, cookware, and bodies changed dramatically during the 20th century. At the beginning of the century, a Saturday night bath using water heated on the stove and soap was the norm for bodily cleanliness. Clothes were washed by hand in a wash-tub using soap and a scrub-board. Dishes, pots, and pans, too, were washed by hand. During the first third of the century machines for washing clothes and dishes arrived. Personal care products expanded from soap to include many more specialized and/or versatile products. Convenience of use became a major factor in product preference. By the end of the century, the aisles of drug, grocery, and other retail outlets offered consumers thousands of products to help them keep clean and look good. We'll talk about cleaning products for clothes, dishes, body, and hair. These are the high-volume household and personal care products. Also considered are skin and oral care products, as well as cosmetics. For the most part, we will be talking about exterior appearance in this chapter. We'll also look a bit into the materials used in medical treatments for cosmetic purposes.

A component of nearly all cleansers and personal care products is a *surfactant*. Surfactant is a word created by contraction of the phrase "surface active agent." In Europe an equivalent word *tenside*, derived from *tension-active*, is often used. The original surfactants were soaps. We will start by discussing what surfactants are and how they work. We will learn that there are now many synthetic surfactants, often grouped together as *detergents*. The usage of various soaps, detergents and other surfactants is in the millions of tons per year. Worldwide, about 20 million tons are produced annually. The United States market for surfactants was valued at about $60 billion in 2000. There are four major international producers, Procter & Gamble, Unilever, Henkel, and Colgate-Palmolive. In the developed countries, the industry is highly sensitive to consumer preferences and the success of advertising in forming them.

6.1   SURFACTANT TYPES

Surfactants are a component of most consumer products that involve cleaning, such as dishwashing and laundry detergents, shampoos and body soaps. Surfactant molecules are called *amphipathic*, meaning they can interact with both water (hydrophilic, polar) and nonpolar (hydrophobic, lipophilic) phases. The hydrophilic portions are charged or polar and interact favorably with water and are often called "head groups." The hydrophobic portions are usually hydrocarbon chains of 8–20 carbons in length and are called the "tail." All surfactants have a common mechanism of action. They reduce surface tension at phase boundaries, and thus permit formation of high surface area systems such as emulsions and foams. Surfactants can be classified as *anionic, cationic, amphoteric, zwitterionic*, or *neutral*, depending on their charge characteristics. Amphoteric compounds have both acidic and basic groups and their charge depends on the pH of the medium. Zwitterionic surfactants are neutral but have separate anionic and cationic sites. Neutral surfactants have no charge but have polar substituent groups. One example is short polyethylene glycol (PEG) chains. Other types of compounds that have a long chain and a polar functional group, including alcohols, amides, and amine oxides exhibit surfactant properties.

Surfactants can acquire supramolecular structure. *Micelles* have more or less spherical shapes with the polar groups on the exterior and the nonpolar chains in the internal space. In aqueous environments, surfactants form micelles at the *critical micelles concentration* and above. A balance of size between the polar and nonpolar groups is necessary to optimize micelles formation. Small micelles tend to adopt spherical shapes to compensate between charge repulsion on the surface and hydrophobic attraction in the interior. There are also *inverted micelles* in which the polar groups are facing inward with an encapsulated pocket of water. One material that tends to form such inverted micelles in sodium *bis*-(2-ethylhexyl)sulfosuccinate.

sodium *bis*-(2-ethylhexyl)sulfosuccinate

The micelles can also aggregate into larger ordered assemblies. They can form cylinders and worm-like structures. Surfactants can also form *lamellar sheets* that consist of stacks of bilayers of polar surfaces separated by the nonpolar chains, as shown in Figure 6.1. Liquid crystal phases, which consist of fluid but organized structures, are formed by many surfactants.

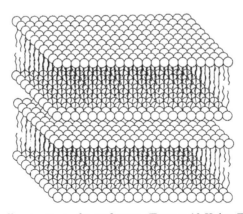

FIGURE 6.1 Lamellar structure of a surfactant, (From p 10 Kaler, E. W. *Detergents and Cleaners: A Handbook for Formulators*; Lange, K. R., Ed., 1994; pp 1–42; Used with permission from Hanser Publications.)

The concentration at which the various shapes form and are stable is determined by surfactant structure. The nature of the micelles and assemblies that are present affect the physical properties (e.g., viscosity) and appearance (e.g., opalescence) of the mixture. Maintenance of the correct micelle structure is an important factor in the formulation and manufacturing of cleansers and shampoos. The detailed behavior of any particular surfactant can be summarized by a phase diagram, showing regions of solubility, simple micelles, and more complex vesicles, such as in Figure 6.2. The *Kraft point* is the intersection between the micelle, soluble, and insoluble phases.

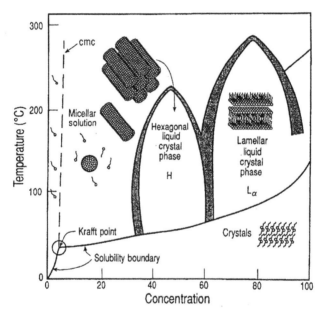

FIGURE 6.2 Schematic phase diagram showing micelles and micellar assemblies (From Kaler, E. W. *Detergents and Cleaners: A Handbook for Formulators*; Lange, K. R., Ed., 1994; pp 1–42; Used with permission from Hanser Publications.)

Surfactants offer an interesting counterpoint between fossil and renewable sources because either can serve as the starting material for the hydrophobic hydrocarbon portions of the molecules. The long-chain constituents of soaps and detergents can be obtained from either *petrochemicals* or *oleochemicals*. The latter are derived from long-chain fatty acids found in substances such as palm oil, coconut oil, soybean oil, and animal tallow. Extensive cultivation of palm and coconut trees for use in various surfactants occurs in Southeast Asia. In the recent past, there has been considerable shift from petroleum-based to plant-based starting materials. Several reasons for this can be cited: (1) increased prices for petrochemicals; (2) growing supply of vegetable oils; (3) consumer preferences for plant-based materials; (4) pressure for efficient energy and resource utilization; and (5) global economic trends. We will begin by discussing the molecular structure, source, and properties of some of the most common surfactants. These appear in most of the major product categories that we will discuss, along with materials that serve other purposes.

6.1.1 SOAPS

Soaps are sodium, potassium, or ammonium salts of long-chain carboxylic acids. Soaps have been used for at least 4500 years, with the earliest examples prepared from vegetable oil or animal fat and wood ash (potash), essentially the same process that is used today. Soap-making was practiced as a craft in medieval Europe. Soap-making was one of the first industrial-scale chemical processes and began in Europe in the early 1800s based on the LeBlanc soda process (see Section 3.1). The fatty acids in soaps can be obtained either from animal fats or vegetable oils. The former are primarily C_{16} (palmitic) and C_{18} (stearic) compounds, while the latter contain mainly C_{12} (lauric) and C_{14} (myristic) acids. Saturated acids are used for soaps so if the starting material contains unsaturation, a hydrogenation step is included. The natural fats and oils are esters of glycerin (triglycerides) and the acids are obtained either by high temperature or alkaline hydrolysis and then isolated as the sodium salts. Alternatively, more soluble materials can be obtained using potassium or ethanolamine salts. The triglycerides can also be converted to fatty acid methyl esters and glycerin by ester exchange with methanol. This process is used when the fatty acids are to be converted to long-chain alcohols. Table 6.1 gives the names of the most common fatty acids. There are several ways of locating the position of unsaturation (see Section 9.3.1). In the table, the symbol Δ, with a superscript number indicates the position of the double bond. The corresponding alcohols use the beginning of the acid name, followed by the suffix "-yl" and the word "alcohol."

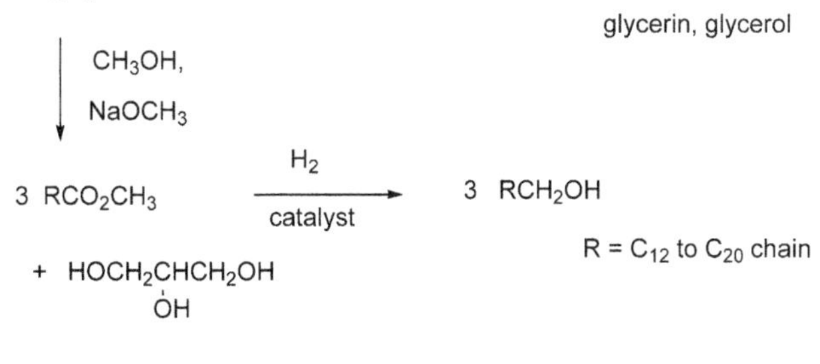

TABLE 6.1 Names of Common Fatty Acids and Alcohols.

	Saturated acids	Unsaturated acids	Alcohol
C_{12}	Lauric	Lauroleic	Lauryl
C_{14}	Myristic	Myristoleic	Myristyl
C_{16}	Palmitic	Palmitoleic, $\Delta^{9,10}$	Palmityl or cetyl
C_{18}	Stearic	Oleic, $\Delta^{9,10}$	Stearyl
C_{20}	Arachidic	Gadoleic, $\Delta^{9,10}$	Arachadyl

Soaps tend to form precipitates with cations such as Mg^{2+} and Ca^{2+} and therefore are not effective in "hard" water, which contains these ions. Soaps give alkaline solutions because they are salts of weak acids and cause partially hydrolysis, increasing pH. This property makes soaps irritating to eyes, mucous membranes and, on extended exposure, the skin.

$$RCO_2^-M^+ + H_2O \rightleftharpoons RCO_2H + M^+ + {}^-OH$$

6.1.2 ANIONIC DETERGENTS

The desired function of detergents is the same as for soaps, but they have improved properties for some applications. Most of the anionic detergents contain either *sulfate* or *sulfonate* groups. Being salts of strong acids, they have no tendency to increase pH. Also, the Ca^{2+} and Mg^{2+} salts of sulfates and sulfonates are more soluble, making these detergents less susceptible to precipitation in hard water. Work on synthetic detergents began in Germany during WWI, partly because of a shortage of fats for soap-making. Commercial products called *Gardinol* and *Igebon* were introduced in the early 1930s and quickly migrated to the United States, where the initial use was in the textile industry. *Gardinol* is a sulfate ester of long-chain alcohols obtained by hydrogenation of fat. *Igepon* T is the oleoyl amide of taurine.[1] These were originally manufactured in the United States by subsidiaries of German companies, but became the property of the government at the outbreak of WWII and were eventually sold to commercial firms.

$CH_3(CH_2)_{14}CH_2OSO_3Na$

Gardinol

$CH_3(CH_2)_7CH=CH(CH_2)_7\overset{\displaystyle O}{\overset{\|}{C}}NCH_2CH_2SO_3Na$

Igepon T CH_3

6.1.2.1 ALCOHOL SULFATES

The first semi-synthetic detergents to be sold as consumer products in the United States were alcohol sulfates, which were introduced in the 1930s. Long-chain alcohols are converted to monoalkyl sulfate esters by reaction with $H_2S_2O_7$ or SO_3. The latter is preferred and can be carried out as a continuous gas–liquid phase process. The mono esters are then neutralized by NaOH and isolated as the sodium salts. Because of their origin in straight chain fatty acids they are easily biodegradable. Early examples of commercial products were Dreft (1933) and the original version of Tide (1947).

$$R(CH_2)nCH_2OH \xrightarrow{SO_3} R(CH_2)nCH_2O-SO_3H \xrightarrow{NaOH} R(CH_2)nCH_2O-SO_3^-Na^+$$

In addition to hydrogenation of esters from fats or oils, there are three major synthetic routes to the long-chain alcohols. One is based on the Ziegler–Natta polymerization reaction (see Section 5.3.2.3). The starting material is ethylene and the product is entirely straight chain and 100% terminal (primary alcohol). The disadvantage is that the process produces a relatively wide molecular weight distribution so that other uses must be found for products outside the detergent range.

$$n\ CH_2=CH_2 \xrightarrow[\text{catalyst}]{(C_2H_5)_3Al} [CH_3CH_2(CH_2CH_2)n]_3Al \xrightarrow{O_2} 3\ CH_3CH_2(CH_2CH_2)nOH$$

The same type of alcohols can be produced using Fischer–Tropsch technology from synthesis gas (see Section 2.2.1.3). In the early 1960s, Shell Chemicals developed a process for oligomerization of ethylene combined with hydroformylation to prepare straight chain alcohols based on nickel and cobalt catalysts. These materials became competitive with the alcohols prepared from fatty acids. This group of compounds is called the "Neodols."[2] Depending on the catalyst used, some of the C-2 secondary alcohol may be produced.

$$CH_2=CH_2 \xrightarrow{\text{catalyst}} CH_3CH_2(CH_2CH_2)_nCH=CH_2 \xrightarrow[\text{catalyst}]{CO,\ H_2} CH_3CH_2(CH_2CH_2)_{n+1}CH_2OH$$
$$+\ CH_3CH_2(CH_2CH_2)_{n+1}\underset{OH}{CHCH_3}$$

Recent trends toward use of less detergent, less water, lower temperature, and faster cycles in washing machines have led to the development of another modification of the structure of alkyl sulfates. It has been found that introduction of a single methyl substituent along the linear chain improves low-temperature solubility. These materials are called "highly soluble alkyl sulfates." They are produced by a modification of the Shell process for linear alcohols.

6.1.2.2   ALKYL ETHOXY SULFATES

The alkyl ethoxy sulfates are made by adding a PEG chain to the end of a long-chain alcohol, followed by formation of the sulfate ester as for alkyl sulfates. The number of ethylene glycol units can range from 5 into the 20s. The products are a distribution over some average number, depending on the details of production. The suffix "eth" on a fatty acid name indicates the presence of a PEG chain (the eth coming from ether). A number associated with these products such as -9, means the average number of ethylene glycol units is nine. Thus the designation laureth-9 refers to a lauryl ($R = C_{12}$) ether of PEG containing an average of nine PEG units. Properties such as viscosity and solubility are affected by both the length of the alkyl chain and the number of ethylene glycol units that are added.

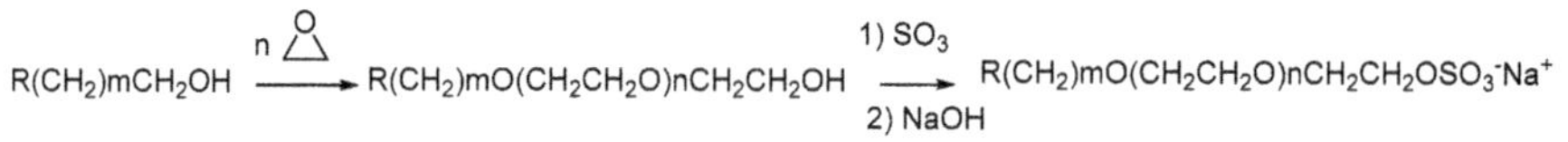

$$R(CH_2)mCH_2OH \xrightarrow{\quad n \triangle \quad} R(CH_2)mO(CH_2CH_2O)nCH_2CH_2OH \xrightarrow[\text{2) NaOH}]{\text{1) }SO_3} R(CH_2)mO(CH_2CH_2O)nCH_2CH_2OSO_3^-Na^+$$

6.1.2.3   ALKYLBENZENE SULFONATES AND LINEAR ALKYLBENZENE SULFONATES

The *alkylbenzene sulfonates* (ABS) detergents were introduced immediately after WWII, but were removed from the market in 1961. They were prepared from propylene oligomers and contained highly branched alkyl groups attached to a benzene ring. They were poorly biodegradable and caused severe foaming problems in water treatment plants and even in streams and rivers.[3] They were replaced by *linear alkylbenzene sulfonates* (LAS). The linear reactants are prepared by monodehydrogenation of straight-chain hydrocarbons from petroleum fractions. The alkenes are then used in Friedel–Crafts alkylation of benzene (see Section 2.1.5). This was originally done using reagents such as $AlCl_3$ or HF, but more recently zeolite catalysts have been used. Besides avoiding use of corrosive materials, the zeolite catalysts have the advantage of tending to attach the benzene ring near one terminus of the alkene. One type is the "Detal" catalysts developed by UOP.[4] This process produces a mixture of position isomers, but minimizes chain-branching. The 2-phenyl isomers are the most desirable and constitute 20–30% of the product mixture, depending on the process. The selectivity is the result of pore size and shape effects of the zeolite catalyst.

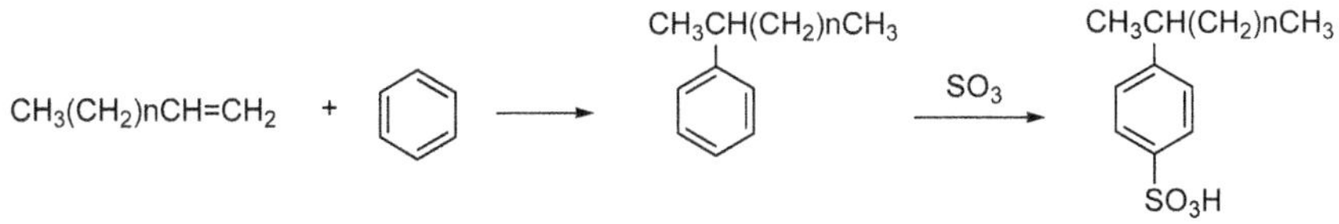

$$CH_3(CH_2)nCH=CH_2 \quad + \quad \bigcirc \quad \longrightarrow \quad \overset{CH_3CH(CH_2)nCH_3}{\bigcirc} \quad \xrightarrow{SO_3} \quad \overset{CH_3CH(CH_2)nCH_3}{\underset{SO_3H}{\bigcirc}}$$

6.1.2.4 α-OLEFIN SULFONATES

α-Olefin sulfonates are produced from linear terminal alkenes in the C_{11} to C_{14} range. Reaction with SO_3 gives sultones, which then form a mixture of alkene sulfonates and hydroxy sulfonates. These are neutralized to provide the sodium salts.

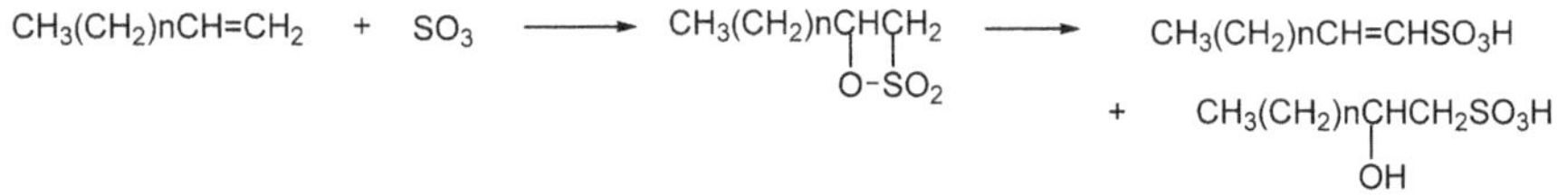

$$CH_3(CH_2)nCH=CH_2 \quad + \quad SO_3 \quad \longrightarrow \quad CH_3(CH_2)n\underset{O-SO_2}{CHCH_2} \quad \longrightarrow \quad CH_3(CH_2)nCH=CHSO_3H$$
$$+ \quad CH_3(CH_2)n\underset{OH}{CHCH_2SO_3H}$$

6.1.2.5 ACYL ISETHIONATES

Isethionates are esters of 2-hydroxyethanesulfonic acid, which is known as isethionic acid. Isethionates are used in formulations where mildness is important. They were, for example, used in the original formulation of the Dove brand of skin cleansers.

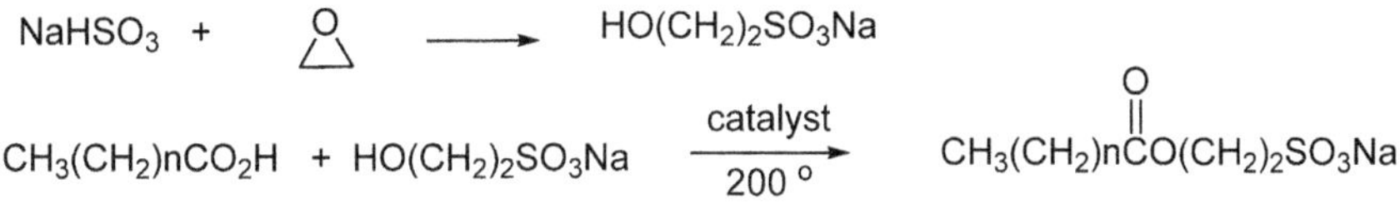

$$NaHSO_3 \quad + \quad \triangle^O \quad \longrightarrow \quad HO(CH_2)_2SO_3Na$$

$$CH_3(CH_2)nCO_2H \quad + \quad HO(CH_2)_2SO_3Na \quad \xrightarrow[200\,°]{catalyst} \quad CH_3(CH_2)n\overset{O}{\overset{\|}{C}}O(CH_2)_2SO_3Na$$
$$\text{acyl isethionate}$$

6.1.2.6 SULFOSUCCINATES

Both mono and diesters of sulfosuccinic acid are used as surfactants. The esters are prepared by reaction of maleic anhydride with long-chain alcohols or ethoxylated alcohols. In the case of the mono esters, both of the isomeric sulfonates are present.

$$ROH + \overset{O}{\underset{O}{\diagup}}^O \quad \longrightarrow \quad RO_2CCH=CHCO_2H \quad \xrightarrow{Na_2SO_3} \quad RO_2C\underset{SO_3^-}{CHCH_2CO_2^-} \quad + \quad RO_2CCH_2\underset{SO_3^-}{CHCO_2^-}$$

6.1.2.7   CHARGED LONG-CHAIN AMIDES

The acyl taurates, introduced in the 1930s as the Igepons, are made by acylation of an N-alkyl-β-hydroxyethanesulfonate. These are made from ethylene oxide and sodium bisulfite.

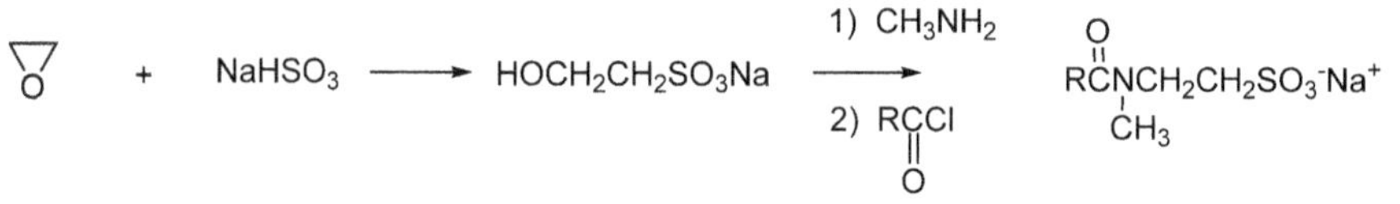

$$\text{(epoxide)} \quad + \quad NaHSO_3 \quad \longrightarrow \quad HOCH_2CH_2SO_3Na \quad \xrightarrow[\text{2) } RCCl]{\text{1) } CH_3NH_2} \quad \underset{\underset{CH_3}{|}}{RCNCH_2CH_2SO_3^-Na^+}$$

Other examples having similar properties are N-acylglutamates, N-acylsarcosinates, and N-acyl arginine esters. Because they are derived from naturally occurring amino acids, in principle, these surfactants could be made entirely from nonsynthetic substances.

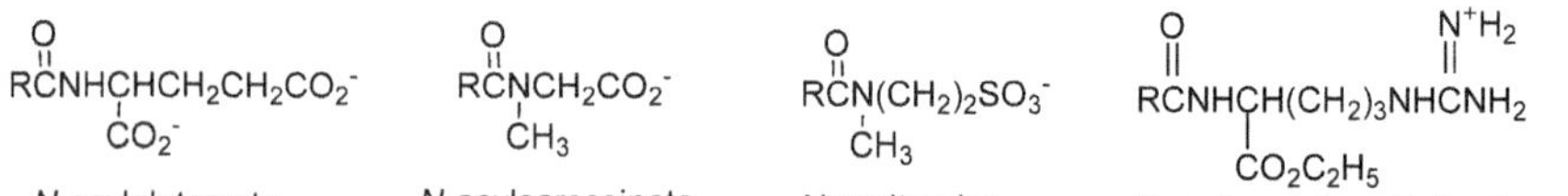

$\underset{\underset{CO_2^-}{|}}{RCNHCHCH_2CH_2CO_2^-}$	$\underset{\underset{CH_3}{|}}{RCNCH_2CO_2^-}$	$\underset{\underset{CH_3}{|}}{RCN(CH_2)_2SO_3^-}$	$\underset{\underset{CO_2C_2H_5}{|}}{RCNHCH(CH_2)_3NHCNH_2}$
N-acylglutamate	N-acylsarcosinate	N-acyltaurine	N-acyl arginine ethyl ester

6.1.3   NONIONIC SURFACTANTS

6.1.3.1   ALCOHOL ETHOXYLATES

Long-chain alcohols can be reacted with ethylene oxide under basic conditions to append a series of ethoxyethanol groups. This reaction is applicable to long-chain alcohols made from hydrolysis of fats or by synthesis.

$$CH_3(CH_2)nCH_2OH \quad + \quad x \;\; \text{(epoxide)} \quad \xrightarrow{^-OH} \quad CH_3(CH_2)nCH_2(OCH_2CH_2)xOCH_2CH_2OH$$

6.1.3.2   ALKYL PHENOL ETHOXYLATES

The alkyl phenol ethoxylates are made from phenol in two steps. In the first step, either propylene trimer or diisobutylene is used to carry out a Friedel–Crafts alkylation of phenol under acidic catalysis. These results in branched chain products called nonyl- and octylphenol, respectively. The mixture called nonylphenol is complex, containing as many as 20 isomers, whereas the octyl product contains mainly the 1,1,3,3-tetramethylbutyl substituent group. In the second step, the alkylated phenol reacts with ethylene oxide to introduce 5–15 ethoxyethyl (PEG) groups. A key feature of these materials is that the branched chain alkyl substituents are not easily biodegradable. This class of detergents has also come into question as being potential sources of endocrine disruptors (see Section 15.7.1).

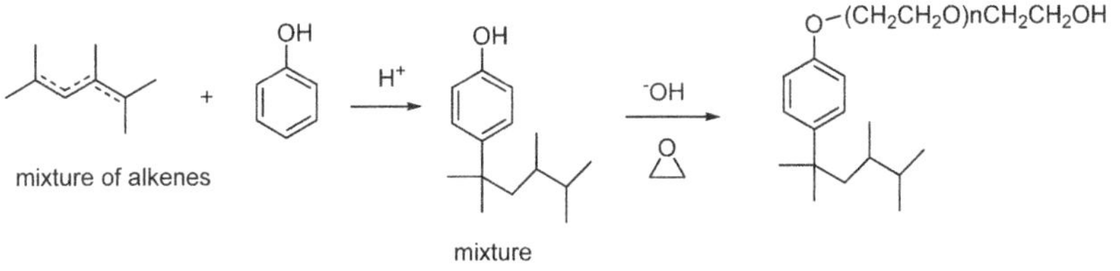

6.1.3.3 ALKYL POLYGLYCOSIDES

The reactions used to synthesize alkyl polyglycosides go back to the carbohydrate pioneer Emil Fischer, who reported examples in 1911. Patents for use as detergents were issued in Germany as early as 1934. Although Rohm & Haas and BASF marketed such products for specialized uses in the 1970s, the introduction into laundry and dishwashing detergents was led by Henkel, which opened plants in the United States and Germany in 1992 and 1995, respectively. The process is based on the same methodology used by Fischer and involves introduction of a long-chain alcohol into the acetal (anomeric) position of glucose. The materials can be made either from monomeric glucose or polymeric forms such as starch or glucose syrup (see Section 9.2.1). When the starting material is of high MW, a preliminary partial depolymerization/acetalization is carried out using a lower MW alcohol such as butanol. The alcohol groups in the alkyl polyglycosides are mainly the C_{12} and C_{14} compounds. The commercial product consists of a mixture of mono- (50–65%), di- (~20%), tri- (~10%), tetra- (5–10%) and higher (2–5%) saccharides. The precise composition is determined by the ratio of glucose to alcohol and the conditions used for the reaction. The alkyl glycosides can be considered to be completely renewable, although the long-chain alcohols can be obtained from either petrochemicals or oleochemicals. They are readily biodegradable, nontoxic and nonirritant. They thus have an especially favorable environmental profile among the major surfactants.

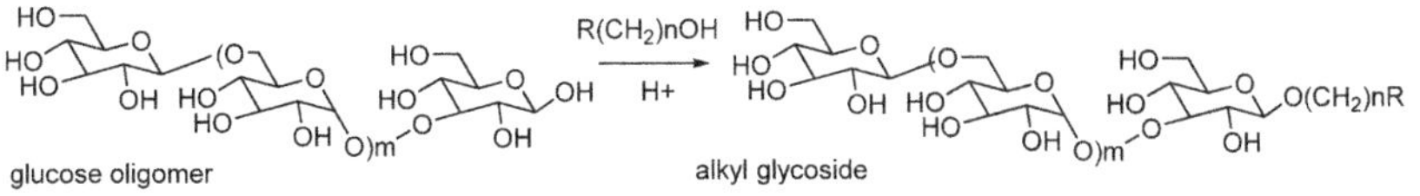

6.1.4 AMPHOTERIC AND ZWITTERIONIC DETERGENTS

Other classes of detergents can have both anionic and cation charges. For *amphoteric* materials, the charge depends on the pH. *Zwitterioinic* detergents contain both a cationic nitrogen and anionic carboxy or sulfonate group but are neutral in the normal pH range. They are usually derived from long-chain fatty acids (or mixtures thereof) and a diamine and are quaternized by a carboxymethyl group. The amphoteric and zwitterionic surfactants are considerably more expensive than the sulfate, sulfonate, and ethoxylate materials. Surfactants such as lauroamphoacetate are mild and are used primarily in shampoos, personal care, and cosmetic products.

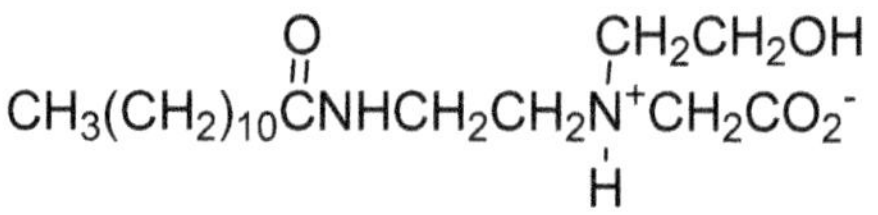

$$CH_3(CH_2)_{10}\overset{\overset{\displaystyle O}{\|}}{C}NHCH_2CH_2\overset{\overset{\displaystyle CH_2CH_2OH}{|}}{\underset{\underset{\displaystyle H}{|}}{N^+}}CH_2CO_2^-$$

lauroamphoacetate

6.1.4.1 ACYLBETAINES

The most commonly used betaines (pronounced "beta-ene") are *N*-acyl betaines. They are prepared from a mixture of fatty acid esters by reaction with *N,N*-dimethylpropylenediamine. The resulting amide is then alkylated with chloroacetic acid. The most widely used example is cocamidopropyl betaine, which contains mainly C_{12} and C_{14} acyl groups.

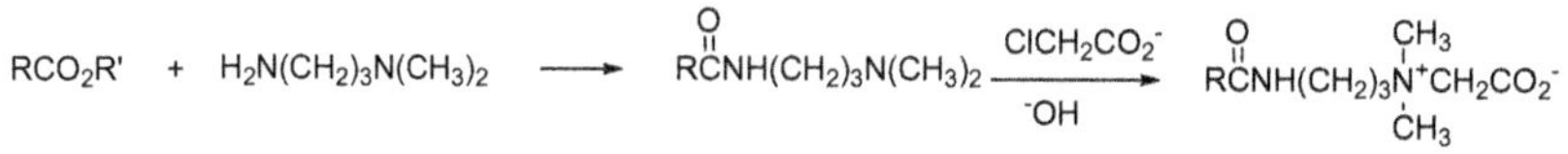

$$RCO_2R' \;+\; H_2N(CH_2)_3N(CH_3)_2 \;\longrightarrow\; R\overset{\overset{\displaystyle O}{\|}}{C}NH(CH_2)_3N(CH_3)_2 \;\xrightarrow[\;^-OH\;]{ClCH_2CO_2^-}\; R\overset{\overset{\displaystyle O}{\|}}{C}NH(CH_2)_3\overset{\overset{\displaystyle CH_3}{|}}{\underset{\underset{\displaystyle CH_3}{|}}{N^+}}CH_2CO_2^-$$

Although cocamidopropyl betaine is used for mildness, it is also considered to be a potential allergen. It is not entirely clear whether it is the final surfactant or a residual impurity, that is the source of the problem.[5] The nitrogen atom can also be substituted by hydroxyethyl groups, as in lauroamphocarboxy glycinate.

$$CH_3(CH_2)_n\overset{\overset{\displaystyle O}{\|}}{C}NH(CH_2)_3\overset{\overset{\displaystyle CH_3}{|}}{\underset{\underset{\displaystyle CH_3}{|}}{N^+}}CH_2CO_2^-$$

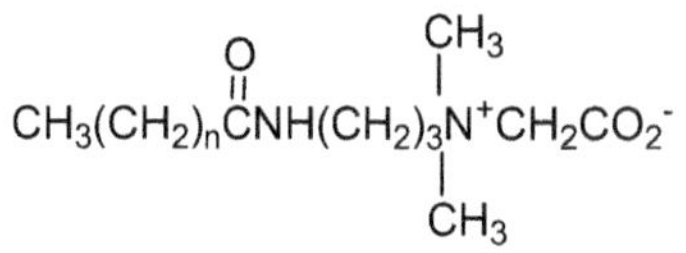

cocoamidopropyl betaine

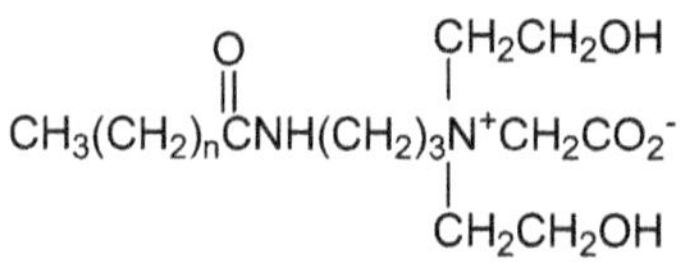

$$CH_3(CH_2)_n\overset{\overset{\displaystyle O}{\|}}{C}NH(CH_2)_3\overset{\overset{\displaystyle CH_2CH_2OH}{|}}{\underset{\underset{\displaystyle CH_2CH_2OH}{|}}{N^+}}CH_2CO_2^-$$

lauroamphocarboxy glycinate

6.1.4.2   AMMONIO SULFOBETAINES

Reaction of tertiary amines with 4- and 5-membered sultones produces ammonio sulfobetaines. Long-chain tertiary amines or pyridines can be used as the tertiary amine. The sultone used is usually propanesultone, $n = 2$, $R = H$.

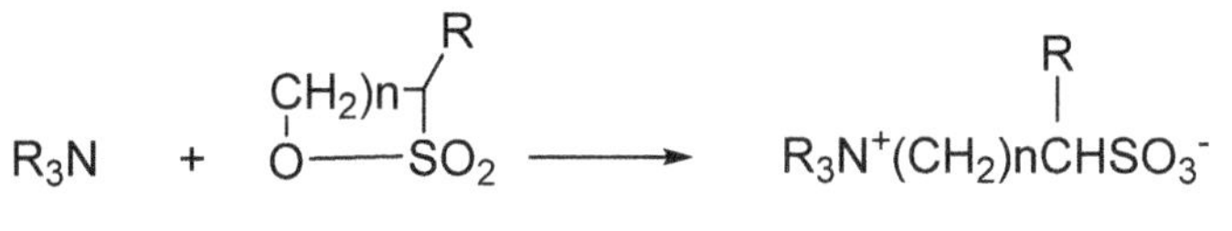

$$R_3N \quad + \quad \underset{O-\!\!-SO_2}{\overset{\overset{\displaystyle R}{\underset{|}{CH_2)n}}{}} \quad \longrightarrow \quad R_3N^+(CH_2)nCHSO_3^-$$

6.1.4.3   AMPHOACETATES AND AMPHOPROPIONATES FROM IMIDAZOLINES

Heating long-chain acids with N-(2-hydroxyethyl)ethylenediamine leads to imidazolines. These are then alkylated with chloroacetic acid, at which point the imidazoline ring is opened by hydrolysis. These compounds are called *acylamphoacetates*.

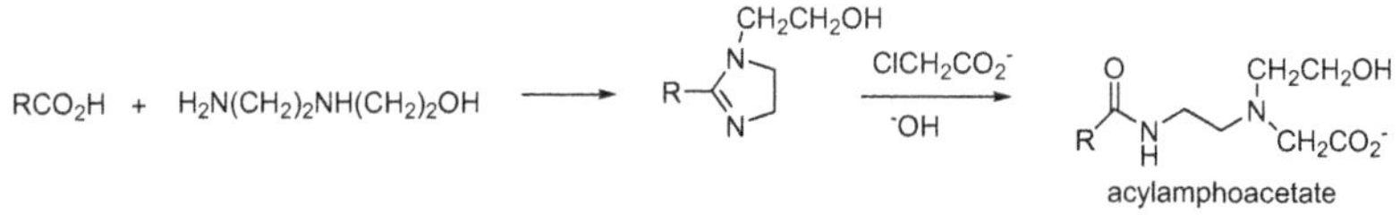

$$RCO_2H \;+\; H_2N(CH_2)_2NH(CH_2)_2OH \longrightarrow$$

acylamphoacetate

Similar compounds can be prepared using acrylic acid as the alkylating agent. These are called *acylamphopropionates*.

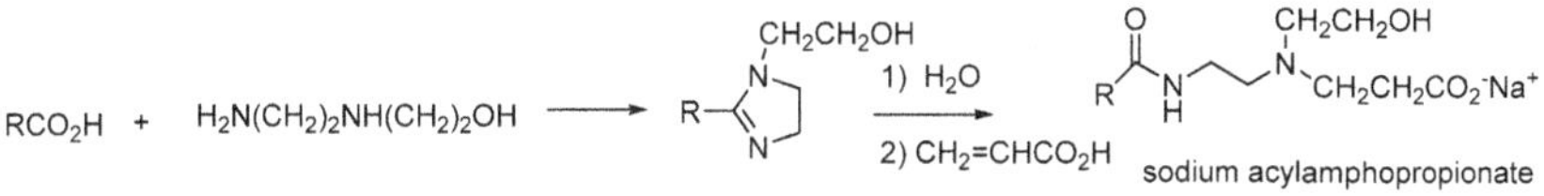

$$RCO_2H \;+\; H_2N(CH_2)_2NH(CH_2)_2OH \longrightarrow$$

1) H_2O
2) $CH_2=CHCO_2H$

sodium acylamphopropionate

There are also products in which two acetate or propionate units have been introduced by alkylation. These are called amphodiacetates and amphodipropionates, respectively. The particular acyl group is designated with the appropriate fatty acid name, such as coco- or lauroyl. Some of the commercial formulations are mixtures of the mono- and dialkyl-derivatives.

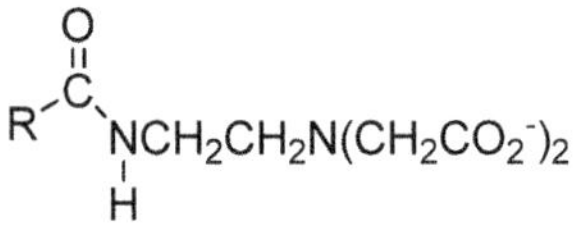

$$R\overset{\overset{\displaystyle O}{\|}}{\underset{}{C}}\underset{H}{\overset{}{N}}CH_2CH_2N(CH_2CO_2^-)_2 \qquad R\overset{\overset{\displaystyle O}{\|}}{\underset{}{C}}\underset{H}{\overset{}{N}}CH_2CH_2N(CH_2CH_2CO_2^-)_2$$

acyl amphodiacetate 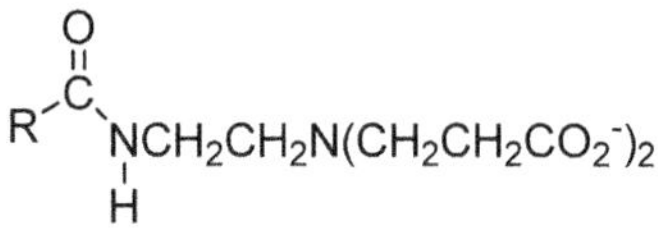   acyl amphodipropionate

6.1.4.4 LECITHINS

The *lecithins* are naturally occurring zwitterionic surfactants. They are phospholipids that contain choline and can be obtained from eggs or soybeans. Lecithins are used extensively in foods (see Section 9.1) but can also be found in cosmetics.

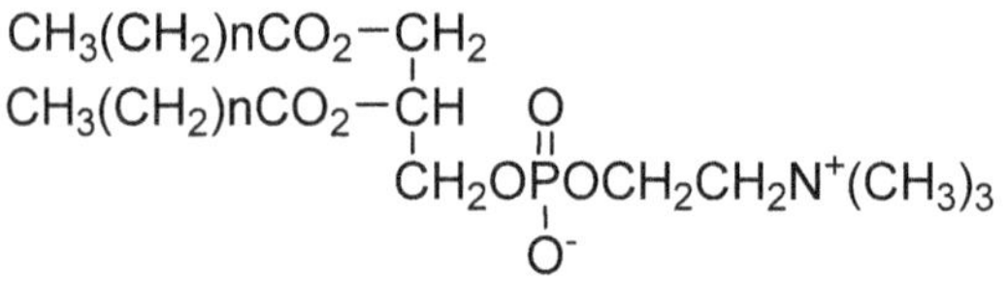

$$CH_3(CH_2)nCO_2-CH_2$$
$$CH_3(CH_2)nCO_2-CH \quad O$$
$$CH_2OPOCH_2CH_2N^+(CH_3)_3$$
$$O^-$$

6.1.5 CATIONIC SURFACTANTS

The cationic surfactants have positively charged head groups, usually quaternary nitrogens. They can be monoquaternaries with one charged group or they can have several charged groups (polyquaternaries). Monoquaternaries are long-chain hydrocarbons terminating in a quaternary nitrogen group. The polyquaternaries are various polymers that have been modified to incorporate cationic nitrogen groups. In some cases, the quaternary groups are introduced on silicones rather than hydrocarbon structures.

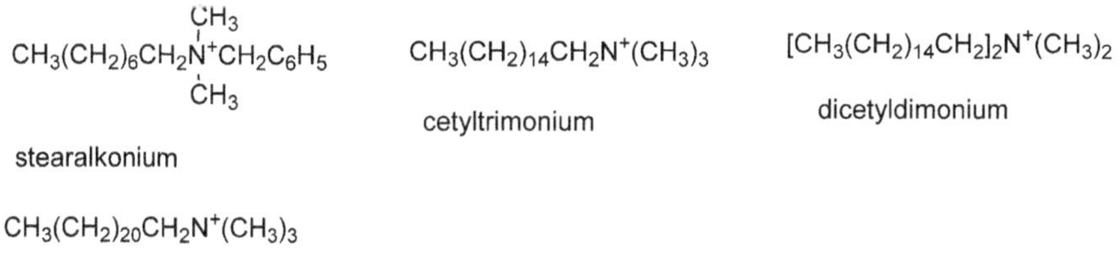

$$CH_3$$
$$CH_3(CH_2)_6CH_2N^+CH_2C_6H_5 \qquad CH_3(CH_2)_{14}CH_2N^+(CH_3)_3 \qquad [CH_3(CH_2)_{14}CH_2]_2N^+(CH_3)_2$$
$$CH_3$$

stearalkonium   cetyltrimonium   dicetyldimonium

$$CH_3(CH_2)_{20}CH_2N^+(CH_3)_3$$

behentrimonium

SCHEME 6.1 Structures found in cationic surfactants.

6.2 HOUSEHOLD CLEANING PRODUCTS

6.2.1 LAUNDRY DETERGENTS

The function of laundry detergents is to dislodge and suspend grease, oil, and grit and to remove stains. The products usually contain both anionic and nonionic surfactants. Another major constituent of laundry detergents are called "builders" and function to sequester calcium and other "hard" metal cations. Laundry detergents often contain "antideposition" agents, whose purpose is also to prevent precipitation of calcium salts and a "greying" effect. Carboxymethylcellulose is an example. Polyester–polyether block copolymers can serve the same function. Bleaches and enzymes are often included, to remove stains. The recent trend has been to decrease both energy and water use in domestic washing machines. This requires increased surfactant concentration and improved effectiveness at lower temperatures. Reduced water usage requires that detergents must rinse out easily.

The first synthetic laundry detergents were introduced in the United States after WWII. They were based on ABS surfactants and phosphate builders. The original ABS surfactants were branched chain and not very biodegradable and led to foaming in sewage treatment plants and eventually streams and rivers. They were replaced in the 1960s by the more biodegradable LASs. A second major change began in the 1970s, first in Japan and then in the United States and Europe. Bulky low-density detergents were replaced by more compact agglomerated powders. These materials have a density of 600–900 g/L and contain higher concentration of surfactants. Several factors drove this change. One was the convenience of the smaller packaging and the reduced amount required per wash. The amount of energy consumed in manufacturing was also reduced. At the same time, phosphates were being replaced by other builders. This required somewhat more complex formulation of the detergents because phosphates had other functions, including buffering. New ingredients to control pH and to reduce precipitation, such as silicates and polycarboxylates were introduced.

Powdered laundry detergents are prepared by spraying a suspension into heated towers where a fluffy powder is formed by rapid drying as the particles fall through the tower. The switch from low-density to high-density detergents required several modifications of this basic process. High-intensity mixing can provide some increase in density by rounding the particles and decreasing the amount of void space. The powders can be agglomerated into more dense material by using a binder, for example, water. Alternatively, they can be compacted and then repulverized into optimal size and shape. These methods all start with the spray-tower product and thus do not save the thermal energy required by this process. An alternative is to carry out the neutralization of the acidic sulfate or sulfonate with base under very high-concentration conditions, thus minimizing the need for drying. This method requires very efficient mixing and temperature control and the use of specialized process equipment.

Many laundry detergents contain a bleaching agent, usually sodium perborate, which functions by release of hydrogen peroxide. However, at lower wash temperatures, <60°C, hydrogen peroxide is not a very effective bleach. Low-temperature conditions require "activators" that react with the hydrogen peroxide to form more active bleaches. Two of the most common are tetraacetyl ethylenediamine and nonanoyloxybenzenesulfonate. Both of these compounds act as acylating agents forming peroxy acids, which are more reactive than hydrogen peroxide at the lower temperatures. A limiting property of a bleach is the extent of attack of dyes and other potential susceptible compounds.[6]

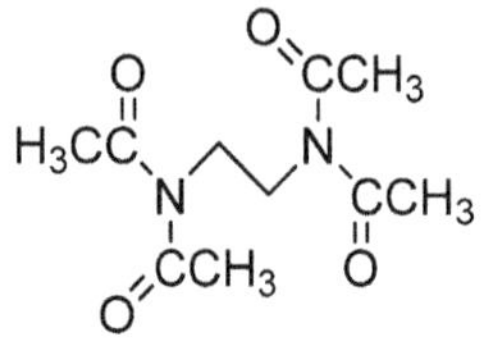

tetraacetyl ethylenediamine

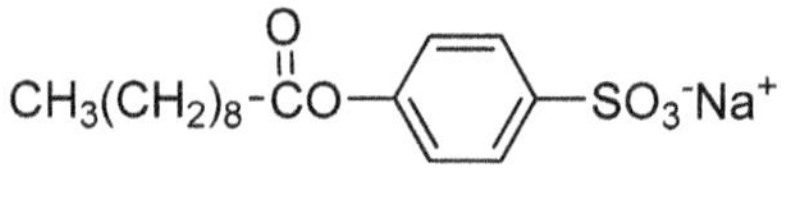

sodium nonanoyloxybenzenesulfonate

Special soaps and detergents for hand-washing of wool and delicate fabrics tend to have a greater portion of nonionic and cationic surfactants. They are designed to give a soft feel to the washed material. Cationic surfactants, usually dialkyl quaternium ammonium salts, are used as fabric softeners and to reduce static charge build-up. They are commonly used as sheets included in the drying process.

Laundry detergents also contain "brighteners" intended to make laundry look "whiter than white." This effect is achieved by adding a colorless material that absorbs in the near UV but fluoresces in the visible, imparting an extra shine to the fabric. The most common compound for this purpose is a substituted stilbene-2,2'-disulfonate.

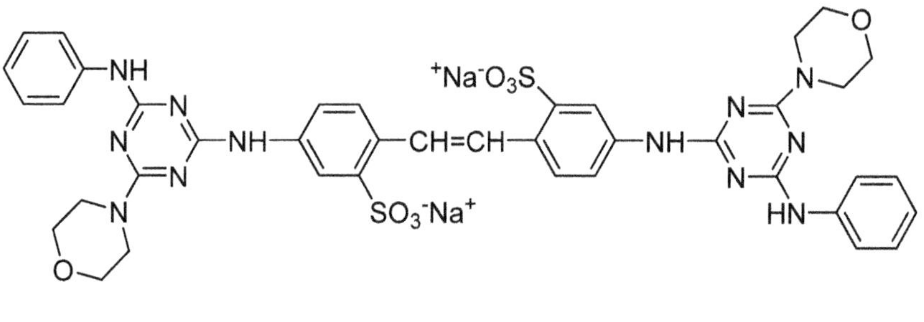

Enzymes made their first appearance in the household laundry as pre-soaks in the 1960s. These products contain proteases isolated from microbiological fermentation, such as *Bacillus licheniformis*. Beginning from the 1980s, enzymes were introduced into powdered laundry detergents and included proteases, amylases, lipases, and in some cases cellulases. (One system for classifying enzymes is to append the suffix "ase" to the name of the substrate. Thus a "protease" degrades proteins, a "lipase" attacks lipids, while amylases and cellulases attack starch and cellulose structures, respectively.) These materials are catalysts and have high turnover and therefore can be present in relatively small concentration. The first enzymes used in laundry products were proteases of the subtilisin group. The lipases are of fungal or bacterial origin. The amylases are produced by *Bacilli*. The cellulases are the most recently introduced enzymes and are used for cotton-containing fabrics. They remove surface material by breaking amorphous regions of microfibrils and loosening the adhered grime. By removing accumulated surface material they also brighten colors. They do not attack crystalline regions in fibers.

The practical requirements for enzymes in laundry detergents are good storage and thermal stability, compatibility with surfactants, optimal function at an alkaline pH, and relative nonselectivity with respect to materials that can be digested. To be compatible with bleaching formulations, the enzymes must also be stable to oxidative conditions. Because the enzymes are proteins, they are potentially allergenic and both in manufacture and use, exposure to dust containing enzymes should be avoided. This is achieved in the final products by coatings such as PEG or one of the nonionic surfactants. Improved enzyme-producing strains have been sought by screening from various locations, such as alkaline lakes, and by mutation. Beginning in the 1990s, genetically engineered subtilisins began to be commercialized. Of the enzymes in use in 2000, about half were "wild type" and half "engineered."[7] In the United States, detergents provide the largest market for industrially produced enzymes, with sales of about $500 million in 2000. Enzymes reduce the overall environmental impact of detergent use. The enzymes are used in relatively small amount, their production is less energy-intensive, they allow reduced wash temperature, and they are rapidly biodegraded.

The most well-known environmental issue associated with laundry products involves the phosphate builders. Phosphates are associated with eutrophication, which is caused by excess algae and plant growth that leads to excess oxygen consumption and fish die-off. Although detergents, at most, accounted for 10–15% of phosphate input, restrictions were adopted in a number of states in the United States and in certain countries in Europe in the 1980s. There have been two results. Presumably to reduce the complexities of dealing with nonuniform regulations, phosphate use was greatly reduced by the manufacturers and waste water treatment facilities were often modified to improve phosphate removal.[8] This has resulted in reduced phosphate levels, depending on local sewage treatment conditions and regulations. Zeolites are the most common replacement for phosphates as builders in laundry detergents. The zeolites function as ion-exchange agents, adsorbing Ca^{2+} and releasing Na^+. They can constitute significant amounts by weight of laundry detergents (see Table 6.2). They are in the form of fine particulate solids and are rinsed away and accumulate as sludge in the water treatment process. No significant environmental hazards are associated with this use of zeolites.

TABLE 6.2   Typical Solid Laundry Detergent Composition (% by Weight).[a]

A. Anionic detergent Powder		B. Nonionic detergent Agglomerated powder		C. Compact Agglomerated powder	
Surfactant		*Surfactants*		*Surfactants*	
Alkylbenzenesul-fonate	12–20	Alcohol ethoxylates	6–10	Neodols	20
		Alkylbenzenesulfonate	0–4		
Builders		*Builders*		*Builders*	
Na_2CO_3	15–30	Na_2CO_3	15–30	Na_2CO_3	27
Zeolite A	15–25	Zeolite A	15–25	Zeolite A	45
Sodium silicate	3–6	Sodium silicate	3–6	Sodium silicate	6
Antideposition agent		*Antideposition agent*		*Antideposition agent*	
Polycarboxylate	1–4	Carboxymethylcellulose	1	Carboxymethyl-cellulose	2

[a]The remainder is usually water.

(Adapted from Raney, O. In *Surfactants—A Practical Handbook*; Lange, K. R., Ed.; Hanser Publishers: Munich, 1999; pp 171–203).

Laundry detergents are regulated in the United States by the Consumer Protection Safety Commission, which requires warning labels of hazards and safe packaging. Very little information about the contents is available on the packages for laundry detergents. If a laundry product makes disinfectant claims, they are regulated by the EPA and the claims must be supported by data. New detergent ingredients must be registered with the EPA.

6.2.2   DISHWASHING DETERGENTS

Dishwashing detergents can be designed for either hand or machine washing. Machine washing is dominant in the United States and Western Europe. In the United States, about equal amounts of solid and liquid product are used. The liquids are aqueous suspensions of varying viscosity, stabilized by thickening agents. A smaller portion of the US market, but more of the European market, is in the form of prepackaged unit doses. A significant difference in design affects the United States, as opposed to European, automatic dishwashers. The latter contain built-in water softeners that reduce the burden on the detergent formulation for removal of "hard" cations. As a result, American dishwater formulations have higher surfactant and builder concentrations than is typical in Europe.

The primary function of dishwashing detergents is to break up and emulsify oil and grease. The detergent concentration is low for machine washing and most of the cleaning is accomplished by the mechanical force of the water and the effect of the elevated temperature. Higher temperatures improve cleaning by softening greasy material, but the advantage of high-temperature needs to be balanced against energy consumption. The alkalinity is also important to cleaning effectiveness. The basic solutions help to dissolve fatty acids and hydrolyze fats, making them more soluble. Machine detergents also usually contain a chlorine source to assist in chemical degradation of protein, remove stains, and sanitize the dishes. Some dishwashing products contain enzymes, in particular proteases, amylases, and lipases. Their inclusion, however, restricts the other agents that can be present, especially bleaches.

The builder is a major constituent of dishwasher detergents and functions to sequester calcium ions and prevent deposition of calcium salts, either as $CaCO_3$ or as salts of organic acids. The builders also function synergistically with the detergent's surfactant action. Detergents without builders have much higher levels of surfactants. The most effective builder is sodium triphosphate, $Na_3P_3O_{10}$. However, because phosphates are believed to contribute to eutrophication, their use is restricted in many jurisdictions. In the United States, the regulations are at the state and local level and are not consistent. European regulation also differs from country to country. Sodium triphosphate can be replaced by chelating carboxylates such as citrate and nitrilotriacetate. Recently, the use of zeolites and carboxylate polymers has become common. The zeolites function by ion exchange, adsorbing Ca^{2+} into the pores and exchanging Na^+ into the water phase. Polyanionic materials such as carboxymethyl cellulose function by stabilizing emulsions. Dishwashing products also contain sodium silicates to provide alkalinity and function as an anticorrosion agent.

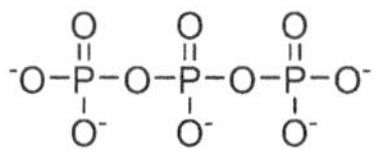

triphosphate (also
called polytriphosphate)

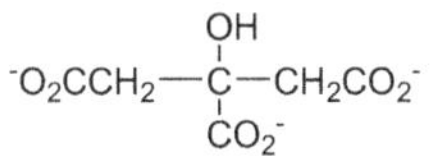

citrate

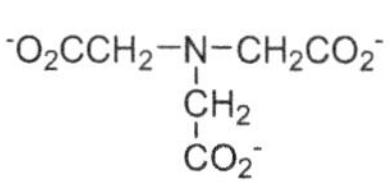

nitrilotriacetate

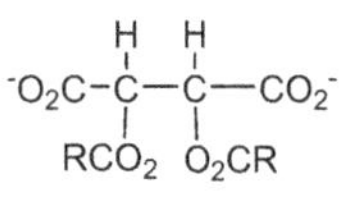

tartrate mono-
and di-succinate

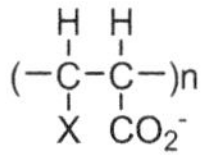

polycarboxylate

Liquid detergents often contain xylene- or cumene-sulfonates, which are called "hydrotropes" and serve to increase the solubility of the more hydrophobic surfactants. Foam control is also important. Too much can lead to overflow and machine damage, while too little may be perceived as ineffective washing. Hard water tends to reduce foaming. Foam stabilizers, usually alkanolamides or amine oxides, are used. Alkyl diphenyloxide disulfonate surfactants meet the low foam requirement. Most products also have coloring and fragrances, but these account for only a small fraction of the total weight.

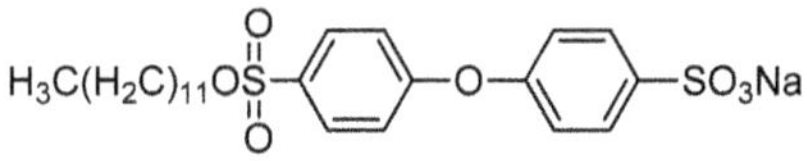

$$H_3C(H_2C)_{11}O\overset{\displaystyle O}{\underset{\displaystyle O}{\overset{\|}{\underset{\|}{S}}}}\!\!-\!\!\langle\ \rangle\!\!-\!\!O\!\!-\!\!\langle\ \rangle\!\!-\!\!SO_3Na$$

sodium dodecyldiphenyl oxide disulfonate

Table 6.3 gives the composition of several types of dishwashing detergents.

TABLE 6.3 Typical Dishwashing Detergent Formulations (% by Weight).[a]

A. Hand liquid		B. Machine powder		C. Machine gel	
Surfactant		*Surfactant*		*Surfactant*	
Alcohol ethoxysulfate	18	Nonanionic detergent	3	*n*-Decyldiphenyl oxide disulfonate	1.0
Alkylbenzenesulfonate	30				
Hydrotrope		*Chlorine source*		*Chlorine source*	
Na xylenesulfonate	8.5	Sodium dichloroisocyanurate	2	NaOCl	7.5
Foam booster		*Water softener and pH*		*Builder*	
Fatty acid diethanolamide	4.0	Na_2CO_3	9	$Na_5P_3O_{10}$	13
Thickener		$Na_3P_3O_{10}$	35	$K_4P_2O_7$	15
NaCl	3.0	*Anticorrosion*		*pH adjustment*	
		Sodium silicate	12	NaOH	2.5
		Filler		*Thickener*	
		Na_2SO_4	22	Polycarboxylate	1.0
				Anticorrosion	
				Sodium silicate	21

[a]The remainder is usually water.

(Adapted from Raney, O. In *Surfactants—A Practical Handbook*; Lange, K. R., Ed.; Hanser Publishers: Munich, 1999, pp 189–192).

6.3 PERSONAL CARE PRODUCTS

6.3.1 HAND AND BODY SOAPS

The true soaps used in hand and body bars are made from shorter chain (C_{12}/C_{14}) fatty acids obtained from vegetable sources such as coconut oil. These tend to produce better foaming action than the longer chain acids. The ratio of Na and K cations can be adjusted. Many hand and body bars are not technically soaps in that they contain surfactants other than fatty acid salts. The alkyl sulfates and alcohol ethoxy sulfates are commonly used in such products. The technical term "syndet bar" derived from "synthetic detergent" is sometimes used, but has negative advertising cachet.

Skin and body care products are intentionally applied to the skin and have the potential to damage skin in the case of excessive use or skin sensitivity. The surfactants can loosen outer cells and remove lipid material and can increase the permeability of the skin. Mildness is a major objective in hand and body soaps. One combination used for mildness is fatty acid isethionates with fatty acid salts. Disodium laurylsulfosuccinate is also used. In Japan, acyl glutamates are popular surfactants in body soaps (see Section 6.1.4 for structures). These ingredients are more expensive than laundry and dish-washing detergents.

One of the requirements of a hand and body soap is a smooth feel after cleansing. This is achieved by moisturizers (humectants, emollients) and fat re-placement agents. Among the compounds used are glycerin and other polyols, long-chain esters of lactic acid, polyquaternaries, as well as peptides and pro-teins. Natural oils or mineral oil may be included as "refatting" agents. Liquid hand and body soaps usually include pearlescent agents such as glycol stearate. Plasticizers and binders are used to permit stamping of bars and maintenance of shape and feel. Typical plasticizers are long-chain alcohols, polylol esters such as glycol or glycerin stearates and PEG ethers. The fatty acid salts themselves also function as plasticizers. Some soaps are cast in molds and this process can achieve translucence if fairly high concentration of polyols are included. Table 6.4 gives some representative compositions.

TABLE 6.4 Typical Bar Soap Composition (% by Weight).

Surfactant	30–70
Binder, emollients	20–50
Foam enhancers	0–5
Mildness enhancers	0–5
Fillers	5–30
Water	5–12

(Adapted from Friedman, M. *SODEOPEC*, 2004, 147–188).

6.3.2 HAIR CARE PRODUCTS

6.3.2.1 SHAMPOOS AND CONDITIONERS

The scalp normally has 100–150 thousand hair follicles corresponding to a density of 200–250/cm². Figure 6.3 is a magnified image of the scalp. The scalp also contains sebaceous and sweat glands that produce lipids, proteins, and salts. This environment supports a considerable growth of microflora. As a result, hair becomes sticky and greasy and rapidly accumulates grime and dirt. Shampooing frequency averages 3–7 times/week in Europe and the United States, but twice per day is not unknown. Washing of hair typically removes some of the natural lipids associated with the outer (cuticle) layer of the hair. Shampoo formulations seek to remove the excess sebum and soil, but without excessive drying of the scalp and hair. Hair normally is anionic with a pI of 3.67, because the protein has more acidic than basic amino acids. The anionic character increases with hair damage as a result of oxidation of cysteine residues to sulfonic acids. Bleached hair is particularly strongly anionic.

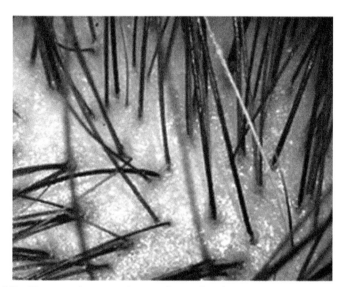

FIGURE 6.3 Electron microscope image of normal scalp.

Hair is composed mainly of the protein keratin which is structurally similar to the collagen on the skin surface. The outer layer is composed of flattened cells that can absorb moisture. The hair is not connected to the skin surface so that by pushing the skin down, the strand of hair can be cut below the surface. Hair shaft thickness and strength varies with individuals and depends on ethnic background. Thin, fragile hair can result from excess lipid removal by shampoos, bleaching, excessively alkaline treatments, swimming, or exposure to sun and wind. Both fatty acids and glycerides, as well as other lipids such as cholesterol and ceramides, are present in natural sebum and these compounds may be present in hair care products to replace materials lost by shampooing. Because of the anionic character of hair, cationic materials are held by electrostatic attraction. Hair care products may also include protein hydrolysates from wheat, rice, oats, almonds, etc.

The primary surfactant in shampoos is a detergent such as an alkyl sulfate or an alcohol ethoxy sulfate. An example of the former is sodium lauryl sulfate and the latter would be the laureth sulfates. All of these compounds are salts of strong acids, so they are anionic but are neutral in pH.

$$CH_3(CH_2)_{10}CH_2O\overset{\overset{O}{\|}}{\underset{\underset{O}{\|}}{S}}O^-\ Na^+$$

sodium lauryl sulfate

$$CH_3(CH_2)_{10}CH_2O\text{-}(CH_2CH_2O)_n\text{-}\overset{\overset{O}{\|}}{\underset{\underset{O}{\|}}{S}}O^-\ Na^+$$

laureth sulfate

laureth-6, average n = 6
laureth-9, average n = 9

$$CH_3(CH_2)_{14}CH_2O\text{-}(CH_2CH_2O)_{20}\text{-}\overset{\overset{O}{\|}}{\underset{\underset{O}{\|}}{S}}O^-\ Na^+$$

ceteth sulfate

ceteth 20,
average n = 20, cetyl = C_{16}

Shampoos also contain viscosity modifiers and foam stabilizers. The physical form is usually an emulsion containing various micelles and vesicles. The fundamental function of the detergent is to remove soil that is imbedded in the sebum. The shampoo detergents must have other properties such as being "sudsy" and minimally irritating to the eyes. Foaming serves as an "end-point" indicating sufficient cleaning, because residual sebum is a foam destabilizer. Cosurfactants are used to promote vesicles, in particular rods and worm-like structures. Cocoamide MEA, cocoamide DEA, and cocoamidopropyl betaine are commonly used as cosurfactants in shampoos. Polymers such as carboxymethyl cellulose, hydroxyethyl cellulose and [acrylate-steareth-20 methyl acrylate]-copolymer are also used. Thickening is achieved with a variety of long-chain molecules that can form organized micelles and/or liquid crystals. These include long-chain esters of ethylene glycol or glycerin and long-chain amides of ethanolamine. Natural cellulosic gums such as guar, tragacanth, karaya, or xanthan are sometimes used. Some of these materials also provide the opalescence effect that is desired for many products. Small particulate material, such as silica, titania, or mica may be included to promote thickening. Table 6.5 gives some typical shampoo compositions.

$$CH_3(CH_2)_{10}\overset{\overset{O}{\|}}{C}NHCH_2CH_2OH$$

cocamide MEA

$$CH_3(CH_2)_{10}\overset{\overset{O}{\|}}{C}N(CH_2CH_2OH)_2$$

cocamide DEA

TABLE 6.5　Hair Shampoo Compositions (% by Weight).[a]

A. Typical shampoo		B. Mild shampoo	
Surfactants		*Surfactants*	
Lauryl ether sulfates	12–16	Alcohol ethoxy sulfate	7.5
Cocoamidopropyl betaine	2–5	Alcohol ethoxycarboxylate	7.5
Fatty acid polyglycol esters	0–2	Cocoamidopropyl betaine	7.5
Foam booster		*Foam booster*	
Fatty acid ethanolamides	0.5–2.5	Fatty acid ethanolamides	2.1
Refatting agent			
Ethoxyethylated lanolin alcohol	0.2–1.0		
Conditioner			
Quaternized hydroxyethylcellulose	0.4–1.0		

[a]The remainder is usually water.

(Adapted from Raney, O. In *Surfactants—A Practical Handbook*; Lange, K. R., Ed.; Hanser Publishers: Munich, 1999; pp 171–203).

Conditioners can be formulated to be used immediately after shampooing, left in few minutes, then rinsed out. Conditioners function by replacing the lipids removed by shampoos and by promoting attraction between hair strands. The conditioners provide lubrication and diminish surface friction by providing a hydrophobic surface layer. They also reduce the build-up of electrostatic charge that leads to "frizzing." Cationic conditioners include stearalkonium chlorides, hydroxyethyl cetyldimonium chlorides and polyquaternaries (see Scheme 6.1 for structures). The quaternary salts are deposited most heavily on the hair toward the ends of the strands, because these strands are the most anionic as the result of oxidative damage to the hair. van der Waals attractions also play a significant role in the attractions and quats that contain more than one long chain provide a heavier level of conditioning. An important group of conditioners are polyquaternaries in which a polymer backbone such as cellulose, polyacrylamide, or poly-*N*-vinylpyrrolidone is modified by incorporation of cationic groups. These cationic polymers form thin films on the hair shaft as the result of electrostatic attraction. Conditioners also often contain long-chain alcohols such as cetyl and stearyl alcohols that help to stabilize emulsions. The formulations are intended to maximize conditioning without build-up. The long-chain monoquats have poor compatibility with anionic soaps and detergents and are used mainly in stand-alone conditioners. The compatibility with shampoos can be improved by using "ethoquats," which contain PEG groups. A relative new type is the "ester quats," which can be attracted to the hair but then undergo hydrolysis, leaving the fatty acid component.

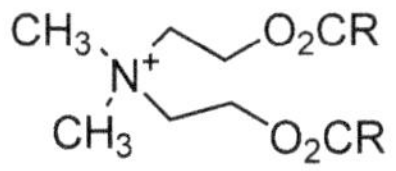

$$CH_3 \underset{CH_3}{\overset{}{\underset{|}{N^+}}} \begin{matrix} \diagup \diagdown O_2CR \\ \diagdown O_2CR \end{matrix}$$

ester quat

Silicones are also important examples of conditioners and are especially useful for combined shampoo conditions. The basic silicone polymer *dimethicone* (see Section 7.1) is usually present and *amodimethicone*, which contains 2-aminoethylaminopropyl side chains, is also used. Another silicone used in conditions is *dimethicone copolyol*, which contains PEG side chains. Silicones reduce surface friction and also provide sheen. A silicone elastomer with PEG cross-linking that assists in styling and manageability has been introduced.[9] Volatile cyclic silicones are often included as carriers for the high MW material. The silicones and quaternary type surfactants appear to have synergistic effects. Table 6.6 illustrates some compositions of conditioners.

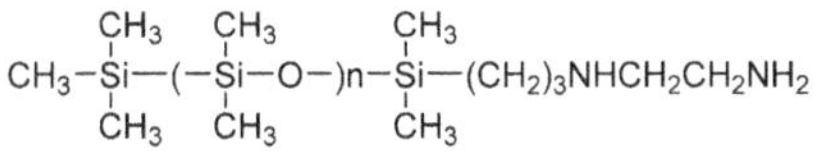

$$CH_3-\underset{\underset{CH_3}{|}}{\overset{\overset{CH_3}{|}}{Si}}-(-\underset{\underset{CH_3}{|}}{\overset{\overset{CH_3}{|}}{Si}}-O-)n-\underset{\underset{CH_3}{|}}{\overset{\overset{CH_3}{|}}{Si}}-(CH_2)_3NHCH_2CH_2NH_2$$

amodimethicone

$$CH_3-\underset{\underset{CH_3}{|}}{\overset{\overset{CH_3}{|}}{Si}}-(-\underset{\underset{CH_3}{|}}{\overset{\overset{CH_3}{|}}{Si}}-O-)n-\underset{\underset{OSi(CH_3)_3}{|}}{\overset{\overset{CH_3}{|}}{Si}}-(CH_2)_3O(CH_2CH_2O)xH$$

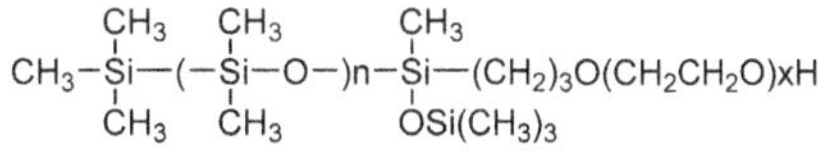

dimethicone copolyol

TABLE 6.6   Hair Conditioner Composition (% by Weight).[a]

Quaternaries, polycations, and silicones	0.5–3.0
Long-chain alcohols	1.0–4.0
Oils and silicone	0.2–1.0
Emulsifier	0–2.0
Thickener	0–1.0
Other	0–5

[a]The remainder is usually water.
(Adapted from Krummel, K.; Chiron, S.; Jachowicz, J. *Chem. Manuf. Cosmet.* **2000,** *2*, 359–396).

6.3.2.2   HAIR-STYLING, WAVING, AND STRAIGHTENING PRODUCTS

The shape of hair is to a considerable extent governed by disulfide bonds. To increase curl, the disulfide bonds are reduced with a reductant, usually ammonium thioglycolate, and the hair is shaped as desired. The disulfide bonds are then reformed with an oxidant. The oxidant is then rinsed out. Hair straightening involves removal of disulfide bonds and requires reduction and some desulfurization. Strongly basic solutions of thioglycolate are used, sometime with heating. The hair is combed under these conditions to promote straightening. Hair styling products can contain polymers such as poly-β-alanine, monosuccinoyl chitosan, acrylates, poly-N-vinylpyrrolidone or PEG-polymers. These materials provide fine coatings that help the hair maintain its shape after styling. The polymers can be applied as sprays, gels, mousses, or creams. In the case of sprays, the polymers are soluble in the solvent, which is usually ethanol–water. Various emulsifiers and foaming agents are used in the products.

6.3.2.3   HAIR-COLORING PRODUCTS

Modern hair-dyeing was introduced early in the 20th century in Paris and London. Hair-dyeing involves mixing certain compounds, usually aromatic amines and phenols with an oxidant, often H_2O_2. The aromatics are oxidized to intermediates that combine to give colored pigments. Particular shades are obtained by mixing different ingredients. About 15 various compounds were in use between 1935 and the 1970s. One of the early compounds, p-phenylenediamine, was recognized as early as 1910 as an allergen in some people. At that point, preliminary skin tests were introduced to detect those who might be allergic. The 1938 revision of the Food, Drug, and Cosmetic Act made preliminary testing obligatory in the United States for any hair dye made from coal tar derivatives. Concern about toxicity began to increase in the 1970s. Although none of the compounds was a direct carcinogen or mutagen, when tested in combination with metabolic activation (Ames Test), several were found to be active mutagens. One of the most active was 2,4-diaminoanisole. As a result, in the 1980s, the EU banned a number of the compounds and in the United States, warning labels were required. These developments moved the industry to use of less highly suspect materials.

6.3.2.4   SHAVING FOAMS

Beard and shaving styles have changed many times. At the turn of the 20th century shaving was done with straight razors and soaps. If done by a barber, it probably included wrapping the face in a warm moist towel to soften the beard. In 1904, King C. Gillette patented a safety razor and introduced it with a "do-it-yourself" advertising campaign. The principle embodied in the patent is depressing the skin, which allows the blade to cut beneath the skin line, resulting in a closer shave. Because the skin surface is uneven, a protective coat of surfactant minimizes cuts and nicks. Several improvements have been made subsequently, including the use of dual and triple blades. Shaving product sales in the United States total several billion dollars annually. Aerosol shave creams were introduced in the late 1940s and quickly became the dominant product. The shaving cream has several purposes. One is to lubricate the skin permitting the blade to slide evenly. Another is to moisten and soften the hair. The cream also serves as a marker to show where shaving has occurred. Table 6.7 shows the composition of a typical shaving cream. Some formulations also include emollients such as propylene glycol or PEG-25-lanolin. The propellants can be low MW hydrocarbons (mainly propane and butane), hydrofluorocarbons, carbon dioxide, or air. Shaving gels were introduced in the 1970s. They are formulated so that rubbing the gel on the skin resulting in foaming and they give thicker foams than do aerosols. Shaving products specifically designed for women are often formulated as gels.

TABLE 6.7   Shaving Cream Composition (% by Weight).[a]

Lauryl methyl gluceth-10	8%
Stearic acid	5%
Ceteth-25	1.5%
Ceteth-5	0.5%
Triethanolamine	2.5%

[a]Remainder is water.

6.3.2.5   HAIR GROWTH CONTROL PRODUCTS

More hair? Less hair? Hair here but not there? For men and to a lesser extent for women, maintaining hair coverage on the head is an important issue. Balding or thinning of hair is commonly observed as *androgenic alopecia* or *male pattern baldness*, that is, hair loss associated with androgenic steroids. This trait is more common in Caucasians than other racial groups and about 35–40% males in the United States and Northern Europe experience hair loss with aging, whereas the figure is in the 20% range in Japan and Taiwan. Women also can experience this form of hair loss, more typically as thinning rather than complete loss. There are estimated to be about 100,000–150,000 hair follicles on the scalp. At any given time, roughly 90% are actively growing hair (*anagen phase*) while 10% are in a resting phase (*telogen phase*). Males tend to produce more "terminal" hairs, which are thicker and darker. Several factors decrease hair growth. One is atrophy of the hair follicle, which is associated with male pattern baldness.

Decreased blood flow and nutritional imbalance can also lead to hair loss. Aging reduces the number of active hair follicles.

Several compounds are marketed as hair growth promoters. In the United States, the best known are minoxidil (Rogaine) and finasteride (Propecia). Minoxidil was first approved in 1988 and became an OTC product in 1996. Finasteride received FDA approval for men in 1998 but is not prescribed for women, because it can cause specific birth defects. In Japan, both 6-(benzyl-amino)purine (cytokinin B) and glyceryl monopentadecanoate are marketed as hair growth stimulants. The structures are given in Scheme 6.2. Minoxidil was discovered accidentally. It was originally developed as an oral antihypertensive drug and during the trials excess hair growth (*hypertrichosis*) was observed. Certain related antihypertensive drugs, such as diazoxide and pinacidil, exhibit similar effects. Minoxidil is used as a topical formulation for promoting hair growth. The exact mechanism of action is uncertain. It has been shown that a sulfated metabolite is the active form and that it is a potassium channel activator. Finasteride is taken orally and is an inhibitor of type II testosterone hydrogenase, thus reducing the androgen level. Other mechanisms have been identified for stimulation of hair growth. One is activation of the follicles and promotion of telogen phase to anagen phase. Cytokinin B and glyceryl monopentadecanoate appear to follow this mechanism. Various natural extracts have been recommended as hair growth promoters.

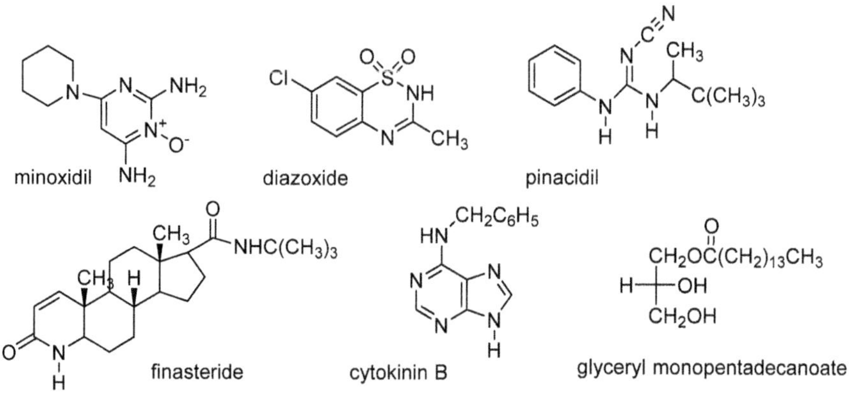

SCHEME 6.2   Hair growth promoters.

Recently, a pharmaceutical product for unwanted facial hair was introduced as *Vaniqa*, by the Gillette Company. As with minoxidil, this potential application was discovered in connection with another study. The active ingredient, eflornithine, is an inhibitor of ornithine decarboxylase, an enzyme important in certain growth regulation processes. Eflornithine has been evaluated both as an anticancer drug and in the treatment of trypanosomiasis (see Section 18.2.1). The product is FDA-approved for topical facial hair removal as a 13.9% solution in an aqueous vehicle containing glyceryl stearate, PEG-100 stearate, cetearyl alcohol, cetareth-20, mineral oil, stearyl alcohol, and dimethicone.[10]

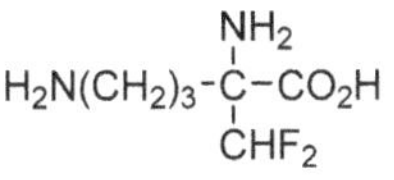

$$H_2N(CH_2)_3-\underset{\underset{CHF_2}{|}}{\overset{\overset{NH_2}{|}}{C}}-CO_2H$$

eflornithine

Approval and regulation of hair growth products varies around the world. In the United States, they are considered drugs and must be approved by the FDA new drug process (See Section 11.4.1.2). In Europe, they are treated as cosmetics and ingredients must be on the approved list of cosmetics. Generally speaking, the US regulations require evidence of efficacy as well as safety, whereas the European regulations are limited to safety. In Japan, products can fall into either cosmetic or drug categories, depending on the ingredients.

6.3.3   TOOTHPASTE

The origin of toothpaste goes back to ancient times. The Greeks used pumice, talc, or other powders, using their finger to rub the teeth. The Romans used powders made from bone, eggshell, or oyster shells. The prophet Mohammed recommended a twig from the tree *Salvador persica*, which turns out to contain $NaHCO_3$ and tannins. Native Americans used plant roots and resins or tobacco mixed with ground mussel shells. The Chinese are thought to have invented the tooth brush around 1500. The modern form of toothbrush was introduced in England around 1780.

The surface or enamel of the tooth is primarily the mineral hydroxyapatite, $Ca_{10}(PO_4)_6(OH)_2$. Under that is a softer material, the dentin, containing calcium phosphate and about 20% collagen. The soft tissue within is called the "pulp." The tooth is connected to the bones of the skull and jaw by the *periodontium*, which consists of a thin bone-like layer (called *cementum*), and the periodontal ligament, a connective tissue. The lower part of the tooth is covered by the gums or *gingival*.

Oral care in the United States greatly improved after WWII. Campaigns to increase tooth-brushing and community water fluoridation contributed. Improved dentifrices were also developed. Tooth decay, called "cavities" or "caries," results from acidic decomposition of the calcium hydroxyapatite in tooth enamel. A main cause of the decay is formation of lactic acid by bacterial metabolism of carbohydrates, so there is an association between tooth decay and sugar use. The function of the fluoride is to promote remineralization of weakened areas of the enamel and Ca^{2+} and PO_4^{3-} also are required for deposition of fluorohydroxyapatite. The first fluoride-containing toothpaste with a demonstrated anticavity effect was *Crest*, introduced by Procter and Gamble (P&G) in 1956. The product was developed in collaboration with Joseph C. Muhler of the University of Indiana and contained stannous fluoride (SnF_2) and calcium pyrophosphate ($Ca_2P_2O_7$). At the time, both the NIH and American Dental Association (ADA) were strongly promoting fluoridation of community water supplies as the best means of decay prevention. The ADA was also leery of the advertising of dental products and had challenged P&G's advertising for its nonfluoridated brand *Gleem*. FDA approval for *Crest* was received in 1955 and sales began in 1956. Tests conducted between 1954 and 1958 established 30–60% reduction in tooth decay. In 1958, a successful ad campaign, "Look Mom, No Cavities" spurred *Crest* sales. The ADA agreed to review the data supporting the claims and in August 1960 concluded that *Crest* was "an effective decay preventive agent." *Crest* became the leading toothpaste by 1961. Major competitors soon introduced stannous fluoride tooth pastes.

Toothpastes contain several ingredients including abrasives, surfactants, humectants, thickeners, and specific varieties also include fluoride, tartar control agents, whiteners, antibacterials, flavors, and colors. The abrasives assist removal of adhering material, particularly a thin proteinaceous film called the *pellicle*. Surfactants disperse plaque and other material loosened by brushing. The humectants keep the paste fluid and contribute to taste. Tartar and calculus are hardened material built up from deposits of calcium salts. Inhibition of tartar build-up is one of the goals of toothpaste and tooth-brushing. The most common agents for this purpose are pyrophosphate salts. They seem to function by inhibiting the hardening of the initial amorphous deposits and this effect can be modeled *in vitro*. A combination of NaF and pyrophosphate can attain about 25% inhibition of tartar build up and a combination of pyrophosphate with (methyl vinyl ether)–maleic acid copolymer (PVM–MA) achieves a 50% reduction.[11] Whitening agents are usually either hydrogen peroxide or the hydrogen peroxide precursor carbamide peroxide.[12] They are applied in such a fashion as to minimize contact with the gum. Table 6.8 lists examples of the types of major ingredients.

Toothpastes frequently contain a mild antibacterial such as benzethonium, chlorhexidine, or triclosan. Time of exposure to antibacterials through toothpaste is limited to one or two minutes. Thus the degree of retention of the antibacterial agent is a significant factor in its effectiveness. The bacteria responsible for caries frequently are imbedded in biofilms consisting of carbohydrate polymers. These films tend to protect the bacteria from transient exposure to antibacterials in products such as toothpaste or mouthwash. For that reason, approaches that lead to controlled release of antibacterials are of interest. These can take the form of varnishes coated onto the teeth or dental appliances such as night-guards. A number of clinical studies on toothpaste formulated with triclosan and PVM–MA have indicated a reduction of the number of sites per patient with plaque or gingivitis.[13] The polymer forms a film that incorporates triclosan and essentially acts as a slow release agent. Results on some of the other methods of application, such as varnishes, are inconclusive.[14]

TABLE 6.8   Toothpaste Ingredients.

Gums	Thickeners	Abrasives	Surfactant	Humectants	Anti-tartar
Carboxymethyl cellulose	Silica	Silica	Sodium lauryl sulfate	Glycerin	$Na_4P_2O_7$
Cellulose ether	Alum	$CaPO_4H·H_2O$	Sodium lauryl sarcosinate	Propylene glycol	$Na_5P_3O_{10}$
Carbopol[a]	Clays	$CaCO_3$	Sodium lauryl sulfoacetate	Sorbitol	Gantrez S-70[f]
Xanthan gum[b]		$NaHCO_3$	Polaxmer[e]	Xylitol	
Carrageenans[c]		$Ca_2P_2O_7$		Polyethylene glycol	
Sodium alginate[d]		Alumina			

[a]Polyacrylic acid.
[b]Natural polymer of glucose, mannose and glucouronic acid.
[c]Partially sulfated polymer of galactose extracted from seaweed.
[d]Polymer of mannouronic acid extracted from seaweed and kelp.
[e]Block copolymer of ethylene glycol and propylene glycol.
[f]Copolymer of methyl vinyl ether and maleic acid.

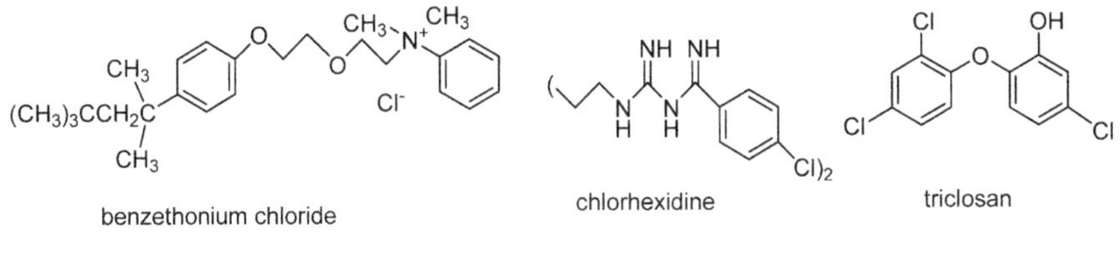

benzethonium chloride   chlorhexidine   triclosan

6.4   SKIN CARE

The outer surface of skin consists of flattened interlocking cells called the *stratum corneum*. The cells are mainly *keratinocytes*, in which the main protein, keratin, is heavily cross-linked by disulfide and glutamyl–lysine amide bonds. The next layer is the *epidermis*, which is connected to the *dermis* by the *basement membrane*. The dermis also has fibroblasts, nerve fibers, blood vessels, and sweat glands. Finally there is the *subcutaneous layer*. The skin has several crucial functions. One is to keep water in, or at least regulate its transport. Another is to keep microbes, toxins and exogenous chemicals out. The skin is also crucial in temperature control. These functions of the skin depend on several components. There are lipid layers, containing ceramides, sphingolipids, glycerides, cholesterol, and fatty acids. The elasticity is provided by the proteins collagen and elastin. Moisture retention depends on glycosaminoglycans such as hyaluronic acid. As with any living system, proper function depends on maintaining the health and balance of each component.

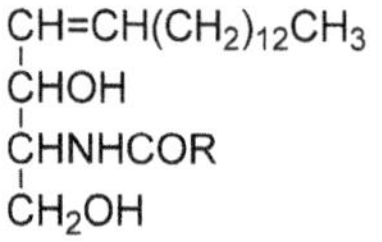

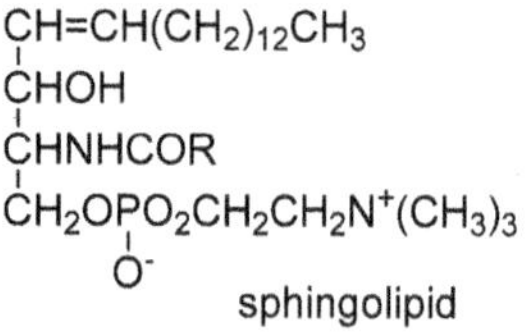

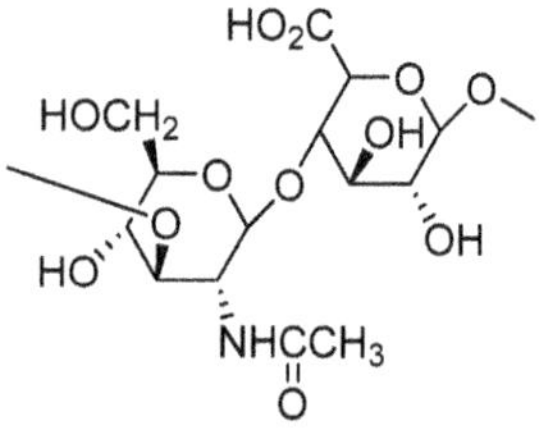

Both normal aging and external factors affect the condition of skin. Sun exposure, smoking, and exposure to certain substances can damage the skin. Soaps and detergents, for example, can remove lipids from the skin, which reduces its ability to retain moisture. Damage is often attributed to reactive oxygen and free radical species. Such reactive molecules are generated as the result of both endogenous metabolism and external factors. The damage from reactive radicals is controlled by a redox system that can intercept the reactive species. The balance between generation and destruction of free radicals determines the extent of damage. As people age, the efficiency of the redox system decreases, increasing the potential for damage. Aging has other effects on the skin as well. The rate of replacement of the epidermis slows, so the damaged cells are replaced more slowly. The skin becomes thinner. Hormone levels decrease, and this is associated with loosening and drying of the skin. Age spots, *melasma*, frequently appear as a result of uneven distribution of the skin pigment *melanin*. Skin care products are designed to prevent, reverse or mask the visible effects of skin damage and aging.

6.4.1 PRODUCTS FOR SKIN PROTECTION AND REJUVENATION

The components of skin care products may serve several purposes. One is to prevent moisture loss. For example, petrolatum, a highly purified hydrocarbon used in many products, can form a protective layer, but it is not a natural membrane component and probably cannot improve membrane function. Glycerol is used as a *humectant*. It holds water and in a cosmetic formulation can maintain water in contact with the skin. The products may also contain an *excipient*, whose purpose is to provide for adhesion of the product to the skin.

Several substances are used in efforts to improve skin appearance by minimizing the effects of aging, sun damage, acne, and other conditions. The materials have been called *cosmeceuticals*, meaning they have a cosmetic effect along with a presumed pharmaceutical benefit. Cosmetics are not classified as drugs in the United States and therefore are not required to prove clinical benefit, unless a specific health claim is made. Many of the ingredients in skin rejuvenation products are "natural" in the sense that they can be found in natural sources. This includes lipids, vitamins, botanicals, and protein hydrolysates. Natural substances have been used for cosmetic purposes from ancient times. Examples are the dyes henna and indigo, and the minerals malachite and stibium.[a] One question that needs to be considered is "What is natural?" There is no legal, chemical, nor widely accepted definition, certainly as applied to cosmetic or personal care products. Perhaps the current consumer definition might be that the substance "comes from something living." If so, how much manipulation prior to use is permissible? Chemical modification is presumably excluded, but what about a physical purification such as filtration, solvent extraction or distillation that might remove some original constituent? The FDA, which regulates cosmetics states: "There is no basis in fact or scientific legitimacy to the notion that products containing natural ingredients are good for the skin."[15] Another authority states that the belief: "natural lipids are good for you because they are natural" has no scientific foundation.[16]

6.4.1.1 FATTY ACIDS AND TRIGLYCERIDES

Oils derived from various plant sources contain unsaturated triglycerides. These oils can be incorporated into skin lipids, replenishing those removed by surfactants. Examples are borage, blackcurrant, and camelina oils. Somewhat heavier oils are called "butters" and include shea, mango, and kokum butter. These triglycerides have two structural variables, the length of the fatty acids and the degree of unsaturation. Examples that are mainly saturated are macadamia,

---

[a]Malachite is $Cu_2(OH)_2CO_3$. Stibium is antimony metal and was used as a cosmetic as a finely ground powder.

coconut, and palm kernel oils. Monounsaturated oils dominate in olive, peanut, canola, sweet almond, apricot (persic oil) and palm oils and shea butter.
Soybeans are a rich source of di-unsaturated oils, as are corn and sunflower oil.
The composition is somewhat variable depending on the precise source and process used. Beeswax contains fatty acid esters of long-chain alcohols and some
of the acids are hydroxylated. It contains excess fatty acid that can be neutralized with borax to produce creams and pastes. Other waxes found in cosmetics
include jojoba and carnauba wax. These, too, consist largely of long-chain fatty
acid esters of long-chain alcohols, such as myristyl cerotrate. The polyunsaturated oils are susceptible to oxidation and this can be accelerated by natural
photo-sensitizers. The susceptibility to oxidation increases with the degree of
unsaturation with di-unsaturated compounds about 10 times more susceptible
than mono-unsaturated compounds. Thus for cosmetic uses, to increase storage
life, both purification and the addition of antioxidants is necessary. One antioxidant that is used is the plant product tocopherol (also known as vitamin E, see
Section 10.1.1.8).[17] Another approach to this problem is to develop plants with
high mono-unsaturation by selective breeding. At least one such product, "high-
oleic sunflower oil" has been commercialized.

6.4.1.2 BOTANICALS

Various plant products are used in cosmetics. Purified apple wax can be obtained by supercritical CO_2 extraction from pomace, which is a byproduct of
apple juice production that contains the skin, core, and seeds. Apple wax's protective properties are readily observable in the storage of apples. It contains a
mixture of esters, free fatty acids, hydrocarbons, and terpenes. Ilex resins are
extracted from leaves of holly-type plants and contain mixtures of terpenes and
terpene esters. The protective properties of the holly-leaf wax are evident from
the leaves ability to stay green for long periods of time. The wax can be extracted with supercritical CO_2. Aloe contains flavones such as aloeresin A. Soy
beans are a source of genistein.

aloeresin A   genistein

6.4.1.3   LANOLIN

Lanolin is purified sheep wax.[18] The word is derived from lana (wool) and oil (oleum) in Latin. Lanolin can be extracted from wool by organic solvents or soapy water. It is a mixture of fatty acid esters and hydrophobic alcohols, diols, and steroids, including large amounts of cholesterol and lanosterol. The alcohols and sterols largely are present as esters prior to hydrolysis. About half the carboxylic acid is hydroxylated. The hydroxyl groups appear at either the α- or ω-position. The fatty acids and alcohols are not entirely straight chain with ω-1 and ω-2 methyl branching being most common. Lanolin has been used in cosmetic preparations since the time of the early Egyptians. It is readily absorbed onto the skin and conveys a smooth feel. It softens the skin and promotes rehydration of dry skin. It is used in such products as lipsticks and glosses. It has the ability to both disperse pigment and form an adhesive bond to the surface of the lip. The lanolin mixture can be fractionated into liquid and waxy components. It can also be subjected to vigorous hydrogenation, in which case the alcohol content is increased by reducing the ester groups to alcohols. Chemically modified forms of lanolin, such as acetylated or ethoxyethylated, are also available as commercial products. These products are considered to be safe in cosmetic applications by the Cosmetic Ingredient Review Expert Panel.[19] Lanolin is also used in pharmaceutical preparations. Lanolin has the potential to cause allergic reactions, although the incidence is estimated to be at most $1/10^6$. In the late 1960s and early 1970s, a "lanolin scare" arose, abetted in part by advertisers of "lanolin free" products. For a time "contains lanolin" warnings were required on cosmetics, but these were lifted in the United States and Europe by 1982. In the meantime, lanolin producers found ways to further refine lanolin, to meet exacting standards for pesticide residues, color, and odor.

6.4.1.4   HYALURONIC ACID

Hyaluronic acid is a natural component of skin. It is a copolymer consisting of alternating units of gluconic acid and N-acetylglucosamine. The acetamido glucose is bound in the polymer through the 3-position. Hyaluronic acid is used as an injectable filler in cosmetic procedures.[20] Chondroitin is a sulfate derivative of hyaluronic acid with the sulfate at either O-4 or O-6 of the acetylglucosamine subunit. At neutral pH, hyaluronic acid is ionized and is a polyanion. It is called hyaluronan or hyaluronate to indicate this ionic character.

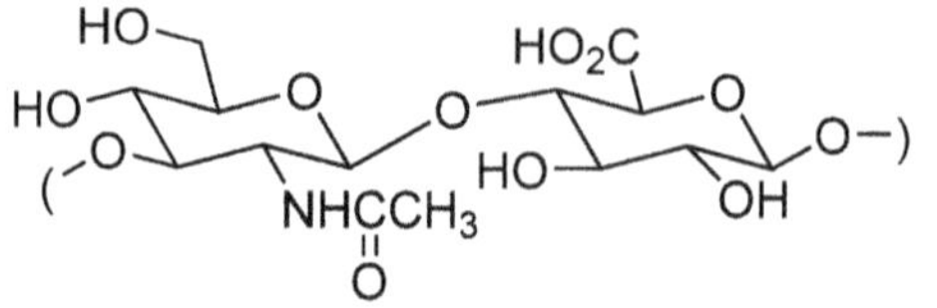

repeating unit in hyaluronic acid

6.4.1.5 EXFOLIANTS AND DEPIGMENTATION AGENTS

The α-hydroxy acids including glycolic acid and lactic acid are used as exfoliants and at higher concentration for skin peels. The FDA concentration limit on glycolic acid is 10% for OTC preparations. Trained cosmetologist can use preparations containing up to 40% and physicians can use >40% concentration for skin peels. Glycolic acid action is terminated at the desired point by washing with sodium bicarbonate, which neutralizes the acid. Glycolic acid peels are usually affected by several applications over a period of weeks. The mechanism of action is uncertain but it appears that the α-hydroxy acids promote the formation of a new layer of epidermis. They also stimulate synthesis of several components of the dermis, including collagen, glycosaminoglycan and hyaluronic acid. One potential problem with α-hydroxyl acid is posttreatment hyperpigmentation. This seems to be more common in patients with darker skin types. Because there is evidence that susceptibility to sun damage is increased immediately after exposure to α-hydroxy acids, the FDA recommends a label caution to that effect. A number of materials containing α-hydroxy acids, including glycolic acid and lactic acid, were used for cosmetic purposes in ancient times. Glycolic acid is present in sugar cane, lactic acid in sour milk and tartaric acid is present in residues of wine fermentation.

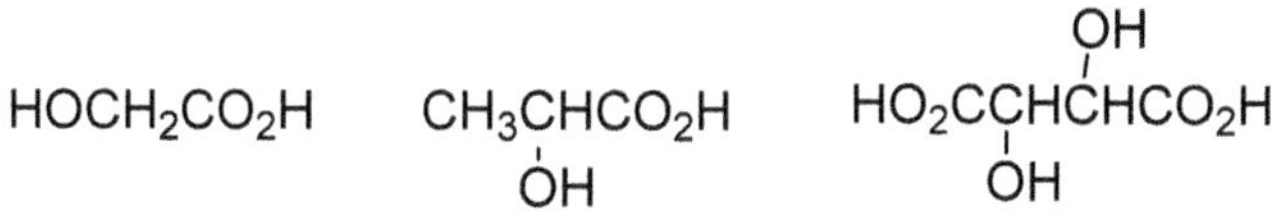

		OH
$HOCH_2CO_2H$	CH_3CHCO_2H	$HO_2CCHCHCO_2H$
	OH	OH
glycolic acid	lactic acid	tartaric acid

Some polyhydroxy acids used for skin exfoliation are carbohydrate derivatives, including gluconolactone and lactobionic acid. As is evident from its structure, gluconolactone is not an acid, but rather an acid precursor. The lactone ring can be hydrolyzed to the corresponding acid. Lactobionic acid is derived from lactose, a disaccharide of glucose and galactose. These products are considered to be less immediately irritating to the skin upon application, with less burning or stinging.

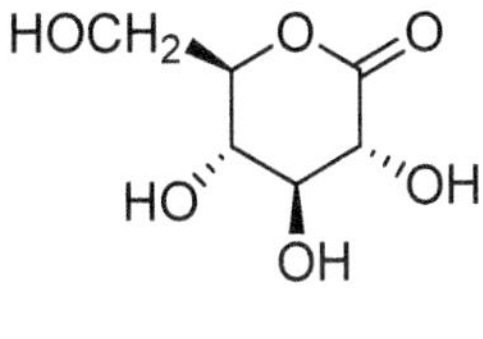

gluconolactone 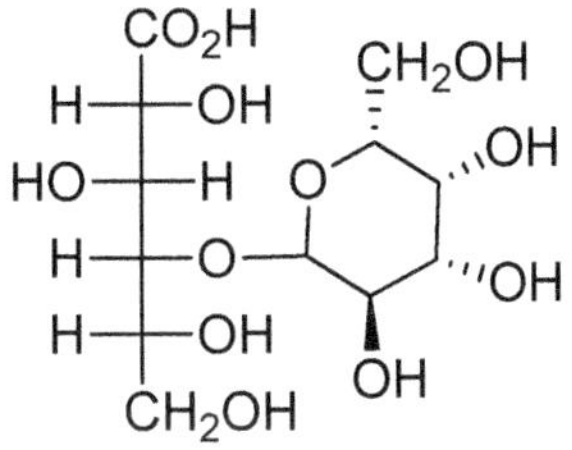   lactobionic acid

Azelaic acid is nonanedicarboxylic acid. It is used in several skin rejuvena-tion products. Kojic acid was used primarily in Japan as a skin-lightener, but has now been banned. Salicylic acid has been used since ancient times to remove corns and calluses. It is currently used in the treatment of acne and some of its effect may be the result of its anti-inflammatory activity.

azelaic acid 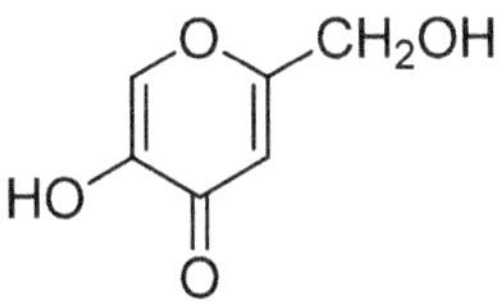   kojic acid 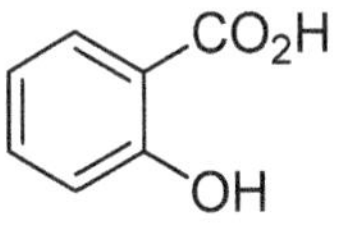   salicylic acid

Phenol was one of the first chemicals to be used for exfoliation. It is a po-tent agent and requires care in usage. Resorcinol, 1,3-benzenediol, is also used. Hydroquinone is used to effect depigmentation. A derivative, called ar-butin, which is the glucopyranoside, has a similar effect, but is less irritating. Hydroquinone is a suspect carcinogen and is banned in some parts of the world. The mechanism and efficacy of these agents are matters of uncertainty.

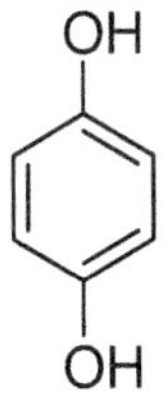

hydroquinone 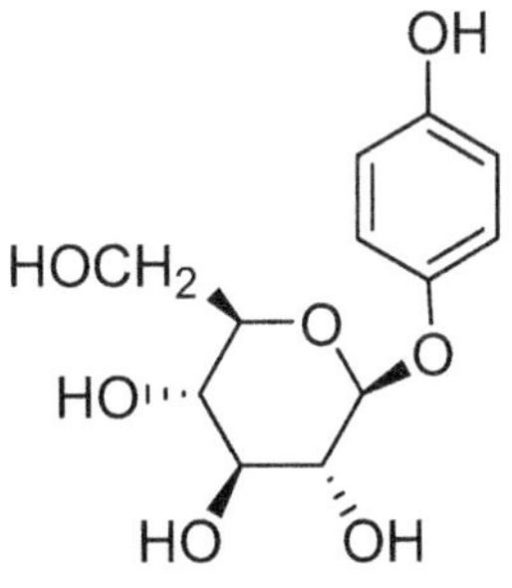   arbutin

Most of the studies evaluating the benefit of these products have been carried out over relatively short periods, for example, a few months. There seems to be very little information on long-term effects. Most of the exfoliants are used on a repetitive basis to maintain the beneficial effects.

TOPIC 6.1 TREATMENT OF ACNE

Acne is a common skin problem caused by several factors. There is hyperkeratinization and obstruction of sebaceous gland follicles, followed by infection with the organism *Propionibacterium acnes*. This results in inflammation and can lead to scarring. The skin eruptions are called *comedones*. The sebaceous glands are under hormonal control and acne is especially common in adolescents. Mild case of acne can be treated with topical retinoids, benzoyl peroxide, azelaic acid, or topical antibiotics. Systemic courses of antibiotics are also used. Hormonal treatments that have estrogenic and antiandrogenic effects (similar to birth control pills—see Section 15.3.3) are sometimes effective for females. The most effective treatment for severe acne is the *cis* isomer of retinoic acid, called *isotretinoin*, but it is a potent teratogen and strict regulation is required for its prescription. The brand name product Accutane, approved by the FDA in 1982, was the largest selling drug of the Swiss drug company Roche for a time. Generic versions have been available since 2001. Because of the powerful teratogenicity of the drug, special controls have been put in place for its prescription. During the 1980s, package warnings and letters to physicians were used to alert patients. Nevertheless, it is estimated that there were 4 instances of pregnancy for every 1000 prescriptions, which were running at 200,000 per year. With the availability of the drug in generic form, the FDA and manufactures agreed to establish an internet-based registry system whereby distributers, pharmacies, physicians, and patients are registered. Only those properly registered can distribute, prescribe, dispense, and receive the drug. Monthly confirmation of a pregnancy test and follow-up of any instance of pregnancy are required. This represents the most thorough implementation of a system for control of a potentially harmful drug yet tried.[21] In the EU, there are similar mandatory pregnancy tests before and during treatment, with immediate discontinuence in the case of pregnancy.[22] In the public media, isotretinoin (as Accutane) has been associated with adolescent depression and suicide. The possible association is complicated by the fact that acne itself can cause social and psychological problems that may lead to depression. The scientific information that is available is contradictory and inconclusive.[23] Some ocular side-effects have also been documented for isotretinoin, most commonly blurred vision, keratitis and opacity, eye dryness, and reduced night vision.[24]

Sources: Williams, C.; Layton, A. M. *Expert Rev. Dermatol.* **2006,** *1*, 429–438; Gancev-
iciene, R.; Zouboulis, C. C. *Expert Rev. Dermatol.* **2007,** *2*, 693–706; Degitz, K.; Ochsendorf,
F. *Expert Opin. Pharmacol.* **2008,** *9*, 955–971.

6.4.2 SUNSCREENS

The radiation from the sun includes ultraviolet radiation that is classified as
UV-B (290–320 nm) and UV-A (340–400 nm). There is also shorter wave-
length radiation (<290 nm) called UV-C, but it is absorbed by the ozone layer.
Exposure to UV-B radiation leads to demonstrable DNA damage. In most individuals, there is a functioning enzymatic repair mechanism. Individuals with
genetic or other impairment of the repair mechanism are very susceptible to skin
cancer. As life expectancy has increased so has the incidence of skin cancer.
Even young skin suffers damage from exposure to sun. The connecting layer
between the epidermis and dermis, known as the *basement membrane*, seems
to be the location of some of the damage. Normally, *angiogenesis* (formation of
blood vessels) is controlled by a protein called thrombospondin-1. In response
to wound healing, angiogenesis occurs. Sun exposure also leads to angiogenesis
and weakens the epidermis–dermis connection.[25] There is a wide variation in
individual sensitivity to sunlight, with more highly pigmented skin being both
less susceptible to sunburn and less efficient at vitamin D synthesis (see Section
10.1.8). It has been speculated that as humans moved from tropical regions to
higher latitudes, vitamin D photosynthesis became more important and resulted
in reduced skin pigmentation.

Sunscreens are formulated to absorb, reflect or scatter UV light and thus prevent photo-reactions with components of the skin. Several requirements must
be met: (1) the product must provide a thin uniform level of protection; (2) the
product should be easily applied by the user; (3) the product should not leave
a sticky or whitish residue; (4) the product should be durable, for example, for
protection during swimming, it must be water resistant. Sunscreens are rated by
the *sunscreen protection factor* (SPF), which represents the time of exposure
required to produce skin reddening. Thus, a factor of 30 means that 30 times
as long an exposure is required with the protecting agent. The testing is done
with volunteers. The measurement is made relative to UV-B and so the direct
effectiveness of the agents relative to UV-A is unknown. The tested rate of application is higher than typically achieved in normal use, so the SPF number
may overestimate the degree of protection. Australia, Japan, and the EU also
have UV-A tests and criteria, but they are not uniform. There are also standards
for water resistance that must be met. Existing data provide convincing evidence that sunscreens reduce the development of skin cancers in animals. The

epidemiological data for humans suggests the same effect, but limited data on sunscreen use and the long-term nature of the effect make conclusive results difficult to obtain.

The United States, Canada, and Australia regulate sunscreens as drugs, while they are considered cosmetics in the EU and Japan. In the United States, approval as an OTC product requires use of approved active components and can claim to protect against sunburn. A list of approved active ingredients is maintained by the FDA. Approval of a new ingredient must follow the New Drug Application process (see Section 11.4.1.2). Although there are dozens of UV-protective materials, only a few are approved world-wide. These include avobenzone, zinc oxide, and titanium dioxide. Other common screening ingredients are octinoxate, octisalate, octocrylene and ensulizole. Their structures are shown in Scheme 6.3. There are 23 approved sunscreens in the United States.

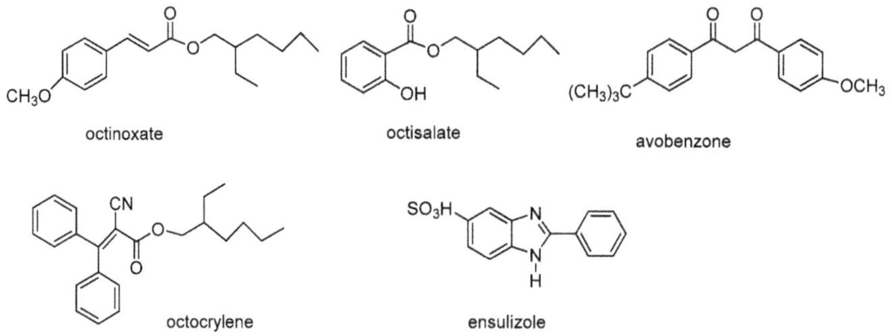

SCHEME 6.3   Representative Sun screen ingredients.

The formulation of sunscreen products can be based on oils, water, ethanol–oil or emulsions. They can range in viscosity from liquids to creams or even waxes. Purely oil or water formulations are rare. The oil-based products are difficult to formulate in ways that ensures even coverage. Water-based formulas are difficult to make durable. Ethanol–oil mixtures are fairly common, but the ethanol has the tendency to cause dry skin. Emulsions are the most common form and like other emulsions require the use of a surfactant to ensure good performance. Depending on the thickness desired, thickeners may also be required. Many products also contain humectants and emollients such as glycerin.

Barbara Gilchrist, a dermatologist at the Boston University Medical School, has been particularly active in supporting the American Dermatological Association recommendations for minimal UV exposures. She sums up her arguments as follows: "common sense, clinical observation, animal experiments, and mechanisitic studies at the molecular level all support reducing sun exposure over the entire lifetime as the preferred means of avoiding the inter-related problems of skin cancer and photoaging."[26] The "tanned look" emerged as a sign of well-being in the early 20th century. Ricketts was identified as a vitamin D deficiency disease associated with insufficient exposure to sun light (see section 10.1.1.8). Gilchrist attributes the public perception of a "healthy tan" to Coco Chanel and subsequent cosmetic industry image-making. More specifically, she regards the $5 billion tanning parlor industry as a source of much of the media interest and misinformation. Gilchrist points out that the tanning parlor industry is targeted at the population least likely to need sun exposure for vitamin D synthesis, namely young Caucasian women, whereas the most likely candidates for vitamin D deficiency are dark-skinned urban dwellers in high latitudes and the elderly. Furthermore, the kinetics of vitamin D formation are such that extended exposure to UV radiation is not beneficial. As shown in Figure 6.4, the maximum vitamin D level is already reached very near the onset of tanning.

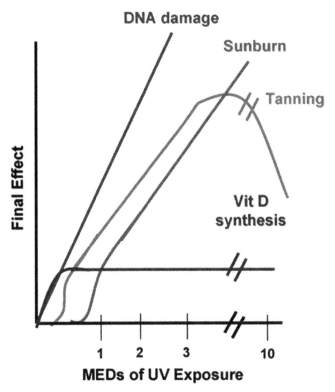

FIGURE 6.4 Relationships between sun exposure and skin damage. DNA damage is linearly related to exposure, while vitamin D synthesis reaches a plateau <0.5 MED. MED is the level at which skin-reddening is observed. (From Gilchrist, B. A. *Steroid Biochem. Mol. Biol.* **2007,** *103*, 655–663 by permission of Elsevier.)

6.5   COSMETICS

Cosmetics are considered to be substances that are externally applied to the body for appearance, scent, or feel. In the United States, cosmetics do not require prior approval for marketing, but they must list ingredients in order of descending concentration. The FDA maintains a list of several thousand ingredients that can be used. An *International Cosmetic Ingredients Dictionary* is published by the Cosmetics, Toiletries, and Fragrances Association. Colorants are regulated and must also be drawn from an approved list. Materials that are intended for some specific biological effect, such as antidandruff, anti-acne or sun-screen are considered drugs and require approval by the FDA in the United States. The regulatory system for cosmetics in Europe is similar. An inventory of about 6000 compounds that are approved is maintained. In the United States, new cosmetic ingredient testing is carried out under the auspices of the Cosmetics, Toiletries, and Fragrances Association, which is industry-sponsored. Reports contain: (1) information on physical and chemical properties; (2) effects on biological systems, including aquatic species, plants, and certain insects (e.g., honeybees); (3) biodegradation products and rates; and (4) toxicology, including acute toxicology, irritation, allergic sensitization, mutagenicity or carcinogenicity, etc. The Federal Hazardous Substances Act, administered by the Consumer Product Safety Commission, requires warning of any hazardous aspects of products. For some ingredients, but by no means all, acute toxicity data exists, but there is little information on the eventual environmental impact of the products.

In cosmetic formulations, the "active ingredients" are the substances that are believed to have beneficial effects. Other ingredients that typically are included are such things as surfactants, antioxidants, preservatives, and stabilizers. The stabilizers include materials such as hydroxypropylcellulose or xanthan gum. Partially esterified polymers of acrylic acids are also used. The products are supplied as solutions, emulsions, creams, waxes, pastes, or powders. Sticks and pastes can be made from a variety of waxy materials and usually are mixed with other ingredients in a molten state, then cooled in a container or mold. The materials contain binding agents that are called *excipients*, which include natural polymers such as chitin, chitosan, and hyaluronic acid. Powders can be made using a number ingredients such as talc (magnesium silicate = steatite), starch, clays, or silica.

6.5.1   COSMETIC FILLERS AND IMPLANTS

A number of materials can be used as injectable fillers in cosmetic applications, such as filling wrinkles. The materials that are used include bovine collagen, cultured human collagen, hyaluronic acid from rooster comb, as well as hyaluronic acid and hyaluronan produced by microorganisms. Other materials that are used include poly-L-lactic acid, fine particulate calcium hydroxyapatite and poly(methyl methacrylate) microspheres. Silicones have also been used for cosmetic purposes, as discussed in Section 7.3.1. These materials differ in potential for allergic response, ease of injection, and duration of effect. Most are FDA-approved, at least for some aesthetic applications. Some of the fillers contain microparticles that remain at the site of the injection more or less permanently. They tend to become enmeshed in tissue which, if properly controlled, accomplishes the cosmetic objectives, but sometimes give rise to nodules. It has been suggested that the nodules may result from low-grade bacterial infections maintained in biofilms associated with the particles.[27]

6.5.2 NAIL POLISHES AND EXTENDERS

Nail polishes contain a soluble resin, a platicizer and pigments in a relatively volatile solvent. Nitrocellulose, acrylates, and polyamides are the most common resins. Camphor and dibutyl phthalate are common platicizers. Nail extenders can be nylon film that is built up using a lacquer.

6.5.3 LIP AND EYE MAKEUP

Lipsticks and lip liners contain a pigment dispersed in a lipophilic material. High-gloss lip makeup often contains castor oil (which is a triglyceride containing mainly ricinoleic acid) or 2-octyldecanol. Eye shadow contains mascara dispersed in a lipid such as glyceryl monostearate. Eye liners contain mainly talc, with pigment, aluminum stearate, and lipid.

6.5.4 PRESERVATIVES IN COSMETICS

Many cosmetics and similar personal care products contain preservatives to control bacterial growth. Some of these have become controversial. One group is called *parabens* and are alkyl esters of *para*-hydroxybenzoic acid. The parabens have been considered to possibly be exogenous estrogens and endocrine disruptors, based mainly on *in vitro* studies[28] (see Section 9.7). Another type that has elicited concern are formaldehyde generators such as 1,3-*bis*-(hydroxymethyl)-5,5-dimethyl hydantoin (DMDM hydantoin), quaternium-15, and imidazolidinyl urea. The structures are given in Scheme 6.4. Also used is a mixture of methylisothiazoline and methylchloroisothiazolidine. These compounds have some potential for allergenic sensitization. The incidence of sensitivity to the parabens is <0.5% and for the formaldehyde-release agents it ranges around 1–2%. Both the United States and EU require that the preservatives be listed and there are maximum concentrations that are permitted. New preservatives must be approved by appropriate safety committees. In the United States, this is done through the Cosmetic Ingredient Review process, which includes representatives of the industry, government, and consumers.

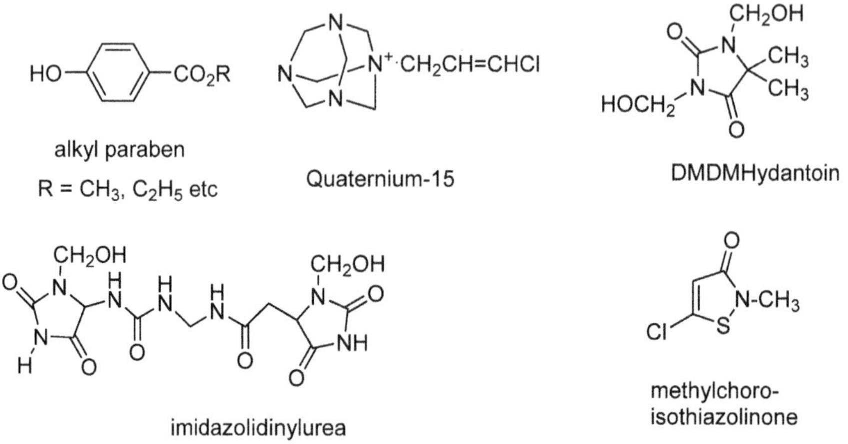

alkyl paraben
R = CH$_3$, C$_2$H$_5$ etc

Quaternium-15

DMDMHydantoin

imidazolidinylurea

methylchoro-
isothiazolinone

SCHEME 6.4   Preservative found in personal care products.

6.6   ADVERTISING OF HOUSEHOLD AND PERSONAL CARE PRODUCTS

Household and personal care product usage has been intimately connected with advertising throughout the 20th century. The national market in the United States began in the late 1800s with the development of transportation and communication systems and the beginning of urbanization. At the same time, the increase of productive capacity and income meant that consumption could move beyond bare necessity. Advertising was critical to bridging the gap between necessity and productive capacity. The first means of mass advertising was national magazines, which blossomed in the early 1900s. Titles such as *Cosmopolitan* and *The Saturday Evening Post*, still claim recognition, but dozens more existed. Magazines such as *Women's Home Companion, Good Housekeeping, Ladies Home Journal*, and *Better Homes and Gardens* appealed particularly to homemakers. The nationally circulated magazines depended on advertising to keep the cost of publication and distribution affordable.

After WWI, soap-makers faced declining sales, at least in part because life had become cleaner. Automobiles had replaced horse-drawn vehicles. Factory work relied more on machinery and less on manual labor. The population was shifting to urban from rural. At the same time, the increasing availability of electricity and household appliances was changing the way personal care was done. In 1927, the soap industry founded the "Cleanliness Institute" to raise the standard of cleanliness and hygiene (and to sell soap). The Institute campaigned for cleanliness in schools, promoting hand-washing, and frequent bathing. The widening recognition of the role of germs in causing disease made soap a "*life-preserver*" as in advertisements for Lifebuoy a "*sanitary disinfectant*" soap sold by Lever Brothers. Magazines, newspapers, and the emerging radio media featured advertisements for cleanliness. Radio introduced the "soap operas" largely supported by soap-company advertising. Procter & Gamble sponsored *Ma Perkins*, Lever Brothers, "Lux Radio Theatre," etc. In 1935, combined advertising in magazines for cleansers and toiletries amounted to about $20 million, ranking behind only automobiles and food in national publications.[29] Several themes dominated the advertising. Mildness to the skin has long been one of the advertising themes for soaps and related products. Ivory's "99.44% pure" was introduced in 1882.[b] In 1926, P&G introduced a perfumed soap, Camay, following extensive market research on both the shape and perfume for the bar. Advertising phrases included "You'll be lovely every day with fabulous pink Camay." Advertising also connected products with "loveliness" and beauty as in "The skin you love to touch" appeal of Woodbury's Facial Soap. Oil of Olay's slogan for a time was "Love the Skin You're In." The 10 years following WWII saw the introduction of TV into American households. Between 1949 and 1956, the ratio of TV:radio advertising dollars went from 1:10 to 2.5:1. In 1952, P&G launched the long-running TV daytime dramas, *Guiding Light* and *Search for Tomorrow*. Its 1957 budget for advertising was $57 million, with more than 80% devoted to TV. By 1970, P&G was spending $200 million annually on TV advertising.

---

[b]James Norris Gamble had studied chemistry and made contact with Campbell Morfit, a leading expert on soap-making. Morfit had authored an 1847 book entitled *A treatise on chemistry applied to the manufacture of soap and candles: being a thorough exposition, in all their minutiae, of the principles and practice of the trade, based upon the most recent discoveries in science and art*. Morfit was the first chemistry professor at the Maryland Institute for the Promotion of Mechanic Arts, located in Baltimore. During the 1870s, Gamble experimented with various combinations, with one goal being a soap that would float. According to one story, the first floating soap was produced accidentally when a batch was stirred longer than the normal time. If so, the "accident" was nevertheless part of Gamble's effort to produce a unique product. The product was given the name "Ivory" and soon (1882) the mottos "*It floats*" and "*99 and 44/100th pure*" were introduced. The soap was distinctive in being sold as individually packaged bars. Up to that time, most soap was sold as large slabs that were cut by the retailer. Ivory was priced between this form and more expensive imported bar soaps.

6.7   ENVIRONMENTAL EFFECTS AND REGULATION

The environmental profile of detergents and other cleansers is very complex. A comparison of the energy use, gaseous emissions and solid waste generation found relatively small differences among the various sources of raw materials and processes, but the analysis is extremely complex. Energy and the attendant CO_2 emissions occur throughout raw material production and processing. Emissions are associated with "natural" and well as "synthetic" products. For example, methane is produced by anaerobic treatment of the byproducts of palm oil mills. Despite the huge quantities of cleansers produced, the total usage of material and generation of CO_2 is very small relative to the amount generated by transportation and industry. Of the 10% of petroleum that is used for petrochemicals, less that 3% is used in surfactant production. On the other hand, as much as 15% of oleochemicals go into surfactants.[30]

The first environmental problem to surface connected with consumer products was excessive foaming in water treatment plants, which sometimes extended to streams and rivers. The foaming caused by the ABS detergents was addressed by voluntary industry change to more easily degradable material (see Section 6.1.2.4). The issue of phosphate eutrophication was addressed in the United States by a variety of local and state regulations that led manufacturers to reduce phosphate use. In Europe, there are also differing limitations in place, but phosphate use has been reduced as the product mix shifted to compact detergents. Because of the very large amounts used, the environmental impact of household detergents is of continuing interest. Furthermore, most are used in such a way that they go "down the drain" to the sewage treatment system. In general, the current major surfactants appear to have low environmental impact.

There have been studies of the eventual disposition of surfactants. The LAS occur at very significant levels in the initial sludge of water treatment plants, evidently precipitated as Ca^{2+} and Mg^{2+} salts. Concentrations in the range 5–15 g/L are observed in initial settling sludge. They are not degraded rapidly under anaerobic conditions but are rapidly biodegraded in aerobic conditions. Because there is significant use of water treatment sludge in agriculture, studies have also been conducted on this use. It is found that the compounds are readily biodegraded in soil (half life of 15–30 days), with no evidence of accumulation by either plants or animals.[31] In the soil, LAS are degraded oxidatively, first at the alkyl chain and then the benzene ring.

The alkylphenol ethoxylates (Section 6.1.3.1) have come under particular scrutiny. The issue is the biotransformation, including in water treatment facilities, to alkylphenols, particularly octyl- and nonylphenol, which exhibit both toxicity to aquatic organisms and weak estrogenic effects.[32] While the alkylphenol ethoxylates are effectively biodegraded under aerobic conditions, they can accumulate in anaerobic sludge. The phenolic metabolites have been detected in significant quantity downstream of plants that use the surfactants in manufacturing, such as wool and paper processers. Among the degradation products are octyl and nonylphenol. Since the side chains of these phenols are highly branched, they are quite resistant to oxidative degradation. Both octyl and nonyl phenols show estrogenic activity, which is not surprising given their general structural similarity to estradiol. One of the problems is that the nonyl phenols is a mixture of as many as 20 different isomers, making it difficult to identify and study specific degradation products. The products were voluntarily removed from many European household products beginning in 1995 and from industrial products in 2000.[33] In contrast, these materials remain available in the United States. An industry-sponsored review of data available in 2008 concluded that neither the products nor their metabolites meet the criteria for persistent, bioacumulative, and toxic compounds.[34]

Some European countries have introduced "eco-labeling" as a means of providing incentives for manufacturers to produce and consumers to select, products with low potential for environmental harm.[35] This requires more information than is usually available about the relative characteristics of potential ingredients. However, it also suggests use of the minimal number of ingredients necessary for satisfactory performance. For example, ingredients whose sole purpose is appearance of the product might be eliminated if consumers were convinced of their lack of benefit and their potential environmental significance.[36]

KEYWORDS

- **surfactants**
- **soaps**
- **detergents**
- **micelles**
- **oleochemicals**
- **petrochemicals**
- **personal care products**
- **skin care products**
- **cosmetics**

BIBLIOGRAPHY BY SECTION

Section 6.1.1: Scheibel, J. J. *J. Surfactants Deterg.* **2004,** *7*, 319–328.

Section 6.1.2: Kaler, E. W. *Detergents and Cleaners: A Handbook for Formulators*; Lange, K. R., Ed., 1994; pp 1–42; Porter, M. R. *J. Surfactants Deterg.* **1997,** 579–608; Edwards, C. L. *Surfactant Sci. Ser.* **1998,** *72*, 87–121; Adami, I. *Surfactant Sci. Ser.* **1999,** *142*, 83–115; Scheibel, J. J. *Surfactant Sci. Ser.* **1999,** *142*, 117–137.

Section 6.1.3.1: Edwards, C. L. *Surfactant Sci. Ser.* **1998,** *72*, 87–121.

Section 6.1.3.3: von Rybinski, W.; Hill, K. *Angew. Chem. Int. Ed.* **1998,** *37*, 1328–1345; Grover, A. R. *Surfactant Sci. Ser.* **1999,** *142,* 49–67; Schmid, K.; Tesmann, H. *Surfactant Sci. Ser.* **2001,** *98*, 1–69.

Section 6.1.4: Koeberle, P. *R. Chem. Soc. Spec. Publ.* **1999,** *230*, 213–226; Otterson, R. *Chem. Technol. Surfactants* **2006,** 170–185.

Section 6.2.1: Showell, M. S. Ed.; *Powdered Detergents*, Marcel Dekker: New York, 1998; Olsen, H. S.; Falholt, P. *J. Surfactants Deterg.* **1998,** *1*, 555–567; Gupta, R.; Beg, Q. K.; Lorenz, P. *Appl. Microbiol. Biotechnol.* **2002,** *59*, 15–32; Scheibel, J. J. *J. Surfactants Deterg.* **2004,** *7*, 319–328; Hargraeves, T. *Chem. Rev. (Deddington, UK),* **2005,** *14*, 17–20.

Section 6.2.2: Cahn, A. *Surfactant Ser.* **1997,** *67*, 1–19; Raney, O. *Surfactants—A Practical Handbook*; Lange, K. R., Ed.; Hanser Publishers: Munich, 1999; pp 171–203.

Section 6.3.1: Reever, R. *Surfactant Sci. Ser.* **1997,** *67*, 409–432; Friedman, M. *Soaps, Deterg., Oleochem., Pers. Care Prod.* **2004,** 147–188.

Section 6.3.2: Lochhead, R. Y. *The Chemistry and Manufacture of Cosmetics*, Vol. 2, 3rd ed.; Allured Publishing Corp., 2000; pp 277–326; Zviak, C.; Vanlerberghe, G. In *The Science of Hair Care,* 2nd ed.; Bouillon, C.; Wilkinson, J., Eds.; CRC Press: Boca Raton, FL, 2005; pp 83–127; Dubief, C.; Mellul, M.; Loussouarn, G.; Saint-Leger, D. In *The Science of Hair Care*; Bouillon, C.; Wilkinson, J., Eds.; CRC Press: Boca Raton, FL, 2005; pp 129–187; Reich, C.; Su, D. T. *Handbook of Cosmetic Science and Technology*, 2006; 331–346; Tarng, J. J.; Reich, C. *Surfactant Sci. Ser.* **2006,** *129*, 377–449.

Section 6.3.2.2: Lochhead, R. Y.; Huisnga, L. R. *Cosmet. Toiletries* **2005,** *120*, 59–68; Gomes, A.; de Almeida, D. F.; Abbehasen, C.; Johnson, B. K. *Cosmet. Toiletries* **2006,** *121*, 57–66.

Section 6.3.2.3: Corbett, J. F. *Dyes Pigments* **1999,** *41*, 127–136; Hefford, R. J. W. *Household Pers. Care Today* **2008** (Suppl. 1), 35–38.

Section 6.3.2.4: Jaynes, Jr., E. N. In *The Chemistry and Manufacture of Cosmetics*, 3rd ed., Vol II; Schlossman, M. L., Ed.; Allured Publishing Corp.: Carol Stream, IL, 2000; pp 205–244.

Section 6.3.2.5: Trancik, R. J. *Cosmet. Sci. Technol. Ser.* **2000,** *23*, 57–72.

Section 6.3.3: Prencipe, M.; Masters, J. G.; Thomas, K. P.; Norfleet, J. *ChemTech* **1995,** *25*, 38–42; Gaffar, A. *Handbook of Cosmetic Science and Technology*, 2001; pp 619–643; Davies, R. M. *Dental Update* **2004,** *31*, 67–71; Sellou, L.; Harrison, T. *Chem. Rev. (Deddington, UK)* **2007,** *17*, 17–20.

Section 6.4: Dykes, P. *Int. J. Cosmet. Sci.* **1998,** *20*, 53–61; Menon, G. K.; Duggan, M. In *Skin Care Delivery Systems*; Wiley, J. J., Ed.; Blackwell Publishing: Ames, IA, 2006; pp 25–41.

Section 6.4.1: Kretz, A.; Moser, U. *Handbook of Cosmetic Science and Technology*, 2001; 463–472; Huang, C. K.; Miller, T. A. *Aesthetic Surg. J.* **2007,** *27*, 402–412; Tsai, T. C.; Hantash, B. M. *Clin. Med.: Dermatol.* **2008,** *1*, 1–20; Shahi, S.; Athawale, R.; Ghadge, S. *SOFW J.* **2008,** *134*, 2–16.

Section 6.4.1.2: Krautheim, A.; Gollnick, H. P. M. *Dry Skin and Moisterizers*, 2nd ed., 2006; pp 375–390.

Section 6.4.1.3: Rieger, M. M. *Cosmet. Toiletries* **1994**, *109*, 57–68; Kripp, T. C. *Biorefineries—Ind. Processes Prod.* **2006**, *2*, 409–442.

Section 6.4.1.5: Murad, H.; Shamban, A. T.; Premo, P. S. *Cosmet. Dermatol.* **1995**, *13*, 295–307; Monheit, G. D. *Skin Ther. Lett.* **2004**, *9*, 6–11; Brinden, M. E. *Cutis* **2004**, Suppl. 2, 18–24; Fabbrocini, G.; De Padova, M. P.; Tosti, A. *Facial Plast. Surg.* **2009**, *25*, 329–336.

Section 6.4.2: Gasparro, F. P.; Mitchnick, M.; Nash, J. F. *Photochem. Photobiol.* **1998**, *68*, 243–256; Tanner, P. R. *Dermatol. Clin.* **2006**, *24*, 53–62.

Section 6.5: Rieger, M. M. *Kirk–Othmer Encyclopedia of Chemical Technology*, 5th ed., 2004, Vol. 7; pp 820–865; Eppley, B. L.; Dadvand, B. *Plast. Reconstr. Surg.* **2006**, *118*, 4 (Suppl.) 98e–106e; Broder, K. W.; Cohen, S. R. *Plast. Reconstr. Surg.* **2006**, *118*, 3(Suppl) 7s–14s; Brandon, R. *Ugly Beauty, Helena Rubenstein, L'Oreal and the Blemished History of Looking Good*; Harper: New York, 2011.; Vaughan, H. *Sleeping with the Enemy: Coco Chanel's Secret War*; Knopf: New York, 2011.

Section 6.6: Lundov, M. D.; Moesby, L.; Zachariae, C.; Johansen, J. D. *Contact Dermatitis* **2009**, *60*, 70–78.

Section 6.7: Nimrod, A. C.; Benson, W. H. *Crit. Rev. Toxicol.* **1996**, *26*, 335–364; Ying, G.-G.; Williams, B.; Kookana, R. *Environ. Int.* **2002**, *28*, 215–226; Montgomery-Brown, J.; Reinhard, M. *Environ. Eng. Sci.* **2003**, *20*, 471–486; Stalmans, M.; Sabaliunas, D. *SOFW J.* **2004**, *130*, 28–40.

REFERENCES

1. For a contemporary summary see: Kastens, M. L.; Ayo, Jr., J. J. *J. Ind. Eng. Chem.* **1950**, *42*, 1626–1638.

2. von Fenyes, C. K.; Johnson, R. C.; Norton, D. G. *Drug Cosmet. Ind.* **1970**, *107*, 36–47, 116–121.

3. For a contemporary account see Anonymous. *Popul. Mech.* **1962**, *118*(1), 90–94.

4. The Detal process is based on a solid bed catalyst that replaced alkylation reactions based on hydrogen fluoride and aluminum chloride catalysts. Vora, B.; Pujado, P.; Imai, T.; Fritsch, T. *Chem. Ind.* **1990**, 187–191; Kocal, J. A.; Vora, B. P.; Imai, T. *Appl. Catal. A* **2001**, *221*, 295–301.

5. Jacob, S. E.; Amini, S. *Dermatitis* **2008**, *19*, 157–160.

6. Milne, N. J. *J. Surfactants Deterg.* **1998**, *1*, 253–261.

7. Bryan, P. N. *Biochim. Biophys. Acta* **2000**, *1543*, 203–222; Maurer, K.-H. *Curr. Opin. Biotechnol.* **2004**, *15*, 330–334.

8. Matzer, E. A. *Surfactant Sci. Ser.* **1998**, *71*, 314–344; Windd, T. *Tensides, Surfactants, Detergents* **2007**, *44*, 19–24.

9. Gomes, A.; De Almeida, D. F.; Abbehausen, C. *Cosmet. Toiletries* **2006**, *121*, 57–66.

10. Shander, D.; Ahluwalia, G. S.; Morton, J. P. *Cosmet. Sci. Technol. Ser.* **2005**, *27*, 489–510.

11. Gaffar, A.; Afflitto, J.; Nabi, N. *ChemTech* **1993**, 38–42; Panagakos, F. S.; Volpe, A. R.; Pettone, M. E.; DeVizio, W.; Davies, R. M.; Proskin, H. M. *J. Clin. Dentist.* **2005**, *16*(Suppl 1), S1–S19.

12. Joiner, A. *J. Dentist.* **2007**, *35*, 889–896. Carbamide peroxide is a 1:1 complex of urea and hydrogen peroxide.

13. Davis, R. M.; Elwood, R. P.; Davis, G. M. *J. Clin. Periodontol.* **2004**, *31*, 1029–1033; Ciancio, S. G. *Compend. Continuing Educ. Dentistry* **2007**, *28*, 178–183.

14. Twetman, S. *Caries Res.* **2004**, *38*, 223–229.

15. Lewis, C. *FDA Consumer*, May 1998.

16. Rieger, M. M. *Cosmet. Toiletries* **1994**, *109*, 57–68.

17. Arquette, D. J.; Cummings, M.; Dwyer, K.; Kleinman, R.; Reihnardt, J. *Cosmet. Toiletries* **1997**, *112*, 67–72.

18. Barnett, G. *Cosmet. Toiletries* **1986**, *101*, 21–44; Thewlis, J. *Agrofoodind. Hi-Tech*, **1997**, *8*(May–June), 14–20; *8*(July–August) 10–15; *8* (September–October), 37–40; Harris, I.; Hoppe, U. *Dry Skin and Moisturizers: Chemistry and Function*, 2nd ed., 2006; pp 309–317.

19. Pang, S. N. J. *Int. J. Toxicol.* **1999**, *18*(Suppl. 1), 61–68.

20. Rohricjh, R. J.; Ghawami, A.; Crosby, M. A. *Plast. Reconstr. Surg.* **2007**, Suppl., 41S–54S; Klein, A. W.; Fagien, S. *Plast. Reconstr. Surg.* **2007**, Suppl., 81S–88S.

21. Honein, M. A.; Lindstrom, J. A.; Kweder, S. L. *Drug Saf.* **2007**, *30*, 5–15.

22. Dreno, B.; Bettoli, V.; Ochsendorf, F.; Perez-Lopez, M.; Mobacken, H.; DeGreef, H.; Layton, A. *Eur. J. Dermatol.* **2006**, *16*, 565–571; Layton, A. M.; Dreno, B.; Gollnick, H. P. M.; Zouboulis, C. C. *J. Eur. Acad. Dermatol. Venerol.* **2006**, *20*, 773–776.

23. Jacobs, D. G.; Deutsch, N. L.; Brewer, M. *J. Am. Acad. Dermatol.* **2001**, *45*, S168–S175; Magin, P.; Pond, D.; Smith, W. *Br. J. Gen. Pract.* **2005**, *55*, 134–138; Marqueling, A. L.; Zane, L. T. *Semin. Cutan. Med. Surg.* **2007**, *26*, 210–220.

24. Fraunfelder, F. W. *Drugs Today* **2004**, *40*, 23–27.

25. Amano, S. *SOFW J.* **2008**, *134*, 10–14.

26. Gilchrist, B. A. *Steroid Biochem. Mol. Biol.* **2007**, *103*, 655–663.

27. Christensen, L.; Breiting, V.; Janssen, M.; Vuust, J.; Hogdall, E. *Aesth. Plast. Surg.* **2005**, *29*, 34–48.

28. Harvey, P. W.; Dabre, P. *J. Appl. Toxicol.* **2004**, *24*, 167–176.

29. Vinikas, V. *Soft Soap, Hard Sell;* Iowa State University Press: Ames, IA, 1992; p 16.

30. Stalmans, M.; Berenbold, H.; Berna, J. L.; Cavalli, L.; Dillarstone, A.; Franke, M.; Hirsinger, F.; Janzen, D.; Kosswig, K.; Postlethwaite, D.; Rappert, T.; Renta, C.; Sharer, D.; Schick, K.-P.; Schul, W.; Thomas, H.; Van Slotten, R. *Tenside, Surfactants, Detergents* **1995**, *32*, 84–108.

31. Scott, M. J.; Jones, M. N. *Biochim. Biophys. Acta* **2000**, *1508*, 235–251; Jensen, J.; Smith, S. R.; H. Krogh, P.; Versteeg, D. J.; Temara, A. *Chemosphere* **2007**, *60*, 880–892.

32. Renner, P. *Environ. Sci. Technol.* **1997**, *31*, A316–320.

33. Hoehn, W. *Textilveredlung* **2006**, *41*, 8–12.

34. Klecka, G. M.; Staples, C. A.; Naylor, C. G.; Woodburn, K. B.; Losey, B. S. *Hum. Ecol. Risk Assess.* **2008**, *14*, 1007–1024; Staples, C. M.; Klecka, G. M.; Naylor, C. G.; Losey, B. S. *Hum. Ecol. Risk Assess.* **2008**, *14*, 1025–1055.

35. Klaschka, U.; Liebig, M.; Moltmann, J. F.; Knacker, T. In *Pharmaceuticals in the Environment*; Kuiemmerer, K. Ed.; Springer: Berlin, 2004; pp 411–428.

36. Herber, R. M.; Duffus, J. H.; Christensen, J. M.; Olsen, E.; Park, M. V. *Pure Appl. Chem.* **2001**, *73*, 993–1031; Fuijtier-Poelloth, C. *Arch. Toxicol.* **2009**, *83*, 23–35.

CHAPTER 7

SILICONES: FROM CONTACT LENSES TO PAVEMENT SEALANTS

ABSTRACT

Work on silicones began in the 1930s as an effort to combine the properties of glass $(SiO_2)_n$ with the newly developed synthetic polymers. The silicones are polymers with the repeating unit $(SiR_2O)_n$ where R is a hydrocarbon group, usually methyl. The starting material is $(CH_3)_2SiCl_2$ but $(CH_3)_3SiCl$, $CH_3Si(Cl)_3$, and $SiCl_4$ can be used to introduce terminal and branched positions. The linear polymers are relatively non-viscous and can be used as lubricants, while branched materials form rubbers, caulks, and coatings. The properties of silicones can be further modified by including other groups such as phenyl, fluoroalkyl, polyethylene glycol, or amino. Silicones with modified structures have a number of biomedical uses in catheters, implanted devices and contact lenses.

7.1 PREPARATION AND PROPERTIES OF SILICONES

Silicon is one of the major elements in the earth's crust, where it occurs as the dioxide silica in sand and quartz and as silicates in rocks and minerals. Nevertheless, it is not a major element in biochemistry, with examples of biological function being rare. One reason for this is presumably the great thermodynamic stability of the unreactive oxidized forms of silicon. An exception is diatoms and sponges which accumulate structures based on oxidized silicon. This is the origin of the kieselguhr (also called *diatomaceous earth*) that Alfred Nobel used to formulate dynamite (see Section 1.2.2.1). However, there is no evidence of biological capacity to produce the more reduced forms of silicon found in silicones and other synthetic organosilicon compounds.

There are several structures possible for SiO_2 and silicates. The most symmetrical for the former is an adamantane-like structure, in which silicon occupies each of the "corner" positions with oxygen bound between adjacent silicons. In silica, the structure continues in all directions, just as diamond is continuous three-dimensional adamantane structure. The silicates are also built up from SiO_4 tetrahedra that can share one, two, or three corners of a tetrahedron.

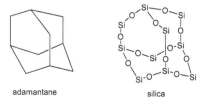

adamantane   silica

The Si–O bond is very strong (108 kcal/mol) and the Si–C bond moderate (76 kcal/mol) but the Si–Si bond is rather weak (46.4 kcal/mol). Unlike carbon, silicon does not form long chain compounds containing Si–H bonds (silanes) nor does it form double or triple bonds. The Si–H bond is more reactive than the C–H bond toward most reagents. The fundamentals of organosilicon chemistry were worked out between 1900 and 1940 by Prof. F. S. Kipping at the University of Nottingham in the UK. He characterized many organosilicon compounds and recognized the fundamental differences between carbon and silicon compounds. His work was plagued by frequent formation of gums and glasses, which eventually turned out to be useful polymers based on –Si–O–Si–O–Si–O– chains. Kipping called such compounds "silicones" and the word has entered the common vocabulary. A more technically correct name is *polysiloxanes*.Figure 7.1 shows a representation of the polysiloxane chain.

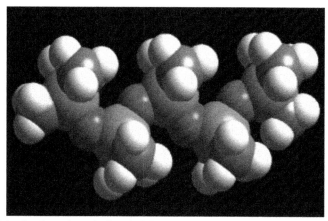

FIGURE 7.1   Portion of a polydimethylsiloxane chain. (From de Buyl, F. *Int. J. Adhesion Adhesives* **2001**, *21*, 411–422 by permission of Elsevier.)

Work on potential commercial applications of silicon compounds began at the Mellon Institute in the 1930s with financial support from Corning Glass and soon after at the General Electric Company (GE). J. F. Hyde and G. R. McGregor at Mellon and E. Rochow and others at GE found ways to prepare useful materials. There were contemporaneous studies in Russia. Metallic silicon can be obtained from silica and then converted to methylsilyl chlorides by reaction with methyl chloride.

$$SiO_2 + C \xrightarrow[\text{furnace}]{\text{electric}} Si + CO_2$$

$$CH_3Cl + Si \xrightarrow[250\text{-}300\ °C]{\text{copper}} (CH_3)_xSiCl_y$$

The silicones are prepared from dichlorodimethylsilane with the Si–O bonds being introduced by hydrolysis. Usually the compounds contain two methyl groups attached to silicon and are called *polydimethylsiloxanes*, abbreviated PDMS. Chains are terminated by trimethylsilyl groups and cross-linked by monomethyl silyl groups so that molecular weight and structure can be controlled by the addition of these linkages. If cross-linking between the chains is introduced, the materials can become resins and glasses, depending on the degree of cross-linking. There is a system for designating the type of silicon, based on the number of oxygens to which it is bound. A terminal R_3Si is referred to as M (mono, one oxygen), a dialkyl group is D (two oxygens), and cross-linking silicones with one or zero organic groups are T and Q (for tertiary and quaternary, three and four oxygens). There are also cyclic siloxanes. The most important cyclic monomer is the tetramer. These cyclic siloxanes can also be polymerized by ring-opening polymerization that can be catalyzed by either acid or base.

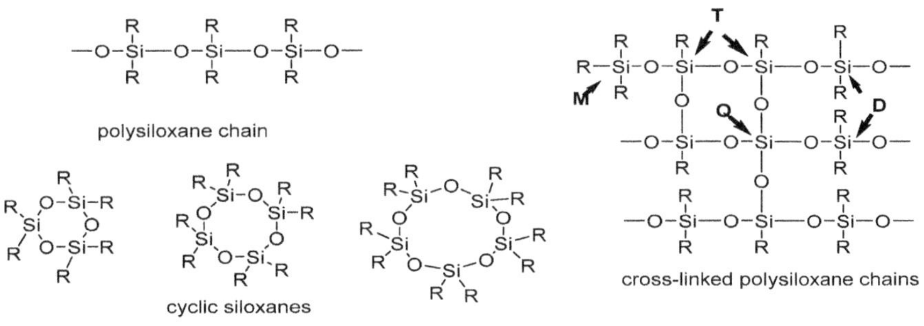

polysiloxane chain

cyclic siloxanes

cross-linked polysiloxane chains

The introduction of branching and cross-linking gives a three-dimensional structure as shown in Figure 7.2.

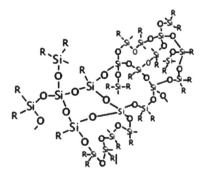

FIGURE 7.2 Structure of cross-linked silicone resin. *Technical Background Silicone Resins.* www.teg.de.

The Si–O–Si bond is very flexible and as a result the silicones have a low thermal coefficient of viscosity. At surfaces the polymers present the nonpolar methyl groups, which results in weak interactions so the PDMS are hydrophobic and have low surface tension. The polymers remain fluid liquids to quite high MW. The low surface tension allows silicone liquids to "creep" and coat other materials with a thin film. Because of their inherently low viscosity, silicone elastomers can be formulated without plasticizers. The relatively open structure also results in easy diffusion of small molecules such as water, carbon dioxide, or oxygen. The materials also have high compressibility. Silicones are quite thermally stable and resistant to flame. The chemical stability results in good weatherability of external silicone coatings.

The first applications of silicones were in insulation of electrical materials, especially in high-altitude airplanes, and provided an important technological advance during WWII. Corning and Dow Chemical formed a joint company, Dow–Corning, in 1943 and General Electric began commercial production in 1947, followed by Union Carbide in 1949. Production was about 1000 t in 1950 but by 2000, that amount had grown to 350,000 t. Dow–Corning remains a producer, but in 2006, General Electric divested its silicone and quartz product lines to Apollo Management, a private equity firm. Other companies including Wacker (Germany), Shin-Etsu (Japan), Rhone-Poulenc (France), OSi Specialties (United States), Bayer (Germany), Rhodia (France), Degussa (Germany), and Huls (Germany) contribute to the world's supply. There are multicompany industry organizations in the United States, Europe, and Japan coordinated through the Global Silicone Council. There are as many as 15,000 different products on the market containing silicones.

Silicones can be formulated to provide for "curing," that is cross-linking in place. For certain sealants, the silicones terminate with acetoxysilyl groups. After application moisture diffuses into the sealant, hydrolyzing the acetoxysiloxanes to hydroxysiloxanes, which then cross-link by condensation. Another cross-linking method involves a two-component mixture. One component contains vinyl-substituted siloxanes and a platinum catalyst. The other component is a siloxane oligomer terminating in silane (Si–H) groups. The catalysts effect addition of the silane to the vinylsiloxane, forming a covalent cross-link. A third method for cross-linking is by peroxide-initiated chain reaction, which requires heating.

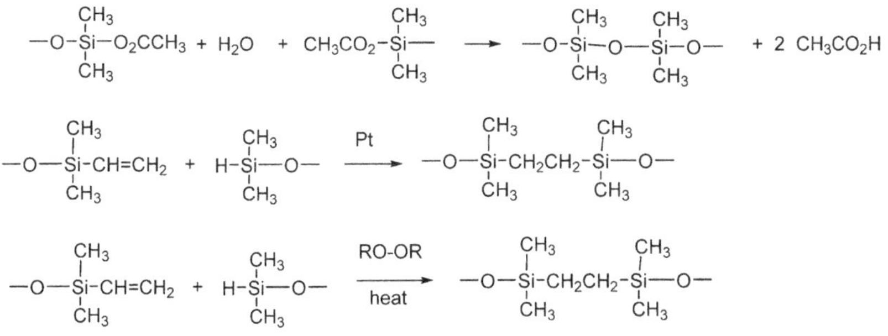

Cross-linking reactions for silicone polymers

The physical properties of silicones can be modified by substitution of specific groups. Use of phenylsiloxane units increases the hydrophobicity and solubility in organic solvents. The aryl-substituted silicones are less susceptible to degradation in the environment. Use of glycol or PEG substituents or linkers can impart properties that can range from complete solubility in water to emulsifying and foaming action. Long chain groups derived from fatty acid alcohols, such as stearyloxy silicone units provide waxy substances with applications in personal care products. Silicones with aminoalkyl substituents provide cationic materials used as hair conditioners (see Section 6.3.2.1).

Silicones that contain polyfluoroalkyl substituents are referred to a *fluorosilicones*. Fluorinated silicones were first prepared by the group of Prof. Paul Tarrant at the University of Florida and further developed at Dow–Corning. Although CF_3, C_2F_5, and C_3F_7 materials are used to some extent, the fluorinated group is usually 3,3,3-trifluoropropyl, $CF_3CH_2CH_2$. The fluorosilicones combine the high thermal and chemical stability of fluorocarbons (see Section 4.3) with the characteristic properties of silicones. Together they result in high chemical and thermal stability, resistance to attack by oil or solvents, and very low surface tension. Fluorosilicones are used in a number of high performance applications. The polymer chain is prepared from 3,3,3-trifluoropropene via the cyclic trimer.

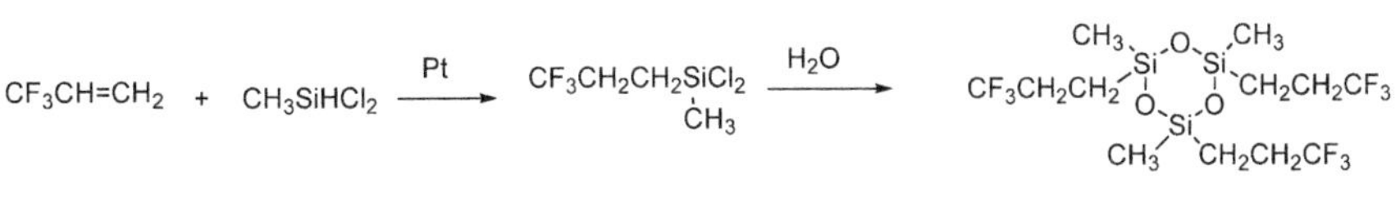

The trimer is in equilibrium with other oligomers, but can be selectively distilled from the mixture. It is then polymerized by base. Cross-linking can be achieved by including vinyl groups or by using two component systems based on hydrosilylation, as described above. Sealants based on terminal acetoxyfluorosilicones are available. The fluorosilicones are more expensive than corresponding polymethylsiloxanes.

7.2 APPLICATIONS OF SILICONES

Silicones have many applications. One is as anti-foaming agents. In this application, silicone-treated fine particulate silica is used. The silicone material spreads into and destabilizes the foam. In the manufacture of polyurethane foams (see Section 5.4), copolymers of silicones and polypropylene oxide are used to promote uniform cell formation. Silicones are used as backing for transferable labels such as stamps and address labels. The adhesive on the label adheres only weakly to the silicone backing but then under pressure sticks to the surface, for example, paper, being used. For personal care products, the silicones spread easily but are not absorbed through the skin, resulting in a soft smooth feel. The low surface attraction of silicones is also important to their biomedical applications. Proteins and lipids have little tendency to adhere to silicone surfaces.

Fluid silicones are used as heat transfer and hydraulic fluids. Silicones are used as greases and lubricants in specialized applications and may include fillers such as graphite, molybdenum disulfide, and silica. Among the applications of fluorosilicones are in highly resistant tubing and for lubrication in corrosive environments, such as the presence of acids and bases. Fluorosilicones have a wide temperature range for applicability, from −40 to 200°C, as lubricants. Fluorosilicone rubber is used for O-rings, gaskets, and bellows in critical hydraulic systems, such as aircraft.

7.2.1 USE AS SEALANTS AND COATINGS

Silicones are used extensively as sealing and caulking materials. Silicones are used in construction as structural glazing for bonding to glass, ceramic, metal, or composite material in buildings. The silicones have several features that make them useful in this application. They provide strong binding that, for example, can hold glass panes in windows against the force of wind. They provide sufficient flexibility to resist cracking or tearing under physical stress or thermal strain. Silicones are also very resistant to moisture, UV light, and common pollutants such as ozone. Pavement sealants must be able to resist both the physical force of traffic and prevent seepage of moisture through the seal into the subsurface. Another application is for coating of underwater concrete structure, such as cooling-water intakes for power plants, to prevent fouling by algae, snails, barnacles, oysters, or mussels. The non-adherent surface impedes attachment of the organisms.[1]

Silicon rubbers are not as strong as either natural or synthetic SBR rubber, but they can be strengthened with fillers such as silicates, aluminates, alumina, titania, or similar minerals. One such filler is "fumed silica" which is an aggregate of small spheroidal particles of silica formed by reaction of $SiCl_4$ with H_2 and O_2 at high temperature. This material has extremely high surface area. Furthermore, it provides S–OH bonds that can covalently bond to silicones having reactive acetoxy or alkoxy groups.

$$SiCl_4 + 2 H_2 + O_2 \xrightarrow{1800^\circ} SiO_2 + 4 HCl$$

The effectiveness of a sealant or coating depends upon adhesion to the surface that is being protected. This adhesion can be the result of several kinds of forces. One is London or van der Waals forces, which are relatively weak and depend on very close contact. Other potential attractions are hydrogen bonds or electrostatic attraction between oppositely charged groups. Materials with hydroxyl groups exposed on surfaces, such as cellulose, can be coated with a silicone layer that covalently attaches to the surface. The same is true of silica-derived glass that has exposed Si–OH groups. Some surfaces require chemical or physical treatment to permit covalent bonding. For example, hydrocarbon polymers can be subjected to surface oxidation by plasma or corona discharge to introduce hydroxyl groups. Silicones are widely used as fabric finishes and add softness as well and wrinkle and tear resistance. Alkyl-terminated silicones are nondurable, but incorporating groups that can covalently bond to the fibers make the coating durable. For fabric softeners, aminosilicon groups are used. These bond by electrostatic forces resulting from the positive charge on the amino groups.

7.2.2 USES AS SURFACTANTS

Silicones have many uses in personal care products, that depend on their surfactant properties and these are discussed in Section 6.3. A review article about silicones in personal care products opens with these words "…silicone polymers are not only among the safest ingredients, they are also a unique category of chemicals. They are environmentally friendly, and improve the application, functionality and aesthetics of personal care formulations. Thus, they should not be perceived as belonging to the same classes as the synthetic polymers or organic chemicals, but as an independent category."[2] Silicones are used in skin and hair care, makeup, fragrances, and deodorants. For skin care products, the favorable properties include excellent spreadability and a smooth nongreasy feel. Silicones are also a major component of hair conditioners, where they impart smoothness and sheen. The dimethiconecopolyols contain both silicone and polyether chains. These polymers enhance the tendency of silicones to bind at surfaces as a result of the attraction of the polyether chains for hydrophilic surfaces.

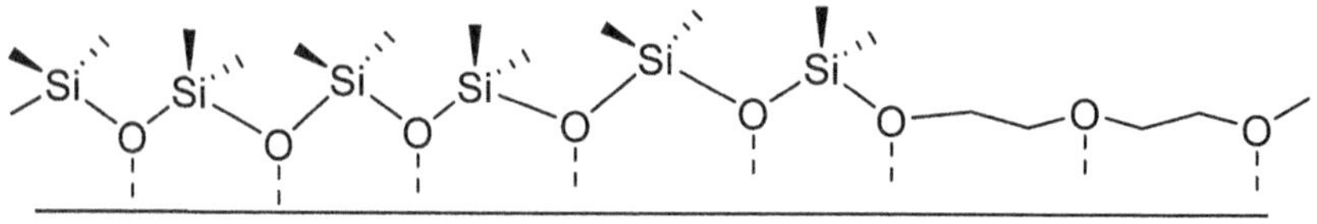

In the consumer products industry, the silicones often appear under a common name.

Structure	Common name
Polydimethylsiloxane	Dimethicone
Cyclic polydimethylsiloxanes	Cyclomethicone
α,ω-Hydroxypolydimethylsiloxanes	Dimethiconol
Polydimethylsiloxane polyoxyethylene copolymers	Dimethiconecopolyol
Phenyl polydimethylsiloxane	Phenyl trimethicone

Another example of the application of surfactant properties is the use with the hydrophilic herbicide glyphosphate (see Section 8.2.3.3) for treatment of mature woody plants. Plants with thick waxy cuticles adsorb glyphosphate poorly. Formulation with a polydimethylsiloxane polyoxyethylene copolymer (Silwet L-77) gives much better results and lower application rates.[3] The surfactant properties of silicones can be enhanced by adding charged groups. Quaternary ammonium groups can be introduced at terminal epoxide functions. Anionic phosphate, sulfate, sulfonate, and carboxylate groups can also be introduced. Zwitterionic betaines are introduced by alkylating a terminal tertiary amino group with chloroacetate.

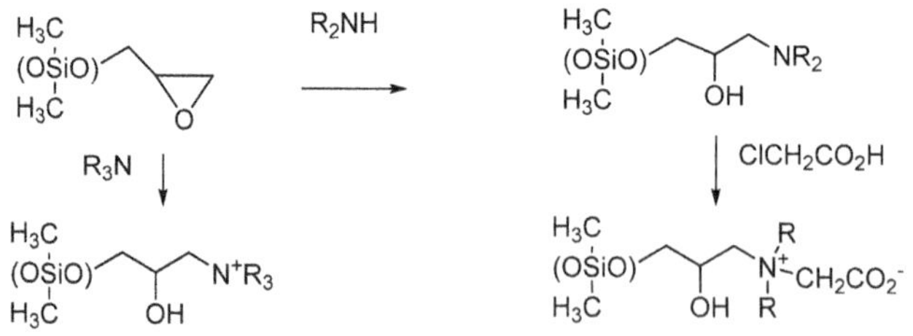

7.2.3 PRESSURE-SENSITIVE ADHESIVES

The properties of silicones are also responsible for their use as pressure-sensitive adhesives. This application is encountered in a number of familiar items, such as "no-lick" postage stamps, "post-it" labels, and in somewhat more sophisticated form, in patches for transdermal delivery of pharmaceuticals. Industrial applications include masking during fabrication of printed circuit boards and similar electronic devices. The adhesive characteristics are the result of interaction between a silicone layer laid down on the label or tape and the material that is being covered.

7.3 BIOMEDICAL APPLICATIONS OF SILICONES

7.3.1 MEDICAL DEVICES

The physical characteristics of silicones are superficially similar to hydrocarbon oils. The silicones are less susceptible to biological degradation than hydrocarbons. Silicones attracted attention as biomedical materials because their low tendencies to cause blood clotting or allergic reactions. They have weak interactions with proteins. Another valuable property is that they allow easy diffusion of small molecules such as O_2, H_2O, and CO_2. One early medical application was coating of syringes and needles with methylchlorosilanes to "silanize" them with a hydrophobic coating. This prevents blood clotting in needles and syringes and also makes the injection less painful. Silicones are inherently flexible and do not require platicizers. Platicizers are undesirable in biomedical applications because of their tendency to leach from the polymer. Silicones are used in many catheters and valves. An early use was the Holter shunt for hydrocephalic infants.[4] Silicones are used as parts of some joint implants. They are also used in implanted devices such as pace-makers and cardiovascular bypasses.

7.3.2 CONTACT LENSES

Contact lenses provide a particularly good example of how chemical composition and physical properties relate to material usage in a biomedical application. Glass was used for contact lenses to a limited extent from 1900 through the 1950s but they were expensive to make and uncomfortable to wear. The first contact lenses to be made from polymers were poly(methyl methacrylate) (PMMA) (see Section 5.4.7.1), but they had low gas permeability and hydrophobic surfaces. They could be worn for only limited times and required careful cleaning. Some wearers of these lenses experienced corneal irritation and inflammation. These symptoms were associated with long-distance air travel and the reduced oxygen level of high-altitude flight. Improvement in oxygen permeability of PMMA contact lenses was achieved by including either silicone or polyfluoro-acrylates as copolymers.

There are several crucial requirements in addition to oxygen permeability. The material must be sufficiently hydrophilic to be wettable, it must be resistant to deposits of both lipids and proteins, and it must have adequate strength and durability. Contact lenses must also be permeable to water and ions and move freely on the surface of the eye. The original material for most soft contact lenses was poly(2-hydroxyethyl methacrylate), commonly abbreviated HEMA. The material was developed by Otto Wichterle, a polymer chemist at the Czechoslovak Academy of Science in Prague. They were commercialized in the United States by Bausch and Lomb. FDA approval was obtained in 1971. These lenses were eventually produced sufficiently cheaply to allow for disposable use. Current versions often include either acrylic acid or N-vinylpyrrolidone in a copolymer. Their oxygen permeability is not high enough to permit extended wear.

For extended wear, contact lenses must have improved permeability to oxygen, water, and ions. For the purposes of soft contact lenses, hydrogels are attractive. These are polymer networks that have high water content (>35%) but nevertheless have adequate strength. These are generally copolymers of a silicone or fluorosilicone with a more hydrophilic monomer containing polar groups such as hydroxyl, polyether or amide. These materials provide a surface that is hydrophilic and yet resistant to accumulation of biomaterials. The first silicone hydrogel received FDA approval in 1998. The current materials for silicone hydrogel lenses include fluorosilicones copolymerized with various vinyl monomers such as N,N-dimethylacrylamide, vinyl carbonate, and vinyl carbamate. Another polymer used for hydrogels is polyvinyl alcohol, which is water soluble. This combination has several advantages in the molding and processing of the lens.[5] The materials are also sometimes subjected to surface treatment with a plasma. Lotrafilcon, for example, is coated with a 25-nm layer of polymer modified by the plasma. Balafilcon is oxidized to convert some of the silicone groups on the surface to silicate. The FDA post-approval study on one of the available silicone hydrogel 30-day lenses found an incidence of 1.9% for inflammatory events in the first year of wear. This was lower than in the pre-approval study.[6]

Intraocular lenses (IOL) which are used to replace opaque lenses in cataract surgery also have gone through evolution of the materials that are used. The original IOLs introduced by a British opthamologist, Sir Harold Ridley, in the 1950s used PMMA. This remains one of the main materials, although it has some disadvantages in terms of rigidity and tendency to cause adhesion of biomaterials and inflammation. Currently, most IOLs are inserted in a folded form that is unfolded during the implantation. This permits use of a smaller incision with less trauma and faster recovery. These can be made from silicone materials, but also from alternative methacrylate esters including the 2-phenylethyl and 2,2,2-trifluoroethyl esters. The IOLs also now often incorporate a material to shield the retinal from blue/violet radiation.[7]

7.3.3 TRANSDERMAL PATCHES

Transdermal patches for delivery of pharmaceuticals were introduced in 1979. One of the major advantages of this means of drug delivery is that it can provide a steady, controlled release of the medication. Examples of drugs that can be delivered are analgesics, steroids, nitroglycerin, and nicotine, the latter for smoking-cessation programs. Silicones are often components of both the matrix or reservoir for the drug and the membrane that controls release. The crucial feature of a transdermal drug delivery system is the diffusion of the drug from the reservoir or matrix through the outer layer of the skin. The silicones are often used in conjunction with other polymers to adjust the polarity and porosity of the matrix. Figure 7.3 indicates some of the designs that are used.

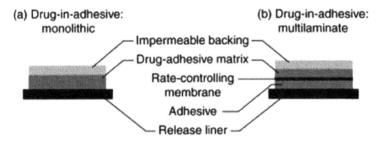

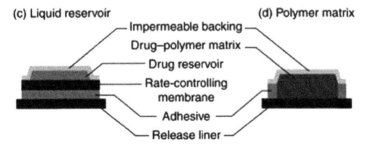

FIGURE 7.3 Various designs for transdermal drug delivery patches. (From Tan, H. S.; Pfister, W. R. *Pharm. Sci. Technol. Today* **1999**, *2*, 60–69 by permission of Elsevier.)

TOPIC 7.1 SILICONES IN COSMETIC AND RECONSTRUCTIVE SURGERY

The public's acquaintance with silicones includes their use as lubricants and sealants, but the most controversial association is with their use in cosmetic and reconstructive plastic surgery, particularly breast implants. Use of silicones injections for breast augmentation began soon after WWII, but impure industrial material was used and frequently led to disastrous results. It was not until 1963 that a product meant for medical use was available. In the meantime, numerous adverse results from direct injection had been observed. In 1964, Dow–Corning filed an "investigational exemption for a new drug" application with FDA. Over 1300 patients were treated at several centers, usually for amelioration of facial deformities or scars. A significant incidence of thickening or hardening of the treated area was observed and reported and the study was suspended by the FDA in 1976. In 1987, Dow–Corning filed an amended protocol for severe facial deformities, and a limited but inconclusive study was performed and no FDA approval followed. Other studies in this time frame also reported incidences of thickness or nodules at the silicone injection sites. During the 1970s, several clinicians reported satisfactory treatment of facial deformities by using modified techniques, specifically smaller spaced injections and use of higher viscosity silicones. These microinjection techniques are thought to be followed by growth of surrounding tissue that limits migration of the silicone. Few complications were reported in these studies; however, they were not designed for extended follow-up periods. High-purity/high-viscosity silicones are currently FDA-approved for treatment of severe retinal detachment and for prevention of retinal detachment during eye surgery. However, considerable off-label use for treatment of facial deformities continues. This is legal under FDA regulations that allow physicians to use approved materials in unapproved application if it is considered to be of medical benefit.

Contained silicone breast implants were introduced by Dow–Corning in the early 1960s. These devices used a pure medical grade silicone fluid encased in a silicone or fluorosilicone pouch. These implants were much more satisfactory than injections, but were not without problems. The scar tissue that formed around the implants tended to deform them. Also it was found that small amounts of silicone could diffuse through the envelope. Some versions of implants contained silicone in polyurethane foams, which tended to cause tissue adherence. In 1976, the medical device amendment to FDA authority brought breast implants under regulation (see Section 11.4.1.1). Because the contained silicone implants had already been in use for some time, they were exempt from the requirement of demonstration of safety. Between 1982 and 1988, FDA announced and eventually formalized a requirement for evidence of safety, prompted in part by increasing reports of problems and a few early lawsuits. The manufacturers, however, did not conduct suitable studies. Ralph Nader's Public Citizen Group took up the issue and its Health Research Director, Dr. Sidney Wolfe, began to press the FDA for action. In a 1990 TV show, Connie Chung brought the matter to wide public attention. David Kessler, the FDA Commissioner at the time, set a deadline for a decision and the FDA convened an expert panel in November 1991. Although finding the lack of evidence of safety "appalling," they recommended leaving the devices on the market for 2 years, pending further study. The AMA, American Cancer Society, and American Society of Plastic and Reconstructive Surgeons favored this decision in their testimony. Another panel was convened in February 1992. This time, Dow–Corning internal documents containing various indications of the company's concern with leakage were part of the evidence and the panel voted to remove the devices from the market. Dow–Corning withdrew them from the market in March 1992. Kessler imposed a ban in April 1992, except for reconstructive surgery and with a requirement for follow-up study.

Between 1980 and 1990, it is estimated that 100,000–150,000 implants were performed annually. By 1990, it was estimated about 1% of women had implants and surveys indicated about 90% were satisfied with the outcome. After the ban, only saline-filled implants were available for cosmetic purposes and the number of such implants dropped to about 40,000 annually. The publicity surrounding the ban of silicone implants accelerated the spread of information about them. Support groups arose and liability attorneys actively recruited potential claimants. Following the ban, the number of liability suits claiming injury or illness from breast implants exploded from 200 in 1991 to 10,000 in 1992. A class action suit included 400,000 women, of whom 248,500 were reported to be ill. A settlement of this suit in April 1994 set up a fund of $4.5 billion, of which $1 billion was for lawyer's fees. However, in May 1995, Dow–Corning filed for bankruptcy and the agreement collapsed. In the meantime, several studies failed to find any link between silicone breast implants and immune disorders or any other disease. In 1999, the Institute of Medicine, a branch of the National Academy of Sciences, issued a broad report on breast implants, concluding that there was no direct relationship between silicone breast implant devices and specific medical conditions. The study also noted that there were often effects such as tissue hardening associated with the surgical procedure, regardless of the material used. These include encapsulation of the implant by scarring and scleroderma (hardening of the skin). Currently, saline-filled implants are in wide use, with over 100,000 breast augmentation surgeries per year. In 2006, two silicone-filled devices were approved for both reconstruction and augmentation. These are manufactured by Allergan Corp. and Mentor Corp. A 10-year follow-up study is underway.

Sources: Angell, M. *Science on Trial.* W. W. Norton: New York, NY, 1996; Zimmerman, S. *Silicone Survivors: Womens' Experience with Breast Implants.* Temple University Press, 1998; Sarwer, D. B.; Nordmann, J. E.; Herbert, J. D. *J. Women's Health Gender-Based Med.* **2000**, *9*, 843–856; Brown, S. L. *Epidemiology (Cambridge, MA)* **2002**, *13*(Suppl. 3), S34–S39.

7.4   ENVIRONMENTAL ISSUES

Large quantities of silicones are introduced into the environment, mainly in the form of sealants or coatings and as personal care products. The latter end up in municipal waste water systems. Silicones that are used as sealants and coatings are typically disposed of as solid waste and are usually incinerated or placed in landfills. Incineration of silicone wastes leads to CO_2 and SiO_2, and the latter is dispersed as fine particulate silica or found in the ash. The direct environmental impact of silicones appears to be fairly small, but they are now widely dispersed in the environment. The polysiloxanes in aqueous environments can hydrolyze. The ultimate hydrolysis product of polydimethylsiloxane is $(CH_3)_2Si(OH)_2$ and it is believed to be mainly lost by evaporation to the atmosphere. Small oligomers are also quite volatile. In the atmosphere, the methylsiloxanes are degraded to SiO_2 and CO_2, mainly by reaction with hydroxyl radicals. Because the small siloxane oligomers are not believed to contribute to smog formation, they are exempt from VOC regulations.[8]

Silicones are generally regarded as rather benign in terms of adverse health and environmental effects. Their use is regulated by the Toxic Substances Control Act of 1976, which requires that EPA be notified of "new chemical substances" to be manufactured or imported and of "significant new uses" for existing chemical substances. Impending use of such materials requires 90 days notice and the requirement that "available data" on health, ecological, and environmental effects be provided. New studies, such as required for new drugs, are not normally done, but typically EPA and the manufacturer agree through a consent order to perform acute toxicity studies and to identify any aquatic toxicity. Subsequently, manufacturers are required to keep records of any reports of adverse effects. In 1989, testing was initiated on the cyclic tetramer, which is widely used in personal care products. These studies determined that likely aquatic concentrations were about 0.005 of the "no observable effect level" concentration and no further studies were recommended. In the 1990s, the EPA reviewed a list of silicones as potential targets for Priority Testing List for health effects. All but 14 were deemed not to require priority testing.[9]

Personal care products contain significant amounts of low MW silicones, including the cyclic tetramer and pentamer. These cyclic oligomers do exhibit acute toxicity at high doses. The tetramer has an LD_{50} of 6–7 g/kg by oral administration, similar to carbon tetrachloride and other halogenated solvents.[10] On the other hand, exposure of rats to cyclomethicone vapors at up to 540 ppm, 30 h/weeks for 4 weeks showed little acute toxicity, but did show elevated liver weights.[11]

Higher molecular weight polymers from consumer products appear to be concentrated mainly in sludge from waste water treatment plants. In the aqueous environment, they would be expected to accumulate in oily surface films and microlayers, and there is some evidence for this. The highest concentrations of aqueous siloxanes are found in areas with a history of sludge disposal by dumping at sea. Higher molecular weight silicones are poorly absorbed by aquatic organisms and they have not been observed to have adverse effects on a variety of marine organism.[12] In contrast to organic halogen compounds, there is no evidence of bioaccumulation. Most of the studies reported have been done on polydimethylsiloxanes. Some sludge from water-treatment plants is used as fertilizer and soil conditioners, and in this case, transient concentrations of polysiloxanes are observed. However, hydrolysis to smaller units and volatilization reduces the concentration fairly rapidly. The silanols produced by hydrolysis have significant water solubility. Also, because the reaction with water and other hydroxyl groups is reversible, silanols can become covalently bound to solids.

KEYWORDS

- silicone
- commercial applications
- physical properties
- adhesion
- cross-linking
- coatings and sealants
- lubricants
- surfactants
- biomedical applications

BIBLIOGRAPHY BY SECTION

Section 7.1: Hardy, D. V. N. *Endeavor* **1947,** *6*, 29–35; Miller, D. C. R. *Can. Chem. Process Ind.* **1949,** *33*, 744–767, 858–870; Owen M. J. *ChemTech* **1981,** 1–11; Lane, T. H.; Burns,

S. A. *Cur. Top. Microbiol. Immunol.* **1996,** *210,* 3–12; Moretto, H. H.; Schulze, M.; Wagner, G. *Industrial Polymers Handbook*; Wiley-VCH: Weinheim, Germany, 2001, Vol. 3; pp 1349–1408; Owen, M. J. *Chim. Nouv.* **2004,** *85,* 27–33; Romenesko, D.; Chorvath, I.; Olsen, Jr., C. W.; Tonge, L. M. *Kirk–Othmer Encycl. Chem. Technol.* **2006,** *20,* 239–248.

Section 7.2.1: Lower, L. D.; Klosowski, J. M. *Handbook of Adhesive Technology*; Marcel Dekker: New York, NY, 2003; pp 813–821; Parbhoo, B., O'Hare, L.-A.; Leadley, S. R. *Adhesive Science and Engineering*; Elsevier: Amsterdam, 2002; pp 677–709.

Section 7.2.2: Demby, D. H.; Stoklosa, S. J.; Gross, A. *Chem. Ind.* **1993,** *48,* 183–203; Floyd, D. T. *Surfactant Sci. Ser.* **1999,** *86,* 181–207; Butts, M.; Cella, J., Wood, C. D.; Gillette, G.; Kerboua, R.; Leman J.; Lewis, L.; Rubinsztajn, S.; Schattenmann, F.; Stein, J.; Wicht, D.; Rajaraman, S.; Wengovius, J. *Kirk–Othmer Encycl. Chem. Technol. 5th edition* **2006,** *22,* 547–626.

Section 7.2.3: Lin, S. B.; Durfee, L. D.; Ekeland, R. A.; McVie, J.; Schalau II, G. K. *J. Adhes. Sci. Technol.* **2007,** *21,* 605–623; Lin, S. B.; Durfee, L. D.; Knott, A. A.; Schalau II, G. K. *Technology of Pressure Sensitive Adhesives and Products*; CRC Press: Boca Raton, FL, 2009, pp 6-1–6-26.

Section 7.3.1: Yoda, R. *J. Biomat. Sci., Polym. Ed.* **1998,** *9,* 561–626; Colas, A.; Curtis, J. In *Biomaterials Science,* 2nd ed.; Ratner, B. D., Hoffman, A. S., Schoen, F. J., Lemons, J. G., Eds.; Elsevier: San Diego, CA, 2004, pp 80–86, 697–707.

Section 7.3.2: Narins, R. S.; Beer, K. *Plast. Recon. Surg.* **2006,** *118*(Suppl. 3), 77S–84S; Chasan, P. E. *Plast. Recon. Surg.* **2007,** *120,* 2034–2040.

Section 7.3.3: Smith, J. M.; Thomas, X.; Gantner, D. C.; Lin, Z. *Adv. Contr. Drug Deliv.: Sci., Technol. Prod.* **2003,** *846,* 113–127.

Section 7.3.4: Tighe, B. *Silicon Hydrogels: The Rebirth of Continuous Wear Contact Lenses*; Oxford Press: Boston, MA, 2000; pp 1–21; Foulks G. N. *Am. J. Opthamol.* **2006,** *141,* 369–373; Dillehay, S. M. *Eye Contact Lens* **2007,** *33,* 148–155; Chapoy, L. L.; Lally, J. M. *Key Eng. Mater.* **2008,** *380,* 149–166.

Section 7.3.5: Bruner, S.; Freedman, J. *Drug Deliv. Technol.* **2006,** *6,* 48–52.

Section 7.4: Hirner, A. V.; Flassbeck, D.; Gruemping, R. *Organometallic Compounds in the Environment*, 2nd ed.; Craig, P. J., Ed.; Wiley: Chichester, UK, 2003; pp 305–351; Chandra G.; Maxim, L.; Sawano, T. *Handbook of Environmental Chemistry*, Vol. 3, Part H; Chandra, G., Ed.; Springer Verlag: Berlin, 1997; pp 295–319.

REFERENCES

1. Wiebe, D.; Connor, J.; Dolderer, G.; Riha, R.; Dyas, B. *Mater. Perform.* **1997,** *36,* 26–31.

2. Floyd, D. T. *Surfactant Sci. Ser.* **1999,** *86,* 181–207.

3. Stevens, P. J. G. *Pestic. Sci.* **1993,** *38,* 103–122; Zabkiewicz A. *Chem. N. Z.* **2008,** *72,* 8–12.

4. Baru, J. S.; Bloom, D. A.; Murasko, K.; Koop, C. E. *J. Am. Coll. Surg.* **2001,** *192,* 79–85.

5. Buehler, N.; Haerri, H.-P.; Hofmann, M.; Irrgang, C.; Muehlenbach, A.; Mueller, B.; Stockinger, F. *Chimia* **1999,** *53,* 269–274.

6. Donshik, P.; Long, B.; Dillehay, S. M. *Eye Contact Lens* **2007,** *33,* 191–195.

7. Bozukova, D.; Pagnoulle, C.; Jerome, R.; Jerome, C. *Mater. Sci. Eng. Rep.* **2010,** *R69,* 63–83.

8. Graiver, D.; Farminer, K. W.; Narayan, R. *J. Polym. Environ.* **2003,** *11,* 129–136.

9. Hatcher, J. A.; Slater, G. S. In *The Handbook of Environmental Chemistry*, Vol. 3, Part H; Chandra, G., Ed.; Springer Verlag: Berlin, 1997; pp 241–266.

10. Lieberman, M. W.; Lykissa, E. D.; Barrios, R.; Ou, C. N.; Kala, G.; Kala, S. V. *Environ. Health Perspect.* **1999,** *107,* 161–165.

11. Klykken, P. C.; Galbraith, T. W.; Kolesar, G. B.; Jean, P. A.; Woolhiser, M. R.; Elwell, M. R.; Burns-Naas, L. A.; Mast, R. W.; McCay, J. A.; White, Jr. K. L.; Munson, A. E. *Drug Chem. Toxicol.* **1999,** *22,* 655–677.

12. Stevens, C.; Powell, D. E.; Makela, P.; Karman, C. *Mar. Pollut. Bull.* **2001,** *42,* 536–543.

PART III
Chemistry and Food

CHAPTER 8

CHEMISTRY AND AGRICULTURE: HELPING TO FEED THE WORLD

ABSTRACT

Agricultural production increased greatly during the 20th century, freeing much of the world from food shortages. Chemical methods contributed to this increase, especially through production of nitrogen fertilizer, herbicides, and insecticides. The Haber–Bosch process for conversion of nitrogen from air to ammonia by reaction with hydrogen was introduced in 1915 and now accounts for >99% of commercial nitrogen fertilizer. Synthetic herbicides and insecticides were commercialized soon after WWII, based on the discovery of 2,4-dichlorophenoxyacetic acid (2,4-D) and 1,1,1,-trichloro-2,2-*bis*-(4-chlorophenyl)ethane (DDT) during and immediately before the war. Many different types of herbicides were subsequently developed by research at chemical companies. More than 20 different classes have been recognized, based on mechanism of action. Agrochemical companies, the Department of Agriculture and the Environmental Protection Agency have developed information on optimal use and regulations that establish specific instructions for use. Nevertheless, two problems persist. One is the development of herbicide-resistant weeds and the other is herbicide residues on food crops. There are also several groups of insecticides, but most target the insect central nervous system. Some have substantial toxicity to other species, including humans. As with herbicides, regulations establish the specific conditions for use. Biotechnology has permitted the development of plant varieties that incorporate tolerance to specific herbicides and/or toxicity to certain insects. These have been widely adopted for major agricultural crops such as corn, cotton, and soy beans. Considerable public concern about the use of pesticides in food production has given rise to increased demand for "organic" or "natural" foods.

In this chapter, we will discuss substances that are used in the production of food. These compounds rank in importance with petroleum and pharmaceuticals in sustaining the current human population. Agricultural production has increased dramatically in almost all parts of the world with a doubling of productivity per acre from 1965 to 2000. This increase was based on several factors: (1) breeding of improved plants and animals; (2) use of fertilizers to increase crop yield; (3) improved agricultural methods, including irrigation; (4) use of pesticides to control insects, weeds, and fungi; (5) intensified production of livestock and poultry, including use of antibiotics; (6) mechanization that increased the productivity of individual farmers. Three of these factors, use of fertilizers, antibiotics and pesticides, have strong chemical components. During the last decade of the 20th century, methods of crop improvement and protection based on molecular biology and genetic engineering became important (see Section 21.2). Table 8.1 provides an indication of the increases in productivity.

TABLE 8.1   Worldwide Production of Cereals and Oilseeds (in Millions of Tons).[a]

Cereals	1965	1975	1985	1995	2005
Corn	227	342	486	517	715
Rice	254	357	468	547	632
Wheat	264	356	500	627	629
Barley	93	135	173	141	138
Sorghum	47	62	76	55	60
Millet	23	27	27	26	31
Oilseeds	30	42	64	92	143

([a]From Food and Agricultural Organization Statistics).

The crop protection industry in the United States and Europe largely evolved out of the chemical industry. Currently, major companies that are involved in manufacture and distribution of chemicals for agriculture include Aventis, BASF, Bayer Crop Science, Dow Agrochemicals, DuPont, Monsanto, and Syngenta. In many ways, the development of the agricultural chemicals industry parallels that of the pharmaceutical industry. The discovery of herbicides and pesticides has progressed from relatively inefficient random screening to more targeted approaches based on biological rationales. In the case of crop protection, because of the relatively large areas to be covered, the cost of the materials is a more significant factor than in human health-care products. Another major aspect of use of agricultural chemicals is their environmental and ecological consequences.

8.1   FERTILIZERS

8.1.1   UNDERSTANDING THE NITROGEN CYCLE

In the late 1700s, it was realized that only about 1/5 of the gas in the atmosphere (oxygen) supported combustion. Carl Wilhelm Scheele, a Swedish chemist,

Joseph Priestly, and Antoine Lavoisier were instrumental in this discovery. The remaining 4/5 was called azote by Lavoisier, the name being derived from Greek for *without life*. Later, the English word nitrogen arose from the recognition that *nitre* (nitrate salts) could *generate nitrogen*. The German word for nitrogen, *stickstoff*, comes from *suffocating substance*. The Russian αζστ is a transliteration of azote. The understanding of the role of nitrogen in agriculture began with Justus von Liebig, one of the founders of the discipline of organic chemistry. His book, *Die organische Chemie in ihren Anwendung auf Agricultur und Physiologie* was published in 10 editions between 1853 and 1876 and translated into the major European languages. Liebig recognized that certain minerals such as potash and lime were beneficial to plant growth and crop yield. Liebig also demonstrated that "fixed" nitrogen, that is in compounds such as ammonia or nitrate salts, is essential to plant growth. In 1840, Liebig formulated his "laws" pertaining to the role of soil nutrients in agriculture. They can be summarized as follows:

1. Soil fertility requires the presence of nutrients in proper amount and form.
2. When crops are harvested, some of the nutrients are removed and must be replenished.
3. Animal husbandry also results in removal of nutrients in the form of milk, meat, etc.
4. If sufficient nutrients are returned, for example, in the form of compost or manure, soil fertility is maintained.

Liebig also formulated a "law of the minimal," which stated that whichever nutrient is present in the minimal amount (relative to need) is the limiting factor in plant growth. At the time, Liebig incorrectly believed that the major source of fixed nitrogen was from the atmosphere, where it is formed by lightning. Liebig also incorrectly believed that nitrogen's principle function was to allow plants to absorb minerals such as potassium and calcium from the soil.

The next advance in understanding the role of nitrogen came from experiments begun in the 1840s by John Bennett Lawes, an English landowner, and Joseph Henry Gilbert, a chemist trained in Liebig's laboratory. Their experiments demonstrated the need for nitrogen, which could be supplied as either manure or ammonium salts. They also demonstrated that a year's growth of clover supplied nitrogen for a subsequent year's crop. The practice of crop rotation with legumes, such as peas, soybeans, or clover, had been known for hundreds of years. Lawes and Gilbert clearly demonstrated the need for nitrogen, but could not explain the role of legumes. Liebig, still convinced that minerals were the limiting factor, called their results "humbug." The role of legumes was clarified by Jean-Baptiste Boussingault, who after experience as a mining engineer and chemistry professor, began experiments on a farm in Alsace.

He showed that clover and peas were able to increase the nitrogen content of the medium in which they were growing, that is, they could somehow "fix" nitrogen from the inert form present in the atmosphere. In 1888, two Germans, Hermann Hellriegel and Hermann Wilfarth, reported that nitrogen levels in legumes were irregular and not correlated with the amount of nitrogen fertilizer supplied, while grains and grasses showed a direct relationship. Pursuing this result further, they were able to associate the nodules on the roots of legumes with nitrogen fixation. Soon several microorganisms capable of fixing nitrogen were found in the nodules. In addition to legumes, algae and cyanobacteria, as well as certain *Archea*, can fix nitrogen.[1]

The remaining step in the nitrogen cycle is the relationship between ammonia and nitrates. It was found that microorganisms capable of oxidizing ammonia to nitrate existed in the soil, while others can reduce the nitrate to dinitrogen, completing the cycle. The main components in the nitrogen cycle, then, are ammonia (NH_3), nitrate (NO_3^-), and dinitrogen (N_2). Microorganisms are the mediators of interconversion. Decaying plant and animal matter also provides NH_3. In recent years, other contributions of compounds of nitrogen to the overall cycle have been recognized. Nitrous oxide (N_2O) is mainly inert, but is a significant greenhouse gas. Nitric oxide (NO) contributes to ozone formation in the biosphere and nitrogen dioxide (NO_2) leads to acid deposition.[2] The oxides of nitrogen are also produced by fossil fuel combustion. Figure 8.1 summarizes the chemical reactions that contribute to the nitrogen cycle.

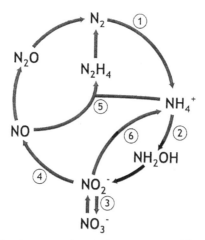

FIGURE 8.1   Steps in the nitrogen cycle. (1) Fixation; (2) aerobic oxidation of ammonia; (3) aerobic oxidation of nitrite; (4) reductive denitrification; (5) anaerobic ammonia oxidation; (6) reduction of nitrite to ammonia. (From Jetten, M. S. M. *Environ. Microbiol.* **2008,** *10,* 2903–2909 by permission of Wiley. © 2008, Society for Applied Microbiology and John Wiley & Sons, Ltd.)

Farmers in many parts of the world had learned to recycle waste and/or rotate legumes to maintain soil productivity. These methods could support up to about 10 people/ha, depending on climate. The recycled waste included human and animal excrement, plowed under stubble from grain crops and legumes, sometimes grown as a cover crop. By the 1800s, crop rotation was widely practiced in England. In 19th-century Paris, suburban garden plots called "marais" were fertilized by the waste from horses used for transportation in the city. In South China, year-round agriculture was based on recycled plant and animal waste and nitrogen fixation by cyanobacteria in rice paddies. Another form of intensive agriculture practiced in China involved fish ponds in conjunction with grains and fruits. These more intensive forms of agriculture could support somewhat larger population densities.

8.1.2   SOURCES OF NITROGEN FERTILIZERS

The first commercial source of nitrogen fertilizer was guano, the accumulated deposits of bird droppings found mainly on a few islands off the Peruvian coast of South America. A sample was brought to Europe by Alexander von Humboldt in 1804. Commercial shipments commenced in 1840. The guano was dug by hand by indentured Chinese workers under hellacious conditions. Production peaked at 70,000 t/year in the 1860s. By 1872, the supplies were nearly depleted and Peru stopped the exportation. Charles Darwin had observed deposits of a white mineral called "caliche," in the desserts on shore in Peru during his voyage on the Beagle in 1835. These deposits were rich in $NaNO_3$ and remained in place because of the arid climate. Exports began in the early 1840s and by the 1870s were 70,000 t/year. In 1883, Chile defeated Peru and Bolivia in a war for the deposits and assumed control of the nitrate export trade. The crude deposits were leached to extract the relatively soluble $NaNO_3$ and the solution was then allowed to evaporate, resulting in the crystallization of the $NaNO_3$. It was about 95% pure. By 1900, exports were more than 200,000 t/year. The nitrate was used not only as a fertilizer but was also as a basic starting material for the growing explosive industry (see Section 1.2). The importance of access to Chilean nitrate for production of explosives was an issue of military concern, especially for Germany, which because of its limited access to the sea was particularly vulnerable.

There were three other conceivable sources of nitrogen fertilizer at the start of the 20th century. One was ammonia gas, which was produced during the pyrolysis of coal to make coke used in blast furnaces. The gases from the process could be sprayed with water and the dissolved ammonia precipitated as the sulfate by sulfuric acid. This method was practiced mainly in the United Kingdom and Germany. Two other methods were feasible after electric power was available on an industrial scale. One was based on calcium carbide (CaC_2), which was produced from lime and coke in high temperature electric furnaces. The CaC_2 could then react with nitrogen, again at high temperatures, to produce calcium cyanamide and urea, either of which could be used as fertilizers. This method was economically feasible only where electric power was cheap and was practiced mainly in Norway and the United States.

$$2\ Ca + 2\ C \longrightarrow Ca_2C_2$$

$$Ca_2C_2 + 2\ N_2 \longrightarrow 2\ Ca^{2+}\ {}^-N{=}C{=}N^- \xrightarrow{\ 3\ H_2O\ } H_2N\overset{\overset{O}{\|}}{C}NH_2 + Ca(OH)_2$$

The other method dependent on electricity was based on oxidation of N_2 by O_2 in an electric arc. The reactions involved formation of N_2O_3, which was converted to HNO_3 and isolated as the calcium salt.

$$N_2 + O_2 \longrightarrow N{=}O + NO_2 \rightleftharpoons N_2O_3$$

$$N_2O_3 + H_2O \longrightarrow 2\ HNO_2 \xrightarrow{\ O_2\ } 2\ HNO_3$$

It was in this context that William Crookes[a] warned in 1898, that the European nations would eventually face starvation if new nitrogen sources could not be found. Crookes believed chemists could do just that.

8.1.3   THE HABER–BOSCH PROCESS FOR NITROGEN FIXATION

As one can imagine, Crookes' proclamation of impending starvation attracted a good deal of attention. There had been several attempts to synthesize NH_3 from N_2 and H_2 in the late 1800s but none had succeeded in producing more than trace amounts. One such report was made in 1900 by Wilhelm Ostwald, a leading German chemist, who reported to BASF that he had made ammonia from N_2 and H_2 over an iron catalyst. Carl Bosch (see Topic 8.2 for biographical information) a relatively new BASF employee, was assigned the task of confirming

---

[a]William Crookes was a chemist and physicist who began his studies at the Royal College of Chemistry under A. W. Hofmann in 1847. After holding two brief academic appointments, an inheritance permitted him to establish a private laboratory in London. He also founded and edited a journal, *Chemical News*. He was highly respected in scientific circles and served as president of the Chemical Society, the Society of Chemistry and Industry, and the Institute of Electrical Engineers. He was knighted in 1897. Among his principal scientific accomplishments were the identification of the element thallium, the invention of the cathode ray tube (Crookes tube), and the observation of scintillation when particles of radioactive decay impinge on certain surfaces. His prediction of impending nitrogen deficiency was made in his presidential address to the British Association for the Advancement of Science.

the experiment, but he found that the NH_3 actually arose from nitride present in the iron catalyst.

When Fritz Haber (see Topic 8.1 for biographical information) arrived at the Technical University in Karlsruhe as a laboratory assistant in 1894, he had had some exposure to organic chemistry, as well as short-term stays in several industrial organizations and a year of compulsory military service. At Karlsruhe, Carl Engler, the director of the chemistry department, introduced him to hydrocarbon combustion. Haber became particularly interested in electrochemistry and published a book on the subject in 1898. In 1902, he toured the United States as a representative of the German Electrochemical Society and reported his findings. In 1905, he published another book on the thermodynamics of gas reactions. He became a professor at Karlsruhe in 1906. Haber was aware of work by Ramsey and Young in England which indicated that the amount of NH_3 at equilibrium with N_2 and H_2 would be small. He studied the reaction and concluded that the equilibrium concentration would be in the range of 0.005–0.0125% under standard conditions. He also recognized that higher pressure and lower temperature would increase the amount of NH_3, but at the time the prospects for a commercial process seemed dim. Haber reported his values. Hermann Nernst had also conducted experiments on the position of the $N_2 + 3\ H_2 \rightleftharpoons 2\ NH_3$ equilibrium and had found a value somewhat lower than Haber (by about 1/4). Nernst reported these results to Haber who, with the assistance of Robert La Rossignol, an English student, arrived at a revised value of 0.0048%, just below his original range. Nernst publicly criticized Haber's value and seems to have gone out of his way to do so. Haber was offended and responded by renewing his studies. La Rossignol showed that operating at 500°C and 3 MPa, increased the NH_3 concentration 28-fold. The thermodynamics of the system required a catalyst that would operate at as low a temperature as possible. For example, at 20 MPa and 600°C, an 8% conversion to NH_3 was to be expected.

BASF was actively researching the electrochemical routes to nitrogen fixation. Carl Engler was a member of BASF's board and encouraged Haber to enter into a research agreement with the company. The agreement called for any patents to be assigned to BASF, with Haber to receive a 10% royalty. Although BASF's main interest was in improving the electrochemical processes, Haber was more interested in the $N_2 + 3\ H_2$ reaction. Two things were required for success: an apparatus that could function at the required pressure and temperature and a catalyst that would be active at as low a temperature as possible. There also had to be provision for rapid cooling of the equilibrium mixture to recover the ammonia. The necessary apparatus was constructed by La Rossignol. With an osmium catalyst, a conversion of 6% was realized. Haber reported the result to BASF in March 1909 and a patent was filed on March 31, 1909. After the successful demonstration to BASF representatives on July 2, 1909, Carl Bosch was assigned the task of developing a commercial scale process. A second patent was filed on Sept 14, 1909, emphasizing the temperature and pressure conditions. The first public announcement of Haber's work was made in 1910. In 1911, Hoechst filed a suit claiming that the Haber–BASF patents were invalid and that Nernst's earlier work had disclosed the process. Nernst, however, testified for BASF that the Haber process indeed merited a patent and it was sustained.

The operation of a commercial plant required solution of several problems. One was a source of the starting materials N_2 and H_2. Another was finding a cheaper catalyst, and the third was the design and construction of a large-scale system. The N_2 problem was solved by access to the Linde process for air lique-faction and separation. Hydrogen was made using coke obtained by the pyroly-sis of coal. In the first stage, glowing coke reacts with water to give CO, which is then reacted again with water giving a second equivalent of H_2.

$$\text{Pyrrolysis:} \longrightarrow H_2, CO_2, CH_4, C_2H_4, \text{coke}$$

$$\text{Steam Reforming: coke} + H_2O \longrightarrow CO + H_2$$

$$\text{Water-Gas Shift: } CO + H_2O \longrightarrow CO_2 + H_2$$

Early laboratory work showed that the H_2 rapidly caused carbon-containing steel to become brittle by removing the carbon. This problem was solved by al-lowing the hydrogen to diffuse out of a soft iron liner and be flushed out before it could damage the steel. By 1912, production of 1000 kg/day was reached and the claims of patent infringement had been resolved. By 1914, a full-scale plant had been constructed at Oppau near the BASF plant at Ludwigshafen and was producing 20 ton/day of NH_3.

But WWI also broke out in 1914 and soon the NH_3 was needed to make nitrate for munitions. German ammunition supplies were thought to be adequate for about 6 months and the German military command expected a short war. Because of the British control of the sea, however, Germany had lost access to the Chilean source of nitrate. When the war bogged down into trench warfare, Germany's military and industrial leaders recognized that the lack of nitrates would be catastrophic. Carl Bosch and Carl Duisberg of Bayer began urging development of other sources of nitrates. BASF found a catalyst for oxidation of ammonia to nitrate and was in production by May 1915. The government want-ed to expand ammonia production, but BASF refused unless the government underwrote the construction costs, fearing the plants would be useless at the end of the war. The government agreed and a huge new plant was built at Leuna in central Germany. Haber was becoming wealthy as the result of his royalties on NH_3 production. These activities were an early example of the science–mili-tary–industry collaborations that would continue throughout the 20th century.

The treaty of Versailles at the end of WWI required that France be licensed to construct a 100,000 ton/year plant. A Haber–Bosch type plant began operating in Britain in 1919. US production began in 1921. In the 1920s, the BASF ammonia plants accounted for two-thirds of IGF's profits (see Section 3.2.2). Modified processes were also developed. In France, the Claude process which operated at higher pressures of 90–100 MPa was introduced. This process was adopted by DuPont for its plant at Belle WV in 1924 (see Section 3.2). In the 1920s, the source of the hydrogen began to be switched to natural gas. The process, developed at IGF, was licensed to SO(NJ) and a plant began operating in Baton Rogue, LA, USA in 1931. Shell began production of ammonia using methane at Martinez CA, near San Francisco, in 1931. After 1950, natural gas became the primary source of H_2 worldwide.

$$CH_4 + H_2O \longrightarrow CO + 3H_2 \quad \Delta H = +206 \text{ kJ/mol}$$

$$CO + H_2O \longrightarrow CO_2 + H_2 \quad \Delta H = -41.2 \text{ kJ/mol}$$

Figure 8.2 is a schematic outline of the Haber–Bosch process base on methane.

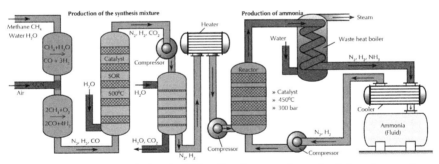

FIGURE 8.2 Schematic design for Haber–Bosch ammonia synthesis based on methane. http://commons.wikimedia.org/wiki/File:Haber-Bosch-En.svg. By Francis E Williams (Own work) [GFDL (http://www.gnu.org/copyleft/fdl.html) or CC BY-SA 4.0-3.0-2.5-2.0-1.0 (http://creativecommons.org/licenses/by-sa/4.0-3.0-2.5-2.0-1.0)], via Wikimedia Commons.

At the beginning of WWII, the United States rapidly expanded NH_3 capacity and it reached 880,000 ton/year by 1945, while much of the European capacity was destroyed during WWII. Expansion continued in the 1960s, especially in Russia, Asia, and South America. Plant sizes grew in most countries with many plants having 1000–2000-ton/day capacity. Global production was about 120 million tons in 1990 and 160 million ton/year in 2000. During this time, energy consumption was cut to about two-thirds of the original process.

TOPIC 8.1 FRITZ HABER

Fritz Haber was born in Breslau in 1868. His mother never recovered from childbirth and died shortly thereafter. Haber studied chemistry in Berlin and received a Ph. D. in organic chemistry. Haber returned to Breslau for a short time, working in his father's dye business. His interests were turning to electrochemistry and in 1894 he obtained a position at the Karlsruhe Technical University, which was not far from the BASF factory in Ludwigshafen. In 1901, Haber married Clara Immerwahr, like Haber a native of Breslau. She was the first woman to receive a Ph. D. in chemistry from the University of Breslau, where she worked with Richard Ahegg, a friend of Haber. Although she and Haber were from Jewish families, both had been baptized as Christians. In 1902, shortly after the birth of their son, Hermann, Haber left for an extended trip to the United States as a representative of the German Electrochemical Society. He visited many factories and industrial facilities. Haber was impressed with the efficiency of the machines he saw in America. In 1906, after publication of a book entitled *Thermodynamics of Technical Gases*, Haber became a full professor at Karlsruhe. He had a group of 30 assistants and students. Haber worked incessantly, except when colitis or exhaustion sent him for stays in a sanatorium. Clara is said to have missed the opportunity to pursue her own studies and research and to not have adapted well to life as a Professor's wife. Haber's obsession with his work led his wife to concern for his health and despair for their marriage.

The idea that Germany's place in the world depended on scientific strength took hold and Kaiser Wilhelm was persuaded to establish the Kaiser Wilhelm Institutes. One, in Berlin, was largely financed by Leopold Koppel, a wealthy banker and businessman who arranged for Haber to be its director. Haber moved to Berlin in 1911 to take the leadership of the newly created Institute for Physical Chemistry. Nationalism was taking hold, embraced and encouraged by the Kaiser, and eventually led to war with Russia and France in August 1914. This was followed by German invasion of Belgium and war with Britain as well. By virtue of his directorship of the Institute, Haber was part of the structure of governmental advisers. The idea of gas warfare had occurred in the late 19th century and in 1899 a conference of European governments had banned the use of "asphyxiating and deleterious gases" delivered by munitions. With the war at a stalemate, however, the Germans, British, and French all began to explore the possibilities of using noxious gases. Haber began work on gas warfare at his Institute. Haber focused on Cl_2 and he began collecting scientifically trained soldiers to man a "gas brigade." It included meteorologists, because the plan was to use wind to deliver the gas. In March 1915, the brigade arrived at Ypres in Belgium. The chlorine cylinders were dug in on a hill overlooking French and Algerian positions. The wind patterns were generally unfavorable, however, and weeks passed. The commanding officer became impatient. On April 22, a favorable wind arose and some 400 tons of Cl_2 was released. It drifted into the Algerian positions and the troops either died or fled in panic. A 4-mile wide hole opened in the French line, but the Germans advanced only one mile. The Germans used Cl_2 four more times during the next week. Work on both defensive and offensive counter-measures began immediately in Britain, France, and the United States.[3]

Haber returned to Berlin and was promoted from sergeant to captain in the reserves. On the night of May 1 1915, Clara Haber shot herself. Whether it was the direct result of Haber's role in the gas attacks or the state of their marriage is unclear. Haber returned to duty and the Institute began working on other gases. At the front, phosgene $O=CCl_2$ was used by both the Germans and French. The Germans introduced mustard gas in 1917 and the British and American used it as well. Haber considered the use of poison gas as an inevitable consequence of scientific progress. He believed it would be used eventually and that whatever advantage could be had should be taken. He was perplexed at moral objections, seeing no qualitative differences in the means of warfare. He considered gas to be a psychological weapon and was quoted as saying in 1925 that it was "more frightening and less destructive" than traditional weapons.

German hopes for victory began to unravel. The Germans turned to submarines, but their use brought the United States into the war. By September 1918, the military leaders were asking the government to negotiate for peace, but before any agreement was reached, mutinies and strikes led to the Kaiser's abdication and exile on Nov. 9. A Social Democratic government signed the armistice on Nov. 11. A tenth of the German population had been killed or disabled in the war, and the casualties in France were even higher. Over all of Europe, 20 million people had been killed or wounded. The Bolsheviks had taken control in Russia. In 1919, Haber's work on nitrogen fixation was awarded the Nobel Prize in Chemistry. There was a furor over Haber's selection and American and French recipients boycotted the awards.

The peace treaty of Versailles imposed heavy reparations on Germany. In 1923, France occupied the coal-producing Rhineland to force payment of reparations. Germany's economy began to collapse and inflation soared out of control. One US dollar, worth 14 marks in 1919 was worth 4 billion marks by the end of 1923. Haber's personal fortune and his Institute's endowment melted away. Haber hoped to solve the reparations demands by finding a way to extract gold from the ocean. He took two ocean voyages to collect samples, but when they were analyzed they showed much less gold than had been anticipated, making the scheme impossible.

Haber's Institute was a vigorous and productive place immediately after the war and Haber was a respected figure in Germany. He was instrumental in placement of many students and he worked for the reintegration of Germany into international scientific organizations, such as the IUPAC. Members of his former gas brigade began to rise to prominence, including three who would win Nobel Prizes in Physics, James Franck, Gustav Hertz, and Otto Hahn. Although they were of opposite political outlooks, Haber maintained a friendship with Albert Einstein, who was located at the nearby Institute for Physics. While Haber was an ardent German patriot, Einstein was a pacifist and detested German militarism.

In 1926, Hermann, Haber's son with Clara, married against his wishes and moved to the US. Late that year, his second wife, Charlotte and their two children left the household and they were divorced in late 1927. The stock market crash of 1929 and the ensuing economic crisis again seriously depleted Haber's finances. In 1932, Haber stated "The great technical accomplishments the past 50 years have granted us, when controlled by primitive egoists, are like fire in the hands of small children." The next year Adolf Hitler became Chancellor of Germany and was voted absolute power in March 1933. On April 1, the Minister of Justice of Prussia ordered that all Jewish judges take a leave of absence. A week later, the government ordered the removal of all Jewish civil servants, except those who had served in the military in WWI. The Nazi's took particular interest in Haber's institute, but he began to resist, dismissing only two of the most prominent members, Herbert Freundlich and Michael Polyani, both of whom were in a position to immediately relocate to England. Haber attempted to assist more junior Jewish staff members to find positions. Haber himself soon resigned. Max Planck tried to dissuade the authorities, including Hitler himself, from accepting Haber's resignation, but the only outcome was to provoke Hitler's rage. Haber began a search for a new position that would allow him to leave Germany in honor. He met Chaim Weizmann in England, who urged him to relocate to Palestine. Former English gas warfare adversaries of WWI arranged a position at Cambridge, but Haber worried about the financial consequences for himself and his younger children, then teenagers. Late in 1933, he arrived in Cambridge with his half-sister, Else, who had managed his household since his divorce from Charlotte. His lifelong faith in Germany was disintegrating and his health was poor. Early in 1934, he contacted his son Ludwig[b] and, with Else, returned to Basel. Exhausted by the trip, he died there of heart failure on January 30, 1934.

---

[b]When the Nazis came to power, Charlotte and her son (Ludwig Fritz) and daughter (Eva) left for Switzerland and then England. Ludwig F. (Lutz) Haber was interned as a "friendly foreigner." He eventually received a degree from the London School of Economics and in 1971 published a history of the chemical industry, *The Chemical Industry 1900–1930.* He became a Reader at the University of Surry and published a second book *The Poison Cloud. Chemical Warfare in the First World War*, in 1986.

Sources: Smil, V. *Enriching the Earth: Fritz Haber, Carl Bosch, and the transformation of world food production.* MIT Press: Cambridge, MA, 2001; Charles, D. *Mastermind, the Rise and Fall of Fritz Haber, the Nobel Laureate Who Launched the Age of Chemical Warfare.* Harper-Collins Publishers: New York, 2005; Hager, T. *The Alchemy of Air, A Jewish Genius and a Doomed Tycoon and the Scientific Discovery that Fed the World but Fueled the Rise of Hitler.* Harmony Books: New York, 2008.

TOPIC 8.2 CARL BOSCH

Carl Bosch was born in Cologne in 1874 and grew up there. He first studied metallurgy and engineering at Charlottenburg near Berlin, but then switched to organic chemistry, receiving his degree from Leipzig. He joined BASF in 1899 and worked on the nitrogen fixation problem. He was instrumental in BASF's decision to pursue the commercial development of Fritz Haber's nitrogen fixation reaction and it became Bosch's responsibility to convert the reaction into a workable industrial process. Early experiments using strong carbon–steel resulted in rupture of the vessels after some time in use. It turned out that the H_2 diffused into the steel, reacted with the carbon, removing it and making the steel brittle. No suitable type of steel could be found. To solve the problem, a soft iron liner was used inside a strong carbon–steel outer container. The H_2 diffused through the soft iron and was vented out without damaging the strong outer container. Alwin Mittasch was put in charge of finding a better catalyst. The successful catalyst was a type of magnetite mined in Sweden. It contained aluminum and calcium oxides that evidently acted as promoters of the catalyst. Pure N_2 had become available via the liquefaction and separation from air, the Linde process. The hydrogen was made from water by the reforming reaction with hot coke. By 1911, prototype reactors were producing up to 30 kg of NH_3 per day. In May 1911, construction started at Oppau. The ammonia plants began operating in September 1913, and by 1914 were producing 20 t/day. The plants represented a new level of sophistication. In addition to the unprecedented pressure and temperatures used, the valves and other control systems were specifically designed for the plant.

With the outbreak of WWI and the realization that Germany's nitrate supplies for munitions were inadequate, BASF turned to finding ways to convert NH_3 to nitric acid. Again Mittasch explored a number of catalysts, finding an iron–bismuth–calcium mixture to be the best. In May, 1915, the plant began producing nitric acid from ammonia. Oppau, however, was close to the frontier and French planes dropped small bombs in May 1915. It was decided to build an even larger plant at Leuna, in central Germany, far from the frontier. Production of ammonia at Leuna began in April 1917. Bosch became the director of BASF in 1919.

After the end of the war, the French occupied the Oppau plant. The terms of the armistice required access to munitions facilities, but not to civilian commercial enterprises. The French and others were very interested in learning the details of the BASF ammonia process. Bosch guarded, as best he could, the technological information. Bosch stalled for time, hoping the peace conference of 1919 would establish BASF's right to the patented technology. Bosch was a representative of German industry at the peace conference. The French wanted both Oppau and Leuna shut down, but Bosch argued that further destabilization of Germany would lead to a German version of the Russian revolution. Bosch secretly met with a French industrialist and offered to license a 100,000 ton/year plant in exchange for leaving the Oppau and Leuna plants in operation. This proposal was part of the final agreement reached in early 1920. Meanwhile, the British had obtained drawings of the Leuna plant and began construction of their own ammonia plants, with Brunner–Mond as the operator.

During 1923, Bosch visited many industries in the United States. He saw first-hand the dramatic growth in automobiles and the many chemical products they used, including rubber tires, lacquer coatings, upholstery, and above all gasoline. He was familiar with the work of Friedrich Bergius on coal hydrogenation (see Section 2.2.1.1). He began to promote development of synthetic gasoline, based on the then current assumption that declining crude production would lead to increased prices. The formation of the IGF combine was in part motivated by the need to assemble capital for the synthetic gasoline project (see Section 3.2.2). The plan was to coordinate research and marketing, while maintaining the operating independence of the companies. The combine included both Bayer and Hoechst, as well as several other smaller companies. IGF began operating in 1925, with Bosch as its first director. Bosch was awarded the 1931 Nobel Prize in chemistry, along with Bergius, for their development of high-pressure chemistry.

In order to finance the synthetic fuel project, Bosch again traveled to the United States and interested SO(NJ) and Ford Motor Company in the project. IGF and SO(NJ) signed a technology sharing agreement and SO(NJ) paid IGF $35 million for access to the technology that developed. Bosch pushed the work hard, but it went slower than expected. Soon, new oil was discovered in Texas and nearby states. The likelihood of an imminent oil shortage disappeared. Bosch lobbied the German government for increased tariffs on imported petroleum, hoping to keep the synthetic fuel competitive. The stock market crash of 1929, setting off the worldwide depression, brought other problems. Unemployment soared in Germany and the politics fractured into extreme left (Communists) and right (Nazis). Business evaporated and by 1933, IGF's income was less than a fifth of that in 1929. By then synthetic gasoline production was nearing 100,000 gal/year, but prices were low. In early 1933, IGF executives met with Hitler and he was favorably inclined toward the synthetic fuel concept, because it conformed with his "autarky" (national self-sufficiency) policy. IGF, in turn, contributed to the Nazi political campaign. But, in a separate meeting, when Bosch tried to dissuade Hitler from purging of Jews from the civil service, Hitler, as he had with Max Planck, flew into a rage. In December 1933, IGF signed an agreement with the government, funding the synthetic fuel project. The government agreed to buy any fuel not sold on the civilian market and encouraged the expansion of the plant. Bosch and IGF had made their deal with the devil. IGF later negotiated with the government to develop Buna synthetic rubber (see Section 5.3.2.3). In 1938, all remaining Jews were removed from IGF.

It is said that German military leaders asked Bosch to intervene in Hitler's plan to invade Czechoslovakia on the basis that the industry could not support war. Bosch agreed to do so and tried to meet with Hermann Goering, but without success. In 1939, Bosch was asked to speak at the Deutches Museum in Munich, where he was on the board. He appeared at the meeting drunk, but in his speech he defended independence in science. The ardent Nazis in attendance walked out. Bosch was removed from the Museum board. Soon, he retreated to his Heidelberg villa and sank into alcoholism and depression. He burned all his correspondence. In his last days, he warned his son of the impending disaster for Germany. He died on April 26, 1940.

Beginning in May 1944, the US Air Force began intensive bombing of the Leuna plant. It was heavily defended by anti-aircraft weapons. In the early raids, severe losses were suffered. But the raids continued, 22 in all. By November 1944, the plant had largely been destroyed and with it, much of the German capacity for making gasoline and aircraft fuel. After the war ended, 23 IGF executives were tried at the Nuremberg war crimes trials. Thirteen were convicted and given prison terms. The charges included use of slave labor. The IGF combine was dismantled, giving rise to Bayer, BASF, and Hoechst. The Leuna works, located in the eastern zone occupied by Russia, were repaired and put back into operation.

Sources: Smil, V. *Enriching the Earth: Fritz Haber, Carl Bosch, and the transformation of world food production.* MIT Press: Cambridge, MA, 2001; Hager, T. *The Alchemy of Air; A Jewish Genius and a Doomed Tycoon and the Scientific Discovery that Fed the World but Fueled the Rise of Hitler.* Harmony Books: New York, 2008.

8.1.4   TRENDS IN WORLD NITROGEN FERTILIZER USE

During the 20th century, nitrogen fixation by the Haber–Bosch process became the dominant source for nitrogen fertilizer in agriculture. Table 8.2 shows the changes in source and amount the occurred in the period 1850–2000.

TABLE 8.2 Nitrogen Fertilizer Production from Various Sources—150–2000 (in kt).

Year	Chilean nitrate	Guano	Coke oven [$(NH_4)_2SO_4$]	Calcium cyanamide	Calcium nitrate	Synthetic ammonia	Total
1850	5	–	–	–	–	–	5
1860	10	70	–	–	–	–	80
1870	30	70	–	–	–	–	100
1880	50	30	–	–	–	–	80
1890	130	20	–	–	–	–	150
1900	220	20	120	–	–	–	360
1910	360	10	230	10	–	–	610
1920	410	10	290	70	20	150	950
1930	510	10	425	255	20	930	2150
1940	200	10	450	290	–	2150	3100
1950	270	–	500	310	–	3700	4780
1960	200	–	950	300	–	9540	10,990
1970	120	–	950	300	–	30,230	31,600
1980	90	–	970	250	–	59,290	60,600
1990	120	–	550	110	–	76,320	77,100
2000	120	–	370	80	–	85,130	85,700

(From Dawson, C. J. *Proc. Int. Fertilizer Soc.*, **2008**, 623, 2-39 based on data in V. S mil, *Enriching the Earth,* 2001, courtesy of MIT Press.)

There are five predominant forms for application of synthetic nitrogen fertilizer. Three are as solids, ammonium sulfate $(NH_4)_2SO_4$, ammonium nitrate (NH_4NO_3) and urea. The fourth is pure liquid ammonia and the fifth is aqueous solutions of ammonia or urea. The salts account for about 3 and 10%, respectively, of current use. The sulfate has a lower N content (21%) than the nitrate (35%). The most significant feature of the nitrate is its capacity to act as an explosive when mixed with a fuel, as discussed in Section 1.2.5. Urea currently accounts for nearly half of synthetic nitrogen fertilizer. It is slightly more expensive to prepare than the sulfate or nitrate salts. It is hydrolyzed in soil to produce ammonia. Pure liquid NH_3 is used primarily in the United States, where a pipeline delivers the material from the Gulf Coast to the Midwest. It must be handled under pressure and is inserted below the surface of the soil with suitable machinery. Nitrogen fertilizers can be used as water solutions containing either urea or ammonium nitrate that can be applied by ground or aerial sprayers.

Fertilizer use increased worldwide during the 20th century. By 1950, the global annual consumption was about 3.6 million tons. This rose to 11.0 by 1960, 31.6 by 1970, 60.6 in 1980, about 80 in 1990, and close to 90 million tons in 2000. This translates to per capita use of about 2 kg in 1950, 15 kg in 1990, and 14 kg in 2000. In the United States, consumption has been at about 11 Mt/year since about 1980. China is the world's largest producer and consumer of nitrogen fertilizer. Use in Africa, except for Egypt, remains very low. In fact, in much of Africa, the rate of nitrogen application is insufficient to maintain maximum soil productivity. Table 8.3 breaks down nitrogen fertilizer consumption by areas of the world. It includes phosphorus and potash fertilizer usage as well (see Section 8.1.5).

TABLE 8.3 Share of Worldwide Fertilizer Consumption by Region.[a]

Region	Nitrogen share %	Phosphorus share %	Potash share %
Africa	3.4	2.5	1.6
North America	13.5	12.0	17.1
Latin America	6.3	13.0	17.5
West Asia	3.5	3.3	1.4
South Asia	19.6	20.5	10.9
East Asia	38.3	36.1	35.2
Central Europe	2.7	1.5	2.4
Western Europe	8.4	5.6	9.5
Eastern Europe	3.0	2.0	3.1
Oceania	1.4	3.5	1.3

([a]From current world fertilizer trends and outlook to 2011/12, Food and Agricultural Organization).

The role of nitrogen fertilizer in maintaining agricultural production has been analyzed by V. Smil.[4] The current nitrogen requirements for maintenance of soil productivity are about 220 kg/ha for corn or wheat and 300 kg/ha for double crop rice. Smil calculates that about 60–80% of this must come from synthetic fertilizer. Currently, worldwide, about 85% of food protein comes from crops and domesticated animals, with the rest coming from fishing and aquaculture. Smil calculates that about 40% of total protein intake originates in synthetic nitrogen fertilizers. The actual daily nitrogen consumption ranges from around 100 g/day in affluent countries to 66 g/day in poorer ones. Another approach to the analysis indicates that food for about 2.2 billion people is dependent on synthetic nitrogen fertilizers. Even in an idealized world of equalized protein consumption of 75 g/day, about a third of the nitrogen required would need to come from synthetic fertilizer. It is believed that the level of population that can be supported has risen to as high as 35–45 people/ha, perhaps even 50/ha under optimum conditions and a mainly vegetarian diet. The highest current population density is in Japan, at 25 people/ha, but this level depends on imported food.

The degree of dependence of individual countries on synthetic nitrogen fertilizer varies with the state of agriculture and population density. In the United States, if food exports were eliminated and a lower meat diet adopted, adequate food could probably be produced without synthetic nitrogen fertilizers. Highly populated and less agriculturally rich countries such as China have a high dependency on synthetic fertilizer. In 1949, when the communists came to power, China had only two fertilizer plants and a production of 27,000 ton/year of ammonium sulfate. In 1963–1965 several ammonia plants were purchased from the United Kingdom and the Netherlands. But in 1966, the Cultural Revolution and its chaos were launched. That and mismanagement brought food supplies back to 1950 levels, while the population had increased by nearly 25%. Food was rationed in all cities and the need for nitrogen fertilizer was drastic. Following the Nixon–Kissinger opening in 1972, China purchased 13 of the largest ammonia plants in the world. By 1979, China became the largest consumer of synthetic nitrogen fertilizer, and in 1990, it became the largest producer as well. China's population is projected to grow to 1.5 billion by 2050 and China will require increased nitrogen fertilizer production. Most of the rest of the low-income world is also dependent on synthetic fertilizer for a minimally adequate diet. Although the rate of population growth is slowing, it is estimated that world population will reach 8.9 billion by 2050. Most of the growth will be in relatively food-poor countries such India, China, Pakistan, Nigeria, and Ethiopia. Most of these countries have no prospect for increase in arable land. As a result, arable land per capita will decrease. Altogether, it seems likely that the requirement for synthetic nitrogen fertilizer will be about 140 Mt/year and that it will be indispensable to feed 60% of the world's population.[5]

The increase in world population has been supported by a number of changes in addition to synthetic nitrogen fertilizer. Among them are expanded land use in South America, Russia, and Australia, improved crop varieties, increased mechanization and energy input, and use of pesticides and weed control agents. Together, these have increased yield by about 4-fold over the century. The improved yield also reduced pressure to convert forested and marginal land to agricultural use. Improvement in yields slowed in the 1990s. Annual increases dropped to 1%/year from 3%/year. Population growth was about 1.4%/year. Grain production per capita reached a maximum around 1990. As a result, there is an increasing demand for agricultural production. This demand is also accelerating as the result of increasing living standards, particularly in Asia. Limits are being reached, not only in productive land, but also in the water supply.

8.1.5 PHOSPHORUS AND POTASSIUM FERTILIZERS

The first commercial production of phosphate fertilizers was by treatment of bone meal in the early 1800s. In the mid-1800s procedures were developed for production of phosphate fertilizers by treatment of phosphate minerals with acid. The main constituents are the calcium salts of the dianionic, $CaHPO_3$, and monoanionic, $Ca(H_2O_2PO_2)_2$, forms of phosphate. This process produces large amounts of $CaSO_4(H_2O)_2$ as a byproduct known as "phosphogypsum". Some phosphate is also produced as diammonium hydrogen phosphate $(NH_4)_2HOPO_3$. As with nitrogen, microorganisms in the soil are needed to convert phosphorus into forms that can be assimilated by plants.[6] The typical phosphorus cycle in modern agriculture is summarized in Figure 8.3.

The Phosphorus Cycle

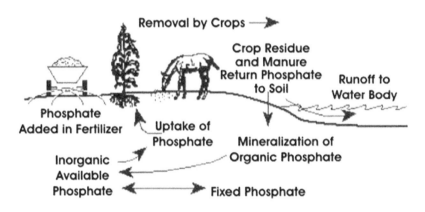

FIGURE 8.3 Schematic diagram representing phosphorus cycle in agricultural production. (From www.epa.gov.)

Plants store phosphate largely in the form of inositol hexaphosphate, also known as *phytic acid*. Phytic acid is considered to be an "anti-nutrient" for humans and other nonruminants, because it is nondigestible, and it can complex metal ions such as Al^{3+}, Zn^{2+}, Ca^{2+}, and Fe^{2+}. Ruminants can digest phytates because of the presence of suitable enzymes (phytases).[c]

---

[c]Interestingly, the triphosphate of inositol plays a crucial role in the animal kingdom as the messenger molecule inositol triphosphate, which transmits signals for muscle relaxation or contraction.

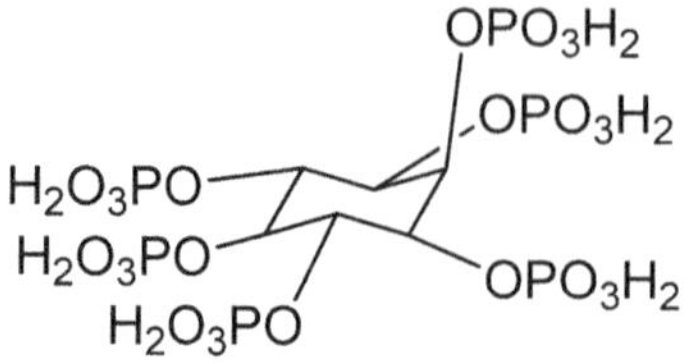

$$OPO_3H_2$$
$$OPO_3H_2$$
$$H_2O_3PO$$
$$H_2O_3PO$$
$$OPO_3H_2$$
$$H_2O_3PO$$

phytic acid; inositol hexaphoshate

The original form of potassium used for fertilizer was potassium carbonate, K_2CO_3 (potash) leached from wood ash. Potash was produced in the United States and Canada from wood ash in the 18th and 19th centuries and was exported to Europe for use in manufacture of soap and glass. Discovery of mines in Germany, shifted the source to imports from Europe. Subsequent exploration for North American sources located them first in Carlsbad NM and later much larger deposits were found in Canada, primarily in Saskatchewan. The largest known deposits are in Russia. The deposits contain both NaCl and KCl laid down in ancient seabeds. The salts are separated by differential flotation. Potash use in the United States has shown a growth pattern similar to that for nitrogen fertilizer, rising rapidly between 1950 and 1980, but then leveling off.

8.1.6 ENVIRONMENTAL AND ECOLOGICAL EFFECTS OF FERTILIZERS

What are the ecological effects of the large-scale use of fertilizers? Current synthetic nitrogen fertilizer production is roughly 100 Mt/year, while biological processes provide between 150 and 200 Mt. Less than half of the nitrogen, however, is incorporated into crops. Some ends up in rivers and aquifers as nitrate. Nitrate levels have risen substantially in many of the world's rivers, but as of now, there is no evidence of serious health effects. On the other hand, many bodies of water, particularly the Gulf of Mexico, Chesapeake Bay, the Baltic, Black, East China, Mediterranean, and North Seas have shown evidence of eutrophication, the excessive growth of algae, which leads to oxygen depletion.[7] Phosphate fertilizers suffer from some of the same problems as nitrogen sources, for example, relatively low efficiency of use and substantial run-off into water. There are other important sources of phosphate including household products. A large dead zone exists and is growing in the Gulf of Mexico at the outlet of the Mississippi River, which drains the largest area of agricultural land in the United States, as shown in Figure 8.4.

FIGURE 8.4   Dead zone in the gulf of Mexico off the Mississippi river delta.

Several factors in addition to the undesirable environmental effects are driving efforts to improve the efficiency of nitrogen fertilizer utilization, including the cost of the fertilizer. The synthetic fertilizers tend to be rapidly available and subject to inefficient utilization. One approach to improving nitrogen utilization is by matching nitrogen availability with the crop's demand for nitrogen on both a spatial and temporal basis, by using precise GPS-controlled application. Another issue is effective utilization of the applied fertilizer. One of the key differences between synthetic nitrogen and phosphorus fertilizers, and manure or compost is the presence of microorganisms in the latter materials. The system of microorganisms and fungi that coexists with plants, particularly in the vicinity of the root system, plays a key role in the absorption of nutrients from the soil. Optimization of nutrient utilization may benefit from soil management methods that promote healthy communities of microorganisms. Use of cover crops that can both prevent erosion and facilitate long-term retention of nutrients may be particularly beneficial. Plant breeding or genetic manipulation may also provide variants that can improve efficiency of nutrient use.

8.2 HERBICIDES

8.2.1 DISCOVERY OF HERBICIDES

The first synthetic herbicides introduced to agriculture were 2,4-D (2,4-dichlorophenoxyacetic acid) and MCPA (4-chloro-2-methylphenoxyacetic acid). The discovery of 2,4-D was made more or less simultaneously by several groups working in both the United States and United Kingdom just before and during WWII.[8] The origins of the discovery can be traced to the recognition in the 1920s that indole-3-acetic acid is the major natural plant growth regulator (also called *auxin*). A structurally similar compound naphthalene-1-acetic acid was also active and it was found that both compounds could act as either promoters or inhibitors of plant growth, depending on concentration and the plant species. Working at ICI from 1936 onward, William G. Templeman, a plant physiologist, found that both 2,4-D and MCPA were very active. He also noted that they were more toxic to broad leaf plants than to cereal crops. A patent was filed in April 1941. At the time publication of scientific results potentially relevant to the war effort was restricted and the results were not published until 1945 and 1946. The discovery was considered to have significance both for crop destruction in enemy territory or for saving labor in domestic agriculture.

indole-3-acetic acid   naphthalene-1-   2,4-dichlorophenoxy-   4-chloro-2-methyl-
auxin   acetic acid   acetic acid   phenoxyacetic acid

In the same time period, P. S. Nutman, M. G. Thornton, and J. H. Quastel, working at the Rothamstead Agricultural Research Station in the United Kingdom also discovered that indole-3-acetic acid and naphthalene-1-acetic acid could exhibit selective toxicity to certain plants. They learned from P. W. Zimmerman and A. E. Hitchcock, who worked at the Boyce Thompson Institute, Yonkers NY, that 2,4-D had auxin activity and included it in their studies. Zimmerman and Hitchcock published information on the growth-promoting activity of 2,4-D in 1942. The herbicidal activity of 2,4-D was also discovered independently by J. Lontz at DuPont. Another line of investigation was initiated by Ezra J. Kraus, a plant physiologist at the University of Chicago. In 1941, he suggested to two of his former students working at the USDA lab in Beltsville, MD, that auxins applied at high levels might act as herbicides. A cooperative program was initiated and when Zimmerman and Hitchcock reported on the auxin activity of 2,4-D, it was included in the studies. Kraus also suggested the potential for military use for crop destruction and a program with that goal was established by the Chemical Warfare Service at Ft. Detrick, MD. Again, because of the military interest, access to the work became restricted and it was not published until 1947. Yet another independent discovery was also made by Franklin D. Jones at the American Chemical Paint Company. He had joined the company in 1938 as manager of agricultural chemicals. The company was interested in plant growth regulators. Jones was particularly interested in killing poison ivy and found that 2,4-D was very effective. He filed Canadian and US patents in 1944 and 1945. Because of the various security and commercial restrictions on these groups, the first open publication (1944) of the herbicidal activity of 2,4-D and 2,4,6-T (2,4,6-trichlorophenoxyacetic acid) came from yet another laboratory, the New York Agricultural Experimental Station. That work was initiated by Charles L. Hamner, who had been one of the participants in the studies at the USDA in Beltsville. Immediately after the war, 2,4-D was commercialized by the American Chemical Paint Company (since renamed AmChem and now part of Aventis). In the United Kingdom, ICI marketed MCPA. The patent situation was unclear, however. Lawsuits were filed, but eventually AmChem agreed to licenses for modest royalties.

The variety and effectiveness of herbicides increased rapidly in the 1960s and became a \$3 billion market by 1970. Annual growth was around 6% in the 1970s and 2% to the 1990s, at which point the market became more or less static at about \$30 billion/year. There are several reasons for the leveling off of the growth. Maximum coverage was reached, at least in the developed world. New herbicides increased competition and decreased prices. Also, agricultural commodities were in over-supply by the 1990s, putting economic pressure on producers. Fewer new types of herbicides were discovered and development-approval time and costs increased. During the 1990s, increased consumer and environmental concerns also led to pressures against herbicide use. In the mid-1990s, herbicide-resistant crops were introduced as the result of genetic engineering, drastically reshaping herbicide-use patterns. This aspect of herbicide use will be considered in Section 21.2.1. In 2000, the major companies producing herbicides were global chemical companies such as Aventis, BASF, BayerCropScience, Dow AgroScience, DuPont, Monsanto, and Syngenta.

8.2.2   GENERAL CHARACTERISTICS OF HERBICIDES

Herbicides fall into two broad groups, those applied to the foliage and those applied to the soil. The latter generally function as *pre-emergence herbicides* and are intended to kill weeds when they sprout. The other category, *post-emergence herbicides*, is applied to foliage. The herbicide must be absorbed by the plant and transported to the site of activity. Plant foliage is normally covered by a waxy cuticle that protects the plant from moisture loss. The use of surfactants is often necessary to ensure adherence and transfer through this cuticle. Other chemicals that assist the process, called *adjuvants*, are also used. Because herbicides are used over relatively large areas, even at low dosages, they must be relatively cheap to produce.

The fact that there are many biochemical and physiological differences between plants and animals makes it possible, in general, to minimize animal toxicity. On the other hand, herbicides must select between desirable (crop) and undesirable (weed) plants. As with drugs, herbicides have a mechanism or mode of action directed against some particular biochemical process. It has been estimated that there might be perhaps 25 ideal targets for herbicide action in the biochemistry of a typical plant.[9] Currently, about 20 such targets are known, so if the estimate is correct, the majority have been found. On the other hand, other estimates of potential targets are as high as 50,[10] but there have been no entirely new mechanisms of action discovered since about 1980. There are two critical negative aspects of herbicide use that have long-term implications. One is environmental or food contamination and its potential ecological and health effects. The second is the development of weeds that are resistant to herbicides, which, in the long term, can reverse the yield and economic benefits of herbicide use.

8.2.3 MAJOR CLASSES AND STRUCTURES OF HERBICIDES

The major classes of herbicides are grouped according to their mode of action in the sections that follow. Table 8.4 shows the 10 most widely used herbicides in American agriculture for 2001–2007.

TABLE 8.4 Most Widely Used Herbicides in US Agriculture.[a]

Herbicide	Target	2001	2003	2005	2007
Glyphosate	EPSPS	85–90	128–133	155–160	180–185
Atrazine	PS-II	74–80	75–80	70–75	73–78
Metochlor-S	VLCFAS	20–24	28–33	27–32	30–35
Acetochlor	VLCFAS	30–35	30–35	26–31	28–33
2,4-D	H⁺-ATPase	28–33	30–35	24–28	25–29
Pendimenthalin	MTI	15–19	9–12	9–12	7–9
Simazine	PS-II	5–7	6–7	5–7	5–7
Trifluralin	MTI	8–10	6–7	7–9	5–7
Propanil	PS-II	6–9	5–7	4–6	4–6
Diuron	PS-II	3–6	4–6	4–6	2–4

[a]In millions of pounds. ESPS (Section 8.2.3.3); PS-II (Section 8.2.3.10); VLCFAS (Section 8.2.3.7); MTI (Section 8.2.3.8). (Data from Gruba, A.; Donaldson, D.; Kiely, T.; Wu, L. *Pesticide Industry Sale and Usage Market Estimates,* US Environmental Protection Agency, 2011).

8.2.3.1   ACETOLACTATE SYNTHASE INHIBITORS

There are several dozen herbicides that are targeted at acetolactate synthase (ALS), also known as acetohydroxyacid synthase (AHAS). This enzyme is critical for the synthesis of branched chain amino acids in plants. Several of the ALS-inhibitors are believed to act by binding at the channel leading to the active site of the enzyme. Scheme 8.1 outlines the function of AHAS in biosynthesis of amino acids and Figure 8.5 shows the binding of two herbicides to the enzyme.

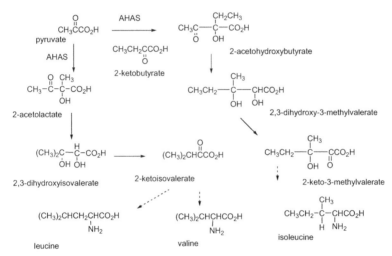

SCHEME 8.1   Biological function of acetolactate synthase.

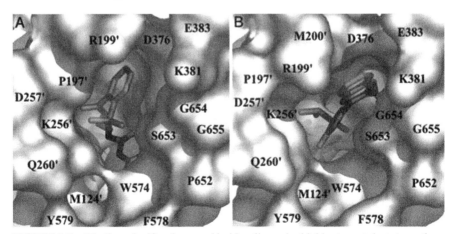

FIGURE 8.5   Binding of sulfonylurea and imidazolinone herbicides to acetolactate synthase of *Arabidopsis thaliana* (From McCourt, J. A.; Pang, S. S.; King-Scott, J.; Guddat, L. W.; Duggleby, R. G. *Proc. Natl. Acad. Sci. USA* **2006,** *103*, 569–573, © 2006, National Academy of Sciences, USA.).

The ALS inhibitors fall into four structural categories including sulfonyl-ureas, imidazolinones, triazolopyrimidines, and sulfonylaminocarbonyltriazoli-nones. The sulfonylureas were discovered at DuPont in the 1970s by screening and structural optimization.[11] Among the commercialized products were chlor-sulfuron (Glean) and sulfometuron-methyl (Oust). The imidazolinones were de-veloped at American Cyanamid.[12] The triazolopyrimidines were developed at DowElanco in Indianpolis IN.[13] The sulfonylaminocarbonyltriazolinones were discovered at Bayer.[14] Representative structures are shown in Scheme 8.2.

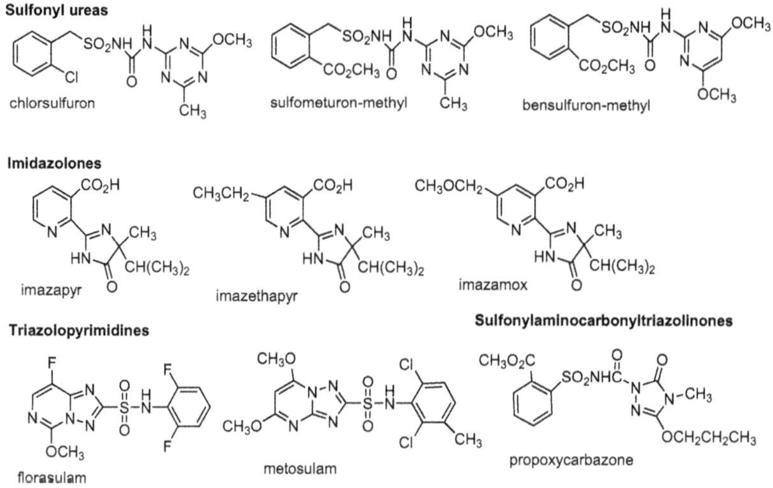

SCHEME 8.2   Herbicides targeted at acetolactate synthase.

The ALS inhibitors are sprayed as post-emergence herbicides and therefore must be selective for the weeds, while causing minimal damage to crops. The basis for selectivity is usually metabolic differences in the plant species. The more resistant species can detoxify the herbicide more rapidly. In some cases, metabolic activation of the herbicide is necessary and protected plants may be less effective at this activation. Among the attractive features of most ALS-inhibitors is that they are used at low application level and have no toxicity to fish, amphibians, birds, or mammals. A negative is that development of resis-tance is fairly rapid.[15]

8.2.3.2   ACETYL COENZYME-A CARBOXYLASE INHIBITORS

The enzyme acetyl coenzyme-A (Co-A) carboxylase (ACAC) is a critical en-zyme in fatty acid biosynthesis and is present in prokaryotes, plants and animals.

It is a biotin-dependent enzyme that catalyzes carboxylation, decarboxylation, and carboxy-transfer reactions. In the first step of fatty acid biosynthesis, it catalyzes carboxylation of acetyl Co-A to malonyl Co-A. In plants, fatty acid biosynthesis takes place in plastids. Plants have two different forms of the enzyme, both consisting of several subunits. The plastid ACAC enzyme in grasses is particularly susceptible to the ACAC inhibitors.[16] Inhibition of the enzyme results in disruption of fatty acid biosynthesis and secondarily membrane disruption and generation of peroxides. There are two major classes of herbicides directed at the ACAC enzyme. These are aryloxyphenoxy propionates (fops) and cyclohexanediones (dims). These herbicides show selectivity toward grasses over broad leaf crops and are called *gramicides*. Structures are shown in Scheme 8.3.

Metabolic Functions of Acetyl Coenzyme-A Carboxylase

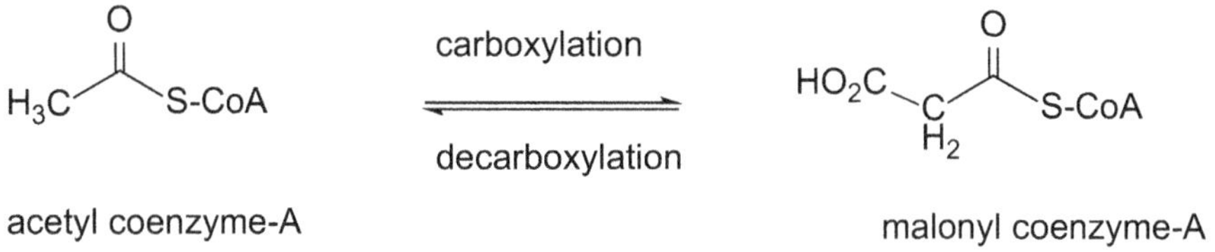

acetyl coenzyme-A

malonyl coenzyme-A

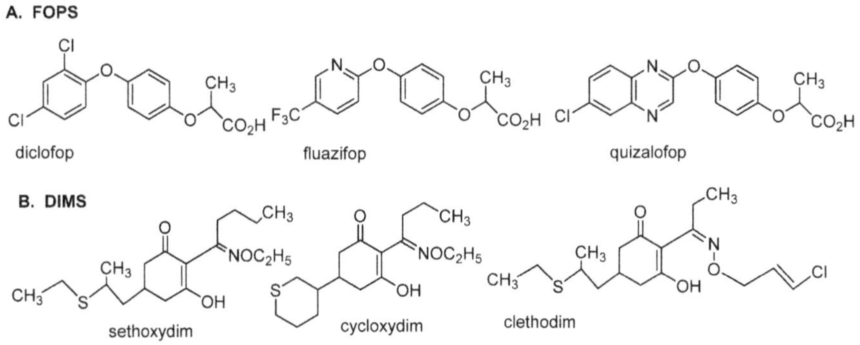

SCHEME 8.3   Herbicides targeted at acetyl Co-A carboxylase.

8.2.3.3   GLYPHOSATE, AN ENOLPYRUVYLSHIKIMATE SYNTHASE INHIBITOR

Glyphosate is the prime example of this group and is currently the most widely used herbicide. Sales in 2001 were around $3 billion. The discovery of glyphosate resulted from a program for synthesis of water treatment chemicals. Monsanto had in place a systematic screening of all new compounds for potential agricultural applications. The lead compounds showed significant, but relatively weak, herbicidal activity. Follow-up was undertaken by John E. Franz, a chemist in the Agricultural Chemicals group.

$(H_2O_3PCH_2)_2NCH_2CO_2H$ $H_2O_3PCH_2N(CH_2CO_2H)_2$ $H_2O_3PCH_2NHCH_2CO_2H$

lead compounds   glyphosate

The investigation of glyphosate was undertaken on the hypothesis that it might be a metabolite of the two lead compounds. Glyphosate proved to be an extremely potent but nonselective herbicide. Originally, it was used for weed control in areas such as orchards, where it could be sprayed on selected areas. Later, it became the first herbicide used with genetically protected crops (see Section 21.2.1). Glyphosate is an inhibitor of the enzyme 5-enolpyruvyl shikimate-3-phosphate synthase (EPSPS), which catalyzes synthesis of chorismic acid. Chorismic acid is an intermediate in the synthesis of the aromatic amino acids, phenylalanine, tyrosine, and tryptophan. There is no comparable enzyme in animals, so the toxicity of glyphosate is very low. The mechanism is shown in Scheme 8.4.

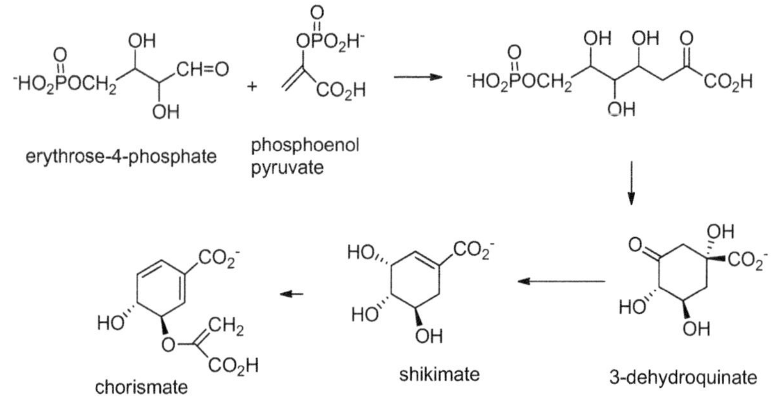

erythrose-4-phosphate   phosphoenol pyruvate

chorismate   shikimate   3-dehydroquinate

SCHEME 8.4   Biological function of enolpyruvylshikimate synthase.

8.2.3.4   GLUFOSINATE, A GLUTAMATE SYNTHASE INHIBITOR

Glufosinate is derived from a tripeptide antibiotic called bialaphos. The herbicidal activity was discovered at Hoechst in Germany. Glufosinate is a structural analog of glutamic acid and is an inhibitor of glutamate synthesis, which is critical to ammonia metabolism in plants.

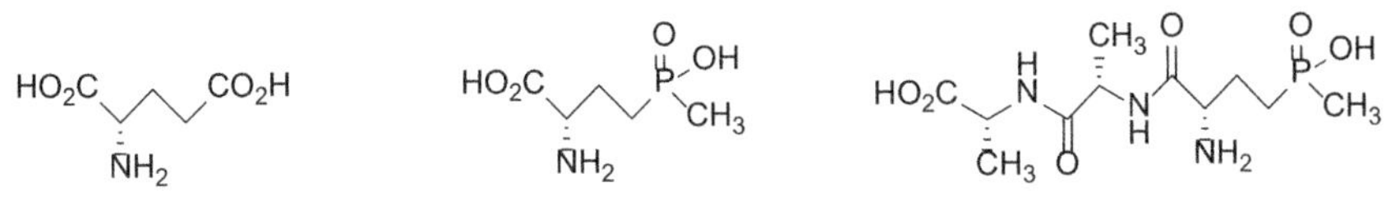

glutamic acid   glufosinate (phosphinothricin)   bialaphos

Glufosinate has low toxicity to animals ($LD_{50} \sim 100$ mg/kg in rodents) and is nonpersistent. It is used as a postemergence herbicide, but has low selectivity among plants and is used in situations where it can be applied selectively to weeds. It is also used in certain genetically modified crops (see Section 21.2.1).

8.2.3.5   AUXIN ANALOGS

In plants, 2,4-D and related compounds act as auxin analogs, disrupting the auxin–cytokinin system. The natural auxin is indole-3-acetic acid. The function of auxin and its analogs is to control biosynthesis of a plasma membrane H^+-ATPase that balances H^+ and K^+ levels in the cell. The ATPase is essential to plant growth as it allows the cell structure to loosen and elongate during growth.[17] The process requires tight control and over-activation by auxin analogs is lethal to the plant. Other examples of this class are fluroxypyr and dicamba. The structures are shown in Scheme 8.5.

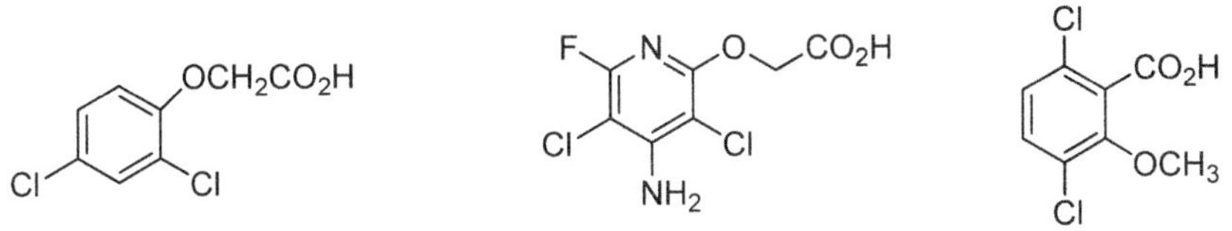

2,4-dichlorophenoxyacetic acid   fluroxypyr   dicamba

SCHEME 8.5   Structures of auxin analogs.

There has been extensive study on the acute and chronic toxicity of 2,4-D and the NOAEL is 5 mg/kg/day.[d] At high doses, kidney toxicity and neurological effects are observed. There is no evidence of genotoxicity or carcinogenicity.[18] The detailed mechanism(s) of mammalian toxicity has not been established.

The auxin class of herbicides had a time of notoriety during the Viet Nam war when 2,4-D and its analog 2,4,5-trichlorophenoxyacetic acid (2,4,5-T) were

---

[d]NOAEL = no observed adverse effect level. This is a general parameter used to define acute toxicity, not only in herbicides but also in preclinical studies of potential drugs.

used in military operations. Between 1961 and 1971, large amounts of mixtures of the two compounds were used for defoliation and crop destruction. The most famous of these was known as *Agent Orange*, but there were also *Agents Purple, Green, Pink*, and *White*, depending on the manufacturer and source. All were mixtures of esters of 2,4-D and 2,4,5-T containing variable amounts of the contaminant 2,3,7,8-tetrachlorodibenzo-1,4-dioxin (see Section 4.2.2). Controversy in the United States arose, first over the morality of defoliant use, and later on the health effects on personnel exposed to the agents. Extensive records exist documenting the amount and location of the agents sprayed in Viet Nam, but there have been no studies of the long-term ecological or health effects.[19] In 1991, congress authorized the Department of Veterans Affairs to establish a list of presumptive health consequences attributable to exposure and to provide compensation and treatment. A number of cancers and other conditions, including *spina bifida* in children are eligible.

8.2.3.6   BIPYRIDINIUMS

The bipyridiniums paraquat and diquat were discovered at ICI in the United Kingdom in the 1950s. They are active in the leaves of green plants and involve both light and oxygen. The pyridinium salts intercept electrons generated in the photosynthetic process and are then oxidized by oxygen, producing superoxide ion and hydroxyl radicals. These reactive species damage the plant membrane. The above-ground parts of the plant are rapidly killed but roots can survive. The bipyridinium salts are used for killing a broad spectrum of annual weeds. They can be applied prior to emergence of the desired crop to kill weeds or cover crops as part of no-tillage systems. Bipyridinium salts are also used as defoliants, for example, in cotton, to facilitate harvest. The pyridinium salts are strongly absorbed by soil, and this both decreases biological activity and slows biodegradation. Depending on soil type the half-life can be 5 years or more and the binding to certain types of clays is nearly irreversible.

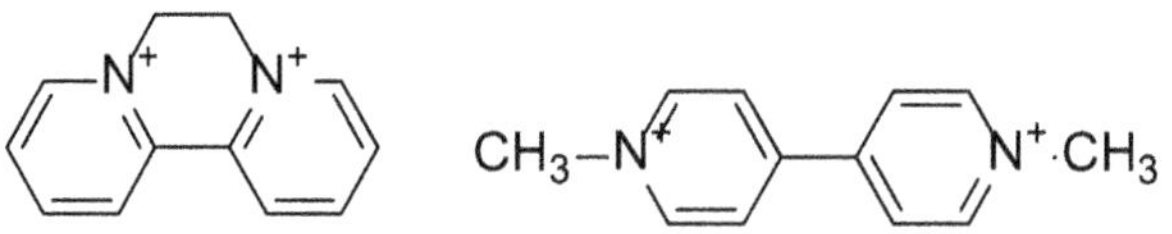

diquat   paraquat

8.2.3.7 CHLOROACETANILIDES

The chloroacetanilides appear to be inhibitors of enzyme systems that are involved in synthesis of very long-chain fatty acids from the C_{16} and C_{18} fatty acids.[20] Examples are shown in Scheme 8.6.

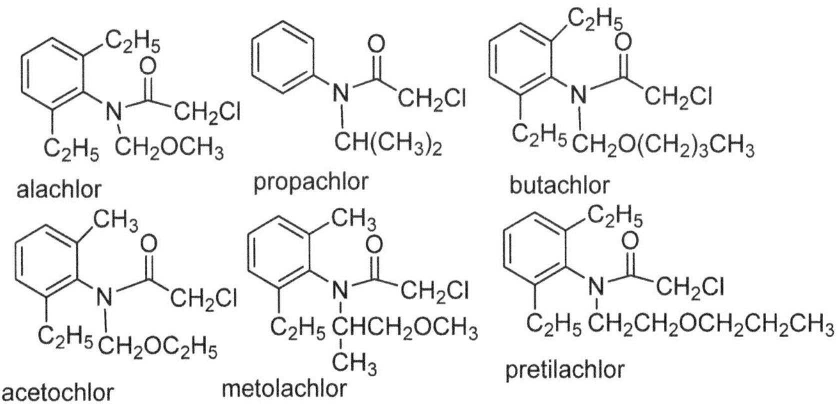

SCHEME 8.6 Chloroacetanilide herbicides.

The chloroacetanilides, along with some other herbicides, are sometimes used with "safeners" or "protectants" that are applied either prior to or simultaneously with the herbicide. They increase the relative tolerance of the crop to the herbicide. For example, two oxime ethers, cyometrinil and oxabetrinil, are used with alachlor. Another protectant is flurazole, which is used with alachlor in sorghum. Fenclorim is used with pretilachlor in rice. Not a lot is known about how these safeners work, but one possibility is that they trigger a stress response mechanism in the crop, which in turn allows it to detoxify the herbicide.[21] Scheme 8.7 shows the structures of some of these protectants.

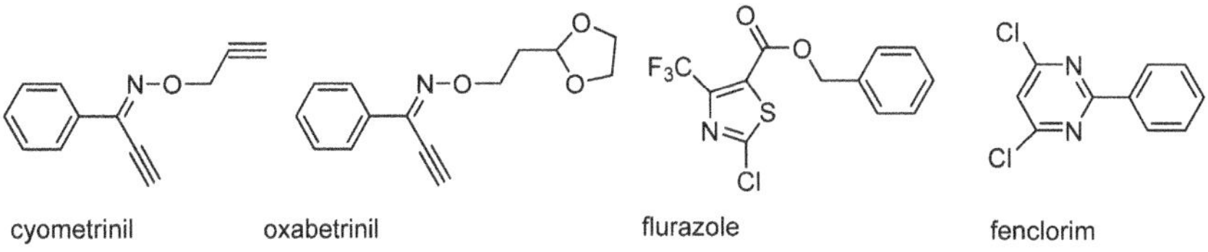

SCHEME 8.7 Structures of herbicide safeners.

8.2.3.8 MICROTUBULE DISRUPTERS

There are several herbicides that act as microtubule disruptors. These include the dinitroanilines, such as oryzalin and trifluralin, aryl amides and carbamates such as propamide, propham, and possibly other compounds such as dithiopyr. Like mammalian microtubule inhibitors, they interfere with mitosis and result in stubby malformed roots. The compounds are used as pre-emergence herbicides by incorporation into a thin soil layer. Monocots are more sensitive than dicots and large-seeded plants can normally grow through the inhibited zone in the soil. The main targets are small-seeded annual grasses. Representative structures are shown in Scheme 8.8.

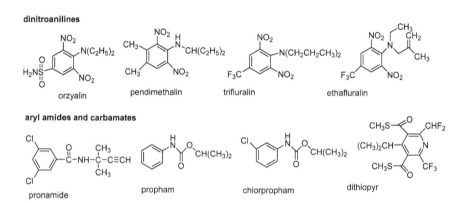

SCHEME 8.8 Herbicidal microtubule disrupters.

8.2.3.9 TRIAZINES

The triazines were originally synthesized and investigated at Geigy in Switzerland. The lead compound was 2,4-*bis*-(diethylamino)-6-chloro-1,3,5-triazine. The first two compounds to be commercialized were atrazine and simazine. Simazine has primarily pre-emergence activity, while atrazine shows both pre- and post-emergence activity. Atrazine was quickly adopted as a major corn herbicide. As of 2005, atrazine was used mainly in corn (85%) with other uses in sorghum (9%) and sugar cane (3%). The level of production is about 75 million lbs/year. It has a fairly long half-life and is a widespread contaminant in surface and ground water. Simazine is used mainly on citrus, fruit and nut trees, and grapes. The starting material for the triazine herbicides is hydrogen cyanide, which reacts with chlorine to give the key intermediate, cyanuric chloride.

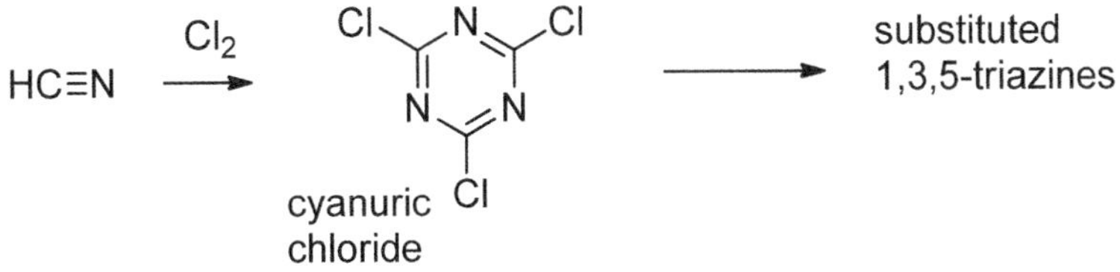
HC≡N
Cl2
Cl
N
Cl
N
N
Cl
cyanuric chloride
substituted 1,3,5-triazines

Other classes of triazines soon followed including 6-methoxy and 6-methylthio derivatives. Other companies also developed triazines, including compounds with cyano and azido groups. Unsymmetrical 1,2,4-triazinones, such as metribuzin and metamitron, were also found. Some of these proved to be useful in broad-leaf crops such as soybeans.

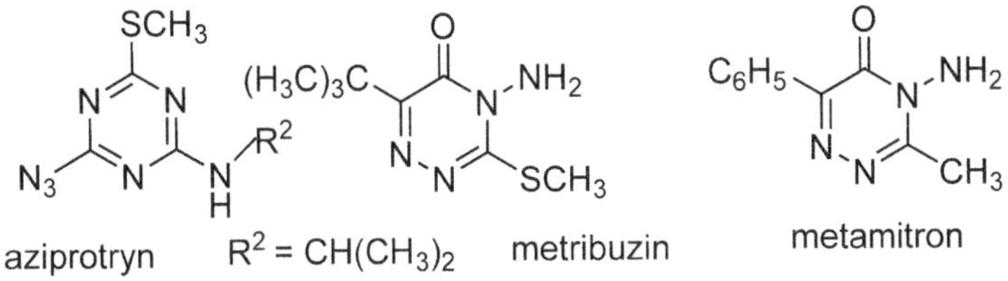
SCH3
N3
R2
aziprotryn
R2 = CH(CH3)2
(H3C)3C
NH2
SCH3
metribuzin
C6H5
NH2
CH3
metamitron

Atrazine and the related triazines are targeted at photosystem II in plant chloroplasts, where they displace plastoquinone and disrupt the electron flow. The selectivity among plant types is usually the result by different rates and extent of metabolic detoxification. The basic patents expired in the 1970s, and there are currently many producers of triazine herbicides around the world. The structures of some of the triazine herbicides are shown in Scheme 8.9.

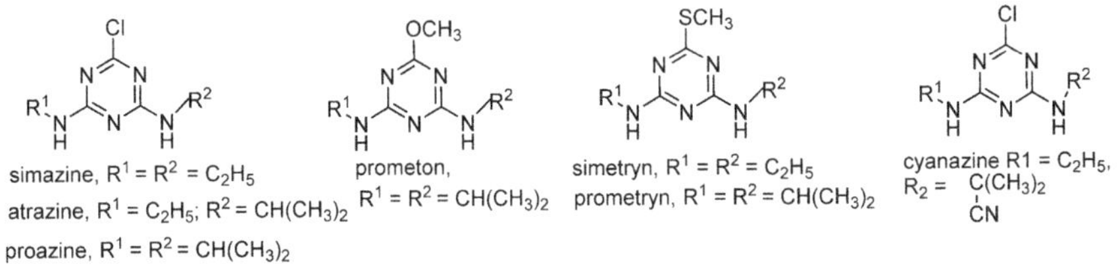
simazine, R1 = R2 = C2H5
atrazine, R1 = C2H5; R2 = CH(CH3)2
proazine, R1 = R2 = CH(CH3)2
prometon,
R1 = R2 = CH(CH3)2
simetryn, R1 = R2 = C2H5
prometryn, R1 = R2 = CH(CH3)2
cyanazine R1 = C2H5,
R2 = C(CH3)2 CN

SCHEME 8.9 Triazine herbicides.

The European Union banned atrazine in October 2003 on the basis that it exceeded the <0.1 ppb level in drinking water that applies to all pesticides, although there is no evidence that atrazine is harmful at that level. In the United States, the permissible level was set at 3 ppb in 1991. The National Research Defense Council sued the EPA in 2003, charging that the EPA had improperly negotiated with the producer and had failed to maintain transparency in its deliberations on atrazine.[22] An EPA review in 2006 concluded that there was "reasonable certainty that no harm will result to the general US population, infants, children or other major subgroups" from continued atrazine use. Most studies on the potential toxicity of atrazine have been focused on its role as a possible endocrine disruptor (see Section 15.6). Several studies have indicated that atrazine has an effect on sexual development in amphibians. On the basis of various studies on potential carcinogenicity, the International Agency for Research on Cancer (IARC) in 1999 classified atrazine as "not classified as a carcinogen in humans." Atrazine does have carcinogenic effects in rodents, but the mechanism is thought not to be relevant in humans.

8.2.3.10   HYDROXYPHENYLPYRUVATE DIOXYGENASE INHIBITORS

The enzyme 4-hydroxyphenylpyruvate dioxygenase (HPPD) is a non-heme iron protein that catalyzes conversion of the substrate to homogentisate. Homogentisate is a key biosynthetic precursor of plastoquinone, a critical component of the photosynthetic electron transport system. HPPD-inhibitors also interfere with biosynthesis of tocopherols and carotenoids. A characteristic manifestation of this class of herbicides is bleaching of the leaves from the loss of chlorophyll and carotenoids. This class also tends to be synergistic with other herbicides directed at photosystem II, such as the triazines.

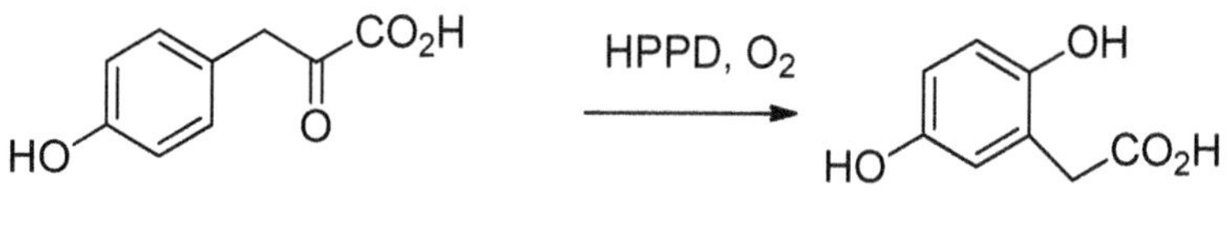

4-hydroxyphenylpyruvic acid   homogentic acid

There is a corresponding mammalian enzyme that mediates the catabolism of tyrosine. However, there is little evidence of mammalian toxicity with this class of herbicides. Rats develop corneal lesions, associated with elevated tyrosine levels, but most other mammals, including humans, evidently metabolize the excess tyrosine without deleterious effect. In fact, there is a genetic condition, tyrosinemia, that is the result of defective tyrosine metabolism. The HPPD inhibitor 2-[2-nitro-4-trifluoromethyl-benzoyl]-1,3-cyclohexanedione (NTBC) has beneficial effects in this condition.[23]

The original example of a HPPD inhibitor was identified at Stauffer Chemical in California, which is now a part of Syngenta. Reed Gray had noted that few weeds grew under bottlebrush plants (*Callistemon citrinus*). He examined extracts and found that one component stunted and bleached germinating grasses. The active compound turned out to be a known substance, leptospermone. The level of activity was not economically attractive, however. Other synthetic cyclohexadiones were investigated and activity was seen in various 2-acyl cyclohexane-1,3-diones. One such compound was sulcotrione. Further optimization led to mesotrione, currently sold under the trade name Callisto. It has good activity against broad-leaf and some annual grasses in maize.[24]

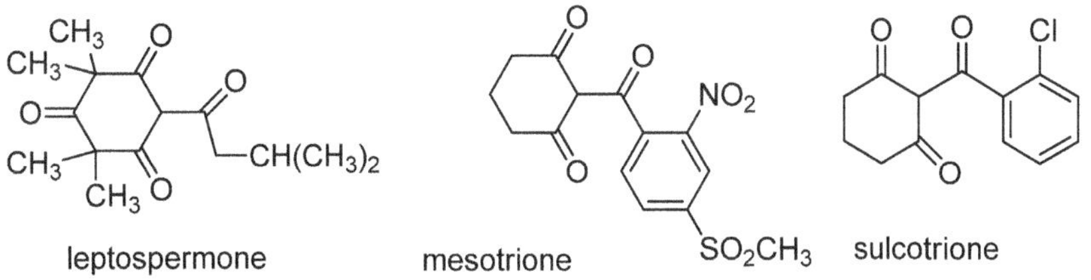

leptospermone   mesotrione   sulcotrione

Related compounds were developed at Hoechst (now part of Bayer Crop Sciences). In Japan, pyrazole derivatives were developed at Sankyo and Ihshihara Kaishi. These included pyrazolynate and pyrazoxyfen, both of which act by being converted to the same pyrazolone. A closely related pyrazolinone, pyrasulfotole[25] is used as a post-emergence herbicide against broad-leaf weeds in cereal crops.

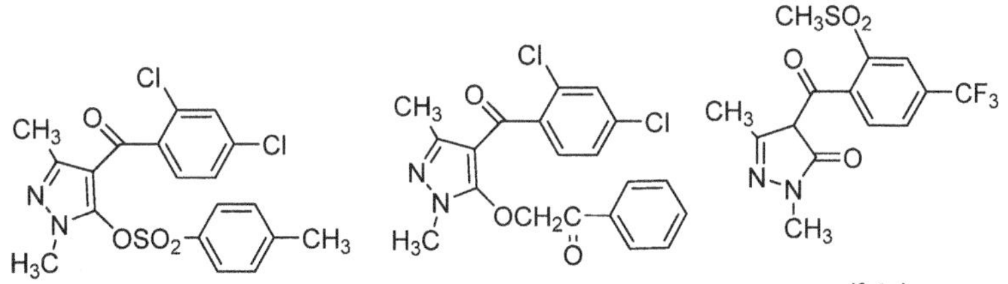

pyrazolynate   pyrazoxyfen   pyrasulfotole

Another HPPD inhibitor is benzobicyclon. It was synthesized at SDS Biotech in Japan.[26] A program of synthesis of analogs of the cyclohexane-1,3-dione class led to various candidates that were then evaluated for safety in rice, low aqueous solubility and mobility, and low fish and mammalian toxicity. The compound passed the usual toxicity, carcinogenicity and teratogenicity screens. The rice grains produced contain no residue of the herbicide or its metabolites. The compound causes the characteristic bleaching in susceptible plants. The herbicidal action is due to hydrolysis of the phenylthio enol ether to the corresponding trione.[27]

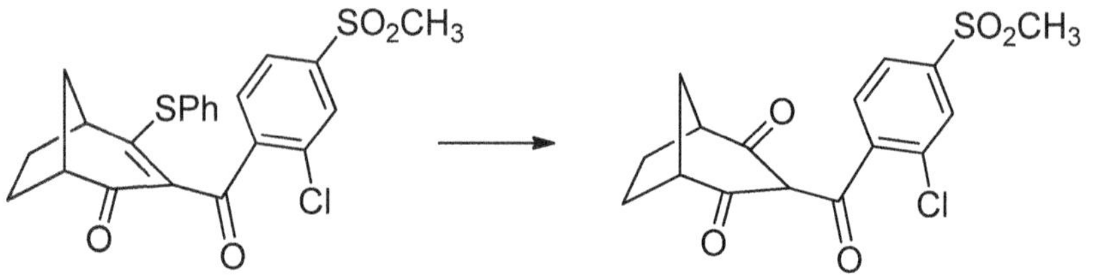

benzobicyclon

Another compound in the class is isoxafutole.[28] The compound evolved from a lead compound that contained a piperidine-2,4-dione structure synthesized at Rhone–Polenc's research station in Ongar, United Kingdom. However, a patent issued to Stauffer covered the most interesting analogs. When work was resumed in the 1980s, one of the substitution made was replacement of the piperidine-dione ring by an isoxazole. This led to the identification of isoxaflutole. It is believed to be metabolized to the active diketo nitrile derivative.

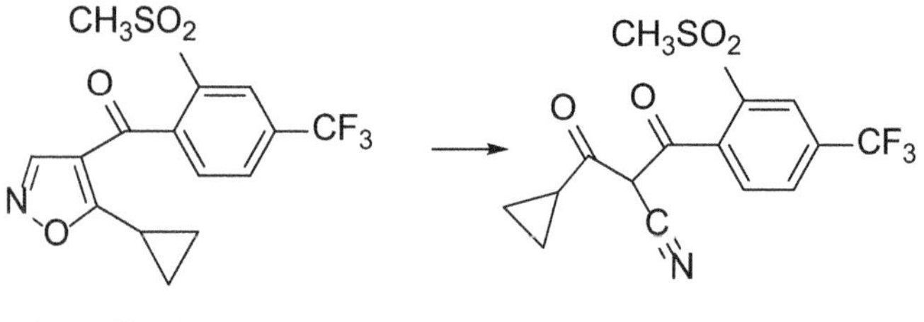

isoxaflutole

A common structural feature of the HPPD inhibitors is a 1,3-dione structure, either in the original compound or the active metabolite. Such compounds are good chelators of iron. A crystal structure of the complex with HPPD shows pyrasulfotole participating in octahedral chelation at the iron of HPPD.[29]

8.2.3.11 PROTOPORPHYRIN OXIDASE INHIBITORS

Protoporphyrin oxidase (PPO) is a key enzyme in the biosynthesis of both chlorophyll and hemes. It oxidizes protoporhyrinogen IX to protoporphyrin IX. Inhibition not only disrupts chlorophyll biosynthesis but also heme-based cytochromes, oxygenases, peroxidases, and catalazes. This results in active oxygen intermediates and membrane damage and rapid killing of the weed. Some of the herbicides in this group are shown in Scheme 8.10.

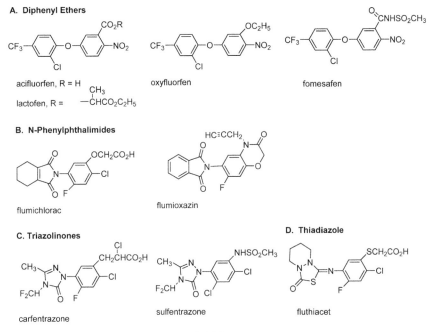

SCHEME 8.10 Protoporphyrin oxidase inhibitors.

8.2.4 *HERBICIDE-RESISTANT WEEDS*

There are now many herbicide-resistant weed species around the world.[30] Repeated use of a particular herbicide selects for the strains that are most resistant and eventually results in resistant variants. There are several mechanisms by which resistance can arise. The most common mutations inhibit the binding or block access of the herbicide to the binding site. Other resistant lines have enhanced detoxification activity, for example, through P-450 type monooxygenases. There are other examples where resistance seems to be the result of reduced absorbance or transport.[31] It has recently been demonstrated that resistant to glyphosate, the world's most important herbicide, can develop by gene amplification.[32] The occurrence of herbicide-resistant weeds has serious implications for agricultural economics. One strategy for slowing resistance is to keep weed numbers and the resulting seed numbers low. This reduces the chances that a random mutation conveying resistance will occur. It is also possible to rotate among herbicide types. Another strategy is the use of combinations of herbicides with different targets of action. If the genes for resistance are independent, the chances of a plant being resistant to both are the product of the fractional frequency of the resistance gene. One of the characteristics of many weeds, however, is production of very large numbers of seeds for each plant, so a single resistant plant can produce many resistant offspring.

8.3 INSECTICIDES

Insecticides have several important roles in maintaining human well-being. One is control of insects that are carriers of human disease, such as malaria, which is transmitted by mosquitoes (see Section 18.1.1). Insecticides are also important for both animal care and crop protection in agriculture. Prior to WWII, only a few substances were available for protection of crops from insects. One was sulfur, a relative ineffective material. Another was natural pyrethrin insecticide, but it was too expensive for agricultural use. Then in the late 1930s, Paul Mueller, at the Geigy Company in Switzerland, discovered the insecticidal properties of DDT. DDT had an immediate and powerful impact. Late in WWII, when lice-borne typhus threatened Naples, treatment with DDT eradicated the infestation. Immediately after the war, malaria was virtually eliminated from Italy and other parts of Europe, as a result of mosquito control. The impact of the discovery is indicated by the speed with which Mueller was recognized by the Nobel Prize in Medicine or Physiology in 1948. Many other chlorinated insecticides followed but their drawbacks soon became apparent (see Section 4.2).

The next major group of insecticides to be developed was organophosphates such as parathion and malathion. Some of the background for these compounds came from chemical warfare research in Russia and Germany prior to WWII, which gave rise to the so-called nerve gases such as sarin, soman, and tabun. Another series of nerve agents, such as amiton and VX, are aminoalkylthio esters and were developed in the United Kingdom. These are extraordinarily toxic compounds.

$$
(CH_3)_2CH-O-\overset{\overset{\displaystyle O}{\|}}{\underset{\underset{\displaystyle CH_3}{|}}{P}}-F
\qquad
(CH_3)_3C\overset{\overset{\displaystyle CH_3}{|}}{C}H-O-\overset{\overset{\displaystyle O}{\|}}{\underset{\underset{\displaystyle CH_3}{|}}{P}}-F
\qquad
CH_3CH_2\cdot O-\overset{\overset{\displaystyle O}{\|}}{\underset{\underset{\displaystyle CN}{|}}{P}}-N(CH_3)_2
\qquad
(C_2H_5O)_2\overset{\overset{\displaystyle O}{\|}}{P}S(CH_2)_2NR_2
$$

sarin	soman	tabun	amiton R = C_2H_5
			VX R = $CH(CH_3)_2$

Another major group of insecticides, the carbamates, was introduced in the late 1950s, with carbaryl being the initial example. The pyrethrins gave rise to a host of synthetic analogs. Another group of compounds, synthetic analogs of nicotine, were introduced in the 1990s. As of 2000, the organophosphates (25%), pyrethroids (20%), nicotinoids (15%), and carbamates (10%) were the most important insecticides in terms of market share. One of the striking features of these insecticides is that their mode of actions is highly concentrated on one target system, neurotransmission.

8.3.1 ORGANOCHLORINE INSECTICIDES

1,1,1-Trichoro-2,2-*bis*-(4'-chlorophenyl)ethane (DDT) was first synthesized in 1873, but its insecticidal properties were not recognized, nor hardly conceivable, at that time. Paul Mueller, working at Geigy in Switzerland was systematically looking for contact insecticides when he resynthesized DDT in 1939 and observed its potent insecticidal properties. Soon, the world was at war and the dislocations of war encouraged the spread of insect-borne disease such as typhus. After the end of the war, DDT became the basis of extensive antimalaria programs by mosquito control, and in 1972, the WHO reported that malaria had been eradicated in 37 countries and reduced in 80 others.[33] In India, it was estimated that the annual death rate from malaria was reduced from 750,000 to 1500 by 1966.[34] DDT also had many agricultural applications. Production peaked in the United States at 160 million pounds in 1961.

Some of the environmental consequences of the chemical stability and long half-life (years) of DDT soon were recognized. The problem is exacerbated by concentration in the food chain by fish and birds. Studies in the 1950s began to detect toxic effects in amphibians, crustaceans, birds, and fish. Bird species differ in their sensitivity, both because of inherent metabolic differences and factors related to concentration through the food chain. Brown pelicans on the Pacific Coast were one of the first species noted to be affected and concentrations of >1000 µg/g were observed and traced to dissemination from a manufacturing plant. Other sensitive species include peregrine falcons, the bald eagle and osprey. Domestic chickens, on the other hand, are relatively insensitive. The public's attention was galvanized by publication of Rachel Carson's *Silent Spring* in 1962 (see Section 4.2). Many studies demonstrated the widespread occurrence of DDT and related chlorinated compounds in the 1960s and 1970s. These included compounds such as methoxychlor, lindane, and aldrin (see Section 4.2 for structures). The use of DDT was banned in the US in 1972. Nevertheless, nearly all residents of the United States have detectable levels of DDT or its major metabolite DDE. DDT is one of 12 *persistent organic pesticides* listed in the Stockholm convention of 2001, and about 180 countries have agreed to limit its use to insect vector control where no other agent is practical. DDT is still used for malaria control under these restrictions in India and Africa.

8.3.2 ORGANOPHOSPHATE INSECTICIDES

The organophosphates are directed toward *acetyl choline esterase* (AChE), an enzyme that is critical in neurological function in both insects and mammals. The compounds phosphorylate the active site serine in the enzyme, resulting in accumulation of the neurotransmitter acetyl choline. In addition, some of the agents undergo a process called *"aging,"* involving hydrolysis of one of the phosphorus substituents. This leads to irreversible inactivation of the enzyme. Most of the organophosphates are very toxic to mammals and a significant number of fatalities have occurred as the result of accidental exposure. Sarin was used in the Tokyo subway attack of 1995, which killed 12 people and sickened several hundred others. Sarin was reportedly also used in civil conflicts such as in Iraq in 1988 and Syria in 2013. Most of the insecticides are thionophosphates. These compounds show somewhat improved selectivity for insects, although they are still very toxic to mammals. Scheme 8.11 shows some of the organophosphate structures.

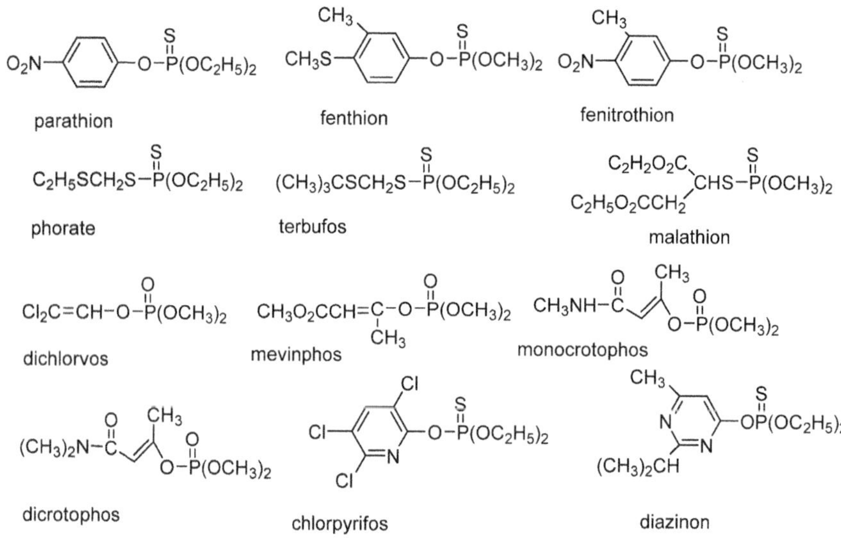

SCHEME 8.11   Organophosphate insecticides.

There is some evidence that low-level repeated exposure is detrimental to human neurophysiological performance. For example, a study of English sheep-dippers with a history of exposure to diazinon indicated lower performance in areas such as fine motor control and higher levels of anxiety and depression.[35] Banana plantation workers in Costa Rica with a history of organophosphate poisoning also showed deficits in motor skills and increased neuropsychiatric symptoms.[36] Concern about possible toxicity of chlorpyriphos and diazinon led to considerable restriction on their use as of 2000 in the United States. They are no longer permitted for indoor use, residential termite treatment or lawn use. The remaining usages are restricted to professional applicators.

8.3.3   CARBAMATE INSECTIDES

The recognition that certain aryl carbamates are inhibitors of acetylcholine esterase can be traced back to the isolation and investigation of a natural product, physostigmine. The origin of physostigmine is the west coast of Africa. One of these tribes, the Efik, was located in the province of Calabar, in what is now Nigeria used ordeal by a poison in judicial proceedings, trials for witchcraft and executions. The British civil authorities in the 1870s suppressed these uses of the calabar bean. The plant was first characterized by John Hutton Balfour, a professor of Botany at Glasgow and Ediburgh, who was also trained as a physician. He described the plant and gave it the name *Physostigma venenosum* in 1861. One of his students, Robert Christison, investigated the toxin's effect, including trying the extract on himself. He isolated the active toxin in 1863 and called it "eserine." The first entirely pure sample was isolated a few years later by Jobst and Hesse in Germany, who called it "physostigmine." One of Christison's students, Thomas Richard Fraser, continued study on the calabar bean. He found that atropine acted as an antidote and quantified the relationship between the dosage of poison and the antidote. The structure of physostigmine was determined in 1925 and it was synthesized by the American chemist Percy Julian in 1935.[e]

---

[e]Percy L. Julian was the first African-American chemist elected to membership in the National Academy of Science. His father, a railway mail clerk, was self-educated and saw to it that each of his six children received higher education. Although Julian's education in his native Alabama ended at the 8th grade, he was able to enter DePauw University in Greencastle, IN, USA. He graduated as valedictorian of the class in 1920. Eventually, the entire family moved to Greencastle and all of Julian's brothers and sisters received degrees there. Julian was a chemistry instructor at Fisk University until he entered graduate school at Harvard in 1923. He received an M. S. but, for lack of financial support, was unable to complete a Ph. D. He served as an instructor at West Virginia State and Howard University and then received a fellowship from the Rockefeller Foundation to do graduate work at the University of Vienna, where he received his Ph.D. in 1931. He returned to Howard to teach but was forced to resign after a confrontation with a colleague led to release of embarrassing personal letters. He returned to DePauw as an instructor and with Josef Pikl, who had come from Vienna to work with him, synthesized physostigmine in 1935. There was a discrepancy in properties with the report of the eminent English chemist, Sir Robert Robinson, but Julian and Pikl's data proved to be correct. Julian later moved to Glidden Soya Products in Chicago where he worked on proteins and oils from soy bean. He developed several successful commercial products. He also devised a method to isolate stigmasterol from soybean oil and this led to methods for synthesis of steroid hormones, including progesterone (see Section 15.3). In 1953, he left Glidden to found his own company, Julian Laboratories, which was engaged in production of steroid intermediates. Competition with Syntex and other producers of steroid intermediates was fierce but his company was successful. He eventually sold the company to Smith Kline French for $2.3 million. Throughout his career, Julian met racial obstacles, including fire-bombing of his home in the Chicago suburb Oak Park. Nevertheless he remained optimistic and committed to improvement of race relations. He served on many civic boards and was eventually the recipient of 19 honorary degrees and numerous other awards. A documentary on his life was prepared by the PBS NOVA series in 2009.
Sources: Witkop, B. *Biogr. Mem., Natl. Acad. Sci. U.S.A.* **1980**, *52*, 3–46; Tischler, M. *Chemist* **1965**, *42*, 105–106.

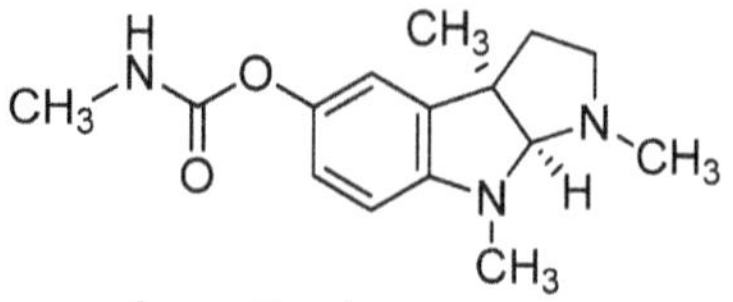

physostigmine

The role of acetylcholine as a neurotransmitter was recognized by Henry Dale in 1914, and in 1926, Otto Loewi demonstrated that physostigmine was an AChE inhibitor. The mechanism of inhibition, covalent carbamoylation of the active site serine, was established by I. B. Wilson and associates in 1960. The leap to a commercial insecticide was evidently serendipitous. Carbamates synthesized by D. H. Gysin at Geigy as insect repellants were observed to have insecticidal activity and led, eventually to compounds such as carbaryl and propoxur, as shown in Scheme 8.12.

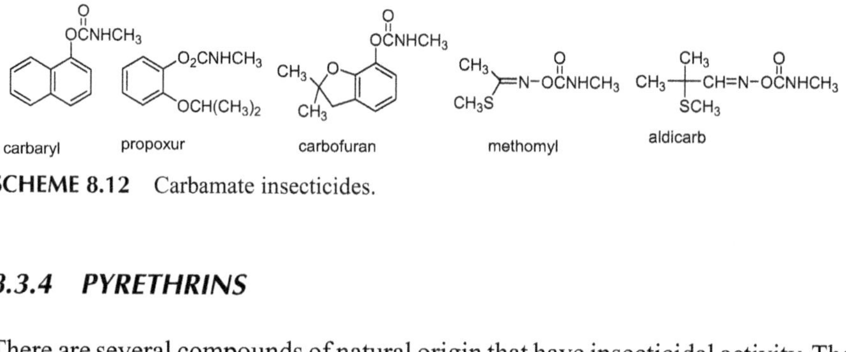

carbaryl   propoxur   carbofuran   methomyl   aldicarb

SCHEME 8.12   Carbamate insecticides.

8.3.4 PYRETHRINS

There are several compounds of natural origin that have insecticidal activity. The most important of these are the *pyrethrins*, most of which are now made by synthesis. The pyrethrins are found in flowers of *Chrysanthenum cinerariaefolium*. The dried powdered flowers can be used as an insecticide. The original source of the insecticidal preparation seems to have been the Dalmatian coast on the Mediterranean. From there pyrethrins spread to the rest of Europe, the United States, and Japan in the second half of the 19th century. Japan became the major commercial producer and just before WWII the production was 13,000 t/y. After WWII, production of natural pyrethrins shifted to East Africa. There are four major constituents of natural pyrethrins, as shown in Scheme 8.13.

A number of synthetic analogs of pyrethrins have been introduced as insectidies. The first synthetic compound, allethrin, was prepared at the USDA Agricultural Research Station in Beltsville MD in 1948.[37] Extensive work was also carried out at the Rothamsted Experimental Station in the United Kingdom[38] and at the Sumitomo Chemical Company in Japan.[39] As structural investigations continued, many active analogs with only general resemblance to the original compounds were found. The advantages of the synthetic compounds included price, activity and field stability. The pyrethrins generally have fairly low toxicity to mammals (rats, LD_{50} >50 mg/kg) but most are quite toxic to fish, although some of the more recent analogs have reduced fish toxicity. Pyrethrins are hydrophobic. Fluorinated analogs such as tefluthrin have enhanced vapor pressures.

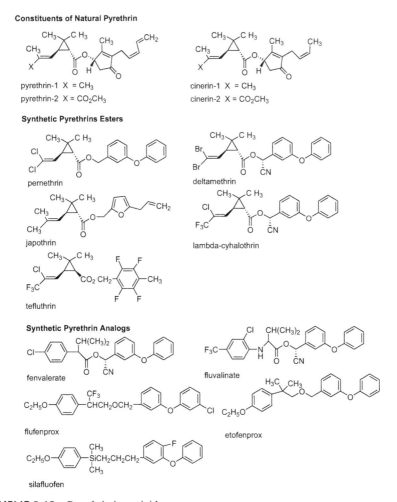

SCHEME 8.13 Pyrethrin insecticides.

The main target of insecticidal activity by the pyrethrins is a voltage-sensitive channel that controls sodium permeability of cells in the nervous system. The compounds cause prolonged excitation leading to tremors, convulsion, and death. This is also the site of action of DDT and there is cross-resistance between the two classes of insecticides. Mammalian toxicity is associated with Ca^{2+} and Cl^- channels. The fact that there are many different pyrethrins, with somewhat different profiles of activity, makes generalization of the toxic effects difficult.[40] Pyrethrins are nonpersistent in the environment. Pyrethrins are used extensively in household insect control and are important in malaria control in Africa, where pyrethrin-treated mosquito netting is used. Household use of pyrethrins can leave residues in rugs, fabrics, etc., and this has given rise to concern that they might lead to harmful chronic exposure. This issue has received particular media attention in Germany.[41] Most pyrethrins decrease neuromotor functions at sub-acute doses, but there is little evidence that they accumulate as a result of chronic exposure.

8.3.5   NICOTINOIDS AND NEONICOTINOIDS

The alkaloid nicotine has insecticidal activity and extracts from tobacco plants have been used in agriculture since the 1700s. However, nicotine is highly toxic to mammals (LD_{50}, IV, mice 0.3 mg/kg) and must be used with care. A number of synthetic nicotine derivatives have attained importance as insecticides. The major development of the neonicotinoids was done by a Japanese firm, Nihon Tokushu Noyaku Seizo that is now part of Bayer Crop Sciences. The successful introduction of imidacloprid was followed by other neonicotinoids such as thiamethoxam (Syngenta), acetamiprid (Nippon Soda), nitepyram and clothianidin (Takeda Agro), and dinotefuran (Mitsui Chemical). The neonicotinoids are systemic insecticides that are absorbed by plants and are most effective against piercing and sucking insects such as aphids. They can also be used on pets for control of fleas, lice, mites, and ticks. The advantage of the new synthetic neonicotinoids is that they have much higher selectivity for the insect AChE receptor. One of the chemical characteristics of the neonicotinoids is the neutrality at physiological pH.[42] The structures of several of the neonicotinoids are shown in Scheme 8.14.

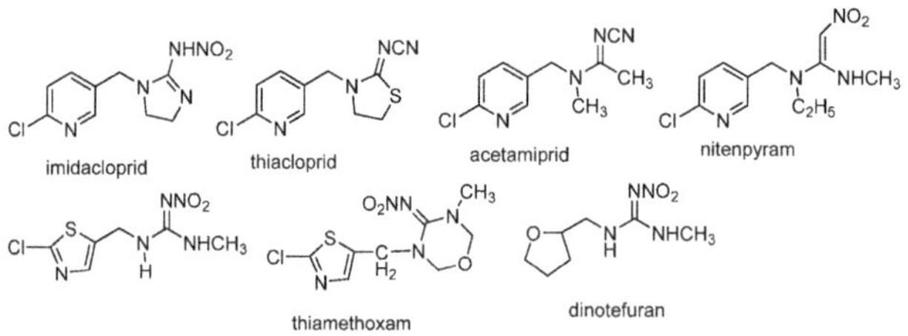

SCHEME 8.14   Neonicotinoid insecticides.

8.3.6 CALCIUM FLUX REGULATORS

The most recently developed insecticides are aromatic amides that interfere with Ca^{2+} flux regulation.[43] A natural compound ryanodine was the first insecticidal material discovered to have this mode of action.[44] Ryanodine is a complex highly oxygenated diterpene alcohol esterified by pyrrole-2-carboxylic acid. Among the synthetic compounds with activity are flubendiamide and DP-12. The structures are shown in Scheme 8.15.

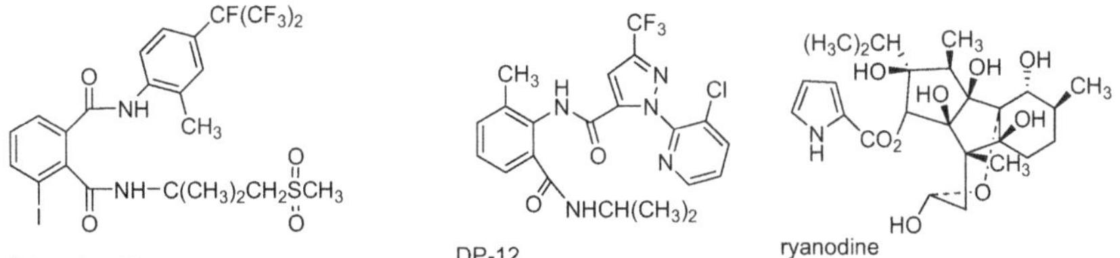

SCHEME 8.15 Calcium flux regulators.

8.3.7 NATURAL SUBSTANCES WITH INSECTICIDAL ACTIVITY

In addition to the pyrethrins and nicotine, other natural products such as rotenone, deguelin, and azadirachtin have insecticidal activity and are the basis for a number of natural or "organic" insecticide products. Rotenone functions by inhibiting one of the proteins in the mitochondrial electron transport complex.[45] Lest we fall into the fallacy that "natural means safe," it is reported that chronic exposure to rotenone induces Parkinson-like symptoms in rats.[46]

Azadirachtin is isolated from the seeds of the neem tree (*Azadirachtin indica*), which is native to India, but now grows in many parts of the semitropics. Neem leaves have been used for centuries in India as an insect repellant for stored grains. It was also observed that locust swarms completely avoided Neem trees, even when devouring everything else. Study of the active constituent was undertaken at the University of Keele in the United Kingdom in the 1960s and the antifeedant component was isolated. It was also demonstrated that the compound inhibits the normal development of immature insects. Commercialization of azadirachtin was undertaken in the 1990s. The neem fruit are harvested and the dried seeds are crushed. An inedible oil is obtained and the azadirachtin is present to the extent of 12–26% in the residual powder, which can be dried and stored. This is the material used for formulation of insecticides. The oil can be used in products such as paints. The US EPA approved azadirachtin formulations under relaxed requirements that pertain to natural substances and various products are now available in the United States and Europe, as well as India.

At the present time, however, the product does not compete economically in large-scale applications. The structures of these natural insecticides are shown in Scheme 8.16.

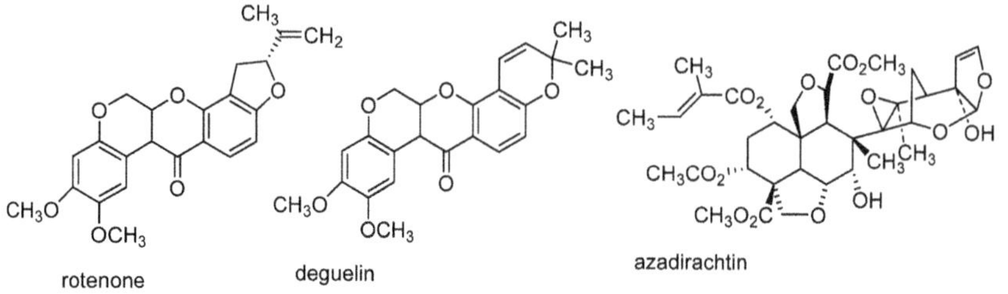

SCHEME 8.16   Natural insecticides.

8.3.8 OTHER APPROACHES TO CONTROL OF INSECTS

The environmental and ecological disadvantages of insecticides have spurred efforts to develop other approaches to control of insect pests. This is called *integrated pest management*, implying that several aspects of insect biology and behavior are targeted in a way that maximizes effective control of pests, while minimizing environmental impact. The most successful examples of integrated pest management are based on the pheromones that insects use for sexual attraction or as alarm, aggregation, and trail signals. These are typically mixtures of volatile compounds including hydrocarbons, alcohols, and esters. Many pheromones have been isolated from insects and characterized by structure determination. In many cases, it has been possible to synthesize the individual components of the pheromone mixtures. There are several ways that pheromones can be used in agriculture. The most widely used application is for monitoring. By use of pheromone-baited traps, the level of insect infestation can be determined. This information can be used to optimize the application of conventional insecticides. Pheromones can also be used for baiting traps in combination with lethal insecticides. The sex pheromones can generate confusing signals that reduce the likelihood of successful mating.

Pheromone-based approaches have several advantages over conventional insecticides. They are usually species-specific so that only the targeted insect pest is affected. The compounds are usually nontoxic and frequently can be used in quite small amounts. They are nonpersistent and unlikely to appear as residues in food products. Unfortunately, they also have significant disadvantages. The compounds are generally quite expensive to make. In some cases, the activity of the pheromones depends on precise composition of several components. Their effectiveness under field conditions depends on the concentration that can be obtained and the temperature. Because of these limitations, pheromones have been most useful in applications such as orchards and vineyards, where fairly small areas can be treated under well-controlled conditions. Overall, only about 1% of the total insect control market for commercial crops production goes to pheromone-based systems. Among the most successful examples is control of the codling moth (*Cydia pomonella*) in fruit orchards by the pheromone *codlemone*. The moths were attacked with a succession of insecticides beginning with DDT in the 1950s followed by carbamates and organophosphates, but by the 1970s, resistance required high levels of application. A group of growers, with advice and support from the University of California Extension Service and USDA, began an integrated management program-based primarily on the organophosphate azinphos-methyl. In the 1990s, mating disruption based on a synthetic codling moth pheromone consisting of 53% E,E-dodeca-8,10-dien-1-ol, 30% dodecanol and 6% tetradecanol was introduced.

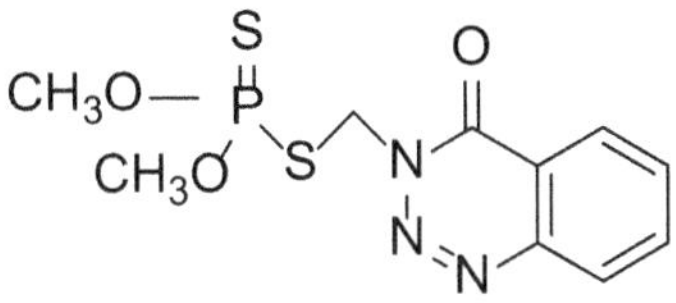

azinphos methyl

Several economically important palm species, including those that produce dates, coconuts, and palm oil, are subject to attack by weevils of the species *Rhynchophorus*. These insects use aggregating pheromones that include compounds such as 4-methyl-5-nonanol and 2-methylhept-5-en-4-ol. The pheromones, typically mixed with decaying plant material, are used for mass trapping. A combination of pheromone and insecticide was successfully used against oil palm beetles in Central and South America. In European vineyards, the grape vine moths *Lohesia botrana* and *Eupoecilia ambiguella* are significant pests. Pheromones have been used in control by mating disruption.[47] Eggplant is a major vegetable crop in India and Bangladesh. The borer *Leucinodes orbonalis* is trapped by a pheromone mixture that is 1% hexadec-11-enol and 99% hexadec-11-enyl acetate. The reduced borer population also minimizes secondary damage by other insects.[48] Mass trapping was successfully used in the late 1970s and early 1980s in spruce forests in Scandinavia. Some 4.5 billion bark beetles were trapped after distribution of almost a million traps. Tree damage was reduced significantly.

In the United States, a program intended to slow the spread of the gypsy moth uses pheromones as well as insecticides and insect pathogens. Since accidental release in New England in 1869, the gypsy moth has spread as far south as North Carolina and as far west as Minnesota. It defoliates many species of trees and plants. Flakes treated with the female gypsy moth pheromone are distributed by air for mating disruption. The program also uses the insecticide diflubenzuron and the insect pathogen *Bacillis thuringiensis.* It is considered that the program slows the rate of spread of the gypsy moth from 12.5 mi/year to 7.5 mi/year.

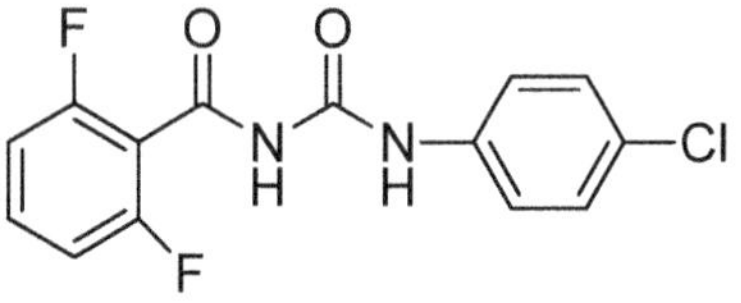

diflubenzuron

Another example is control of the cotton boll weevil. The boll weevil arrived from Mexico late in the 19th century. It spread throughout the cotton-growing regions of the United States, including California, the Southwest and Southeast. The USDA in cooperation with cotton-growers began a program of regional eradication in 1978, starting in Virginia and North Carolina. The program involved application of insecticides (mainly malathion), soil management (stalk-shredding and plowing), and, eventually, pheromone-baited monitoring traps. The program is judged to have been highly successful, achieving eradication in all of the United States except small parts of Texas and New Mexico. The program maintains traps to monitor for reappearance of the boll weevil. The successful eradication over large areas has greatly reduced the need for insecticide application during the growing season.

8.4 ENVIRONMENTAL AND ECOLOGICAL EFFECTS OF FOOD PRODUCTION

8.4.1 *GENERATION OF GREENHOUSE GASES*

Agriculture has major environmental impacts. It is a significant source of greenhouse gases, particularly CO_2, methane, and N_2O. As much as half of methane emission is attributed to animal husbandry, both directly from animals and by microbial metabolism of animal waste.[49] Microbial degradation of both synthetic nitrogen fertilizer and plant and animal waste generates N_2O. It is estimated that agriculture accounts for about 10% of total manmade greenhouse gas emissions. Agriculture also plays an important role in the overall carbon balance because crop growth uses CO_2 converting it to food and fiber. Roughly speaking, agriculture, including fuel use, accounts for about 5% of the net emission of CO_2.

Various agricultural practices have the potential to reduce these emissions. No-till farming decreases energy use and favors carbon retention in the soil. Grasslands sequester CO_2, and under optimal management including fertilization, the extent of carbon capture can increase. Restoration of degraded land, including improved water management, can also increase carbon retention. Precision nitrogen fertilization can decrease N_2O emission and fertilizer runoff. Possible means of reducing methane generation include capture from waste. It may also be possible to reduce methane generation directly from animals by feeding practices, or biochemical (drug) manipulation. The largest factor will probably be the extent of biomass conversion for energy. Biomass conversion reduces net generation of CO_2. The balancing factor here, however, is the competing use of land to generate food supplies (see Section 2.2.3).

8.4.2 PESTICIDES

8.4.2.1 ORGANOCHLORINE COMPOUNDS

Many studies have demonstrated the effect of the organochlorine insecticides on wildlife[50] (see Section 4.2). DDT and closely related compounds have long half-lives in the environment, can accumulate in certain species, and generally are only slowly metabolized. Many specific connections have been established between DDT use and bird deaths. In the assessment of one writer (Blus ref. 50 p. 314): "It is likely that no other group of contaminants of anthropogenic origin has exacted such a heavy toll on the environment as have the organic chlorine pesticides." North America and Western Europe were the first areas of the world to use large amounts of organochlorine insecticides and the first to strongly restrict their use. The same problems arose in other parts of the world. Agricultural use was banned in China in 1983, but mirex and chlordane are still used in termite control.[51] DDT is still used in parts of the world where there is no practical alternative for mosquito control.

8.4.2.2 PESTICIDE RESIDUES IN FOOD

There are both national and international guidelines for the amount of pesticide residues in foods. There are several concentration levels that are considered significant in evaluating pesticide toxicity. These are based on toxicological and related studies that establish the "No Observed Adverse Effect Level" (NOAEL), the level at which no evidence of toxicity is observed over a relatively short period of time (e.g., 30 days). The "Acute Reference Dose" is the lowest dose at which a rapid adverse response in noted. The "Maximum Residue Level" (MRL) is the highest permissible level of the contaminant. The "Allowable Daily Intake" is the daily amount of a substance over a total lifetime that is considered to have no appreciable health risk. In the United States, each MRL is set by the EPA, but the responsibility for monitoring food additives and contaminants rests with the FDA. These laws have been updated from time to time. The most recent legislative revision was the Food Quality Protection Act of 1996. A USDA study in 2003, found that about 0.3% of samples of fresh fruits and vegetables exceeded the MRL. The EU and Japan, have similar regulatory systems, with default levels of 0.01mg/kg for substances where inadequate data is available to establish a MRL. An international organization, the *Food and Agricultural Organization* (FAO), maintains recommended levels for many pesticides and some countries use these as the permissible levels. The FAO promotes implementation of agricultural and processing procedures that minimize residue levels.

8.4.2.3 PESTICIDES AND HUMAN HEALTH

The case of DDT and other highly chlorinated insecticides clearly showed that pesticides can have environmental effects. There is widespread public concern about many aspects of pesticide use. Are there pesticide residues in the food we eat and if so, how much and what effects do they have? Do pesticides in food, water, or other sources of exposure lead to such consequences as cancer,[52] reproductive,[53] or developmental[54] problems? Do pesticides contribute to neurological diseases such as Alzheimer's or Parkinson's disease? In all cases except Alzheimer's, one can point to pesticides that do have specific adverse effects. Those cases where clear connections exist between exposure and health effects have generally been detected by clusters of cases that could be directly related to an exposure. No one knows, however, exactly how important pesticides might be in the etiology of various diseases and other health issues. There are many complicating factors. (1) There may be wide variations in individual susceptibility to harmful effects based on genetics, age, gender, immunological status, etc. (2) Related, or even unrelated, chemical species might have cumulative or synergistic effects. (3) Detecting effects in the broad population by epidemiological studies faces many potential confounding factors, and in the case of pesticides, there is usually great uncertainty in extent of exposure. (4) Some diseases such as cancer have long latency periods so there is often a long lapse between exposure and occurrence of cancer, increasing the difficulty in establishing a causative link. New biotechnological techniques may provide some data that will help address these kinds of issues. *Biomarkers* for example, can be used to detect effects of exposure, even in the absence of overt effects. A biomarker is some measurable function that can be related to pesticide exposure. For example, AChE levels can be used to indicate exposure to organophosphates.

One of the broad public concerns about pesticides is that they might cause cancer. In the United States, the CDC, NIH, and EPA all conduct studies to determine the extent of the risk of pesticides. An international organization, the IARC publishes studies on the cancer risk associated with various chemicals, including pesticides. Unfortunately, drawing definitive conclusions from these studies is often difficult. The IARC assigns cancer risk on the basis of epidemiological and/or animal studies. Specific compounds are given one of four designations with respect to carcinogenicity: (1) sufficient evidence; (2) limited evidence; (3) inadequate evidence; (4) no evidence. The highest category is assigned on the basis of epidemiological studies that indicate a relative risk of >2.0 for the material. Category 2 is assigned on the basis of epidemiological studies indicating a relative risk of > 1.3 or suggestive animal studies. Category 3 pertains to compounds for which there is insufficient evidence, either positive or negative, to make a judgment. Category 4 requires reliable evidence that the substance in noncarcinogenic. Organizations such as IARC also take into account the "coherence" of the evidence, that is, does it fit together and is it biologically plausible. Of pesticides that are currently in wide use, the phenoxyacetic acids, such as 2,4-D and its analogs are classified 2. Atrazine, with some evidence of increased rate of ovarian cancer, also falls in category 2, although this classification is controversial. Most organophosphates are classified 3, although in some cases there is epidemiological association with leukemias or lymphomas.

There have been a number of epidemiological studies of the possible relationship between DDT exposure and breast cancer. Most of the studies show no evidence of a relationship.[55] There have been more limited studies of other potential adverse effects, but no definitive associations exist as of now.[56] In 2009, a group of knowledgeable scientists reviewed the most recent (past 5 years) studies on the health effect of DDT exposure. There is some suggestion that adolescent exposure to DDT increases the risk of breast cancer. There is also some evidence of association of DDT with type-II diabetes and possibly pancreatic cancer.[57] Nevertheless, some 70 years after, DDT was introduced into the environment, it remains largely uncertain as to whether it has had a significant adverse effect on human health.

8.4.2.4 PUBLIC PERCEPTIONS AND ATTITUDES TOWARD PESTICIDES

There is considerable public confusion about the health effects of pesticides. In many cases, the confusion is justified because there is no compelling evidence one way or the other. For example, there is very little scientific information on possible synergistic effects among pesticides. On the other hand, the news media promote the confusion by sensationalizing anecdotal information, which is often without any scientific basis. There is a range of opinion on the nature of the problem. For example, R. Krieger, an agricultural toxicologist, in his 2005 address upon receiving the International Award for Research in Agrochemicals, focuses on the public's tendency to over-estimate the risk and under-appreciate the benefits of pesticides.[58] The interplay between the benefits and risks of pesticide use are very complex. On one hand, agricultural production is highly dependent on their use. On the other hand, the possible adverse effects are difficult to pin down. At the current time, regulatory practices are largely based on data from animal studies. These studies, in general are of relatively short duration, and more likely to detect acute toxicity than long term effects.[59]

KEYWORDS

- **agricultural production**
- **ammonia**
- **nitrogen cycle**
- **soil microorganisms**
- **herbicides**
- **mechanism of action**
- **insecticides**
- **integrated pest management**
- **pheromones**
- **environmental effects**

BIBLIOGRAPHY BY SECTIONS

Section 8.1.3: Appl, M. *Nitrogen* **1976,** *100*, 47–58; Appl, M. *Ammonia: Principles and Industrial Practice.* Wiley-VCH: Weinheim, New York, 1999; Smil, V. *Enriching the Earth*; MIT Press: Cambridge, MA, 2001.
Section 8.1.5: Durst, R. C. *Fert. Res.* **1991,** *38*, 103–107.
Section 8.1.6: Atkinson, D.; Black, K. E.; Dawson, L. A.; Dunsiger, Z.; Watson, C.A.; Wilson, S. A. *Ann. Appl. Biol.* **2005,** *146*, 203–215; Drinkwater, L. E.; Snapp, S. S. *Adv. Agron.* **2007,** *92*, 163–186.

Section 8.2.2: Cobb, A. H.; Kirkwood, R. C. *Herbicides and their Mechanisms of Action*; Sheffield Academic Press: Sheffield, UK, 2000; pp 1–24; Shaner, D. L. *Pest Manage. Sci.* **2003**, *60*, 17–24; Smith, K.; Evans, D. A.; El-Hiti, G. A. *Philos. Trans. R. Soc., B* **2008**, *363*, 623–637.

Section 8.2.3.1: Shaner, D. L.; Singh, B. K. In *Herbicide Activity: Toxicology, Biochemistry and Molecular Biology*; Roe, R. M.; Burton, J. D.; Kuhn, R. J., Eds., 1997; pp 69–110.

Section 8.2.3.2: Burton, J. D. In *Herbicide Activity, Toxicology, Biochemistry and Molecular Biology*; Roe, R. M.; Burton, J. D.; Kuhn, R. J., Eds.; IOS Press, 1997; pp 187–205.

Section 8.2.3.3: Franz, J. E.; Mao, M. K.; Sikorski, J. A. *Glyphosate; A Unique Global Herbicide*, ACS Monograph *189*; American Chemical Society: Washington, DC, 1997; Baylis, A. D. *Pest Manage. Sci.* **2000**, *56*, 299–308; Magin, R. W. *Pest.Formul. Appl. Syst.* **2003**, *23*, 149–157.

Section 8.2.3.4: Hoerlein, G. *Rev. Environ. Contam. Toxicol.* **1994**, *138*, 77–145.

Section 8.2.3.6; Bromilow, R. H. *Pest Manage. Sci.* **2003**, *60*, 340–349.

Section 8.2.3.8: Vaughn, K. C. In *Plant Microtubules; Potential for Biotechnology*; Nick, P. Ed.; 2000; pp 193–205.

Section 8.2.3.9: LeBaron, H. M.; McFarland, J. E.; Burnside, O. C. *The Triazine Herbicides*; Elsevier: San Diego, CA, 2008.

Section 8.2.4: Shaner, D. L. In *Weed and Crop Resistance to Herbicides*; DePrado, R.; Rafael, J. J.; Garcia-Torres, C., Eds., 1997; pp 29–38; Owen, M. D. K.; Zelaya, I. A. *Pest Manage. Sci.* **2005**, *61*, 301–311; Powles, S. B.; Yu, Q. *Ann. Rev. Plant. Biol.* **2010**, *61*, 317–41; Shaner, D. L. *Weed Sci.* **2014**, *62*, 427–431.

Section 8.3: Cassida, J. E.; Quistad, G. B. *Annu. Rev. Entomol.* **1998**, *43*, 1–16; Cassida, J. E.; Quistad, G. B. *Agric. Chem. Biotechnol.* **2000**, *43*, 185–191.

Section 8.3.2: Marrs, T. C.; Maynard, R. L.; Sidell, F. *Chemical Warfare Agents, Toxicology and Treatment.* John Wiley: Chichester, UK, 1996; Marrs, T. C. In *Organophosphates and Health.* Karalliedde, L. Ed.; Imperial College Press, 2001, pp 1–36; Amnenta, F.; Di Tullio, M. A.; Parnetti, L.; Tayebati, S. K. *Curr. Enzyme Inhib.* **2006**, *2*, 249–259.

Section 8.3.3: Homstedt, B. In *Plants in the Development of Modern Medicine*; Swain, T. Ed.; Harvard University Press, 1972, pp 303–360; Proudfoot, A. *Toxicol. Rev.* **2006**, *25*, 99–138.

Section 8.3.4: Elliott, M. *Pestici. Sci.* **1989**, *27*, 337–351; Elliott, M. *L'Actualite Chim.* **1990**, 57–70; Katsuda, Y. *Pestic. Sci.* **1999**, *55*, 775–782.

Section 8.3.5: Tomizawa, M.; Casida, J. E. *Annu. Rev. Pharmacol. Toxicol.* **2008**, *45*, 247–268; Jeschke, P.; Nauen, R. *Pest. Manage. Sci.* **2008**, *64*, 1084–1098.

Section 8.3.7: Crouse, G. D. *Chemtech* **1998**, *28*, 36–46; Mordue, A. J.; Morgan, E. D.; Nisbet, A. J. *Compr. Mol. Insect Sci.* **2005**, *6*, 117–135; Copping, L. G.; Duke, S. O. *Pest Manage. Sci.* **2007**, *63*, 524–554; Morgan, E. D. *Bioorg. Med. Chem.* **2009**, *17*, 4096–4105.

Section 8.3.8: Rosell, G.; Queno, C.; Coll, J.; Guerno, A. *J. Pestic. Sci.* **2008**, *33*, 103–21; Witzgall, P.; Kirsch, P.; Cork, A. *J. Chem. Ecol.* **2010**, *36*, 80–100.

Section 8.3.11: Hao, G.-F.; Zuo, Y.; Yang, S.-G.; Yang, G.-F. *Chimia* **2011**, *65*, 961–969.

Section 8.4.2.2: Racke, K. D. In *Pesticide Chemistry, Crop Protection, Public Health, Environmental Safety*; Ohkawa, H.; Miyagawa, H.; Lee, P. W., Eds.; Wiley-VCH: Weinheim, 2007, pp 29–41; Jackson, L. S. *J. Agric. Food Chem.* **2009**, *57*, 8161–8170.

Section 8.4.2.4: Goldsmith, D. J. *Rev. Toxicol.* **1998**, *2*, 17–36; Kamel, F.; Hoppin, J. A. *Environ. Health Perspect.* **2004**, *112*, 950–958.

REFERENCES

1. Bhattacharjee, R. B.; Singh, A.; Mukhopadhyay, S. N. *Appl. Microbiol. Biotechol.* **2008,** *80*, 199–209; Berg, G. *Appl. Microbiol. Biotech.* **2009,** *84*, 11–18.
2. Vitousek, P. M.; Aber, J. D.; Howarth, R. W.; Likens, G. E.; Matson, P. A.; Schindler, D. W.; Schlesinger, W. H.; Tilman, D. G. *Ecol. Appl.* **1997,** *7*, 737–750.
3. Fitzgerald, G. J. *Public Health: Then Now* **2008,** *98*, 611–625.
4. Smil, V. *Enriching the Earth.* The MIT Press: Cambridge, MA, 2001.
5. Smil, V. *Enriching the Earth*; MIT Press: Cambridge, MA, 2001, p 173.
6. Gyaneshwar, P.; Kumar, C. N.; Parekh, L. J.; Poole, P. S. *Plants Soils* **2002,** *245*, 83–93.
7. Diaz, R. J.; Rosenberg, R. *Science* **2008,** *321*, 926–929.
8. Peterson, G. E. *Agric. Hist.* **1967,** *41*, 243–254; Troyer, J. R. *Weed Sci.* **2001,** *49*, 290–297.
9. Gressel, J. *Molecular Biology of Weed Control*; Taylor and Francis: London, UK, as cited by Kudsk, P.; Streibig, J. C. *Weed Res.* **2003,** *43*, 90–102.
10. Berg, D.; Tietjen, K.; Wollweber, D.; Hain, R. *Brighton Crop Protection Conference—Weeds* **1999,** *2*, 491–500.
11. Levitt, G. *ACS Symp. Ser.* **1991,** *443*, 16–31; Brown, H. M.; Kearney, P. C. *ACS Symp. Ser.* **1991,** *443*, 32–49.
12. Shaner, D. L.; Singh, B. K. In *Herbicide Activity: Toxicology, Biochemistry and Molecular Biology*; Roe, R. M.; Burton, J. D.; Kuhn, R, J., Eds.; IOS Press, 1997; pp 69–110; Los, M. *ACS Symp. Ser.* **1998,** *656*, 8–16.
13. Kleschick, W. A. *Chem. Plant Prot.* **1994,** *10*, 119–143.
14. Mueller, K.-H.; Gesing, E.-R. F.; Sandel, H.-J. *Modern Crop Protection Compounds;* 2nd ed., 2012; pp 142–162.
15. Tranel, P. J.; Wright, T. R. *Weed Sci.* **2002,** *50*, 700–712.
16. Incledon, B. J.; Hall, J. C. *Pestic. Biochem. Physiol.* **1997,** *57*, 255–271; Nikolau, B. J.; Ohlrogge, J. B.; Wurtele, E. S. *Arch. Biochem. Biophys.* **2003,** *414*, 211–222; Sasaki, Y.; Nagano, Y. *Biosci. Biotech. Biochem.* **2004,** *68*, 1175–1184.
17. Hager, A. *J. Plant Res.* **2003,** *116*, 483–505.
18. Bus, J. S.; Hammond, L. E. *Crop Protect.* **2007,** *26*, 266–269.
19. Stellman, J. M.; Stellman, S. D.; Christian, R.; Weber, T.; Tomasallo, C. *Nature* **2003,** *422*, 681–687.
20. Schmalfuss, J.; Matthes, B.; Knuth, K.; Boger, P. *Pestic. Biochem. Physiol.* **2000,** *67*, 25–35.
21. Abu-Qare, A. W.; Duncan, H. J. *Chemosphere* **2002,** *48*, 963–974; Harrios, K. K.; Burgos, N. *Weed Sci.* **2004,** *52*, 454–467.
22. Sass, J. B.; Colangelo, A. *Int. J. Occup. Environ. Health* **2006,** *12*, 260–267.
23. Moran, G. R. *Arch. Biochem. Biophys.* **2005,** *433*, 117–128.
24. Lee, D. L.; Prisbylla, M. P.; Provan, W. M.; Fraser, T.; Mutter, L. C. *Weed Sci.* **1997,** *45*, 601–609; Mitchell, G.; Bartlett, D. W.; Fraser, T. E. M.; Hawkes, T. R.; Holt, D. C.; Townson, J. K.; Wichert, R. A. *Pest Manage. Sci.* **2001,** *57*, 120–128; Beaudegnies, R.; Edmunds, A. F. J.; Fraser, T. E. M.; Hall, R. G.; Hawkes, T. R.; Mitchell, G.; Schaetzer, J.; Wendeborn, S.; Wibley, J. *Bioorg. Med. Chem.* **2009,** *17*, 4134–4152.
25. Schmitt, M. H.; van Almsick, A.; Willms, L. *Pflanzenschutz-Nachr. Bayer, Engl. Ed.* **2008,** *61*, 7–14.
26. Komatsubara, K.; Sekino, K.; Yamada, Y.; Koyanagi, H.; Nakahara, S. *J. Pestic. Sci.* **2009,** *34*, 113–114.

27. van Almsick, A. *Outlooks Pest Manage.* **2009,** *20,* 27–30.
28. Pallett, K. E. *Herbicides and Their Mechanism of Action;* In Cobb, A. H.; Kirkwood, R. C. Eds.; Sheffield Academic Press, 2000; pp 215–238; Pallett, K. E.; Cramp, S. M.; Little, J. P.; Veerasekaran, P.; Crudace, A. J.; Slater, A. E. *Pest Manage. Sci.* **2001,** *57,* 133–142.
29. Freigang, J.; Laber, B.; Lange, G.; Schulz, A. *Pflanzenschutz-Nachr. Bayer, Engl. Ed.* **2008,** *61,* 15–28.
30. Heap, I. M. *Pestic. Sci.* **1997,** *51,* 235–243.
31. Preston, C.; Mallory-Smith, C. A. *Herbicide Resistance and World Grains*; CRC Press: Boca Raton, FL, 2001, pp 23–60.
32. Gaines, T. A.; Zhang, W.; Wang, D.; Bukun, B.; Chisholm, S. T.; Shaner, D. L.; Nissen, S. J.; Patzoldt, W. L.; Tranel, P. J.; Culpepper, A. S.; Grey, T. L.; Webster, T. M.; Vencill, W. K.; Sammons, R. D.; Jiang, J.; Preston, C.; Leach, J. E.; Westra, P. *Proc. Natl. Acad. Sci. U.S.A.* **2010,** *107,* 1029–1034.
33. Anonymous, *World Health Organization Chronicle* **1972,** *26,* 485.
34. World Health Organization, *World Health*, April, 1968.
35. Ross, S. J. M.; Brewin, C. R.; Curran, H. V.; Furlong, C. E.; Abraham-Smith, K. M.; Harrison, V. *Neurotoxicol. Teratol.* **2010,** *32,* 452–459.
36. Wesseling, C.; Kiefer, M.; Ahlbom, A.; McConnell, R.; Moon, J.-D.; Rosenstock, L.; Hogstedt, C. *Int. J. Occup. Environ. Health* **2002,** *8,* 27–34.
37. Schecter, M. S.; Green, N.; La Forge, B. L. *J. Am. Chem. Soc.* **1948,** *71,* 3165–3173.
38. Elliott, M. *Pestic. Sci.* **1989,** *27,* 337–351.
39. Yoshioka, H. *Rev. Plant Protect. Res.* **1978,** *11,* 39–52.
40. Ray, D. E.; Fry, J. R. *Pharmacol. Ther.* **2006,** *111,* 174–193; Sonderlund, D. M.; Clark, J. M.; Sheets, L. P.; Mullin, L. S.; Piccirillo, V. J.; Sargent, D.; Stevens, J. T.; Weiner, M. L. *Toxicology* **2002,** *171,* 3–59.
41. Kolaczinski, J. H.; Curtis, C. F. *Food Chem. Toxicol.* **2004,** *42,* 697–706.
42. Tomizawa, M.; Talley, T. T.; Maltby, D.; Durkin, K. A.; Medzihradszky, K. F.; Burlingame, A. L.; Taylor, P.; Casida, J. E. *Proc. Natl. Acad. Sci. USA* **2007,** *104,* 9075–9080; Tomizawa, M.; Casida, J. E. *Ann. Rev. Pharmacol. Toxicol.* **2005,** *45,* 247–268.
43. Nauen, R. *Pest Manage. Sci.* **2006,** *62,* 690–692.
44. Rogers, E. F.; Koniuszy, F. R.; Shavel, Jr. J.; Folkers, K. *J. Am. Chem. Soc.* **1948,** *70,* 3086–3088.
45. Casida, J. E.; Quistad, G. B. *Agricul. Chem. Biotech.* **2000,** *43,* 185–191.
46. Betarbet, R.; Sherer, T. B.; MacKenzie, G.; Garcia-Osuna, M.; Panov, A. V.; Greenamyre, J. T. *Nature Neurosci.* **2000,** *3,* 1301–1306.
47. Anfora, G.; Baldessari, M.; De Cristofaro, A.; Germinara, G. S.; Ioriatti, C.; Reggggiori, F.; Vitagliano, S.; Angeli, G. *J. Econ. Entomol.* **2008,** *101,* 444–450.
48. Cork, A.; Alam, S. N.; Das, A.; Das, C. S.; Ghosh, G. C.; Farman, D. I.; Hall, D. R.; Maslen, N. R.; Vedham, K.; Pythian, S. J.; Rouf, F. M. A.; Srinivasan, K. *J. Chem. Ecol.* **2001,** *27,*1867–1877; Cork, A.; Alam, S. N.; Rouf, F. M. A.; Talekar, N. S. *Bull. Entomol. Res.* **2005,** *95,* 589–596.
49. Broucek, J. *J. Environ. Protect.* **2014,** *5,* 1482–1493.
50. Blus, L. J. *Handbook of Ecotoxicology,* 2nd ed., 2003; pp 319–339.
51. Hu, J.; Zhu, T.; Li, Q. *Dev. Environ. Sci.* **2007,** *7,* 159–211.
52. Goldsmith, D. F. *Rev. Toxicol.* **1998,** *2,* 17–36.
53. Perry, M. J. *Hum. Reprod. Update* **2008,** *14,* 233–242.

54. Weselak, M.; Arbuckle, T. E.; Foster, W. *J. Toxicol. Environ. Health, Part B Crit. Rev.* **2007,** *10*, 41–80.
55. Snedeker, S. M. *Environ. Health Perspect. Suppl.* **2001,** *109*, 35–47.
56. Beard, J. *Sci. Total Environ.* **2006,** *355*, 78–89.
57. Eskenazi, B.; Chevrier, J.; Rosas, L. G. et al., *Environ. Health Perspect.* **2009,** *117*, 1359–1367.
58. Krieger, R. *Outlook Pest. Manage.* **2005,** *16*, 244–248.
59. Moser, V. C. *Hum. Exp. Toxicol.* **2007,** *26*, 321–331.

CHAPTER 9

CHEMISTRY, FOOD AND THE MODERN DIET: WHAT'S IN FOOD BESIDES FOOD?

ABSTRACT

Understanding of nutrition at the molecular level began in the 19th century with the determination of the structures of the main food components, carbohydrates, fats, and proteins. The role of enzymes in the digestive process also began to be recognized. Urbanization and changed lifestyle patterns have resulted in the need for a complex system of food processing, storage, and distribution. This, along with changes in dietary preferences, has resulted in the use of many materials in food processing. These include natural gums and emulsifiers, chemically modified forms of cellulose, biologically modified substances such as high fructose corn syrup and xanthan gum, fat substitutes, low-calorie sweeteners, and preservatives. Food additives are subject to regulation by the FDA in the United States and most have been in use for many years. Other examples of chemically based food processing include coffee decaffeination, controlled fruit ripening, and fermentation process such as in the preparation of yogurt. The final quarter of the century saw increased incidence of obesity, type-2 diabetes and other conditions associated with the metabolic syndrome. Excess caloric intake and sedentary lifestyle appear to be the main causes, but specific food additives or contaminants have also been considered as possible contributing factors.

Modern systems for food production, processing, and distribution depend heavily on input from chemistry. As we have seen in Chapter 8, most nitrogen fertilizers are produced synthetically. Protection of crops from insects and weeds is managed chemically. Many foods are now distributed in forms that include chemically modified ingredients. Modern food distribution depends on long-distance shipping and storage that requires physical stability and extended shelf life. We will look at some of the molecules involved in food processing, storage, and distribution.

9.1   SOME COMMON FOOD INGREDIENTS

Glancing at the label of a prepared or packaged food will provide a list of ingredients by name. Some of the ingredients most likely to be found are described briefly below.

Agarose is a gelable fraction of the polysaccharide agar, which is extracted from *Rhodophyceae* algae, found mainly in the Pacific and Indian Oceans and the Sea of Japan. It contains a polymer of a disaccharide called *agarobiose* derived from D-galactose and 3,6-anhydro-L-galactose. It is accompanied by a sulfated form called *agaropectin*.

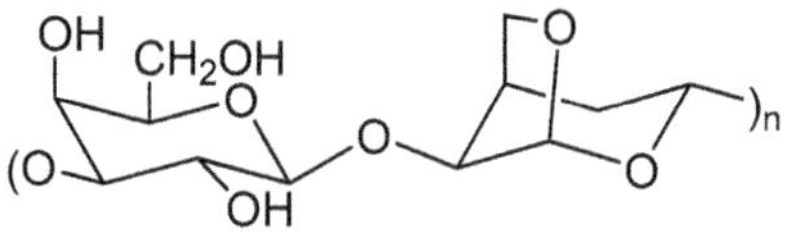

D-galactose   3,6-anhydro-L-galactose

Baking powder is a mixture of three salts, sodium bicarbonate ($NaHCO_3$), calcium hydrogen phosphate ($CaHPO_4$), and sodium hydrogen pyrophosphate ($Na_2H_2P_2O_7$). Baking powder was invented by a Harvard chemistry professor, E. N. Horsford,[a] in the mid-1800s. Its function is to release CO_2, forming bubbles and causing "rising" in baked goods, substituting for the action of yeast.

Baking soda is sodium bicarbonate, $NaHCO_3$. In the United States it is manufactured from a mineral, trona ($Na_3H(CO_3)_2$), that was discovered in the Green River Basin in Wyoming in the 1950s. The mineral is combined with water and CO_2 from nearby natural gas wells.

$$Na_3H(CO_3)_2 + CO_2 + H_2O \longrightarrow 3\ NaHCO_3$$

In most of the rest of the world, $NaHCO_3$ is made by the Solvay process in which salt, ammonium chloride, and lime are heated together (see Section 3.1).

$$NaCl + NH_3 + CO_2 + H_2O \longrightarrow NaHCO_3 + NH_4Cl$$

---

[a]Not coincidentally, Horsford held the Rumford Chair at Harvard. Count Rumford's will (see Topic 9.1) establishing the chair stated that the purpose should be "to teach by regular courses of academic and public lectures, accompanied by proper experiments, the…physical and mathematical sciences, to the improvement of the useful arts, and for the extension of the industry, prosperity, happiness and well-being of society." Horsford and George Wilson established a chemical company that was eventually called the Rumford Chemical Works. Their first product was calcium sulfite, used to neutralize chlorine in bleached fabrics. Horsford and Wilson began the manufacturing of baking powder in the 1860s. It remains on the market today under the brand name Rumford Baking Powder.

Butylated hydroxy toluene and *Butylated hydroxy anisole* are used as antioxidants. They are made from 4-methylphenol and 4-methoxyphenol, respectively, by Friedel–Crafts alkylation with isobutene (see Section 2.1.5).

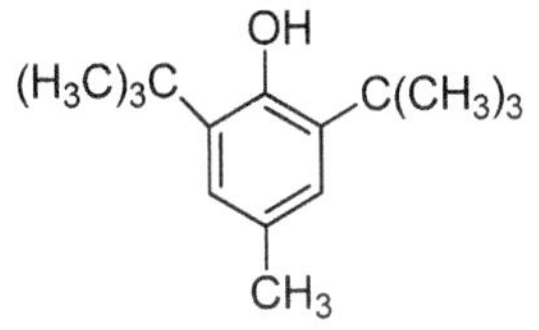

Carrageenan is a partially sulfated polysaccharide containing mostly D-galactose units. It is isolated from a red seaweed and used as a gelling, emulsifying, and viscosity-enhancing agent in both foods and nonfoods. Carrageenan is also used as a fat substitute in low-fat meat products such as patties and sausages.

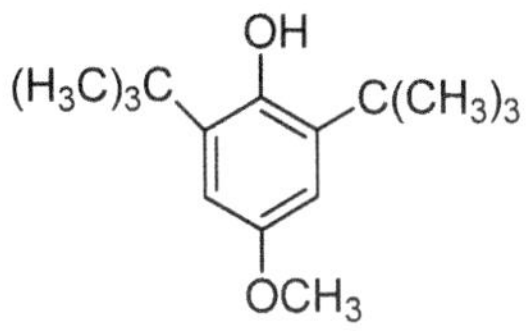

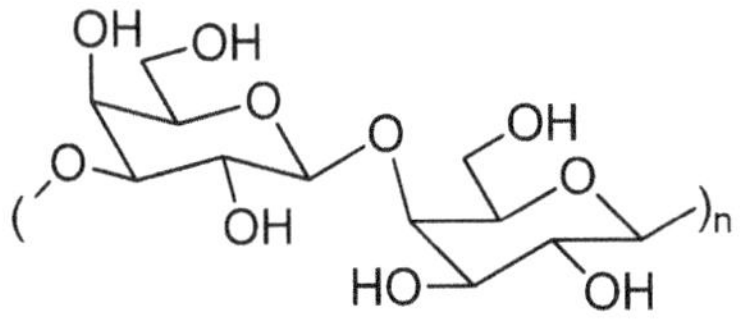

there are a variable number of sulfate
groups found at the 2, 3, 4 and 6 positions

Gelatin is a denatured form of the protein collagen, which consists mainly of the amino acids glycine, alanine, proline, and hydroxyproline. The collagen is obtained commercially by extraction from animal skin and tendons. The material is first treated with either acid (HCl) or base (lye or lime) and then extracted with hot water. The extracted material has an MW range of 40–90 kDa. The acid-treated material is called type A, while the base-treated material is called type B. Gelatin is soluble in hot water and has unique gel-formation ability. In the United States, about half is used in gelatin-based desserts. The remainder is used in a variety of foods, including candies, bakery goods, and ice cream, as well as meat and dairy products.[1]

Gelatinized (modified) starch is made by reaction of corn starch with propylene oxide and phosphorus oxychloride. This introduces short polyether chains terminating in phosphate groups.

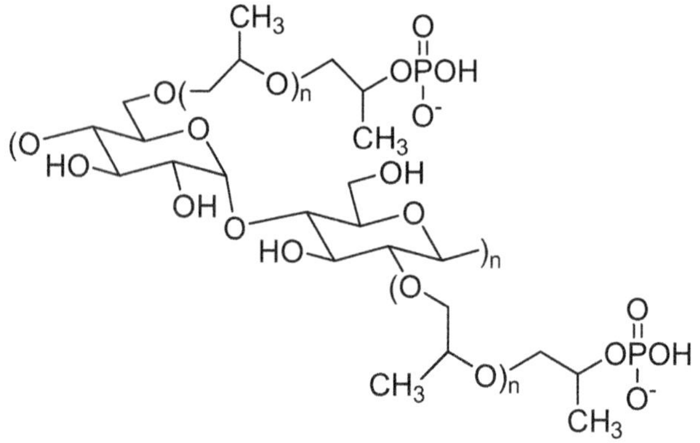

Gums. There are various gums, such as alginates, guar, and xanthan that are used in processed foods. They are also called *hydrocolloids*. The gums include both natural and chemically modified carbohydrate polymers. They form extended structures that incorporate large amounts of water and increase viscosity. *Alginate* is made from kelp. It is a polymer of β-1,4-D-mannuronic acid and L-guluronic acids in varying ratios. Because of the carboxy groups, it is a polyanion at neutral pH and can be obtained as sodium, calcium, or magnesium salts.

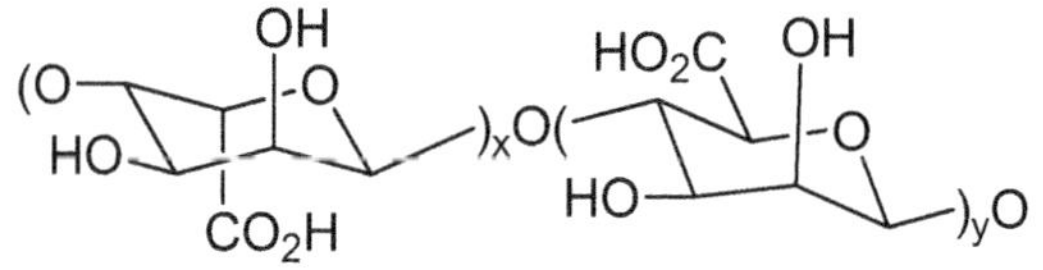

L-guluronic acid   D-mannuronic acid

alginic acid

Cellulose gum is made from partially hydrolyzed cellulose. It is alkylated with chloroacetic acid to give "sodium carboxymethyl cellulose." This material voraciously adsorbs water and gives baked goods a fatty texture. Because it can replace fat, it is classified as a "fat-reducer." It also helps capture air in baking and gives fluffiness and flakiness to the product. *Gellan gum* is a partially acetylated polysaccharide consisting of repeating units of β-D-glucose: β-D-glucuronic acid: β-D-glucose: α-L-rhamnose. It is produce by fermentation of *Psuedomonas elodea*. It is also available as the deacetylated form. It is used as a nonviscous gelling agent that helps suspend particulate components. It is used in foods, personal care products, and pharmaceuticals. *Guar* is isolated from a legume, *Cyamaopsis tetragonoloba*, that is grown mainly in India. It is a polymer of β-D-mannose with galactose side chains. It is frequently used in salad dressings and soups as a thickener. Guar gum has applications in the textile, pharmaceuticals, and oil and gas drilling industries, as well.

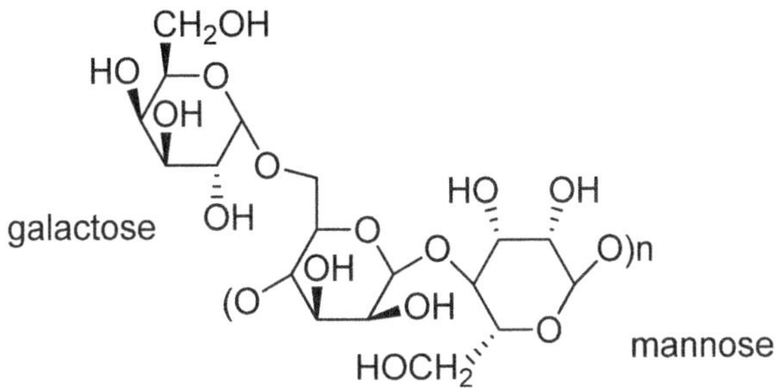

Xanthan gum was developed at the Northern Regional Laboratory of the USDA at Peoria, IL. It is produced by a strain of the bacterium *Xanthomonas campestris*. Currently, it is manufactured by companies such as Aventis, Merck, Pfizer, and Sanofi. Xanthan gum is a branched polysaccharide. The main chain is the same as cellulose. Side chains are attached at alternating glucose units and consist of a mannose, a glucuronic acid, and a terminal mannose. The overall ratio of carbohydrates is about 2.8 glucose, 2 mannose, and 1 glucuronic acid. About half of the terminal mannose units on the side chain have 4,6-acetals formed from pyruvic acid. The first mannose is partially acetylated at the C-6 OH. The presence of the glucuronic acid and the pyruvate groups makes the material a polyacid, and it is polyanionic at neutral pH. Aqueous solutions are very viscous and mixtures with other carbohydrate polymers show synergism in increasing viscosity. Xanthan gum is used as a thickening agent in many foods and also has industrial application for the production of slurries.[2]

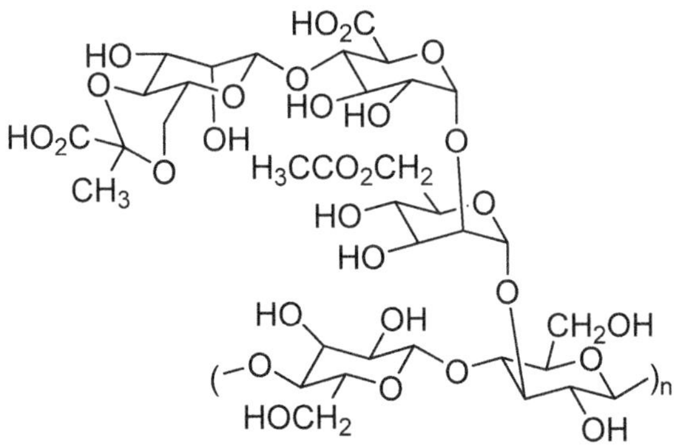

High-fructose corn syrup (HFCS) is made from corn starch by partial hydrolysis, followed by treatment with the enzyme glucose isomerase, which converts about half of the glucose to fructose. HFCS is available in two commercial grades HFCS-42 and HFCS-55, containing 42% and 55% fructose, respectively, the remainder being mainly glucose.

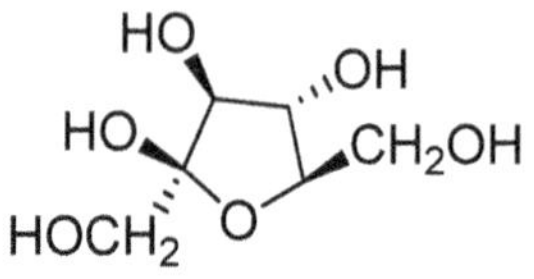

D-Fructose

HFCS was introduced in 1970s and in the United States rapidly became a widely used sweetener for many prepared foods and drinks. The favorable factors include its ability to be handled as a fluid and its ready solubility in water. Its cost and availability are also not as sensitive to the weather and economic conditions that can cause fluctuation in the price of sucrose. HFCS was adopted more rapidly in the United States than most of the rest of the world, either for economic or technological reasons. In recent years, there has been a controversy concerning the possible role of HFCS in contributing to the increase of incidence of obesity in the United States and much of the rest of the world (see Section 9.2.1).

Lecithins are phosphatidyl cholines that have two fatty acid chains. One form, "soy lecithin," is extracted from soy bean oil. Lecithin can also be isolated from egg yolks. Lecithins are used in foods as emulsifiers. Lecithins also have uses as surfactants in cosmetics.

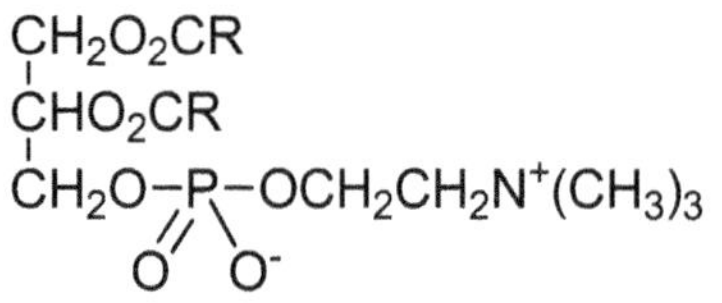

Monocalcium phosphate (MCP) is obtained from the mineral "apatite" or calcium fluoroapatite, mined in the west. It is the monobasic calcium salt of phosphoric acid.

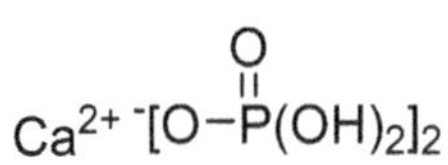

Monosodium glutamate is the sodium salt of the amino acid, L-glutamic acid. It is made by microbial fermentation (see Section 21.1.1).

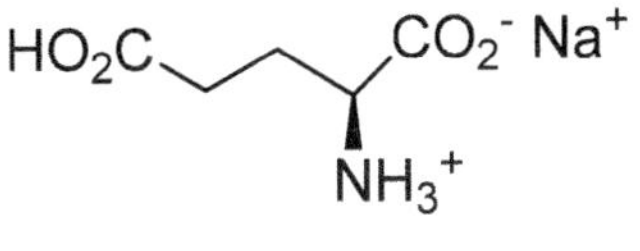

Pectin is a polysaccharide found in plant tissue, especially citrus rind. It is a partial esterified (methyl)poly-D-galacturonic acid with some intervening L-rhammose units. Other saccharides are attached as side chains.

Polysorbate (20, 40, 60, 80) are fatty acid monoesters of sorbitan, a sugar alcohol derived from glucose by reduction and cyclization. The monoesters can be further modified by reaction with ethylene oxide to introduce about 20 PEG on free alcohol sites, similarly to the related PEG surfactants. The number indicates the fatty acid that is used, 20 for lauric, 40 for palmitic, 60 for stearic, and 80 for oleic acid. These materials are used as emulsifiers. Brand names associated with polysorbates include, Emsorb, Liposorb, Sorlate, and Tween.

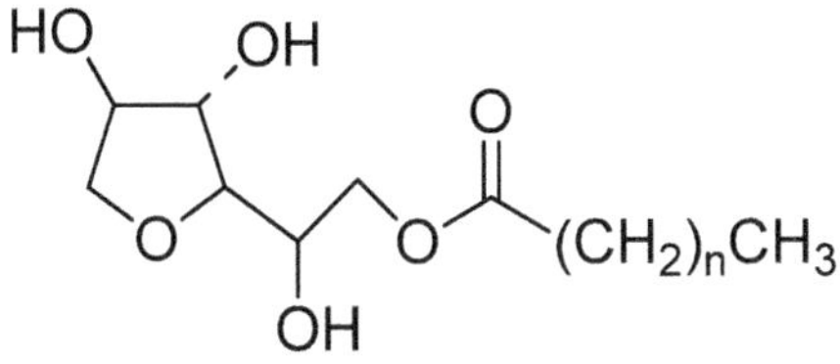

sorbitan ester, n = 12, 14, 16, 18

Soy protein is isolated from the solid product remaining after the oil is pressed from soy beans. The solid is subjected to partial alkaline hydrolysis, followed by neutralization, which precipitates the protein as "curd." One use of soy protein is to make veggie-burgers.

Stearoyl Lactate, the stearoyl ester of lactic acid, used as the sodium salt, is an emulsifier.

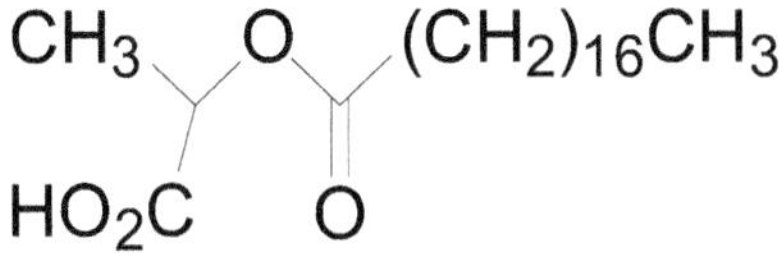

TOPIC 9.1 FOUNDATIONAL UNDERSTANDING OF THE CHEMICAL ASPECTS OF NUTRITION

Among the origins of European medical thinking were the writings attributed to Hippocrates. He regarded balanced food intake and exercise as the basis for good health. Curative properties were attributed to various salt solutions. Galen (Claudius Galenus, 131–201 CE) was trained in the Hippocratic system of medicine at Pergamun in Anatolia, then part of the Roman Empire. He also studied at Smyrna, Corinth, and Alexandria. He returned to Pergamun, where he was physician to the gladiators. He eventually became physician to the Emperor Marcus Aurelius. Galen had many incorrect concepts, but he did recognize that air interacted with venous blood to turn it bright red. He considered health to be governed by four "humors": blood, phlegm, green bile, and black bile. Food was given attributes of hot, cold, moist, and dry. Interaction among the humors was considered to determine health. These ideas became the foundation of European medical thought and remained in place into the 1600s.

Gradually, other ideas began to take hold. Andreas Vesalius, a 16th century Dutch physician taught medicine at the University of Padua in Italy. He published an illustrated medical text *De Humanis Corpore Fabrica* in 1543, followed by a second edition in 1564. He envisioned the circulation of food, digestive juices, blood, and waste. To this concept, Johannes Baptisa van Helmont, another Dutch physician, added the idea that "ferments" (which we now call *enzymes*) accomplish the transformation of food. van Helmont saw the process orchestrated by an "immortal mind" and "sensible soul" that had religious connotations. But he was skeptical of the ability of religious practice to affect any cure, and as a result came to the attention of the Inquisition. He was forced to recant, imprisoned for a year and subjected to house arrest, rather lenient treatment for the time. In the late 1500s, Santorio Santorio at Padua, added some experimental data to the picture. He weighed everything he ate and drank, what he excreted and his own weight. Something was missing. Santorio called it "insensible respiration."

Herman Boerhaave became professor of medicine, botany, and chemistry at Leiden early in the 1700s.[b] By this time, William Harvey had demonstrated that the heart circulated blood by pumping and Antoni van Leeuwenhoek, by his invention of the microscope, had discovered that blood was a suspension of corpuscles, especially red blood cells. Boerhaave considered digestion and nutrition to be a mechanical and chemical processing of food that served to repair

---

[b]Boerhaave was a professor at Leiden from 1703 to 1727. At the time training in Chemistry was primarily associated with the preparation of medications by mixing various ingredients. Boerhaave introduced quantitative concepts into Chemistry. In 1732, he published *Elementa Chemiae*. It was one of the books that began the formation of chemistry as a scientific discipline.

Source: J. C. Powers, *Inventing Chemistry, Herman Boerhaave and the Reforming of the Chemical Arts.* University of Chicago Press, 2012.

wear and tear on the organs that did the processing. He reasoned that a perfectly balanced diet would lead to a long healthy life. One of his students, Albrecht von Haller, from Berne in Switzerland, further elaborated these ideas and published in 1766, an eight volume treatise, *Elementa Physioligie Corporis*. But the role of oxygen was still unrecognized.

Antoine Laurent Lavoisier wrote on geology, agriculture, and improved the process for making saltpeter (see Section 1.2.1). In 1768, at the age of 25, he was made a member of the Academie des Sceances in Paris. In 1771, he married Marie-Anne Pierette Paulze, who subsequently helped him in many ways, including translation from English and illustrating his work. In 1775, Lavoisier was made responsible for gun powder production at the Arsenal of Paris. In his laboratory there, he established the role of oxygen in combustion and in 1789 published *Traite Elementaires de Chimie*. Lavoiser also studied the role of oxygen in metabolism. He considered the sequence of plant growth, animal consumption of plant material, death and regeneration to be a cycle, and that the growth of plants was the reverse of combustion. Lavoisier demonstrated that CO_2, then called "fixed air," was produced not only by combustion and decay of vegetable matter, but also by respiration. He quantitatively measured the amount of CO_2 and heat produced by metabolism of a guinea pig and compared it with that produced by combustion. Lavoisier, using his collaborator Armand Sequin as the subject, measured the rate of CO_2 production at various levels of activity, showing that CO_2 production increased with exercise. Measurements of the amount of water produced by respiration were also made. Lavoisier proposed that many other measurements be made on the chemical character of blood and bile, but the French revolution intervened and both he and his father-in-law were guillotined on May 4, 1794.[c]

---

[c]Lavoisier's widow eventually married Benjamin Thompson, otherwise known as Count Rumford. Rumford had a curious history. He was born in Massachusetts in 1753. Just prior to the Revolutionary War, he was appointed a major in the New Hampshire militia by the royalist governor. He became the target of local discontent and abandoned his wife and young child. He fled to Massachusetts where he was accused of acting as a British spy and escaped with them, first to Canada, and eventually to London. In London, he made measurements on the power of exploding gun powder, became a consultant to the Royal Navy and was elected to the Royal Society. He invented a device for measuring the intensity of light and conducted important experiments on the heat generated by boring cannons, which laid the foundation for the modern understanding of heat. Among his other inventions were a bright illuminating lamp and the drip coffee pot. Eventually, Rumford ended up in Bavaria in the employ of the Elector of Bavaria and devised "Rumford's Soup" as a means of feeding the Bavarian army. Besides his government and scientific activities, Rumford organized workhouses for the poor in which they made military uniforms. In a book, *Of Foods and Particularly of Feeding the Poor*, Rumford advocated a mixture of pearl barley, peas, potatoes, wheat bread, vinegar, salt, and water as an inexpensive nutritious food. Rumford held various posts in the Bavarian government, eventually being named a Count of the Holy Roman Empire. When war broke out between Austria and France in 1794, with Bavaria located between them, Rumford was made commander of the Bavarian army. He managed to avoid any military confrontation and was acclaimed a hero. He returned to London, where he founded the Royal Institute, which would soon house Humphry Davy and then Michael Faraday. He soon outwore his welcome and returned to Munich, where he died rich in 1814. His estate endowed the Rumford Chair at Harvard, which was eventually held by Eben Horsford, the inventor of baking powder (see Section 9.1).

During the 1800s, the analysis of carbohydrates, fats, and proteins began to yield correct chemical compositions and, eventually, the nature of the chemical changes involved in the digestion and metabolism of food. Among the contributors were Justus von Liebig, Jean-Baptiste Dumas, and Jean-Baptiste J. D. Boussingault (see Section 8.1.1). By the end of the 19th century, there was correct information on the chemical structure of carbohydrates, fats, and proteins and this information formed the basis for working out many biochemical pathways and mechanisms during the 20th century. From the point of view of food production, the correct understanding of the source and role of nitrogen permitted successful strategies for fertilization that resulted in large increases in production (see Section 8.1.4).

Liebig was interested in practical application of the new knowledge on the nature of food. He formulated fertilizers on the basis of his ideas. Of more dubious merit was an excursion into meat processing based in Argentina, through a company called Fray Bentos. Beef was boiled with water, and the water collected and canned. It was shipped to Europe and sold as *"extractis carnis Liebig."* It was thought to have curative and restorative properties, but chemical analysis indicated that it contained little of nutritional value. Liebig also marketed an infant formula called *"Liebig's neue Suppe fur Kinder"* made from wheat and malt flour, cow's milk, and $KHCO_3$. It was one of the precursors of modern infant formulas. Liebig became very rich as the result of these endeavors. Many other infant formulas were developed, among the most successful being those introduced in the 1860s by Henri Nestle in Switzerland.

Source: Gratzer, W. *Terrors of the Table: The Curious History of Nutrition.* Oxford University Press: Oxford, UK, 2005.

9.2   SUGARS AND LOW-CALORIC SWEETENERS

9.2.1   SUGARS

The sugars are carbohydrates. The sweet taste of certain sugars seems to appeal to most people. Sweet-tasting sugars that are present in many foods include fructose (fruits), lactose (milk), maltose (grains), in addition to common table sugar sucrose (isolated from sugar cane or sugar beets). Historically, the most available source of a sweet taste was honey, which consists mainly of fructose (~38%), glucose (~31%), and maltose (~7%). Sugars are found in many processed foods. In addition to providing a sweet taste, they have other functions. They influence the texture of baked goods and also give the characteristic brown coating, such as in bread. They also act by retaining moisture and, at high concentration, have a preservative function. Sucrose, lactose, and maltose are *disaccharides* made by linking two individual carbohydrate molecules. Sucrose contains fructose and glucose, lactose contains galactose and glucose, while maltose contains two glucose molecules. Small polymers of glucose having the same subunits as maltose are called *dextrins* and are also present in many foods. They are prepared by partial hydrolysis of starch.

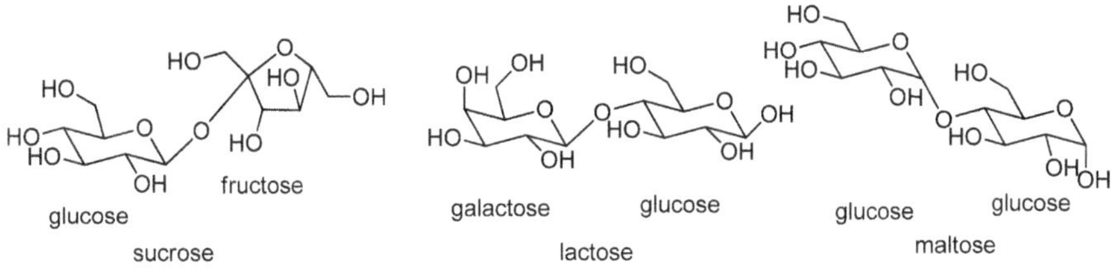

glucose   fructose   galactose   glucose   glucose   glucose

sucrose   lactose   maltose

A major new source of carbohydrate sweetener appeared in the 1970s in the form of *HFCS*. HFCS is commercially produced by enzymatic treatment of corn starch. There are two major commercial forms of HFCS, one containing 42% fructose and the other 55% fructose. The rest of HFCS is mainly glucose. By comparison, common table sugar, sucrose, is 50% fructose and 50% sucrose. *Corn syrup*, sometime confused with HFCS, on the other hand is mainly small oligomers of glucose.

Glucose is the most prevalent carbohydrate and is a fundamental energy source for animals. There are two polymeric forms of glucose called *cellulose* and *starch*. They differ in the stereochemistry of the connection between the monomer units, with starch having the α-configuration and cellulose the β-configuration. Humans and most other mammals cannot digest carbohydrates having the β-configuration. As a result, carbohydrates derived from cellulose-like materials do not supply energy. Ruminants, on the other hand, can convert cellulose to carbohydrates and fatty acids through the action of microorganisms.

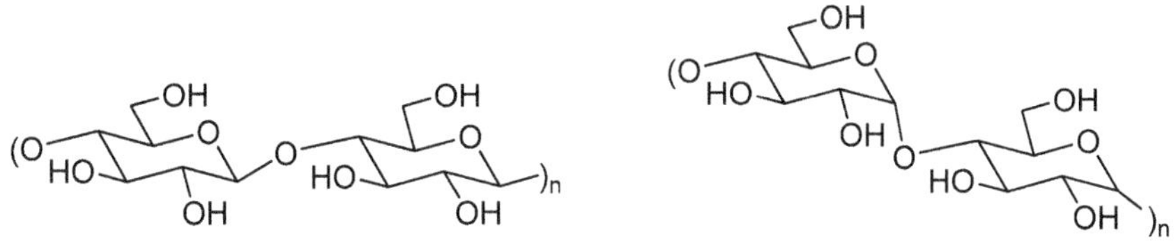

cellulose, glucose with β-connection   starch, glucose with α-connection

There has been a worldwide increase in related medical conditions including obesity, type-2 diabetes, cardiovascular disease, and fatty liver disease that are grouped together as the *metabolic syndrome*. These factors, in turn, are correlated with overall caloric intake and metabolism (see Section 9.8). The issue is whether HFCS is a specific causative factor. The temporal pattern of introduction of HFCS beginning in 1970, largely coincides with the increase in the disorders associated with the "metabolic syndrome." This correlation, of course, does not prove a cause-and-effect relationship. Careful chemical distinctions are needed to consider this controversy. As mentioned above, HFCS contains the same carbohydrates as sucrose or honey, in only slightly different ratio. Thus, if there are specific effects due to HFCS, they must result from the relatively small differences in composition. There is some evidence that pure fructose is converted more rapidly to fatty acid lipids than is glucose, although the metabolic pathways eventually converge to the same intermediates.[3] To date, there is no strong evidence that HFCS is significantly different from sucrose and other carbohydrates in its metabolic effects.[4] The usage of HFCS is higher in the United States than most of the rest of the world, primarily because the economics of production and use are more favorable. The prevalence of obesity and the related disorders does not show the same pattern, suggesting that total caloric intake may be the major factor in causing the metabolic syndrome.[5] Another part of the puzzle has to do with the difference in caloric intake from beverages as opposed to solid foods. It has been suggested that beverages do not lead to satiation, and as a result are more likely to increase caloric intake[6] (see Section 9.8 for more discussion of the metabolic syndrome).

Total caloric intake from carbohydrate sweeteners, including HFCS, increased substantially in the United States until about 2000. Since then, it has shown a noticeable decrease. Figure 9.1 shows estimates of *per* capita consumption from the USDA.

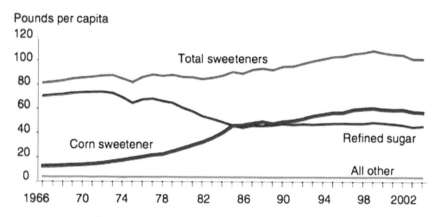

FIGURE 9.1 Estimated per capita annual consumption of sucrose and corn sweeteners. (From USDA Report SSS-23-01, August, 2005.)

9.2.2 LOW-CALORIC SWEETENERS

Such terms as "diet," "light" or "lite," and "sugar-free" appear on many food products. These products use *non-nutritive sweeteners* or *low-calorie sweeteners* that provide little or no caloric energy. The sweeteners function by interaction with one or more receptor proteins located in the "taste buds."[7] The main justification for use of non-nutritive sweeteners is to reduce caloric intake and thus, presumably, reduce obesity and its consequences. As food additives, sweeteners must be approved in the United States by the FDA. The approval process involves setting the *acceptable daily intake* (ADI), which is the amount that a person could consume per day over a lifetime and expect no adverse result. It is expressed in mg/kg body weight and is usually set at 1/100 the level at which no adverse effect (NOAEL) was noted in animal tests. Sometimes, a precise NOAEL cannot be set, because non-specific effects (e.g., weight-loss because of poor palatability) occur at very high doses in animal tests. Table 9.1 gives the generic name, some brand names, and approximate sweetness relative to sucrose for the sweeteners that we will discuss.

TABLE 9.1 Low-Calorie Sweeteners.

Generic name	Brand names	Relative sweetness	ADI (mg/kg)
Saccharin	Sweet n Low	300	5
Acesulfame K	Sunette, Sweet One	200	15
Aspartame	Nutrasweet, Equal	200	50
Neotame		8000	18
Sucralose	Splenda	600	5
Stevioside	Truvia	300	4

In some products, the non-nutritive sweeteners are used with *sugar alcohols* that provide physical properties resembling sucrose. These are called *polyols*, and examples include sorbitol (a.k.a., glucitol), mannitol, xylitol, lactitol, and maltitol. These names are related to the corresponding sugars whose name ends in "-ose" and are prepared from the sugars by chemical reduction. Related substances may also be referred to as "hydrogenated glucose syrups" or "hydrogenated starch hydrolysates." The polyols are less completely absorbed by the digestive tract than sucrose or other carbohydrates and as a result provide less caloric input. They can cause gastrointestinal disturbances and this is indicated on labels of certain products with high content of sugar alcohols.

9.2.2.1   SACCHARIN

Saccharin was first made at Johns Hopkins University in 1878. It is made by oxidation of *o*-toluenesulfonamide. Its sweetness was presumably discovered as a result of the custom at the time of tasting new chemicals. It has no structural resemblance to the carbohydrates.

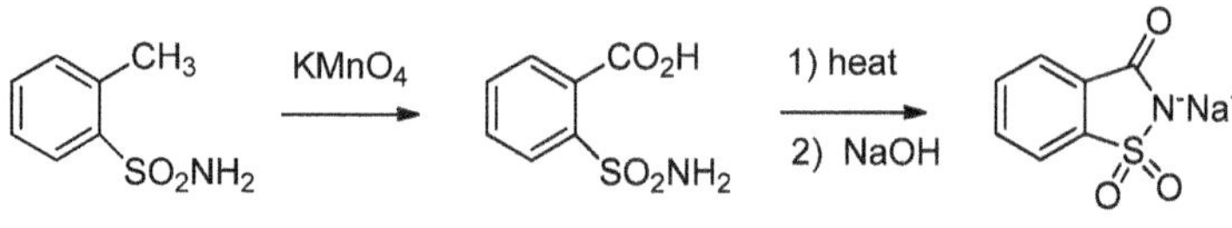

saccharin

For many years, it was the only available noncaloric sweetener, although used alone, it can have a bitter after-taste. Its use was especially high during WWII when sugar was rationed. When regulation of food additives was mandated in 1958, saccharin was approved as "generally recognized as safe" on the basis of its extensive previous use. The 1958 law, however, also stated that any substance found to "induce cancer in man or animal" should not be deemed "safe" (Delaney Amendment). In the 1970s, several studies showed that high doses of saccharin can cause bladder cancer in rats. The FDA announced its intention to ban saccharin, but Congress intervened with a series of moratoria. Further study showed that the mechanism of induction of cancer was not specific to saccharin but also applied to other sodium salts, including table salt, and was due to precipitation of calcium phosphate crystals that causes irritation and cell proliferation. There is not a comparable mechanism for tumor induction in humans. Finally, in 2000, FDA removed saccharin from the list of suspected carcinogens and Congress mandated the removal of the warning label.

9.2.2.2   ACESULFAME K

Acesulfame K was synthesized at Hoechst in Germany in 1967. It is manufactured by cyclization of acetoacetamide-*N*-sulfonic acid with sulfur trioxide, followed by conversion to the potassium salt by potassium hydroxide.[8] It is structurally similar to saccharin in being a cyclic acyl sulfonamide.

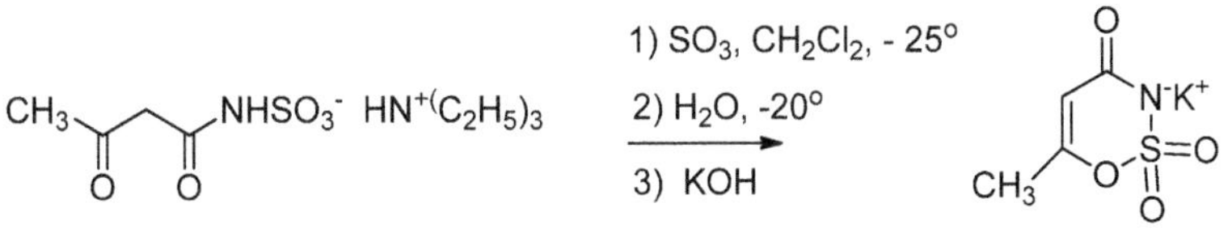

$$CH_3 \quad NHSO_3^- \ HN^+(C_2H_5)_3$$

1) SO_3, CH_2Cl_2, - 25°

2) H_2O, -20°

3) KOH

It is marketed under the names *Sunette* and *Sweet One*. Acesulfame K is about 200 times sweeter than sucrose. It is highly stable both chemically and biologically. It is pharmacologically inert and excreted unchanged. It is thermally stable and can be used in cooking. Acesulfame K has a quick-acting sweetness and is often used in blends with other sweeteners. Products that may contain acesulfame include low-calorie yogurt, ice cream, candy, canned fruits and jams, and sauces such as ketchups. It is also used in some toothpastes. The various toxicological tests submitted for approval indicate no acute or chronic toxicity, nor is there evidence of carcinogenicity or genotoxicity. The ADI is 15 mg/kg in the United States and 9 mg/kg in Europe. Using a factor of 200 for sweetness relative to sucrose, the latter value translates to an equivalent of 108 g/day of sucrose for a 60 kg individual, indicating that exceeding the ADI is quite unlikely.

9.2.2.3 ASPARTAME

The sweetness of aspartame was originally recognized at the pharmaceutical company, G. D. Searle.[9] The compound was synthesized in connection with an effort to develop inhibitors of the peptide gastrin, which is involved in generation of stomach acid. Aspartame is a methyl ester of a dipeptide phenylalanyl-glutamate. The dipeptide is a natural substance that is formed from gastrin.

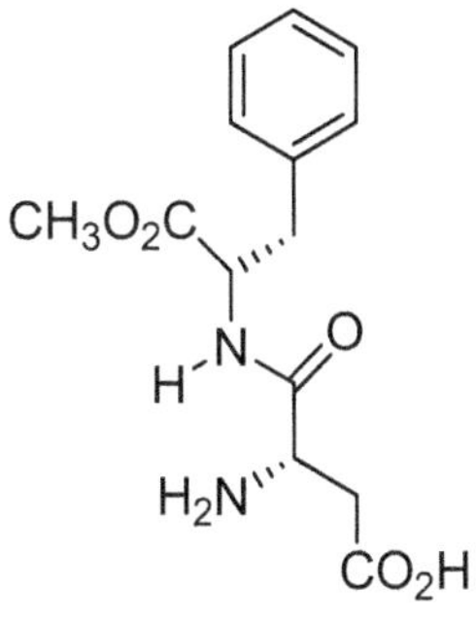

aspartame

Aspartame is produced by several companies. The major current method was developed at Toyo Soya in Japan and is known as the Tosho process. It involves enzymatic formation of the peptide bond using methyl phenylalaninate and racemic aspartic acid. The reaction is selective for the α-carboxy group and uses only the L-amino acid, generating the correct stereoisomer. The enzyme used is thermolysin, a thermally stable enzyme originally isolated from a bacteria found in hot springs in Japan.[10] Und er the conditions of operation, the aspartame separates as a complex with the unreacted D-enantiomer of aspartic acid, which further facilitates the process. The D-enantiomer can be recovered, racemized, and recycled. The process is outlined in Figure 9.2.

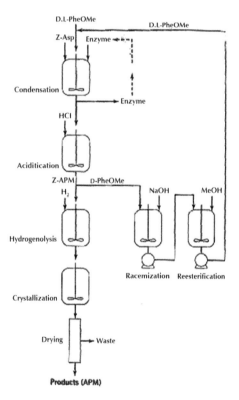

FIGURE 9.2 Transformations and flow diagram for enzymatic production of aspartame. (From Hanazawa, S. *Encycl. Bioprocess Technol.: Ferment., Biocatal. Bioferment.* **1999,** *1,* 201–210 by permission of Wiley. © 1999, John Wiley & Sons.)

Aspartame was approved for use in dry foods in 1981 and for carbonated beverages in 1983. The ADI initially approved by the FDA was 20 mg/kg/day. This was increased to 50 mg/kg/day in 1983. In terms of sweetness, this is equivalent to consumption of about 1.3 lb of sugar daily. Since 1996, aspartame has been approved for all categories of foods and beverages. It is unstable to heating, however, and cannot be used in baking and similar processes. Because it is a source of phenylalanine, aspartame is not appropriate for use by people with phenylketonuria, a genetic condition caused by inability to metabolize phenylalanine.

The ester bond in aspartame is susceptible to hydrolysis and this is fastest at acidic and basic pH, as is the case for most esters. The pH–rate profile, shown in Figure 9.3, shows maximum aqueous stability at about pH 4.5, where the half-life is about 500 days. Thus, aspartame does have some tendency to hydrolyze in solution, such as in soft drinks.

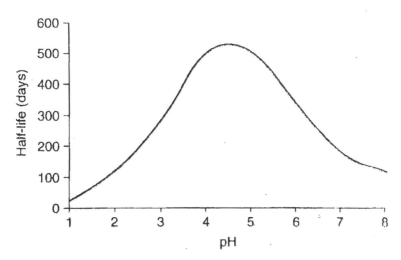

FIGURE 9.3 pH Profile for aspartame in aqueous solution. (From p 90, O'Donnell, K. *Sweeteners and Sugar Alternatives in Food Technology*; Mitchell, H., Ed.; © 2006 John Wiley & Sons. pp 86–102 by permission of Wiley.)

A unique feature of aspartame is that its components are all found in natural foods and it is converted back to them in the process of digestion, so that aspartame "disappears" and its pieces are diluted into the considerably larger pool of its components. To illustrate, a glass of milk contains several times more phenylalanine and aspartic acid than a glass of aspartame-sweetened drink. Even methanol, which is toxic in high doses, is present in some natural foods at levels higher than are obtained by ingesting aspartame. Because its components are ultimately digested and metabolized, aspartame is not strictly "non-nutritive," but the amount consumed is very small and the caloric input is negligible.

Aspartame is used in many "diet" foods to reduce caloric intake. Table 9.2 shows the amount from several common sources. Extensive food survey data suggest that at the 90th percentile of usage, the daily intake is in the range of 2–3 mg/kg. Because of lower body weight, children consume, on average, somewhat higher amounts. Aspartame is sometimes used in combination with saccharin or acesulfame K and is synergistic with these compounds and tends to diminish their metallic aftertaste. An equimolar salt of aspartame and acesulfame is marketed as "*TwinSweet*."[11] Table 9.2 gives the approximate amount of aspartame present in some representative foods.

TABLE 9.2   Approximate Aspartame Content of Some Common Foods.

Food (diet)	Serving size	Content (mg)
Beverage	12 oz	180
Yogurt	8 oz	125
Gelatin	4 oz	95
Hot chocolate	6 oz	50
Table sweetner	1 packet	35
Pudding	4 oz	25

Despite the presumably innocuous nature of aspartame and its components, numerous reports of adverse effects have appeared in the popular media. The NutraSweet Company, the initial marketer of aspartame, initiated a post-marketing surveillance program as did the CDC. No clinical syndrome that could be associated with aspartame was found. A 1995 follow-up study indicated a declining number of adverse reaction reports. The number of such reports peaked in 1985 and appears to be related to the degree of media coverage (see Fig. 9.4). Despite the number of individual reports of adverse reactions, the weight of the evidence indicates that aspartame is safe at and well beyond the approved levels.

The debate on the safety of aspartame is on-going. Much of the information, such as found on the internet, is anecdotal and impossible to evaluate. However, some qualified scientists also have expressed concern. A 2005 study reported that long-term exposure of rats led to significantly increased levels of cancer.[12] Most of the results pertained to very high doses but significant trends for lyphomas–leukemias were reported at doses >400 ppm (i.e., 0.04% of food intake). This study differed from standard studies in following all the animals until death occurred. (Normally studies are terminated at 24 months.) The researchers concluded that the observation of the tumors resulted from this extended study.

Both the FDA and European Food Safety Authority reviewed this study and maintained the existing level of safety for aspartame.[13] More recently, the same group of researchers studied exposure of rats beginning *in utero* and followed throughout their life-span.[14] The data again suggested an increased tumor incidence. On the basis of the most recent study, a new appeal has been made to the FDA to review the safety of aspartame.[15]

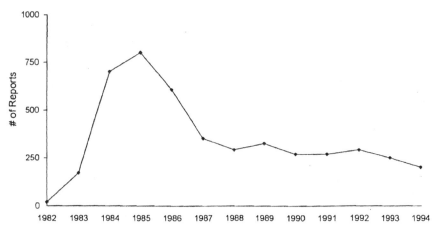

FIGURE 9.4   Reports of adverse effects attributed to aspartame: 1982–1994. (From Butchko, H. H.; Stargel, W. W. *Regul. Toxicol. Pharmacol.* **2001**, *34*, 221–233 by permission of Elsevier.)

9.2.2.4   NEOTAME

Neotame is related structurally to aspartame. Neotame was developed on the basis of a structural hypothesis for the sweetness binding site and synthesized by Jean-Marie Tinti and Claude Nofre in France. It differs from aspartame in the inclusion of an alkyl (3,3-dimethylbutyl) group on the basic nitrogen. It is about 40 times sweeter than aspartame and 8000 times sweeter than sucrose. It is also thermally more stable than aspartame and can be used in cooking. The ADI is 18 mg/day. Neotame passed all of the various toxicological tests and was approved by the FDA in 2002. The fate of neotame in humans is somewhat more complex than aspartame, which is simply hydrolyzed to its components. About half of neotame is eliminated without any change. Most of the rest is hydrolyzed at the methyl ester and excreted in the urine. A minor pathway involves oxidation of the 3,3-dimethylbutyl group to 3,3-dimethylbutanoic acid. Several other derivatives are even sweeter, but have not been commercialized.

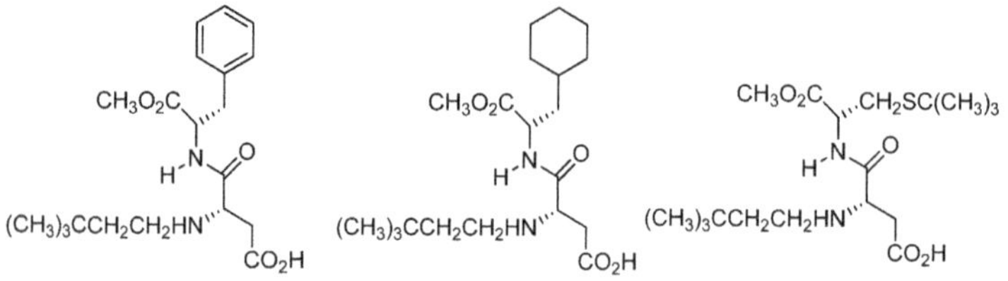

neotame (8000)   (13,500)   (23,000)

9.2.2.5 SUCRALOSE

Sucralose was first synthesized in England in 1976. It was synthesized as part of a systematic program to examine sucrose derivatives for sweetness.[d] One series of compounds replaced hydroxy groups by chlorine. Successive increases in sweetness were noted as the number of chlorines was increased. Sucralose differs from sucrose in the replacement of two primary hydroxy groups by chlorine and the replacement and inversion of one secondary hydroxy group. Sucralose is distributed under the brand name *Splenda*.

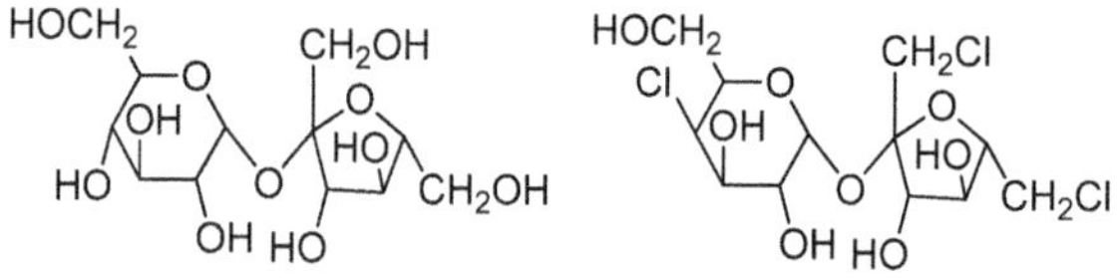

sucrose   sucralose

Sucralose is prepared from sucrose by a series of five chemical steps that involve distinguishing between the primary and secondary hydroxy groups by use of protecting groups. This justifies the "made from sugar" statement that appears on certain labels. The initial petition for approval as a food additive was submitted to the FDA in 1987 and it was granted in 1998. Sucralose is a general purpose sweetener and is used in a number of products. The concentrations are usually in the 100–300 ppm level, but are higher in products such as syrups used as sweeteners with other foods (1000–1500 ppm). Sucralose is about 600 times sweeter than sucrose so that a concentration of 20 mg/L is equivalent in sweetness to about a 2% solution of sucrose. Its stability in aqueous solution and its compatibility with high temperatures in cooking and sterilization give it certain advantages over aspartame.

---

[d]The research was done at the University of London and sponsored by Tate & Lyle, a firm whose history in sugars and carbohydrates traces back to the mid-nineteenth century. Tate & Lyle currently has worldwide interests in sugar, sweeteners, and other related products.

The extensive physical characterization and toxicological studies on sucralose have been summarized.[16] Despite the replacement of three hydroxyls by chlorine, sucralose remains water soluble. The stability of the glycosidic linkage to hydrolysis is also enhanced. Sucralose is not digested or metabolized and is therefore noncaloric. Most of the sucralose is excreted rapidly in animals and man but a fraction (15% in man) is absorbed and excreted more slowly. About 2% is accounted for as glucuronide conjugates. In rats, enlargement of the cecum (pouch at the end of the large intestine) is observed. This effect is not observed in other species and is caused by an osmolytic effect that does not suggest toxicity. Human volunteers administered 1000 mg/day of sucralose in a 12-week study reported no gastrointenstinal effects. In high dose studies (270 mg/kg) with female rats, there was evidence of small developmental changes in offspring. The ADI for sucralose is 5 mg/kg/day. The estimated average consumption by users is about 1 mg/kg/day. There is no current indication of adverse effects associated with the allowable human intake.

In 2006, the Sugar Association filed a complaint with the FTC that the slogan "Made from Sugar" used by McNeil Nutritionals, the US marketer of *Splenda*, was misleading advertisement. Producers of other low-calorie sweeteners have also filed suits claiming that the slogan implied that Splenda was more "natural" than the competing products and was false advertising. One such suit was settled out of court, evidently with McNeil paying damages to the complainant, but no definitive legal ruling seems to have been made.

9.2.2.6   STEVIOSIDE

Stevioside is a naturally occurring glycoside of the terpene kaurene. It is isolated from a plant, *Stevia rebaudiana*. It contains three glucose units and is considered to be about 300 times sweeter than sucrose. One commercial brand is *Truvia*. Stevioside differs from the other noncaloric sweetener in being nonsynthetic.

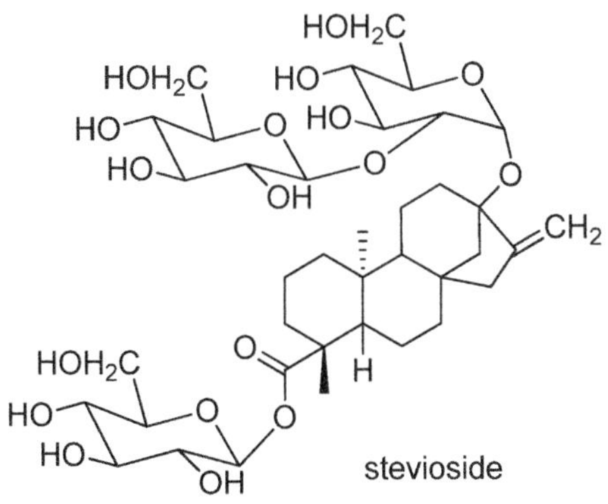

stevioside

9.3   FATS AND OILS

9.3.1   STRUCTURE OF FATS AND OILS

Fats are esters in which fatty acids are attached to 1,2,3-trihydroxypropane, also known as glycerol or glycerin. Fats are also called *triglycerides*. The three fatty acids need not be the same. In *diglycerides*, only two fatty acids are present.

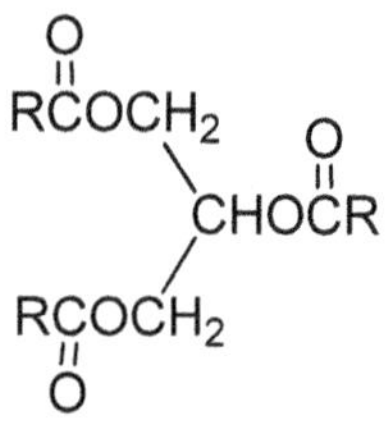

The fatty acids have an even number of carbons and are unbranched. The detailed description of fats can get complicated. We hear the terms "omega-3," "*trans*," and "polyunsaturated" in connections with fats and fatty acids. Exactly what do these terms mean and what are the implications? To completely describe the structure of a fatty acid, its chain length, the position, and stereochemistry of double bonds must be specified. One system for doing this is to give the length of the chain followed by the positions and stereochemistry—c (*cis*) or t (*trans*) —of the double bonds. For example, linolenic acid, a major constituent of soy bean oil, can be designated 18:9c,12c,15c.

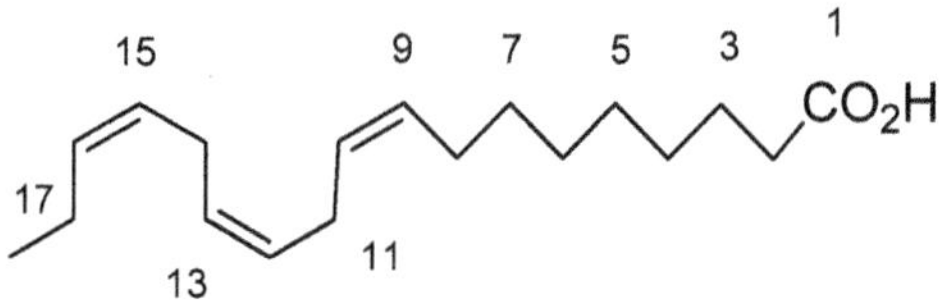

Omega-3 refers to a double bond terminating 3 carbons from the methyl end of the chain, that is, between C-15 and C-16 in a C-18 acid. The location of the double bond can also be indicated by the symbol Δ, followed by superscript numbers for the position of the double bond(s). The fats found in most vegetable and fish oils include unsaturated fatty acids. Natural fats usually have *cis* double bonds. Animal fats contain higher proportions of saturated fatty acids. Another significant feature of the structure is whether or not the double bonds are "conjugated," that is separated only by a single bond rather than by intervening C-atoms. Figure 9.5 gives the composition of saturated and unsaturated fats in some of the common fats and oils.

Comparison of Dietary Fats

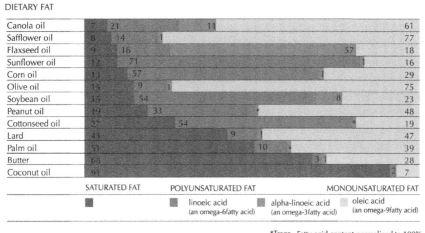

FIGURE 9.5 Composition of fats and oils.

9.3.2 TRANS-FATS

Consumers have learned to be on the look-out for *trans*-fats in food. So, where do *trans*-fats come from? In the United States, the principal source is the hydrogenation of soybean oil. In addition to linolenic acid shown above, soybean oil contains linoleic (18:9c,12c) and oleic (18:9c) esters. The objective of hydrogenation is to increase the oleic acid content, while avoiding the formation of the saturated stearic acid. The hydrogenation increases the melting point of the material, giving it the butter-like consistency found in "margarines" and "spreads." It also decreases the tendency to react with oxygen and become rancid. The position of the double bond is variable with $\Delta^{9,10}$ and $\Delta^{10,11}$ predominating. The ratio of *cis:trans* fats produced by hydrogenation can be controlled to some extent by the methodology. The conventional hydrogenation method involves a nickel catalyst and elevated temperature (>120°C). These conditions can give as much as 25–50% of *trans* product. Use of other metals, especially Pd, permits lower temperature reaction and reduces the amount of *trans*-fat.[17]

Hydrogenated vegetable oil was first used in Europe as a substitute for butter around 1900. The term "margarine" was introduced at that time. The first example of a hydrogenated vegetable oil in the United States was *Crisco*, introduced in 1911 by Procter and Gamble. Butter substitutes called margarines became popular after WWII. Dairy industry interests fought the inclusion of yellow dyes, but eventually lost the battle. Partially hydrogenated vegetable oils are used extensively in prepackaged foods because they provide increased shelf-life. Initially it was thought that *trans* fatty acids (TFA) would primarily replace saturated fatty acids in the diet and thus might be beneficial in terms of reducing cardiovascular disease. Several epidemiological studies, however, have associated consumption of TFA with increased risk of cardiovascular disease. Also, high intake levels of TFA increases the LDLP/HDLP ratio (low/high-density lipoprotein, see Section 15.4.1), which is also associated with cardiovascular risk. There may be other mechanisms by which TFA increases risk for cardiovascular disease. With the concern about *trans* fat, changes are occurring in the composition of spreads and prepared foods. Commercial spreads can achieve the texture of hydrogenated spreads by incorporating more saturated fats.

In the United States, legislation was passed in 2003 requiring the amount of *trans* fat to be included on food labels as of January 1, 2006. The lower limit for reporting is 0.5 g per serving, so the appearance of "zero *trans* fats" on a food label can be misleading. In the United States, products with as much as 0.49 g per serving legally can be labeled as "zero *trans* fat." Effective January 1, 2004, TFA was restricted to a maximum of 2% of total fat content in Denmark. A study of TFA content in typical fast food and packaged food products conducted by Danish researchers found a wide variation over the world, even in food provided by international companies. Some products, such as packaged French fries and microwave popcorn had fat that was as much as 50% TFA, making the potential ingestion as high as 40 g/d. Similar products sold in Denmark contained less than 1 g TFA.[18] The implication is that TFA can be substantially reduced, at least for products in a specific market.

9.3.3   FAT SUBSTITUTES

The general purpose of fat substitutes is to replace a fat and its caloric content with a material that provides similar physical appearance and texture, but with reduced caloric content. Fats are the highest of the three basic food components in caloric content (9 cal/g), being somewhat more that twice both carbohydrates and proteins (4 cal/g). The modified materials can also have reduced absorption or metabolic conversion, further reducing caloric input. There seems to be little evidence on the nutritional consequences of fat replacement, beyond the reduction in caloric intake.[19] Several specific fat substitutes are described in the paragraphs that follow.

Modified diacyl and triacyl glycerides. Salatrim is a mixture of triglycerides that have one long-chain acid (usually stearic) and two short-to-medium acids (C_2–C_4). Although chemically similar to fats, they have only about half the caloric value, in part because of relatively low metabolism. It is used in spreads, salad dressings, dairy products, candies and desserts.

β-*Glucans* include materials such as *Oatrim* that are prepared by chemical or enzymatic modification of cellulose from grains, usually oats or barley.[20] Its caloric capacity is considered to be equivalent to that of digestible carbohydrates, that is, 4 cal/g. On nutritional labels it is listed as hydrolyzed oat fiber (or hydrolyzed oat flour). Mixed with water it forms a cream with fat-like physical attributes. Related products go by the names *Nu-Trim* and *Z-Trim*.

Inulin is a polymer of fructose ranging up to 50–60 subunits, terminated by a single glucose. It is present in several plants but is usually prepared from chicory root (*Cichorium intybus*). It also occurs in onions, leeks, and garlic and in Jerusalem artichokes. It is considered to be a food, as opposed to an additive, and has the "generally recognized as safe" status in the United States and is also accepted as an ingredient in Canada, Europe, and Japan. Inulin can be processed by partial enzymatic digestion to obtain a mixture of shorter oligomers (<10) that is called *oligofructose*. When mixed with water, inulin or oligofructose can form spreadable gels and creams. Inulin itself is tasteless, but the smaller oligomers are about a third as sweet as sucrose. Like other carbohydrates, inulin has humectant and preservative functions.[21] The fructose–fructose bonds are not readily hydrolyzed in the human digestive system. As a result, only about a third of the caloric content is obtained, and this is the result of metabolism to fatty acids in the digestive tract. Inulin acts as a dietary fiber, and because it can replace fat, it can be used in "reduced fat" and "fat free" foods. Inulin and oligofructose are used as a fat substitute in spreads, processed cheeses, frozen desserts, baked goods and bread, and salad dressings, among others. It is also considered to be a "prebiotic" in that it alters the distribution of microorganisms in the digestive tract, favoring healthful bacteria and reducing potentially harmful ones.

Maltodextrin contains enzymatically digested glucose oligomers. They are digestible and have a *caloric* content equivalent to sugars, about 4 cal/g. Maltodextrin can be used as a fat substitute in the form of aqueous gels.

Methylcellulose and hydroxypropylcellulose are chemically modified versions of cellulose in which some of the cellulose hydroxy groups have been substituted by methyl and/or hydroxypropyl (introduced using propylene oxide). They are noncaloric but impart some fat-like qualities to foods such as salad dressings and frozen desserts.

Microcrystalline cellulose is partially hydrolyzed cellulose that has been subjected to mechanical sheer. It forms a colloidal dispersion that has a cream-like consistency. It is not caloric and is usually used with one of the gums listed in Section 9.1.

Microcrystalline protein is prepared from precipitated milk or egg protein. It is processed into microparticles that provide a dispersion that has the physical texture of fat. On a weight-by-weight basis, it has about 40% of the caloric content of fat. It is used in salad dressings, mayonnaise, and frozen desserts.

Polydextrose is a polymer of glucose modified with sorbitol (10%) and citric acid (1%). It is incompletely digested and qualifies as a fat reducer (1 cal/g) and as a dietary fiber. It was developed at Pfizer and is currently marketed by Danisco under the brand name *Litesse*.

Propoxylated glycerol esters are made from glycerol by reaction with propylene oxide to introduce some 2-hydroxypropyl groups. The material is then esterified with long-chain fatty acids. The physical properties are similar to normal fats, but the materials are nondigestible.

9.4   COFFEE DECAFFEINATION

9.4.1   A LITTLE BACKGROUND ON COFFEE USE

Coffee is widely consumed in the western world, with US per capita consumption averaging 4 cups per day. Two species are grown commercially, *Coffea arabica*, and *Coffea caneplora (robusta)*. After harvest, the beans are dehusked and roasted, which causes many chemical changes and gives coffee its distinctive flavors. At some point prior to use, the roasted beans are ground. In the United States, coffee is usually consumed as a filtered extract of either ground regular or decaffeinated coffee. In Scandanavia, the eastern Mediterranean and the Middle-East, it is often served as a decanted boiled extract of ground coffee. The means of preparation influences the chemical constituents present in the brew. Regular coffee contains about 30–175 mg of caffeine per cup (~150 mL). Decaffeinated coffee contains from 3% (United States) to <0.1% (the European Union) of this amount. The lethal dose of caffeine in humans is about 10 g, the equivalent of roughly 50–100 cups of regular coffee. Boiled coffee contains terpenoid oils, mainly cafestol and kahweol, that are not present in filtered coffee. Depending on the specific brewing process, these can be present up to ~25 mg/cup. The other biological active substances present in coffee are phenolic antioxidants, particularly chlorogenic acid. These are present in up to 300–400 mg/cup.

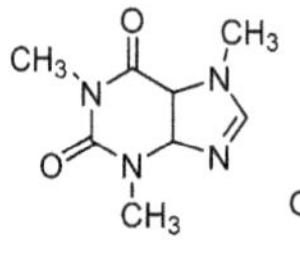

caffeine 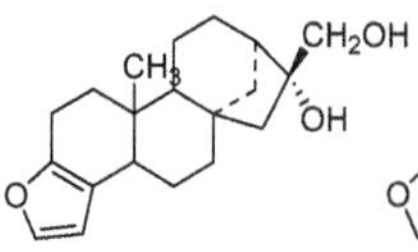   cafestol 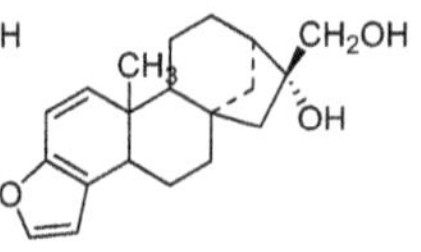   kahweol 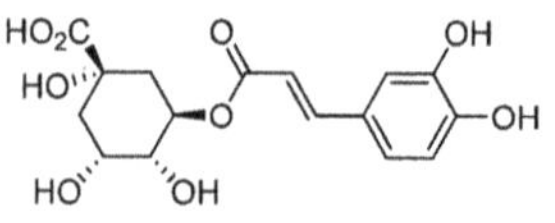   chlorogenic acid

There have been numerous studies of the pharmacological effects of coffee and its constituents. The short-term effects of caffeine include an increase in blood pressure, diuretic activity, and an increase in gastric secretion. Caffeine increases energy consumption in humans. It is estimated that at 6 cups/day, energy consumption is increased by about 100 cal/day. This appears to result from accelerated fat metabolism. The details are not clear, but it may be that caffeine and its metabolites affect adenosine's role in metabolic control. Several studies indicate that coffee consumption reduces the risk of type 2 diabetes, perhaps by promoting weight loss. There have also been epidemiological studies aimed at detecting relationships between coffee consumption and other health outcomes. In most of these studies, there is little evidence of either beneficial or harmful effects. The diterpenes found in boiled unfiltered coffee appear to have negative effects on cardiovascular health, especially at high consumption levels. The antioxidants may have beneficial effects.

Caffeine is found in a number of other beverages, including tea (10–50 mg/cup), colas, and other carbonated soft drinks (40–70 mg/12 oz can) and energy drinks (30–150 mg/250 mL). The latter also contain herbal stimulants including taurine and glucuronolactone. There is little information on the short or long-term effects of consumption of this combination of substances.

$H_3N^+CH_2CH_2SO_3^-$

taurine

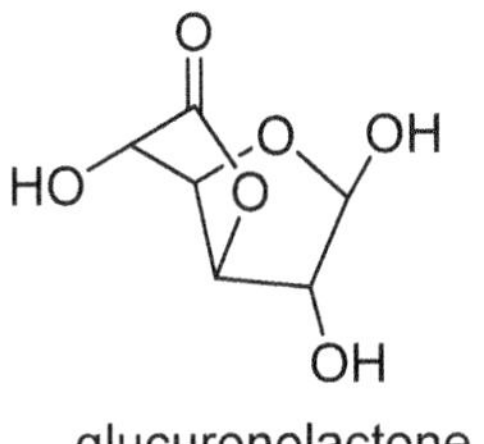

glucuronolactone

9.4.2 DECAFFEINATION

Several processes have been developed for decaffeination of coffee. One is solvent extraction. Methylene chloride was originally used, but ethyl acetate is now used. The beans are treated with hot water, which extracts much of the caffeine and also flavor components. The caffeine is extracted from the water by the solvent, and the flavor components in the extracted water are returned to the decaffeinated coffee. The success of this process depends on reconstitution of the flavor. The main concern with this process is removal of any residual solvent. The limit was specified at 10 ppm for methylene chloride, but there is no specified limit for ethyl acetate, which is a natural constituent of some fruit flavors.

There are also processes in which the flavor and caffeine are extracted into the hot water, and then the caffeine is removed from the water by an adsorbing agent such as activated charcoal. The water containing the flavor components is then used in successive extractions, retaining the flavor components. About half of the caffeine that is removed from coffee is recovered. This amounts to about 2500 t/year. Total consumption of caffeine is four times that, mostly for formulation of soft drinks, and the remainder of the caffeine is prepared by synthesis.

Another decaffeination process uses supercritical carbon dioxide for extraction. This process has been in commercial use since 1978. At pressures of 300 atm and about 150°C, the supercritical CO_2 selectively extracts caffeine. The supercritical CO_2 is particularly effective at penetrating the beans. The caffeine is then removed from the CO_2 by passage through water. The CO_2 can easily be recycled, and being a natural substance and a gas, it causes no residue problems in the decaffeinated coffee. Figure 9.6 gives a schematic diagram of the process.

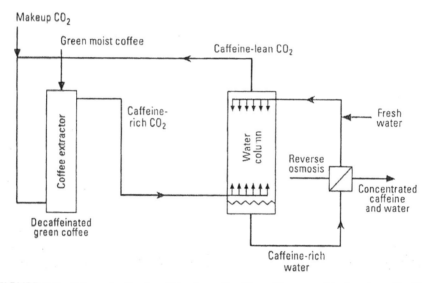

FIGURE 9.6 Schematic for decaffeination of coffee with supercritical carbon dioxide. (From Beckman, E. *Environ. Sci. Technol.* **2002,** *36,* 347A–353A. © 2002 used with permission of the American Chemical Society.)

9.5   CHEMICAL BASIS OF FRUIT RIPENING

It turns out that one of the most fundamental petrochemicals, ethylene, which is used to make tons of plastics and many other products, is also the natural plant hormone responsible for turning tomatoes and other fruits ripe. And, of course, the ethylene from either source causes fruits to ripen. This information has been put to use in the produce industry. The demands of transportability and storage impose conditions on agricultural produce. While it is no problem moving ripe tomatoes a few miles, moving them a thousand miles and through several distribution facilities is another matter. There have been several ways of addressing this problem. An important one is plant breeding; selecting varieties for their durability and storability. Another is a chemical method. The discovery that ethylene promotes ripening of certain fruits led to technology in which fruit is harvested prior to ripening, when it is typically more durable, and then initiating the ripening process with ethylene during storage.[e] The effect of ethylene also can be deleterious. If fruit stored in close quarters begins to ripen, the evolved ethylene may promote ripening at an undesirably fast rate. Ethylene-controlled ripening is used extensively in the storage–distribution cycle for avocados, bananas, mangos, melons, tomatoes, and other fruits. In the case of bananas, cultivars that do not ripen naturally have been developed, permitting even more precise control of the ripening process. One process that evidently takes advantage of ethylene's ripening activity has been employed since ancient time. Figs are lightly coated with olive oil, which advances ripening. Modern investigation of this technique indicates that the olive oil stimulates ethylene production.[22]

Certain fruits are classified as *climacteric*, which means that their ripening is associated with a burst of ethylene production. These fruits then progress to senescence rather quickly, at which point they are no longer commercially attractive. Apples are a climacteric fruit with an annual harvest cycle, which mean that year-round availability depends on long-term storage. Apples respond quite well to cold storage, but this can be improved by chemical inhibition of ethylene-mediated ripening. An *inhibitor* of ethylene-mediated ripening is commercially available.[23] The compound is 1-methylcyclopropene (MCP). It is sold as "*SmartFresh*" by AgroFresh, Inc., a subsidiary of Rohm and Haas. This product is a powder of MCP absorbed in cyclodextrin. When the powder is dissolved in water, the MCP is released. As of 2002, treatment of apples with MCP has been permitted in the United States, with a maximum allowable concentration of 1 μg/L. In Topic 21.1, we will learn how biotechnology can be used to control tomato ripening.

---

[e]An early description of ethylene treatment of tomatoes can be found in E. F. Kohman, **1931**. *Ind. Eng. Chem.*, *23*, 1112–1113. Kohman reports that if tomatoes have already started to turn red, the ethylene has little effect. However, the reddening process is accelerated in green tomatoes. Kohman recognized the potential of the method, but was not enthusiastic about its prospects, stating: "Tomatoes picked so green that ethylene treatment is necessary to develop red color are not suitable for [commercial] canning because of low yield and poor quality." Despite Kohman's pessimism, "gassed" tomatoes are currently a major product and the underlying biochemistry and molecular biology of the process and its inhibition have been extensively studied (see Topic 9.2). Kohman researched many other aspects of food preservation and the nutritional content of canned foods. He published an autobiography, *Canorama of a Chemist: Memories of Fifty Years in the Food-Processing Industry,* in 1965.

TOPIC 9.2 THE BIOCHEMISTRY OF ETHYLENE PRODUCTION IN PLANTS

The process by which plants form ethylene is very intriguing from a chemical point of view. Like nearly all biological processes, it is catalyzed by enzymes. The starting point is the amino acid methionine, and other important players are bicarbonate ion and ascorbate (vitamin C). A very interesting compound, 1-aminocyclopropanecarboxylate, is the key intermediate. In the first step, methionine is converted to *S*-adenosylmethionine, a process that occurs in many other biological reactions of methionine. The cyclopropane ring is formed in the second step, by the enzyme *1-aminocyclopropanecarboxylate synthase*. This enzyme uses pyridoxal phosphate as a coenzyme (see Section 10.1.1.5). The mechanism is similar to others in amino acid metabolism, but the formation of the cyclopropane ring is unique. The modified sulfur in methionine is displaced to form the cyclopropane ring.[24] This mechanism is summarized in Scheme 9.1.

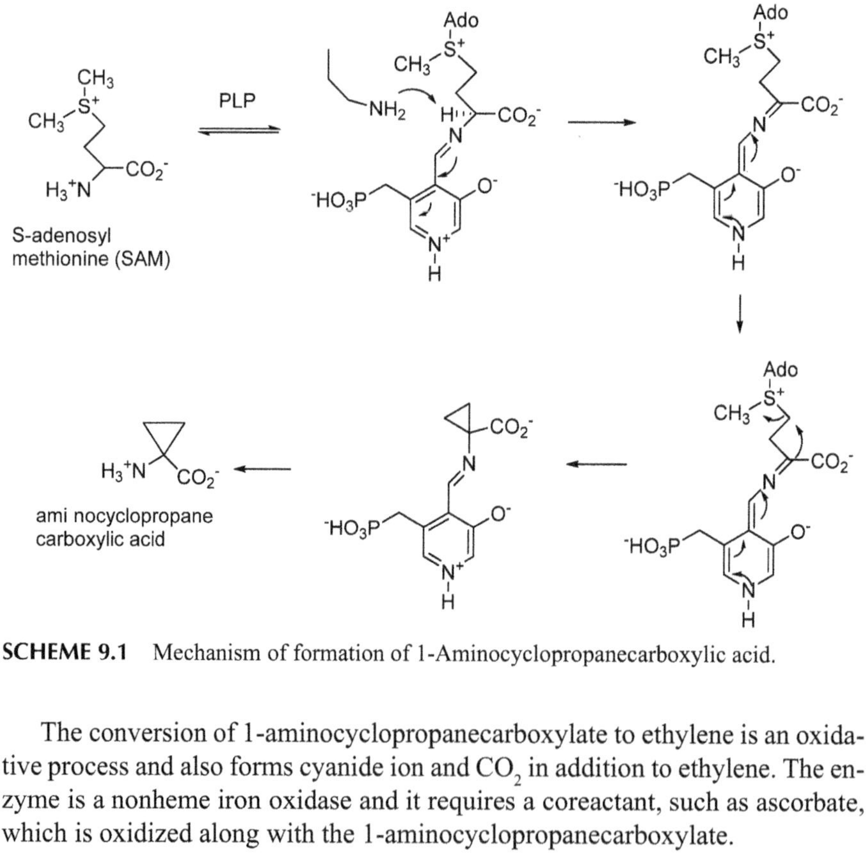

SCHEME 9.1   Mechanism of formation of 1-Aminocyclopropanecarboxylic acid.

The conversion of 1-aminocyclopropanecarboxylate to ethylene is an oxidative process and also forms cyanide ion and CO_2 in addition to ethylene. The enzyme is a nonheme iron oxidase and it requires a coreactant, such as ascorbate, which is oxidized along with the 1-aminocyclopropanecarboxylate.

The key step appears to be a one-electron oxidation of the amino group in the substrate.[25] Because ethylene formation controls several crucial stages in the plant's development, the process is under the control of several gene families.

Sources: Pech, J.-C.; Puirgatto, E.; Bouzayen, M.; Latche, A. *Ann. Plant Rev.* **2012**, *44*, 275–304; Serrano, M.; Zapata, P. J.; Gullen, F.; Martinez-Romero, D.; Castillo, S.; Valero, D. In *Tomatoes and Tomato Products*; Preedy, V. R., Watson, R. R., Eds., 2008; pp 67–84.

9.6 ECONOMICS AND POLITICS OF FOOD PROCESSING

During the second half of the 20th century, first in the United States, and then in much of the rest of the world, the supply of food came to exceed the basic nutritional needs of the population. The surplus of food engendered competition among producers and processers, resulting in more effort at marketing and advertising. As a result, during the late 20th century, for much of the world, the problem shifted from under-nutrition to over-nutrition. In this environment, food producers and processors are enemies of any message to "eat less." Generally speaking, they are also enemies of "eat better," because profit margins tend to increase with the degree of processing. (Compare the price of a potato with that of a bag of potato chips.) Food content can also be manipulated to help establish preferences. The human palate seems to find sugar, salt, and fat especially appealing. The government is the primary source of information about nutrition and regulation of food safety. As a result of the role of financing of political campaigns, the federal agencies are subject to the influences of the agricultural and food industry. For example, during the Reagan administration, the Public Health Service produced the first (and only) *Surgeon General's Report on Nutrition and Health*, but political pressure ensured that the report did not recommend reduction of any specific category of food.[26] Since then, the industry has completely stymied any update, despite Congress having legislated in 1990, that such a report be produced biennially. The result is considerable confusion and skepticism about dietary recommendations on the part of the public.

The primary responsibility for development of nutritional guidelines rests with the Department of Agriculture (USDA), which also has responsibility for the economic well-being of the agricultural industry. This frequently puts the agency in a position of internal conflict of interest. For example, in 1991, the Secretary of Agriculture, Edward R. Madigan blocked the release of the "Eating

Right Pyramid" because of complaints from the meat and dairy industry. The places of dairy and meat products were smaller than those for cereals, vegetables, and fruits. Groups with an interest in public health joined the fight in support of the Triangle. Eventually USDA, along with the Department of Health and Human Services, contracted for a study of the effectiveness of the triangle. The results favored the Triangle and in April 1992, Secretary Madigan approved release of the Eating Right Triangle. In 2005, the meaning of the food triangle was obscured by removal of the food representations. It could be interpreted only by visiting the USDA web site. A new symbol for nutritional guidelines was introduced in 2011. It is called "My Plate" and shows four food groups as proportional parts of a diet represented on a plate. Dairy products are represented by an accompanying glass. Figure 9.7 shows this new figure as well as some of its predescessors.

Choose Your Food Wisely

STUDY THESE FIVE FOOD GROUPS

Every food you eat may be put into one of these groups. Each group serves a special purpose in nourishing your body. You should choose some food from each group daily.

1. VEGETABLES AND FRUITS.

2. MILK, EGGS, FISH, MEAT, CHEESE, BEANS, PEAS, PEANUTS.

3. CEREALS—CORN MEAL, OATMEAL, RICE, BREAD, ETC.

4. SUGAR, SIRUPS, JELLY, HONEY, ETC.

5. FATS—BUTTER, MARGARINE, COTTONSEED OIL, OLIVE OIL, DRIPPINGS, SUET.

You can exchange one food for another *in the same group.* For example, oatmeal may be used instead of wheat, and eggs, or sometimes beans, instead of meat; but oatmeal can not be used instead of milk. Use both oatmeal and milk.

YOU NEED SOME FOOD FROM EACH GROUP EVERY DAY—DON'T SKIP ANY

1916 - 1943

1943 to 1956

1957 to 1993

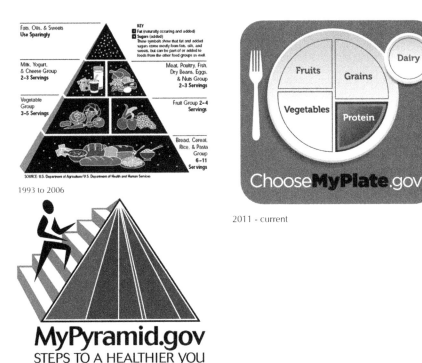

FIGURE 9.7   Evolution of USDA symbols for food groups.

The down side of food abundance is occurrence of the so-called life-style diseases that are exacerbated by diets high in calories, fats, and sugars. Adult obesity in the United States increased from 12% to nearly 20% over the last quarter of the 20th century. Caloric intake increased from about 3200/person to 4000/person at the end of the century, a 25% increase. The source of the calories also changed significantly. For example, per capita whole milk consumption decreased by 2/3 and low-fat milk increased by 2.5 times, presumably a beneficial result of the public perception of the value of low-fat milk. On the other hand, annual soft-drink consumption more than doubled, from 24.3 gallons/person to 53 gallons/person.

Economic and demographic changes have added pressure favoring prepared and convenience food. The large shift of women to the work–force has been a major factor, as has the pressures of time on individuals and families. By the end of the century about half of the total food intake was either processed food or food eaten outside the home, and therefore subject to and dependent on advertising. Roughly $33 billion was spent in 2000 on advertising by food companies.

The expenditures for nutritional education by the government and private interest groups are on the order of 1/100 of this amount.

9.7   EXAMPLES OF PROCESSED FOODS

9.7.1   BREAKFAST CEREALS

Breakfast foods, cereals in particular, have changed profoundly in the 20th century. Cereals were used primarily in the form of porridges by Europeans and Native Americans. Oatmeal has been a common breakfast food since colonial times. It was considered to be a curative agent for digestive problems and was sold mainly by druggists in the early part of the 19th century. In 1881, Henry Parsons Crowell bought an oatmeal mill in Ravena, OH, USA that had rights to the name "Quaker Oats." Crowell managed to establish an "oatmeal trust" and make Quaker Oats the dominant brand. Crowell promoted the still iconic cylindrical package as more wholesome than the then customary distribution from bulk. Crowell distributed Quaker Oats worldwide and it was the leading breakfast food in America at the turn of the century. A confluence of several factors resulted in great interest in new kinds of food, particular breakfast cereals in the early 20th century. One source of these trends was Sylvester Graham, a New England clergyman, who lectured widely and published a book, *Lectures on the Science of Human Life*. His views were not scientific, however, and he was scornful of the developing scientific information on the composition of foods. (He, died, a rather unhealthy man, at 57.) One of Graham's followers, James Caleb Jackson, successfully marketed Graham's flour and crackers, both based on whole grain flours. Jackson also introduced an early form of breakfast cereal called *Granula*, a form of bran nuggets.

Battle Creek MI, however, became the home for many American breakfast cereals. The Seventh-Day Adventists, a religious group with a strong interest in abstentious living, established a Sanitarium in Battle Creek. One of its leaders, Ellen G. White, encouraged a protégé, John Henry Kellogg, to obtain medical training, which he did at the University of Michigan and Bellevue Hospital in New York. He returned to Battle Creek to become Director of the Sanitarium. Dr. Kellogg and his brother, William K. Kellogg, discovered how to make flakes from a paste of cereal flour. J. H. Kellogg also invented peanut butter. W. K. Kellogg was the more financially oriented of the brothers and eventually founded what is currently the Kellogg Cereal Company. Presaging battles to follow, the brothers had a falling out over whether to include sugar in the flakes, and eventually battled in court about the ownership of the company and its patents. W. K. Kellogg won the fight. The industry now uses a billion pounds of sugar a year. Breakfast cereals rank third behind soft drinks and milk in total volume of grocery sales and are second only to automobiles in advertising expenditures.

Box cereal products proliferated in the early years of the 20th century. *Corn Flakes* were invented by the Kellogg brothers. Battle Creek was also the site of the Post cereals, including *Post Toasties* and *Grape Nuts. Cream of Wheat* was made from dehusked wheat. It was an immediate success and gave rise to the Ralston Purina Company, headquartered in St. Louis, MO, USA. The first puffed cereal *Puffed Rice* was patented in 1902, by Alexander P. Anderson.[f] The process was scaled up using surplus artillery and was commercialized by Quaker Oats, using the motto "The Grain Shot from Guns." Shredded wheat type products were invented by Henry D. Perky by passing cooked wheat through grooved rollers. Perky named his company Natural Foods. In 1902, he built a large factory at Niagara Falls, NY, USA. It became the first show place of food technology. The company became part of the National Biscuit Company (Nabisco) in 1928. *Rice Krispies* were developed at Kellogg in 1927 and promoted by the slogan "Snap, Crackle and Pop," and cartoon characters with those names. Many others were added to the cast of characters including *Sonny, the Cocoa Puff Cuckoo Bird*, *Tony the Tiger* of *Frosted Flakes* fame and *Cap'n Crunch*. The first cereal to incorporate vitamins was *Kix*, introduced by General Mills.

The first pre-sweetened cereal was invented by James Rex, a Philadelphia heating equipment salesman, who came up with the idea, hoping to cut the amount of sugar his children put on their cereal. He manufactured and sold his product as *Ranger Joe Puffed Wheat Honnies* in the Philadelphia area, but they tended to become sticky and then hard as the coating of sugar, honey and corn syrup adsorbed moisture. The problems were solved by Moses Berger, who bought out Rex's company and then sold the brand to Nabisco in 1954. Post became the first of the major companies to introduce sugared flakes with *Sugar Crisp* in 1949. Kellogg countered in the early 1950s with *Sugar Frosted Flakes*, *Sugar Corn Pops*, and *Sugar Smacks*. General Mills joined the fray with *Sugar Jets*, *Trix*, and *Cocoa Puffs*. The Quaker Oats company had largely avoided the pre-sweetened cereal market, but in the early 1960s they worked with the Arthur D. Little Company to develop *Cap'n Crunch*, corn-oats particles coated with butterscotch. It quickly became one of the top 10 cereal brands. Some of the healthy food tradition reemerged at Kellogg with the introduction of *Special K* in 1955. It was a mixture of grains, supplemented with vitamins. The *Total* brand

---

[f]Alexander P. Anderson grew up in rural Minnesota and received a B. S. from the University of Minnesota in 1894. He then did work for an M. S. at the New York Botanical Garden and received a Ph.D. in Munich for work on starch particles. It was there that he did his first experiments on "exploding" grains. After teaching briefly at Clemson and the University of Minnesota, he returned to the New York Botanical Gardens and did extensive experiments on exploding grains. He then went to work at Quaker Oats and stayed there during the remainder of his career.

was introduced by General Mills in 1961, with supplements that met 100% of most cereal-based nutritional requirements.

Sugar back-lash began in the 1970s. Robert F. Choate, served as President Richard Nixon's representative on the White House Conference on Nutrition. He became convinced that children were being "programmed to demand sugar and sweetness in every food." He ranked cereals by nutritional value, with *Special K*, *Kaboom* and *Total* coming out at the top. The FTC was under the gun at the time, criticized by Ralph Nader as totally ineffective. The cereal industry looked like a good target. In 1972, the FTC launched a complaint about "unfair advertising" but after 10 years in court, lost the battle. As concern for the nutritional consequences of cereal content and consumption grew, legislation was passed to restrict some of the marketing methods. The Consumer Product Safety Acts of 1972 and 1976 put limits on items such as premiums. The Nutrition Labeling and Education Act of 1990, required standardized food labeling. In 1992, limits were placed on the amount of advertising on children's television programs.

Health and nutritional studies also began to recognize the value of fiber. Several studies suggested that oat bran reduced cholesterol level. The turn to more healthful products resulted in a "back to nature" trend in the cereal industry. The pioneer in this area was called *Crunchy Granola*, which was later renamed *Heartland Natural Cereal*, sold by a small company. The majors followed with "*Country Morning*" (Kellogg), and "*Nature Valley*" (General Mills).

Studies of the relationship between breakfast habits and nutritional status indicate that children and adolescents who regularly eat breakfast have better nutritional status and body-weight index, even though they consume more calories. The improved nutritional status is at least in part due to the vitamin and mineral supplements associated with prepared cereals and the fact that cereals usually are consumed with milk and sometimes fruit.[27] There seem to be no studies of prepared cereals *per se* on the health status of either children or adults.

9.7.2 SALAD DRESSINGS

Salad dressing is another example of prepared food products. Salad dressings have two purposes. One is to enhance taste but they also function to help absorb lipophilic components, for example, vitamins, from the vegetable material. Like cereals, there are a number of competing brands, albeit without quite the same array of colorful promotional characters. Salad dressings are typically oil in water mixtures or emulsions. A look at the label of a typical salad dressing will reveal the presence of several of the food ingredients listed in Section 9.1. They often contain emulsion-stabilizing agents such as carboxymethyl cellulose, methyl cellulose, or hydroxypropylmethyl cellulose. They may also contain gums. Another common component is vinegar (dilute acetic acid), which not only gives a particular taste component, but also retards growth of microorganism. Reduced caloric salad dressing must replace digestible fat with noncaloric materials that nevertheless provide the smooth texture of the fat component. Several are so-called fat substitutes described in Section 9.3.3. An example is the β-glucans, which are prepared from plant fiber, particularly from oats or barley, by partial chemical or enzymatic degradation to short glucose oligomers. The limited data on the effect that the fat substitutes suggest they might reduce adsorption of lipophilic vitamins.[28]

9.7.3 YOGURT AND RELATED DAIRY PRODUCTS

Yogurt is produced by fermentation of milk by bacteria that produce a thickened product containing lactic acid. Yogurt originated in India and the Middle-East and has been a major food in that part of the world for hundreds of years. It was introduced to the United States primarily by immigrants. Yogurt is a "probiotic" and has been recommended as a dietary aid. John H. Kellogg, for example, was an early advocate. During the last quarter of the 20th century, yogurt became widely regarded as a healthy food. At the same time, the trend to "low-fat" and "low-sugar" diets gave rise to various forms of yogurt and yogurt-like products. Table 9.3 indicates some of the combinations of dairy source, calorie content, fat and fat substitutes, and sweeteners that can be found.

TABLE 9.3 Contents of Some Yogurts and Yogurt-Like Products.

Name	Dairy source	Calories	Fat(g)	Carbohy-drate (g)	Fat subst.	Sweetener
A: Grade A lite	Non-fat milk	80	0	22	Modified corn starch	Sucralose
A: Cultured dairy Blend	Non-fat milk solids	60	1.5	33	Modified corn starch	Sucralose, acesulfame
A: Fruit on bottom	Low-fat milk	170	2.0	57	Modified corn starch	Sucrose
B: Light	Non-fat milk	100	0	20	Modified corn starch	HFCS, aspartame
B: Greek non-fat	Non-fat milk	120	0	19	Locust bean gum	Sucrose
C: Light and fit	Non-fat milk	80	0	14	Modified starch	Fructose, sucralose, acesulfame
C: Fruit on bottom low fat	Low-fat milk	150	2.5	26	Modified starch	Sucrose

9.8   METABOLIC SYNDROME, CARDIOVASCULAR DISEASE, AND TYPE-2 DIABETES

Metabolic syndrome is a term used to describe a cluster of conditions that is associated with increased risk of cardiovascular disease and type-2 diabetes. The factors that are widely recognized include obesity, dyslipidemia (decreased HD-lipoprotein), hypertension (high blood pressure), and insulin resistance (see Section 16.3.3). Other factors also appear to be related, including chronic inflammation, as indicated by elevated levels of cytokines and c-reactive protein and hypercoagulability of blood. In the United States, the National Cholesterol Education Program has provided specific test readings that define the metabolic syndrome. The WHO and other organizations have related definitions of the condition. In the last quarter of the 20th century, obesity and the factors associated with the metabolic syndrome have increased substantially. It is estimated that about one-third of middle-aged Americans exhibit the metabolic syndrome and the incidence increases with age. This poses an extraordinarily serious problem for future demands on the health-care system. Current annual medical costs in the United States related to the metabolic syndrome diseases are on the order of $50 billion. Furthermore, it is estimated that perhaps 400,000 premature deaths annually are related to obesity. The metabolic syndrome is related to the level of stored fat in the body. Fat is stored in cells called *adipocytes*. The fundamental evolutionary purpose of fatty deposits is as a hedge against starvation, but excessive fat storage has adverse consequences. Dietary and lifestyle approaches can reduce the occurrence of the metabolic syndrome. Exercise and other physical activity reduce the levels of the markers for and the risk of both cardiovascular disease and type-2 diabetes. Decreased caloric intake and increased unsaturated fats and fiber in the diet are beneficial. Specific dietary patterns such as the "Mediterranean diet" and moderate alcohol consumption also seem to be helpful.

The causes of the metabolic syndrome are under vigorous investigation and discussion. Excess caloric intake and sedentary life style clearly contribute. The laws of conservation of energy and thermodynamics apply and any caloric energy that is taken in but not expended as physical work or heat is stored. The modern diet tends to be high in refined carbohydrates. In addition to amount and type of food, other factors may contribute. These include environmental contaminants such as *bis*-phenol-A, phthalate esters, halogenated aromatics, and arsenic. The term *obesogen* has been applied to such materials, although it remains to be shown that such compounds contribute to the obesity epidemic.[29] Other compounds that are under consideration are called *advanced glycation end-products*, AGEs for short. These are molecular combinations of amine and carbohydrates that are formed on cooking foods, especially at higher temperatures. A variety of products can be formed by the *Maillard reaction*. These materials contribute to the taste of cooked food. Some of the structures, such as pentosidine, are quite complex.

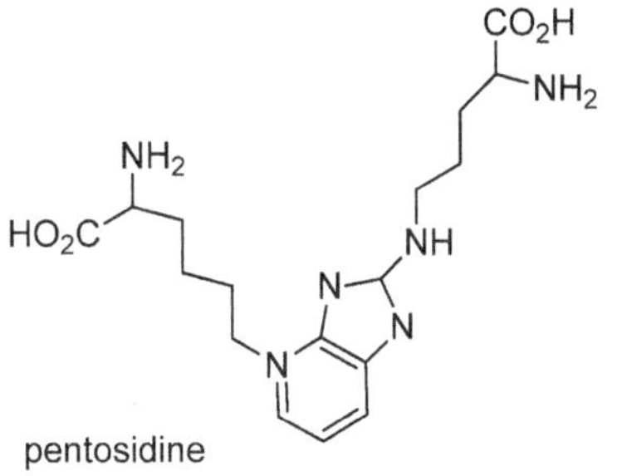

pentosidine

The suggestion is that the expanded use of processed food may have increased the exposure to AGEs. Some of the AGEs have been shown to have adverse metabolic effects, for example, increase in oxidative stress. There are natural defensive mechanisms, but the suggestion is that they may have been overwhelmed by the increased level of AGEs.[30]

Many other factors may also be involved. Maternal nutrition is believed to influence the metabolic pattern of offspring and the metabolic syndrome seems to be associated with low birth weight.[31] Stress may contribute.[32] Socioeconomic changes such as urbanization and immigration may also be factors.[33] A list of 10 such possible contributing factors has been compiled[34] and includes: (1) sleep deprivation; (2) environmental endocrine disruptors and/or obesogens; (3) temperature control; (4) decreased smoking; (5) increase use of pharmaceuticals, some of which may induce weight gain; (6) change in population age and ethnicity; (7) increased maternal age at birth; (8) nutritional and other influences *in utero*; (9) correlation of fecundity with weight; and (10) selectivity for higher weight in reproduction. For each factor, there is suggestive evidence and a plausible explanation, but to date no definitive results demonstrating cause and effect are available.[35]

KEYWORDS

- **gums**
- **emulsifiers**
- **glucose**

- **high fructose corn syrup**
- **low-calorie sweeteners**
- **fat substitutes**
- **coffee decaffeination**
- **fruit ripening**
- **processed foods**

BIOBLIOGRAPHY

Kessler, D. *The End of Overeating: Taking control of the insatiable American appetite,* Rodale, New York, 2009.

Ettinger, S. *Twinkie, Deconstructed: My Journey to Discover How the Ingredients Found in Processed Foods are Grown, Mined (Yes, Mined), and Manipulated into What America Eats.* Hudson Street Press: New York, 2007.

Nestle, M. *Food Politics*, University of California Press: Berkeley, CA, 2007.

Estabrook, B. *Tomatoland: How Modern Industrial Agriculture Destroyed Our Most Alluring Fruit.* Andrews McMeel: Kansas City, MO, 2011.

BIBLIOGRAPHY BY SECTIONS

Section 9.1: Gomez, M. *Food Engineering Aspects of Baking Sweet Goods*; CRC Press; Boca Raton, FL, 2008; pp 248–273.

Section 9.2.2: Kroger, M.; Meister, K.; Kava, R. *Compr. Rev. Food Sci. Food Saf.* **2006**, *5*, 35–47; Weerasinghe, D. K.; DuBois, G. E., Ed.; *ACS Symp Ser.* **2008**, *979*; Lemus-Mondacca, R.; Vega-Galvez, A.; Zura-Bravo, L.; An-Hen, K. *Food Chem.* **2012**, 1121–1132.

Section 9.2.2.2: Rathjen, S.; von Rymon Lipinski, G.-W. In *Sweeteners*, 3rd ed.; Wilson, R., Ed.; Blackwell Publishing, 2007; pp 3–19.

Section 9.2.2.3: Kotsonis, F. N.; Hjelle, J. J. In *The Clinical Evaluation of a Food Additive*; Tschanz, C.; Butchko, H. H.; Stargel, W. W.; Kotsonis, F. N., Ed., 1996; pp 23–41; Butchko, H. H.; Stargel, W. W.; Comer, C. P.; Mayhew, D. A.; Andress, S. E. *Food Sci. Technol.* **2001**, *112*, 41–61; Lovett, R. *New Sci.* **2006**, *190*, 41–43; Huff, J. J.; Ladou, J. *Int. J. Occup. Environ. Health* **2007**, *13*, 446–448; Meyer, H. In *Sweeteners*, 3rd ed.; Wilson, R. Ed.; Blackwell Publishing, 2007; pp 31–46.

Section 9.2.2.4: Nofre, C.; Tinti, J.-M. *Food Chem.* **2000**, *69*, 245–257; O'Donnell, K. *Sweeteners and Sugar Alternatives in Food Technology.* Blackwell Publishing: Oxford, UK, 2006, pp 95–102; Amino, Y.; Mori, K.; Tomiyama, Y.; Sakata, H.; Fujieda, T. *ACS Symp. Ser.* **2008**, *979*, 463–480.

Section 9.2.2.5: Jenner, M. R. *Adv. Sweeteners* **1996**, 253–262; Goldsmith, L. A.; Merkel, C. M. *Alternative Sweeteners* **2001**, *112*, 185–207; Quinlan, M. In *Sweeteners*, 3rd ed.; Wilson, R., Ed.; Blackwell Publishing: Oxford, UK, 2007; pp 113–125.

Section 9.2.2.6: Lemus-Mondaca, R.; Vega-Galvez, A.; Zura-Bravo, L.; An-Hen, K. *Food Chem.* **2012**, *132*, 1121–1132.

Section 9.3.2: Mozaffarian, D.; Katan, B. K.; Ascherio, A.; Stampfer, M. J.; Willett, W. C. *N. Engl. J. Med.* **2006**, *354*, 1601–1613; Mozaffarian, D.; Willett, W. C. *Curr. Atheroscler. Rep.* **2007**, *9*, 486–493; Crupkin, M.; Zambelli, A. *Compr. Rev. Food Sci. Food Saf.* **2008**, *7*, 271–279.

Section 9.3.3: Cho, S. S.; Prosky, L. *Complex Carbohydrates in Foods, Food Science and Technology*; Marcel Dekker: New York, 1998; pp 411–429; Sandrou, D. K.; Arvantoyannis, I. S. *Crit. Rev. Food Sci. Nutr.* **2000**, *40*, 427–447; Artz, W. E.; Mahungual, S. M.; Hansen, S. L. *Lipid Analysis and Lipomimics*; AOCS Press: Champaign, IL, 2006; pp 379–397.

Section 9.4.1: Geenberg, J. J.; Boozer, C. N.; Geliebter, A. *Am. J. Clin. Nutr.* **2006**, *84*, 682–693; Bonita, J. S.; Mandarano, M.; Shuta, D.; Vinson, J. *Pharmacol. Res.* **2007**, *55*, 187–198; Smith, B. D.; White, T.; Shapiro, R. *Caffeine and Activation Theory: Effects on Health and Behavior*; Smith, B. D.; Gupta, U.; Gupta, B. S. Ed.; CRC Press, 2007; pp 9–42.

Section 9.4.2: Lack, E.; Seidlitz, H. *Extraction of Natural Products using Near-Critical Solvents*; King, M. B.; Bott, T. R., Ed.; Blackie Professional Publishing: New York, 1993; pp 101–139; Ramalakshmi, K.; Raghavan, B. *Crit. Rev. Food Sci. Nutr.* **1999**, *39*, 441–456; Heilmann, W. In *Coffee—Technology II*; Clarke, R. J.; Vitzthun, O. G., Ed.; Blackwell Science Ltd: Oxford, UK, 2001; pp 108–124.

Section 9.5: Saltveit, M. E. *Postharvest Biol. Technol.* **1999**, *15*, 279–292.

Section 9.7.1: Bruce, S.; Crawford, B. *Cerealizing America: The Unsweetened Story of American Breakfast Cereal.* Faber and Faber: Boston, MA, 1995.

Section 9.8. Laaksonen, D. E.; Niskanen, L.; Lakka, H.-M.; Lakka, T. A.; Uusitupa, M. *Ann. Med. (London, UK)* **2004**, *36*, 332–346.

REFERENCES

1. Djagny, K. B.; Wang, Z.; Xu, S. *Crit. Rev. Food Sci. Nutr.* **2001**, *41*, 481–492.
2. Rees, D. A.; Walsh, E. J. *Angew. Chem., Intern. Ed.* **1977**, *16*, 214–224; Garcia-Ochoa, F.; Santos, V. E.; Casas, J. A.; Gomez, E. *Biotechnol. Adv.* **2000**, *18*, 549–579.
3. Ouyang, X.; Cirillo, P.; Sautin, Y.; McCall, S.; Bruchette. J. C.; Diehl, A. M.; Johnson, R. J.; Abdelmalek, M. F. *J. Hepatol.* **2008**, *48*, 993–999.
4. Melanson, K. J.; Angelopoulos, T. J.; Nguyen, V.; Zukley, L.; Lowndes, J.; Rippe, J. M. *Am. J. Clin. Nutr.* **2008**, *88*(suppl), 1738S–1744S.
5. White, J. S. *Am. J. Clin. Nutr.* **2008**, *88*(suppl.), 1716S–1721S.
6. Wolf, A.; Bray, G. A.; Popkin, B. M. *Obes. Rev.* **2007**, *9*, 151–164.
7. Nelson, G.; Hoon, M. A.; Chandrashekar, J.; Zhang, Y.; Ryba, N. J.; Zuker, C. S. *Cell* **2001**, *106*, 381–390; Margolskee, R. F. *J. Biol. Chem.* **2002**, *277*, 1–4; Li, X.; Staszewski, L.; Xu, X.; Durick, K.; Zoller, M.; Adler, E. *Proc. Natl. Acad. Sci. USA* **2002**, *99*, 4692–4696.
8. Linkies, A.; Reuschling, D. B. *Synthesis* **1990**, 405–406.
9. Mazur, R. H. *J. Toxicol. Environ. Health* **1976**, *2*, 243–249.
10. Oyama, K. In *Chirality in Industry*; Collins, A. N.; Sheldrake, G. N.; Crosby, J. Eds.; Wiley, 1992; pp 237–247.
11. Fry, J. C.; Hoek, A. C. *Food Sci. Technol.* **2001**, *112*, 481–498.
12. Soffriti, M.; Belpoggi, F.; Delgi Esposti, D.; Lambertini, L. *Eur. J. Oncol.* **2005**, *10*, 107–116; Soffritti, M.; Belpoggi, F.; Delgi Esposti, D.; Lambertini, L.; Tivaldi, E.; Rigano,

A. *Environ. Health Perspect.* **2006**, *114*, 379–385; Belpoggi, F.; Soffritti, M.; Padovani, M.; Esposti, D. D.; Lauriola, M.; Minardi, F. *Ann. N. Y. Acad. Sci.* **2006**, *1076*, 559–577.

13. Magnuson, B. A.; Burdock, G. A.; Doull, J.; Kroes, R. M.; Marsh, G. M.; Pariza, M. W.; Spencer, P. S.; Waddell, W. J.; Walker, R.; Williams, G. M. *Crit. Rev. Toxicol.* **2007**, *37*, 629–727.

14. Soffritti, M.; Belpoggi, F.; Tibaldi, E.; Esposito, D. D.; Lauriola, M. *Environ. Health Perspect.* **2007**, *115*, 1293–1297.

15. Abdo, K. M.; et al. *Int. J. Occup. Environ. Health* **2007**, *13*, 449–450.

16. Grice, H. C.; Goldsmith, L. A. *Food Chem. Toxicol.* **2000**, *28* (Suppl. 2), S1–S129.

17. Jung, M. Y.; Min, D. B. *Healthful Lipids* **2005**, *2005*, 65–77; Belkacemi, K.; Harmoudi, S.; Arul, J. *Recent Devel. Catal.* **2005**, *3*, 43–65.

18. Stender, S.; Dyerberg, J.; Astrup, A. *Scand. J. Food Nutr.* **2006**, *50*, 155–160.

19. Cheung, S.-T. *Nutr. Health* **2000**, *14*, 271–280.

20. Morgan, K. *Food Sci. Technol. Nutr.* **2009**, *173* (Hydrocolloids) 287–307; Stevenson, D. G. *Food Sci. Technol. Nutr.* **2009**, *173* (Hydrocolloids) 615–652.

21. Meyer, D.; Blaauwhoed, J.-P. *Food Sci. Techol. Nutr.* **2009**, *173* (Hydrocolloids), 829–848.

22. Ben-Yehoshua, S.; Iwahori, S.; Lyons, J. M. *Isr. J. Agric. Res.* **1970**, *20*, 173–177; Koshio, K. H.; Takahashi, H.; Ota, Y. *Plant Cell Physiol.* **1995**, *36*, 1511–1517.

23. Watkins, C. B. *Biotechnol. Adv.* **2006**, *24*, 389–409.

24. Capitani, G.; Hohenester, E.; Feng, L.; Storici, P.; Kirsch, J. F.; Jansonius, J. N. *J. Mol. Biol.* **1999**, *294*, 745–756.

25. Zhang, Z.; Ren, J.-S.; Clifton, I. J.; Schofield, C. J. *Chem. Biol.* **2004**, *11*, 1383–1394.

26. Nestle, M. *Food Politics*; University of California Press: Berkeley, CA, 2007; p 3.

27. Rampersaud, G. C.; Pereira, M. A.; Girard, B. L.; Adams, J.; Metzl, J. D. *J. Am. Diet. Assoc.* **2005**, *105*, 743–760; Ritchie, L. D.; Welk, G.; Styne, D.; Gerstein, D. E.; Crawford, P. B. *J. Am. Diet. Assoc.* **2005**, *105*, S79–S90.

28. Brown, M. J.; Ferruzzi, M. G.; Nguyen, M. L.; Cooper, D. A.; Eldridge, A. L.; Schwartz, S. J.; White, W. S. *Am. J. Clin. Nutr.* **2004**, *80*, 396–403.

29. Grun, F.; Blumberg, B. *Mol. Endocrinol.* **2009**, *23*, 1127–1134.

30. Vlassara, H.; Striker, G. E. *Nat. Rev. Endocrinol.* **2011**, *7*, 526–539.

31. Bruce, K. D.; Byrne, C. D. *Postgrad. Med. J.* **2009**, *85*, 614–621; Bruce, K. D.; Hanson, M. A. *J. Nutr.* **2010**, *140*, 648–652.

32. Kyrou, L.; Tsigos, C. *Hormone Metab. Res.* **2007**, *39*, 430–438.

33. Candib, L. M. *Ann. Family Med.* **2007**, *5*, 547–556.

34. Keith, S. W.; Redden, D. T.; Katzmarzyk, P. T.; Boggiano, M. M.; Hanlon, E. C.; Benca, R. M.; Ruden, D.; Pietrobelli, A.; Barger, J. L.; Fontaine, K. R.; Wang, C.; Aronne, L. J.; Wright, S. M.; Baskin, M.; Dhurandhar, N. V.; Lijoi, M. C.; Grilo, C. M.; De Luca, M.; Westfall, A. O.; Allison, D. B. *Int. J. Obes.* **2006**, *30*, 1585–1594.

35. McAllister, E. J.; Dhurandhar, N. K.; Keith, S. W.; Aronne, L. J.; Barger, J. Baskin, M.; Benca, R. M.; Biggio, J.; Boggiano, M. M.; Eisenmann, J. C.; Elobeid, M.; Fontaine, K. R.; Gluckman, P.; Hanlon, E. C.; Katzmarzyk, P.; Pietrobelli, A. O.; Redden, D. T.; Ruden, D. M.; Wang, C.; Waterland, R. A.; Wright, S. M.; Allison, D. B. *Crit. Rev. Food Sci. Nutr.* **2009**, *49*, 868–913.

CHAPTER 10

VITAMINS AND NUTRITION

ABSTRACT

While prevention or treatment of certain conditions such as scurvy and night-blindness had been associated with specific foods for centuries, the association of these conditions with specific chemical substances called vitamins began late in the 19th century and was largely complete by the 1940s. Physicians, biochemists, chemists, and nutritionist identified specific molecules associated with the vitamin deficiency diseases. They were isolated and structurally characterized and their biological functions determined. By the 1950s, commercial production by synthesis or biological preparation made the vitamins widely available and relatively inexpensive. Vitamins are relatively loosely regulated in the United States, with only a minimum content required. On packaged foods the percent of the daily requirement is stated on the basis of a 2000 calorie diet, but the information is required for only vitamins A, C, calcium, and iron. Enriched and fortified products list other nutrients. Since the 1990s, various "dietary supplements" have been exempt from FDA approval, but the FDA can litigate to remove hazardous materials from the market. The effect on vitamin content of food preparation and cooking is an important factor in the level of vitamins obtained through the diet but the information available is limited. In general, measurement of vitamin content is relatively difficult. The huge economic scope of the food production industry has made nutritional labeling and regulation a source of continuing controversy.

The role of vitamins in human health came into focus in the first third of the 20th century as a result of studies in medicine, nutrition, chemistry, and biochemistry. A vitamin is defined as a substance that is required for proper function, but that cannot be synthesized by the organism and must be taken in from the diet. There is some variation in the particular set of vitamins for each species. For humans, 13 substances are recognized as being essential. Vitamins have a number of biological functions including as coenzymes, mediators of redox reactions, gene regulators, and antioxidants. Deficiency of a vitamin can result in a recognizable set of symptoms, called a *vitamin deficiency disease*. As with any chemical substance, excessive amounts of a vitamin presumably can have deleterious effects, including toxicity. Within the limits of deficiency and toxicity, there are two other significant points, often imprecisely defined. One is the *minimum required level* and the other is the *optimum level*. Furthermore, there are surely individual variations in these levels. Both national and international organizations have published, and from time to time revised, minimum and recommended levels. In the United States, *Recommended Dietary Allowances* (RDA) were developed during WWII. These are revised from time to time with the Department of Agriculture as the lead government agency. Currently, foods are labeled with the percentage of the RDA contained in a serving. In 1997, the Institute of Medicine of the US National Academy of Sciences, developed an alternative scale called the *Dietary Reference Intakes* (DRI). It provided recommended levels that vary somewhat with age and gender. The DRI are expressed in units of weight (μg or mg per day). So far these DRI values have not been incorporated into food labeling laws or regulations. For many vitamins there are also *International Units* (IU) that have been defined historically and vary from vitamin to vitamin.

The discussion of vitamins and supplements brings up the words, natural, organic, and synthetic, and their meaning. To chemists, the word "organic" usually means "derived from carbon," even though its original meaning was closer to "from living things." The "from living things" definition became untenable when Friedrich Woehler in 1828 found that urea, an "organic" compound, could be made from ammonium isocyanate, which was not "organic." This, plus innumerable other observations, disproved the "vital force" theory: the idea that molecules from living things are fundamentally different from those derived from nonliving substances. The concept lives on, however. Currently it is exploited primarily in marketing, where the term "natural" is used. Its counterpoint is "synthetic." This fundamentally flawed concept is also ensconced in US law. While "drugs" (whether natural or synthetic) must receive FDA approval, "herbs and supplements" can be sold without approval, unless they are known to be toxic. This applies even if the "supplement" is *synthetic*! All that is required is that the substance exist somewhere in nature.

10.1   VITAMIN DEFICIENCY DISEASES AND THE DISCOVERY OF THE VITAMINS

Several cases of diseases that could be treated by specific foods were known prior to the discovery of vitamins. How this was possible, was not understood. At the end of the 19th century two concepts dominated medical thinking about the relationship between diet and health. One was derived from the studies of Justus von Liebig and others, which taught that carbohydrates, fats, and proteins constituted the essentials of a healthy diet (see Topic 9.1). The other was the newly recognized role of germs as the cause of many diseases. The idea that a lack of specific nutrients could cause disease was not recognized. The broad concept of vitamins was formulated by Frederick Gowland Hopkins.[a] Between 1910 and 1930, several vitamin-deficiency diseases and conditions were described and associated with particular vitamins. By the end of the 1930s, most of the major vitamins had been identified, but widespread use of vitamin supplements was just beginning. Writing in 1940, W. H. Sebrell, a physician with the USPHS, estimated that about one-third of the American public was unable to afford an entirely adequate diet and that many more were still unfamiliar with how to achieve an adequate diet.[1] He estimated that around $25 million was being spent on vitamins, but more than half of this was on products that did not conform to the then current pharmaceutical standards. Since then, use of vitamins and other nutrition supplements has increased dramatically. Let's start this story by looking at the essential vitamins and nutrients, with the "alphabetical" vitamins listed first.

10.1.1   THE ESSENTIAL VITAMINS

10.1.1.1   VITAMIN A (RETINOL)—NIGHT-BLINDNESS

As long ago as 1500 BC, the Egyptians realized that certain food could improve night vision. Liver was beneficial, as was known to Hippocrates, Galen, and Chinese traditional medicine. Late in the 1880s, several observations were made that a diet of carbohydrate, precipitated milk protein and fat did not maintain the health of laboratory mice, even though all the major food categories were present. On the other hand, a small amount of whole milk added to this diet sustained the mice. The first effort to isolate the substance responsible was by Wilhelm Stepp, working in Strassburg. He cooked milk and bread together and extracted the product with both ethanol and ether. Both extracts, if used with the otherwise inadequate diet, kept the mice healthy. This work was followed up by Elmer V. McCollum, an American chemist. McCollum joined the Wisconsin Agricultural Experimental Station in 1907, to work on nutrition of cattle. He set up a mouse colony to use as a model system. A diet of milk protein (casein), starch, lard, and salt resulted in early death, preceded by blindness. Following

---

[a]Frederick Gowland Hopkins was born in 1861. He was a prize-winning student in chemistry at the Institute of Chemistry of University College, London. For a time he was an associate of a prominent forensic chemist. In 1888, he entered medical school at Guy's Hospital. He graduated in 1894 and continued as a lecturer in physiology and forensics. In 1898, he went to Cambridge University to develop a program in chemistry and physiology that eventually became the department of biochemistry. He became chair of biochemistry in 1914 and the department expanded in to the William Dunn Institute of Biochemistry in 1925. Hopkins formulated the general theory of vitamins. Among his other discoveries were the isolation of the amino acid tryptophan and the important tripeptide glutathione. He shared the 1929 Nobel Prize in Physiology or Medicine with Christiaan Eijkmann, another vitamin pioneer.

the lead of Stepp, McCollum extracted the lipid component of milk and egg yolk and showed that the extract maintained the mice in a healthy state. He was also able to show the factor was not a fat, because it was unaffected by alkaline hydrolysis. McCollum named the material "fat-soluble factor A." Later on, material that could be extracted from wheat bran and prevented beriberi, was called "water-soluble factor B." Other water soluble B vitamins received numerical subscripts. Eventually, beginning with vitamin C, new vitamins got the next letter in the alphabet.

One of McCollum's students, Harry Steenbock (see also Section 10.1.1.8), noted that the same favorable response was obtained with yellow extracts of certain plants, carrots, and sweet potatoes in particular. Thomas Moore, working at the Dunn Nutritional Laboratory in Cambridge, UK, isolated a yellow compound, carotene, from carrots. On further work, he showed that mice converted carotene to a material with the same properties as McCollum's "fat-soluble factor A," a.k.a. vitamin A. The structures of the compounds were determined by Paul Karrer at the University of Zurich, who shared the 1937 Prize in Chemistry with W. N. Haworth (see Section 10.1.1.7). The connections were now becoming fairly clear. The relationship with blindness is that retinol is the precursor of the corresponding aldehyde, retinal, which is an essential component of the light-detection system of the retina. Vitamin A deficiency remains a problem, especially among children in impoverished areas of the world. The first symptom, called *xerophthalmia*, results from reduced tear formation and leads to lesions in the eye.

carotene, present in yellow
vegetables and fruits

retinol X = CH_2OH; retinal X = $CH=O$
vitamin A, retinol, is formed in the
liver and is present in whole milk

10.1.1.2   VITAMIN B1 (THIAMINE)—BERIBERI

The symptoms of beriberi include weight loss, muscle weakness, fatigue, and neurological disturbances. Beriberi was especially prevalent in Asian societies in which polished rice was the dietary staple. It was, for example, common among Japanese seaman. By 1882, Dr. Kanehiro Takaki of the Japanese Medical Service had noted that its incidence was reduced by a more varied diet. He arranged to test this concept on two similar ships destined for New Zealand. One, the Ryujo, was provided a traditional rice–fish diet, while on the other, the

Tsukuba, the diet was supplemented with wheat, milk, and meat. On the Ryujo, after 9 months, nearly half the crew was ill and 25 had died. On the Tsukuba, less than a tenth were ill, and none had died. Similar improvements in health were noted in prisons when some rice in the diet was replaced by barley (to save money).

The disease was also a problem in colonial Dutch Indonesia. A small institute was set up to investigate the cause. Experimenting with chickens, Christiaan Eijkman discovered that use of unpolished rice prevented beriberi-like symptoms.[b] He believed, nevertheless, that the disease itself was caused by microorganisms (germs), but was somehow counteracted by the unpolished rice. Eijkman shared the 1929 Nobel Prize in Physiology or Medicine with Frederick G. Hopkins, although he retained an erroneous concept of the role of vitamins and believed that they functioned to prevent infectious diseases. His successor, Gerrit Grins, carried out further experiments and solidified the idea that some factor present in the husk of the rice was curative. Grins also demonstrated that other foods, mung beans, for example, could prevent and cure beriberi.

The British were also investigating the cause of beriberi in Malaysia, where it primarily affected Chinese workers. While the Chinese preferred polished (white) rice, other ethnic groups used other (brown) forms. Among the theories proposed was that the storage of the white rice led to contamination by a toxin. William Fletcher, the District Surgeon in Kuala Lumpur, conducted experiments at the local insane asylum in 1905 and 1906. The inmates were divided into two groups and housed separately. Their diets were identical except that one group received white rice and the other brown. To check the possibility that location (contamination of some sort) was the cause of the disease, the groups exchanged housing at mid-year. Patients who became ill with beriberi were given the brown rice diet. At the end of the experiment 20 of 220 on the white rice had developed beriberi, but none of the 273 on brown rice had become ill. Most of the affected patients recovered when switched to the brown rice diet. A similar experiment was done by Henry Fraser and Thomas Stanton involving a crew of Chinese rail workers, with the same results. The United States encountered beriberi in the Philippines. It was especially common in infants and young children. Two physicians, Edward Vedder and Weston Chamberlain, found that an alcohol extract of rice hulls was an effective treatment. Gradually, the common conclusion that the brown rice bran contained some protective substance was accepted.

---

[b]Christiaan Eijkman was born in 1858 in the Netherlands. In 1875, he entered the military medical school at the University of Amsterdam. He studied physiology from 1879 to 1881 and received a doctorate in 1883. He was then dispatched to colonial Indonesia as a medical officer. He contracted malaria and returned to Holland, during which time he studied bacteriology. On his return to Indonesia, he was assigned to study beriberi. He also became director of a medical school for Javanese students. During this period, he observed and confirmed the effect of substituting unpolished rice for polished rice on beriberi. In 1898, he returned to Holland as Professor of Hygiene and Forensic Medicine in Utrecht. His subsequent research was primarily on bacterial metabolism and pathology.

Several groups around the world worked on the isolation of the protective factor. Casimir Funk, working at the Lister Institute in London, obtained crystalline material containing the factor, as did Umetaro Suzuki, working at the University of Tokyo. B. C. P. Jansen, working with A. P. Dornath, in Dutch Indonesia isolated 100 mg of material starting with 100 kg of rice husk. They conducted an elemental analysis, but failed to recognize the presence of sulfur. Robert R. Williams, an American chemist, working with Edward Vedder in the Philippines, was able to extract the active substance, but not in pure form. He continued his studies, largely on his own time, subsequently at the USDA Bureau of Chemistry and at Bell Labs. He eventually obtained pure material and determined the correct formula $C_{12}H_{17}N_4OSCl$. Finally, Adolf Windaus, using material isolated by chemists at IGF, determined the structure of the active material, by then given the name *thiamine* and the designation Vitamin B_1. Thiamine can also be obtained with the alcohol group in the form of a diphosphate ester. Thiamine functions as a coenzyme for enzymes involved in carbohydrate metabolism.

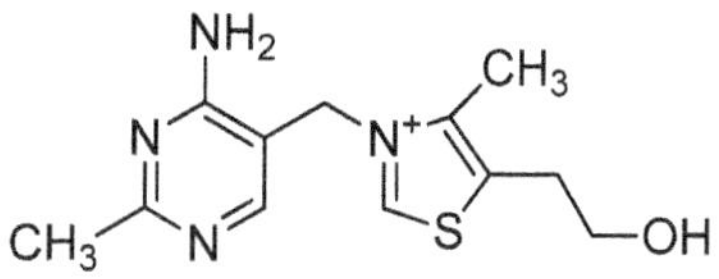

thiamine, vitamin B_1

10.1.1.3   VITAMIN B2—(RIBOFLAVIN)

Riboflavin was first found as a component of whey in 1879, but no nutritional connection was made then. In 1932, its reversible oxidation–reduction properties and role as a coenzyme were recognized. In 1933, several groups established the identity of riboflavin with the material designated vitamin B_2.[2] The correct structure was proposed in 1934 by Richard Kuhn at Heidelberg. It was synthesized by both Kuhn and Paul Karrer in Zurich.

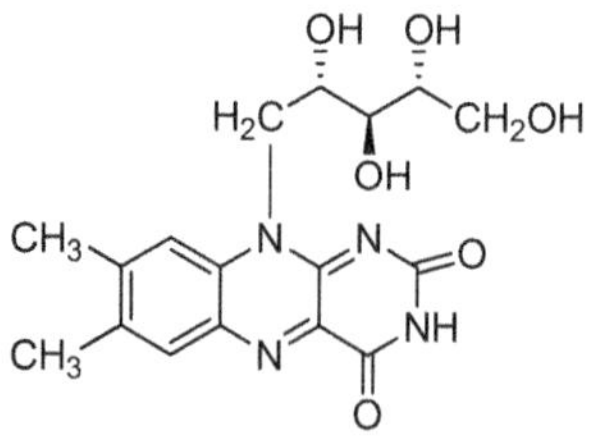

riboflavin, Vitamin B_2

Riboflavin is a *flavin ribonucleotide*. It functions as a coenzyme in various redox processes. Natural food sources include egg yolk, milk, fish, and green vegetables. Riboflavin deficiency in humans is rare but typically involves epithelial lesions (lips, tongue) and neurological disorders. Humans also show decreased white blood cells and platelets. Deficiency in domesticated animals, especially poultry, is more common. Riboflavin is used extensively as a supplement in breads and cereals. For this purpose it can be made either by synthesis or produced by microorganism. The latter method is subject to improvement by genetic manipulation.

10.1.1.4   VITAMIN B₃ (NIACIN)—PELLAGRA

The name Pellagra comes from Italian for "angry skin." Pellagra causes skin rashes, oral lesions, diarrhea, and, if left untreated, leads to mental deterioration. The disease was associated with corn-rich diets and poverty in Europe in the late 1800s. The disease suddenly increased in the US South early in the 20th century. Dr. Joseph Goldberger, a Public Health Service (PHS) physician, was assigned in 1914 by the Surgeon General to investigate. Goldberger recognized the connection between poverty, poor diet and the disease, which was especially prevalent in orphanages, prisons, and asylums. He persuaded the state of Georgia to provide a more balanced diet and the incidence of pellagra decreased. Using prisoners, he was able to demonstrate that a poor diet led to the symptoms of pellagra and that a more varied diet cured the condition. In this era, however, the idea that germs cause disease was dominant and much of the medical community was skeptical. Dr. Goldberger persisted and with Edgar Sydenstrikecker, a PHS statistician, was able to demonstrate a geographical correspondence between diet and the level of pellagra. In 1927, floods reduced the food supply and the incidence of pellagra increased. Goldberger recommended treatment with yeast, which was successful. It is suspected that one cause of the sudden increase in pellagra in the US South was a changed method of milling that removed most of the germ of the corn.[3] One of Goldberger's observations provided the starting point for another connection. In a single case, he noted that treatment with the amino acid tryptophan led to improvement.

Several groups had isolated niacin (nicotinic acid) from rice hulls in the course of study of beriberi, but it had no effect on that condition. Then a German biochemist, Otto Warburg, showed that the amide, nicotinamide, was part of the crucial coenzyme nicotinamide-adenine-dinucleotide (NAD). Conrad Elvehjem[c]

---

[c] Conrad Elvehjem was educated at the University of Wisconsin, receiving a PhD. in 1927. He taught agricultural chemistry and biochemistry and eventually became the University president in 1958. He received the Albert Lasker Award in 1952 and the Willard Gibbs Medal of the American Chemical Society in 1943. *Source*: Kline, O. L.; Baumann, C. A. *J. Nutr.* **1971**, *101*, 569–578.

showed that both nicotinic acid and its amide prevented and cured pellagra in dogs. Successful trials on humans soon followed. Eventually pellagra was eliminated in the United States by use of vitamin-enriched flour. Goldberger's observation that tryptophan is beneficial derives from a metabolic route from tryptophan to nicotinic acid.

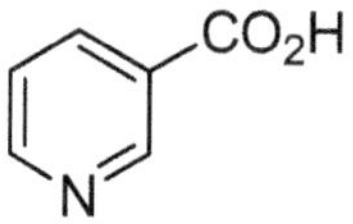

nicotinic acid
niacin

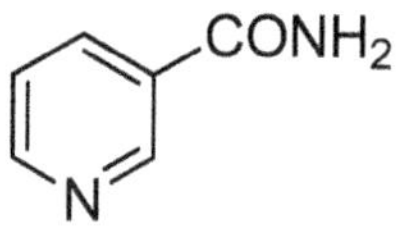

nicotinamide
niacinamide

Nicotinic acid and nicotinamide are made from readily available synthetic pyridines such as 3-methylpyridine or 5-ethyl-2-methylpyridine. The 3-methyl-pyridine is oxidized to 3-cyanopyridine and then hydrolyzed to nicotinamide. Nicotinic acid can be made from 5-ethyl-2-methylpyridine by oxidation and decarboxylation.

10.1.1.5   VITAMIN B$_6$ (PYRIDOXINE, PYRIDOXAL, AND PYRIDOXAMINE)

The B$_6$ vitamins are pyridoxine, an alcohol and the related aldehyde (pyridoxal) and amine (pyridoxamine). The compounds were isolated during the search for vitamin B$_1$, but their identity as vitamins was established by Paul Gyorgi in 1938.[d] The compounds serve as coenzymes for several metabolic enzymes, particularly transaminases. Vitamin B$_6$ is widely distributed in foods, including whole grain cereals, nuts, meats, fish, and vegetables. Overt vitamin B$_6$ deficiency is rare but it can be induced by artificial diets low in B$_6$ sources.

---

[d]Paul Gyorgi was born in Hungary and received an M. D. at Budapest in 1915. After WWI, he studied at Heidelburg, where he became a professor of pediatrics. While there, he participated in studies on riboflavin. In 1933, he moved to Cambridge, UK and then in 1935 to Western Reserve in Cleveland, OH, USA. It was there that he isolated vitamin B$_6$. He moved to the University of Pennsylvania in 1944. He was interested in many aspects of infant nutrition and was a strong advocate of breast feeding. He was also involved in training of health care workers in Indonesia and Thailand.

Source: Barness, L. A.; Tomarelli, R. M. *J. Nutr.* **1979,** *108,* 19–23.

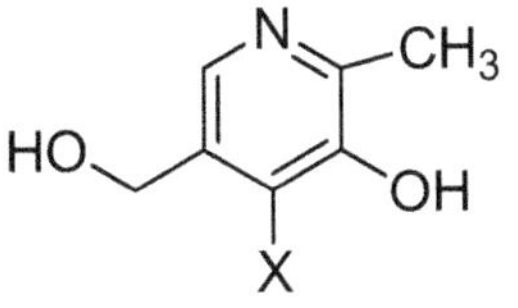

pyridoxine X = CH_2OH
pyridoxal X - CH=O
pyridoxamine X = CH_2NH_2

10.1.1.6   VITAMIN B$_{12}$ (COBALAMIN)—PERNICIOUS ANEMIA

Vitamin B$_{12}$ is also known as *cobalamin*. Cobalamin functions as a cofactor for two critical enzymes, methionine synthase and L-methylmalonyl Co-A mutase. The former also requires folic acid and is important in biosynthesis of nucleosides. The latter converts methylmalonyl Co-A to succinoyl Co-A, which is a crucial step in metabolism of fats and in the biosynthesis of hemoglobin. The best food sources of B$_{12}$ include meat, seafood, cheese, and eggs. Clams, mussels, and crabs are especially high. Plants do not produce B$_{12}$ and the source in other foods is evidently the ingestion of material synthesized by bacteria. Several diseases are associated with vitamin B$_{12}$ deficiency. A form of anemia, called *chlorosis*, because of the green palor of patients, was found to be resistant to treatment with iron. In 1892, William Oster, then at the Johns Hopkins University, noted that young women who were over-worked and poorly nourished were most likely to fall victim. The condition came to be called *pernicious anemia*, because it was almost always fatal. In the 1920s, George H. Whipple discovered the value of eating liver and other organs in treating dogs with experimental anemia. A Boston physician, William P. Murphy, working mainly with poor Irish patients, noticed that one of his male patients seemed to be surviving longer than usual and inquired about his diet. It turned out the man ate a lot of liver and ate it raw. Murphy turned for help to George Minot, a physician who was affiliated with a Harvard teaching hospital. They found that indeed eating raw liver improved patients and they were able to isolate an extract that had the same effect. Whipple, Murphy, and Minot received the 1934 Nobel Prize in Physiology or Medicine.

Vitamin B$_{12}$ has a much more complicated chemical structure than any of the other vitamins. It contains a cobalt ion surrounded by a ring system that is structurally related to hemoglobin and chlorophyll. The structure was determined by

Dorothy Crowfoot Hodgkin[e] at Oxford University, using X-ray crystallography. Hodgkin won the 1964 Nobel Prize in chemistry for her work on penicillin and vitamin B_{12} and later completed the crystallographic determination of the structure of insulin.

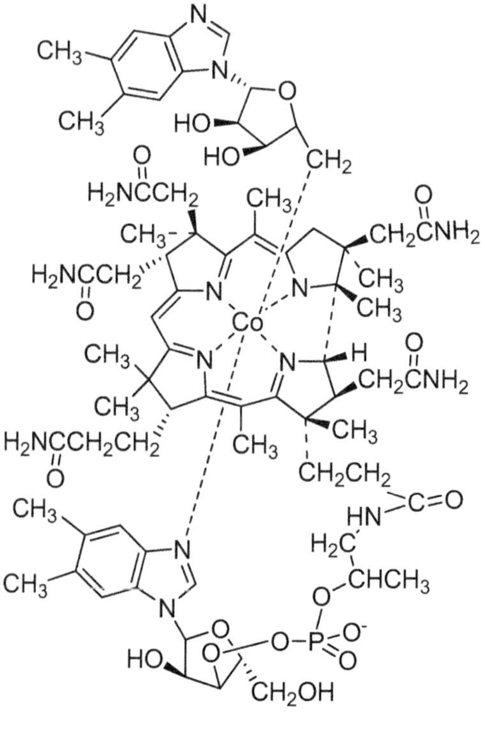

Cobalamin, Vitamin B_{12}

In the elderly, B_{12} deficiency can develop because of poor absorption by the digestive tract. In addition to anemia, there are also possible associations with cardiovascular disease, breast cancer, Alzheimer's disease, and depression. Vitamin B_{12} for supplements is produced commercially using microorganisms. The microorganisms have been selected from mutated strains that are high producers. In recent years, further improvement in production has been achieved by

[e]Dorothy Crowfoot Hodgkin's parents lived in the Sudan, where her father was Director of Education and Antiquities. She was educated in England, but returned to the Sudan regularly. She became interested in chemistry and crystals as a youngster. She began a serious study and research in crystallography as an undergraduate at Somerville College in Oxford and then at Cambridge. She returned to Somerville College and remained at Oxford throughout her career. Among the most important structures she solved were cholesterol, penicillin, vitamin B_{12}, and, after many years of work, insulin.

gene amplification. Currently, the main producer is Aventis, using technology originally developed at Rhone-Polenc.[4]

10.1.1.7   VITAMIN C (ASCORBIC ACID)—SCURVY

The symptoms of scurvy are bleeding gums, loose teeth, internal hemorrhaging and diarrhea and are the result of weakened blood vessels caused by lack of vitamin C, ascorbic acid. The disease has appeared several times in recorded history, as for example during the crusades in the 12th century. During his explorations of North America, Jacques Cartier lost many men to scurvy, but learned from Native Americans that a tea made from pine bark and needles could treat the condition. The undertaking of long duration sea journeys in the 1600s made the disease more common. English seaman recognized the value of lemons and limes in preventing and treating the disease. Nevertheless, on a voyage around the world in 1741–1744, Commodore George Anson lost 90% of his 2000 men to scurvy and typhus.

James Lind, trained as a surgeon's apprentice at the Medical School in Edinburgh, became a surgeon's mate in the Royal Navy in 1739. In 1746, Lind was assigned as the surgeon on HMS Salisbury. He conducted a series of experiments, perhaps the first controlled medical experiments ever performed, in which he treated six pairs of scurvy patients with different treatments. Those treated with citrus juice fared the best. In 1753, he published his observations and opinions on scurvy in a work entitled *Treatise on Scurvy*. He revised the work again in 1772. While he had confirmed the curative effect of citrus, his explanation remained mired in the medical theory of "humors" that was dominant at the time.

Commodore Anson was appointed First Lord of the Admiralty in 1751 by William Pitt. Anson, in 1758, appointed Lind as physician at the Haslar Naval Hospital in Portsmouth. Anson, seeing the value of citrus fruit, required it as provision, but the fruit usually spoiled on long voyages, and scurvy remained a problem. There were also competing explanations and recommended treatments for scurvy. Scurvy had long been associated with the "bad air" of the cramped crew quarters on ships. This was the era of Priestley's and Lavoisier's discovery of oxygen. The idea took hold that scurvy was somehow associated with depleted oxygen content. Pumps to improve ventilation were installed. It was not until the early 1800s that the British Admiralty finally required stocking of sufficient lemon or limes for all sailors. Louis H. Roddus writing in his *A Short History of Nautical Medicine* says "In the 200 years from 1600 to 1800 nearly 1,000,000 men died of an easily preventable disease. There are in the whole of human history few more notable examples of official indifference and stupidity producing such disastrous consequences to human life."

Scurvy reappeared mysteriously on Arctic expeditions in the late 1800s; mysteriously because the expeditions had adequate supplies of citrus. The puzzle was solved at the Lister Institute. It turned out that the source of limes had been shifted from the Mediterranean to the Caribbean and that the Caribbean fruit had much lower ascorbic acid concentration. During the Crimean War and the Siege of Paris in 1870, scurvy again appeared as the result of limited food supplies. Scurvy also frequently occurred in asylums, orphanages, and prisons, where diets were poor. Frozen orange juice concentrate was developed during WWII to serve as a transportable food source of vitamin C for US military forces.

In 1907, two Norwegian scientists, Axel Hoist and Theodor Froehlich, produced scurvy in guinea pigs by limiting diet to cereals. Inclusion of cabbage or fruits in the diets cured the condition. Most animals can synthesize adequate amounts of vitamin C, but guinea pigs, like humans, cannot. This provided a workable assay in the search for the vitamin. A Hungarian biochemist, Albert Szent-Gyorgyi, working at the University of Szeged with Joseph Svirbely an American of Hungarian descent, first isolated the active substance, which was called hexouronic acid.[f] They found that local paprika peppers were a rich source. The same compound was also isolated by Glenn King, a biochemist at the University of Pittsburgh and shown to be ascorbic acid. The structure was determined by W. N. Haworth, a famous carbohydrate chemist, who worked at the University of Birmingham, UK. Vitamin C is a cyclic oxidation product of glucose. Haworth received the 1937 Nobel Prize in Chemistry and Szent-Gyorgyi the Prize in Physiology and Medicine. The slight of King was a source of controversy in the United States.

ascorbic acid,
vitamin C

---

[f] Albert Szent-Gyorgi was born into a privileged Hungarian family in 1893. One of his uncles was Professor of Anatomy in Budapest, and Szent-Gyorgi began studies there that were interrupted by WWI. After the war, he studied pharmacology and electrophysiology at several locations including Berlin, Leiden, and Groningen. In 1926, he went to Cambridge as a Rockefeller Fellow, where he worked in the laboratory of F. G. Hopkins. He received a Ph. D. in 1927. His studies included oxidative processes and he first isolated ascorbic acid from adrenaline glands, although its structure and identity with vitamin C were not immediately recognized. In 1930, he became professor of medicinal chemistry at the University of Szeged and it was there that he and Joseph Svirbely isolated ascorbic acid. He also made major discoveries in the carboxylic acid metabolic cycles and in the characterization of actin and myosin in muscles. Szent-Gyorgyi left Hungary for the United States after WWII and the Russian occupation of Hungary. He worked first at Woods Hole and then for at time at NIH. He eventually established a laboratory supported by a private group, The National Foundation for Cancer Research. Szent-Gyorgi developed a number of broad-ranging theories on the biology of cancer.

The Nobel Prize winning American chemist Linus Pauling (Chemistry 1954, Peace 1962) was a proponent of high doses of vitamin C. In 1970, he published *Vitamin C and the Common Cold* in which he advocated doses of vitamin C well in excess of those recommended at the time. He based his recommendation on his interpretation of the published literature in the area, but it is a matter of controversy as to whether there is any beneficial effect. The medical profession was, in general, dismissive. Large segments of the population did however heed his advice and vitamin C sales increased. It remains a question as to whether there is any benefit, none having been conclusively demonstrated.

10.1.1.8   VITAMIN D, THE SUNSHINE VITAMIN (CHOLECALCIFEROL AND ERGOCALCIFEROL)—RICKETS

Rickets was described in the 17th century, notably by Francis Glisson at Cambridge. Rickets began to become prevalent in urban Europe, especially in London, in the 1800s. An English physician made the connection with lack of sunlight, noting that the incidence was less in Southern Europe. The high incidence in England was also related to diet, which consisted mainly of white bread and porridge. A Polish doctor, Jedrzej Sniadecki,[5] noted in 1822 that rickets was rare in rural Poland and that when urban children got more sunlight they improved. People around the Baltic Sea had learned the value of cod liver oil in preventing and treating the disease. Several physicians in Germany confirmed that cod-liver oil had anti-rickets activity.[6] In 1922, the British physician and physiologist Edward Mellanby showed that dogs kept out of sunlight and fed a diet of porridge developed rickets. He was able to cure the rickets by treatment with cod liver oil or egg yolk and deduced that rickets must be caused by the absence of a substance that could be obtained from these sources.

Between 1924 and 1927, the research groups of Alfred Hess,[7] Harry Steenbock,[8] and S. Otto Rosenheim[9] showed that UV-C radiation of several foodstuffs gave them anti-ricket activity. Steenbock worked in the School of Agriculture at the University of Wisconsin and was particularly interested in the well-being of the state's dairy industry. Although patenting of academic research was unusual at the time, Steenbock felt there were several reasons to patent the process. He was familiar with the contemporary development of insulin (see Section 16.1) and felt a patent would inhibit unscrupulous or incompetent exploitation of his discovery. He was also interested in using the process for the benefit of the dairy industry. The margarine industry had recently added vitamin A to its product, and Steenbock wanted to block incorporation of vitamin D into margarine, which would have strengthened its competition with butter. A strong sentiment existed against patenting the results of publicly funded research and many of Steenbock's faculty colleagues also opposed the idea. The

Board of Regents equivocated and delayed. At that point, Steenbock and others conceived the idea of the independent Wisconsin Alumni Research Foundation (WARF) and the patents were eventually assigned to the Foundation. Along with several other valuable patents, the Steenbock patents became a source of research funds. Both food companies and major suppliers of vitamins such as Abbot, Mead-Johnson, Parke-Davis, Winthrop Laboratories, and Squibb, were among the early licensees. The irradiation patents were eventually declared invalid when a federal court ruled that irradiation was a "natural" process and therefore not patentable. Nevertheless, the establishment of WARF and its financial success set a pattern for post WWII development of research commercialization by universities in the United States.

The chemical structure of vitamin D was determined by collaboration between Hess, Rosenheim, and Adolph Windhaus. Windhaus and Heinrich Wieland were leading the efforts to determine the structure of steroids. When it was discovered that irradiation of certain steroids, including cholesterol, resulted in the development of anti-rickets activity, the connection between vitamin D and the steroids was recognized. It was eventually found that the actual vitamin D precursors were 7-dehydrocholesterol and 7-dehydroergosterol, which were present as minor impurities in the steroids. These substances are converted to cholecalciferol and ergocalciferol by UV light. Cholecalciferol and ergocalciferol then undergo further metabolic activation by oxidation, first at C-25 and then at C-1 to be converted to the biologically active calcitriol. The hydroxylations occur mainly in the kidney, but also in the skin. The 1,25-dihydroxy compound is a transcription factor that regulates genes that are involved in maintaining the calcium level. Vitamin D also appears to have a function in the immune system, where it controls the level of a natural peptide antibiotic, *cathelicidin*. This could explain the efficacy of rest and *sunshine* as a treatment for tuberculosis (see Section 12.2). Also intriguing is a correlation between vitamin D levels and the occurrence of autoimmune diseases such as multiple sclerosis.

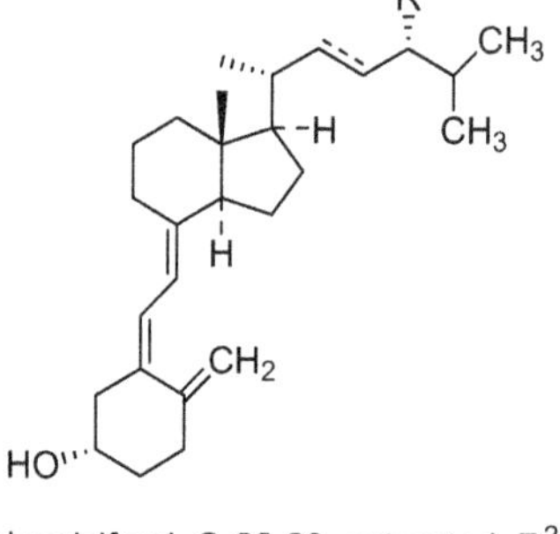

cholecalciferol, C-22,23, saturated, R^{24} = H
ergocalciferol, C-22,23, unsaturated, R^{24} = CH_3 calcitriol

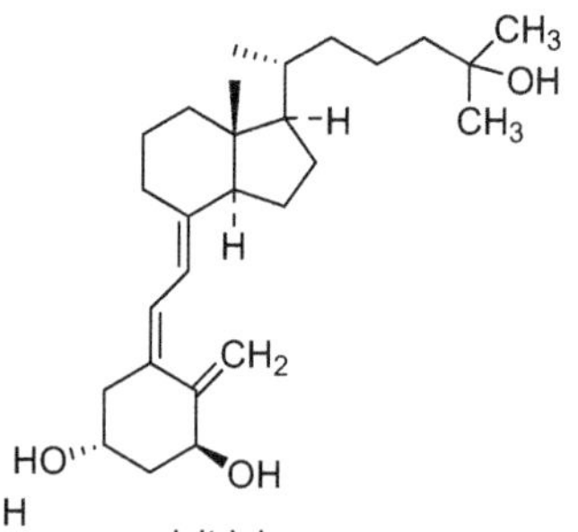

The level of vitamin D produced in humans by sunlight depends on many factors. The amount of UV reaching the earth's surface decreases with the solar angle, which depends on the time of day, season, and distance from the equator. The extent of cloud cover and atmospheric quality also affects the UV level. In recent years, evidence has accumulated that humans need considerably more vitamin D than had previously been believed and that vitamin D levels below optimum are common.[10] There are thought to be several reasons for this, including less exposure to sun as a result of life-style changes and lower intake of vitamin D from the diet because of avoidance of certain D-rich foods (e.g., skin of poultry, fish). Because of its role in control of calcium level, excessive levels of vitamin D can be harmful leading to osteoporosis and bone fractures. There is also some indication of increased lung cancer risk for smokers. Recently, a large study has been started in an attempt to determine if regular supplementation by vitamin D reduces illness from a number of causes.[11]

The connection between light, steroids, and vitamin D also has a special place in pure chemistry. The investigation of the relationship between the cyclic and open forms of vitamin D analogs revealed that the thermal reaction and photochemical reaction produce different stereoisomers. This was one of the experimental observations that led to the formulation of the orbital symmetry rules by R. B. Woodward and R. Hoffmann, for which Hoffmann shared the 1981 Nobel Prize in Chemistry.[12]

10.1.1.9   VITAMIN E (TOCOPHEROLS AND TOCOTRIENOLS)

Effects attributed to vitamin E were first reported by Herbert McLean Evans and Katharine S. Bishop of the Anatomy Department at UC, Berkeley. The substance is present in vegetable oils, especially wheat germ oil. It was found to be necessary for successful reproduction of rats. A pure sample was isolated in 1936. Subsequently, it was found to contain three related compounds, α, β, and γ-tocopherol. The most biologically active constituent is α-tocopherol. The related compounds β-tocopherol and γ-tocopherol, have only two methyl groups. There are also related tocotrienols, which have unsaturation in the side chain. The activity of the various isomers range down to <1% of the standard that is used, α-tocopherol acetate.

tocopherols   α–R⁵,R⁷, R⁸ = CH₃
                β–R⁵, R⁸ = CH₃
                γ–R⁷, R⁸ = CH₃

tocotrienols   α–R⁵,R⁷, R⁸ = CH₃
                β–R⁵, R⁸ = CH₃
                γ–R⁷, R⁸ = CH₃

Vitamin E is synthesized only in plants. Human food sources of vitamin E are mainly vegetable oils, seeds, nuts, and cereal grains. Wheat germ oil is particularly rich in vitamin E. Milk can also be a source if the animals' diet contains sufficient forage. There is no specific human disease condition associated with vitamin E deficiency, but in rats it leads to diminished reproduction. Humans who are genetically deficient in the ability to utilize vitamin E suffer from neurological disorders.[13] Some epidemiological studies have suggested reduced rates of cardiovascular disease in groups with higher vitamin E levels or intake. Vitamin E is an antioxidant and is believed to function as a chain-breaker in lipid oxidation by various oxygen radicals. Vitamin E is frequently used as a supplement both by humans and in animal feed. Vitamin E is also a common constituent of cosmetics. Vitamin E is a case where excess amounts may cause adverse results. Vitamin E can slow blood coagulation, especially if vitamin K level is low, and affect the high-density lipoprotein/low-density lipoprotein (HDLP/LDLP) ratio unfavorably. Retrospective statistical analysis of a number of studies of use of vitamin E alone or in combination with other antioxidants have suggested slight increases in all-cause mortality.[14] Long-term clinical studies have been carried out to see if vitamin E supplements reduce the risk of cancer or cardiovascular disease. The result is that they do not.[15]

An important aspect of the structure of vitamin E is its stereochemistry. There are three chiral centers and therefore eight stereoisomers of the tocopherols. The natural α-tocopherol has the R configuration at each center, but the synthetic material is generally racemic at all three centers. So far, synthesis of the pure natural enantiomer has not been economically feasible.[16] About 90% of the vitamin E supplements available for both humans and animals is the synthetic material. The main component, α-tocopherol, is synthesized from its two components 2,3,5-trimethylhydroquinone and the alcohol phytyl alcohol. The latter is obtained by hydrogenation of a precursor from soybean oil. These synthetic steps are non-stereospecific, which means that eight separate stereoisomers of natural α-tocopherol are present in more or less equal amount. Only about one-eighth of the material has exactly the same structure as actual α-tocopherol. So, a more accurate description of commercial vitamin E would be 12.5% Vitamin E. Does it make any difference?

Biological systems usually discriminate between the mirror image R and S forms of chiral molecules. That seems to be the case for the tocopherols. The all R- isomer is thought to be 1.5–2.0 times more active than the racemic mixture. The 2R-isomer is preferentially absorbed and the 2S-isomer preferentially excreted. There are also stereogenic centers at C4' and C8' but these probably have smaller effects because of the flexible nature of the chain. But the story may be even more complicated. Some members of the vitamin E family seem to have biological activity beyond being antioxidants. For example, some can inhibit the cyclooxygenase system (see Section 13.1.1) and therefore have anti-inflammatory activity. This activity is associated mainly with the γ-structure, which has an unsubstituted 5-position. Here is the kicker! It is conceivable that the increased level of S-tocopherol present in the synthetic supplement might *reduce by competition* the activity of the R-enantiomers. Nobody knows for sure and it would be a monumental effort to compare the various stereoisomers to find out. The important point is that because synthetic vitamin E is not a *pure* substance, but a mixture of stereoisomers, it is quite conceivable that the natural and synthetic materials may have different properties.

10.1.1.10 *VITAMIN K (MENAQUINONES AND PHYLLOQUINONES)*

Vitamin K was identified by Henrik Dam, a Danish physiologist. He received the 1943 Nobel Prize in Physiology or Medicine, along with Edward Doisy of St. Louis University Medical School, who synthesized the compound. Working with chicks fed with low-lipid diets, Dam noted hemorrhages and slow blood-clotting. The designation K in this case came from the German *koagulation*. There are at least two natural forms of vitamin K. One, called vitamin K_1 or phylloquinone, has four isoprene units, the first being unsaturated, attached to a methylnaphthoquinone ring. The other forms, called menaquinones, have a variable number of unsaturated isoprene units. The dietary sources of vitamin K_1 include green leafy vegetables, such as spinach, broccoli, and cabbage. Vitamin K_1 is also found in some (canola, soybean, olive) but not all (corn, peanut) vegetable oils. The menaquinones are primarily bacterial in origin.

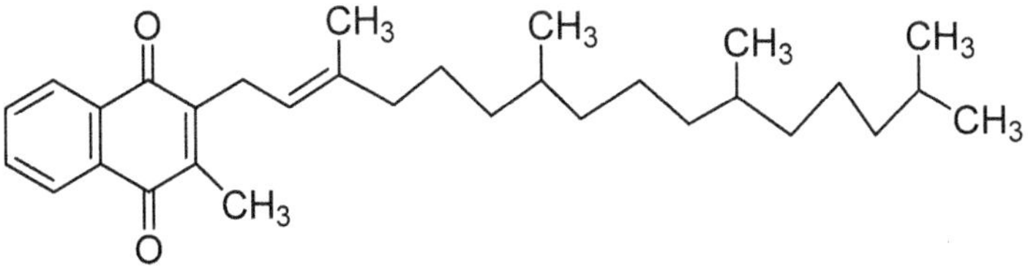

vitamin K₁

Vitamin K_1 is a cofactor for γ-glutamyl carboxylase. This enzyme is involved in both calcium deposition in bones and in the carboxylation of prothrombin. The latter function is presumably the cause of K_1 deficiency in certain newborns that can cause excessive bleeding. The deficiency occurs in newborns because of the low transplacental transfer of K_1 and the relatively low level in milk.

10.1.2 OTHER ESSENTIAL NUTRIENTS AND COFACTORS

10.1.2.1 BIOTIN

Biotin, sometimes called vitamin B_7 or H, is present in most foods, but in very small amounts. The richest natural sources are "royal jelly," which is produced by honeybees and induces the reproductive ability of queen bees, and brewer's yeast. Milk, liver, and egg yolks are the most important sources in the human diet. Another likely source is absorption in the gut of biotin produced by microorganisms. Biotin is strongly bound by *avidin*, a protein found in egg white, and biotin deficiency can be produced by use of avidin to remove biotin. Biotin is sensitive to oxidation and its level is reduced by many types of food processing. Biotin is a coenzyme for a family of enzymes that catalyze carboxylation, decarboxylation, and transcarboxylation. One of these enzymes, pyruvate carboxylase, is critical in carbohydrate biosynthesis.

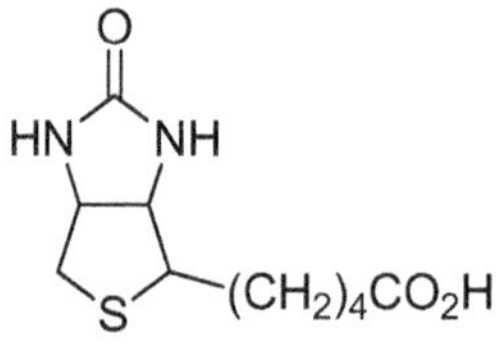

biotin

10.1.2.2 FOLIC ACID

The search for folic acid, also called vitamin B_9, was initiated by Lucy Wills, an English pathologist. In 1930, Margaret Balfour, head of the Indian Medical Service, asked Dr. Wills to investigate a form of anemia that afflicted pregnant textile workers. Dr. Wills found that Marmite, a spread prepared from yeast, was an effective treatment. Several research groups isolated a factor having anti-anemia activity from sources such as liver and spinach leaves. Sometime later, Robert Stokstad, working at the Lederle Labs in the United States, determined the structure as folic acid. Folate deficiency is particular harmful in pregnant women because it can lead to neurological defects in the child. Folic acid plays a role in biosynthesis of two very important molecules, the amino acid methionine and the nucleic acid base thymine.[17]

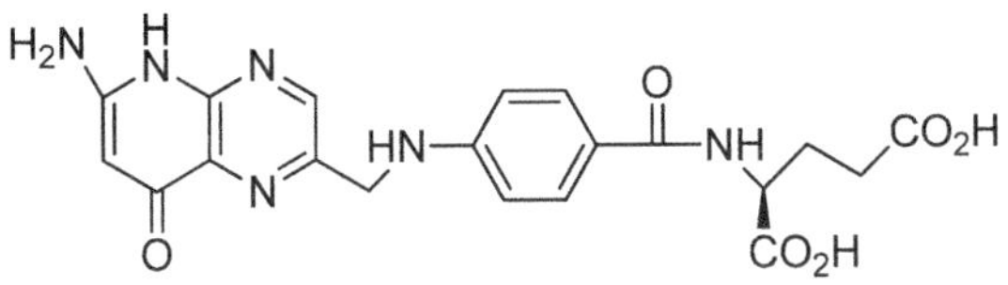

folic acid

10.1.2.3 LIPOIC ACID

Lipoic acid is involved in acyl transfer reactions in which it undergoes reversible reduction to the corresponding dithiol. It is involved in the biosynthesis of acetyl coenzyme A from pyruvate. Animals, including humans, can synthesize lipoic acid and no external source is required. Thus, it does not meet the definition of a vitamin.

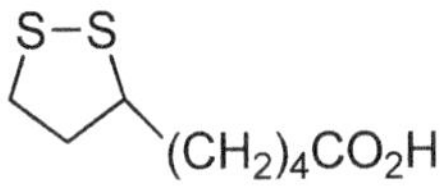

lipoic acid

10.1.2.4 PANTOTHENIC ACID

Pantothenic acid, sometimes called vitamin B_5, was isolated from yeast and a variety of other sources by Roger J. Williams, a biochemist at the University of Texas in Austin. It is a structural component of coenzyme A, which is a critical component in metabolism of fatty acids, carbohydrates, and amino acids. Pantothenic is widely distributed in foods and is therefore not associated with any particular deficiency disease.

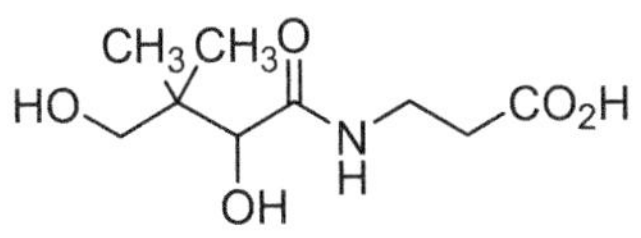

pantothenic acid

Pantothenic acid can be synthesized from pantenolactone. The crucial step is enantioselective reduction of the oxidation product.[18]

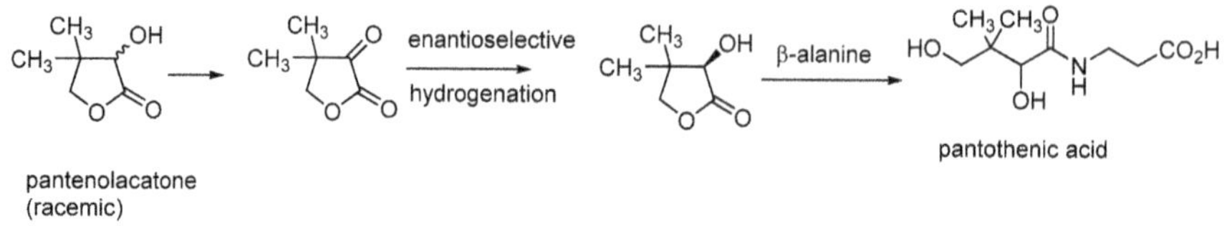

pantenolacatone
(racemic)

Chart 10.1 gives a timeline summarizing the discovery of the vitamins.

CHART 10.1 TIMELINE OF VITAMIN DISCOVERY

1905, Chritiaan Eijkman and William Fletcher observed that unpolished rice prevented beriberi.

1912–1914, Elmer V. McCollum and M. Davis discovered fat soluble factor A (vitamin A).

1912, Casimir Funk isolated vitamin B_1, thiamine from rice hulls.

1922, Edward Mellanby discovered vitamin D.

1922, Herbert Evans and Katharine S. Bishop discovered vitamin E.

1928–1932, Vitamin C was isolated and structurally characterized.

1933, Several groups identified riboflavin as vitamin B_2.

1933, Lucy Wills discovered folic acid.

1933, Vitamin D was first added to milk in the United States.

1937, Conrad Elvehjem identified vitamin B_3, niacin.

1938, Paul Gyorgy isolated vitamin B_6.

1941, B Vitamins were added to flour in the United States.

1996, Folic acid supplementation was mandated in the United States.

10.2   THE USE AND VALUE OF VITAMIN SUPPLEMENTS

As knowledge about vitamins and vitamin deficiencies developed, the medical community gradually came to recognize the importance of nutrition and diet. WWI brought on food shortages throughout Europe. Flour was rationed in Germany beginning in 1915. During the winter of 1915–1916, turnips were a prime food source. The German U-boat blockade of Britain also reduced supplies of imported foods there. One of the centers of nutritional research in Britain was the Lister Institute in London. Following WWI, Harriette Chick and Elsie Dalyell travelled to Vienna to investigate malnutrition in post-war

Austria.[g] The dominant medical opinion at the time was that diseases such as rickets were caused by germs, the victims being predisposed by poor nutrition. Chick was convinced that nutritional deficiencies were the direct cause. She received the support of Professor K. E. Wenkebach, a Dutch cardiologist who was highly respected in Vienna. Working with Clemens Baron von Pirquet, Director of the Kinderklinik in Vienna, she was able to demonstrate the beneficial effect of lemon juice in infantile scurvy and butterfat in treatment of rickets. In 1920, the Medical Research Council approved a study, supported by the International Red Cross, which demonstrated the value of full-cream milk powder and cod liver oil in treatment of rickets.

Nutritional deficiencies were prevalent in England, especially among the poor. As late as 1935, 60% of potential military recruits failed to meet standards. Among the leaders in promoting nutrition was John Boyd Orr, a Scottish physician, who, after WWI, became director of the Rowert Institute and initiated a research effort in nutrition. His book *Food, Health and Income*, published in 1936 called attention to the role of poverty and malnutrition in health. He was awarded the Nobel Peace Prize in 1949. Elsie Widdowson and Robert A. McCance,[h] undertook comprehensive studies in food analysis that resulted in the publication in 1940 of *The Chemical Composition of Foods*. When WWII broke out, a War Rations diet was developed to insure basic nutrition. Among other actions taken, calcium phosphate was added to flour and mothers of new infants received rations of orange juice, milk, and cod liver oil.

Vitamin use increased greatly in the United States during and immediately after WWII, as vitamin supplements became readily available and the benefits were recognized. Many people in the United States and the rest of the developed world now regularly use vitamin supplements, including multivitamin mixtures.

---

[g]Harriette Chick was born in 1875. She enrolled in the University College of London in 1894 and eventually earned a doctorate in bacteriology. She received an appointment at the Lister Institute of Preventive Medicine, a private research institution originally funded by the Guinness family of brewing fame. She was engaged in research on the denaturation of proteins. During WWI, she worked on nutritional requirements of military rations. After the completion of the studies on rickets in Vienna, she continued to work at the Lister Institute for more than 50 years and was a leading figure in nutritional research. She died in 1977 at the age of 102. Elsie Dalyell was born in Sydney Australia in 1881. She received a medical degree in 1910 and was awarded a fellowship to study at the Lister Institute. She served in the Royal Medical Corps during WWI, serving in Malta and Greece. At the completion of her studies in Vienna with Harriette Chick, she returned to Australia where she worked in the Department of Public Health.

[h]Elsie Widdowson studied chemistry at Imperial College (one of only 3 women in her class of 100) and received a Ph. D. in 1928. She undertook extensive nutritional studies in collaboration with Robert A. McCance at the Medical Research Council Laboratories in Cambridge. McCance was born in Northern Ireland and served in the Royal Navy Air Service during WWI. He then studied at Cambridge, receiving a Ph. D. in biochemistry in the laboratory of F. G. Hopkins. He also received a medical degree at Kings College, London. He conducted studies on the nutritional content of food first at Kings College and then at the MRC laboratory in Cambridge. Widdowson and McCance formulated the nutritional basis of WWII food rationing in Britain, partly based on self-experimentation. After the war, Widdowson was head of Infant Nutritional Research at the Dunn Nutritional Laboratory at Cambridge. McCance undertook studies on malnutrition in Germany immediately after WWII and for a time served as director of an MRC program in infant nutrition in Kampala, Uganda.

There have been many studies aimed at determining whether this is beneficial. One retrospective study examined the "anti-oxidant" vitamins including A, C, and E. The combined results of 68 trials involving a total of 232,606 participants that met criteria for randomization, blinding, and adequate follow-up were analyzed by the technique of meta-analysis (see Section 13.1.2). Of these 21, were entirely aimed at preventive effects, while 47 were related to some specific condition, such as digestive or cardiovascular. The numbers of reported deaths in the studies were examined. The supplements had been given in widely varying doses and often in combination with one another, or with other supplements such as zinc or selenium. When the results were analyzed statistically, each of the antioxidant vitamins showed an *increase in mortality.* For β-carotene, the increased risk of mortality was 1.09, with a range of 1.06–1.13; vitamin A: 1.20 (1.12–1.29); vitamin C: 1.06 (0.99–1.14); and vitamin E 1.06 (1.02–1.10).[19]

In 2006, the NIH convened an expert panel in an attempt to assess the value of vitamin supplements.[20] They found little reliable evidence for positive effects in areas of major health concern such as cardiovascular disease and cancer. Negative effects were also limited, except for those mentioned above for specific vitamins. Why is there so little clear evidence, one way or the other? Most people probably get adequate vitamins from their diet, in which case supplements may provide no added benefit. Relatively few well-controlled placebo-based studies are available, partly because of the wide use of multivitamins and the difficulty in accurately assessing actual vitamin usage. Studies that deliberately limited vitamin intake, of course, would be unethical. The answer to the question "Are multivitamin supplements good for you?" appears to be "Nobody knows!" The most general clinical advice at the end of the 20th century was that many in the population, especially among the elderly, might be receiving less than the optimal level (as opposed to minimal level) and should take a general purpose multivitamin supplement.[21]

10.2.1 CAN THERE BE DIFFERENCES BETWEEN NATURAL AND SYNTHETIC VITAMINS?

At the molecular level, the answer to this question is absolutely "NO!" Samples of a *pure substance*, no matter what their origin, are identical. This axiom applies to all substances, including vitamins. But the emphasized words "pure substance" are critical and raise significant issues. (1) Are the substances, both natural and synthetic, really pure? (2) How about stereochemistry, is it the same in both substances? (3) The definition of *pure* excludes *anything else* being present; is this true for *both* natural and synthetic samples? (4) Are there any significant differences in physical form that might influence biological effect, that is, are the substances *functionally equivalent*? Those who argue that there are or could be differences in natural and synthetic vitamins must consider these questions. The effects of vitamins obtained from foods might very well be different than vitamins in the form of pills, not as a result of chemical composition, but because there may be differences in the physical form of the substance, or interactions with other components present in the food. A specific example of a difference between natural and synthetic sources occurs with vitamin E. The natural material is the pure *R,R,R*-enantiomer, but the synthetic material is a mixture of eight stereoisomers (see Section 10.1.1.9). There is evidence of differing bioavailability and adsorption among the stereoisomers.[22]

On the other hand, critics of synthetic vitamins sometime resort to erroneous distinctions. Most vitamins are acids or bases and thus can exist as salts. Some synthetic vitamins are supplied as salts. Vitamins taken orally are subject to the pH of the digestive tract and the protonated status of the vitamin will be controlled by the local pH, not the form in which it was ingested. A more significant factor may exist when vitamins are provided as close chemical derivatives. An example is vitamin A, retinol, which is frequently supplied as the acetate ester. This substitution requires that the ester be biologically equivalent, which is reasonable, but requires verification. Critics sometimes resort to listing the origin of synthetic vitamins as "coal tar" or "petroleum."[23] This is an invalid argument, because the origin of the synthetic material is not relevant, as long as the substance in question is *pure*. The origin of a particular sample can be relevant only if it leads to some significant impurity or difference in physical form. There is no "vital force" that distinguishes natural from synthetic.

10.2.2 MARKETING, ECONOMICS, AND REGULATION OF VITAMIN SUPPLEMENTS

The discovery of the relationship between vitamins and deficiency diseases led to efforts to improve public nutrition by use of vitamin supplements. Harry Steenbock, working under the auspices of the Wisconsin Alumni Research Foundation (see Section 10.1.1.8) promoted the irradiation of milk to increase its vitamin D content. Natural sources of vitamins such as wheat germ and cod liver oil were also promoted. Initially the medical community was skeptical. There seems to have been several elements to this skepticism. (1) As mentioned above, the medical community was slow to accept the concept that a disease could result from vitamin deficiency rather than from "germs". (2) Use of vitamin supplements probably seemed like a form of back-sliding, because the AMA had only recently won some of the battles against patent medicines (see Section 11.4.1). (3) Vitamins were a form of "self-medication" outside the control of the profession.

By the early 1940s, the vitamins had moved from being mysterious "fac-tors" present in minute amount to well-characterized compounds, most of which could be produced commercially by synthesis. The prices dropped dramatically in several cases.[24] Vitamin C from natural sources cost $200/oz in 1935 but in 1940, the synthetic material cost $2.00/oz. Similarly, vitamin B_1 dropped from $350/g in 1935 to around $1/g in 1940. Niacin dropped from $35/lb to $10/lb. During WWII, vitamin use gained momentum. Rationing had made certain foods scarce, increasing the possibility of vitamin deficiencies. Some believed that vitamins promoted efficiency and decreased absenteeism in the work force, both directly and indirectly. Major pharmaceutical companies including Abbott, Merck, Pfizer, Squibb, and Upjohn became producers of vitamin supplements. Most vitamins were initially sold in drug stores, so pharmacists also became interested in promotion of vitamins. During the 1940s and 1950s, extensive ad-vertising appeared. It was usually directed at mothers and aimed to convince them of the benefits of vitamin use for their children's health. The ads promoted "sturdy young bodies" but also warned of the consequences of neglect, such as the specter of weak or bent bones. Pharmacists and pharmaceutical compa-nies were not the only promoter of vitamin's benefits. Food producers, such as the Pineapple Producers Cooperative, promoted the vitamin content of their products.

In 1939, the Kroger grocery chain began to sell a combination vitamin called A–B–D–G.[i] The drug store industry objected and filed various suits claiming that vitamins were drugs and should be sold only by licensed pharmacists. Most of the cases were never clearly resolved, but gradually a distinction be-tween "prescribed for a particular condition" and "not for treatment of illness" evolved. The share of vitamin sales in drug stores dropped throughout the 1960s and 1970s, from about 90% in 1950 to about 33% in 1980. One of the lon-gest commercial successes in marketing of vitamins was the *One-A-Day* brand of Miles Laboratory. Miles was a maker of proprietary products, most nota-bly *Alka-Seltzer*. *One-A-Day* was launched under the leadership of Walter A Compton, a grandson of one of the company founders, who had a chemistry degree from Princeton and an MD (1937) from Harvard. Compton worked to build a reputation of reliability. He coordinated with the FDA on labeling and packaging information, although not required to do so. The *One-A-Day* brand was aggressively advertised on the basis of "extra insurance" and "year-round" usages. Another successful brand of multiple vitamins was *Unicap*, produced by the Upjohn Company. At the time, the FDA's general position was that vitamin supplements were unnecessary. Miles Laboratory and other major vitamin mak-ers founded the Nutritional Vitamins Foundation in 1946, which supported re-search in nutrition and the distribution of educational information. It is estimated

---

[i]Vitamin G was an alternative name for vitamin B_2, riboflavin.

that between 33 and 50% of American adults currently use vitamin supplements. The sales in 2005 are believed to have been about $23 billion.

10.2.3 TOO MUCH OF A GOOD THING? CAN VITAMINS USE HAVE ADVERSE CONSEQUENCES?

The laws on vitamin content in the United States do not require that the content be stated exactly, only that the concentration meets the specified minimum level. It is therefore conceivable that the vitamin content is actually higher than stated. Overdose, however, is more likely to result from use of several different products, in addition to the vitamins present in enriched foods. Upper recommended daily limits have been adopted for seven of the 13 essential vitamins, as shown in Table 10.1.

TABLE 10.1   Recommended Upper Daily Limits for Certain Vitamins.

Vitamin	Dose
A	3000 µg
E	1000 mg
B_6	100 mg
C	2 g
D	50 µg
Folic acid	1000 µg
Niacin	35 mg

In a few cases, there are indications that excessively high levels of vitamin intake can be harmful. There is some evidence that high (>1500–3000 µg/day) vitamin A intake level may increase the risk of hip fracture in post-menopausal women. The risk was reduced in women using estrogen replacement therapy. This effect might be caused by antagonism of vitamin A toward vitamin D in maintaining calcium level.[25] Folic acid represents an example of the potential benefits and risks of mandatory supplementation. Mandatory supplementation was introduced in the United States and Canada in the 1990s as a result of evidence that folate deficiency led to a neural tube birth defect. As shown in Figure 10.1, there has been an upward shift in the serum levels of folate. Smith and co-workers, considering the advisability of mandatory folate supplementation, have discussed some of the possible mechanisms by which increased levels might have adverse effects.[26] The supplement is in the form of the monoglutamate, folic acid. Most dietary folate is in the form of polyglutamates that may have different rates or levels of absorption. The active form of the coenzyme is the reduced tetrahydro derivative. It has been suggested that increasing folate levels may activate certain enzymes, while inhibiting others. There are also some data suggesting that there is an optimum folate level. In the case of colorectal cancer, it was found that incidence increased both below and above a certain level. There is also evidence that the effect may depend upon individual genetic factors. It remains an open question as to whether supplemental folate has any deleterious effects.

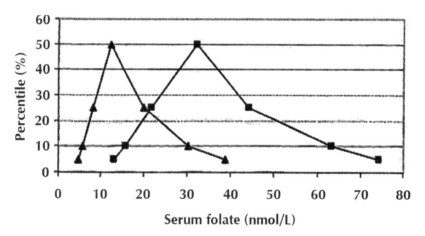

FIGURE 10.1   Upward shift in serum folate levels between 1988–1994 (triangles) and 1999–2000 (squares). (From C. M. Pfeiffer, S. P. Caudill, E. W. Gunter, J. Osterloh and E. J. Sampson, 2005. *Am. J. Clin. Nutr.* **82**, 442–450 by permission of Wiley. © 2005, American Society of Clinical Nutrition.)

10.3   DIETARY SUPPLEMENTS

10.3.1   BAD LAW–BAD MEDICINE?

In the 1960s, the FDA attempted to impose labeling on vitamin supplements that read "Vitamins and minerals are supplied in abundant amounts by the foods we eat. The Food and Nutrition Board of the National Research Council recommends that dietary needs be satisfied by foods. Except for persons with special medical needs, there is no scientific basis for recommending routine use of dietary supplements." This labeling was never imposed. The FDA tried to tighten regulation of vitamins again in the 1970s based on the newly established *Recommended Daily Allowances.* The basic proposal was to regulate products that contained more that 150% of the RDA. The FDA believed that "a typical well-balanced American diet does not lead to deficiencies that need to be made up by taking dietary supplements, vitamins, etc."[27] Opponents objected to regulation and government interference and were afraid that regulation might lead to the requirement for prescription of vitamins. The National Health Federation, an affiliation of the producers of nutritional supplements, was particularly active in generating opposition. The public outcry was heard in Congress and in 1974 Sen. William Proxmire (D, WI) offered legislation to the affect: The FDA…"shall not limit the potency, number, combination, amount or variety of any synthetic or natural vitamin, mineral or other nutritional substance…" This amendment was not accepted in the House, but in 1976, after additional public and press engagement, the language adopted was "…may not classify any natural or synthetic vitamin or mineral (or combination thereof) as a drug solely because it exceeds the level of potency which…is nutritionally rational or useful." In the 1990s, FDA Commissioner David Kessler again raised the issue of advertisements and claims for vitamins and supplements. The health supplement industry again mounted opposition. The FDA lost again. This time the Hatch Amendment, known as the "Health Freedom Act" exempted supplements from FDA regulation if they occurred as natural substances, *even if the product is made synthetically or by using a genetically engineered microorganism.* The Dietary Supplement Health and Education Act of 1994 includes the following provisions:

1. Dietary supplements comprised of herbals and botanicals are not required to demonstrate either safety or efficacy.
2. New products require no testing for safety. Notice of a new product to be marketed gives the FDA 75 days to prove the product is unsafe and this can be done only through court action.
3. Dietary supplements can make claims of *functional benefits* such as "strengthens the immune system" or "enhances energy" without any evidence.
4. An all-purpose disclaimer is required. "This statement has not been evaluated by the Food and Drug Administration. This product is not intended to diagnose, treat, cure, or prevent any disease."

Under this law, Congress exempted the makers and sellers of supplements from any obligation to show that the product is safe, effective, or even for that matter, "natural." Dietary supplements need not meet any specific requirements of either safety or efficacy. Herbal and dietary supplements require no studies of metabolism or long term toxicology. This means that any toxic effects will be discovered only as the result of use. Manufacturers of dietary, nutritional, and herbal supplements are not required to report adverse events. The law established no criteria of purity or quality and prohibited only products with "significant or unreasonable risk." Thus, anyone who purchases a "nutritional supplement" assumes the risk that the product is safe, let alone effective. These regulations, and other factors, have given rise to extensive marketing and use of any number of materials as dietary or nutritional supplements. By 2000, it was estimated that 29,000 different products were on the market. Some specific examples of nutritional supplements are given below. Dietary and nutritional supplements cannot make claims for treatment of a disease or its symptoms. However, "health claims" are permitted and frequently used in advertising for supplements.

The legislation exempting dietary supplements from purity and efficacy standards assigned to the Office of Dietary Supplements of the National Institutes of Health the responsibility for developing analytical methods and reference standards for evaluation of dietary supplements. There are many practical problems that need to be solved. There is considerable variation in the content of different samples. This can result from different plant varieties, climate, parts of the plant harvested and differences in processing. In some cases, the active ingredient is unknown. There is a lack of standard methods for analysis. Compared to synthetic drugs, which are subject to quality control and usually consist of a single pure substance, it is much more difficult to determine the quality and purity of herbal preparations. Many studies have shown wide variation in the contents of materials sold under identical or similar labels. Adulteration by other substances is a common problem. Because there is no requirement for proof of efficacy before an herbal or dietary supplement is marketed, there is often little basis for knowing whether the product is beneficial. Many of the trials that have been reported fall short of the standards of control, blinding, and so on, that are applied to synthetic drugs.[28] The lack of information about pharmacology and mechanism of action also makes it more difficult to recognize the possibilities for interaction with other medicinal substances, including prescription drugs.

Use of herbal and dietary supplements is quite wide spread. A 2002 phone survey of 2743 people throughout the United States found that about 73% used a vitamin, multivitamin, or other supplement. Of these, about 85% reported the use of multivitamins. About 42% reported using a non-vitamin supplement. The most commonly used non-vitamin supplements were echinacea (19.5%), garlic (16.6%), ginko (14.6%), ginger (11.7%), and glucosamine (10.9%). Respondents were also asked to report if they had had adverse reactions that they attributed to vitamins or supplements. About 4% reported such events, 42% of which applied to non-vitamin supplements. About half consulted a doctor about the adverse effect and 12% reported going to an emergency room. The products most commonly associated with adverse events were xenadrine, ginko, metabolife, ginseng, and St. John's wort.[29]

One type of supplement that is fairly widely used are those intended for weight loss. A recent survey indicated usage level of about 45% among women and as high as 50% in women aged 25–34. Half of those surveyed believed that supplements have been approved for safety by the FDA and a third believed that they were safer than OTC or prescriptions medicines.[30] Some of these supplements contained ephedrine and/or caffeine as the principal ingredients. Various studies indicated a 2–3.6 increased risk of side effects of several types, including agitation, anxiety, sweating, tremors, insomnia, hyperactivity, nausea, gastrointestinal upset, irregular heartbeat, and hypertension.[31] The FDA removed ephedrine-based products from the market in 2004.

There is a category of drugs, *botanical drugs*, established by the FDA. These drugs must meet the same criteria of safety and efficacy as synthetic drugs, but may contain several components. One example is *psylium*, which is approved as a source of fiber. One of the types of evidence that can be submitted in support of botanical drugs is prior use, as in traditional medicines. Canadian and European regulatory agencies also have categories for "natural health products" that take extended history of use into account.[32] The increasing role of supplements is also reflected in the *United States Pharmacopeia and National Formulary*. This publication dates to 1820 and is intended to provide information about reliable drugs. Although produced by a nongovernmental agency, it was recognized by both the 1906 and 1938 Pure Food and Drug laws as a source of standards of drug quality and purity. The revisions are currently made annually. Although it originally contained many botanical preparations, the number gradually dropped as botanical preparations were replaced by synthetic drugs. In 1995, the sponsoring United States Pharmacopeial Convention authorized the preparation of purity standards and tests for dietary supplements. In 2005, 35 botanical preparations were listed. The compilation for 2010 contains entries pertaining to botanical preparations under such titles as "garlic," "milk thistle," "saw palmetto," and "St. John's wort."[33]

10.3.2 EXAMPLES OF COMPOUNDS FOUND IN DIETARY SUPPLEMENTS

Many individual chemical compounds have been identified in dietary and nutritional supplements. They can be grouped into several large categories, such as polyenes, flavanoids, polyphenols, and unsaturated fatty acids. Many compounds have also been tested in *in vitro* or animal models to demonstrate potentially beneficial pharmacological activity. Among the types of activity that can be demonstrated are anti-oxidant, anti-inflammatory, cell cycle regulation, apoptosis, and antiangiogenic. Each of these activities, in turn, can be associated with potentially beneficial effects in prevention, or even treatment, of diseases. There is a final step that must be taken, however, to demonstrate that these compounds or supplements containing them, have beneficial effects. The final test is randomized, controlled studies showing beneficial outcome such as decreased incidence or mortality associated with the supplements. Because these kinds of data are not required to market dietary or nutritional supplements, they are seldom available. Some illustrative structures are gathered in Scheme 10.1.

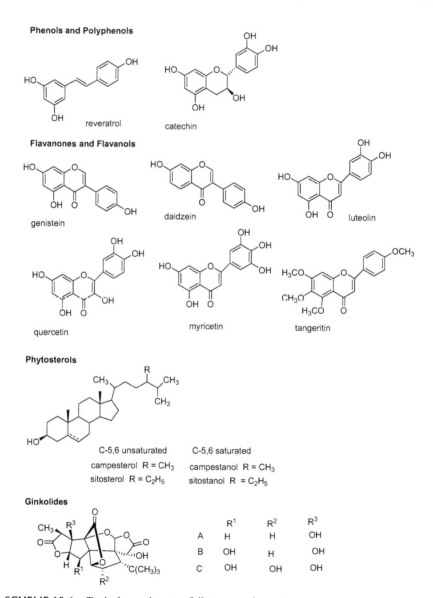

SCHEME 10.1   Typical constituents of dietary supplements.

10.3.3 EXAMPLES OF ADVERSE EFFECTS FROM DIETARY SUPPLEMENTS

Several examples of severe adverse effects due to natural supplements are outlined in the following paragraphs. Of course, there are also examples of adverse effects to FDA-approved drugs, as will be discussed in Section 11.4.2.6.

10.3.3.1 RENAL FAILURE FROM MISIDENTIFIED HERBS

In the 1990s, a cluster of renal failures occurred in Belgium and France, associated with women using a Chinese herbal preparation for weight loss. Eventually, about 100 patients were involved. They were treated for kidney failure by dialysis and/or kidney transplant. Investigation of the contents of the particular preparations revealed that a marker component of the supposed plant material from *Stephania tetrandra* (tetrandrine) was absent, suggesting that some other plant had been used. It was determined that the plant used was *Aristolocholia fangchi*, which contains a nephrotoxin and carcinogen, aristocholic acid. The names of the two plants in Chinese are nearly identical and it is suspected that the toxic plant was inadvertently used.[34]

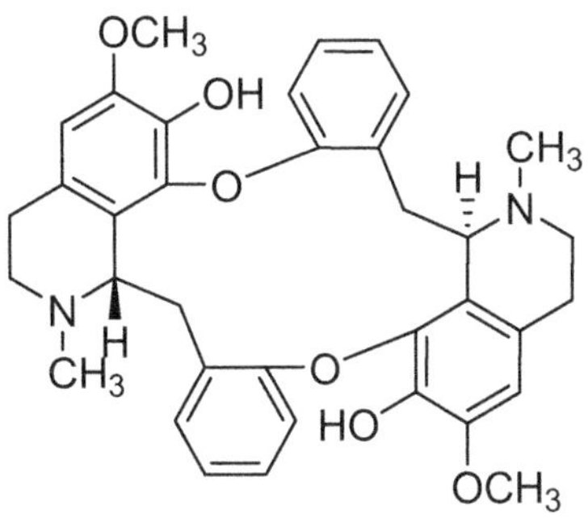

tetrandrine

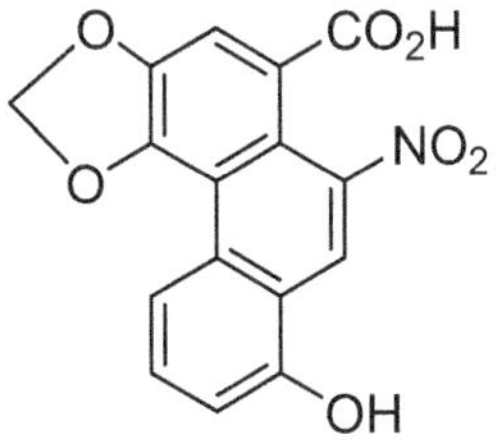

aristocholic acid

10.3.3.2 L-TRYPTOPHAN AND EOSINOPHILIA–MYALGIA SYNDROME

In 1989, physicians in New Mexico noted and reported several cases of elevated white count with muscle pain (eosinophilia–myalgia syndrome) and associated it with use of the amino acid L-tryptophan as a dietary supplement. In November, the CDC issued a warning and advised consumers to discontinue use of L-tryptophan. By 1993, a total of about 1500 cases and 37 deaths had been reported to CDC. There were six producers of L-tryptophan at the time, all in Japan. Investigation showed that only lots from Showa Denka manufactured after 1988 were associated with the syndrome. Showa Denka manufactured the amino acid using a recombinant strain of *Bacillus amyloliquefaciens* and had modified their process just before the outbreak. They had also modified the purification process, reducing the amount of adsorbent charcoal used. Although the material was 99.6% pure, it contained numerous minor impurities, one of which was evidently the cause of the syndrome.[35] L-Tryptophan was removed from the market, despite the evidence that an impurity was the cause of the toxicity. It has been replaced by 5-hydroxytryptophan, a compound one step further down the metabolic chain in the production of the neurotransmitter serotonin.[36]

10.3.3.3   PENNYROYAL TEA

Pennyroyal is an herbal supplement that can be used as a tea or oil. It is made from plants in the mint family and contains pulegone as a major component. It is reputed to be an abortifacient agent, although there are no scientific studies supporting this belief. The pulegone is metabolized to menthofuran, which is believed to be the toxic agent. Fatalities have been reported for both intentional and accidental ingestion of pennyroyal preparations.[37]

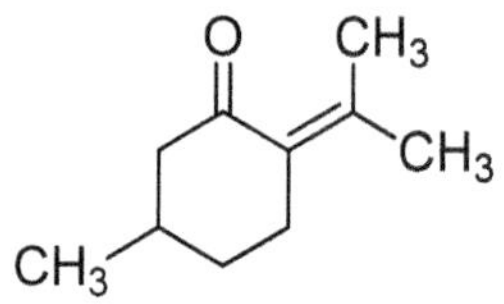

pulegone 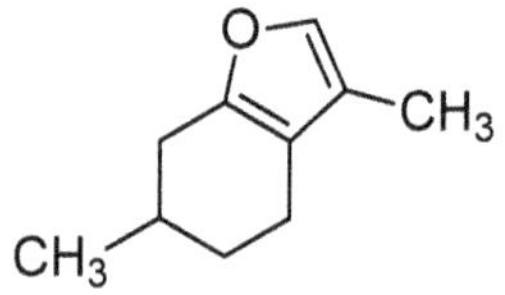   menthofuran

10.4   NUTRITIONAL CONSEQUENCES OF FOOD PRODUCTION AND PROCESSING

The availability of vitamins and nutrients in foods is a complicated matter. There are two distinct aspects that need consideration; one is *content* and the other is *bioavailability*. Content pertains to how much of the substance is present and bioavailability is the availability for adsorption. Bioavailability can be strongly influenced by other materials present and the physical form of the food. Many factors influence the vitamin and nutrient content of any particular food sample. These include the particular variety used and growing conditions, the stage of harvest, the length and nature of storage, the degree of susceptibility of the nutrient to degradation either by enzymes in the plant or by processes such as washing, chopping, canning, or freezing.

Vitamin and nutrient content is also determined by the extent of processing and cooking. Much of what has been learned corresponds to what might be called chemical "common sense." Several fundamental chemical factors are involved. One is heat; some vitamins and antioxidants are destroyed by heating. Another is cooking in water. This tends to reduce the amount of water-soluble vitamins, because they are leached out. Exposure to oxygen depends on length of storage and nature of packaging. Oxygen sensitive materials, notably vitamin C, decrease on extended exposure to air. Even physical slicing can have an effect because it breaks down compartmentalization and, in general, accelerates the action of internal enzymes that are redistributed in the process. Physical damage in harvest and shipping can have the same effects. In some cases, processing increases bioavailability of nutrients. For example, carotenoids levels tend to increase on cooking, presumably the result of release from more complex structures.[38]

10.4.1 EFFECTS OF FOOD PROCESSING

To what extent do preservation, storage, and/or cooking lead to depletion of nutritional value? There are several aspects of food production and processing that affect nutrient level and food quality in general. An obvious one is contamination by pathogenic organisms as a result of poor handling or spoilage. Another is the level of the major classes of food components, proteins, carbohydrates, and fats, as well as vitamins and other nutrients. Another aspect of quality is the presence of preservatives or residues from pesticides (see Section 8.4.2). As the role of vitamins in nutrition became known, several strategies for improved nutrition came into play. Vitamin supplements (pills, etc.) or fortification (addition to foods) have been used to increase the intake of vitamins. Education has been aimed at improved diet and the overall level of nutrition. Methods of food processing and cooking have been evaluated.

By the late 1800s, several kinds of food processing had been introduced that were potentially detrimental to the nutritional content of the diets of many people. In the mid-1800s, especially in England, a preference for white bread from refined flour arose. The bran was removed and the grain pulverized to a very fine powder that was sieved. Most of the germ of the wheat was removed by the sieving process and used in animal feed. Finally, the flour was bleached with SO_2, O_3, or Cl_2, eliminating other nutrients, including any remaining vitamin A. The baking of bread was also modified by using baking powder rather than yeast. In some cases, bread was raised using CO_2, giving especially light bread. The result of these processes was the reduction of the content of the bread to almost entirely carbohydrate and a little fat.

There have been several comparative studies of nutrient concentration after processing. Ascorbic acid can be analyzed relatively accurately and it is one of the more sensitive nutrients both to extraction by water and oxidation so it serves as a useful marker for estimating nutrient loss. Figure 10.2 shows the extent of ascorbic acid loss that occurs by cooking spinach under several conditions.[39]

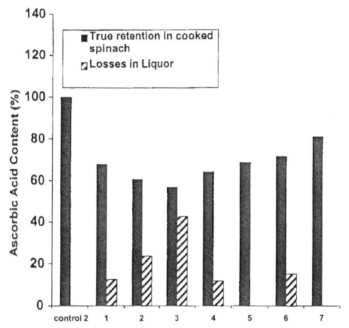

FIGURE 10.2   Retention of ascorbic acid in spinach after several cooking regimes: (1) boiling in water (1:1), 10 min; (2) boiling in water (2:1), 10 min; (3) boiling in water (5:1), 10 min; (4) boiling and frying in water (2:1), 12 min; (5) stir-fry, 10 min; (6) pressure cooker with water (2:1) 6 min; and (7) microwave in water. (From Davey, M. W.; Van Montagu, M.; Inze, D.; Sammartin, M.; Kanellis, A.; Smirnoff, N.; Benzie, I. J. J.; Strain, J. J.; Favell, D.; Fletcher, J. *J. Sci. Food Agric.* **2000**, *80*, 825–860 by permission of Wiley. © 2000, Society of Chemical Industry.)

A study at the Unilever Company compared ascorbic content of several vegetables immediately after harvest, quick freezing and under conditions typical for market or supermarket distribution. The study indicated that ascorbic acid content in quick-frozen vegetables was 70–120% of fresh-picked and that stored vegetables (either chilled or ambient temperature) retained 40–60% of ascorbic acid over 7 days.[40] Spinach was an exception, losing nearly all its vitamin C. The results are given in Figure 10.3.

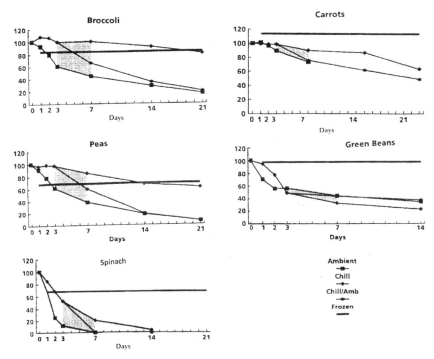

FIGURE 10.3 Comparison of ascorbic acid content of several vegetables after quick-freezing and chilled or ambient temperature storage. Shaded area corresponds to approximate delivery time from harvest to market. (From Favell, D. J. *Food Chem.* **1998**, *62*, 59–64 by permission of Elsevier.)

Puupponen-Pimia and co-workers examined the effect of blanching and freezing.[41] Most of the losses occurred during blanching, presumably the result of either leaching or oxidation. Folate was the most sensitive, with losses in the range of 75%. Vitamin C and other anti-oxidants were reduced on the order of 20–30%. Carotenoids were not strongly affected. Of the long-term storage methods, freezing appears to retain the highest levels of vitamins and antioxidants, which is not surprising from a chemical point of view. Dehydration to provide material for reconstitution appears to lead to the largest losses.

10.4.2 ANALYSIS OF VITAMIN CONTENT

Evaluation of the vitamin content of a food or supplement, of course, depends on an analytical procedure. Most of the early analyses were based on rate of growth of microbiological cultures and these remain in place for several of the vitamins. These assays have relatively large error ranges (e.g., ±20%). More accurate analyses are available, but their introduction into official methods has been rather slow. Liquid chromatography is an excellent method for several of the water-soluble B vitamins. Isotope dilution analysis is also very accurate. The recent interest in vitamin D levels (see Section 10.1.1.8) led to efforts to develop standards and compare methods of analysis. Initial results showed unacceptably high standard deviations (35–50%), but careful investigation of the causes reduced this to 7–12%. Nevertheless, the existing methods are cumbersome and expensive.[42-]

10.4.3   ARE ORGANICALLY GROWN FOODS MORE NUTRITIOUS?

One of the reactions to extensive use of chemical insecticides and herbicides in food production has been the development of organically grown produce. The assumption is that organically grown foods are less likely to have residues of pesticides. But are they more nutritious? Is there any difference in vitamin or other nutritional components? The best answer is probably "Nobody Knows." There have been very few studies that have made systematic comparisons.[43] There is no theoretical basis for believing that organically grown foods should be more nutritious.

10.4.4   CAN NUTRITIONAL CONTENT BE IMPROVED BY GENETIC MODIFICATION?

Agriculturalists practiced cross-breeding for centuries without understanding its genetic basis. Rather, desirable qualities were recognized by trial and error and retained and propagated. This process accelerated rapidly in the 20th century as its genetic basis became clearer. Explicit science-based efforts to *increase the nutrient content* are of relatively recent origin. One such method is traditional cross-breeding (hybridization). This is relatively slow because several generations are required to optimize the overall changes, because the combination of individual genes in the hybrid is more or less random. The techniques of molecular biology offer the prospect of greatly accelerating the process. It is possible to identify the genes associated with desirable traits and focus study on lines containing the most promising combination of genes.

Genetic modification, in which a specific set of genes is introduced with the intent of improving the nutrient content of the plant, is also being applied. One example of a food that might be genetically altered is rice. We have already seen how removal of rice husk during processing was a factor in vitamin B_1 deficiency (see Section 10.1.1.2). Polished rice is also low in vitamin A, and people who obtain most of their caloric intake from polished rice can be vitamin A deficient. One solution is a more varied diet. It is also possible to produce "golden rice" that has been genetically modified to include the vitamin A precursor carotene. Other examples are the "high oleic oils" that can be produced by modifying the relative proportions of the fatty acids in oils. Modified forms of both canola and soy bean oil are available. These modified oils are more stable to storage and thermal treatment, as is the case with partially hydrogenated vegetable oils (see Section 9.3.2). However, they do not have the *trans* fats present in the latter. We will discuss genetically modified crops further in Section 21.2.4.

KEYWORDS

- vitamins
- recommended daily allowance (RDA)
- co-factors and coenzymes
- antioxidant
- dietary supplements
- effects of cooking and freezing

BIBLIOGRAPHY

Gratzer, W. *Terrors of the Table, the Curious History of Nutrition.* Oxford University Press: Oxford, UK, 2005.
Ball, G. F. M. *Vitamins in Foods: Analysis, Bioavailability and Stability.* Taylor Francis: Boca Raton, FL, 2006.
Pollan, M. *The Omnivore's Dilemma.* Penguin Press: New York, 2006.
Frankenburg, F. R. *Vitamin Discoveries and Disasters: History, Science and Controversies.* Praeger: Denver, CO, 2009.

BIBLIOGRAPHY BY SECTIONS

Section 10.1.1.2: Carpenter, K. J. *Bereiberi, White Rice and Vitamin B: A Disease, a Cause and a Cure.* University of California Press: Berkeley, CA, 2000.
Section 10.1.1.5: Mooney, S.; Leuendorf, J.-E.; Hendrickson, C.; Hellmann, H. *Molecules* **2009,** *14,* 329–351; Hellmann, H.; Mooney, S. *Molecules* **2010,** *15,* 442–459.

Section 10.1.1.6: Pacholok, S.; Stuart, J. J. *Could it be B₁₂? The Epidemic of Misdiagnosis.* Quill Driver Books: Sanger, CA, 2005.

Section 10.1.1.7: Carpenter, K. J. *Ann. Nutr. Metab.* **2012**, *61*, 259–264.

Section 10.1.1.8: Bjorn, L. O. *Photobiology* **2002**, 265–280; Wolf, G. *J. Nutr.* **2004**, *134*, 1299–1302; De Luca, H. F. In *Vitamin D*, 2nd ed.; Feldman, D.; Pike, J. W.; Glorieux, F., Eds.; Elsevier, 2005; pp 3–12; Holick, M. F. *The Vitamin D Solution: A 3-Step Strategy to Cure Our Most Common Health Problems.* Hudson Street Press: New York, 2010.

Section 10.1.1.9: Ricciarelli, R.; Zingg, J.-M.; Azzi, A. *Biol. Chem.* **2002**, *383*, 457–465; Yang, B. *Lipid Technol.* **2003**, *15*, 125–130; Jiang, Q. In *The Encyclopedia of Vitamin E*; Preedy, V. R., Watson, R. R., Eds.; AABI Publishing: Wallingford, UK, 2007; pp 807–818; Kosowski, A.; Clouatre, D. L. In *Tocotrineols*; Watson, R. R.; Preedy, V. R., Eds.; OACS Press: Urbana, IL, 2009; pp 61–75.

Section 10.1.2.4: Bramley, P. M.; Kafatos, E. A.; Kelly, F. J.; Manios, Y.; Roxborough, H. E.; Schuch, W.; Sheehy, P. J. A.; Wagner, K.-H. *J. Sci. Food Agric.* **2000**, *80*, 913–938.

Section 10.2: Apple, R. D. *Vitamania: Vitamins in American Culture.* Rutgers University Press: New Brunswick, NJ, 1996.

Section 10.3: Betz, J. J.; Fisher, K. D.; Saldahna, L. G.; Coates, P. M. *Anal. Bioanal. Chem.* **2007**, *389*, 19–25; van Breenen, R. B.; Fong, H. H. S.; Farnsworth, N. R. *Am. J. Clin. Nutr.* **2008**, *87*(Suppl.), 509S–513S; Fu, P. P.; Chiang, H.-M.; Xia, Q.; Chen, T.; Chen, B. H.; Yin, J.-J.; Wen, K.-C.; Lin, G.; Yu, H. *J. Environ. Sci. Health, Part C*, **2009**, *27*, 91–119; Rogovik, A. L.; Vohra, S.; Goldman, R. D. *Ann. Pharmacother.* **2010**, *44*, 311–324.

Section 10.4: Breene, W. M. *J. Foodservice Syst.* **1994**, *8*, 1–45.

REFERENCES

1. Sebrell, W. H. *J. Am. Med. Assoc.* **1940**, *115*, 851–854.
2. The isolation and identification of vitamin B₂ is summarized by Booher, L. E. *J. Am. Med. Assoc.* **1938**, *110*, 1105–1111.
3. Bollet, A. J. *Yale J. Biol. Med.* **1992**, *65*, 211–221.
4. Martens, J.-H.; Barg, H.; Warren, M. J.; Jahn, D. *Appl. Microbiol. Biotechnol.* **2002**, *58*, 275–285.
5. Mozolowski, W. *Nature* **1939**, *143*, 121.
6. Guy, R. A. *Am. J. Dis. Child.* **1926**, *26*, 112–116; Rajakumar, K. *Pediatrics* **2003**, *112*, e133–e135.
7. Hess, A. F.; Weinstock, M. *J. Biol. Chem.* **1924**, *62*, 301–313.
8. Steenbock, H.; Block, A. *J. Biol. Chem.* **1924**, *64*, 263–298.
9. Rosenheim, S. O.; Webster, T. A. *Biochem. J.* **1926**, *19*, 537–544.
10. Hypponen, E.; Power, C. *Am. J. Clin. Nutr.* **2007**, *85*, 860–868.
11. Maxmen, A. *The Scientist* March 1, 2012.
12. Berson, J. A. *Tetrahedron* **1992**, *46*, 3–15.
13. Ouachi, K.; Arita, M.; Kayden, H.; Hentati, F.; Hamida, M. B.; Sokol, R.; Arai, H.; Inoue, K.; Mandel, J.-L.; Koeinig, M. *Nat. Genet.* **1995**, *9*, 141–145.
14. Miller III, E. R.; Pastor-Barriuso, R.; Dalal, D.; Riemersma, R. A.; Appel, L. J.; Guallar, E. *Ann. Intern. Med.* **2005**, *142*, 37–46; Bjelakovic, G.; Nikolova, D.; Gluud, L. L.; Simonetti, R. G.; Gluud, C. *J. Am. Med. Assoc.* **2007**, *297*, 842–857.
15. Brown, B. G.; Crowley, J. *J. Am. Med. Assoc.* **2009**, *293*, 1387–1390.

16. Schmid, R.; Scalone, M. In *Comprehensive Asymmetric Catalysis*, Vol. III; Jacobsen, E. N.; Pfaltz, A.; Yamamoto, H., Eds.; Springer-Verlag: Berlin, 1999; pp 1439–1449.

17. Smith, A. D.; Kim, Y. I.; Refsum, H. *Am. J. Clin. Nutr.* **2008**, *87*, 517–533.

18. Schmid, R.; Scalone, M. In *Comprehensive Asymmetric Catalysis*; Vol. III; Jacobsen, E. N.; Pfaltz, A.; Yamamoto, H., Eds.; Springer: Berlin, 1999; pp 1439–1449.

19. Bjeladovic, G.; Nikoova, D.; Gluud, L. L.; Simonetti, R. G.; Cluud, C. *J. Am. Med. Assoc.* **2007**, *297*, 842–857.

20. NIH Panel, *Ann. Intern. Med.* **2006**, *145*, 364–371.

21. Fletcher, R. H.; Fairfield, K. M. *J. Am. Med. Assoc.* **2002**, *287*, 3127–3129.

22. Schmidt, K.; Nikoleit, D. A. *STP Pharma Pract.* **1993**, *3*, 259–270.

23. For example, see Thiel, R. J. *Med. Hypotheses* **2000**, *55*, 461–469.

24. Major, R. T. *Chem. Eng. News* **1942**, *20*, 517–523; Major, R. T. *Chem. Ind.* **1943**, *62*, 19–23.

25. Melhus, H.; Michaelsson, K.; Kindmark, A.; Bergstrom, R.; Holmberg, L.; Mallmin, H.; Wolk, A.; Ljunghall, S. *Ann. Intern. Med.* **1998**, *129*, 770–778; Feskanich, D.; Singh, V.; Willets, W. C.; Colditz, G. A. *J. Am. Med. Assoc.* **2000**, *287*, 47–54.

26. Smith, A. D.; Kim, Y.-I.; Refsum, H. *Am. J. Clin. Nutr.* **2008**, *87*, 517–533.

27. From a response by then FDA Commissioner Alexander Schmidt, see page 149, Apple, R. D. *Vitamania; Vitamins in American Culture.* Rutgers University Press: New Brunswick, NJ, 1996.

28. Wolsko, P. M.; Solondz, D. K.; Phillips, R. S.; Schachter, S. C.; Eisenberg, D. M. *Am. J. Med.* **2005**, *118*, 1087–1093; Gagnier, J. J.; De Melo, J.; Boon, H.; Rochon, P.; Bombardier, C. *Am. J. Med.* **2006**, *119*, 800 e1–11.

29. Timbo, B. B.; Ross, M. P.; McCarthy, P. V.; Lin, C.-T. J. *J. Am. Diet. Assoc.* **2006**, *106*, 364–371.

30. Pillitteri, J. L.; Shiffman, S.; Rohay, J. M.; Harkins, A. M.; Burton, S. L.; Wadden, T. A. *Obesity* **2008**, *16*, 790–796.

31. Pitler, M. H.; Schmidt, K.; Ernst, E. *Obes. Rev.* **2005**, *6*, 93–111.

32. Chen, S. T.; Dou, J.; Temple, R.; Agarwal, R.; Wu, K.-M.; Walker, S. *Nat. Biotechnol.* **2008**, *26*, 1077–1083.

33. Schiff, Jr., P. L.; Srinivasan, V. S.; Gianaspro, G. I.; Roll, D. B.; Salguerro, J.; Sharaf, M. H. M. *J. Nat. Prod.* **2006**, *69*, 464–472.

34. Vanherweghem, J.-L. *J. Altern. Complem. Med.* **1998**, *4*, 9–13; De Pierrtreux, M.; Van Damme, B.; Van der Houte, K.; Vanherweghem, J.-L. *Am. J. Kidney Dis.* **1994**, *24*, 172–180.

35. Mayeno, A. N.; Gleich, G. J. *Trends Biotechnol.* **1994**, *12*, 346–352; Kilbourne, E. M. *Epidemiol. Rev.* **1992**, *14*, 16–36; Kilbourne, E. M.; Philen, R. M.; Lamb, M. L.; Falk, H. *J. Rheumatol.* **1996**, *23*(Suppl. 46), 81–91.

36. Das, Y. T.; Bagchi, M.; Bagchi, D.; Preuss, H. G. *Toxicity Lett.* **2004**, *150*, 111–122.

37. Bakerink, J. A.; Gospe Jr., S. N.; Dimand, R. J.; Eldridge, M. W. *Pediatrics* **1996**, *98*, 944–947; Anderson, I. B.; Mullen, W. H.; Meeker, J. F.; Khojastch-Bakht, S. C.; Oishi, S.; Nelson, S. D.; Blanc, P. D. *Ann. Intern. Med.* **1996**, *124*, 726–734.

38. Granado, F.; Olmedilla, B.; Blanco, I.; Rojas-Hidalgo, E. *J. Agric. Food Chem.* **1992**, *40*, 2135–2140; Mangels, A. R.; Holden, J. M.; Beecher, G. R.; Forman, M. R.; Lanza, E. *J. Am. Diet. Assoc.* **1993**, *93*, 284–296.

39. Davey, M. W.; Van Montagu, M.; Inze, D.; Sammartin, M.; Kanellis, A.; Smirnoff, N.; Benzie, I. J. J.; Strain, J. J.; Favell, D.; Fletcher, J. *J. Sci. Food Agric.* **2000**, *80*, 825–860.

40. Favell, D. J. *Food Chem.* **1998,** *62*, 59–64.
41. Puupponen-Pimia, R.; Hakkinen, S. T.; Aami, M.; Suortti, T.; Lampi, A.-M.; Eurola, M.; Piironen, V.; Nuutila, A. M.; Oksman-Caldentey, K.-M. *J. Sci. Food Agric.* **2003,** *83*, 1389–1402.
42. Byrdwell, W. C.; DeVries, J.; Exler, J.; Harnly, J. M.; Holden, J. M.; Holick, M. F.; Hollis, B. W.; Horst, R. L.; Lada, M.; Lemar, L. E.; Patterson, K. Y.; Phillips, K. M.; Tarrago-Triani, M. T.; Wolf, W. R. *Am. J. Clin. Nutr.* **2008,** *88*(Suppl.), 554S–557S; Phillips, K. B.; Byrdwell, W. C.; Exler, J.; Harnly, J.; Holden, J. M.; Holick, M. F.; Hollis, B. W.; Horst, R. L.; Lemar, L. E.; Patterson, K. Y.; Tarrago-Trani, M. T.; Wolf, W. R. *J. Food Compos. Anal.* **2008,** *21*, 527–534.
43. Williams, C. M. *Proc. Nutr. Soc.* **2002,** *61*, 19–24.

PART IV
Molecules for the Treatment of Illness

CHAPTER 11

DRUG DISCOVERY, DEVELOPMENT, AND DISTRIBUTION

ABSTRACT

The discovery of drugs can be traced to two origins. One is natural products that have been observed to have pharmacological activity. These include compounds such as morphine (from opium), quinine (from Cinchona bark), cocaine (from Coca leaves), and digoxin (from foxglove). The second is chemical synthesis, which began in the late 1800s and led to drugs such as antipyretics and analgesics (antipyrine, aspirin, acetanilide, phenacetin) and the first antimicrobial agent (arsphenamine for treatment of syphyllis). Further impetus came from the discovery of penicillin and the sulfa drugs by 1940. The approaches were merged in the synthetic modification of natural substances, as in the production of semisynthetic antibiotics. After WWII, the pharmaceutical industry developed rapidly based on screening of both natural and synthetic compounds involving bioassays. Later, crystal structure determination of target enzymes and computational modeling permitted drug design based on structural analysis. High throughput screening based on targets identified by gene sequencing became possible in the 1990s. Regulation gradually developed during the century, usually in response to well-publicized incidents that spurred action. These include requirements of (1) labeling of ingredients (1906); (2) prohibition of addictive substances (1912); (3) evidence of safety (1938); (4) evidence of efficacy (1962). Medical devices came under similar regulation in 1976. The current process involves: (1) discovery and preclinical studies leading to an Investigational New Drug (IND) application; (2) Phase I, II, and III clinical studies designed to determine dose, detect side effects and demonstrate efficacy; (3) approval of a New Drug Application (NDA) including indications for use and labeling. Drug approvals of new chemical entities (NCE) have averaged 20 $\pm$ 5 annually since 2000. About 10–15 new biologic drugs have been approved annually during this period. Controversies have continued since the 1970s about the effectiveness of the regulatory process and drug pricing.

At the beginning of the 20th century, there were only a handful of effective drugs available. The synthetic analgesics antipyrine (1885), phenacetin (1887), and aspirin (1898) had been introduced for treatment of fever and pain. Several important drugs from plant sources were also available, including digitalis from foxglove, quinine from *Cinchona*, morphine from the opium poppy, and cocaine from the *Coca* plant. The concept of treating bacterial infections with chemicals came from Paul Ehrlich, a German histologist working at the Institute of Experimental Therapy in Frankfurt. He was engaged in staining bacteria and conceived the idea that toxic substances might be delivered by dyes that would selectively attach to the bacteria. By systematically investigating hundreds of compounds, in 1909 he found arsphenamine (Salvarsan), which successfully treated syphilis.[a] The discovery of the sulfa drugs by Gerhard Domagk (1932) was a major advance. Synthetic analogs of quinine also were introduced in the 1920s and 1930s. The development of penicillin during WWII was a premier achievement of the first half of the century. Drug discovery accelerated rapidly after WWII and resulted in drugs for treatment of many cardiovascular, metabolic, and similar disorders.

At least four factors played a key role in enabling the advances that occurred during the second half of the 20th century. One was the recognition that infectious microorganisms were the cause of many diseases. This occurred in the period 1875–1890 as Pasteur, Koch, and others identified microorganisms associated with specific diseases by using improved microscopes and staining methods. The recognition of the causes of vector-borne parasitic diseases such as malaria and yellow fever occurred soon afterward (see Section 16.1.1). A third factor was expanded understanding of cellular processes as the structure and function of biological molecules such as proteins and carbohydrates were unraveled. This provided insight into systemic disorders such as cardiovascular disease. Another facet was the ability to synthesize and test potential therapeutic molecules using bioassays. The basic question we want to ask is "How do molecules prevent, ameliorate or cure disease?" A related question is "How are such molecules found?" The detailed answers are specific to the disease and the drug. In this chapter, we will examine the history of drug discovery and describe the regulations and procedures that apply to approval of drugs. In later sections, we will look as specific

---

[a]Paul Ehrlich was born in 1854 and studied at several German universities. He received a Doctorate in Medicine based on work on staining tissues with dyes. In 1878, he began his independent work in Berlin. In 1890, he joined Robert Koch at the Institute for Infectious Diseases, and in 1896, became director of a separate institute devoted to study of vaccines and antisera. He established some of the fundamental aspects of vaccines and antitoxins. Ehrlich formulated the "side-chain theory" suggesting that cells had specific receptors for antibodies, which encompassed, at least in essence, the major aspects of immunology as the story unfolded in the twentieth century. In 1899, he became director of the Royal Institute of Experimental Therapy in Frankfurt. It was there that he began to pursue his ideas about chemotherapy. When the organism responsible for syphilis was identified, Ehrlich began a search among hundreds of compounds, eventually leading to the arsenic compound salvarsan. He was awarded the Nobel Prize in Physiology or Medicine in 1908, recognizing his work in immunology.

classes of therapeutics and the mechanisms by which they act. Chart 11.1 gives a timeline for the introduction of some of the drugs that we will discuss.

CHART 11.1 TIMELINE FOR DISCOVERY OF REPRESENTATIVE DRUGS

1640   Anti-malaria effects of quinine recognized.
1763   Analgesic effects of willow bark reported in England.
1785   Cardiotonic effects of digitalis from foxglove described.
1840   Commercial production of pure morphine began.
1898   Aspirin introduced.
1900   Benzocaine, a synthetic analog of cocaine introduced.
1909   Salvarsan is the first drug used to treat an infectious disease.
1922   Insulin isolated and use in treatment of diabetes began.
1924   First synthetic anti-malaria drug.
1935   Sulfa antibiotics discovered.
1940   First successful treatment with penicillin.
1945   Antituberculosis activity of streptomycin discovered.
1948   Cortisone synthesized and used for treatment of arthritis.
1953   First use of anti-psychotic drugs.
1958   Introduction of β-blockers for treatment of cardiovascular disease.
1960   Steroid contraceptives introduced.
1960   Benzodiazepine tranquilizers introduced.
1982   Recombinant human insulin made by genetic engineering.
1987   AZT, the first anti-HIV drug approved.
1987   First selective uptake inhibitor antidepressant introduced.
1989   First statin for cholesterol control introduced.

11.1   THE ORIGINS OF CHEMOTHERAPY

For most of humankind's existence there has been no rational means of treatment of illness. The writings of the Greek and Roman physicians, Hippocrates and Galen, had some basis in observation and experience, but no fundamental understanding of the nature of illness. Various natural substances, mainly from plants, were used to treat illnesses. Such information had developed on the basis of experience over many generations. Particularly detailed information had been collected and recorded in China and India, but had relatively little influence in Europe. In Europe, an itinerant alchemist and healer, known as Paracelsus (1493–1541) promoted various mixtures including "laudanum," an opium-based material, but it is unclear if any of his treatments were effective.

Paracelsus did have one crucial concept correct: That the *dose* of a particular drug is a critical variable. *Cure or poison – it can depend on the amount!*

The antimalarial compound quinine was introduced into Europe in the 1600s, after discovery in Peru (see Section 18.1.2). Willow bark, which contains salicylic acid derivatives, came into use for treatment of fevers in England during the 1700s. Similarly, digitalis from the plant foxglove, was used to treat fluid build-up from circulatory problems, then called "dropsy." Its heart-stimulating action was beneficial (see Section 15.4.4). The circumstances of the discovery of these substances are not entirely clear and occurred before there was a scientific basis for understanding their effects. As scientific methods developed, many other important drugs were found from natural sources, especially plants and microorganisms.

The understanding that microorganisms can be the causes of illness opened the way to effective preventions and treatments. Edward Jenner, a physician in England, pursuing folklore that infection with cowpox prevented smallpox, devised a means of vaccination for smallpox. This showed that the development of the disease could be prevented by the vaccine. In 1847, Ignaz Semmelweiss in Vienna demonstrated that the incidence of childbed fever could be reduced if medical attendants cleaned their hands and took other measures to prevent patient-to-patient transfer. This indicated that some "organism" must transmit the disease. Improvements in the magnifying power of microscopes enabled Pasteur to directly observe microorganisms associated with disease. When he investigated childbed fever, Pasteur was able to observe microorganisms, not only in sick women, but in bed clothes and on the hands of practitioners. Pasteur went on to develop immunizations for cholera in chickens and anthrax in sheep. He also developed an antitoxin against rabies in humans that could be used as a treatment if given between exposure and development of the disease. Combining microscopy with selective staining, Robert Koch was able to identify the causative agent of tuberculosis in 1882. The question then became how to fight these organisms in order to prevent and treat infectious diseases.

11.1.1 SYNTHETIC DRUGS

The role of synthetic chemistry in the discovery of drugs dates to the late 19th century. Amazingly, one of the first drugs, aspirin, remains one of the most widely used today. It was synthesized at Bayer in Germany in 1898. The Bayer lab also developed phenacetin, a predecessor of acetaminophen, now widely marketed as tylenol. More will be said about aspirin and acetaminophen in Chapter 13 along with other analgesic and anti-inflammatory drugs. At the beginning of the 20th century, there were no effective treatments for bacterial diseases such as pneumonia, tuberculosis, or for *Staphylococcus* and *Streptococcus* infections.

A great flu pandemic began near the end of WWI in1918. More than 50 million people died worldwide, mainly from pneumonia, over the next 2 years.

Following Ehrlich's discovery of Salvarsan, Wilhelm Roehl, who worked at Bayer, in 1916 synthesized Germanin (now known as Suranim), which could be used to treat African sleeping sickness, a disease caused by parasitic trypanosomes. These represented the first cases of treatment of pathogenic organisms by chemicals. By modern standards, these were miserable drugs, but they demonstrated the fundamental feasibility of chemotherapy. Arsphenamine was difficult to prepare and purify and was quite toxic. It had to be administered by a long series of intravenous injections, a technique that was familiar to relatively few physicians at the time. Suranim was also highly toxic and required repeated intravenous administration.

arsphenamine
Salvarsan

Suranim

A major advance occurred in the 1930s. Gerhard Domagk,[b] a physician, was working on the immune process at the Bayer laboratory in Wuppertal. He noted that immunized animals had increased numbers of killer cells (phagocytes). Domagk had worked in field hospitals with the German army in WWI and was dedicated to finding a way to treat infectious diseases. He hypothesized that if the bacteria could be weakened chemically, they would be particularly susceptible to the immune response. In 1929, Wilhelm Roehl, one of Domagk's coworkers, who had synthesized both Suranim and the antimalarial drug, plasmoquine, died of a blood-borne *Streptococcus* infection. Domagk decided to focus his search on a drug that would treat *Streptococcus*. Domagk chose to use infected animals as the test system, believing this to be a more realistic than isolated bacteria in culture. In the course of events, this turned out to be a crucially important decision. Domagk, in conjunction with chemists Josef Klarer

---

[b]Gehard Domagk began medical studies in 1914, but this was interrupted by WWI where he served as a soldier, and after being wounded, as a medic. He completed his medical studies in 1921. Between 1921 and 1927, he held posts in anatomy in several universities. In 1927, he began working at the Bayer laboratory in Wuppertal, where, along with chemists Josef Klarer and Fritz Mietsch, he synthesized and tested many dyes, eventually finding Prontosil. Among the first patients he treated was his own daughter. He was awarded the 1939 Nobel Prize in Physiology and Medicine. But because Hitler was angry at the Nobel Committee for an earlier award of the Peace Prize to Carl von Ossietzky, an anti-Nazi activist, Domagk was forbidden to accept the prize and was briefly arrested. He later found the antituberculosis drug isoniazid.

and Fritz Mietzch, began the synthesis of compounds related to dyes. In 1932, a compound, called *Prontosil*, was found that successfully treated mice infected with *Streptococcus* strains.

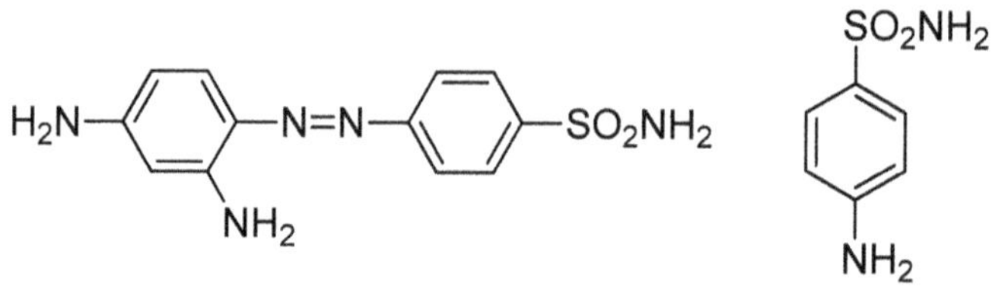

prontosil   sulfanilamide

The first human trial was on an 18-year-old girl with a severe *Streptococcus* infection. She was cured. Subsequent trials showed that the drug was also effective against *Staphylococcus*. The medical community was initially skeptical, but Dr. Leonard Colebrook at Queen Charlotte Medical Hospital in London demonstrated success in treatment of postpartum infections. In 1936, the President's son, Franklin D. Roosevelt, Jr., developed severe tonsillitis. Eleanor Roosevelt feared he was dying. At her insistence, his doctor used Prontosil and he recovered. The resulting publicity increased the awareness of the drug, both within the medical community and the general public.

Eventually, it was found that the active form of Prontosil is generated by *in vivo* metabolism and is *sulfanilamide*. This discovery opened the door to multiple manufacturers, because sulfanilamide had been patented as a dye intermediate in 1909 and the patent had expired. But it did not take long to recognize the need for caution. In 1937, the S. E. Massingil Company of Bristol, TN, USA began selling a liquid form called "Elixir Sulfanilamide" in the South. It was formulated in a toxic solvent, diethylene glycol, and within a month, 107 users were dead and the inventor had killed himself. This led to revision of the Food and Drug Administration (FDA) laws (see Section 11.4.1).

A large number of other sulfa drugs were developed, all containing the 4-aminobenzenesulfonamide structure. Among the most useful are sulfadiazine, sulfathiazole, sulfapyridine, and sulfaguanidine.

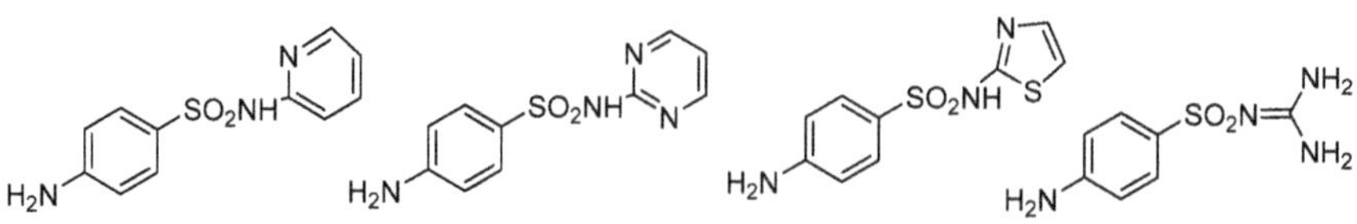

sulfapyridine   sulfadiazine   sulfathiazole   sulfaguanidine

Near the end of WWII, Domagk's group synthesized many compounds and tested them against tuberculosis. The most promising was thiacetazone, called Conteben. At about the same time, a Swedish physician, Jorgen Lehmann, on the basis of structural analogy, predicted that *para*-aminosalicylic acid would have antitubercular activity. The compound was made by the Swedish company Ferrosan and proved to be effective in human trials. These successes led to many new compounds being synthesized by the emerging pharmaceutical industry. One antitubercular compound, isoniazid, was particularly promising. It was synthesized by Dogmak's group, but also at two drug firms, Squibb and Hoffmann-LaRoche. Like sulfanilamide, it had been previously disclosed and could not be protected by a patent. It became widely used. But it also provided an indication of problems to come. While many patients were cured, those who relapsed often failed to respond to a second course of treatment. Evidently, tuberculosis could develop resistance to isoniazid. Further study showed that concurrent use of *p*-aminosalicylic acid reduced the incidence of resistance to isoniazid.

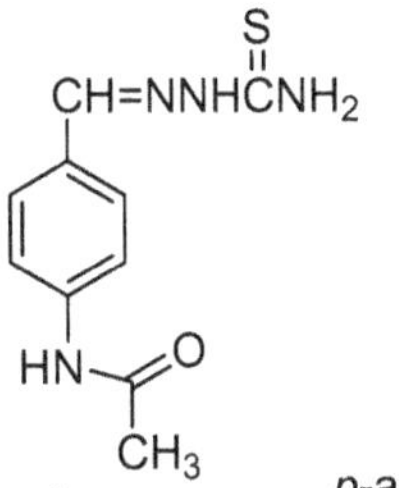

thiacetazone
Conteben

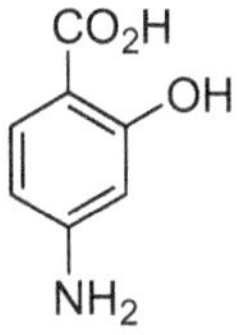

p-aminosalicylic acid

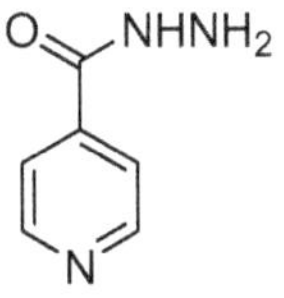

isoniazid

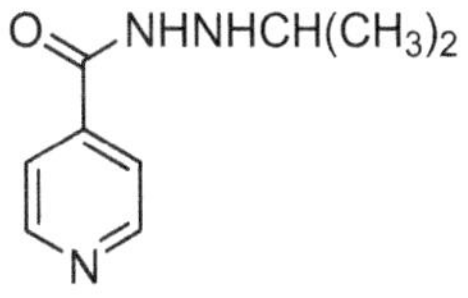

ipronazid

About 15 years later, observation of the effects of an analog of isoniazid, ipronazide, led to the discovery of the first antidepressant drugs (see Section 17.6.1).

11.1.2 NATURAL SOURCES OF DRUGS

Advances in chemical techniques also permitted the isolation of pure substances from natural extracts. The purification of cocaine and its structural determination was followed by the synthesis of the local anesthetics benzocaine, procaine, and novocaine (see Section 13.7). Ergonovine was found to be the active principle in Ergot extract, which had long been used by midwives to enhance contractions and reduce bleeding during birth. Its use, however, was dangerous because the dose was difficult to control. The firm Burroughs–Wellcome in London, founded by two expatriate American pharmacists, Silas Burroughs and

Henry Wellcome, developed a standardized extract. The company came to be one of the leading pharmaceutical firms in Britain.

Two other classes of drugs, steroids and antibiotics, achieved importance in the 1940s and 1950s. Several steroids had been isolated from animal tissue in minute amounts in the 1930s. They were extremely potent biologically. As we will see in Chapter 15, they eventually led to drugs such as cortisone and to birth-control pills. Penicillin, originally discovered by Alexander Fleming in 1928, was developed during WWII by Howard Florey and others under the impetus of the military need to treat infections. Near the end of WWII, Selman Waksman and Albert Schatz discovered streptomycin, which was the first antibiotic that successfully treated tuberculosis. We will pursue these and other antibiotics further in Chapter 12.

As we examine some of the specific types of drugs, we will discover that many had their origin in naturally occurring materials. Many of the antibiotics were both isolated from and are commercially produced by growing bacteria. Other drugs are isolated from plants. These include the antimalarials quinine and artemisinin, the morphine analgesics and some steroids, such as digitonin. In many cases, the original natural materials have been chemically modified. An analysis of the 1031 *new chemical entities* approved as drugs by the FDA between 1981 and 2002 indicates that about 23% were synthetically modified derivatives of natural products. Another 24% were discovered on the basis of structural similarity to natural substances.[1] At the molecular level, of course, the action of drugs is determined by molecular structure, not the source of origin.

11.2   BEGINNINGS OF THE MODERN PHARMACEUTICAL INDUSTRY

Ehrlich's concept of chemotherapy, Domagk's success with the sulfa drugs and the introduction of penicillin and streptomycin set the stage for the expansion of the pharmaceutical industry. A source of chemical compounds coupled with *in vitro* and *in vivo* screens held the promise of finding drugs for the treatment of many pathologic conditions. One other early observation pointed to the importance of a *biochemical rationale*. Paul Fildes and Donald D. Woods at the MRC (see Section 11.4.3.4) had suggested that the basis of activity of sulfanilamide was its structural similarity to *p*-aminobenzoic acid, which is required by bacteria to synthesize folic acid. It was from this suggestion that Jorgen Lehmann postulated that *p*-aminosalicylic acid would have antitubercular activity. The understanding of the nature of diseases began to permit rational approaches to medicines. Chemists played a key role by isolating active compounds from natural sources, determining their structure, and by synthesizing new molecules.

Chemical synthesis, based on a biochemical rationale and pharmacological screening, became the basis of drug discovery in the nascent pharmaceutical industry.

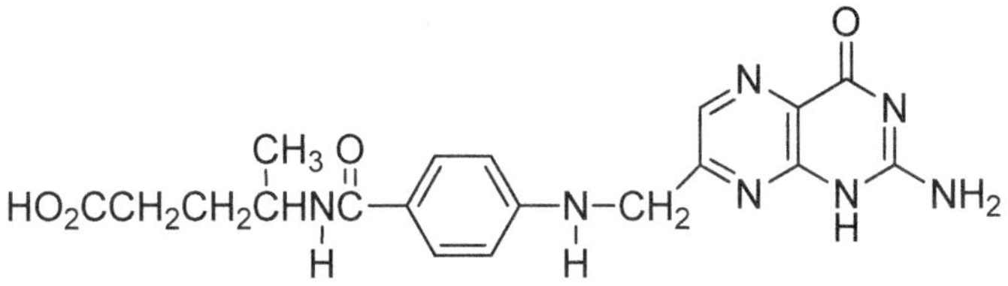

folic acid

The pharmaceutical industry developed first in Germany at the Bayer Company, which began research on analgesics and antipyretics in the late 1800s. The Hoechst Company also began producing drugs and antitoxins late in the 19th century. Other companies began in England, France, Switzerland, Italy, and Scandinavia. Many of the US pharmaceutical firms originated in the late 19th century, when the country was under the thrall of patent medicines of dubious value. Some were formed by physicians or pharmacists appalled by the low quality and lack of effectiveness of these products. Others developed from companies engaged in wholesale distribution of drugs made in Europe. For the most part, the American companies were distinct from the large chemical manufacturers, although many of the latter have at one time or another acquired pharmaceutical divisions. At the present time most major pharmaceutical companies are international enterprises, usually with American, European, or Japanese origins.

At the beginning of the 20th century there was a clear distinction between "ethical" and "proprietary" drugs. The former were sold only through pharmacists and physicians by prescription and were not advertised to the general public. Proprietary drugs, also called *patent medicines*, were formulations of various, usually undisclosed, ingredients. They were sold by vigorous public advertising, including travelling road shows. Both ethical and proprietary drugs could be sold under brand names protected by trademark laws. The term aspirin, for example, was originally a protected trademark of the Bayer Company. One distinction of "ethical" drugs disappeared in the United States in 1996, when direct public advertising of prescription drugs was approved by the FDA. In most of the rest of the world, general advertising of prescription drugs is not allowed. Drugs protected by patents can be sold as *generic drugs* after the patent expires. The generic drugs are sold under a name assigned to the particular chemical compound or formulation. The successors to the "patent medicines" of earlier days are the "dietary and nutritional supplements" that can be sold without evidence of effectiveness or FDA approval (see Section 10.3).

11.3 DRUG DISCOVERY AND DEVELOPMENT

Successful chemotherapy requires some functional interaction of the drug with a biological mechanism or system. In the case of pathogenic organism, efficacy requires *selective toxicity* between the pathogen and the human host. The pathogen must be more sensitive to the drug than the host. This may be the result of differing mechanisms in the two species (a qualitative difference) or different sensitivities to a related mechanism (a quantitative effect). Conditions due to dysfunction in some cardiovascular, metabolic, or digestive process require that the defective function be suppressed or enhanced so that it is restored to its normal level. Similarly, the complex endocrine system may be the target of pharmacological intervention. The same is true of neurological and psychological conditions such as depression, anxiety, or mental illness. In order to regulate these processes, the drug must interact with an enzyme or receptor that is part of the biological process. We will consider examples of such enzymes and receptors, as we discuss representative types of drugs.

Beginning in the 1950s, systematic searches for drugs were done using screens based on either animal models or enzymatic and other pharmacological assays designed to detect the desired activity. These screens were used to find *lead compounds*, which were then subjected to further structural modification to optimize properties. The most promising compounds were submitted for preclinical and then clinical studies necessary for final approval. The first successful drug is often followed by modified versions referred to as "second-generation," "third-generation," etc. These may be introduced by the original company or by competitors. This process usually results in improvement of the overall effectiveness of the drugs. In some cases, unanticipated activity was found for compounds being tested for other purposes. Examples include Furosemide (Section 14.2) and Rogaine (Section 6.3.2.5).

Beginning in the 1970s, *rational drug design* came within reach as the result of advances in several areas. One was deeper understanding of biological mechanisms. In order to intelligently design a drug, it must be targeted at a relevant function. Another advance that was essential was in structural biology. The structural features of potential drug targets, especially proteins, became available from X-ray structure determinations. Another crucial element was computational modeling, which advanced rapidly during the 1980s. This information—the role of the biological mechanism, the structure of the target enzyme, and the tools to visualize the interaction of the drug with the receptor—provided the synthetic chemists with ideas about specific molecules that might be made. Later, we will see specific examples of these methods in the case of drugs for treatment of conditions such as high blood pressure, cardiovascular disease, diabetes, and AIDS.

Molecular biology and biotechnology introduced other means of drug discovery. It became possible to identify, clone, and sequence the genes related to specific biological functions. These genes could then be used to prepare the corresponding proteins for bioassays. These efforts, including the completion of the sequence of the human genome in 2003, permitted identification of many more functional biological molecules. This genetic information provided the basis for identifying a wide variety of potential protein targets. For example, the effect of specific genes can be studied using "knock-out" or "knock-in" mice in which particular genes have been either inactivated or added. Beginning in the mid-1980s, it became possible to use genetic engineering to produce protein-based drugs. Examples include insulin, human growth hormone, and granulocyte stimulating factor (see Section 21.3).

Combinatorial synthesis and *high-throughput screening* were applied to drug discovery beginning in the 1990s. In combinatorial synthesis, a large number of related molecules are synthesized by using several different, but related, compounds in each step of a synthesis. For example, if 10 sets of related molecules are used in all possible combinations over four steps of synthesis, 10^4 products are generated. These compounds can then be screened against enzymes or receptors of interest using a sensitive analytical technique to detect binding or other evidence of interaction. Bioassays were also adapted to rapid screening of many compounds, including those from archives of samples prepared previously.

Despite increasing knowledge and new techniques and tools, the rate of new drug approvals has remained relatively constant since the 1950s. A total of about 1250 new drugs were approved between 1950 and 2010. The total expenditures on the other hand have increased dramatically. This trend is reflected in Figure 11.1 which shows (on a logarithmic scale!) the number of drugs per billions of dollars of investment in the US drug industry. Several factors have been considered as possible causes of this trend. One may be a higher barrier to success, in the sense that the early discoveries were in areas where discoveries were most likely to occur, the "low-hanging fruit" hypothesis. This factor may be especially important in well-explored areas such as antibiotics. It suggests that as successful drugs are developed it becomes more difficult to improve upon them. Another factor is probably more demanding regulation (see Section 11.4), including the more extensive studies and larger clinical trials required for drug approval. Higher costs may also reflect the increasingly complex organizational structure of drug development.

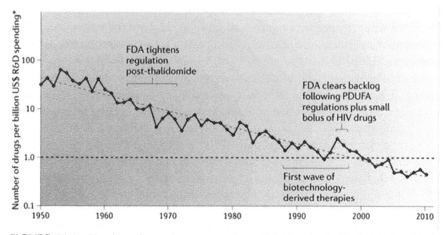

FIGURE 11.1 Number of new drug approvals per $ billion in the United States, 1950–2010. (From Scannell, J. W.; Blankley, A.; Boldon, H.; Warington, B. *Nat. Rev. Drug Discov.* **2012**, *11*, 191–200, by permission of Nature Publishing Group.)

11.4 REGULATION, RESEARCH, POLITICS, AND PUBLIC RELATIONS

11.4.1 *THE FOOD AND DRUG ADMINISTRATION*

The FDA had its origin in the Pure Food and Drug Act of 1906. This law was in response to pervasive problems in both the patent medicine and food industries at the end of the 19th century. A period of *laissez faire* economics following the Civil War had resulted in widespread sale of worthless and even dangerous patent medicine and adulterated or spoiled agricultural products. Patent medicines were promoted as having secret formulas and sold by vigorous advertising campaigns. Industrialization and urbanization required means of large-scale food preservation and shipment, but there was limited knowledge on how to do this safely. Upton Sinclair's book, *The Jungle*, dramatized the practices of slaughter houses. Politically, it was the time of the Progressive Movement and its outrage at fraud and exploitation. It was also a time at which the underlying causes of illness, such as the discoveries of Koch and Pasteur, were coming to be widely recognized. The 1906 legislation set some basic standards for labeling of drugs and prevention of contamination of food. It prohibited "adulteration" or "misbranding and mislabeling" but provided little enforcement authority. The first FDA commissioner was Harvey Wiley, who had been trained in medicine and chemistry at Harvard and in Germany. He was a carbohydrate chemist and had demonstrated that there was widespread adulteration of products such as honey and syrup by sugar water. He became the Chief Chemist at the Laboratory of Chemistry in the Department of Agriculture in 1883 and in 1907 became the first Commissioner of the FDA, serving until 1912.[c] Thereafter, he founded The Good Housekeeping Institute for Consumer Protection.

In 1938, largely in response to the "sulfanilamide elixir" tragedy (see Topic 11.2.1), the FDA's mandate was expanded to include proof of safety. In 1962, the requirement that drug sponsors provide substantial evidence that drugs are effective was added. This greatly expanded the need for clinical studies and introduced the *Investigational New Drug Application* (IND), a process required to begin clinical studies. These procedures are described in more detail in Section 11.4.1.2. The 1962 law also required that the efficacy of drugs introduced between 1938 and 1962 be established. This was a major undertaking, completed in 1969 with the assistance of the National Academy of Science. In the 1970s, in response to revelations of lax record-keeping in clinical trials, the *"Good Laboratory Practices"* and *"Good Clinical Practices"* manuals were developed. The FDA also became responsible for biologics and medical devices. The latter was precipitated by the case of the Dalkon Shield, an intrauterine contraceptive device that resulted in many cases of infection, infertility, and a number of deaths (see Topic 11.2.2). In 1982, as a result of fatal poisoning of several

---

[c] Harvey W. Wiley became Professor of Chemistry at the newly founded Purdue University in 1874. He was also state chemist of Indiana. In 1883, he moved to Washington DC to become chief chemist of the Department of Agriculture. He vigorously promoted laws against adulterated food and, later, dubious drugs. These efforts, along with considerable press interest, catalyzed goverment action. In 1906, after the first pure food and drug legislation was passed, he was responsible for enforcing the legislation. Wiley chose as his first target a product sold as "Curforhedake Brane-Fude," a ludicrous misspelling of "Cure for Headache Brain Food." The material was marketed by Robert N. Harper, a pharmacist, and prominent citizen of Washington, DC, USA, who among other posts had been the president of the retail drug association and the Chamber of Commerce. The mixture contained acetanilide, antipyrine, caffeine, potassium and sodium bromide, and ethanol. Acetanilide, a forerunner of acetaminophen, was widely used as an analgesic at the time, but was also recognized as having substantial toxicity. The label of the product said it contained "no...poisonous ingredients." The charges were brought on the basis of this statement and on the amount of ethanol in the remedy. Harper mounted a vigorous defense, including testimonials from prominent citizens. The judge gave the jury very literal instructions regarding the law, which stated that if the label was misleading in "any particular," it would be a violation. The jury found Harper guilty. Pres. Theodore Roosevelt pressed for jail time, but the judge ordered a fine of $700. It was estimated that Harper had made the then enormous sum of $2 million from the product. Nevertheless, the law and its effect had been established. Even the manufacturers of patent medicines and the druggists associations, which had strongly opposed legislation at the outset, decided to cooperate in the endeavor to improve the quality and reliability of their products. Regulations under the law were the responsibility of a 3-person commission chaired by Wiley. The commission wrote quite stringent regulations interpreting the word "label" in the law. The law was especially effective in reducing the level of opiates in patent medicines. In 1911, however, the Supreme Court ruled that the law's prohibition of "false labeling" did not pertain to claims of therapeutic benefits. Congress soon passed an amendment making "false and fraudulent" claims illegal. The staff available at the Bureau of Chemistry remained small and inadequate to address the many possible violations. Eventually, in 1912, Wiley left and founded *Good Housekeeping Magazine*, where he continued to promote safe foods and medicines and hygienic practice. Wiley was succeeded at the FDA by Carl L. Alsberg, who was an M. D., with training in chemistry and biology. At this point the American Medical Association was taking the leadership in promoting of improvement of all aspects of medical care, including the purity and effectiveness of drugs. Alsberg used the provisions of the 1912 amendment to vigorously attack materials making false claims, with considerable success.

people by cyanide added to tylenol, requirements for tamper-proof containers were introduced.

The "drug lag issue" is the proposition that FDA drug approval was (or is) slower than in much of the rest of the world and that as a result, availability of new drugs is delayed in the United States. The matter has been in dispute since the 1970s. A 1980 report of the Government Accounting Office indicated that the main causes of delayed approval were lack of FDA staff, inadequate industry data, and ineffective communication. Ongoing studies in the 1980s often indicated unfavorable comparison, especially with the United Kingdom. These conclusions were contested by FDA Commissioners Donald Kennedy (1978)[2] and David Kessler (1996)[3] but their contentions were, in turn, vigorously rebutted.[4] The perception of a "drug lag" was largely accepted by the pharmaceutical industry and exploited by politicians.

During the 1980s, the issue of safety of silicone breast implants arose. Although implants were covered by the laws requiring demonstration of safety, the requirements had never been enforced, and no reliable information on safety was available. The FDA moved to require safety studies, but both the manufacturer, Dow–Corning, and the plastic surgeons claimed that experience with use constituted proof of safety. In fact, when data were collected, the rate of rupture and other adverse events was quite high, and received extensive publicity and resulted in many lawsuits claiming injuries. In the end, silicone implants were not completely banned, but safety studies were required for future implants (see Section 7.3.3).

By the early 1990s, the FDA was overwhelmed by applications for new drug approvals and the AIDS crisis. Congress had assigned many responsibilities over the years, but budgets had been cut during the Reagan years. In 1990, David Kessler was appointed FDA Commissioner. The pharmaceutical industry agreed to user fees for financing the drug approval process in 1992. The laws authorizing these fees also set performance goals for the FDA. With the new revenues, the budget and staff of the FDA were expanded. The user fees were extended in 1997 and 2002. As of 2004, these fees were about $575,000 per application and provided more than $250,000,000 of the FDA budget. The user fee seems to have been successful in that approval times were reduced from an average of 35.6 months in 1984–1986 to 16.8 months in 1996–1998. The 1996–1998 period saw the approval of 110 new chemical substances as drugs, the highest ever.[5]

Kessler also launched an effort to invigorate FDA oversight in the area of health claims in food labeling and advertising. These efforts resulted in the Nutrition Labeling and Education Act of 1990. The implementation of this act led to the system of labeling food content on the basis of an average daily input of 2000 calories. A particular struggle arose with regard to labeling of the fat content of meats, in which the interests of the meat industry were advocated by the Department of Agriculture. In 1992, President George H. W. Bush approved the regulations favored by the FDA.

A strange feature of current law is the special status of "dietary and nutritional supplements" and "herbal" products. The "Dietary Supplement and Health Control Act of 1994" exempted such products from the standards applied to drugs. There are no requirements for concentration or purity of these materials. Naturally occurring materials, even if they are made by synthesis, can be sold without any evidence of efficacy or safety. Efforts to require proof of health claims were scuttled by a vigorous public relations campaign mounted by the industry. In an extreme example, an ad portrayed Mel Gibson being accosted by a SWAT team confiscating his vitamin C. This situation has had its occasional negative consequences, including fatalities (see Section 10.3).

11.4.2   THE CURRENT DRUG APPROVAL PROCESS

The responsibility for drug approval in the United States lies with the Food and Drug Administration (FDA) and is carried out by the *Center for Drug Evaluation and Research* (CDER). Other divisions monitor biologicals, medical devices and food safety. The FDA's goal is to achieve a balance between speed and safety in the drug approval process. Recently, it has placed additional emphasis on risk recognition and post-marketing surveillance, intended to ensure that experience with drugs after approval continues to be safe and effective. There is an international group, the *International Conference on Harmonization* (ICH) that works to provide consistency in the drug approval process in the United States, EU, and Japan. In 2003, an internationally acceptable format of drug information (Common Technical Document) became available.

The approval process begins with an IND application, which proposes small-scale human studies. These applications include the results of *preclinical studies* providing preliminary evidence of potential value and safety, including animal toxicology, genotoxicity and carcinogenicity, chemical composition and stability, as well as pharmacological studies on cardiovascular, respiratory, nervous, and endocrine systems. The collection and submission of the data for the preclinical studies are governed by *Good Laboratory Practices*, which indicate acceptable procedures for these studies. These studies provide the NOAEL (*no observed adverse effect level*) in animals, on which the test dosage for human subjects are based. The INDs specify the protocol for the proposed human studies. The FDA review is conducted by teams typically including medical, chemical, and pharmacological members. These teams are organized to review drugs in particular areas of application such as cardiovascular, oncological, neurological, gastrointestinal, anti-infective, and anti-viral. The FDA can issue a "hold" on a study within 30 days if it does not meet the standards for protection of subjects, but if this is not done the study may proceed with tacit FDA approval. About 8–10% of INDs are placed in the hold status.

Human trials typically fall into three successive phases. *Phase 1* determines toxicity and side effects in response to escalating doses in healthy volunteers or patients. These studies typically involve 20–100 subjects. Phase 1 studies provide the first human dosage information and the subjects are monitored closely for observation of side effects. Phase 2 studies examine dose–response, efficacy, toxicity, and side effects in subjects having the condition. These typically involve 50–200 patients. An important goal of these studies is to refine the dosage requirements. Phase 3 involves larger scale, usually multicenter, treatment of patients with quantitative measures of effectiveness, toxicity, and side effects. They involve several hundred to several thousand subjects. If feasible and ethical, the studies are "double-blinded" and involve control groups, with randomized placement of patients in treated and control groups. Normally, at least two independent studies are required. Recently, provision for single studies has been made, but such studies must usually involve multiple test sites and investigators.

The conduct of clinical trials is governed by *Good Clinical Practices*, which are intended to minimize risks to participants and to insure that the data provide the basis for adequate interpretation. During clinical studies, drug sponsors are required to report any adverse experiences and any new information that may become available from other sources. Clinical studies are also under the supervision of *Institutional Review Boards* at the participating institutions. These boards are responsible for the welfare of the participants, including insuring their *informed consent*. The trial sponsor must also select a monitor for clinical studies. The monitor must be qualified to evaluate the ongoing study. Their reports become part of the application for approval of the drug.

Phase 4 involves post-introduction monitoring and reporting of any adverse events. Trials of new formulations or dose regimen studies are also considered Phase 4 studies. *Supplemental New Drug Approvals* are issued for significance changes in an approved drug, including changes in dosage or formulation, changes in the manufacturing process, addition of new medical indications, and changes in labeling or packaging.

The approval of a new drug is obtained by submission of a *New Drug Application* (NDA). This document must provide evidence that the drug is both safe and effective for the use(s) proposed. The information is contained in a voluminous document (many thousands of pages usually), which addresses preclinical, clinical, biopharmaceutical, and statistical analysis. The application is reviewed by a review team. If there are significance issues, the approval may be submitted to one of the drug advisory committees, which are organized in parallel with the "system-oriented" review teams, for example, cardiovascular, endocrine, antiviral, etc. The approval of marketing by FDA is based on conclusions in several areas: (1) Is the drug safe and effective? (2) Is the risk–benefit ratio appropriate? (3) Does the method of manufacture insure adequate purity and stability? (4) Does the label information provide for safe distribution and use?

Several factors may influence, directly or indirectly, the likelihood and speed with which a drug is approved. One is the drug-maker's reputation, resources, and experience. Another is patient and disease advocacy groups. The FDA is strongly motivated to protect its reputation for drug safety. In 1975, drug approval was estimated to take 7–10 years from discovery to launch and to cost $50 million. By 1985, this was 10–12 years and $125 million and by 1995, 12–15 years and $500 million. The time of approval is particularly significant in relation to recouping of research and development costs, because the period of patent protection dates from the date of the patent, not the date of drug approval. There has been significant reduction in review/approval times since 1995. Figure 11.2 shows the number of new drugs and biologics approved between 2001 and 2011.

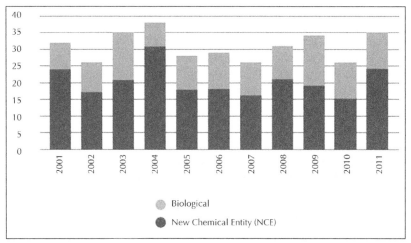

FIGURE 11.2   Approvals of drugs (new chemical entities [NCE]) and biologics by the FDA from 2001 to 2011. (From The Pharmaceutical Industry and Global Health, Facts and Figures 2012. Used with permission of the International Federation of Pharmaceutical Manufacturers.)

11.4.2.1   GENERIC DRUGS

Drugs are assigned a *generic name* that applies to a particular chemical entity or combination. *Brand names* are assigned by the individual companies to their products, including products sold under license from another patent holder. The brand names are usually registered trademarks and are capitalized. After patent protection has expired, manufacturers of generic drugs can enter the market. The critical aspect in the comparison of brand and generic drugs is that of *biological equivalence*. The requirement of "biological equivalence" was established in 1983. The manufacturers of a generic drug need only demonstrate that the product has bioequivalence with the approved drug, and thus generally do not encounter the research and clinical trials cost associated with a new drug. The approval process also requires inspection of production facilities. Generic drugs typically have a different color, shape, etc., because the trademarks of brand-name drugs protect such distinctive features. These differences in appearance cannot have any effect on the drug's activity. Government policy in the United States has encouraged the development of generic drugs as a means of reducing costs. The Hatch–Waxman Act provided a period of 180 days of exclusive market access to the first supplier of a generic drug under certain circumstances.

11.4.2.2   BIOSIMILAR DRUGS

There is a process by which biologically produced drugs can be approved as generic substitutes for biologic drugs, such as those obtained by genetic engineering. A revision of the relevant laws was included in the Patient Protection and Affordable Care Act of 2010. This provides that a biosimilar drug can be approved on the basis of clinical studies demonstrating that the biosimilar drug is "interchangeable" with the original biologic drug. The terminology that applies is somewhat different from pure substances because it is generally impossible to prove that two biologic drugs are "identical" because of the complexity of their structures. The required studies include evaluation of pharmacokinetics and pharmacodynamics and clinical studies showing equivalent "safety, purity, and potency." The European Medicines Agency is developing similar procedures.[6] Experience with biologic drugs has indicated that one of the factors of importance is immunogenicity, as examples of increased immunogenicity following process modification have been observed.

11.4.2.3   SPECIAL CIRCUMSTANCES FOR DRUG APPROVAL

Orphan drugs are those that are potentially useful but for which there is not a sufficient market to make commercial development feasible. For most cases, the orphan drug provisions apply to conditions that affect less than 200,000 people in the United States. The same safety and efficacy requirements are applied as in regular NDA, but certain incentives are offered. The most significant is 7-year-market exclusivity from the date of approval. A tax credit equal to 50% of clinical testing costs is available and can be carried forward for up to 15 years. The FDA also provides assistance in preparation of protocols and a modest amount of funds for costs of testing of potential orphan drugs. As of 2000, about 200 drugs had been approved with orphan status.

Accelerated access and fast track approval. With the appearance of AIDS (see Chapter 19), considerable pressure arose for access to drugs prior to final approval of the NDA. Various procedures were developed whereby drugs in the approval process could be made available to patients not in formal clinical studies. Subsequently, similar provisions were applied to potential anticancer drugs. These procedures were formalized as the *Fast Track Approval Process* in 1997. To be eligible for fast-track consideration, a drug must treat a condition that is considered "serious," that is life-threatening or the cause of significant disability. The drug must also have the potential to significantly improve on existing therapies in efficacy, tolerance, or patient compliance. The fast track approval process permits use of clinical indicators or endpoints that predict clinical benefit, as opposed to the final evidence of clinical outcome. An example of a "surrogate end-point" is the $CD4^+$ cell count in AIDS.

Pediatric drugs. Many drugs that are used in treatment of children have not been subject to specific pediatric studies, and, as a result, information on dosing, efficacy, side effects, etc., and specific pediatric labeling are based on extrapolations. The 1997 FDA Modernization Act specified procedures for improving the quality of pediatric safety and efficacy data. A 6-month extension of patent or other exclusivity can be obtained by conducting a pediatric clinical study. This extension also applies to the expansion of indications for an existing drug on the basis of pediatric studies.

11.4.2.4   OFF-LABEL USE

Drugs are approved for certain specific uses, for example "chronic pain of arthritis in patients >60 years of age." The reason for this is that efficacy must be demonstrated for a specific condition and it is usually practical to carry out studies only for limited conditions and types of patients. However, once approved, physicians are free to prescribe the drug in any circumstances they feel are medically justified, including for other conditions and in different doses. Drug manufacturers, however, are prohibited from advertising or otherwise promoting off-label use. Many recent regulatory actions, including fines and other sanctions, have been against drug companies that have been found in violation of this prohibition.

11.4.2.5   APPROVAL FOR OVER-THE-COUNTER MARKETING

Drugs that are approved for "over-the-counter" (OTC) sale can be purchased without a prescription. In the United States, this means the drug is available on shelves in drug, grocery, etc., stores. Approval of OTC sale requires reasonable evidence that the drug can be safely used without the advice of a medical professional. Labeling information on the active ingredient, recommended dose and other information on use is required. Some drugs have been converted to OTC drugs after extensive use as prescription drugs indicated their safety. Examples are the digestive drugs cimetidine (Tagamet) and omeprazole (Prilosec) and the anti-histamine loratidine (Claritin). In some cases, such as the anti-inflammatory drugs ibuprofen and naproxen, the dose that is available OTC is smaller than the dose available by prescription. Drugs that have abuse potential, such as pseudoephedrine, may not require a prescription, but can only be dispensed by a professional.[7]

11.4.2.6   WITHDRAWAL OF APPROVAL OF A DRUG

Problems with drugs may be encountered after approval of marketing. This is because marketing results in use by much larger groups of patients than in clinical trials. About half of approved drugs undergo label changes and about 20% get "black box" warnings. Historically, about 3–4% of approvals for new drugs are subsequently withdrawn. Publicity associated with drug withdrawal usually treats the decision as the correction of a previous error. It would be more accurately described as inherent limits on the pre-approval information. The primary responsibility for relabeling and withdrawal rests with the *Office of Drug Safety*, whereas the approval process is located in the Center for Drug Evaluation and Research. While this has the positive feature of separating the approval and withdrawal decision, it also sets up some tension in that the decision on withdrawal may reflect adversely on the prior decision to approve a drug. Some have argued that the US drug approval process should put more emphasis on post-marketing surveillance, as summarized in Figure 11.3.

One of the most frequent causes of drug withdrawal or warnings is toxicity associated with liver metabolism, which can result in the generation of toxic metabolites. Around 7–15% of cases of acute liver failure are attributed to adverse drug reactions. Some drugs carry labels that recommend periodic tests of liver function, although these recommendations are not always followed. There can be wide variations in patient sensitivity to some drugs, based on genetic factors.[8] We will consider several specific cases of withdrawal of approved drugs, such as rofecoxib (Section 13.4.2), troglitazone (Section 16.4) and the appetite suppressant fenfluramine (Section 17.6.2).

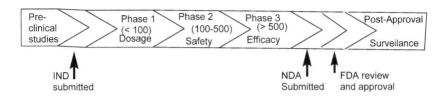

FIGURE 11.3   Summary of FDA drug approval process.

11.4.2.7   CONTROLLED SUBSTANCES

The FDA and Drug Enforcement Agency have the authority to restrict access to various classes of medications based on the potential for abuse. Such drugs are categorized as Schedule I–IV based on their potential for abuse and need for legitimate medical purposes. Schedule I drugs are those that are considered to have high potential for abuse and dependency and for which there is no recognized medical use. The current list includes certain amphetamines, hallucinogens such as LSD and mescaline, and heroin. Marijuana is also currently listed in Schedule I. Schedule II drugs are those with a high potential for abuse but for which there are legitimate medical uses. They can be distributed only by prescription and the prescriptions are monitored and cannot usually be refilled. Current examples include Ritalin and other amphetamines, morphine, morphine derivatives and analogs, and barbiturates used as fast-acting anesthetics. Schedule III drugs are those that have somewhat less potential for abuse and dependency and for which there is an accepted medical use. Examples include anabolic steroids, dihydrocodeine in combinations with other analgesics, certain synthetic marijuana analogs, and amphetamines used as anorexiants. These can only be supplied by prescription and the prescription is limited to five refills over 6 months. Schedule IV drugs have limited potential for abuse and dependency. The restrictions on prescriptions are similar to those in Schedule III. They can only be distributed by prescription. Examples include cough suppressants that include codeine, diarrhea preparations containing small amounts of opiates, benzodiazepines, and certain anticonvulsants.

TOPIC 11.1 CONGRESSIONAL HEARINGS ON DRUG APPROVALS AND COSTS

Throughout the second half of the century, Congress has held hearings aimed at investigating the relationship between drugs, health, and health-care costs. Because of the wide public interest in these topics, the hearings generated extensive media coverage. Unfortunately, these hearings have seldom encouraged, or even permitted, rational discussion. More often, they have been exploited for political purposes.

The Kefauver Hearings, 1959–1962. Estes Kefauver, a Democrat from Tennessee who had campaigned for the presidency in 1952 and 1956, was the party's vice-presidential nominee in 1956. Between 1957 and 1962, his US Senate Anti-Trust and Monopoly Subcommittee investigated the prices of prescription drugs. Kefauver's staff employed a number of distortions. For example, it was claimed that only about 1/3 of drugs were developed in the United States by going back to the 1500s and including highly dubious examples. Misleading information was provided on profit margins and the effect of patent protection on new drug production. A specific charge involved the price mark-up of drugs. The staff produced figures claiming to demonstrate mark-ups of 1000% and more, but these were based solely on raw material costs and ignored research, testing, marketing, and distribution costs. The scheduling of witness and reports to the press were timed to ensure that the most inflammatory charges would receive national news coverage. More responsible members of the committee, such as Philip Hart (D, MI) and Roman Hruska (R, NE), criticized the hearings as efforts to "reinforce the preconceptions of the subcommittee staff." Kefauver eventually drafted proposed legislation that would have limited patent protection on new drugs to 3 years. Witnesses such as Vannevar Bush (see Section 12.1.1) defended the importance of patent protection. Opposition from senators such John McClellan (D, AR) and Sam Irvin (D, NC) eventually side-tracked most of Kefauver's proposals.

During the course of the hearings, the thalidomide tragedy came to light (see Topic 11.2). Dr. Helen Taussig of John Hopkins had noted in a medical meeting in April 1962 that the drug was suspected of causing birth defects. Dr. Frances Kelsey, aware of these concerns, prevented approval of the drug by the FDA, sparing the United States, for the most part, the multiple birth defects caused by the drug that occurred in Europe. Nevertheless, the case brought home to the American public the issue of drug safety and put enormous pressure on the FDA to prevent any such future catastrophe. The thalidomide incident made it clear that the approval of an unsafe drug would have far more visible consequences than failure to approve a safe one and tilted the FDA emphasis toward safety. As a result of the thalidomide case, an update of the FDA legislation passed unanimously in the House and Senate in October, 1962. Its main provision required that drugs be shown to be both effective and safe. The process for NDA (see Section 11.4.2) resulted. The legislation also instituted regulation of manufacturing practice, including inspections and tightened restrictions on advertising. Use of generic and well as brand names in informational material was also required.

The Nelson Hearings (1967–1978). Like Kefauver, Senator Gaylord Nelson (D, WI) conducted a long series of hearing that covered many of the same topics as the Kefauver hearings, and paid particular attention to pricing. Nelson eventually proposed a far-reaching bill to establish federal control of nearly all aspects of drug prescription and marketing. It also proposed government responsibility for clinical trials. No legislation resulted, however.

The Kennedy Hearings, (1973–2007). Senator Edward Kennedy (D, MA) began hearings of the Senate Health Subcommittee in 1973. The original issues included "excessive profits," "excessive prescription of drugs," "marketing practices," and "influence on MD prescriptions." Eventually, the hearings also dealt with such topics as informed consent in clinical trials, FDA effectiveness, the adequacy of clinical trials, and the "drug lag"—the contention that drug development and approval in the United States, was slow compared to Europe and Japan.

These hearings, and others like them that have continued since, have established a wary and often adversarial relationships between the industry, consumer groups, the FDA, and Congress. While many of the issues are technical and legal, some such as drug safety and pricing attract public attention. The arguments continue to this time. At one extreme, the regulatory process is said to slow the development of drugs and reduce the incentive for innovation. Those holding this view argue for relaxing the review process. At the other extreme, some consumer advocates argue for price-reduction schemes, with little regard to the economic consequences. They use any instance of failure of a drug as an occasion to argue for increased regulation. Added to this mix is liability litigation.

Sources: Anonymous, 1961. *The Untold Story of the Drug Hearings*, The Pharmaceutical Manufacturers Association; Cray, W. C. *The Pharmaceutical Manufacturers Association; the First 30 Years*, The Pharmaceutical Manufacturers Association, 1989.

11.4.3 OTHER GOVERNMENT AGENCIES INVOLVED IN DRUG RESEARCH AND DISTRIBUTION

11.4.3.1 THE PUBLIC HEALTH SERVICE

The earliest public health concerns had to do with maritime commerce and included local laws governing the quarantine of ships suspected of having infected crews or passengers. From these concerns, there grew up a group of Marine Service Hospitals for the treatment of sick sailors. By the start of the civil war, there were 27 such hospitals. The origins of the Public Health Service (PHS) and The National Institutes of Health (NIH) can be traced to the establishment in 1887 in New York of a one-man Hygienic Laboratory to research communicable diseases such as cholera, diphtheria, typhoid, typhus, and tuberculosis. The motivation, at least in part, was the great influx of immigrants from Europe, some of whom were carrying these and other diseases. The one man was Joseph James Kinyoun.[d] The Hygenic Laboratory was moved to Washington DC in 1891.

In 1902, Congress passed a law regulating the purity of vaccines. It was originally intended to apply only to the District of Columbia, but the word "national" was inserted and the Hygenic Laboratory suddenly had the authority to set standards throughout the country. This legislation followed a 1901 incident in which diphtheria and smallpox vaccines contaminated with tetanus bacillus led to the deaths of about 20 children in Missouri and New Jersey. The facilities were expanded in 1904 in a building located on the grounds of the Naval Observatory. In 1912, the Hygenic Laboratory became part of the newly formed PHS. Among the investigators who worked under the auspices of the PHS were Joseph Goldberg (pellagra, see Section 10.1.1.4), Dr. Alice C. Evans (brucellosis as the cause of undulant fever) and Dr. Charles Armstrong, who researched many infectious diseases, including malaria, parrot fever (psittacosis), and rabbit fever (tularemia).[9]

11.4.3.2   THE NATIONAL INSTITUTES OF HEALTH

The National Institute of Health (NIH) was established in 1930, when Sen. Joseph E. Ransdell (D, LA) shepherded through Congress a law providing an appropriation for fellowships for study of prevention and cure of disease. The current location of the NIH in Bethesda, MD, USA was established in 1930 on a 45 acre tract donated by Luke Ingalls Wilson, a Chicago-based clothing manufacturer, whose wife, Helen Clifton Woodward, was an heir to the Woodward Lothrop department store in Washington. The first building was an animal research facility. With the support of the Surgeon General Dr. Thomas Parran and President Franklin D. Roosevelt, the decision was made to establish the National Institute of Health at that location. Dr. Lewis R. Thompson became

---

[d]Jospeph J. Kinyoun was born in North Carolina, the son of a physician who eventually moved his practice to Missouri. Joseph J. Kinyoun received his M. D. from Bellevue Medical College in New York in 1882. He also studied pathology and bacteriology with Robert Koch in Berlin and Elie Metchnicoff in Paris. He joined the US PHS and was assigned to establish a Hygiene Laboratory in 1887. He established diagnostic tests for cholera and diphtheria and began production of diphtheria antitoxin. In 1896, he was awarded a Ph. D. from Georgetown University. He held PHS appointments in San Francisco and Detroit before resigning in 1900. After that, he was on the faculty of George Washington University Medical School in Washington, DC, USA and eventually became the Director of the Washington DC Public Health Laboratory.

Source: Winkelstein Jr., W. *Epidemiology* **2007**, *18*, 181.

the Director in 1937. The first of the disease-oriented Institutes, The National Cancer Institute, was authorized by Congress later in 1937. Just days before, Luke I. Wilson had died of cancer, and the family donated additional land as the site for the Cancer Institute. At the end of WWII, some of the wartime research projects related to health, such as those on penicillin (see Section 12.1.1) and cortisone (see Section 15.2) were transferred to the NIH. This was the beginning of the biomedical research programs that became the major activity of NIH. In 1948, new Institutes dedicated to heart, microbiology, experimental biology and medicine, and dental research were established. The hospital center, where studies involving patients are conducted, was constructed in 1953. It treats about 150,000 patients a year as part of clinical studies. The NIH became part of the Department of Human and Health Services (HHS) in 1980. At present, there are twelve institutes, including the National Institute of Environmental Health and Safety, which is located in the Research Triangle, NC. Currently, about 80% of the NIH budget supports extra-mural research. Each Institute also maintains an intramural research program.

11.4.3.3   CENTERS FOR DISEASE CONTROL AND PREVENTION

The Centers for Disease Control (CDC) are located in Atlanta GA, USA and opened in 1946. It is part of HHS. It has responsibility for monitoring infectious diseases, food-borne pathogens, as well as occupational and environmental health. The National Institute of Occupational Safety and Health, located in Washington DC, is part of the CDC. The original purpose of the CDC was prevention of malaria, which accounts for its location in the South where malaria remained endemic after WWII. The Center gradually expanded into other aspects of infectious disease and public health. It operates both in the areas of laboratory investigation and epidemiology. In the 1970s, additional responsibilities for environmental and lifestyle diseases were added. These included such health risks as accidents, violence, and smoking. The environmental areas included issues such as cancer clusters and birth defects.

The "Prevention" part of the name was added in 1992, but the initials CDC were retained. The CDC's role in tracking influenza strains is well known. It also played a major role in detecting new diseases, such as Legionnaires Disease (1976) and AIDS (1981). In August of 1976, a number of people associated with an American Legion convention in Philadelphia became ill and some died from pneumonia. There was concern that the disease might be related to the deadly 1918 swine flu virus. Although tests ruled out swine flu, the cause could not be immediately identified and it was decided to proceed with a nation-wide immunization campaign. About 50 million people were vaccinated, but late in the year evidence of increased risk of Giullain–Barre Syndrome, a neurological condition somewhat like polio, appeared. At that point, the immunization program was stopped. Late in the year, CDC scientist found a bacterium to be the cause of Legionnaires Disease. AIDS was recognized when an unusual number of requests for a drug, pentamidine, used to treat a rare pneumonia, *Pneumocystis carinii*, were received. The requests came to CDC because the drug was not licensed. Clusters of a rare cancer, Karposi sarcoma, were also soon noted and both were concentrated among homosexual men. Blood tests revealed very low T-cell lymphocyte counts. Evidence soon appeared that AIDS was also associated with blood transfusions and intravenous drug use. Chapter 19 provides further information on the AIDS pandemic.

11.4.3.4   THE MEDICAL RESEARCH COUNCIL IN THE UNITED KINGDOM

Interest in public health and infectious disease was an important issue in the United Kingdom in the second half of the 19th century, with much of the focus on tropical diseases that occurred in colonial territories. British military officers, for example, played a major role in the discovery of the cause of malaria (see Section 18.1.1). Private organizations such as the Lister Institute of Preventive Medicine and the Wellcome Research Laboratories also had research program in communicable disease. In 1911, the National Insurance Act established a system of health and unemployment insurance. It also authorized a source of research funding. After the end of WWI, the MRC was established as an entity to support research. Early areas of research focus included tuberculosis and rickets. During WWII, the MRC supported research to find antimalarial agents. The MRC conducts and supports health-related research, in many ways similar to the role of the NIH in the United States. The MRC currently operates through some 25 units distributed in various locations, but primarily in Cambridge, Edinburgh, London, and Oxford. MRC-supported labs have been the site of some of the most momentous advances in biomedical research such as the Watson–Crick model for DNA and the earliest work on protein crystal structures.

11.5   THE PHARMACEUTICAL INDUSTRY

The pharmaceutical industry is a major economic enterprise. The estimated global sales in 2004 were about $600 billion, with about 45% in the United States and 25% in Europe. Annual growth rates are around 5–6% in developed markets but are much higher in developing countries. The pharmaceutical industry has historically enjoyed high rates of return on investment, estimated at

15% during the 1990s. The industry has high research/development costs and profits at individual firms tend to be concentrated in a relatively small number of "blockbuster" drugs.

11.5.1 THE PHARMACEUTICAL MANUFACTURERS ASSOCIATION (BIG PHARMA)

The 1950s was the era of introduction of the antibiotics and their profound, nearly miraculous, impact on public health. But, beginning with thalidomide in 1962, the potential for harmful side effects of drugs became a public concern. Sometimes for its own mistakes, but also because of the efforts of politicians, the media and consumer advocacy groups, the pharmaceutical industry has found itself in the public spotlight for most of the final quarter of the 20th century. Politicians found the pricing of drugs an appealing target.

The Pharmaceutical Manufacturers Association was formed in 1958 by the merger of two predecessor organizations, The American Pharmaceutical Manufacturing Association and the National Association of Manufacturers of Medicinal Products, both of which dated to the early 1900s. The goals of the organization included promotion of education of scientists, funding of basic research, and promotion of rational and ethical claims for pharmaceuticals. It became the center of industry communication with and response to governmental organizations including Congress and the regulatory agencies, that is, it became the industry's major public relations and lobby organization. Among its major objectives is opposition to regulation of drug costs. The nickname Big PHARMA is associated with the Pharmaceutical Manufacturers Association and its member companies.

11.5.2 ECONOMICS OF THE PHARMACEUTICAL INDUSTRY

11.5.2.1 DRUG COSTS

The pharmaceutical industry is characterized by high initial investment costs in research and development, but comparatively low productions costs. Only a fraction of the candidate drugs reach the market as profitable products. The profits of successful drugs must recover costs of the many unsuccessful candidates. The final drug costs, which are set to recover the research, development, and marketing costs, typically show a large "mark-up" relative to production costs. At various times price regulation of drugs has been proposed, but have never been instituted in the United States (see Topic 11.1). On the other hand,

steep discounts of prices are available for large insurance and government programs through either contracting procedures or the establishment of maximum
reimbursement costs. The United States has the highest per capita medical ex-
penditure in the world, roughly double that of most European countries. As of
2000, drugs accounted for about 10% of medical costs. The US government
pays about half of these costs, as opposed to about 80% in Europe and Japan.
Since the 1990s, drug costs have been one of the targets for containment of
health care costs throughout the developed world.

The economic incentives are strongest for pharmaceutical companies to develop drugs for the most extensive markets, for example the so-called life-style
drugs, such as those to control cholesterol levels or to treat anxiety. The greatest effort in the pharmaceutical industry is toward "block-buster" drugs that
can rapidly recoup the investment in the development and approval procedures.
Drugs that treat chronic conditions tend to be more profitable than those for
acute conditions, because they are used over a long term. The competitive process also generates "me-too" or "copy-cat" drugs that are patentable analogs of
other successful drugs, but which may offer only marginal additional benefit.
These are frequently criticized as examples of nonproductive research effort
aimed primarily at enhancing revenues. However, it can also be argued that me-
too drugs provide price competition. As mentioned in Section 11.2, the United
States and New Zealand are the only countries that permit direct advertising
of prescription drugs. It is estimated that the industry spent about $4 billion in
such advertising in 2005. Another $7 billion was spent in promoting drugs to
the medical professions. Table 11.1 lists the top 10 brand name drugs by sales
in 2010. We will consider most of these drugs in subsequent sections on specific
drug types.

TABLE 11.1   Top Ten Brand-name Drugs by Sales for 2010.

Brand name	Generic name	$ Billions	Purpose	Section
Lipitor	Atorvastatin	7.2	Statin	15.4.2
Nexium	Omeprazole	6.3	Antacid	
Plavix	Clopidogrel	6.1	Anticoagulant	13.1.2
Advair Diskus	Fluticasone/salmeterol	4.7	Asthma	15.2/14.2.3
Abilify	Ariprazole	4.6	Antipsychotic	17.4.2
Seroquel	Quetiapine	4.4	Antipsychotic	17.4.2
Singulair	Montelukast	4.1	Asthma	
Crestor	Rosuvastatin	3.8	Statin	15.4.2
Actos	Pioglitazone	3.5	Antidiabetic	16.4.6
Epogen	Erythropoietin	3.3	Anemia	

11.5.2.2   PATENT PROTECTION

The information developed by research in the pharmaceutical industry and other organizations such as universities and independent research laboratories is *intellectual property*. In the United States, Canada, and Europe, long-standing patent law regulates the incentives and rewards associated with a useful invention. More recently, globalization has resulted in widespread international agreements on the protection of intellectual property, such as is involved in the discovery and development of new drugs. The fundamental requirement for a new drug to be patented in the United States is that it be useful, novel and non-obvious. Basic laws of nature and natural substances cannot be patented in the United States. Furthermore, patents are limited to a specific invention. For example, a patent issued to the University of Rochester on all inhibitors of the enzyme cyclooxygenase-II was ruled invalid. Rather only specific drugs for this purpose can be patented. The current maximum period of patent for a drug is 20 years. In actual fact, the period of protection is more likely to be 10–12 years because of time required for clinical studies and FDA approval. The Drug Price Competition and Patent Term Restoration Act of 1984, commonly called the Hatch–Waxman Act, provided for up to a 5-year extension in compensation for the time in clinical trials and the FDA-review process. The overall objective of patent law and drug regulation is to optimize the balance between incentive for research and development of new drugs against the costs and risks to the consumer.

TOPIC 11.2 WORST CASE EXAMPLES

TOPIC 11.2.1 THE "ELIXIR SULFANILAMIDE" INCIDENT

In 1937, soon after sulfanilamide had come to the attention of the American public as a result of the President's son's illness (see Section 11.1.1), salesmen for the S. E. Massengill Company, a pharmaceutical supplier located in Bristol TN, reported a desire for a liquid form of the drug. The company's chief chemist and pharmacist, Harold Watkins, formulated the drug with raspberry flavor and red coloring, using as the solvent diethylene glycol. The company carried out no tests on the stability or toxicity of the formulation. In fact, diethylene glycol, a structural dimer (an ether) of the antifreeze component ethylene glycol, was known to be toxic, but Watkins did not realize this. Beginning in September, 1937, the Massengill Company shipped 240 gallons of the formulation under the name, Elixir Sulfanilamide. The toxicity was first noted by physicians in Tulsa OK and reported to the American Medical Association on October 11.

A physician in New York reported this information to the FDA on October 14. An immediate field investigation confirmed many deaths, and resulted in recall telegrams from the Massengill Company warning about the danger of the product. As the distribution of the product was tracked, 107 deaths were documented. The toxicity was the result of kidney failure induced by the solvent. The only law the company had violated was misbranding because the word "elixir" meant the material contained alcohol, which it did not. Dr. Samuel E. Massengill, the firm's owner, was fined $26,000, the maximum permitted under the existing laws. Harold Watkins committed suicide. The public outcry led to enactment of laws, strengthening the authority of the FDA (see Section 11.4.1.1).

Source: Ballentine, C. *FDA Consumer Magazine*; June, 1981.

TOPIC 11.2.2 THALIDOMIDE

Thalidomide was introduced by a West German pharmaceutical company, Grunenthal, in 1957. The compound was a powerful sedative and hypnotic. At the time, only minimal requirements for drug safety were in place, and while the company demonstrated low acute toxicity, little else was known about its activity or mechanism of action. The drug soon became available without prescription and the company began to promote it as a preventative for "morning sickness," the nausea associated with pregnancy. An Australian obstetrician, William McBride, who began to recommend the drug soon noticed an unusual number of birth defects in newborns he delivered, specifically shortened, malformed or missing limbs. A German physician and geneticist, Widukind Lenz, also suspected the drug. There were soon widespread occurrences of such birth defects in all the countries where the drug had been sold. Altogether about 10,000 children, mainly in Britain and Europe, were born with birth defects attributed to thalidomide.

thalidomide

Although the drug was introduced for clinical trials in the United States, it was never approved for sale. An FDA inspector, Frances Kelsey, refused to approve the drug considering the data on safety to be inadequate. The case received wide publicity and contributed to the passage of the Kefauver–Harris Act of 1962, which required evidence of both efficacy and safety as part of the drug approval process.

Investigation of the mechanism of action of thalidomide indicated that it inhibits angiogenesis and that this is the cause of the failure of limbs to develop normally. The drug was subsequently approved for the treatment of certain forms of leprosy (1998) under conditions of strict monitoring to prevent use by pregnant women. In 2006, the FDA approved the drug, in combination with dexamethasone, for treatment of multiple myeloma.

Source: Stephens, T.; Brynner, R. *Dark Remedy: The impact of thalidomide and its revival as a vital medicine;* Perseus Publishing: Cambridge, MA, 2001.

TOPIC 11.2.3 THE DALKON SHIELD

The Dalkon Shield was an intrauterine contraceptive device developed by Dr. Hugh Davis, a physician associated with the Obstetrics and Gynecology Department at Johns Hopkins University. The Dalkon Corporation was formed 1969 by Davis, Irwin Lerner, an inventor who devised and patented the device, and Robert E. Cohn, Lerner's lawyer. Later, they were joined by Dr. Thad Earl, the director of a family planning clinic in Ohio. Rights to the Shield were purchased in 1970 by A. H. Robins, a pharmaceutical firm located in Richmond VA, for $750,000 and 10% of the future sales. A. H. Robins had begun as a small distributor of drugs, but with the success of *Robitussin* cough syrup and acquisition of products such as *Chap Stick*, it had become a multimillion dollar enterprise. Dr. Davis had published a paper on the effectiveness of one version of the device and had claimed 99% success in prevention of pregnancy. However, he failed to report that a high proportion of the original subjects dropped out. Also, the time of use reported for the study averaged less than 6 months. He also did not disclose that he had a financial interest in the product. After acquiring the product, Robins began a vigorous advertising and public relations campaign. Articles were placed in magazines such as *Redbook, Family Circle* and *Woman's Day*. The device was promoted for use by general practitioners, as well as OB-GYN specialists. As a result, sales rose to 1670,000 in 1971 and eventually the device was used by several million women in the United States.

The Dalkon Shield was not the first such device to have been introduced, but previous devices had been plagued by infections and were more difficult to insert. Soon after its introduction, reports of problems began to surface. These usually stemmed from pelvic inflammations, and resulted in miscarriage, irreversible reproductive damage, and even some deaths. A particular problem associated with the Dalkon Shield was that it used a multistrand filament that was connected to the device and terminated in the vagina. This filament acted as a wick and provided a pathway for bacterial infection of the normally bacteria-free uterus. Furthermore the individual filaments were made of nylon, which was susceptible to degradation in the body.

As additional results from use of the Shield became available, it was found that pregnancy rates were often 5% or more and removal rates were 25% or more. Robins began receiving complaints about the Shield from physicians, but made no effort to investigate. Rather, it continued to rely on the original data provided by Dr. Davis. Reports of severe pelvic infection and septic abortions continued to accumulate in 1972 and 1973. In particular, in June 1972, Dr. Earl himself, one of the original Dalkon Corporation owners and a paid consultant to Robins, reported five cases of severe complications and septic abortions.

In the mid-1970s, Howard J. Tatum of the The Population Council published several papers demonstrating the ability of the Dalkon Shield string to wick and the detection of pathological bacteria in the string.[10] By this time, independent studies had begun to appear that contradicted the 1% pregnancy rate reported by Davis. In June, 1974, the FDA asked Robins to suspend sales of the Shield, which it did. An advisory committee recommended continuation of the morato-rium in October, 1974. Robins terminated sales of Dalkon Shields in 1974, but neither the company nor the FDA initiated efforts to have the devices removed. Lawsuits began in 1974 and the evidence showed that the Robins Company was aware of both the risk of infection and the inaccurate failure rate data. Numerous lawsuits were filed between 1975 and 1980, but Robins continued to deny re-sponsibility. Not until 1980, did Robins send a letter to doctors recommending removal, but refused to pay the cost and continued to obscure the danger of pelvic infection. In 1984, after several deaths attributed to the Shield had oc-curred, Robins issued a formal recall and removal at its expense. By 1985, about 15,000 damage claims had been filed. The rapidly escalating claims led A. H. Robins to file for Chapter 11 bankruptcy protection in 1985. In the approved settlement, the company was sold to American Home Products for $3 billion of which $2.45 billion was set aside for claims. Another $370 million was obtained from insurance. Controversy, delay and litigation persisted for another 15 years. Eventually about 200,000 damage awards were made.

Intrauterine contraceptive devices have had some record of success both before and after the Dalkon Shield. A device called the Lippes Loop, made of polyethylene was used for a time prior to the arrival of the Dalkon Shield. The Population Council and the G. D. Searle Company marketed T-shaped devices that contained copper wire and were more effective, but the Searle product was driven from the market by the threat of litigation after the Dalkon Shield events. An improved version of the Population Council device became available.[11] Currently, about 10% of women in the United States use intrauterine devices for birth control.

Sources: Mintz, M. *At Any Cost*; Pantheon Books: New York, 1985; Perry, S.; Dawson, J. *Nightmare*; Macmillan: New York, 1985; Hicks, K. M. *Surviving the Dalkon Shield IUD*; Teachers College Press: New York, 1994; Hawkins, M. F. *Unshielded*; University of Toronto Press: Toronto, 1997.

KEYWORDS

- **chemotherapy**
- **antibiotics**
- **sulfanilamide**
- **penicillin**
- **tuberculosis**
- **bioassays**
- **drug design**
- **high-throughput screening**
- **investigational new drug (IND)**
- **new drug application (NDA)**

BIBLIOGRAPAHY BY SECTION

Section 11.1: Weathrall, M. *In Search of a Cure*. Oxford University Press: Oxford, UK, 1990.

Section 11.2: Mahoney, T. *Merchants of Life*; Harper: New York, 1959.

Section 11.3: Munoz, B. *Nat. Rev. Drug Discovery* **2009,** *8*, 959–968; Scannell, J. W.; Blanckley, A.; Boldon, H.; Warrington, B. *Nat. Rev. Drug Discovery* **2012,** *11*, 191–200.

Section 11.4.1: Young, J. H. *The Toadstool Millionaires*. Princeton University Press: Princeton, NJ, 1961; Young, J. H. *The Medical Messiahs*; Princeton University Press: Princeton, NJ, 1967; Hilts, P. J. *Protecting America's Health*. Alfred Knopf: New York, 2003; Ceccoli, S. J. *Pill Politics: Drugs and the FDA;* Lynne Riemer Publishers: Boulder, CO, 2004; Hawthorne, F. *Inside the FDA*; John Wiley: Hoboken, NJ, 2005; Borchers, A. T.; Hagie, F.; Keen, C. L.; Gershwin, M. E. *Clin. Ther.* **2007,** *29*, 1–16.

Section 11.4.2: Ng, R. *Drugs from Discovery to Approval*. Wiley-Liss: Hoboken, NJ, 2001; Mathieu, M. *New Drug Development: A Regulatory Overview*, 6th ed. Parexel: Waltham, MA, 2002; Peck, G. E.; Poust, R. *Drugs Pharm. Sci.* **2002,** *121*, 627–643.

Section 11.4.3.3: Etheridge, E. W. *Sentinel for Health*; University of California Press: Berkeley, CA, 1992.

Section 11.5: Williams, T. R. *Sem. Radiat. Oncol.* **2008,** *18*, 175–185.

REFERENCES

1. Newman, D. J.; Cragg, G. M.; Snader, K. M. *J. Nat. Prod.* **2003,** *66*, 1022–1037.
2. Kennedy, D. *J. Am. Med. Assoc.* **1978,** *239*, 423–426.
3. Kessler, D. A.; Haas, K.; Feiden, K.; Lumkin, M.; Temple, R. *J. Am. Med. Assoc.* **1996,** *264*, 2409–2415.
4. Wardell, W. *J. Am. Med. Assoc.* **1978,** *239*, 2004–2011; Goldberg, R. *Untimely Access: An Analysis of America's Drug Lag*; George Washington University Center on Neuroscience, Medical Progress and Society: Washington, DC, 1996.
5. Kaitin, K.; Healy, E. *Drug Inf. J.* **2000,** *34*, 1–14; Berndt, E. R.; Gottschalk, A. H. B.; Philipson, T. J.; Strobel, M. W. *Nat. Rev. Drug Discovery* **2005,** *4*, 545–554.
6. Kay, J. *Arthritis Res. Ther.* **2011,** *13*, 112.
7. Gale, E. A. M. *The Lancet* **2001,** *357*, 1870–1875.
8. Senior, J. R. *Clin. Pharmacol. Ther.* **2009,** *85*, 331–334; Aithal, G. P.; Watkins, P. B.; Andrade, R. J.; Larrey, D.; Molokhia, M.; Takikawa, H.; Hunt, C. M.; Wilke, R. A.; Avigan, M.; Kaplowitz, N.; Bjornsson, E.; Daly, A. K. *Clin. Pharmacol. Ther.* **2011,** *89*, 806–815.
9. Lilienfeld, D. E. *Perpect. Biol. Med.* **2008,** *51*, 188–198.
10. Tatum, H. J.; Schmnidt, F. H.; Phillips, D.; McCarty, M.; O'Leary, W. M. *J. Am. Med. Assoc.* **1975,** *231*, 711–717; Tatum, H. J.; Schmidt, F. H.; Phillips, D. *Contraception* **1975,** *11*, 465–477; Tatum, H. J. *Fertil. Steril.* **1977,** *28*, 3–28.
11. Burnhill, M. S. *Am. J. Gynecol. Health* **1989,** *3-S*, 6–10.

CHAPTER 12

ANTIBIOTICS: THE BATTLE WITH THE MICROBES

ABSTRACT

Both natural and synthetic antibiotics were discovered between the two World Wars. Penicillin was discovered by Alexander Fleming in 1928 and isolated in the late 1930s by a group led by Howard Florey. Gerhard Domagk's group synthesized the first sulfa drug in 1932. After WWII, several important types of antibiotics were discovered including other β-lactams, aminoglycosides, macrolides, tetracyclines, and quinolones. Many semisynthetic derivatives were also introduced. Resistance to most antibiotics was noted within a few years and organisms resistant to multiple drugs appeared as well. The rate of discovery of new classes of antibiotics slowed after 1970.

The term *antibiotics* refers to substances that can be used to treat infectious diseases caused by microorganisms, mainly bacteria. There are several sources of antibiotics. One is isolation by growth of microorganisms and another is synthetic modification of these compounds. Other antibiotics are entirely synthetic in origin. In some cases, the microorganisms used for antibiotic production have been genetically modified to improve yield and efficiency.

The idea that microorganisms, "germs" as we call them, are responsible for infectious diseases took firm hold in the last quarter of the 19th century, aided by improvements in microscopic and staining techniques. Diseases such as typhoid fever, cholera, puerperal (childbed) fever, and tuberculosis were identified with specific microorganism and it was shown that these organisms could cause infection when transferred to a susceptible host. These ideas are now nearly universally understood, but were new at the beginning of the 20th century. Pathogenic bacteria are responsible for a wide variety of illnesses: pneumonia, tuberculosis, respiratory, gastrointestinal and urogenital infections, and so on. Bacteria are broadly classified as Gram-positive or Gram-negative. This classification originally was on the basis of the staining patterns of various bacteria, and is related to the structure of the cell wall. In general, Gram-negative cell walls are more difficult for antibiotics to penetrate.

The idea that chemicals could be used to treat disease is attributed to Paul Ehrlich (see Section 11.1) who conceived the idea as he studied staining bacteria with dyes. He referred to the potential drug as a "magic-bullet," a chemical targeted at the infectious organism. In 1909, he was able to demonstrate the validity of the concept when one of many compounds synthesized in his laboratory, Salvarsan, proved to be active against syphilis. In 1935, Gerhard Domagk discovered that sulfa drugs could be used to treat several common bacterial infections (see Section 11.1). In the 1940s, the remarkable activity of penicillin was demonstrated. From that point on, the development of antibiotics was very rapid. We will consider five classes, the β-lactams, aminoglycosides, macrolides, tetracyclines, and quinolones. There are several other types of antibiotics, but these groups illustrate the sources, mechanism of action, and therapeutic application of antibiotics. The global market for antibiotics was about $30 billion, as of 2006.

12.1 β-LACTAMS

12.1.1 DISCOVERY AND DEVELOPMENT

The β-lactams are so-named because they all contain a four-membered cyclic amide (lactam) ring as shown in the structures in Scheme 12.1. The first penicillin was discovered by Alexander Fleming.[a] The basic story of a Petri dish in which bacterial growth was inhibited by a mold is a familiar one. The background and subsequent progress is less well known. Fleming began research in pathology in London in 1906. At the time, immunization was seen as the main means for combating infectious disease, although only modest progress had been made since Jenner (Smallpox, 1796) and Pasteur (Rabies, 1885) had demonstrated its potential. During WWI, Fleming served in a field laboratory in Normandy and saw first hand the ravages of infected wounds. After the war ended, Fleming returned to the Immunology Department of St. Mary's Hospital and Medical School in London. He conducted experiments on the antibacterial activity in tears. He identified one active substance as the protein lysozyme,

[a]Alexander Fleming was born in rural Scotland in 1881 and moved to London to complete his secondary education. He entered St. Mary's Hospital Medical School in 1903 and received a medical degree in 1906. He then joined the Inoculation Department at St. Mary's where he studied bacteriology and began research under its director, Sir Almoth E. Wright, a leading figure in the field. He earned a B. Sc. degree in bacteriology in 1908. He received an award for his thesis *Acute Bacterial Infections*, which proposed multiple approaches, including Ehrlich's recently proposed chemotherapy. During WWI Fleming served in the Medical Corps and did research as a bacteriologist in a field laboratory in France. He became convinced that the antiseptics being used were doing more harm than good. After the war, he returned to St. Mary's and continued research. He discovered the antibacterial properties of the enzyme lysozyme found in tears and mucus secretions. Fleming succeeded Wright as the Director of the Inoculation Department in 1946 and it was subsequently renamed the Wright–Fleming Institute.

which is also present in blood, especially white cells. Fleming observed that while lysozyme inhibited many kinds of bacteria, it did not affect the most pathogenic ones.

It was in the course of these experiments that Fleming made the famous observation that a mold contaminating one of his cultures had inhibited growth of *Staphylococcus aureus*. He retained a sample of the mold for future culture and tested it against other pathogenic bacteria and found that it inhibited several others. Fleming deduced that the mold must be producing a substance that then diffused through the agar in the culture dish. He prepared an extract from the mold and demonstrated that it was thousands of times more active than lysozyme. The mold was identified as *Penicillium notatum*, and Fleming called the active component *penicillin*. The solubility properties of the substance demonstrated that, unlike lysozyme, it was not a protein. The β-lactam ring is chemically fragile and Fleming was unable to isolate a pure substance. The results were published in the *British Journal of Experimental Pathology* in 1929, but attracted little attention. No efforts to treat infections in experimental animals or patients were made at that time.[1] A few follow-up studies in other laboratories also failed to isolate the active substance.

The isolation of penicillin from the extract was accomplished at the Dunn School of Pathology at Oxford University by a team led by Howard Florey.[b] Florey assumed the Chair of the recently endowed School in 1935 at the relatively young age of 37. Florey was interested in the chemical aspects of pathology and assembled a team of biochemists and chemists that included Ernst Chain,[c] a Jewish refugee from Germany, Norman Heatley, and Edward Abraham. They first began work on lysozyme, following Fleming's original studies, and succeeded in isolating the pure protein. In the course of this work, Chain became interested in Fleming's work on penicillin. With Heatley doing most of the isolation work, they succeeded in obtaining the active component. Heatley found that penicillin could be extracted into base and recovered by neutralization (be-

---

[b]Howard Florey was born in Australia in 1898 and received his undergraduate degree at Adelaide University in 1921. He was awarded a Rhodes Scholarship to Oxford, where he completed B.Sc. and M.A. degrees in 1924. He travelled to the US as a Rockefeller Fellow in 1925 and then returned to Cambridge where he received a Ph.D. in 1927. He held academic posts in pathology at Cambridge and then the University of Sheffield. In 1935, he returned to Oxford as the head of the Dunn School of Pathology, where he took a new approach to the subject based on biochemistry and physiology. He recruited a research team that succeeded in isolating penicillin and reported the first clinical success in 1941. His wife Ethel, a physician, collaborated in the clinical studies. Florey was knighted in 1944 and made a Baron in 1965.

[c]Ernst Chain was born in Berlin in 1906, where his father was a chemist and businessman. Chain became interested in chemistry at his father's factory and received a degree in chemistry from Friedrich–Wilhelm University in Berlin in 1930. He did research in enzymology until 1933, when he left for England after the Nazi's took power. He worked first with Sir Frederick Gowland Hopkins and then moved to Oxford in 1935. He joined Howard Florey's group in 1939. He moved to Rome in 1948, where he was scientific director at the Center for Chemical Microbiology. There, he discovered the enzyme penicillinase that can deactivate penicillin and convey resistance. He returned to England as Professor of Biochemistry at the Imperial College in 1969. He also served on the executive committee of the Weizmann Institute in Israel.

cause of its carboxylic acid group). They used the techniques of counter-current extraction, freeze-drying, and chromatography, all of which were then in their infancy. The concentrated penicillin, though not yet pure, was a million times more active than the original extract. Injections cured mice infected with otherwise fatal *Streptococcus* germs. These results were published in August 1940. The United Kingdom was already at war by this time, and Florey was unable to locate a company willing to undertake the arduous fermentation and extraction process. By 1941, they had tested the toxicity of penicillin and found none and had accumulated enough material for trials involving six massively infected patients. Although the amount of penicillin used was very small by current standards, four recovered and one died for other reasons.

Florey decided to seek support for production of penicillin in the United States. He met first with the Rockefeller Foundation, which had provided financial support for his penicillin work at Oxford. He also saw John Fulton, Sterling Professor of Physiology at Yale, whom he had met on a visit to the United States as a Rockefeller Scholar in the 1920s. Fulton immediately arranged contacts with various government officials. Although the United States was not yet at war, preparations were being made. Office of Scientific Research and Development (OSRD) under Vannevar Bush had assumed responsibility for coordination of research activities, including the "Manhattan Project" for the development of the atomic bomb. Alfred N. Richards, a leading pharmacologist from the University of Pennsylvania, was responsible for medical research. Richards met with Florey in August, 1941. Florey and Heatley visited the USDA laboratories in Peoria, IL, which had expertise in fermentation processes. Heatley stayed in Peoria and began experiments with the scientists there. They soon found a superior strain of the mold and improved the fermentation yield more than ten fold. Other important improvements resulted from using corn steep liquors in the growth medium and use of deep vessels in place of shallow pans.

Alfred Richards began to enlist pharmaceutical companies in the production of penicillin, starting with Merck. Richards had served as a consultant and adviser to Merck, an unusual arrangement at that time. By March 1942, sufficient penicillin had been isolated to begin the first human tests in the United States. The results were profound and penicillin opened a new era in treatment of bacterial infections. The OSRD coordinated research at a number of pharmaceutical companies, and by the end of 1943 about 25 companies were involved. The age of antibiotics manufactured by fermentation had begun, and semisynthetic modification soon followed. Penicillin production increased rapidly in the United States after 1944 and by June, when the invasion of German-occupied Europe began, there was sufficient supply to meet military needs. Civilian testing began in 1944 and by March 1945, the drug began to be available through normal distribution channels.[2] At the present time, β-lactams account for about half of all antibiotics used on a cost basis.

The structure of penicillin remained an enigma until 1945. In 1944, Chain and Abraham had proposed the correct β-lactam structure, but a competing structure was championed by Sir Robert Robinson,[3] an Oxford professor who was then the leading chemist in Britain. An X-ray crystal structure by Dorothy Crowfoot Hodgkin (see also Section 10.1.1.6) showed the β-lactam structure to be correct. This structural feature is responsible both for penicillin's biological activity and its chemical sensitivity. It turned out that the penicillin being produced in Oxford and America were different. The acyl group in the English material was hexenoyl (Penicillin F) while the American material was phenylacetyl (Penicillin G). The difference was presumably due to the composition of the culture medium and the strain of *penicillium* used.

Penicillin G

One of the issues that concerned the pharmaceutical companies during the development of fermentation technology was the possibility that the molecule was simple enough that it might be made easily by synthesis, superseding fermentation. This did not prove to be the case. It was not until 1957 that penicillin was synthesized and synthetic procedures have never competed with fermentation. It is the four-membered ring that makes the molecule a difficult synthetic challenge.

Fleming, Florey, and Chain received the 1945 Nobel Prize in Medicine. In the United Kingdom at the time, there was a strong presumption against patenting the results of academic research and neither Fleming nor Florey had patented their work, so little personal financial gain accrued from the initial work on penicillin. The distribution of public credit for the discovery of penicillin left a residue of animosity among the three prize recipients. As publicity began to appear in 1943 about the powers of penicillin, Florey was reluctant to promote it, because the supply was insufficient to treat patients. Fleming, on the other hand participated in publicity that helped benefit fund-raising by his institution, St. Mary's Hospital. Influential leaders of the press, who were supporters of St. Mary's promoted Fleming as the discoverer of penicillin. As a result, the popular association of Fleming's name with penicillin took hold. Chain, a sensitive, sometimes difficult personality, felt aggrieved that his pursuit and isolation of the active penicillin was not fully credited.

After the war, Ernst Chain relocated to a government laboratory in Italy and continued research on penicillin. One of the key discoveries made there was the isolation of an organism that produced the un-acylated 6-aminopenicillanic acid. This led to the production of the first semisynthetic penicillins as a result of collaboration between the English company Beecham and Bristol-Myers in the United States. The first example was methicillin, but it was followed by ampicillin and amoxicillin (see Scheme 12.1 for structures). These drugs generally provided both oral availability and an expanded range of susceptible organisms.

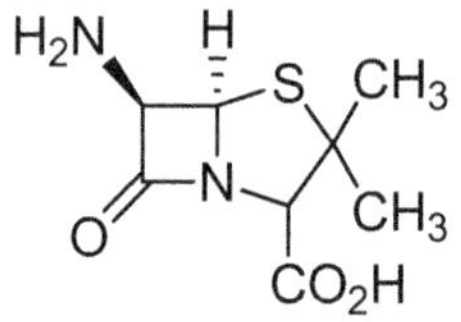

6-aminopenicillanic acid

Near the end of his life in 1979, some 40 years after he began his initial work on penicillin, Chain authored several articles on the history of penicillin.[4] He pointed out two circumstances that were crucial in the chain of events that led to the discovery of the antibiotic activity of penicillin. One had to do with the initial observation of inhibition of *Staphyloccus* growth by penicillin and the fact that this observation could not be easily reproduced by Fleming, nor later, by Chain and his associates. The original Petri dish had been incubated overnight at 37° and then left at a much cooler temperature, around 20°, for some weeks (while Fleming was on vacation). This sequence of temperature was crucial, because it allowed the *Staphyloccus* to initiate growth, but, at the lower temperature, to grow slowly in comparison with the *Penicillium* mold. This allowed the penicillin to reach a concentration that could kill the *Staphyloccus*. Fleming photographed the sample and it appeared prominently in his 1929 publication that later caught Chain's attention. Another aspect of the story is the origin of penicillin-producing mold. The story sometimes goes that the mold "blew in" an open window. In fact, a sample of Fleming's *Penicillium* strain had been maintained in the Immunology Department for unrelated purposes, and was most likely the origin of the contamination of Fleming's sample. Thus, when Chain found Fleming's article and began to follow-up the work, he had a ready source of the mold.

The wartime work on production of penicillin in the United States and United Kingdom was carried out in secret after 1943, because of its military significance. However, some information on isolation, purification, and clinical application was available from the earlier publications. A 1943 publication from the Pharmacological Institute of the University of Berlin summarized the British work.[5] An extensive review was also published in Switzerland.[6] Independent research on isolation of penicillin was carried out at the Swiss Federal Institute of Technology in collaboration with CIBA (originally Chemische Industrie Basel).[7] Production of penicillin was also carried out clandestinely at a privately held company NG&SF in Delft, in the Netherlands. Learning of penicillin through radio from Britain and the Swiss publication, microbiologists there screened a number of *Penicillium* and related species. One, called *Penicillium baculatim*, produced an active extract. In the summer of 1945, near the end of the war, a physician associated with the company obtained penicillin dropped with supplies to the liberated part of Holland, permitting direct comparison with the company's material. By January 1946, penicillin was being produced. The company subsequently became part of DSM, which remains a major penicillin producer.[8] Penicillin production began in several locations in Europe soon after the war, but it was not until 1950 that adequate supplies were available.[9] Japan, of course, was also an enemy of the United States and United Kingdom in WWII, but immediately after the end of the war, penicillin production began with input from American scientists, particularly Jackson W. Foster.[d] The Japanese had considerable prior experience in fermentation technology. A very strong research program developed, particularly in the laboratories of Prof. Hamao Umezawa and Prof. Toju Hata. The Japanese pharmaceutical industry became a major player on the international scene and Japan has remained a source of antibiotics, both from fermentation and synthesis.[10]

12.1.2 ANALOGS AND APPLICATIONS

There are four broad classes of β-lactam antibiotics. In the penicillins (penems), the β-lactam ring is fused to a thiazolidine ring. Some derivatives have a double bond in the ring. In the *cephalosporins*, the thiazolidine ring is expanded to a six-membered thiazine ring. In the *carbapenems*, such as thienamycin, the sulfur is absent from the five-membered ring. Finally, in the *monobactams*, the β-lactam ring exists without a fused component. These structural types are shown in Scheme 12.1.

The cephalosporins originated in a culture isolated from sewage discharge in Sardinia by Guiseppe Brotzu, who provided the culture to Howard Florey and his group at Oxford. Edward Abraham isolated several compounds, the most active being cephalosporin C. Commercial production began at Glaxo in the United Kingdom in the late 1950s. The cephalosporins provided an opportunity

---

[d]Foster, who had worked at Merck, was a Professor of Microbiology at the University of Texas; see J. B. Davis, *Zeitschr. Allgem. Mikrobiol.* **1967**, *7*, 173.

for a wider range of chemical modifications. The goals of the structural modification included oral activity, duration of activity (*in vivo* lifetime), broad spectrum of activity, and avoidance of resistance. Among the more recent examples are cefepime and ceftriaxone. The first of the carbapenems to be discovered was thienamycin, isolated at Merck in the 1970s.[11] A number of semisynthetic and synthetic analogs, such as impenem were developed during the 1980s. However, they, like thienamycin, had considerable nephrotoxicity and needed to be administered with a second drug. The first to be used as a single drug was meropenem, developed at Sumitomo in the 1990s. The first of the monobactams to be introduced was aztreonam. The lead compound was found in soil samples from New Jersey and then modified by synthetic acyl groups.[12]

The chemical modification of penicillin usually involves replacement of the acyl group in the amide side-chain. The acyl groups can be introduced by semisynthesis from 6-aminopenicillanic acid. The acylation can be done by enzymatic coupling.[13] Among the first important derivatives to be discovered were methicillin, amoxicillin, and ampicillin. No single drug is a universal "magic bullet" but matching their particular strengths with the clinical situation allows physicians to treat most bacterial infections successfully.

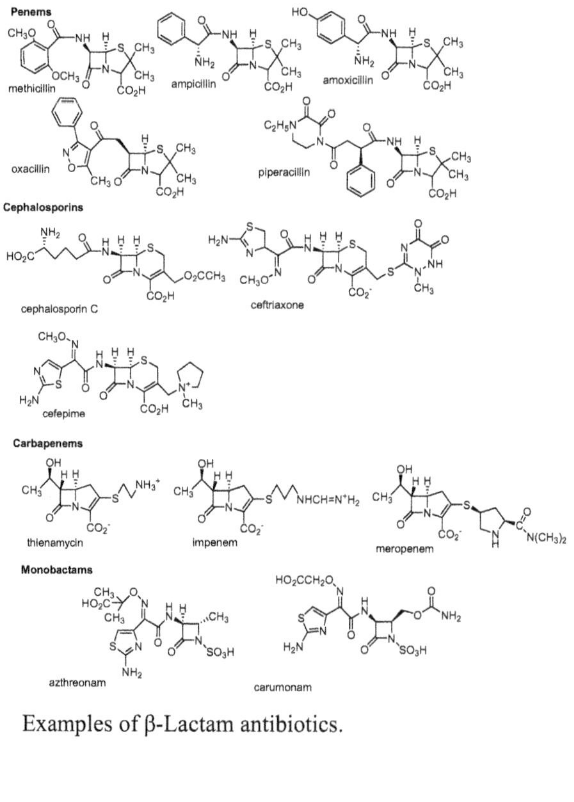

SCHEME 12.1   Examples of β-Lactam antibiotics.

The β-lactams function by inhibiting cell wall formation in bacteria. Because there is not an analogous process in humans, the antibiotics have low toxicity. The purpose of the cell wall in bacteria is to contain the cell's constituents, which are at higher concentration and higher osmotic pressure than the surroundings. The bacterial cell wall is unusual in containing D-amino acids, whereas most functional amino acids are L-, and in having tail–tail bonds. The peptides are linked to carbohydrates, as well as N-acetylglucosamine and N-acetyl muramic acid.[14]

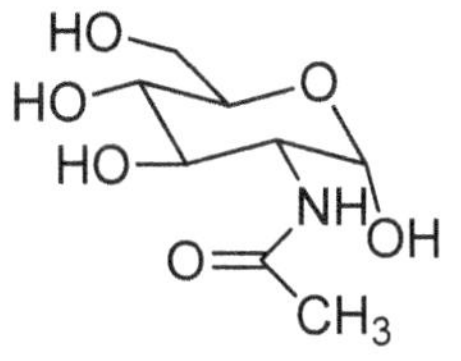

N-acetylglucosamine 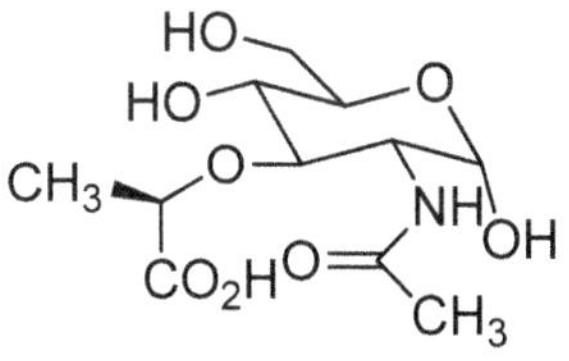   N-acetylmuramic acid

12.1.3 PENICILLIN RESISTANCE

The introduction of penicillin and the other β-lactam was soon followed by the observation of resistance. The rate of resistance development varies with structure. One of the main mechanisms of resistance is *lactamases* that inactivate the β-lactams by hydrolysis. These enzymes evolved in bacteria to control the growth of other competing bacteria. Many bacterial species contain genes that code for β-lactamases. Hundreds of different β-lactamases have been identified.[15] They fall into two broad categories, the serine hydrolases and the metallo-lactamases. The most prevalent of the β-lactamases is called TEM-1. It is found in very high percentages of the population throughout the world. Strains of *Staphylococcus aureus* that are resistant to all β-lactam antibiotics have arisen. Resistance genes can be transferred between bacteria by incorporation of a plasmid containing the gene. Resistance can also arise by mutation of the target cell wall proteins. Other mutations cause changes in membrane permeability or activate *efflux pumps* that expel the drug from the cell. These mechanisms of resistance are also observed in other classes of antibiotics.

The dosing of antibiotics is a critical feature with respect to minimizing resistance. Dosing, that is too low or ends too quickly, permits a few bacteria to survive. The survivors are the most resistant to the particular antibiotic, so inadequate dosing promotes the development of resistance. Some surviving bacteria, called *persisters*, may be dormant but become reactivated as the level of antibiotic drops. The rapid rate of reproduction of bacteria means that mutations leading to resistance can occur within the span of a few hours. The primary strategy to counter resistance has been to produce new derivatives. These, too, of course, eventually encounter resistance. Another strategy is to use inhibitors of the lactamase enzymes. Some β-lactam antibiotics are administered with a second drug that inhibits the β-lactamase. Examples are amoxicillin/clavulanic acid, ampicillin/sulbactam, and piperacillin/tazobactam.

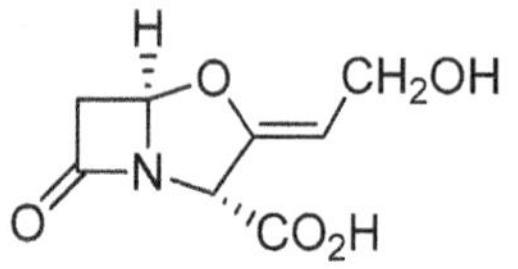

clavulanic acid 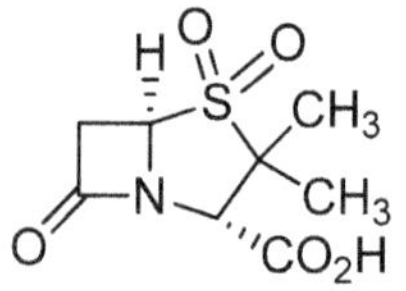   sulbactam 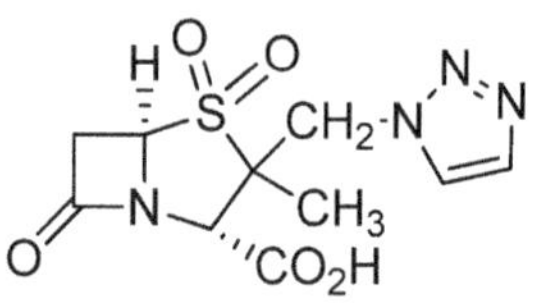   tazobactam

12.2   AMINOGLYCOSIDES AND OTHER ANTITUBERCULOSIS DRUGS

One of the bacterial infections against which penicillin is ineffective is tuberculosis (TB). TB has been present in man throughout recorded history. It is estimated that the disease caused a billion deaths in the nineteenth and 20th centuries. It is a contagious infection that can be transmitted by contaminated materials such as aerosols from a cough or sneeze. The bacterium can survive for some time in dust. Tuberculosis can also be acquired from contaminated food, in particular milk from tubercular cows. Not everyone exposed to the bacterium develops tuberculosis. The estimated rate of infection is 10%. In some individuals exposure results in control of the bacteria by encapsulation into *tubercles*. Such individuals give a positive skin test for exposure, but do not develop the disease unless the latent bacteria are activated. Tuberculosis was a major disease in the United States and the world through the first half of the 20th century. The most common infections were in the lung and in the early 20th century as much as a third of the population had been exposed. The only treatment was isolation and rest. There were sanatoriums for tuberculosis patients throughout the country. The mortality rate in the United States is thought to have been about 50%. The patients who survived often harbored residual pockets of dormant bacteria that could be reactivated, resulting in a relapse years after the initial infection. Currently, there are an estimated 8 million new cases and about 2 million deaths annually, most in the developing world.

The causative agent *Mycobacterium tuberculosis* was identified by Robert Koch in 1882 by using a special staining technique involving heating. The tuberculosis bacillus is particularly difficult to stain because it is surrounded by a waxy coat of polysaccharides that are esterified by extremely long chain (>60) α-branched β-hydroxy fatty acids. Some of the long chains also contain cyclopropane, methyl, and/or methoxy groups. The structure is illustrated in Figure 12.1. This coating is protective against both immunological attack and drug absorption. These structures are produced by a variation of normal fatty acid biosynthesis. An enzyme complex, called *fatty acid synthase II*, affects the α-acylation that branches and elongates the main chain.

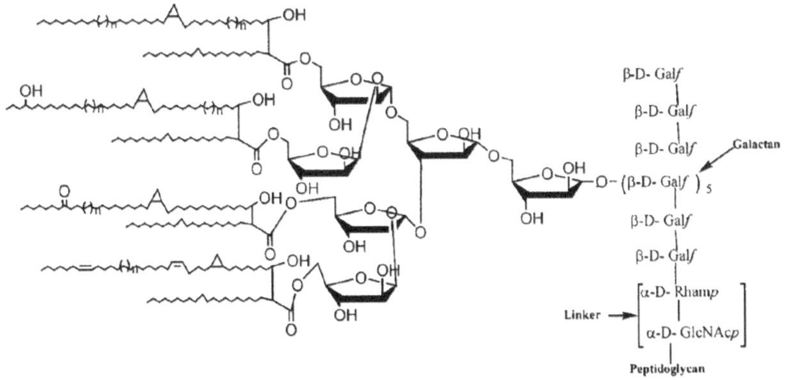

FIGURE 12.1   Structure of the cell wall of *Mycobacterium tuberculosis*. (From Tripathi, R. P.; Tewari, N.; Dwivedi, N.; Tiwari, V. K. *Med. Res. Rev.* **2005**, *25*, 93–131. © 2005, Wiley Periodicals Inc.)

12.2.1   DISCOVERY OF STREPTOMYCIN

The first antibiotic that was found to be active against tuberculosis was strepto-mycin, an aminoglycoside isolated in the laboratory of Selman Waksman,[e] a soil microbiologist at Rutgers University. One of his collaborators, Rene Dubos,[f]

---

[e]Selman Waksman was born in the Ukraine, then part of the Russian Empire in 1888. He was Jewish and op-portunities for education were limited. He emigrated to the United States in 1910 and enrolled at Rutgers and received a B. S. in agriculture in 1915 and an M. S. in 1916. He then received a Ph. D. in biochemistry at UC Berkeley. He returned to Rutgers and eventually became a professor of biochemistry and microbiology. He systematically studied the microbiological species present in soil, specializing in *Acetomyces*. Using the assay techniques that had been developed by his student, Rene Dubos, he discovered a number of antibiotics, includ-ing streptomycin.

[f]Rene Dubos was born in in France in 1901. Unable to attend a university for financial reasons, he enrolled in the Institut Nationale Agronomique and earned a B. S. in 1921. For several years he edited an agricultural magazine published in Rome and from this experience became fascinated with soil microbiology and decided to come to the United States for further study. By coincidence, he met Selman Waksman and joined his laboratory at Rut-gers, earning a Ph. D. in 1927. It was during this work that he developed his assay method. He then moved to the Rockefeller Institute in New York, applying the methodology in a search to find materials that could inactivate the pneumonia organism. Dubos's wife had fallen ill with tuberculosis and died in 1942. After a time away, Dubos returned to the Rockefeller Institute in 1944. He became particularly interested in the environmental and emotional aspects of the immune response. He went on to write a number of books and articles on the holistic nature of humankind and the environment, such as the Pulitzer prize-winning *So Human an Animal* and *Man Adapting*. He is credited with coining the phrase "Think globally, act locally."

was interested in the mechanism by which microorganisms could break down cellulose and proteins in the soil. He devised a technique for finding the most promising organism by isolating those that grew best on the material. In 1927, Dubos moved to the Rockefeller Institute in New York, where he worked with Oswald Avery (see Section 20.1 for biography). Avery was interested in the polysaccharide that encapsulates the organism *Pneumococcus* responsible for pneumonia. Dubos examined soil samples to locate microorganism that might attack the *Pneumococcus* capsule. An extract from one such organism protected mice against *Pneumonoccus*. Collaborating with chemist Rollin Hotchkiss in 1939, they isolated the active compound gramicidin from *Bacillus brevis*. Gramicidin is a modified polypeptide containing several D-Val and D-Leu amino acids. The material was produced in conjunction with Merck. Gramicidin proved too toxic for oral or intravenous treatment, but it could be used on open wounds by topical administration.

In 1939, Waksman turned his efforts to finding a compound active against tuberculosis. His laboratory investigated many microorganisms using Dubos' competitive selection approach. The first two active compounds found, actinomycin and streptothricin, were too toxic for application, but during this period Waksman established a collaboration with Merck. One of his graduate students, Albert Schatz, continued searching for microorganisms with antibiotic activity. After screening several hundred, he found an organism, *Streptomyces griseus*, that produced a substance, called streptomycin, which was active against many microorganisms. It also proved active against tuberculosis bacilli. Waksman did not have facilities for animal tests, but a collaboration was initiated with William Feldman, DVM, and H. Corwin Hinshaw, M. D., Ph. D., of the Mayo Clinic in Rochester MN USA Feldman was an expert in animal TB and Hinshaw was an expert in pulmonary disease. The drug was tested on guinea pigs at the Mayo Clinic in 1944, with promising results. Waksman turned to Merck to produce sufficient material for tests in humans. Merck chemists, led by Karl Folkers and Max Tischler, produced sufficient material. The results were highly promising and led to the first clinically successful drug against TB. Treatment with streptomycin was rapidly adopted, although the drug was very expensive to manufacture. Resistance also began to develop, within months in some patients. Feldman himself became infected, probably through exposure in the course of his laboratory work, but was successfully treated with streptomycin in combination with *para*-aminosalicylic acid (PAS) (see below). Streptomycin had shown that TB was subject to treatment by antibiotics. Although not the final word, it opened the way to more effective treatments. Selman Waksman received the Nobel Prize in Medicine or Physiology in 1952.

Subsequently, a number of other aminoglycosides have been isolated, most from various *Actinomycetes* cultures. Some of the structures are shown in Scheme 12.2. Looking carefully you will see that one of the rings in each compound is *carbocyclic* and the term *aminocyclitol* is often used in naming of the class. As is the case for the β-lactams, a number of synthetically modified aminoglycosides have been investigated.

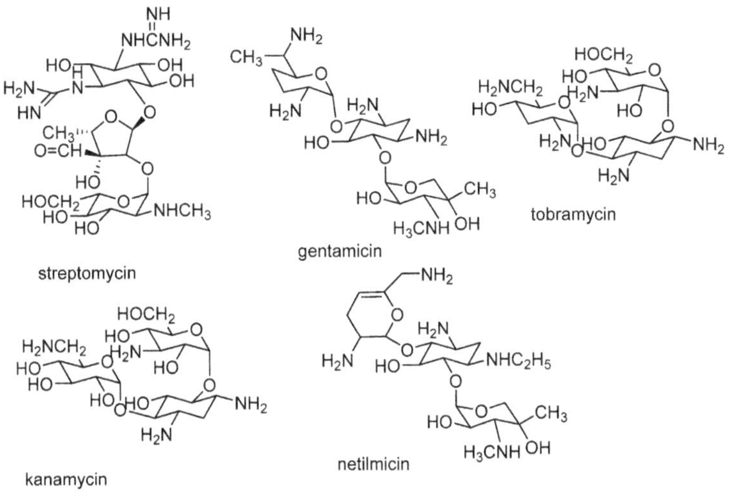

SCHEME 12.2   Aminoglycoside antibiotics.

As is evident from their chemical structures, the aminoglycosides are positively charged compounds and very hydrophilic. They are poorly absorbed from the digestive tract and are usually administered by injection. The cationic antibiotics are electrostatically bound to the anionic lipopolysaccharides on the bacterial outer membrane. The drugs are then incorporated into the bacterial cell by energy-dependent active transport. Eukaryotic cells are resistant to this incorporation and this provides one basis for the antibacterial selectivity. The cellular target of the aminoglycosides seems to be the bacterial ribosomal RNA, which is essential for protein synthesis. Binding of the aminoglycosides leads to mistranslation and synthesis of defective proteins. The lethal effect appears to be the result of defects in membrane permeability. The amino glycosides have two main side effects, kidney toxicity and hearing loss. The latter results from degeneration of the hair cells of the cochlear.

12.2.2   OTHER ANTITUBERCULOSIS DRUGS

Synthetic compounds with antitubercular activity also were discovered. In Sweden, Jorgen Lehmann postulated that PAS would have activity. It had been shown (by Frederick Bernheim at Duke) that aspirin promoted metabolism of *M. tuberculosis.* Lehmann reasoned that it must contribute to the bacteria's growth and that a substituted analog might block that action. He persuaded the Ferrosan Company to synthesize PAS. The compound was tested in several critically ill TB patients with promising results. The results were first published in 1946. Initially, PAS was extremely difficult to synthesize and, like streptomycin, very expensive. This changed when chemists at Ferrosan found that a synthesis analogous to that of salicylic acid, direct carboxylation, could produce the drug inexpensively in good yield. In 1949, it was found that a combination of streptomycin and PAS helped delay development of resistance. PAS remains an important compound in treatment of TB.

In 1942, Gerhard Domagk convinced Bayer's management to begin looking for antitubercular drugs, and his group found that thiosemicarbazones were active. The best of the compounds was thiacetazone (Conteben), the thiosemicarbazone of *p*-acetamidobenzaldehyde. Work was interrupted by extensive bombing and the impending collapse of Germany as the end of the war neared. The research was resumed after the war and led to promising clinical trials. Isoniazid was discovered nearly simultaneously at the US pharmaceutical laboratories of Hoffman–LaRoche and Squibb, as well as by Domagk's group at Bayer. It was considerably more active than the previous drugs, although it too, was susceptible to development of resistance.

The structures of isoniazid and PAS are remarkably simple. There are two other active compounds, pyrazinamide and ethionamide, that resemble isoniazid in structure. Their mechanism of action gives us a glimpse into the relationship between structure and biological activity. All three compounds are "pro-drugs," that is they are metabolically transformed by the target organism. The case of isoniazid is best understood. It is oxidized by a mycobacterial catalase-peroxidase to the isonicotinoyl radical.[16] The radical then forms a series of adducts with NAD^+ and $NADP^+$. Some of the adducts are powerful inhibitors of one of the key enzymes in the formation of the protective coat of the tuberculosis bacilli.[17] Ethionamide is metabolized to 2-ethyl-4-hydroxymethylpyridine and a second, as yet unidentified metabolite. Only the latter material is found inside the bacterial cell. Ethionamide and pyrazineamide are also activated by metabolism, but their precise mechanism of action remains uncertain. Cycloserine interferes with cell wall formation by *M. tuberculosis*. Scheme 12.3 gives the structures of these drugs.

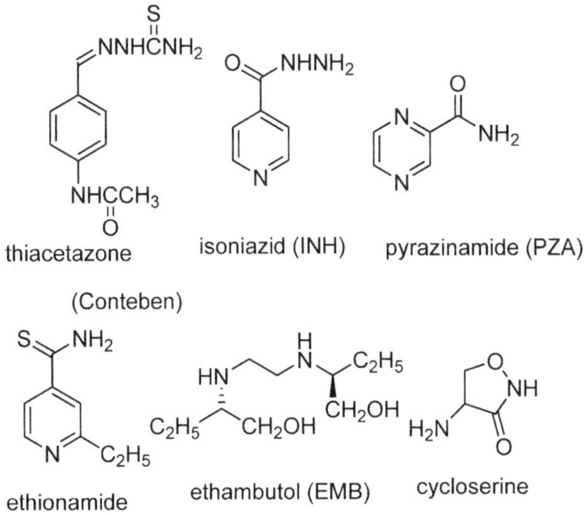

thiacetazone

(Conteben)

isoniazid (INH)

pyrazinamide (PZA)

ethionamide

ethambutol (EMB)

cycloserine

SCHEME 12.3 Antituberculosis drugs.

A number of other antituberculosis drugs were developed in the 1950s and 1960s including derivatives of the antibiotic rifamycin, in particular rifampicin and, the more recently introduced (1998), rifapentine. Rifamycin and its analogs are selective inhibitors of prokaryotic RNA polymerase.

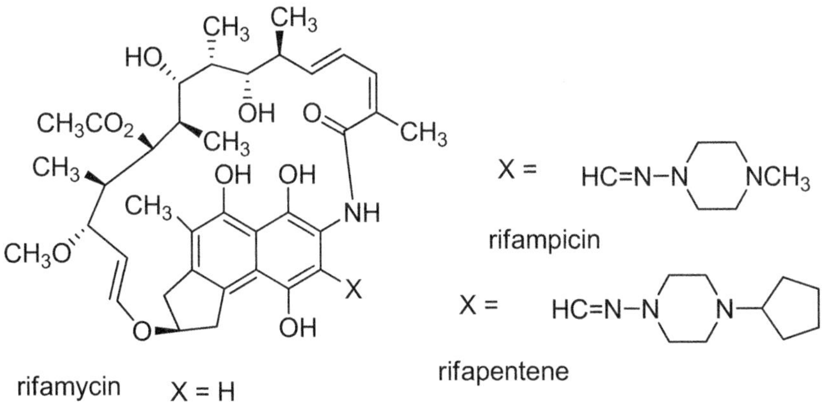

X = HC=N–N⏜NCH₃

rifampicin

X = HC=N–N⏜N–⬠

rifapentene

rifamycin   X = H

12.2.3   TREATMENT STRATEGIES AND DRUG RESISTANCE

As early as the 1950s, three drugs were used in combination, usually streptomy-cin, isoniazid, and PAS. Patients were cultured and if the sample was resistant to one of the drugs, it was removed from the regimen. The combination drug therapy was used along with community-wide screening by chest X-rays in the United States and United Kingdom and rapidly led to reduction of TB incidence. As continental Europe recovered after the war, these approaches were adopted. In 1956, a program supported by the UK MRC, the WHO, and the state govern-ment was instituted in Madras, India. Patients were treated with a combination of isoniazid and PAS. Cure rates of 90% were achieved without hospitalization.

Although streptomycin and other antituberculosis drugs greatly reduced the incidence of tuberculosis in the general population in the United States and Europe, the disease remains a serious health problem, especially in the develop-ing world. It is estimated that 2 billion people worldwide are infected, with as many as 50 million having drug-resistant infections. TB is a common infection in immunocompromised patients, for example, those with AIDS. The current recommended treatment takes 6 months and involves a 2-month treatment with a four-drug combination (INH, RIF, PZA, and EMB), followed by INH and RIF for another four months. This treatment is generally effective, unless the tuber-culosis is multidrug resistant. However, adherence to the regime is a problem. The need for the long treatment derives from factors inherent to *M. tuberculosis*. It is a relatively slowly growing organism and develops dormant forms. These dormant forms are not susceptible to current drugs, which require active metab-olism.[18] A summary of this situation is given in Figure 12.2, which also indicates the drugs that are active against various stages of *M. tuberculosis*.

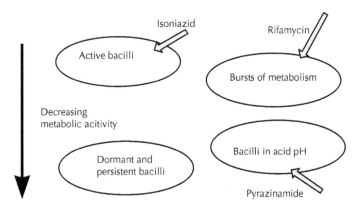

FIGURE 12.2   Schematic summary of the metabolic activity pattern and drug susceptibility of tuberculosis bacilli.

Since the introduction of rifampicin in 1968, there has been a long hiatus in the development of new antituberculosis drugs. There are several reasons. The disease was under control, at least in the developed world, and there was little economic incentive for development of new drugs. Several factors have come into play that made the development of new drugs urgent. These include the emergence of AIDS, the alarming incidence of TB in the underdeveloped world, which is exacerbated by poor access to medical care and poverty, and the widespread development of MDR-TB. In light of these developments, in 1993, the WHO declared TB a "global emergency" and in 2000 a public–private consortium, the Global Alliance for TB Drug Development, was formed. The current goals of drug development for treatment of TB have been defined by the Global Alliance for TB Drug Development. They are (1) drugs that can shorten the length of treatment; (2) drugs that can be used to treat MDR-TB; and (3) drugs to treat latent TB infections.

12.3   OTHER CLASSES OF ANTIBIOTICS

12.3.1   MACROLIDE ANTIBIOTICS

The macrolide antibiotics include erythromycin A, which was discovered in 1952, and several semisynthetic derivatives. They are characterized by large ring lactone structures. One derivative is azithromycin, in which the lactone ring is expanded by inclusion of nitrogen. Erythromycin A and its derivatives are used in treatment of a variety of respiratory tract infections. They are also used in treatment of digestive tract infections including *Salmonella* and *Shigella*. They are also active against *Helicobacter pylori*, the organism responsible for many ulcers. The macrolides bind to ribosomal RNA and interfere with protein synthesis, a mechanism generally similar to that of streptomycin. There is considerable cross-resistance with penicillins, particularly for *Streptococcus pneumoniae*. The resistance mechanisms include enhanced efflux of the drug and modification of the bacterial RNA by methylation. Several of the macrolide antibiotics are shown in Scheme 12.4.

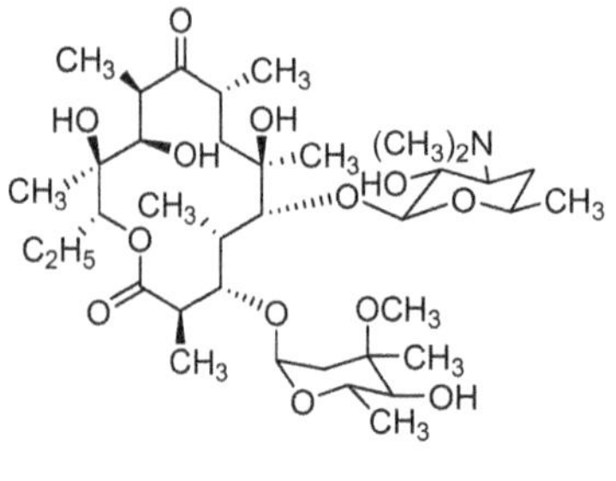

erythromycin A

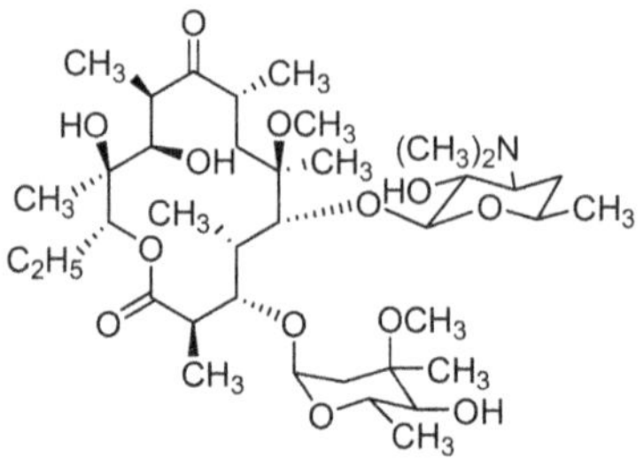

clarithromycin

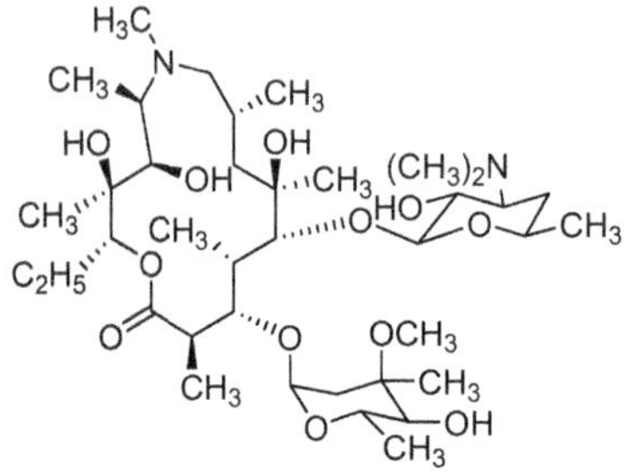

azithromycin

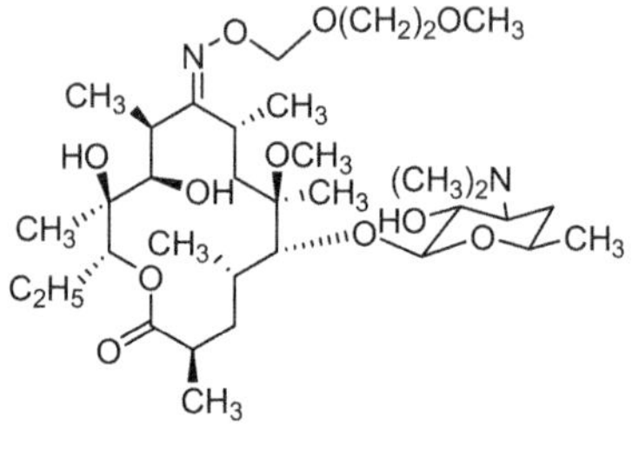

roxithromycin

SCHEME 12.4   Macrolide antibiotics.

12.3.2   TETRACYCLINE ANTIBIOTICS

The tetracyclines were discovered in the 1940s. The first tetracyclines to be introduced were chlortetracycline and oxytetracycline, discovered at Lederle Laboratories and Pfizer, respectively. They were found by screening of soil samples. The tetracyclines have a wide spectrum of activity against both Gram-positive and Gram-negative bacteria. They also show activity against other types of microorganisms, including protozoa and mycoplasma and were the first drugs to show activity against diseases such as Rocky Mountain Spotted Fever and typhus. The original compounds have been followed by a number of natural and semisynthetic analogs, such as shown in Scheme 12.5. Most are active by oral administration. The tetracyclines function by preventing association of aminoacyl transfer RNA with the bacterial ribosome. One of the derivatives, doxycycline, has recently been found to be a preventative drug for malaria (see Section 18.1.3.6).

The tetracyclines have been widely used in animal husbandry. Not only do they prevent infections, they also enhance the animal's rate of growth. The antibiotics have been used in chickens, pigs, and cattle. This use is believed to have contributed to the development of tetracycline-resistant strains. Use in animal feed was banned in Europe in the 1970s but is still practiced in the United States, Australia, and other countries.

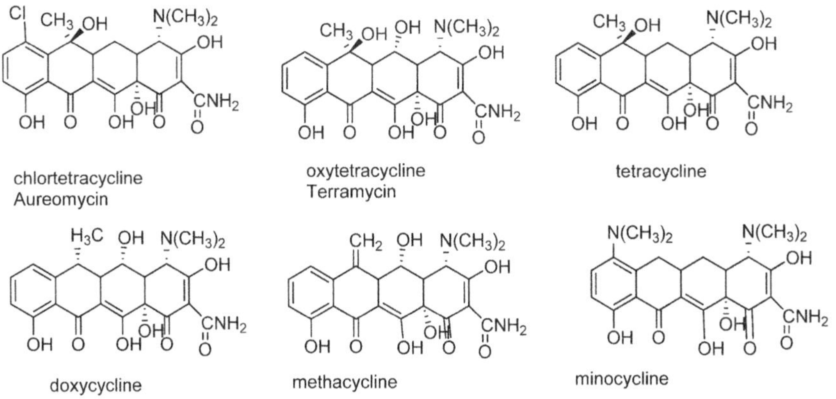

chlortetracycline
Aureomycin

oxytetracycline
Terramycin

tetracycline

doxycycline

methacycline

minocycline

SCHEME 12.5 Tetracycline antibiotics.

The tetracyclines were the first group of antibiotics for which the phenomena of active drug efflux was demonstrated. The process involves proteins that export the antibiotics from the cell, thus negating their effectiveness. There are other mechanisms of resistance, including proteins that prevent the binding of the drugs to the ribosome target.

12.3.3 QUINOLONE ANTIBIOTICS

The quinolone antibiotics had their beginning with the isolation of nalidixic acid[19] in 1962 and oxolinic acid[20] in 1966. These are synthetic compounds. The major importance of this group derives from a number of analogs, particularly fluoro-substituted derivatives such as norfloxacin and ciprofloxacin. Ciprofloxacin is currently the most widely used of these drugs. It is used in the treatment of uro-genital tract infections, gastrointestinal infections, and septicemia. The quinolones function by inhibition of a bacterial topoisomerase and gyrase and prevent proper supercoiling of the bacterial DNA. The quinolones also have activity against tuberculosis and are used as second-line drugs. The structures of several examples are shown in Scheme 12.6.

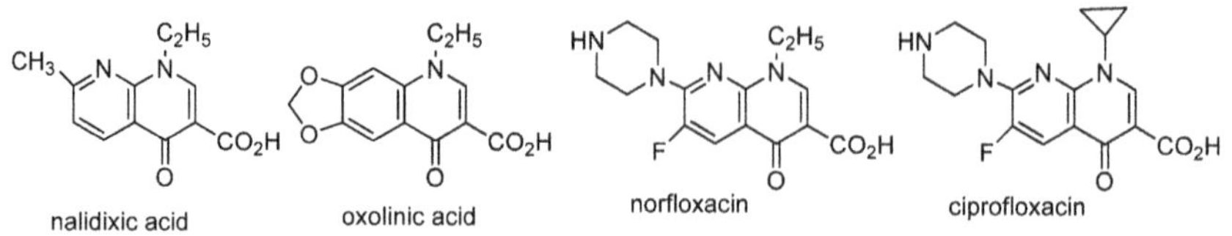

SCHEME 12.6   Quinolone antibiotics.

12.3.4   OLDER CLASSES OF ANTIBIOTICS

There are several other classes of antibiotics. Some of them remain widely prescribed in primary care, especially in the developing parts of the world. Several
diaminopyrimidines were developed in the 1940s, some as antimalarials and
others as antibiotics. The most widely used is trimethoprim.[21] The compound
interferes with folic acid metabolism. This is also the mechanism of the sulfa
drugs (see Section 11.1.1) and trimethoprim and its analogs are sometimes used
in combination with sulfa drugs. Chloramphenicol was first isolated from soil
samples in the 1940s. It is now made by chemical synthesis. It has been used
extensively since then, but is associated with occasional severe side reactions,
especially bone marrow toxicity.[22]

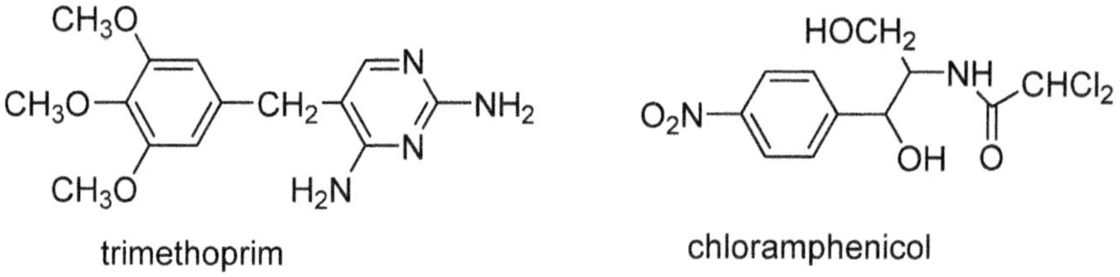

trimethoprim   chloramphenicol

12.4   ISSUES FOR THE FUTURE OF ANTIBIOTICS

The time from 1940 to 1965 is said to have been the golden era of antibiotics. Nearly all of the major classes of both fermentation and synthetic antibiotics were discovered in that period, although many analogs have been prepared
subsequently. The pace of discovery of new types of antibiotics slowed after
1970. It was not until 2003 that a new class of synthetic antibiotics, the oxa-
zolidinones, was introduced. Table 12.1 gives a time line summarizing the date
of discovery, observance of resistance, and mechanism of action for the major
groups of antibiotics.

TABLE 12.1   Timeline for Introduction and Observation of Resistance for Antibiotics.

Antibiotic class First example	Introduction	Resistance seen	Mechanism
Sulfa drugs Prontosil	1936	1942	Folic acid synthesis
β-Lactams Penicillin	1938	1945	Cell wall synthesis
Aminoglycosides Streptomycin	1946	1946	Ribosomal protein synthesis
Phenicols Chlorampenicaol	1948	1950	Ribosomal protein synthesis
Macrolides Erythromycin	1951	1955	Ribosomal protein synthesis
Rifamycins Rifampicin	1958	1962	RNA polymerase
Glycopeptides Vancomycin	1958	1960	Cell wall synthesis
Quinolones Ciprofloxacin	1968	1968	DNA synthesis
Oxazolidinones Linezolid	2003	2001	Ribosomal protein synthesis

(Adapted from Lewis, K. *Nat. Rev. Drug Discov.* **2013**, *12*, 371–387).

The relative lack of new antibiotics is a cause for concern because of the development of resistance to the existing classes. There have been both scientific and economic reasons for the decreased rate of new discoveries. A major scientific reason is the fact that the broad screens used originally are most likely to turn up known antibiotics, making the approach unproductive. This indicated the need to develop new approaches to identifying antibiotic types. In the 1990s, searches for new antibiotics were undertaken based on isolation of specific bacterial genes that coded for essential proteins. While a few leads turned up, so far no new antibiotics have been introduced.[23] The economics of antibiotic are also unfavorable, relative to many other classes of drugs. Because they are generally used in an acute setting, rather than chronically, the market is smaller than for several other major classes of drugs. There is particular concern for so-called *nosocomial infections*, meaning those acquired in a hospital setting, which are especially likely to exhibit multidrug resistance, but represent a still smaller market. As with other drugs, the cost of development has escalated and the approval process slowed, and, as a result, the years of patent protection have dwindled.

Three fundamental ideas underlie the development of antibiotics in the 20th century. One was Paul Ehrlich's concept of the "magic bullet"—a molecule that could attack pathogenic microorganisms. Another was Gerhard Domagk's conviction that such compounds could be synthesized and that they would work in conjunction with the immune system to defeat infection. The third was the recognition, by Fleming, Florey, Chain, Waksman, Dubos, and their many successors that such molecules existed in nature and could be isolated. These fundamental concepts came from disciplines outside of chemistry, but most of the work toward practical application, including isolation, purification, production, synthetic modification, and mechanism of action was done by chemists and biochemists. Indeed, in the introductory chapter of *Antimicrobial Agents*, a wide-ranging summary of chemotherapy of infectious diseases, Andre Bryskier says "The history of contemporary anti-infective chemotherapy is more chemical than bacteriological...therapeutically active substances are obtained by extraction from culture broths, or by synthesis or semisynthesis."[24] Many important concepts came from these studies, including recognition of specific mechanisms of action, the selectivity of certain antibiotics for specific classes of bacteria, synthesis on the basis of structure–activity relationships, and the necessity for appropriate physical properties for adsorption and transport. The progress in these areas has provided a wide array of potential antibiotic drugs. The great nemesis is the capacity of bacteria to survive antibiotics by developing resistance. There is no general solution to this problem, and as a result the pharmaceutical approach to treatment of bacterial infections requires a constant stream of new drugs and combinations. At the beginning of the 20th century, immunization was seen as the best approach to infections. At the end of the century, effective vaccines for tuberculosis, malaria, and HIV remained at the top of the list of goals for prevention of infectious diseases. In the meantime, countless infections have been cured, millions of lives saved, and many options for treatment of bacterial infections have been developed.

KEYWORDS

- antibiotics
- β-lactams
- penicillin

- **cephalosporins**
- **carbapenems**
- **aminoglycosides**
- **macrolides**
- **tetracyclines**
- **quinolones**
- **antituberculosis drugs**
- **multidrug resistance**

BIBLIOGRAPHY

Lax, E. *The Mold in Dr. Florey's Coat*. Holt and Company: New York, 2004.
Bud, R. *Penicillin, Triumph and Tragedy*. Oxford University Press: Oxford, UK, 2007.
Bryskier, A. In *Antimicrobial Agents*, 2nd ed.; Bryskier, A. Ed.; ASM Press, 2005; pp 1–12.

BIBLIOGRAPHY BY SECTION

Section 12.1.1: Richard, A. N. *Nature* **1964,** *201*(4918), 441–445; Amyes, S. G. B. *Magic Bullets, Lost Horizons*. Taylor and Francis: London, Chap. 1–2, 2001; Brown, K. *Pharm. Hist.* **2004,** *34*, 37–43; Quinn, R. *Am. J. Public Health* **2013,** *103*, 426–434.

Section 12.1.2: Hubschwerlen, C. *Comprehensive Medicinal Chemistry II*; Elsevier: Amsterdam, Vol. 7, 2007; pp 479–578.

Section 12.1.3: Bush, K. Bradford, P. A. In *Enzyme-Mediated Resistance to Antibiotics*; Bonomo, R. A.; Tolmasky, M. Ed.; ASM Press: Washington, 2007; pp 67–79.

Section 12.2: Feldman, W. H. *Am. Rev. Tuberculosis* **1954,** *69*, 859–868; Ryan, F. *Tuberculosis: The Greatest Story Never Told*. Swift Publishing: Bromsgrove, UK, 1992; Cegielski, J. P.; Chauhan, L. S.; Chin, D. P.; Granich, R.; Nelson, L. J.; Raviglione, M. C.; Rodrequez, R.; Cruz, E. A.; Talbot, A. B.; Wright; Zaleskis, R. In *New Topics in Tuberculosis Research*; Spiegelburg, D. D., Ed.; Nova Science Publishers: New York, 2007; pp 1–70.

Section 12.2.1: Begg, E. J.; Barclay, M. L. *Br. J. Clin. Pharmacol.* **1995,** *39*, 597–603; Veyssier, P.; Bryskier, A. In *Antimicrobial Agents: Antibacterials and Antifungals*; Bryskier, A. Ed.; ASM Press: Washington, 2005; pp 453–469; Magnet, S.; Blanchard, J. S. *Chem. Rev.* **2005,** *105*, 477–497; Kim, M.-Y.; Nicolau, D. P. *Antimicrobial Pharmacodynamics in Theory and Practice*, 2nd ed.; Informa Healthcare, 2007; pp 147–175; Hermann, T. *Cell. Mol. Life Sci.* **2007,** *64*, 1841–1852.

Section 12.2.2: Street, E. W. *Scott. Med. J.* **1977,** *22*, 279–285; Zellweger, J.-P. *Expert Rev. Respir. Med.* **2007,** *1*, 85–97.

Section 12.2.3: Mitchison, D. A. *Am. J. Respir. Crit. Care Med.* **2005,** *171*, 699–706; Zhang, Y. *Ann. Rev. Pharmacol. Toxicol.* **2005,** *45*, 529–564; Meya, D. B.; McAdam, K. P. W. J. *J.*

Intern. Med. **2007,** *261,* 309–329; Almeida Da Silva, P. E.; Palomino, J. C. *J. Antimicrob. Chemother.* **2011,** *66,* 1417–1430.

Section 12.3: Kaneko, T.; Dougherty, T. J.; Magee, T. V. *Comprehensive Medicinal Chemistry, II,* 2006, Vol. 7; pp 519–566; Jain, S.; Bishai, W.; Nightingale, C. H. *Antimicrobial Phamacodynamics in Theory and Clinical Practice,* 2nd ed., 2007; pp 217–230; Aldred, K. J.; Kearns, R. J.; Osberoff, N. *Biochemistry* **2014,** *53,* 1565–1574.

Section 12.4: Chopra, I.; Roberts, M. *Microbiol. Mol. Bio. Rev.* **2001,** *65,* 232–260; Nelson, M. L.; Projan, S. J. In *Frontiers in Antimicrobial Resistance*; White, D. G.; Alekshun, M. N.; McDermott, P., Eds.; 2005; pp 29–38; Griffin, M. O.; Fricovsky, E.; Ceballos, G.; Villarreal, F. *Am. J. Physiol. Cell Physiol.* **2010,** *299,* C539–C548.

Section 12.5: Emmerson, A. M.; Jones, A. M. *J. Antimicrob. Chemother.* **2003,** *51(Suppl. 1),* 13–20; Andriole, V. T. *Clin. Infect. Dis.* **2005,** *41*(Suppl. 2), S113–S118.

Section 12.6: Fischbach, M. A.; Walsh, C. T. *Science* **2009,** *325,* 1089–1093; Lewis, K. *Nat. Rev. Drug Discovery* **2013,** *12,* 371–387.

REFERENCES

1. Hare, R. *Med. History* **1982,** *26,* 1–24.
2. Richards, A. N. *Nature* **1968,** *4918,* 441–445; Kaufman, G. B. *Chemistry* **1978,** *51,* 11–17; Coghill, R. D. *Chem. Eng. Progress Symp. Ser.* **1970,** *66,* 13–21.
3. Abraham, E. P. *Nat. Prod. Rep.* **1987,** *4,* 41–46; Curtis, R.; Jones, J. *J. Pept. Sci.* **2007,** *13,* 769–775.
4. Chain, E. *Trends Biochem. Sci.* **1979,** *4,* 143–144; Chain, E. *Trends Pharmacol. Sci.* **1979,** *1,* 6–11; Chain, E. *ChemTech* **1980,** *10,* 474–481; Chain, E. *Hist. Antibiot.* 1980, *5,* 15–29.
5. Kiese, M. *Klin. Wochenschr.* **1943,** *22,* 505–511.
6. Wettstein, A. *Schweiz. Med. Wochenschr.* **1944,** *74,* 617–625.
7. Ettinger, L. *Hist. Antibiot.* **1980,** *5,* 57–67.
8. Burns, M.; van Dijck, P. W. M. *Adv. Appl. Microbiol.* **2002,** *51,* 185–200.
9. Friedrich, C. *Pharm. Unserer Zeit* **2006,** *35,* 392–398.
10. Kaumazawa, J.; Yagisawa, M. *J. Infect. Chemother.* **2002,** *8,* 125–133.
11. Papp-Wallace, K. M.; Endimiani, A.; Traxila, M. A.; Bonomo, R. A. *Antimicrob. Agents Chemother.* **2011,** *55,* 4943–4960.
12. Sykes, R. B.; Koster, W. H.; Bonner, D. P. *J. Clin. Pharmacol.* **1988,** *28,* 113–119.
13. Bruggink, A.; Roos, E. C.; de Vroom, E. *Org. Process. Res. Dev.* **1998,** *2,* 128–133.
14. Koch, A. L. *Clin. Microbiol. Rev.* **2003,** *16,* 673–687.
15. Jones, R. N.; Biedenbach, D. J.; Sader, H. S.; Fritsche, T. R.; Toleman, M. A.; Walsh, T. R. *Diagn. Microbiol. Infect. Dis.* **2005,** *51,* 77–84.
16. Zhang, T.; Heym, B.; Allen, B.; Young, D.; Cole, S. *Nature* **1992,** *358,* 591–593.
17. Vilcheze, C.; Jacobs, Jr., W. R. *Ann. Rev. Microbiol.* **2007,** *61,* 3550.
18. Zhang, Y. *Ann. Rev. Pharmacol. Toxicol.* **2005,** *45,* 529–564.
19. Lesher, G. Y.; Froelich, E. J.; Gruett, M. D.; Bailey, J. H.; Brundage, R. P. *J. Med. Pharm. Chem.* **1962,** *5,* 1063–1065.
20. Kanubsky, D.; Metzger, R. I. U.S. Patent 3,287,458; *Chem. Abstr.* **1966,** *66,* 65,399.
21. Then, R. L. *J. Chemother.* **1993,** *5,* 361–381.

22. Fisch, A.; Bryskier, A. *Antimicrobial Agents: Antibacterials and Antifungals*, 2nd ed.; ASM Press: Washington, 2005; pp 925–929.

23. Payne, D. J.; Gwynn, M. N.; Holmes, D. J.; Pompliano, D. L. *Nat. Rev. Drug Discovery* **2007,** *6*, 29–40.

24. Bryskier, A. *Antimicrobial Agents*, 2nd ed., Bryskier, A. Ed.; ASM Press: Washington, 2005; p 1.

CHAPTER 13

ANALGESIC, ANTI-INFLAMMATORY, ANTIPYRETIC, AND ANESTHETIC DRUGS: DEALING WITH PAIN, INFLAMMATION, AND FEVER

ABSTRACT

Analgesic substances from plants that were known from ancient time include salicylates from willow bark and opiates from the poppy *Papaver somniferum*. The analgesic properties of morphine and cocaine (from *Erythroxylon coca*) were used medically from the middle of the 19th century. Purification of the active components occurred in the 19th century and structures were determined by the 1920s. The addictive properties of opiates and cocaine led to their banning by the Harrison Narcotics Act in 1912. The first synthetic analgesics were made in the late 1800s and included antipyrine, aspirin, and phenacetin. Systematic searches for improved analgesics began after WWII and led to the NSAIDS (nonsteroidal anti-inflammatory drugs) including ibuprofen and naproxen. Investigation of the metabolism of phenacetin identified acetaminophen as the active metabolite and led to its introduction in the United States as Tylenol in 1955. Many analogs of morphine were synthesized and shown to have both the analgesic and addictive characteristics of morphine. Some have specific medical application such as short-term analgesics in diagnostic procedures. Investigation of the mechanism of action of aspirin and the NSAIDS identified prostaglandin formation by cyclooxygenase as the target enzyme. Recognition that two forms of the enzyme existed led to the synthesis of selective inhibitors of cyclooxygenase-2, known as coxibs. They showed reduced gastrointestinal side effects but increased the risk of cardiovascular events. Use of cocaine as a topical anesthetic began in the1800s and was followed by the introduction of synthetic analogs including benzocaine and procaine (Novocain) and several others used widely in dentistry. The use of chloroform and ether as inhaled anesthetics began in the mid-1800s. They have been replaced by safer halogenated hydrocarbons such as halothane and isoflurane. Intravenous general anesthetics such as propofol and etomidate were also introduced.

There are a number of both over-the-counter and prescription drugs that are very widely used for a related set of common symptoms. These include analgesics for pain relief, anti-inflammatory and antipyretic (lowering of body temperature) activity. Several drugs, most notably aspirin, have all of these effects. Some are combined with other medications in remedies sold for relief of the symptoms of cold or flu. Others are used in treatment of chronic pain and inflammation, such as occurs with arthritis. Some of these drugs, in particular aspirin and phenacetin, are among the oldest synthetic drugs, having been available since the late 1800s. Because they were marketed long before modern drug regulation and were "grand-fathered," scientifically based study of safety, efficacy, and relative benefit/risk sometimes have gaps.

Aspirin was followed at mid-century by a group of drugs that are now called *nonsteroidal anti-inflammatory drugs* (NSAIDS). It was eventually found that aspirin and the NSAIDS inhibit the formation of *prostaglandins* that are involved in inflammation, fever, and pain. Acetaminophen (tylenol) is another drug with analgesic and antipyretic activity, although it has little anti-inflammatory activity. Inflammation occurs in response to injury, irritation, or infection. The response is mediated by several kinds of molecules, including the monoamine neurotransmitters (histamine, serotonin), cytokinins (e.g., interleukins, tumor necrosis factor), glycoproteins (e.g., fibrinogen), and reactive oxygen species. The physiological purpose of these responses is to remove the cause of the inflammation and repair the local tissue damage. Anti-inflammatory drugs treat the symptoms of swelling, pain, and fever, but they do not directly treat the cause of inflammation. Normally, inflammation is a self-limiting process that removes the infected or damaged cells and is followed by termination of inflammation and healing. Failure to properly regulate the inflammatory process leads to conditions such as arthritis and autoimmune diseases. The anti-inflammatory drugs are also used to treat these chronic conditions, but do not cure them.

There is another group of drugs that includes natural and synthetic analogs of opium that are used for the treatment of chronic severe pain, such as is associated with cancer. These drugs interact with the central nervous system. These drugs are very addictive and subject to abuse. Another group of drugs, anesthetics, are used for temporary blockage of pain in surgery, diagnostic procedures and dentistry. Some of these are related in structure to cocaine, which itself has topical anesthetic properties.

13.1   ASPIRIN

Hippocrates prescribed extracts of willow bark and leaves for treating of inflammation and fever in the fourth century BCE. A publication in 1763 by an English clergyman described the effects of dried willow bark on fever.[1] This study was based on two nonscientific ideas: (1) cures of diseases were thought to be found in proximity to their causes; thus, the association between swamps (bad air, fever) and willows; and (2) quinine (bitter taste) was known to lower fever and willow, too, had a bitter taste. While the theory was wrong, the conclusion was correct. By the mid-19th century, the active ingredient in willow bark had been identified as salicin, a glycoside of salicyl alcohol. It is hydrolyzed and metabolized to salicylic acid. Oil of wintergreen, from the American plant *Gaultheria procumbens*, a source of methyl salicylate, was used in ointments that had anti-inflammatory and analgesic properties. By 1860, salicylic acid was being manufactured as an anti-inflammatory, but it was very irritating to the stomach. Acetyl salicylate, aspirin, was introduced in 1898 by Bayer and has remained in wide use since that time.

salicin   salicylic acid   methyl salicylate   aspirin

But no one knew how aspirin worked until the early 1970s, when John Vane and collaborators showed that it inhibited formation of *prostaglandins*.[2] Vane shared the 1982 Nobel Prize in Medicine for this discovery.[a] The first enzyme in the biosynthetic pathway to prostaglandins is *cyclooxygenase-1* (COX-1), so aspirin and related drugs became known as *COX-1 inhibitors*. The prostaglandins are biosynthesized from *arachidonic acid*, a highly unsaturated 20-carbon fatty acid. The cyclooxygenase enzyme contains a heme group that functions to form peroxides and hydroperoxides and a second site that converts these to hydroxyl and carbonyl groups.[3] The two sites are linked by electron transfer. The key intermediate is *prostaglandin endoperoxide* (shown as PGH_2 in Fig. 13.1), which gives rise to the prostaglandins and two related types of compounds, the *thromboxanes* and *prostacyclin*. As a group, these compounds are called *eicosanoids*, referring to their 20-carbon structures. Several of the prostaglandins are vasodilators and also effect blood clotting. The thromboxanes tend to be

---

[a]John Vane was born in 1927. He obtained B. Sc. degrees in Chemistry (Birmingham, 1946) and Pharmacology (Oxford, 1949) and a Ph.D. in Pharmacology (Oxford, 1953). In 1955, he joined the faculty of Basic Medical Sciences at the Royal College of Surgeons in London. He developed various bioassays to follow physiological effects. In the course of these studies, he discovered that the mechanism of action of aspirin involved inhibition of prostaglandin formation by cyclooxygenase enzymes. He reported these results in 1971. In 1973, he became director of research at the Wellcome Foundation. There, in conjunction with the Upjohn Company, he identified the prostacyclins and their antithrombotic activity. Vane shared the 1982 Nobel Prize in Medicine or Physiology with two Swedish chemists, Bengt Samuelsson and Sune Bergstrom, cited for their work in isolation and characterization of the prostaglandins. In addition to his work on the prostaglandins, Vane was instrumental in the development of the angiotensin-converting enzyme (see Section 14.4) inhibitors, an important group of antihypertensive drugs.

prothrombotic, vasoconstrictors and promote platelet aggregation. Prostacyclin (also called PGI_2) has the opposite effects, acting as a vasodilator and inhibiting platelet aggregation. The balance between these effects is a critical aspect of their biological function. The eicosanoids are involved in several other important biological processes including, kidney function, asthma, maintenance of gastrointestinal mucosa, induction of sleep, and aspects of reproduction.

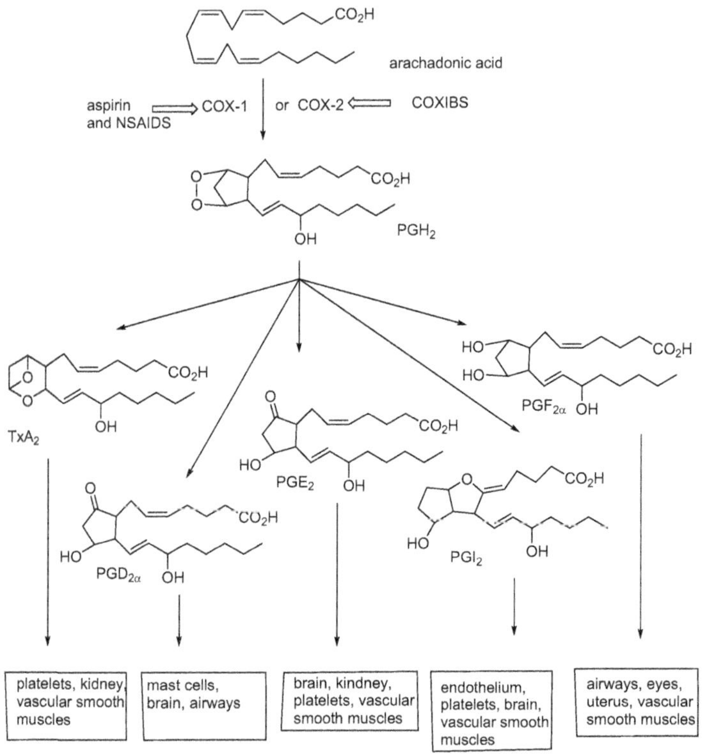

FIGURE 13.1   Summary of conversion of arachidonic acid to prostaglandins, thromboxanes, and prostacyclins and their target organs.

13.1.1 HOW DOES ASPIRIN WORK?

The train of events that led to the discovery of the mechanism of action of aspirin began in the British laboratories of the Parke Davis Company. Henry O. J. Collier, a pharmacologist, found that kinins, such as bradykinin (a nine amino acid polypeptide), were associated with pain. He studied some of the effects of bradykinin, such as its ability to induce an asthmatic response in guinea pigs and found that the effect was prevented if aspirin was given *before*, but not *after*, exposure to bradykinin. He also showed that the action of aspirin was not controlled by the central nervous system. Collier also showed that aspirin was *more potent*, than salicylic acid, which was contradictory to the long-held belief that aspirin was a precursor (prodrug) of the active salicylic acid. Collier began to synthesize his view that aspirin must *influence the body's defensive response to injury or inflammation.* Collier and his assistant Pricilla Piper undertook study of many drugs in an effort to recognize some pattern that would clarify the mechanism. Piper went to work in the laboratory of John Vane in 1968 to learn some of Vane's bioassay techniques. Vane's system was to subject a strip of guinea pig lung to the equivalent of anaphylactic shock by exposure to egg white. He then could study the effect of the substances released on other biological preparations, such as rabbit aorta. Vane called this a "cascade superfusion." With her background in aspirin from Collier's lab, Piper suggested that it be included in the studies. It blocked the response, suggesting that it must have prevented formation of some substance that mediated the response. Vane and Piper identified the substance as one of the prostaglandins, a series of very potent compounds with many physiological effects, including vasodilation and constriction and the regulation of inflammation. Vane recognized that aspirin must be blocking prostaglandin formation and was able to demonstrate that was the case.

Aspirin is unique among the common COX-1 inhibitors. It covalently acetylates a serine in COX-1, while the other NSAIDS are bound noncovalently. As a result, aspirin has a long duration of action, nearly completely suppressing formation of the prostaglandins and thromboxanes on the basis of relatively small dose (30–100 mg) for as long as 48 h. The other NSAIDS shown in Scheme 13.1, most of which are carboxylic acids, block the entry of arachidonic acid to the active site of cyclooxygenase and suppress the formation of the cyclooxygenase product types. Aspirin and the other NSAIDS have a common side effect of causing gastrointestinal upset, intestinal bleeding, and in some cases, ulcers. Individuals vary in their sensitivity to the drugs, but it is a quite common side effect. The cause of this side effect is that some of the prostaglandins have protective functions in the digestive tract and the thromboxanes and prostacyclins regulate blood clotting. Reduction in amounts or imbalance among the various compounds can cause the side effects.

13.1.2   ASPIRIN AND HEART ATTACKS

In 1933, a Wisconsin dairy farmer brought a dead heifer to the University of Wisconsin. Finding the State Veterinarian's office closed for the weekend, he went to the biochemistry department. His herd was suffering from severe fatal hemorrhaging. Prof. Karl Link and a graduate student, Eugen Schoeffel, recognized the cause as "sweet clover disease" and, as was characteristic of the condition, the animal's blood would not clot. The two began to investigate the connection between hemorrhage and spoiled sweet clover hay, the immediate cause of the condition. Link and H. A. Campbell discovered that the condition could be improved with alfalfa extract, which contains vitamin K that promotes blood clotting (see Section 10.1.1.9). In 1939, they solved the second part of the puzzle when they isolated dicoumarol from the spoiled hay. Dicoumarol is a potent anticoagulant. Dicoumarol is metabolized to salicylic acid and Link suspected that the latter might be the active compound. He went on to show that high doses of aspirin did have an anticoagulant effect and suggested that hemorrhage might be a harmful side effect of aspirin. Link also pursued research on synthetic analogs of dicourmarol. From this work came warfarin, which was first used as a rodent poison and then as an anticoagulant in humans.[4]

dicoumarol   warfarin

   The clotting process was fairly well understood, at least at the cellular level. When a blood vessel is damaged, the aggregation of platelets is induced. This is followed by formation of a protective protein network from fibrin. The structure that develops is called a *thrombus* (plural thrombi, and approximate equivalent, blood clot). Accumulation of thrombi can lead to blockage of blood vessels and lead to heart attack, stroke, or phlebitis. The ability to control the coagulation process, therefore, offered promise of being able to prevent or treat these conditions. As understanding of the coagulation process deepened, it was recognized that dicoumarol functioned by preventing formation of fibrin. Harry J. Weiss of the Mt. Sinai Hospital and School of Medicine was the first to note that aspirin also had an inhibitory effect on platelet aggregation. Anticoagulants, heparin in particular, were already in use for treatment of heart attacks. Warfarin quickly attracted attention in the medical community and was patented by the Wisconsin Alumni Research Foudation (WARF) and licensed to Abbott, Lilly, and Squibb (see Section 10.1.1.8 for background on WARF). The clotting process is summarized in Figure 13.2, which also shows potential points of interaction of several kinds of drugs, including aspirin.

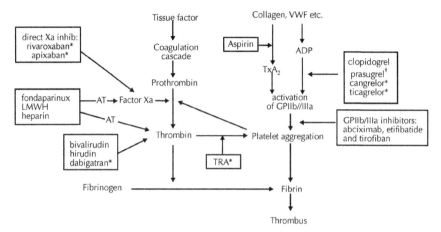

FIGURE 13.2   Points of drug interaction in the platelet aggregation and coagulation cascade. LMWH = low MW heparin; GPIIb/IIa = glycoprotein that induces platelet aggregation; TRA = thrombin receptor antagonist.; TxA_2 = thromboxin A2; ADp = adensosine diphosphate. (From Van de Werf, F. *Eur. Heart J.* **2009,** *30,* 1695–1702, by permission of Oxford University Press.)

In 1950, a California general practitioner, Lawrence L. Craven, noted that when he recommended *Aspergum* after tonsillectomies, some patients suffered severe bleeding. On investigation, he found they were using as many as 20 sticks, rather than the four he had recommended. Craven speculated that because aspirin played a role in blood clotting it might prevent heart attacks. He began to suggest to middle-aged men that they take several aspirins a day. He conducted no scientific study, but formed the impression that taking aspirin lowered heart attack risk. He published the results in minor medical journals. It was also fairly common practice at the time to give aspirin to patients after surgery, although no particular rationale for doing so had been established. Craven eventually collected data on 8300 patients given aspirin as a heart attack preventative. Although not a controlled study, he concluded that the use of aspirin was beneficial. Craven began to popularize his recommendation and claimed that *none* of his patients taking aspirin had died of heart attacks. He also took aspirin daily himself. In 1957, he died suddenly of a heart attack. In retrospect, it seems likely he had exaggerated the beneficial effects, although in the long run, his conclusion was corroborated.

Peter Elwood, M. D., an epidemiologist, was engaged by the MRC in the United Kingdom to investigate of the relationship between aspirin and heart attacks. To make the study manageable, Elwood decided to study the frequency of a second heart attack among patients who had survived a first attack. The study was carried out in "double-blind" fashion, that is neither the patient or physician knew which type of pill was being administered. During the course of the study an American physician, Herschel Jick, also found evidence that aspirin appeared to reduce the risk of heart attack. However, his data was also consistent with the possibility that aspirin might increase the likelihood that a heart attack would be fatal. The Elwood study was expanded. Although a reduction of deaths was not-ed, the difference was not large enough to be statistically significant. A second study also showed a reduction (of 17%), but again was not large enough to be statistically conclusive. In 1975, the US NIH undertook the *Aspirin Myocardial Infarction Study.* The study found a reduction in second heart attacks, but an overall increase in mortality. None of the studies to that point had provided sta-tistically significant evidence that aspirin could prevent heart attacks.

Several other studies on the effect of aspirin were underway in the early 1980s. These included studies of British and American doctors on the pre-ventive value of aspirin. The former was inconclusive but the latter, which included more than 22,000 participants, observed a large (40%) reduction in heart attacks. Studies were also conducted on transient ischemic attacks (which are associated with strokes) and positive results were found for men (but not for women). Based on these studies, the Sterling Drug Company, which sold Bayer aspirin, proposed to the FDA to include an indication on the aspirin label that it was beneficial in preventing second heart attacks. To support its proposal, Sterling presented the results that had been subjected to the newly devised statistical approach of *meta-analysis*, in which trends, even if not statistically significant in their own right, from several related stud-ies can be compared and analyzed as a group. This analysis indicated that there were beneficial results from aspirin in preventing heart attacks. An FDA panel considered the results in March 1983, but was unconvinced by the ar-guments from the meta-analysis and declined to support a label change. A second meeting of the panel in December 1984 concluded that there were beneficial effects; their confidence bolstered by yet another study conducted by the Veterans Administration. Finally, in 1985, the HHS Secretary Margaret Heckler announced approval of a label change for aspirin indicating benefit in preventing second heart attacks. Bayer Aspirin's slogan became "Wonder Drug that Works Wonders." Sellers of products such as Anacin and Bufferin began to emphasize, rather than obscure, the fact that their products contained aspirin. Aspirin sales shot up for a time, but Tylenol remained the leading over-the-counter analgesic (see Section 13.4).

Many of the aspirin-related studies were eventually coordinated by *The Anti-Platelet Trial Collaboration* and included over 200 studies involving some 115,000 patients. Positive results were found for heart attacks, stroke, transient ischemic attacks, pulmonary embolisms, and a variety of other conditions caused by thrombi.[5] Taken together, these studies indicate that daily aspirin reduces the chance of heart attack in the general population by about 1/4 from the low (< 1 per 1000 per year) normal incidence. Similar effects are seen in patients with preexisting cardiovascular conditions, where reduction of myocardial infarction is about 33% and of stroke about 25%.[6] These results have a paradoxical implication. They suggest that several small, cheap aspirin tablets a week could reduce the incidence and/or severity of several large health problems. Yet, there is only weak economic incentive to promote this usage. The relatively small level of profit limits both the value of increased advertising and the incentive to sponsor additional research. The potentially harmful side effect of gastrointestinal bleeding is also a factor.

Aspirin is also used in immediate treatments of people who have suffered heart attacks or are undergoing procedures such as angioplasty as a result of cardiovascular disease. Its function is to prevent platelet aggregation and clotting that can lead to thrombosis. In this application, it may be used with other drugs. The thienopyridines, clopidogrel, ticlopidine, and prasugrel are in this class. They block ADP-induced platelet aggregation (see Figure 13.2). Various studies have compared the drugs alone and in combination with aspirin and the combinations are usually beneficial, but also increase the risk of excessive bleeding.[7] Another problem with clopidogrel is that it requires metabolic activation by a P-450 enzyme and about 25% of patients do not have the particular P-450 variant that is required, making the drug ineffective in these patients.

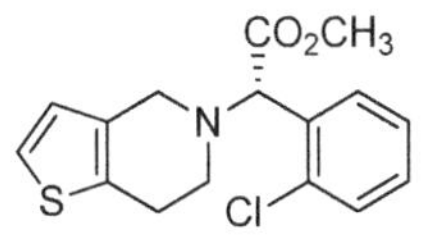

clopidogrel 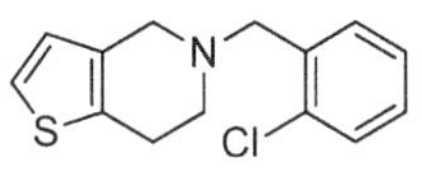   ticlopidine 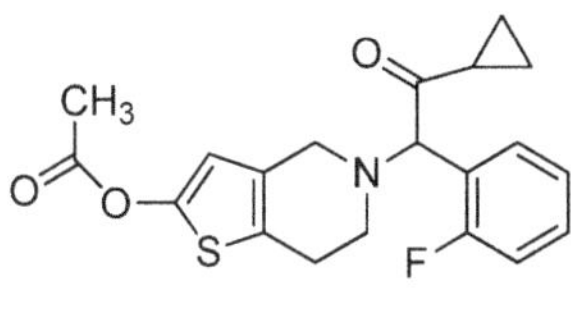   prasugrel

Ticagrelor and cangrelor are ADP analogs that also inhibit the action of ADP in promoting clotting.

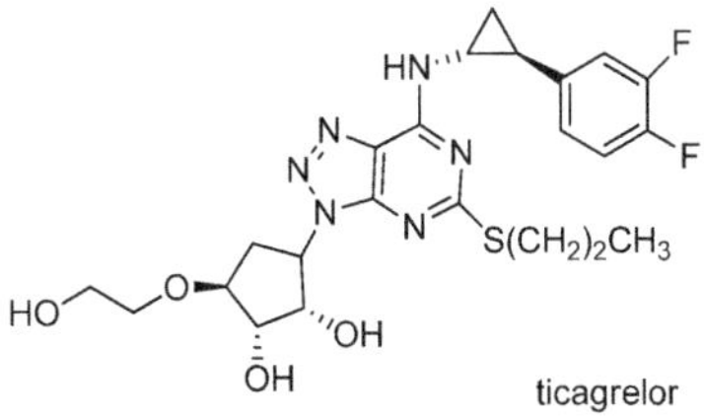

ticagrelor 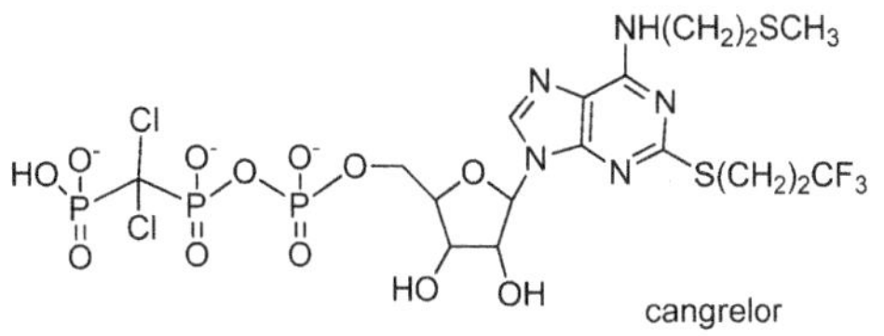   cangrelor

Another point of intervention in the platelet aggregation and coagulation cascade is to interfere with thrombin-mediated formation of fibrin. Among the materials that can inhibit this part of the coagulation process are heparin, *eptifibatide* (a cyclic hexapeptide disulfide), *tirofiban* (a small molecule drug), and *abciximab* (an antibody to a platelet glycoprotein receptor).[8] There are also substances that interact directly with thrombin, including *fondaparinux*, a heavily sulfated pentasaccharide containing aminoglucose, idose, and glucuronic acid unit. This structural feature is also present in heparin. *Hirudin* is a 65 amino acid polypeptide isolated from leeches, while *bivalirudin* is a 20 amino acid polypeptide related to hirudin.

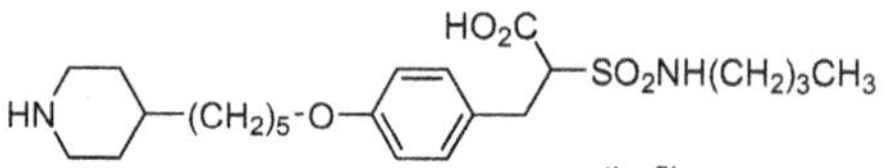

tirofiban

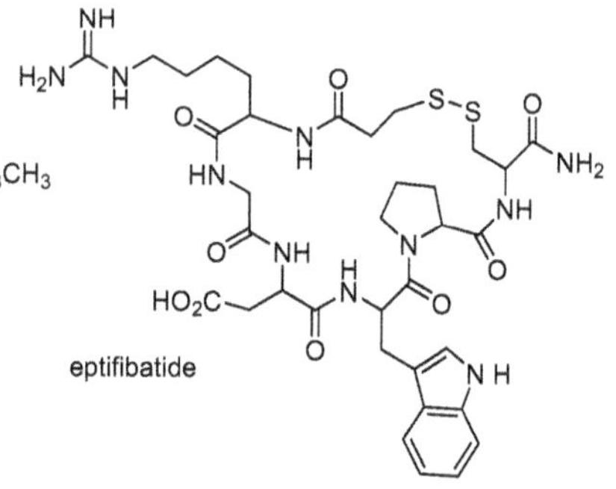

eptifibatide

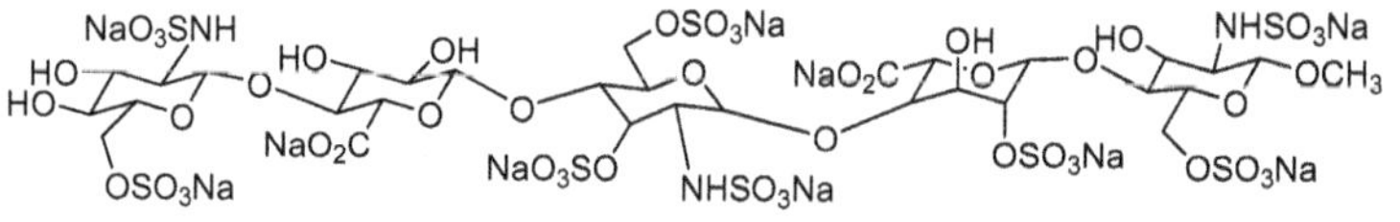

fondaparinux

13.1.3   ASPIRIN AND CANCER

Aspirin and certain other NSAIDS reduce the recurrence of colon polyps, which are associated with development of colon cancer. These and other indications that aspirin might have some beneficial effects, led to the reanalysis of the data obtained in the course of the antiplatelet studies described above. The study provided data on 25,570 patients and 674 cancer deaths during the follow-up study periods. This data was analyzed for effect on all deaths, cancer deaths and type of cancer. The results indicated definite benefits from aspirin, with about a 20% reduction in cancer deaths (odds ratio 0.79 [0.68–0.92]). The effects were strongest in gastrointestinal cancers: odds ratio 0.46 (0.27–0.77). The effects were only evident after fairly long treatment periods (5 years or more). There was also some reduction in cancer deaths from adenocarcinoma of the lung and esophagus. There was no effect on hematological cancers. There was a reduction of 8% (odds ratio 0.92 [0.86–0.99]) in deaths from all causes. Figure 13.3, shows the time course at <5 years, 5–7.4 years, and >7.5 years for all types of solid tumors.

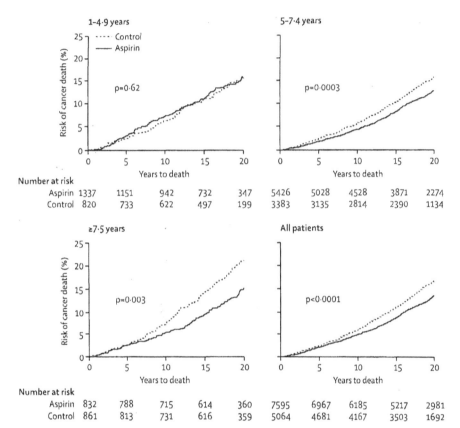

FIGURE 13.3 Effect of aspirin on 20 year death risk from any solid tumor as a function of length of aspirin treatment. (From Rothwell, P. M.; Fowkes, F. G. R.; Belch, J. F. F.; Ogawa, H.; Warlow, C. P.; Meade, T. W. *The Lancet* **2011**, *377*, 31–41 by permission of Elsevier.)

13.2 NONSTEROIDAL ANTI-INFLAMMATORY DRUGS

Beginning in the 1960s, several new synthetic compounds with a profile of action similar to aspirin were introduced. They are called NSAIDS for short. Some of these drugs are shown in Scheme 13.1. It is estimated that in 2000, the total market for NSAIDS was about $20 billion, but at the expense of around $4 billion in costs for side effects. The most common side effect is the same as with aspirin, that is, irritation of the gastrointestinal tract, leading in some cases to ulcers. The first of the NSAIDS to be introduced was ibuprofen, which was developed at the Boots Drug Company in the United Kingdom and introduced in 1969 and in the United States in 1974.[9] The research was led by Stewart Adams and John Nicholson. The goal was to improve on the existing drugs, which included aspirin, phenylbutazone (see Section 13.5) and the corticosteroids (Section 15.2), all of which had significant side effects. A major goal was to find a better drug for the treatment of the chronic pain and inflammation associated with rheumatoid arthritis. The first group of candidate drugs included ibufenac (4-isobutylphenylacetic acid), but it proved to have unacceptable liver toxicity. The research then turned to phenylpropionic acids, of which ibuprofen was the most promising. The recommended dose for arthritis is in the 2–3 g/day range. The drug was later approved for over-the-counter sale at a dose of 1.2 g/day. Another early NSAIDS was indomethacin, which was developed at Merck.[10] It is a more potent drug, but has fairly strong GI side effects as well as some central nervous system effects. Interestingly, the leads to both ibuprofen and indomethacin came from the testing of compounds originally synthesized as herbicides.

Several other related structures were developed. Naproxen was synthesized at Syntex and introduced commercially in 1976. It, as well as ketoprofen, has been available under several brand names without prescription in the United States since the mid-1990s. The intended use for the nonprescription versions of naproxen is for treatment of temporary conditions. Chronic use, as in arthritis, requires prescription versions that are at a higher dosage level. Several other types of NSAIDS have been introduced subsequently. One group is aniline—benzoic acids or phenylacetic acids. These were first recognized at the Parke-Davis Company. Dichlofenac is the most widely used of this group. It was developed at Ciba-Geigy, now part of Novartis. Dichlofenac was introduced in Japan in 1974 and was approved in the United States in 1988. It is one of the most widely used NSAIDS in much of the world, but in the United States, it was not actively marketed until the 1990s. The ester of dichlofenac with glycolic acid, known as acechlofenac, is also used as an analgesic and anti-inflammatory agent. There are other NSAIDS that do not contain the carboxylic acid group, but still have gastrointestinal side effects. Most of these contain sulfonamide structures. One example, meloxicam was approved as an analgesic for treatment of rheumatoid arthritis by the FDA in 2000. These structures are included in Scheme 13.1.

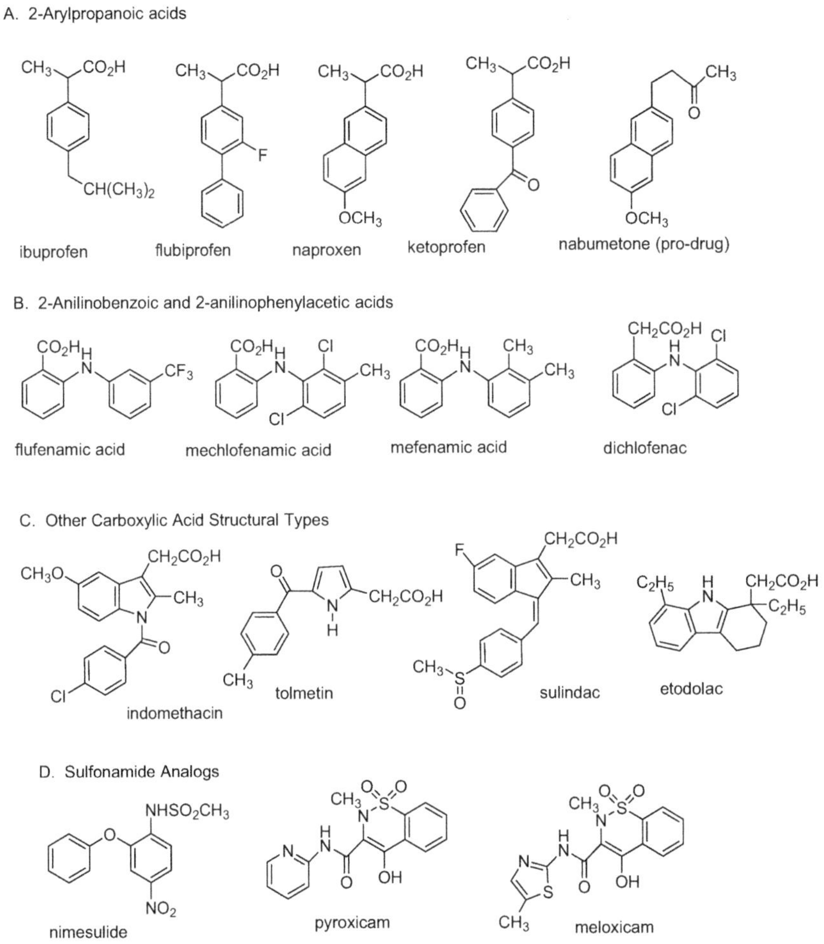

A. 2-Arylpropanoic acids

ibuprofen   flubiprofen   naproxen   ketoprofen   nabumetone (pro-drug)

B. 2-Anilinobenzoic and 2-anilinophenylacetic acids

flufenamic acid   mechlofenamic acid   mefenamic acid   dichlofenac

C. Other Carboxylic Acid Structural Types

indomethacin   tolmetin   sulindac   etodolac

D. Sulfonamide Analogs

nimesulide   pyroxicam   meloxicam

SCHEME 13.1   Structure of nonselective COX inhibitors.

13.3   SELECTIVE CYCLOOXYGENASE-2 INHIBITORS

The next major discovery in the area was a second cyclooxygenase called COX-2. It is rather similar to the COX-1 enzyme, but with somewhat different amino acid sequence, different distribution among cell types, and the property of being induced by inflammation. COX-2 appeared to be more intimately involved in the inflammation process and less so in blood clotting than COX-1, so the hypothesis arose that drugs directed at COX-2 might retain anti-inflammatory activity while reducing side effects resulting from inhibition of blood clotting. A search for such drugs began. In agreement with the hypothesis that the anti-inflammatory benefits and side effects could be separated, several drugs that reduced inflammation but with less evidence of gastric disturbance were found. These drugs are called *cyclooxygenase-2 inhibitors* (COXIBS). The traditional NSAIDS inhibit both COX-1 and COX-2 to a comparable extent. Figure 13.4 summarizes the relationship between cyclooxygenase inhibition and gastro-intestinal side effects. The figure indicates the expected minimization of side effects if only COX-2 is inhibited. It also indicates that inhibition of cyclooxy-genases increases the formation of another class of arachadonic acid metabo-lites, the leukotrienes responsible for leukocyte adhesion.

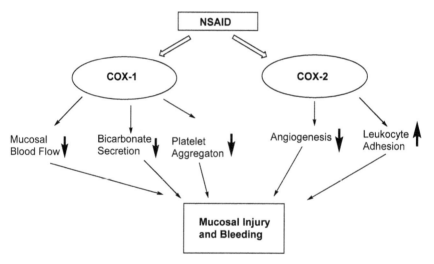

FIGURE 13.4   Overview of the relationships between COX-1 and COX-2 inhibition and gastrointestinal side effects.

Crystal structure determination revealed that the binding pocket in COX-2 is somewhat larger than in the COX-1 enzyme and has a hydrogen-bonding pocket that accounts for the prominence of sulfones and sulfonamides among the COX-2 inhibitors (see Fig. 13.5). Scheme 13.2 gives the structure of some of the drugs that were discovered by these efforts.[11]

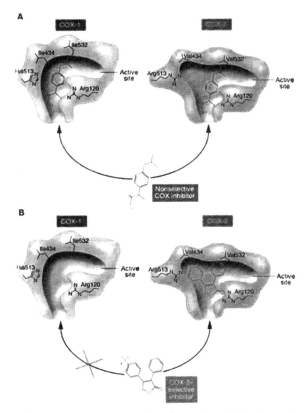

FIGURE 13.5 Schematic diagram identifying the amino acids in the COX-1 and COX-2 binding sites and showing the additional space in COX-2 active site. (From Grosser, T.; Fries, S.; FitzGerald, G. A. *J. Clin. Invest.* **2006,** *116*, 4–15 by permission of the American Society for Clinical Investigation.)

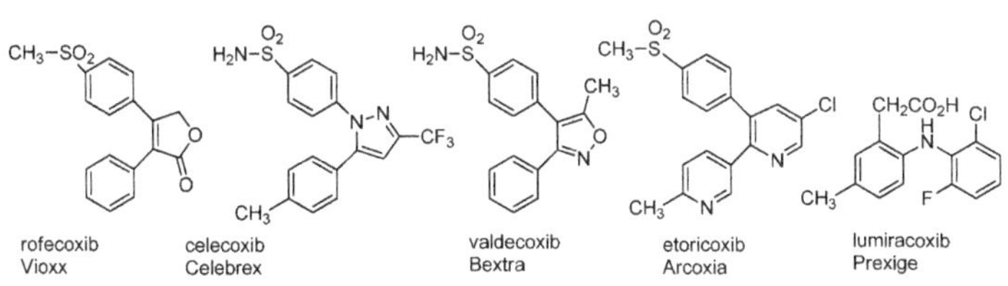

SCHEME 13.2   Structures of selective cyclooxygenase-2 inhibitors.

The first COX-2 inhibitor to be introduced in the United States was celecoxib as Celebrex. It was discovered at Searle, which subsequently had been acquired by Pfizer. Celebrex was introduced in 1999 and had sales of $1.5 billion in its first year. The second COX-2 inhibitor to be introduced was rofecoxib (Vioxx) by Merck in 2000. It had $400 million in sales in its first year and $2.5 billion by 2003. The sales of both drugs were increased by vigorous direct consumer advertising based on the reduced gastrointestinal side effects. The introduction of these drugs was rapid by most standards, barely 10 years between discovery of COX-2 and the introduction of the drugs. A corollary of this rapid introduction was that the longest of the safety studies submitted for drug approval were less than 1 year in duration. For example, the assessment of the gastrointestinal effects was by endoscopic examination searching for abnormalities, rather than the more extended time that would be required for development of symptomatic conditions.

There were some concerns about the safety of the COX-2 inhibitors. It was suggested that COX-2 inhibitors might decrease production of certain beneficial prostaglandins.[12] The first concern about the COX-2 inhibitors based on clinical findings were published in 2001 by physicians associated with the Department of Cardiovascular Medicines at the Cleveland Clinic. They reviewed the studies of rofecoxib and celecoxib and noted that in the case of rofecoxib there was a 2.4-fold excess of thrombotic cardiovascular events, relative to the control drug naproxen. In comparison with a placebo with an incidence of about 0.5%, the rates were about 0.74% and 0.80% for rofecoxib and celecoxib, respectively.[13] This publication triggered a number of letters to the editor of the *Journal of the American Medical Association* criticizing various aspects of the study.[14]

In addition to anti-inflammatory activity, aspirin and other NSAIDS inhibit the reappearance of colon polyps. Because polyps can progress to cancerous lesions, studies were conducted on this effect using the COX-2 inhibitors. These studies ran for longer time periods (at least 3 years) than the original safety studies. The drugs did cause significant reduction (20–30%) in polyp recurrence, but when the data were analyzed, it was found that both rofecoxib and celecoxib had significant increases in cardiovascular events (heart attack, stroke) over the course of the studies.[15] The increase was from a rate of around 1% to 2–3%. These results led Merck to withdraw rofecoxib in September 2004. Soon thereafter, the FDA required stronger warning labels on celecoxib as well as *all NSAIDS*. Numerous lawsuits ensued. Merck won a majority of the individual trials and vowed to defend itself vigorously in all of them. Most of the cases were settled out of court. It is estimated that Merck eventually absorbed nearly $5 billion in these settlements. Pfizer paid nearly $900 million to settle claims pertaining to celecoxib. Both Merck and the FDA were criticized for failing to conduct studies that might have detected the cardiovascular effects as part of the original drug approval process.[16,17] Valdecoxib and lumiracoxib, neither of which were ever marketed in the United States, were also withdrawn because of other side effects. Celecoxib remains available in the United States, with sales in the $2–3 billion range for 2004–2007. Etoricoxib is approved in Europe, but not the United States.[18]

There is a good level of understanding of the origin of the increased cardio-vascular risk.[19] The adverse effects are thought to be due to a decrease in the amount of specific prostaglandins, especially prostacyclin, that act as vasodilators, inhibit platelet aggregation and reduce thrombosis and atherosclerosis. There is also increased production of thromboxanes that are vasoconstrictors and promote platelet aggregation. The COXIBS also tend to increase blood pressure, which may contribute to increased incidence of cardiovascular events. There is now evidence that many of the traditional NSAIDS also increase cardiovascular risk.[20] In retrospect, it appears that there is a continuum of selectivity for COX-1 and COX-2 inhibition that includes both the classical NSAIDS and the COXIBS. As shown in Figure 13.6, some other NSAIDS are somewhat selective for the COX-2 enzyme and thus might be anticipated to have reduced gastrointestinal side effects. These include dichlofenac, nimesulide, etodolac, and meloxicam. Data available to this point are consistent with this expectation.[21]

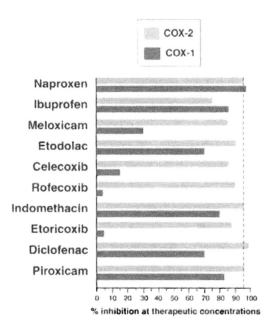

FIGURE 13.6 Comparison of COX-1 and COX-2 inhibition by anti-inflammatory drugs. (From Garcia Rodriguez, L. A.; Tacconelli, S.; Patrignani, P. *J. Am. Coll. Cardiol.* **2008,** *52,* 1628-1638. Used with permission of Elsevier.)

Aspirin has a unique place in the range of activities. It completely inhibits COX-1 while having little effect on COX-2. As a result it allows continued production of the beneficial prostaglandins and prostacyclins, while suppressing the prothrombotic thromboxanes. The various drugs also vary in the time course of their action, which also has implications for their overall effects, as shown in Figure 13.7.

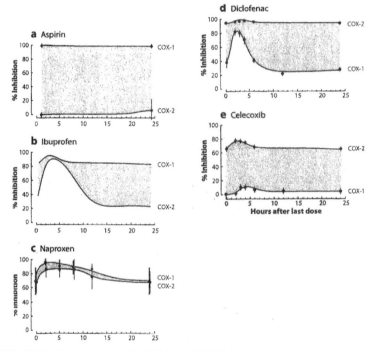

FIGURE 13.7 Time course of inhibition of COX-1 and COX-2 by several of the anti-inflammatory drugs. (From Grosser, T.; Yu, Y.; FitzGerald, G. A. *Ann. Rev. Med.* **2010**, *61*, 17–33. Reprinted by permission of Elsevier.)

Figure 13.8 summarizes the effects of several of the types COX inhibitors and their physiological effects.[22]

COX inhibitor	Platelet TXA$_2$	Whole Body PGI$_2$	Cardiovascular phenotype
Aspirin (50 100 mg)	Decreased by >97%	No significant changes	Cardioprotection
Aspirin (650 1,300 mg)	Decreased by >97%	Decreased 60–80%	Cardioprotection(Effects relative to lower doses are uncertain)
Naproxen (500 mg bid)	Decreased by ~95%	Decreased 60–80%	None/cardioprotection?
Other NSAIDs (high doses)	Decreased by 50–90%	Decreased 60–80%	Increased risk of myocardial infarction
Coxibs (high doses)	No significant changes	Decreased 60–80%	Increased risk of myocardial infarction

FIGURE 13.8 Summary of aspirin, NSAIDS, and COX-2 inhibitors on thromboxane and prostaglandin levels and anticipated physiological consequences. (Adapted from Patrono, C.; Baigent, C. *Mol. Interv.* **2009**, *9*, 31–39.)

13.4 ACETANILIDES, INCLUDING PHENACETIN AND ACETAMINOPHEN

Acetaminophen, also called *paracetamol*, is most familiar to Americans as Tylenol. There are many other brand names, as well. Like aspirin, acetaminophen has a long history. Acetaminophen is a derivative of acetanilide, which was discovered to have antipyretic properties in 1886 when it was accidentally used to treat a patient for intestinal parasites and the patient's temperature dropped. Acetanilide, itself, proved to be quite toxic, but its 4-ethoxy derivative, phenacetin, was also effective and was introduced as a drug by Bayer in the 1890s. Much later, in the 1950s, it was shown that both acetanilide and phenacetin are metabolized to acetaminophen, leading to the introduction of acetaminophen as a drug.[23] Acetaminophen has antipyretic and analgesic properties, but unlike aspirin and the NSAIDS, it has little anti-inflammatory action. It also does not have the gastrointestinal side effects of the NSAIDS. However, as we will see shortly, it has a rather low margin of safety and is the cause of many accidental overdoses.

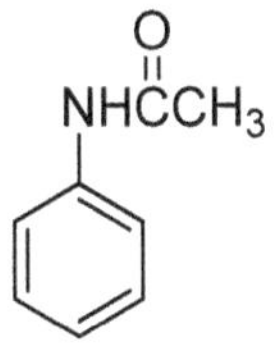

acetanilide

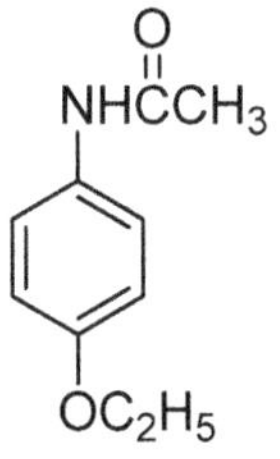

phenacetin
acetophenetidine

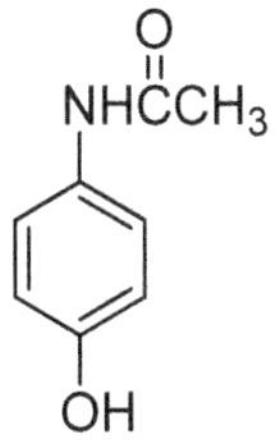

acetaminophen,
paracetamol

Acetaminophen became a widely used analgesic and antipyretic after 1950. It was introduced in the United States by McNeil Laboratories as Tylenol in 1955, and particularly formulated for use by children. It was introduced in the United Kingdom as Panadol, by Frederick Stearns & Company (now sold by GlaxoSmithKline). Currently in the United States, adult dosages of 325, 500, and 650 mg are available with a recommended maximum dose of 4 g/day. For children the dosages are 80 or 160 mg, with the recommended maximum of 75 mg/kg/day. One reason for its use in children is the risk of Reyes syndrome, which is associated with use of aspirin in patients with high fever.

The mechanism of action of acetaminophen is still something of a puzzle and is controversial.[24] One proposal is that there is a third form of COX associated with the CNS that is highly sensitive to acetaminophen. Another possibility is that metabolism of acetaminophen is involved. As a phenol, it is highly susceptible to oxidation, and an oxidative metabolite might be involved in the inhibition of the COX enzymes.

The toxic effects of acetamionphen, primarily liver damage, began to be recognized in the 1960s. It probably would not be approved as an over-the-counter drug under current regulations. In the United States, there are annually about 100,000 calls to poison centers involving acetaminophen, about a quarter pertaining to children under six. There are 56,000 visits to emergency rooms, 2600 hospitalizations and 450 deaths annually associated with acetaminophen overdose.[25] Acetaminophen overdose accounts for about half of acute liver failure cases in the United States. Acetamoinophen is present not only in brand name and generic versions but also in combination medicines for cold, flu, and pain. In a study of healthy volunteers using the 4-g daily maximum recommended dose for 14 days, 39% showed >3times the normal upper limit and 25% >5 time normal upper limit for amino acid transaminase enzymes, an indicator of liver function. Other longer term studies have indicated a much lower incidence of liver function change (2.4%) and no acute liver failure.[26] Children seem to be at lower risk of toxic effects from acetaminophen than adults, and it has been suggested that this may be due to the relatively larger size of the liver in children.[27] Acetaminophen is oxidatively metabolized by CYP-type enzymes, which are in highest concentration in the liver. These enzymes are induced by ethanol, acetone, and other xenochemicals. The toxic metabolite is N-acetyl-p-benzoquinone imine, which is a reactive electrophile that can form covalent bonds with proteins. The metabolite is captured by glutathione and the adduct can be excreted. The toxic effects occur when glutathione is depleted. An antidote, N-acetylcysteine, is available. The antidote is effective, especially if administered within 8 h of overdose, but also carries some risk of anaphylactic response. Because of concern over its side effects, in 2009, the FDA convened a panel to consider several restrictions on access to acetaminophen, including reducing the recommended maximum daily dose to 3.25 g, restricting combinations with other analgesics and making the liquid form available in a single concentration. Except for the latter issue, which the panel strongly recommended, the votes were split on the other proposals.[28]

13.5   OTHER ANALGESIC AND ANTIPYRETIC DRUGS

The first commercially successful synthetic antipyretic/analgesic was antipyrine. It was synthesized at Erlangen by Ludwig Knorr in 1884 and its antipyretic properties investigated in collaboration with Wilhelm Filehne, Professor of Pharmacology. A patent was awarded to Knorr and the commercial development was undertaken at a company that eventually became part of Hoechst. In the

1920s, sales of antipyrine amounted to several hundred thousand of kg. Hoechst also introduced related drugs called dipyrone and pyramidon.

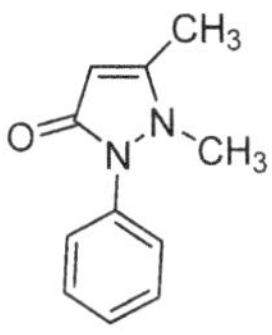

anrtipyrine 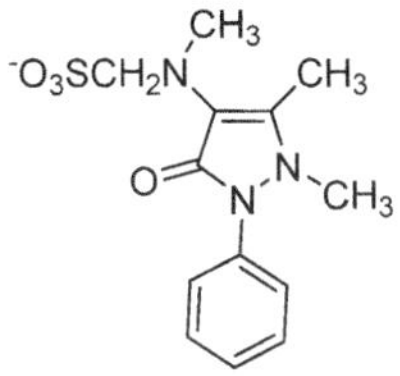   dipyrone 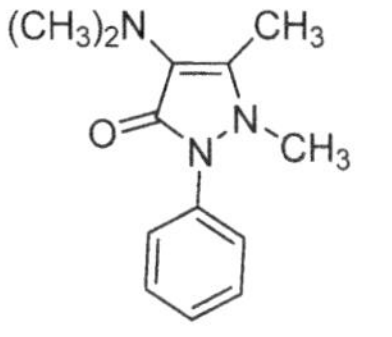   pyramidon 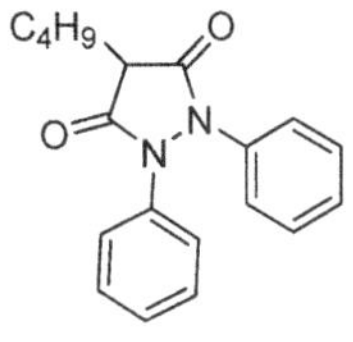   phenylbutazone

Another drug of this group is phenylbutazone, which was developed at Geigy. Phenylbutazone was introduced in 1952. Side effects included intestinal bleeding and ulcers, and there was also a risk of serious blood disorders, including aplastic anemia.[29] In 1971, the FDA added a warning to the label and recommended it be used only for arthritic pain that fails to respond to other drugs.[30] Phenylbutazone is still permitted in treatment of other specific conditions, if no other analgesic is effective.[31] Phenylbutazone is widely used in veterinary practice, especially for treatment of horses. The drug is toxic not far above the recommended dosage, so careful dosing is necessary.[32]

13.6   MORPHINE AND RELATED OPIATES

The isolation of opium from poppy plants has been practiced in the Middle-East for centuries. Immature seed capsules are cut and the liquid that is exuded is collected. Knowledge of opium came to Europe as a result of the crusades in the 11th and 12th century. In the middle ages *laudanum*, a mixture of opium, alcohol, and spices, was popularized by Paracelcus (see Section 11.1). The potential lethality of these preparations was also appreciated. The first recorded modern usage of opium in connection with surgery was by James Moore, a London surgeon in 1784. There was no understanding of the mechanism of action however. By the mid-1800s, opium in the form of laudanum was being used for many ills in both Europe and America. Books popularized opium use, one being Thomas De Quincey's *Confession of an English Opium Eater*, published in 1821. It became fashionable for writers and artists to use opium. The poems and literature from the 1770s through the 1830s suggest images drawn under the influence of opium. It is believed that Edgar Allen Poe wrote some of his works under its influence. In some cases, writers and poets managed their addiction, but in others the vivid dreams and imagination gave way to melancholy and fear. Samuel Taylor Coleridge (1772–1834), who tried many times to overcome his addiction, toward the end of his life wrote: "After my death, I earnestly entreat that a full and unqualified narrative of my wretchedness, and of its guilty cause, may be made public, that at least some little good may be effected by the direful example."

Opium was widely available during the 1800s in unregulated potions and patent medicines. Laudanum could be obtained without a prescription. The United Kingdom imported some 280,000 pounds of opium in 1860 and exported about half of that, much of it to the United States. The material sold by pharmacists was often uneven in concentration and overdoses were common. Among the potions sold were "calming syrups" for children and many children, especially in institutions such as orphanages, were often in a semi-drugged state. In America, in the latter half of the 19th century, alcohol use by women was frowned upon, but not opium use. Opium use was also prevalent among the Chinese laborers who built many of the American railroads in the middle of the 19th century. Addiction was prevalent in other parts of the world, especially China, where it is estimated that a fourth of the adult males smoked opium in the early 1900s.

Several of the active ingredients in opium were isolated in the first half of the 19th century, beginning with morphine by Friedrich Serturner in 1806. Codeine and thebaine are also present in the opium extract (for structures, see Scheme 13.3). Commercial production of pure morphine began at E. Merck & Company (Darmstadt) about 1840. Although the availability of pure morphine improved the quality of the drug, it also resulted in higher doses being administered. Around 1850, the development of the hypodermic syringe, permitted subcutaneous and eventually intravenous administration of morphine. The availability of injectable morphine further expanded the range of conditions for which it was used in medical practice. Morphine was used extensively in treatment of wounded soldiers in both the US civil war and the Franco-Prussian war of 1870–1871 and resulted in many cases of addiction. Some reforms were begun in the 1870s as the problem of addiction became apparent. In the 1890s, aspirin and barbiturates became available, allowing them to be used for more tractable forms of pain. Opium, however, continued to be widely prescribed for a number of ailments.

Diacetylmorphine was investigated at Bayer by the same group that developed aspirin, including Albert Eichengrun and Felix Hoffmann. Bayer registered diacetylmorphine (as heroin) in 1898, incorrectly believing that is was less addictive than morphine. Its recommended use was as a cough suppressant. It became readily available in both the United States and Europe and resulted in a major addiction problem in the United States in the early 1900s. By 1910, it was clear that heroin was highly addictive. Passage of the Pure Food and Drug Act in 1906 led to decreased opium and cocaine use in the United States, as people became aware of the contents of many patent medicines. In 1914, with the passage of the Harrison Narcotic Act, heroin was made a controlled narcotic and was banned in the United States in 1924.[33] It is sometimes used in pain control and the generic name is diamorphine.

The structure of morphine was determined by Sir Robert Robinson at Oxford in the 1920s. This led to the synthesis of a number of analogs. Some were obtained by modification of morphine or other natural constituents of opium. Others were entirely synthetic, such as those derived from the benzomorphan structure and still others from simpler piperidine structures. The synthetic compounds that retain analgesic activity, however, also are highly addictive. A few of the compounds, such as nalorphine, naloxone and naltrexone are antagonists of morphine and can be used for treatment of addiction. Some of the synthetic analogs such as fentanyl and remifenatil are rapid-onset, short duration analgesics and are used during surgery and painful diagnostic procedures. The structures are shown in Scheme 13.3.

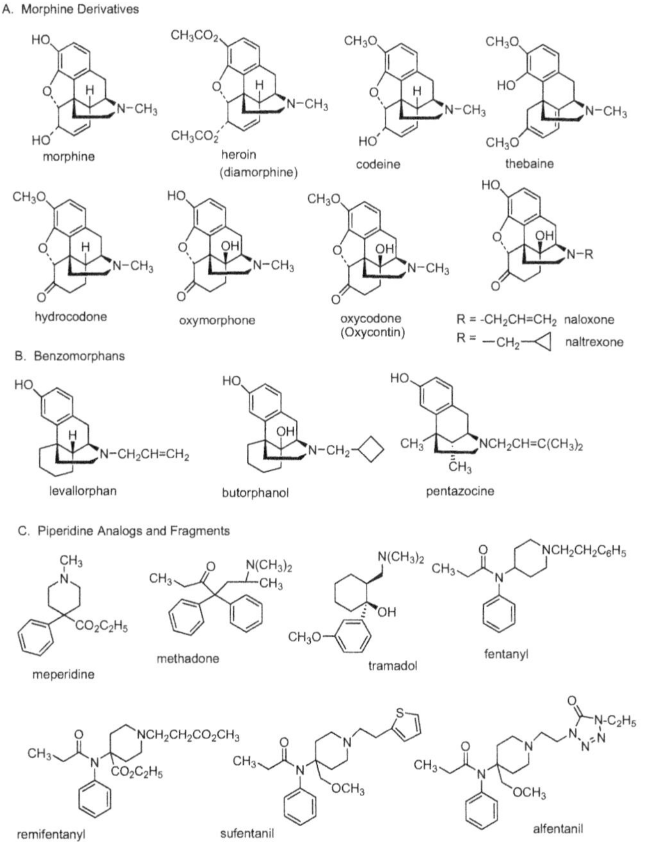

SCHEME 13.3 Morphine derivatives and synthetic analogs.

Key discoveries about the mechanism of action of the opiates were made in the 1970s. It was found that there is a family of CNS receptors and that the natural agonists for these receptors are small peptides called *enkephalins*. There are related peptides called *endorphins* and *dynorphins* that contain the same terminal sets of five amino acids as the enkephalins (Tyr-Gly-Gly-Phe-Met or Tyr-Gly-Gly-Phe-Leu). The receptors, called μ, κ, and δ, are also similar. The receptors can form both homo- and hetero-complexes that have somewhat different profiles of responses in areas such as respiratory depression, heart rate elevation and cough suppression. They also exhibit somewhat different levels of sensitivity to various opiates. The dose-limiting side effect of morphine is respiratory depression. The opiates are highly addictive. The opiates also cause *tolerance*, that is, the need for escalating doses for pain management. The addictive properties of the opiates are believed to result from decreased production of the natural polypeptide pain-killers. Thus, when the opiate "wears off" there is a deficiency of the natural substances and the need to take additional opiate. The addictive properties limit chronic use of opiates, except in the case of terminal conditions.

13.7   COCAINE AND RELATED STRUCTURES

13.7.1   COCAINE

Europeans became aware of cocaine as a result of the exploration of the Andes by the Spanish in the 16th century. Natives of areas which are now Bolivia and Peru chewed the leaves of the coca plant to relieve pain and increase energy and endurance.

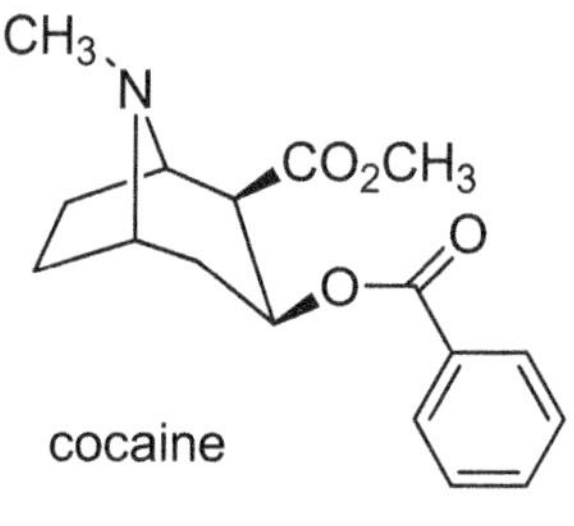

cocaine

The isolation and structure determination of cocaine were carried out in the laboratory of Friedrich Woehler in Gottingen by Albert Niemann and Wilhelm Lossen in 1860. The E. Merck Company began production on a limited scale in 1862. The development of small accurate syringes made possible controlled delivery of cocaine. In the 1880s, medical use of cocaine began to attract attention.

An Austrian physician, Carl Koller, reported the use of cocaine as an anesthetic in eye surgery. In the United States, surgeons such as William S. Halstead and William Hammond began to use it for anesthesia, especially in eye, ear, nose, and throat surgery. Cocaine was also considered useful in treating "nervous diseases," which we would call anxiety, stress, depression, or neuroses. Sigmund Freud was especially interested in the use of cocaine to treat such conditions and hypothesized that cocaine's stimulatory effect was beneficial. The drug was usually administered as an extract of coca leaves, and therefore in relatively low does, although injection of the hydrochloride salt was also sometime used. In the 1880s and 1890s, when opiate addiction was common, cocaine was used as a withdrawal treatment. But, the potential for addiction to cocaine too, soon became apparent. Some physicians became addicted as the result of self-administration. Physicians also observed what was called cocaine toxicity, manifested by delusion and uncontrollable agitation. By the late 1880s, cases of toxicity and respiratory arrest were being reported in the medical literature. The physical and mental deterioration associated with long-term cocaine use, including emaciation, also began to be recognized.

In the 19th century, cocaine was completely legal and required no prescription. Various drug and pharmaceutical companies were engaged in production of pure cocaine. In Germany, Merck and Boehringer were involved. In the United States, in addition to Merck, Parke-Davis, E. R. Squibb, and McKesson & Robbins all produced cocaine. There were also many small firms supplying patent medicines that contained cocaine. The most famous of these was *Coca-Cola* formulated by an Atlanta pharmacist, John Pemberton. The success of *Coca-Cola* spawned many imitators. Most of these colas contained only small amounts of cocaine. Figure 13.9 shows and early Coca-Cola ad.

Coca cigarettes and cigars also appeared, one brand being called *Cocarettes*. Its advertisements said "Coca the finest nerve tonic and exhilarator ever discovered. Coca stimulates the brain to great activity and gives tonic and vigor to the entire system." Cocaine was also available in powders called *catarrh snuffs*, and recommended for treatment sinus and nasal congestion and asthma. These were intended for sniffing into the nose.

Medical use of cocaine peaked around 1900. By that time the medical profession was becoming concerned about addiction due to both medical and non-medical use. Nonmedical use was spreading among laborers, in areas such as the New Orleans docks, railway construction, southern plantations and western mines. In some cases the cocaine was supplied by employers in the belief that it would increase the capacity for work. Cocaine use also began to appear in "red-light" districts. Public concern began to increase. The term "cocaine fiend" began to be used in news accounts to describe those under its influence, but cocaine use was considered to be a personal vice, rather than an addiction. The

Pure Food and Drug act of 1906 required that ingredients such as cocaine be listed, but did not ban them. As a result, cocaine was removed from many patent medicines and tonics, including *Coca-Cola*. Major pharmaceutical firms, wary of its unsavory reputation, also began to stop production. It was banned in 1914 by the Harrison Act. Widespread cocaine use and addiction began to reappear in the 1960s. In the form of the free base, sold on the street as "crack," cocaine is exceedingly addictive. Its distribution, like heroin, was taken over by criminal organizations. Like heroin, cocaine became a target of the "War on Drugs."

FIGURE 13.9 A 1907 advertisement for *Coca-Cola*.

13.7.2 SYNTHETIC ANALOGS OF COCAINE AS ANESTHETICS

The most enduring medical application to come from cocaine is the use of synthetic analogs, such benzocaine, procaine, and lidocaine as local anesthetics, especially in dentistry. Benzocaine and procaine were introduced as topical anesthetics early in the 1900s. Benzocaine is still used in some over the counter products, such as throat lozenges. The brand name for procaine, Novocain, became essentially synonymous with dental anesthesia during the first half of the century. A number of analogs were made, but none seemed better than procaine. Work resumed later and in 1943, lidocaine was synthesized and found to be a potent local anesthetic. Commercial development was undertaken by Astra under the name Xylocaine. After further clinical testing it was launched in Sweden in 1948, including formulations with epinephrine. The subsequent experience was very positive and Xylocaine received approval by the FDA in the United States in 1948.[34]

These and related local anesthetics function by blocking nerve impulses. For local anesthesia, they are used in conjunction with epinephrine, which restricts blood vessels in the affected area and extends the duration of the anesthesia. A large number of analogs were synthesized and evaluated, although most fell by the wayside as the result of toxicity. Several improved analogs including mepivicaine (1957), prilocaine (1960), bupivacaine (1963), articaine (1972), and ropivacaine (1996) have been introduced. Their structures are shown in Scheme 13.4. The pure S-enantiomer of bupivacaine, known as levobupivacaine was introduced in 1996 and has improved effectiveness relative to the racemic material. The "caines" are used extensively for spinal anesthesia, for example, in childbirth. However, this can have some adverse effects on the newborn infant, occasionally fatal, and restrictions were introduced by the FDA in 1983.

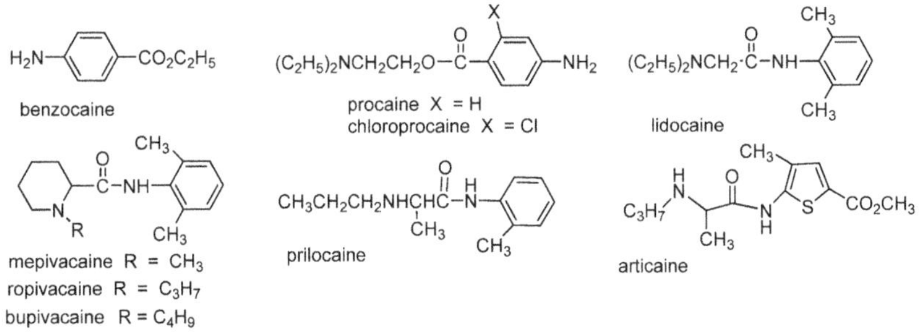

SCHEME 13.4 Cocaine analogs used as local anesthetics.

The "caine" drugs act on the α-subunit of the voltage-gated Na^+ channel. They are protonated at physiological pH, and the charge is evidently important

to the activity. Most of the compounds used as anesthetics have pKa values in
the range that they can exist in both the cationic and neutral form. Quaternary
analogs such as QX-222 and QX-314 have an even more powerful blockade
effect.

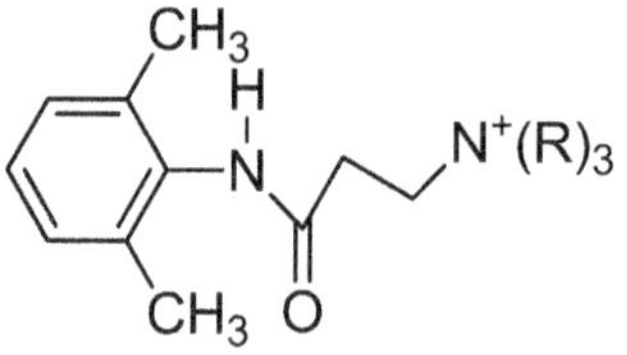

QX-222   R = CH_3

QX-314   R = C_2H_5

13.7.3 ATROPINE

The alkaloid atropine is closely related structurally to cocaine. It is isolated from
plants such as *Atropa belladonna* (deadly nightshade) and *Datura stramonium*
(jimson weed). At one time, atropine was used by women for its cosmetic effect
in dilating the pupils. Atropine derivatives are used for this purpose in ophthal-
mology. An analog homatropine is used because it has a shorter duration of
action. Atropine blocks the effect of acetyl choline at its muscarinic receptors
and is an antidote against the toxic effects of the organophosphorus and carbamate classes of insecticides (see Section 8.2.2 and 8.2.3). Atropine is a racemic
mixture of the enantiomers of the alkaloids D- and L-hyoscyamine. Most of the
pharmacological effects are due to the L-enantiomer.

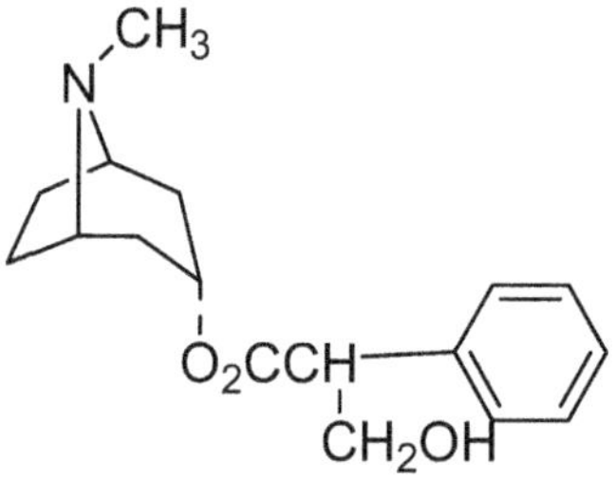

atropine 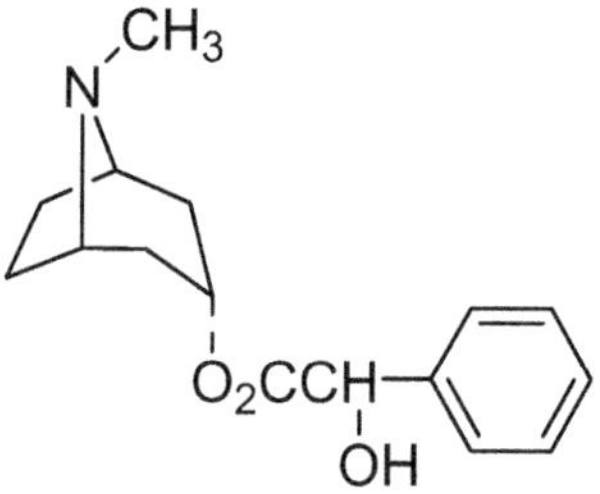   homatropine

13.8   GENERAL ANESTHETICS

The use of diethyl ether as an anesthetic was discovered independently in the 1840s by Crawford Long, a physician in Georgia, and two Massachusetts dentists, Horace Wells and William Morton, the latter at the suggestion of Charles Jackson. A public demonstration of use of ether as an anesthetic occurred in October 1846. Morton had applied for a patent for the process of anesthesia using "ethereal vapors." He had implied that the material, called the "preparation," contained an additional active substance, but this was not claimed in the patent application. A major surgery, amputation at the hip, was carried out by a leading surgeon John Collins Warren, of the Harvard Medical School, on November 7, 1846. The use of Morton's patented "preparation" violated medical ethics, which prohibited the use of secret "nostrums." Warren forced Morton to disclose that the "preparation" was in fact nothing but diethyl ether. News of the discovery spread rapidly and ether was being used in London by December, 1846. The use of diethyl ether as a general anesthetic persisted through the 1950s.[35]

The use of chloroform as an anesthetic was initiated by a Scottish physician, James Young Simpson in 1847. In 1850, Queen Victoria opted to deliver her eighth child under chloroform anesthesia. Use of both ether and chloroform was common during the Civil War. Besides causing respiratory depression, chloroform has the potential to induce fatal ventricular fibrillation. The use of anesthetics in childbirth was somewhat controversial both from a medical and social point of view. In the latter part of the 19th century, advocates of women's rights included anesthetics during childbirth as one of their goals. In the mid-20th century, natural childbirth, aided by relaxation techniques came into vogue. Current medical practice frequently uses a localized block with an anesthetic such as lidocaine or a combination of one of the local anesthetics with an opioid.

The goal of general anesthetics is to induce loss of consciousness, with resulting analgesia, amnesia and muscle relaxation. These states must be followed by relatively rapid, uneventful recovery. The current general anesthetics can be broadly divided into those administered by inhalantion and those administered by injection. The former include halothane, isoflurane, desfulrane, and sevoflurane, while propofol and etomidate are examples of injected anesthetics. Structures are shown in Scheme 13.5. Some of the barbiturates and a benzodiazepine, midazolam, are also used as general anesthetics. The volatile anesthetics are nonpolar compounds and their activity correlates with lipophilicity. There is also a size "cut-off" above which molecules are too large to be effective. The general anesthetics must be carefully administered and monitored as the therapeutic ratios are not very high. Respiratory depression is a specific problem. There is also accumulating evidence that general anesthetics can cause long-term cognitive defects, especially in children and in the elderly.

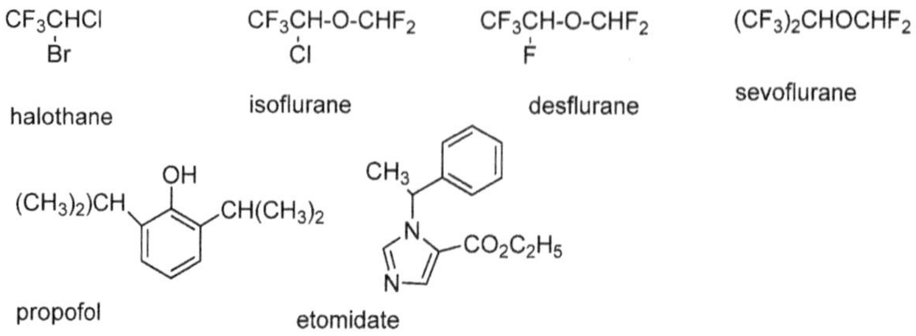

CF$_3$CHCl
Br

halothane

CF$_3$CH-O-CHF$_2$
Cl

isoflurane

CF$_3$CH-O-CHF$_2$
F

desflurane

(CF$_3$)$_2$CHOCHF$_2$

sevoflurane

(CH$_3$)$_2$CH

OH

CH(CH$_3$)$_2$

propofol

CH$_3$

N

N

—CO$_2$C$_2$H$_5$

etomidate

SCHEME 13.5 General anesthetics.

Exactly how anesthetics achieve their effect remains a mystery. Consciousness seems to involve coordinated function of specific parts of the brain, including the cerebral cortex, the thalamus, and the reticular network. With the volatile anesthetics these functions are gradually lost as the concentration of the anesthetic increases. One site of action of the general anesthetics is the GABA$_A$ receptor (see Sections 17.2 and 17.5). The volatile anesthetics activate inhibitory receptors and reduce neuronal activity. The effects may be specific to subsets of receptors. GABA receptors are also the site of action of propofol and etomidate. The most convincing evidence for involvement of GABA receptors is a series of specific mutations in the receptor that have pronounced effect on anesthetic potency.

KEYWORDS

- aspirin
- cyclooxygenase
- prostaglandins
- nonsteroidal anti-inflammatory drugs
- coxibs
- acetanilides
- acetaminophen
- opioids
- cocaine
- topical anesthetics
- general anesthetics

BIBLIOGRAPHY BY SECTION

Section 13.1: Mann, C. C.; Plummer, M. L. *The Aspirin Wars, Money, Medicine and 100 Years of Rampant Competition.* Alfred A. Knopf: New York, 1991; Jeffreys, D. *Aspirin: The Remarkable Story of a Wonder Drug.* Bloomsbury: New York, 2004; Mahdi, J. G.; Mahdi, A. J.; Mahdi, A. J.; Bowen, I. D. *Cell Prolifer.* **2006**, *39*, 147–155.

Section 13.1.1: Botting, R. M. *J. Therm. Biol.* **2006**, *31*, 208–219.

Section 13.1.2: Thompson, R. M.; Anderson, D. C. *Curr. Neurol. Neurosci. Rep.* **2013**, *13*, 327–335*;* Hennekins, C. H.; Dalen, J. E. *Am. J. Med.* **2013**, *126*, 373–378.

Section 13.2: Brune, K.; Beck, W. S. *Agents Action Suppl.* **1991**, *32* 13–25; Hersh, E. V.; Moore, P. A.; Ross, G. L. *Clin. Ther.* **2000**, *22*, 500–548; Vonkerman, H. E.; van de Laar, M. A. F. J. *Semin. Arthritis Rheum.* **2009**, *39*, 294–312; Patrono, C.; Rocca, B. *Pharmacol. Res.* **2009**, *59*, 285–289.

Section 13.3: Flower, R. J. *Nat. Rev. Drug Discovery* **2003**, *2*, 179–191; Whitehouse, M. W. *Curr. Med. Chem.* **2005**, *12*, 2931–2942; Whitehouse, M. *Inflammopharmacology* **2005**, *13*, 403–417; Mitchell, J. A.; Warner, T. D. *Nat. Rev. Drug Discovery* **2006**, *5*, 75–86; Grosser, T.; Fries, S.; FitzGerald, G. A. *J. Clin. Invest.* **2006**, *116*, 4–15; Hinz, B.; Renner, B.; Brune, K. *Nat. Clin. Pract. Rheumatol.* **2007**, *3*, 552–560; Martinez-Gonzalez, J.; Badimon, L. *Curr. Pharm. Des.* **2007**, *13*, 2215–2227; Rao, P. N. P.; Knaus, E. E. *J. Pharm. Pharm. Sci.* **2008**, *11*, 81s–110s; Marnett, L. J. *Ann. Rev. Pharmacol. Toxicol.* **2009**, *49*, 265–299; Patrono, C.; Rocca, B. *Pharmacol. Res.* **2009**, *59*, 285–289; Grosser, T.; Yu, Y.; FitzGerald, G. A. *Ann. Rev. Med.* **2010**, *61*, 17–33.

Section 13.4: Prescott, L. F. *Am. J. Ther.* **2000**, *7*, 143–147; Rumack, B. H. *Clin. Toxicol.* **2002**, *40*, 320; Josephy, P. D. *Drug Metab. Rev.* **2005**, *37*, 581–594; Graham, G. G.; Scott, K. F.; Day, R. O. *Drug Saf.* **2005**, *28*, 227–240; Amar, P. J.; Schiff, E. R. *Expert Opin. Drug Saf.* **2007**, *6*, 341–355; Bateman, D. N.; Dear, J. *Clin. Toxicol.* **2010**, *48*, 97–103.

Section 13.5: Brune, K. *Acute Pain* **1997**, *1*, 33–40.

Section 13.6: Booth, M. *Opium, A History,* St. Martin's Press: New York, 1998; Hamilton, G. R.; Baskett, T. F. *Can. J. Anesth.* **2000**, *47*, 367–374; Sabatowski, R.; Schaefer, D.; Kasper, S. M.; Brunsch, H.; Radbusch, L. *Curr. Pharm. Des.* **2004**, *10*, 701–716; Corbett, A. D.; Henderson, G.; McKnight, A. T.; Paterson, S. J. *Br. J. Pharmacol.* **2006**, *147*, S153–S162.

Section 13.7.1: Flynn, J. C. *Cocaine*; Carol Publishing: Secaucus, NJ, 1991; Weiss, R. D.; Mirin, S. M.; Baartel, R. L. *Cocaine*, 2nd ed. American Psychiatric Press: Washington, DC, 1994; Spillane, J. *Cocaine.* The Johns Hopkins University Press: Baltimore, MD, 2000; Streatfield, D. *Cocaine.* St. Martin's Press: New York, 2002.

Section 13.7.2: Ruetsch, Y. A.; Boni, T.; Bortgeat, A. *Curr. Top. Med. Chem.* **2001**, *1*, 175–182.

Section 13.8: Franks, N. P. *Br. J. Pharmacol.* **2006**, *147*, S72–S81; Lugli, A. K.; Yost, C. S.; Kindler, C. H. *Eur. J. Anesthesiol.* **2009**, *26*, 807–820.

REFERENCES

1. Stone, E. *Philos. Trans. R. Soc. Lond.* **1763**, *53*, 195–200.
2. Vane, J. R. *Nature New Biol.* **1971**, *231*, 232–235; Flower, R. J. *Pharmacol. Rev.* **1974**, *26*, 33–67.

3. Rouzer, C. A.; Marnett, L. J. *Chem. Rev.* **2003**, *103*, 2239–2304.
4. Copeland, C. E.; Six, C. K. *J. Surg. Educ.* **2009**, *66*, 176–181.
5. Anti-Thrombotic Trialist's Collaboration, *Br. Med. J.* **2002**, *324*, 71–86.
6. Patrono, C.; Baigent, C. *Mol. Interact.* **2009**, *9*, 31–39; Patrono, C. *Am. J. Med.* **2001**, *110*, 625–655; Patrono, C.; Garcia-Rodriguez, L. A.; Landolfi, R.; Baigent, C. *N. Engl. J. Med.* **2005**, *353*, 49–59.
7. Patel, J. H.; Stone, J. A.; Owora, A.; Mathew, S. T.; Thadani, U. *Am. J. Cardiol.* **2009**, 1687–1693; Bowry, A. D. K.; Brookhart, M. A.; Choudhry, N. K. *Am. J. Cardiol.* **2008**, 960–966; De Luca, G.; Marino, P. *Drugs*, **2008**, *68*, 2325–2344.
8. Van de Werf, F. *Eur. Heart J.* **2009**, *30*, 1695–1702.
9. Adams, S. S. *Clin. Pharmacol.* **1992**, *32*, 317–323; Rainsford, K. D. *Intern. J. Clin. Pract. Suppl.* **2003**, *135*, 3–8.
10. Shen, T. Y. *Angew. Chem. Intern. Ed. Engl.* **1972**, *11*, 460–472.
11. An overview of the development of COX-2 inhibitors is given by Flower, R. J. *Nat. Rev. Drug Dev.* **2003**, *2*, 179–191.
12. McAdam, B. F.; Catella-Lawson, F.; Mardini, I. A.; Kapoor, S.; Lawson, J. A.; FitzGerald, G. A. *Proc. Natl. Acad. Sci. U.S.A.* **1999**, **96**, 272–277; Penglis, P. S.; James, M. J.; Cleland, L. G. *Intern. Med.* **2001**, *31*, 37–41.
13. Mukherjee, D.; Nissen, S. E.; Topol, E. J. *J. Am. Med. Assoc.* **2001**, *286*, 954–959.
14. Letters Section, *J. Am. Med. Assoc.* **286**, Dec. 12, 2001.
15. Solomon, S. D.; McMurray, J. J. V.; Pfeiffer, M. A.; Witters, J.; Fowler, R.; Finn, P.; Anderson, W. F.; Zauber, A.; Hawk, E.; Bertagnolli, M. *N. Engl. J. Med.* **2005**, *352*, 1071–1080; Bresalier, R. S.; Sandler, R. S.; Quan, H.; Bolognese, J. A.; Oxenius, B.; Horgan, K.; Lanas, A.; Konstam, M. A.; Baron, J. A. *N. Engl. J. Med.* **2005**, *352*, 1092–1022.
16. Topol, E. J. *N. Engl. J. Med.* **2004**, *351*, 1707–1709.
17. For a view indicting aspirin as well as the classical NSAIDS as toxic drugs, along with the COXIBS, see Whitehouse, M. *Inflammopharmacology* **2005**, *13*, 403–417.
18. Cervantes, J. V.-M.; Collantes-Estevez, E.; Escudero-Contreras, A. *Future Rheumatol.* **2007**, *2*, 545–565.
19. Grosser, T.; Fries, S.; FitzGerald, G. A. *J. Clin. Invest.* **2006**, *116*, 4–15.
20. Kearney, P. M.; Baigent, C.; Godwin, J.; Halls, H.; Emberson, J. R.; Patrono, C. *Br. Med. J.* **2006**, *332*, 1302–1308.
21. Ahmed, M.; Khanna, D.; Furst, D. E. *Expert Opin. Drug Metab. Toxicol.* **2005**, *1*, 739–751.
22. Patrono, C.; Baigent, C. *Mol. Intervent.* **2009**, *9*, 31–39.
23. Brodie, B. R.; Axelrod, J. *J. Pharmacol. Exp. Ther.* **1949**, *97*, 58–67.
24. Davies, N. M.; Good, R. L.; Roupe, K. A.; Yanez, J. A. *J. Pharm. Pharm. Sci.* **2004**, *7*, 217–226 ; Kis, B.; Snipes, J. A.; Busija, D. W. *J. Pharmacol. Exp. Ther.* **2005**, *315*, 1–7; Aronoff, D. M.; Oates, J. A.; Boutaud, O. *Clin. Pharmacol. Ther.* **2006**, *79*, 9–19.
25. Lee, W. M. *Hepatol. Res.* **2008**, *38*, S1–S8.
26. Temple, A. R.; Benson, G. D.; Zinsenheim, J. R.; Schweinle, J. E. *Clin. Ther.* **2006**, *28*, 222–235.
27. Tenenbein, M. *J. Toxicol. Clin. Toxicol.* **2004**, *42*, 145–148.
28. Krenzelok, E. P. *Clin. Toxicol.* **2009**, *47*, 784–789.
29. Mauer, E. F. *N. Engl. J. Med.* **1955**, *253*, 404–410.
30. Faich, G. A. *Pharmacotherapy* **1987**, *7*, 25–27.

31. Toussirot, E.; Wendling, D. *Drugs* **1998,** *56*, 225–240.
32. Goodrich, L. R.; Nixon, A. J. *Vet. J.* **2006,** *171*, 51–69.
33. Sneader, W. *Lancet* **1998,** *352*, 1697–1699.
34. Holmdahl, M. H. *Acta Anaesthiol. Scand. Suppl.* **1998,** *42*, 8–12.
35. Moore, F. D. *Ann. Surg.* **1999,** *229*, 187–196; Macdonald, A.; Zuck, D. *Anaesthesia* **2006,** *61*, 553–556.

CHAPTER 14

ANTIHYPERTENSIVE DRUGS: CONTROLLING BLOOD PRESSURE

ABSTRACT

Hypertension, high blood pressure, is associated with an increased risk for heart attack and stroke, making it an important target for pharmaceutical intervention. Blood pressure is regulated by the renin–angiotensin–aldosterone system (RAAS), which presents several specific points for control. Drug types that have been introduced include vasodilators, diuretics, β-blockers, angiotensin-converting enzyme (ACE) inhibitors, angiotensin receptor blockers (ARB), calcium channel blockers (CCB), renin inhibitors, and aldosterone analogs.

Elevated blood pressure, *hypertension*, is an important cause of stroke, being implicated in about 70% of cases. Hypertension is also a significant factor in risk for heart disease, kidney disease, and diabetes. In this section, we will consider drugs that are used to control blood pressure. Treatment with antihypertensive agents to bring blood pressure to acceptable levels is associated with an approximately 30–40% reduction in the risk of stroke. There is also a reduction in myocardial infarction and congestive heart failure. These drugs also slow the pathological progression of diabetes. Elevated blood pressure is associated with certain dietary and life-style factors. Blood pressure can often be reduced by weight loss (if overweight), exercise, decreased sodium intake and reduced use of alcohol. Blood pressure tends to increase with age, both as a result of loss of flexibility in blood vessels and development of atherosclerosis. The significance of treatment of hypertension increased in the last quarter of the 20th century as obesity, diabetes, and cardiovascular disease (metabolic syndrome, see Section 9.8) contributed to increased illness, disability, and death. Besides their importance in treating a major cause of disability and death, the antihypertensive drugs provide an excellent example of the relationship between the understanding of physiology and biochemistry and the successful development of drugs.

14.1   THE RENIN–ANGIOTENSIN–ALDOSTERONE SYSTEM

Normal control of blood pressure is effected by the *renin–angiotensin–aldoste-rone system* (RAAS, sometimes RAS), by which the kidneys regulate sodium, potassium, and calcium levels. The pharmacological understanding of hyper-tension goes back to 1897, when two physiologists, Robert Tigerstedt and Per Bergman working at the Karolinska Institute in Sweden discovered that extracts of kidneys could raise blood pressure when injected into normal animals.[1] They named the presumed active substance *renin*. Not much more happened until the 1930s, when Harry Goldblatt, a pathologist working at Case Western University, showed that clamping of kidney arteries produced hypertension. He also no-ticed that hypertensive patients frequently had arteriosclerosis in the kidney. He showed that the sympathetic nervous system was not directly involved in the hy-pertension induced by blockage of the kidney arteries. Goldblatt proposed that there must be some hormonal substance that was responsible for these effects, but he had no direct evidence beyond Tigerstedt's observations.

Goldblatt's proposal triggered a search for renin. Around 1940, it became clear that the substance was a protease with high specificity. Renin is produced mainly in the kidney, initially as a 406 AA precursor. The active form is made by cleavage to remove 23 and 43 AA segments and glycosylation. Renin is pro-duced in response to several factors, including Na^+ flux and vascular pressure in the kidney. It is an *aspartyl proteinase* that cleaves the protein angiotensinogen to form the actual active substance, *angiotensin-II* (AT-II). In the 1950s, it was shown that there were two steps in the process, leading first to angiotensin-I (AT-I) a decapeptide and then to AT-II, an octapeptide. This implied that there was another enzyme involved in the conversion of AT-I to AT-II, called the *an-giotensin converting enzyme* (ACE). Late in the 1950s, it was shown that AT-II functions by stimulating aldosterone secretion. Aldosterone is a critical factor in maintaining salt and fluid balance, as well as blood pressure.[2] The polypeptides were made by synthesis at Ciba and became available for extensive studies.

Asn-Arg-Val-Tyr-Ile-His-Pro-Phe-HIs-Leu

Angiotensin-I

cleavage point for formation
of Angiotensin-II

Bradykinin has opposing effects to AT-II and promotes vasodilation and so-dium excretion. Bradykinin was discovered by Mauricio Rocha e Silva at the Biological Institute in Sao Paulo Brazil in 1948. It was found as the result of investigations of the effect on blood pressure of the venom of a Brazilian viper,

Bothrops jararaca, which frequently bit agricultural workers in Brazil. At the time, histamine was recognized as a major factor in the hypotension associated with anaphylactic shock. In his studies of the venom, however, Rocha e Silva observed effects that could not be attributed to histamine. He was able to isolate the responsible substance, which he called *bradykinin*. The isolated material was able to induce smooth muscle contraction and a drop in blood pressure. It was shown to be involved in the causation of certain forms of anaphylactic shock. Structural investigation, and, eventually synthesis, showed bradykinin to be a nonapeptide. In the 1930s, it had been observed that extracts of pancreas were hypotensive. The name *kallikrein* from the Greek word for pancreas was assigned to the presumed active substance. When the structure of the active substance was determined, it was found to be a decapeptide, kallidin-II, that has the same sequence as bradykinin with an additional N-terminal lysine.

Arg-Pro-Pro-Gly-Phe-Ser-Arg-Phe-Arg   Lys-Arg-Pro-Pro-Gly-Phe-Ser-Arg-Phe-Arg

bradykinin   kallidin-II

Rocha e Silva and Sergio Ferreira discovered that *B. jararaca* venom also contained small peptides called "bradykinin potentiating factors" that inhibited the proteolytic breakdown of bradykinin. These were small peptides with N-terminal pyroglutamic acid residues. They turned out to also be inhibitors of the important enzyme ACE.

Determination of the structure of these peptides led to the design and synthesis of inhibitors of ACE (see Section 14.2.4). Drugs were also found that could inhibit the effects of AT-II by blocking its receptor. These are called *angiotensin receptor blockers* (ARB). The formation of renin is promoted by the adrenergic amines and this step can be inhibited by drugs called β-blockers (BB). Antagonists that blocked the effect of aldosterone were also identified. These relationships are summarized in Figure 14.1, which includes the points at which various drugs can affect the system.

The evolutionary purpose of the RAAS is to elevate blood pressure and conserve fluid in response to stress. However, the effects of AT-II also promote inflammation and endothelial dysfunction and chronic activation of the RAAS is detrimental to both the cardiovascular and renal systems. More recently other aspects of the system have been uncovered. It has been found that both the nona- and heptapeptides related to AT-I and AT-II have biological activities in their own right. Generally speaking, these peptides tend to counteract and balance the effects of AT-II. Also, a second receptor for AT-II has been discovered and its effects are generally opposed to those of the dominant receptor. Thus, the system includes not only the main components summarized in Figure 14.1, but also a feedback network that maintains control of vasoconstriction and blood pressure. It has been found that both the ACE-i and ARB type of drugs upregulate formation of renin and this tends to counteract the effect of the downstream inhibition. The overall picture that emerges is somewhat reminiscent of the relationships discussed for the prostaglandins in Section 13.1.1. That is, each of the angiotensin peptides may have multiple, sometime opposing, affects.

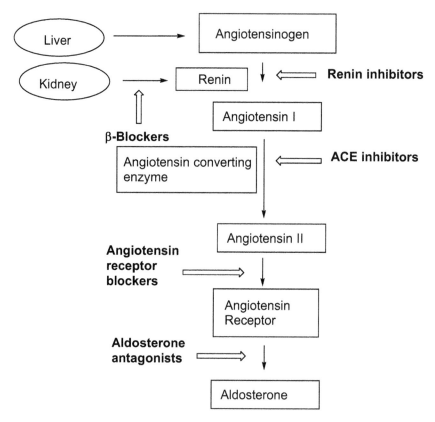

FIGURE 14.1   The renin–angiotensin system and indication of drug targets. AT1 and AT2 = angiotensin receptors. (Adapted from Epstein, B. J.; Gums, J. G. *Ann. Pharmacother.* **2005**, *39*, 471).

14.2 DRUGS FOR TREATMENT OF HYPERTENSION

Several types of treatments have been introduced for hypertension. In the 1940s, low-sodium dietary regimens were developed, but were effective only at extreme levels of sodium reduction. Beginning in the 1960s, effective drugs became available. There are several major categories of antihypertensive drugs. These include vasodilators, diuretics, BB, angiotensin-converting enzyme inhibitors (ACE-i), ARB, calcium channel blockers (CCB), and aldosterone antagonists (AA). Recently, a new class of drugs called direct renin inhibitors (DRI), represented by the first example aliskiren, has been introduced. Frequently, drugs of two different types are required to achieve reduction of blood pressure to the target level. The renin and sodium-volume aspects of hypertension can be treated separately or together. The renin system responds primarily to the ACE-i and ARB drugs, while the sodium-volume system responds to BB, diuretics and CCB. Within each category, there are a number of drugs and they are often structurally related. Several drugs in these categories are among the most widely prescribed drugs, including atenolol (BB), metoprolol (BB), hydrochlorthiazide (diuretic), furosemide (diuretic), lisinopril (ACE-i), lorsartan (ARB), and amlopidine (CCB). Contributing to the wide use of these drugs is the high incidence of hypertension and the chronic nature of the condition. The availability of these drugs has not only had clinical significance, but has also permitted the development of more detailed understanding of the interactions and mechanisms that control the related factors of heart rate and output, ion concentration levels, and vasodilation/constriction that influence blood pressure. Scheme 14.1 gives examples of these drugs.

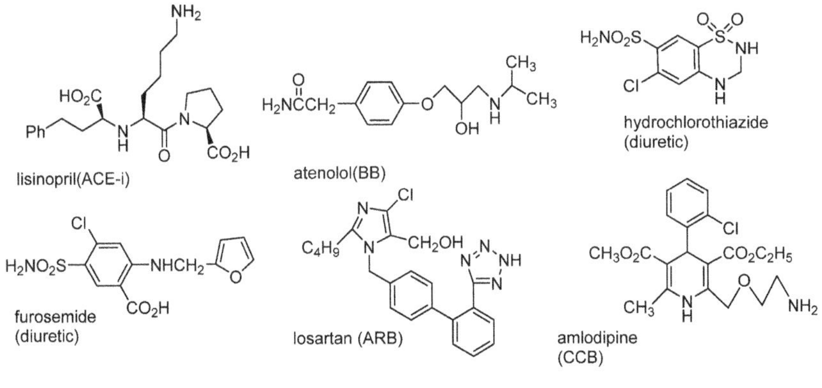

SCHEME 14.1 Representative types of antihypertensive drugs.

14.2.1   VASODILATORS

Nitroglycerine was introduced for treatment of angina pectoris in the 19th century. Its vasodilatory effect is the result of release of the endogenous vasodilator NO. Other nitrite esters have similar effects. The hydrazine derivatives were among the first antihypertensive agents to be introduced. They are not used very much at present, but are sometimes used in combination with other drugs.

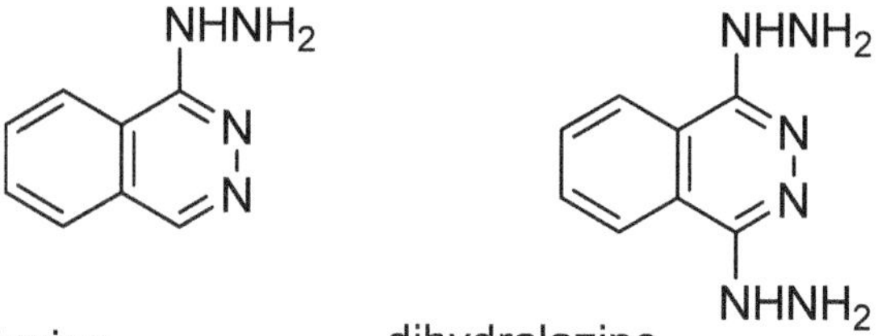

hydralazine   dihydralazine

14.2.2   DIURETICS

The thiazides were the first successful diuretics. The thiazide diuretics promote elimination of Na^+. They inhibit reabsorption of Na^+ in the kidneys and reduce fluid build-up. They may also reduce vascular resistance. The thiazide diuretics can enhance the antihypertensive effects of other types of drugs and are therefore often used in combinations. The details of their mechanism of action are somewhat uncertain but seem to involve both vasodilation and Na^+ and K^+ transport. Urinary loss of K^+ can be a dose-limiting factor. Chlorothiazide and hydrochlorothiazide are examples and other compounds of this class include metolazone, chlorthalidone, and indapamide. All these drugs contain an o-chlorosulfonamide structure.

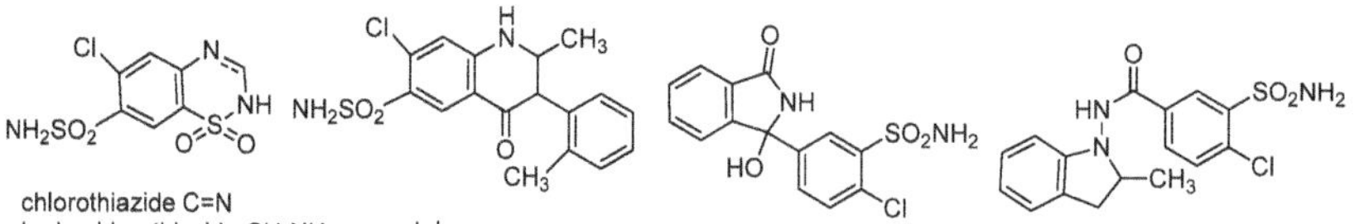

chlorothiazide C=N
hydrochlorothiazide CH-NH   metolazone   chlorthalidone   indapamide

A second group of diuretics, including furosemide, bumetamide, and torsemide, is called "loop diuretics." The first of these compounds, furosemide, was discovered when it was tested in a search for carbonic anhydrase inhibitors. These compounds have very rapid diuretic and natriuretic activity. The locus of action is a region of the kidney where little readsorption takes place. The loop diuretics also have vasodilation effects. The loop diuretics are also used in treatment of acute pulmonary and renal edema and in congestive heart failure.

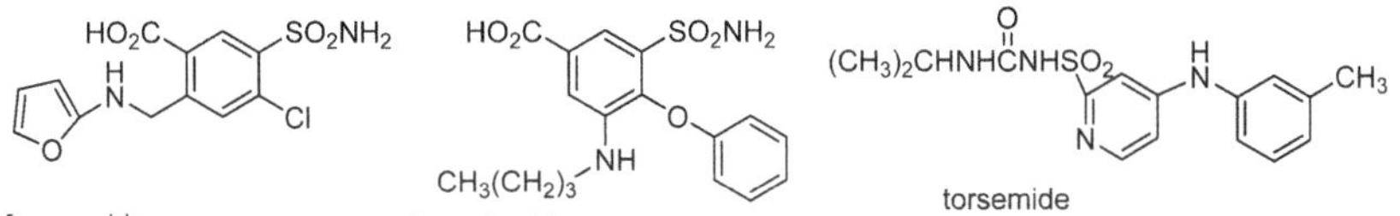

furosemide

bumetamide

torsemide

A third class of diuretics, call "potassium-sparing" diuretics includes amiloride and triamterene.

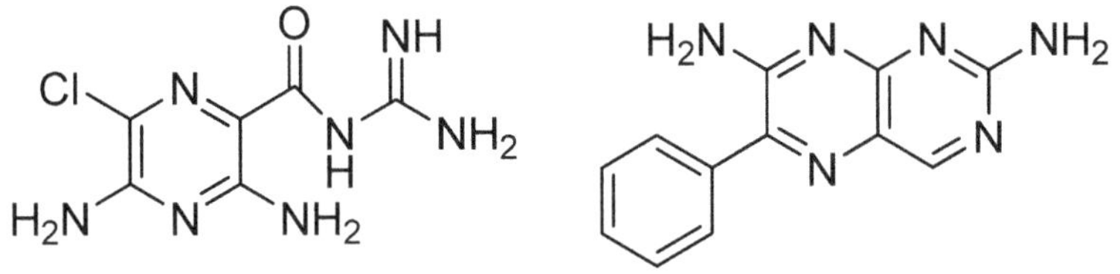

amiloride

triamterene

14.2.3   *ADRENERGIC α- AND β-BLOCKERS*

One of the key neurotransmitters involved in regulation of blood pressure is epinephrine (adrenaline) along with the closely related norepinephrine. These compounds, along with dopamine are called *catecholamines*, so named because of the *ortho*-dihydroxybenzene (catechol) ring they contain (see also Section 17.6).

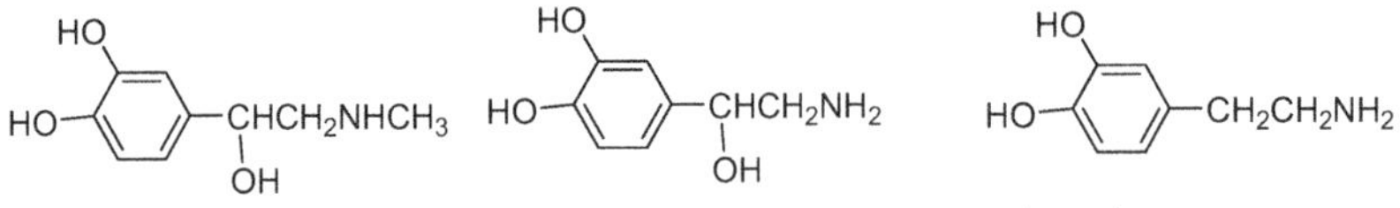

epinephrine

norepinephrine

dopamine

Raymond Ahlquist proposed in 1948 that there were two types of adrenaline receptors, α and β, based on different patterns of response to a series of drugs. Adrenaline increases the force and rate of heart contractions and also causes vasoconstriction and so raises blood pressure. Drugs that counter these effects can lower blood pressure. Sir James Black,[a] a pharmacologist working at ICI, pursued Ahlquist's theory by searching for a selective β-antagonist, on the premise that this would reduce the heart's need for oxygen and alleviate the pain of angina pectoris. Taking a lead from an Eli Lilly compound, dichloroisoprena-line, a compound called pronethalol was synthesized. It proved to be active and decreased heart rate, as well as reducing the occurrence of angina. This com-pound did not pass toxicological hurdles, but the related compound propranolol did. Not only was angina pectoris successfully treated, but as clinical results accumulated, it was also found to lower blood pressure and reduce the incidence of myocardial infarction.[3]

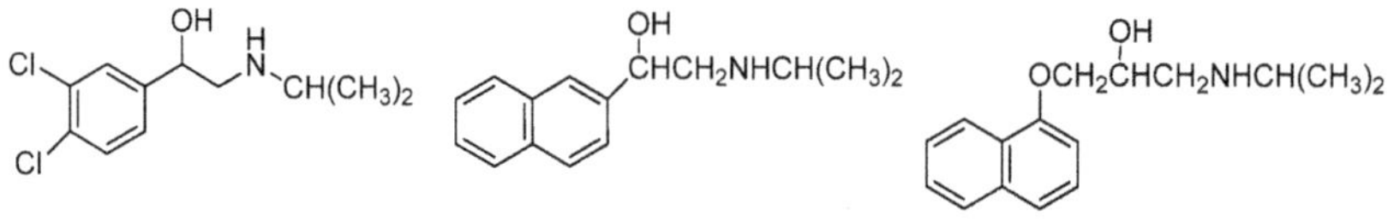

dichloroisoprenaline   pronethalol

propranolol

Subsequently, many other drugs that interact with adrenergic receptors have been found. They can have several effects, depending on their selectivity for the various receptors. The BB, particularly carvedilol, are used in treatment of congestive heart failure.[4] Another important use of the BB is as bronchodilators in treatment of asthma. The structures of several of these drugs are shown in Scheme 14.2.

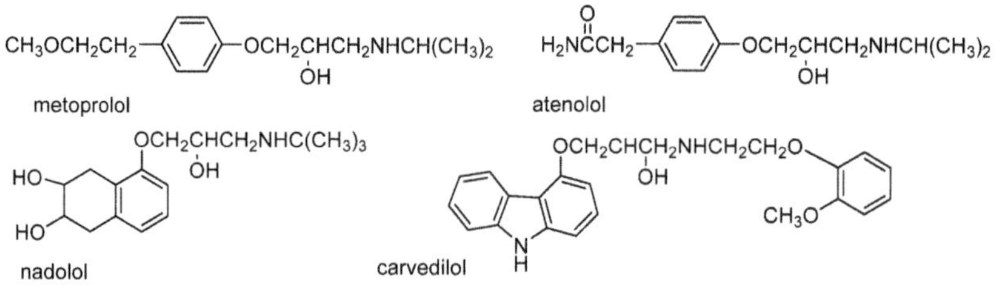

metoprolol   atenolol

nadolol   carvedilol

SCHEME 14.2   Examples of β-blockers.

[a]Sir James Black was born in Scotland in 1924. He received a medical degree from St. Andrew's University, spending part of his time at Dundee. After receiving his medical degree, he taught for a time in Singapore and was then the head of the Physiology Department of the Veterinary School at the University of Glasgow. He began working at ICI in 1958. His discovery of the first β-blocker, propanolol, was one of the earliest cases in which rational structural concepts were used to identify candidates, as opposed to finding "hits" by random screening, and this aspect of the work was cited in the award of the Nobel Prize in Physiology or Medicine for 1988. Later in his career, working at Smith, Kline, & French (now part of Glaxo) he developed a second major drug, the antiulcer drug cimetidine. In 1973, he returned to the academic world as head of the department of pharmacology at University College, London. He then moved to a third pharmaceutical company, Wellcome in 1978. Later, he was a professor of pharmacology at King's College, London. In 1992, he returned to Scotland as chancellor of the University of Dundee, serving until 2006.

Drugs targeted at the α-adrenergic receptor are also used to treat hypertension, such as prazosin, terazosin, and doxazosin. Drugs, such as clonidine and guanabenz are classified as α_2-receptor agonists. They, too, are sometimes used to treat hypertension.

pyrazosin R =

terazosin R =

doxazosin R =

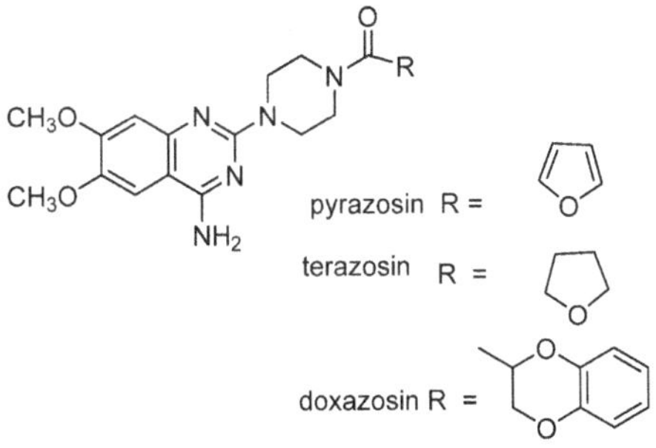

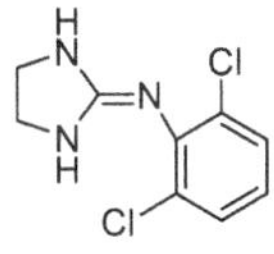

clonidine

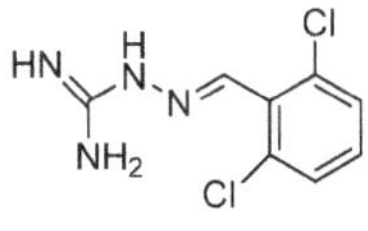

guanabenz

14.2.4 ANGIOTENSIN-CONVERTING ENZYME INHIBITORS

The centrality of the RAAS in hypertension made it an appealing target for pharmacological intervention. The possible effectiveness of angiotensin analogs was first demonstrated in 1972 and 1975 by H. Gavras and H. R. Brunner who hypothesized that inhibition of binding of AT-II would ameliorate renin-dependent hypertension. Dramatic effects were observed in two patients with extreme hypertension by infusion of a synthetic AT-II peptide analog saralasin.[5] Polypeptides present in the venom of the viper *B. jararaca* were found by John Vane (see Section 13.1.1 for biography) and coworkers to be inhibitors of ACE. Vane suggested that related polypeptides be synthesized. A nonapeptide, teprotide (Squibb), was shown to be effective in reducing renin-dependent hypertension. Both saralasin and teprotide, being peptides, had short lifetimes and needed to be administered intravenously, and thus were not suitable as drugs. They did show that inhibition of the binding or generation of AT-II could lower blood pressure.

$CH_3NHCH_2\overset{O}{\overset{\|}{C}}$–Arg-Val-Tyr-Val-His-Pro-Ala

saralasin

–Trp-Pro-Arg-Pro-Gln-Ile-Pro-Pro

teprotide

The first group of ACE inhibitors was designed by taking advantage of deductions about the ACE active site. This was done prior to availability of any crystallographic information. A simple compound, *N*-succinoylproline, showed weak but specific inhibitory activity and was the starting point. Introduction of a thiol group at a point consistent with interaction of a Zn^{2+} ion at the active site led to a dramatic increase in activity. These studies at Squibb in Princeton, NJ, USA eventually led to captopril, the first clinically used ACE inhibitor.[6] It was introduced in 1980.

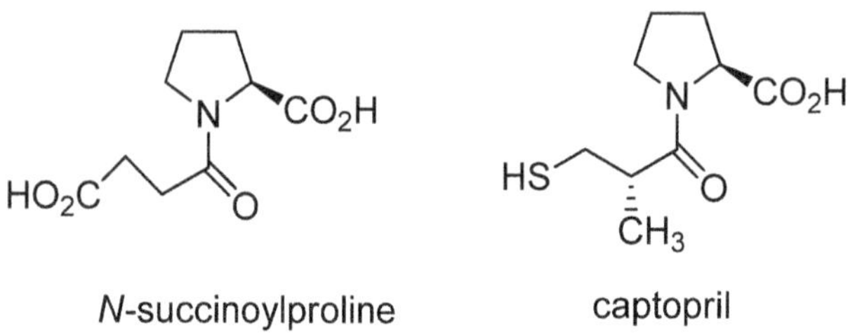

N-succinoylproline   captopril

Captopril and related ACE inhibitors appear to have other beneficial functions. They have vasodilating activity and induce recovery of abnormalities in heart structure associated with hypertension. ACE inhibitors also slow the progression of kidney disease associated with hypertension and type-2 diabetes. Statistical data also indicate a reduction in the risk of heart attacks. Currently, about 10 chemical entities are approved as ACE inhibitors in the United States. Several of the ACE inhibitors are esters that are hydrolyzed to the active forms, including ramipril (to ramiprilat), perindopril (to perindoprilat), and moexipril (to moexiprilat). Representative structures are shown in Scheme 14.3. The ACE inhibitors are often used in combination with a diuretic.

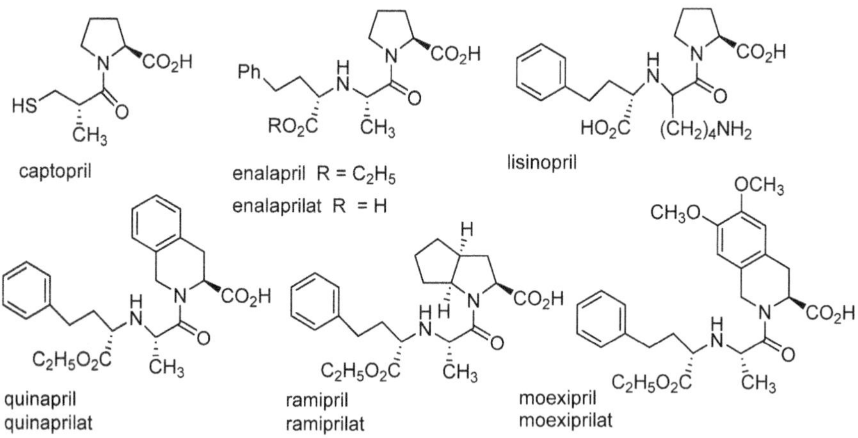

SCHEME 14.3   Examples of angiotensin-converting enzyme inhibitors.

Much later, the crystal structure of the ACE complex with lisinopril was determined and is shown in Figure 14.2. The zinc ion is bound by His-383, Glu-384, and His-387. The structure has a number of α-helices and only a small amount of β-structure. The enzyme is known to be activated by Cl⁻ ions and there are two bound Cl⁻ ions, but they are not directly at the active site. In the complex, a carboxylate binds to the zinc ion. The terminal carboxylate in lisinopril is bound by Lys-514 and Tyr-520.

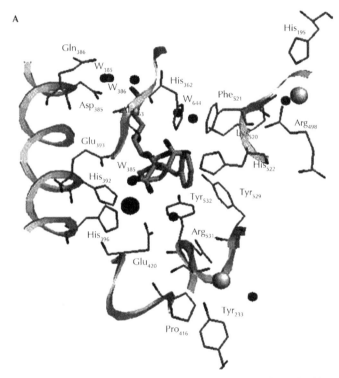

FIGURE 14.2   Protein drug interactions in the complex of lisinopril with ACE. (From Naresh, R.; Schwager, S. L. U.; Sturrock, E. D.; Acharya, K. R. *Nature* **2003**, *421*, 551–554; Fernandez, J. H.; Hayashi, M. A. F.; Camargo, A. C. M.; Neshich, G. *Biochem. Biophys. Res. Commun.* **2003**, *308*, 219–226 by permission of Elsevier.)

14.2.5   ANGIOTENSIN RECEPTOR BLOCKERS

This class is the newest group of widely used hypertensive drugs. Losartan, the first drug of this group, was introduced in 1995. It was developed by the Pharmaceutical Division of Du Pont. The work began with a molecule reported by the Takeda Pharmaceutical Company, which was a weak but selective inhibitor of the AT-II receptor. By synthetic variation, activity (IC_{50}) was improved from the 100 µM to the 0.01 µM range.[7] This study also found that other functional groups such as trifluoromethylsulfonamido and tetrazole, by virtue of their comparable acidity, could replace a carboxy function. However, the compounds were not orally active. Oral activity and the clinical compound losartan were found by inserting the biphenyl rings, with a tetrazole ring as the acidic (negatively charged) group. Its activity (IC_{50}) was 0.019 µM.[8] The active compound is the carboxylic acid formed by metabolism of the hydroxymethyl group. The route to lorsartan is outlined in Scheme 14.4.

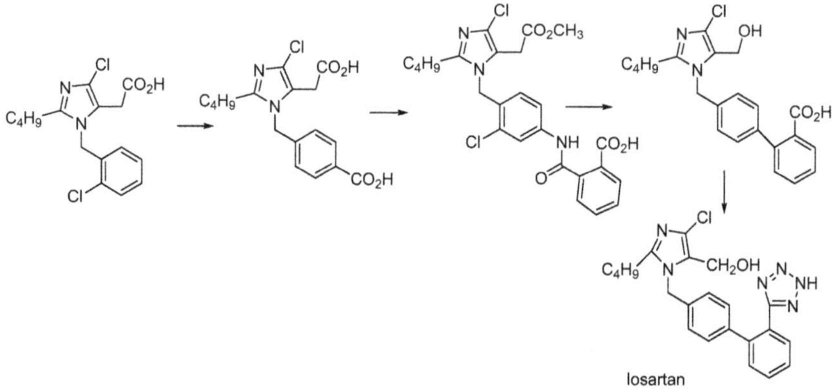

SCHEME 14.4   Structural modification resulting in Losartan.

Several other ARB compounds have been introduced, most of which are structurally similar to losartan. Irbesartan was developed at Sanofi in France. Telmisartan was originally synthesized at Dr. Karl Thomae Gmbh., in Germany.[9] An improved synthesis has been published by the generic drug firm, Dr. Reddy's Laboratory in India.[10] Valsartan originated in the Ciba-Geigy Laboratory in Basel, and is now marketed by Norvatis. Structures such as that found in eprosartan were synthesized at SmithKline.[11] The benzimidazole analogs, such as candesartan and azilsartan, were developed at Takeda in Japan.[12] Examples of these drugs are given in Scheme 14.5.

The ACE inhibitors and the ARB blockers not only lower blood pressure but have several other related beneficial effects. They improve glycemic control and reduce the incidence of type-2 diabetes. These drugs also inhibit and even reverse damage to the heart, kidney, and blood vessels. A major study ONTARGET compared the effects of an ACE inhibitor (ramipril), an ARB (telmisartan) separately and in combination for effectiveness in reducing the incidence of stroke and heart attack. The results indicated the drugs were equally effective, but that no additional benefit was achieved by a combination.

SCHEME 14.5   Other examples of angiotensin receptor blockers.

14.2.6   DIRECT RENIN INHIBITORS

Examination of Figure 14.1 suggests another possible point of control of the RAAS, namely inhibition of renin itself. This objective has been recognized since the 1970s and became especially relevant when it was observed that both the ACE and ARB classes of drugs induce a feedback mechanism that increases renin concentration. As a result there have been extensive efforts to identify drugs that would be *DRI*. Renin is an aspartic acid peptidase and its crystal structure and several complexes with inhibitors became available around 1990. This permitted structurally modeling to identify molecules that might act as inhibitors. Most attention was focused on *peptidomimetics*, which are structures that resemble peptide substrates but are not susceptible to hydrolysis. This is a theme that is applicable to many areas of drug design and several structural motifs that can act as peptidomimetics have been identified.

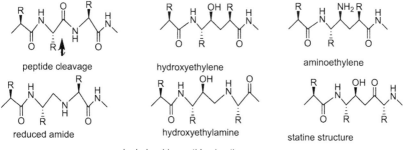

peptide cleavage

hydroxyethylene

aminoethylene

reduced amide

hydroxyethylamine

statine structure

non-hydrolyzable peptidomimetics

Aliskiren was the first DRI to be introduced. However, studies to date have failed to demonstrate significant advantages over the ACE-i or ARB classes of drugs.

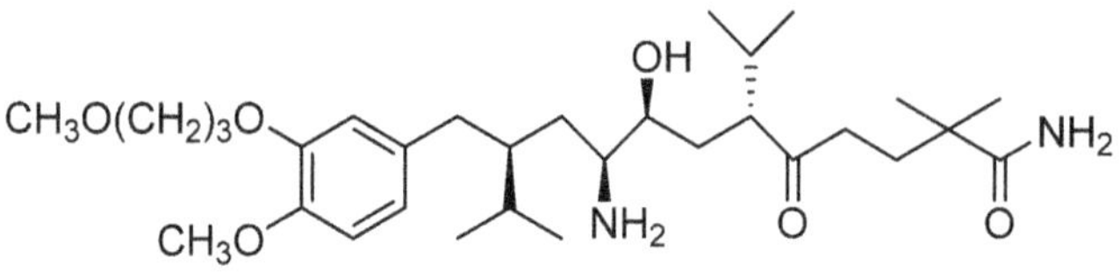

aliskiren

14.2.7   CALCIUM CHANNEL BLOCKERS

The primary effect of the CCB is to reduce calcium entry in smooth-muscle cells by binding to calcium channels in the membranes. The Ca^{2+} channel blockers lower blood pressure by reducing arterial and vascular resistance. There are three types of calcium blockers presently in use. These are phenylalkyl amines, benzothiazepines, and dihydropyridines. The first widely used calcium blocker was verapamil. Other examples of this class are prenylamine, cinnarizine, and flunarizine.

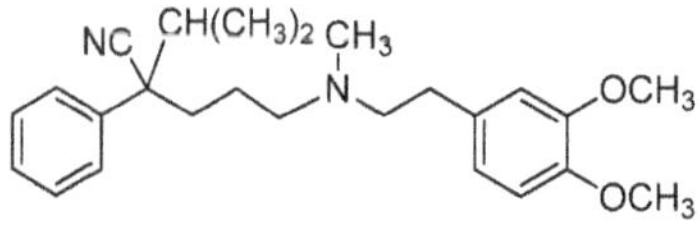

verapamil

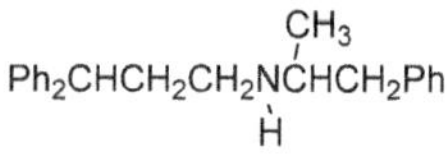

prenylamine

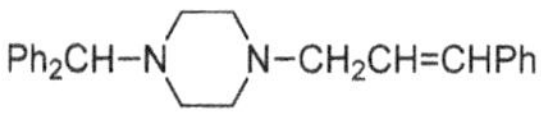

cinnarizine

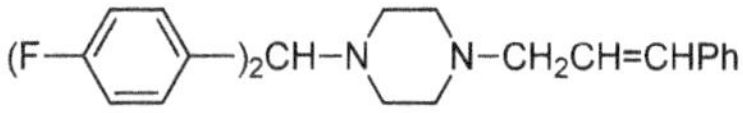

flunarizine

The most important of the benzothiazepins is diltiazem.

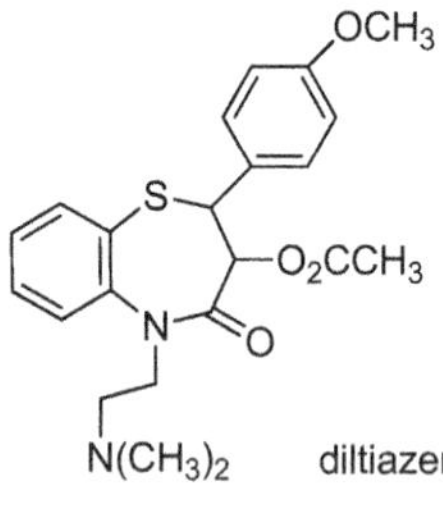

diltiazem

The largest groups of CCB are the dihydropyridines and several examples are shown in Scheme 14.6. These can be classified into first through fourth generation and have achieved improvements in pharmacodynamics and side effects.

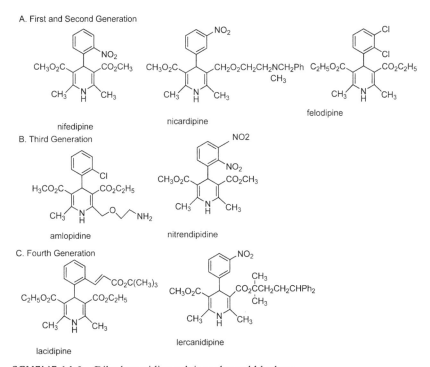

SCHEME 14.6 Dihydropyridine calcium channel blockers.

A large number of dihydropyridines have been investigated as CCBs. The first and second generation, as represented by nifedipine and nicardipine, were fast-acting but with fairly short durations of action. Several studies suggested that the fast-acting dihydropyridines may be associated with increased risk of heart attacks. Controversy arose in the mid-1990s about the studies indicating these relationships.[13] They were reformulated in extended release forms in response to the problems. In 1996, the FDA approved continued use, but required a warning on potential hazards.[14] The third generation drugs including amlodipine and nitrendipine have improved pharmacodynamics profiles. Amlodipine is used extensively in fixed combinations with the other types of antihypertensive drugs. The fourth generation drugs have still longer periods of effectiveness, which was achieved by introducing more lipophilic substituents.[15] Some of the differential effects of the various dihydropyridines may be the result of discrimination between the types and location of calcium channels that are affected.[16]

14.2.8   ALDOSTERONE ANTAGONISTS

As shown in Figure 14.1, aldosterone is one of the key mediators in the blood pressure regulating system. Inhibitors of its action can enhance sodium ion transport and decrease blood pressure and fluid retention. The currently known inhibitors are steroid analogs that have some action on androgen and progesterone receptors and can give rise to side effects in both male and female reproductive systems. The first such agent, spironolactone, exhibited such side effects, but a newer compound, eplerenone, shows blood pressure lowering with reduced side effects. The aldosterone antagonists are used in certain special conditions for treatment of hypertension.

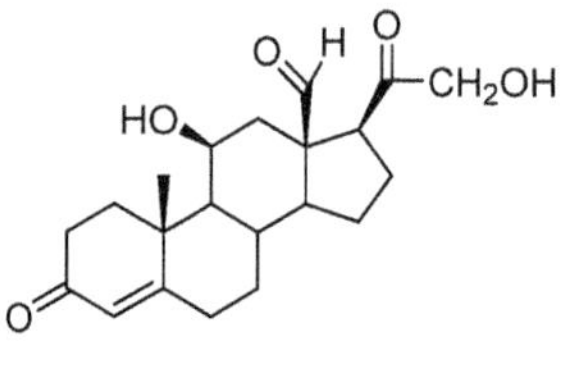

Aldosterone

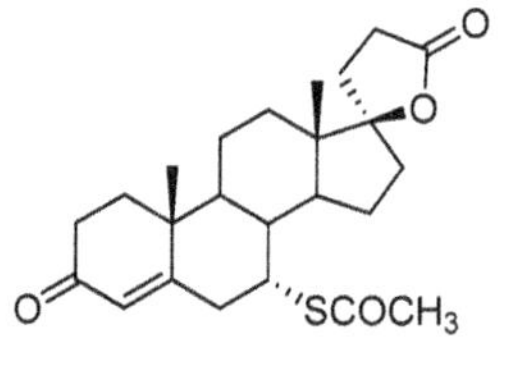

Spironolactone

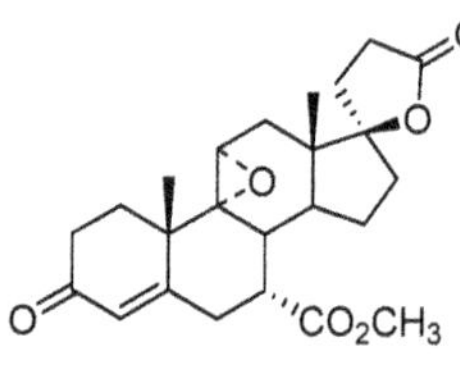

Eplerenone

KEYWORDS

- hypertension
- blood pressure
- renin–angiotensin–aldosterone system
- diuretics
- β-blockers
- ACE inhibitors
- angiotensin receptor blockers
- calcium channel blockers
- aldosterone analogs

BIBLIOGRAPHY

Cleland, S. J.; Reid, J. L. *Heart* **1996**, *76*(Suppl. 3), 7–12; Hawgood, B. J. *Toxicon* **1997**, *35*, 1569–1580; Laragh, J. H. *Am. J. Hypertens.* **1998**, *11*, 170S–174S; Phillips, M. I.;

Schmitt-Ott, K. M. *News Physiol. Sci.* **1999**, *14*, 271–274; Hollenberg, N. K.; Sever, P. S. *J. Renin–Angiotensin–Aldosterone Syst.* **2000**, *1*, 5–10; Prado, G. N.; Taylor, L.; Zhou, X.; Ricupero, D.; Mierke, D. F.; Polgar, P. *J. Cell. Physiol.* **2002**, *275*, 275–286; van Zwieten, P. A. In *Handb. Hypertens. Hypertension in the Twentieth Century: Concepts and Achievements*; Birklenhaeger, W. H.; Robertson, J. I. S.; Zanchetti, A. Eds.; Elsevier, 2004, Vol. 22; pp 457–486; Skrbic, R.; Igic, R. *Peptides* **2009**, *30*, 1945–1950; Campese, V. M.; Park, J. *J. Nephrol.* **2006**, *19*, 691–698; Simoes e Silva, A. C.; Flynn, J. T. *Pediatr. Nephrol.* **2012**, *27*, 1835–1845.

BIBLIOGRAPHY BY SECTION

Section 14.1: Abuissa, H.; Jones, P. G.; Marso, S. P.; O'Keefe, Jr., J. H. *J. Am. Coll. Cardiol.* **2005**, *46*, 821–826; Dazu, V. *J. Hypertens.* **2005**, *23*, Suppl. 1, S9–S17; Brunton, L. L., Ed. In *Gilman and Goodman's The Pharmacological Basis of Therapeutics*, 11th ed., McGraw-Hill, 2006; Chapters 10, 22, 28, 30, and 33; Bornstein, N.; Silvestrelli, G.; Caso, V.; Parnetti, L. *Clin. Exp. Hypertens.* **2006**, *28*, 317–326; Unger, T.; Stoppelhaar, M. *Am. J. Cardiol.* **2007**, *100*, 25J–31J; Boehm, M. *Am. J. Cardiol.* **2007**, *100*, 38J–44J; Perkins, J. M.; Davis, S. N. *Curr. Opin. Endocrinol. Diabetes* **2008**, *15*, 147–152; Weber, M. *J. Clin. Hypertens.* **2008**, *10*, 427–430; Berlaimont, V.; Billiouw, J.-M.; Brohet, C.; Dupont, A. G.; Gazagnes, M.-D.; Heller, F.; Kresinski, J.-M.; Missault, L.; Persu, A.; Pierard, L.; Rottiers, R.; Vanhooren, G.; Van Mieghem, W.; Vervaet, P.; Herman, A. G. *Acta Clin. Belg.* **2008**, *63*, 142–151; Kurkulasuriya, L. R.; Sowers, J. *Contemporary Endocrinology: Cardiovascular Endocrinology: Shared Pathways and Clinical Crossroads*; Fonseca, V. A., Ed.; 2009; pp 121–147; Sica, D. A. *Curr. Hypertens. Rep.* **2010**, *12*, 67–73; Azizi, M.; Menard, J. *Cardiovasc. Drugs Ther.* **2013**, *27*, 145–153; von Lueder, T. G.; Krum, H. *Cardiovas. Drugs Ther.* **2013**, *27*, 171–179; Burnier, M.; Vuignier, Y.; Weerzher, G. *Eur. Heart J.* **2014**, *35*, 557–562.
Section 14.2.1: Chaitman, B. R. *Can. J. Cardiol.* **2005**, *21*, 1031–1034.
Section 14.2.2: Dupont, A. G. *Cardiovas. Drugs Ther.* **1993**, *7*, 55–62.
Section 14.2.3: Wang, J.-G.; Lin, Y. *Expert Rev. Neurotherap.* **2004**, *4*, 1023–1031; Alexander, L. M. *Drugs* **2006**, *6*, 1239–1252; Westfall, T. C.; Westall, D. P. In *Gilman and Goodman's Pharmaceutical Basis of Therapeutics*, 11th ed.; Brunton, L. L. Ed.; McGraw-Hill, 2006; Chapter 10, Chrysant, S. G.; Chrysant, G. S.; Dimas, B. *Clin. Cardiol.* **2008**, *31*, 249–252.
Section 14.2.4: Cleland, S. J.; Reid, J. L. *Heart* **1996**, *76*(Suppl 3), 7–12; Vane, J. R. *Med. Sci. Monitor* **2001**, *7*, 790–800; Brunner, M. *J. Renin–Angiotensin–Aldosterone Syst.* **2007**, *8*, 208–212.
Section 14.2.5: Bauer, J. H.; Reams, G. P. *Arch. Inter. Med.* **1999**, *155*, 1361–1367; Gavras, H. *J. Am. Soc. Nephrol.* **1999**, *10*, S235–S257; Kurtz, T. W. *Nat. Clin. Pract.* **2008**, *5*(Suppl. 1), S19–S26.
Section 14.2.6: Jensen, C.; Herold, P.; Brunner, H. R. *Nat. Rev. Drug Discovery* **2008**, *7*, 399–410; Webb, R. L.; Schiering, N.; Sedrani, R.; Maibaum, J. *J. Med. Chem.* **2010**, *53*, 7490–7520.
Section 14.2.7: Dougal, H. T. McLay, J. *Drug Saf.* **1996**, *15*, 91–106; Ioan, P.; Carosati, E.; Micucci, M.; Cruciani, G.; Brocatelli, F.; Shorov, B. S.; Chiarini, A.; Budriesi, R. *Curr. Med. Chem.* **2011**, *18*, 4901–4922.
Section 14.2.8: Burgess, E. *Expert Opin. Pharmacother.* **2004**, *5*, 2572–2581.

REFERENCES

1. Phillips, M. I.; Schmidt-Ott, K. M. *Physiology* **1999**, *14*, 271–274; J. E. Hall, 2003. *Mol. Biotechnol.* **24**, 27–39.
2. Gross, R. *Klin. Wochnschr.* **1958**, *36*, 693–706.
3. Crowther, A. F. *Drug Des. Delivery* **1990**, *6*, 149–156.
4. Ruffolo, R. R.; Feuerstein, G. Z. *Expert Opin. Drug Discovery* **2006**, *1*, 85–89.
5. Turni, G. A.; Brunner, H. R.; Ferguson, R. K.; Rivier, J. L.; Garvas, H. *Br. Heart J.* **1978**, *40*, 1134–1142.
6. Ondetti, M. A.; Rubin, B.; Cushman, D. W. *Science* **1977**, *196*, 441–444; Cushman, D. W.; Ondotti, M. B. *Hypertension* **1991**, *17*, 589–592; Vane, J. R. *J. Physiol. Pharmacol.* **1999**, *50*, 489–498.
7. Timmermans, P. B. M. W. M.; et al. *Pharmacol. Rev.* **1993**, *45*, 205–251; Wong, P. C.; Timmermans, P. B. M. W. M. *Blood Pressure* **1996**, *5*(Suppl. 3), 11–14.
8. Carini, D. J.; Duncia, J. V.; Aldrich, P. E.; Chiu, A. T.; Johnson, A. L.; Pierce, M. E.; Price, W. A.; Santella III, J. B.; Wells, G. J.; Wexler, R. R.; Wong, P. C.; Yoo, S.-E. Timmermans, P. B. M. W. M. *J. Med. Chem.* **1991**, *34*, 2525–2547; Timmermans, P. B. M. W. M.; Carini, D. J.; Chiu, A. T.; Duncia, J. V.; Price, Jr., W. A.; Wells, G. J.; Wong, P. C.; Wexler, R. R.; Johnson, A. L. *Hypertension* **1991**, *18*(Suppl. 5-III), 136–142; Siegl, P. K. S. *J. Hypertens.* **1993**, *11*, S19–S22.
9. Ries, U. J.; Mihm, G.; Narr, B.; Hasselbach, K. M.; Wittneben, H.; Entzeroth, M.; van Meel, J. C. A.; Wienen, W.; Hauel, N. H. *J. Med. Chem.* **1993**, *36*, 4040–4051.
10. Reddy, K. S.; Srinivasan, N.; Reddy, C. R.; Kolla, N.; Anjaneyulu, Y.; Venkatraman, S.; Bhattacharya, A.; Mathad, V. T. *Org. Process Res. Devel.* **2007**, *11*, 81–85.
11. Keenan, R. M.; Weinstock, J.; Finkelstein, J. A.; Franz, R. G.; Gaitanopoulos, D. E.; Girard, G. R.; Hill, D. T.; Morgank, T. M.; Samanen, J. M.; Peishoff, C. E.; Tucker, L. M.; Aiyar, N.; Griffin, E.; Ohlstein, E. H.; Stack, E. J.; Weidley, E. F.; Edwards, R. M. *J. Med. Chem.* **1993**, *36*, 1880–1892.
12. Kubo, K.; Kohara, Y.; Yoshimura, Y.; Inada, Y.; Shibouta, Y.; Furukawa, Y.; Kato, T.; Nishikawa, K.; Naka, T. *J. Med. Chem.* **1993**, *36*, 2343–2349.
13. Kizer, J. R.; Kimmel, S. E. *Pharmacoepidemiol. Drug Saf.* **2000**, *9*, 25–36.
14. Marwick, C. *J. Am. Med. Assoc.* **1996**, *275*, 423–424 and 1638 (erratum).
15. Karim, A.; Berdeux, A. *Therapie* **2003**, *58*, 333–339.
16. Richard, S. *Drugs* **2005**, *65*(Suppl. 2), 1–10.

CHAPTER 15

STEROIDS: ARTHRITIS, FERTILITY, HEART ATTACKS, AND HOME RUN RECORDS

ABSTRACT

The importance of the steroids was recognized when minute amounts were isolated from animals and shown to have potent biological activities. The structures were determined by chemical transformations in the 1920s and 1930s and confirmed by an X-ray crystal structure of cholesteryl iodide in 1945. The anti-inflammatory corticosteroids were characterized in the1940s and cortisone was synthesized in 1948. Plant sources of steroidal sapogenins provided plentiful starting materials for semisynthetic analogs and corticosteroids and the oral contraceptive pill were soon introduced. The role of cholesterol in cardiovascular disease led to the recognition of the statins, which act as inhibitors of steroid biosynthesis. Other steroids act as cardiotonics. The anabolic–androgenic steroids are used in medical treatments but also were used illicitly to enhance athletic performance. Various phenolic and other synthetic compounds exhibit steroid-like hormonal activity and their potential to act as endocrine disruptors is a subject of research.

15.1 STRUCTURE OF STEROIDS

The isolation and determination of the structure of the steroids was one of the great scientific challenges of the first-third of the 20th century. All steroids share a system of four rings. The rings are referred to as A, B, C, and D and the positions are numbered as shown below. Various substituent groups can be present. There are usually methyl groups at positions 19 and 18, represented by the wedged bonds, which indicate that the methyl groups lie above the approximate plane defined by the four rings. The dashed bonds indicate substituents point toward the lower face of the ring system. These two orientations are designated β- and α-, respectively. The ring junctions also have specific *cis* or *trans* orientations. The correct orientation of the substituent groups and rings is essential to their biological function and must be taken into account in the synthesis of the compounds. The prefixes *nor-* and *de-* (or *des-*) are used to indicate the absence of groups. Thus, deoxychlolic acid lacks a C-7 hydroxy group, while 19-*nor*-progesterone lacks the 19-methyl. Some examples of important steroids are shown in Scheme 15.1.

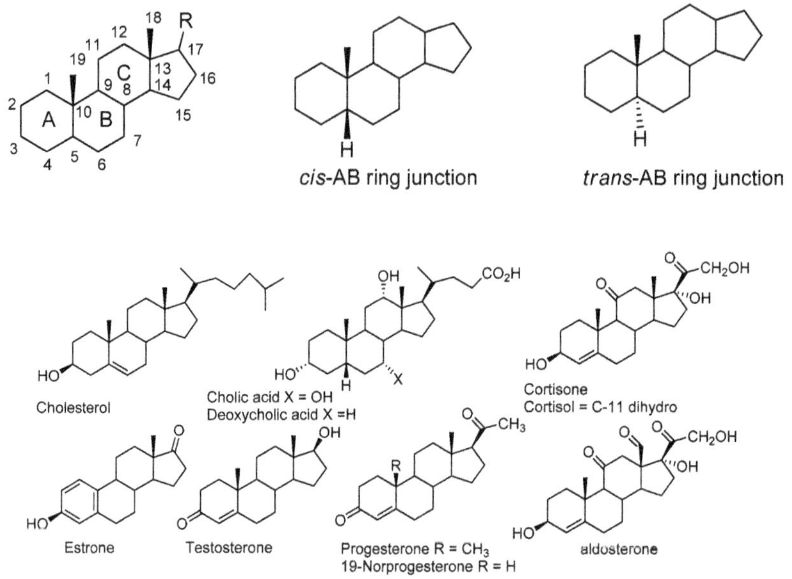

cis-AB ring junction   *trans*-AB ring junction

Cholesterol

Cholic acid X = OH
Deoxycholic acid X = H

Cortisone
Cortisol = C-11 dihydro

Estrone   Testosterone   Progesterone R = CH₃
                             19-Norprogesterone R = H   aldosterone

SCHEME 15.1   Structures of some steroids.

Cholesterol is an important constituent of biological membranes, but at elevated levels, it contributes to arterial plaque and is also the major constituent of gall stones. Deoxycholic acid and cholic acid are liver metabolites of cholesterol and are called *bile acids*. Estrone and testosterone are the primary female and male sex hormones, respectively. The sex hormones are involved in many aspects of the reproductive cycle, including onset of puberty, fertility, and pregnancy. Progesterone is involved in the female reproductive cycle and inhibits ovulation during pregnancy. The steroids function through binding to various receptors called androgen, estrogen, and progesterone, etc. receptors. These receptors in turn function to control activation of genes. Some of the systems appear to be controlled on the basis of a balance between particular steroids. From a pharmacological point of view, it is of interest to have both *agonist* and *antagonists* that can either activate or inhibit a specific receptor and its function. Structures such as cortisone are called *corticosteroids* and are produced by the adrenal glands. They are characterized by an oxygen substituent at C-11 and the hydroxyacetyl group at C-17. The corticosteroids are involved in the hypothala-mus–pituitary–adrenal axis and are critical messenger in the endocrine system. The corticosteroids influence metabolism, cell differentiation and function of the immune system. Aldosterone is an important mediator of blood pressure and salt and water retention (see Section 14.1).

There are completely synthetic materials that can mimic steroids. Two ex-amples are diethylstilbestrol and tamoxifen. More will be said about these com-pounds later.

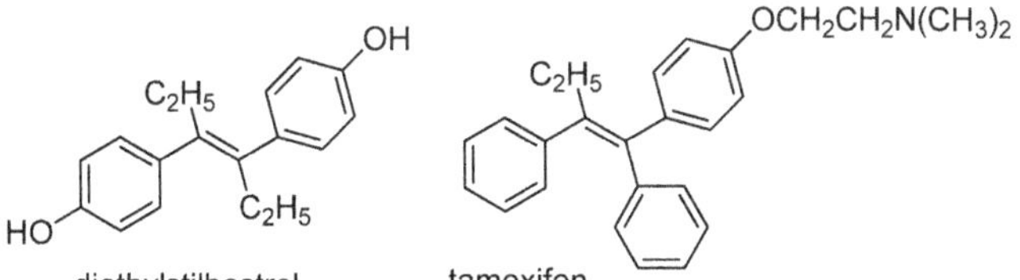

diethylstilbestrol   tamoxifen

Cholesterol is plentiful, but isolation of the steroids with potent biological activity required prodigious efforts. The 1927 Nobel Prize in Chemistry was awarded to Heinrich Wieland for the isolation of the bile acids. The 1928 prize went to Adolf Windaus for isolation of plant sterols and recognition of their connection to vitamin D (see Section 10.1.1.8). The 1939 prize was awarded to Adolf Butenandt, who isolated estrone, androsterone, and progesterone. The steroid structure was confirmed by X-ray crystallography of cholesteryl iodide in 1945 by R. H. Carlyle and Dorothy Crowfoot (Hodgkins) and was one of the contributions cited in the award of the 1964 Nobel Prize in Chemistry to Hodgkins (see Section 10.1.1.6 for biographical information). The structural and synthetic studies on steroids provided the impetus for developing methods to control and determine the orientation of substituents, known as *stereochemis-try*. As is the case with most natural substances, the steroids are *chiral*, having a specific three-dimensional shape. The steroid field was one of the first areas of organic chemistry in which extensive use was made of spectroscopic methods, such as UV–Vis and IR for identification of substituents and of optical rotation for determination of stereochemistry. Somewhat later, NMR spectroscopy be-came an important tool. In 1969, Derek Barton and Odd Hassel were awarded the Nobel Prize in Chemistry for their work on the concept of conformation, which was largely based in steroid chemistry.

15.2 CORTICOSTEROIDS AND ARTHRITIS

The corticosteroids are powerful anti-inflammatory agents. In rheumatoid ar-thritis, they delay the progression of joint destruction. They have favorable effects in other auto-immune pathologies. Cortisol, the 11-dihydro derivative of cortisone, is widely used as an anti-inflammatory, both by injection and topically. Cortisone was one of several steroids isolated from adrenal glands by E. C. Kendall, working in collaboration with P. S. Hench at the Mayo Foundation in Rochester, MN and also by Thadeus Reichstein at the Swiss Federal Institute of Technology (ETH). The amounts isolated were too small for any biological testing. In 1944, Kendall synthesized one of these compounds, and it was re-synthesized at Merck. Testing, however, indicated little activity. Nevertheless, Merck decided to proceed with the synthesis of a second of the compounds. The synthesis of cortisone was completed in 1948.[1] The compound was tried at the Mayo clinic in treatment of severe rheumatoid arthritis. The results were dramatic and interest in steroids as drugs exploded. Kendall, Hench, and Reichstein were awarded the 1950 Nobel Prize in Physiology or Medicine for their work on corticosteroids.[a]

The discovery of biological activity required a more efficient synthesis. The cortisone structure contains an oxygen at C-11 and required an arduous effort for synthesis. One was developed at Merck, which when fully optimized, provided cortisone in slightly over 20% yield from deoxycholic acid in 30 steps. This process remained in production at Merck until 1966. Another approach to synthesis of cortisone was developed at the Upjohn Company. It used a microorganism that could introduce the C-11 oxygen, using progesterone as a starting material.[2] Related semisynthetic corticosteroids, such as prednisolone (Upjohn) and dexamethasone (Merck), were subsequently prepared.

prednisolone   dexamethasone

The discovery of cortisone and the burgeoning interest in steroids required more accessible starting materials, both for manufacturing and for research. Steroid research was greatly advanced by the discovery of plentiful plant sources of steroids such as diosgenin, hecogenin, and stigmasterol that were amenable

[a]Philip S. Hench was the head of the rheumatic disease service and Edward C. Kendall was a Professor of Physiological Chemistry at the Mayo Clinic in Rochester, MN, USA. Kendall had previously isolated the important thyroid hormone thyroxine. They began collaborative research on adrenal hormones and Kendall was able to isolate several active compounds, one of which turned out to be cortisone. Eventually they were able to obtain sufficient material for clinical tests as a result of the synthetic work at Merck.

to synthetic modification. During the 1930s and 1940s, Russell E. Marker,[b] of Pennsylvania State University, had done many synthetic transformation on steroids and had become interested in plants as sources for steroidal starting materials. He carried out a wide-ranging search and found a number of plants containing *sapogenins* in which the steroids are bound to carbohydrates. Eventually he discovered two plants in Mexico that had high concentrations of suitable sapogenins. Marker was convinced that these compounds represented the most promising source of steroids. Unable to convince any US pharmaceutical company to undertake production, he left Penn State for Mexico to achieve the task himself. This eventually (see biographical material) led to the founding of Syntex S. A. In a few years, production of progesterone was raised to tonnage scale in response to the growing demand for cortisone. In the process, Syntex became one of the leading research companies in the steroid field. Eventually, as described in the next section, Syntex became the source of one of the first oral contraceptives. Glaxo in the United Kingdom developed a synthesis of steroids based on hecogenin, obtained from African sisal wastes, which has an oxygen at C-12 in the C ring. Percy Julian (see Section 8.3.3 for biographical sketch), working at Glidden, developed a route to progesterone from stigmasterol isolated from soybeans.

diosgenin   hecogenin   stigmasterol

---

[b]Russell Marker was a brilliant but eccentric person. His graduate training in Chemistry was at the University of Maryland, but unwilling to meet all the course requirements, Marker left without a degree. He worked for a time at Ethyl Corporation and was responsible for development of the octane rating scale. His work came to the attention of F. C. Whitmore of Penn State, who some years later, appointed Marker to the faculty. By this time, Marker had become interested in steroids. He published many papers on transformations and synthesis of steroids. He made an important correction of the structure of the sapogenins and became convinced that they represented ideal starting materials for synthesis of steroids. He undertook examination of a number of plant species, eventually finding Mexican yams that were rich sources of the sapogenin, diosgenin. He had already shown that progesterone could be synthesized from diosgenin. Unable to convince any US pharmaceutical company to invest in Mexican production, in 1943, he decided to leave for Mexico and undertake the task himself. Searching for potential Mexican collaborators, he located a company Laboratorios Hormona S. A. that had been founded in 1933 by two European emigrants, Emerik Somlo and Frederico A. Lehmann. The company was engaged in the extraction of active ingredients from natural material and synthesis of pharmaceuticals. At this point, Marker had been able to prepare kilogram quantities of progesterone from diosgenin, and Somlo and Lehmann were convinced of the potential of this chemistry. With Marker, they founded Syntex S. A. and began the production of progesterone. In 1949, a falling out occurred and Somlo and Lehmann purchased Marker's share of the company. He disengaged from Chemistry and retired to Pennsylvania. Marker had left little documentation of the process and George Rosenkranz, another émigré from Europe, found it necessary to redevelop the procedures.

Source: Lehmann F. P. A.; Bolivar G.; Quintero R. *J. Chem. Ed.* 1973, *50*, 195–199.

15.3   STEROIDS IN FERTILITY CONTROL

The control of fertility was a controversial issue in the United States throughout the 20th century. The Comstock Laws, dating from 1873, made all means of contraception illegal. Margaret Sanger and other early advocates for access to birth control methods were arrested. In 1965, the Supreme Court ruled that remaining state prohibitions on contraceptive methods were unconstitutional and contraceptive drugs became available by prescription. By 1970, nearly 9 million women were using birth control pills in the United States. The total fertility rate dropped from 3.6 births per woman in 1960 to 1.9 in 1982, most of the change occurring in the period 1960–1973. The availability of effective birth control pills was a major factor. There was a decrease in the proportion of "unplanned" births from 65% in 1961–1965 to 37% in 1969–1973.

15.3.1   ORAL CONTRACEPTIVES, A.K.A. THE PILL

The concept underlying the search for a contraceptive drug was to mimic the anti-ovulation effects of the natural hormone progesterone. At Syntex, Carl Djerassi[c] undertook this goal. One critical structural change proved to be the removal of the C-19 methyl group. Djerassi was aware that the C-14 epimer of 19-norprogesterone had been reported in 1944 to have a progestational effect (the experiment had been done using only two rabbits because of lack of material). Djerassi and his group at Syntex synthesized 19-norprogesterone and found

---

[c]Carl Djerassi arrived from Vienna in New York as a teenager in 1939. By 1945, he had a Ph. D. from the University of Wisconsin and had begun work at Ciba in New Jersey. In 1949, he moved to Syntex in Mexico City, attracted by the prospect of freedom to publish his work. Like Djerassi, several of the key scientists at Syntex were Jewish refugees from the Nazis. At Syntex, Djerassi undertook an energetic program of synthetic modification of progesterone and other steroids. Among the compounds synthesized was estrone. A related synthetic route led to 19-norprogesterone, which, in turn, led to the orally active analog norethindrone. Djerassi, in his 1992 review of the history of the Pill, discusses the issue of the similarity of the Searle and Syntex compounds. The Searle compound is converted to the Syntex compound by acid and also by gastric juices. The question for patent law is whether the compound is in fact patentable. This particular case was never resolved from a legal point of view, although subsequent cases indicate the Searle patent might well have been ruled an infringement. Djerassi also commented on the unappreciated contribution of chemistry to advances in biomedical science. He stated that Gregory Pincus, one of the leading figures in the development of oral contraceptives, did not "make the slightest reference in his opus magnum The Control of Fertility, to any chemist or how the active ingredients of the Pill actually arrived in his laboratory." (Some had been supplied by Syntex and some by Searle.) Djerassi goes on: "The 19-nor steroids studied by Pincus and eventually making the Pill a reality did not occur in nature nor did he purchase them in a drugstore. Was this [neglect of the chemists] just a reflection of the low opinion Pincus and other biologists had of the role chemists play in the development of new drugs?"

After leaving Syntex, Djerassi served on the faculty of Wayne State University in Detroit and then at Stanford. He received many awards and honors for his work on steroids. In his autobiography, he recounts an amusing incident at the award of an honorary degree by Columbia in 1975. Cited by the president for his contribution, through the Pill, to the emancipation of women, two cheers arose; first from the female graduates of Barnhard College, but immediately thereafter from the (then) all-male Columbia graduates.

Source: Djerassi, C. *The Pill, Pygmy Chimps and Degas' Horse*; Basic Books: New York, 1992.

it to have high progestational activity. This material had to be administered by injection and so an orally active analog was sought. Following another clue in the published literature, the group introduced the ethynyl group at C-17, giving rise to *norethindrone*. The patent application was filed November 22, 1951 and led to US Patent 2,744,122 issued May 1, 1956.

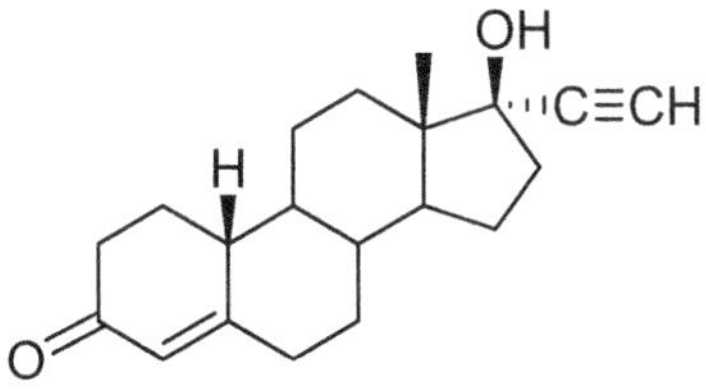

norethrindrone
Ortho-Novum

A second compound originated at the Searle Company in 1953. This compound, norethynodrel, differs only in the location of the double bond at C5–C10. These two compounds became the subject of intense study by several biologists and physicians. The Syntex compound was licensed to Parke-Davis for marketing, because Syntex had no distribution system in the United States. Both the Syntex and Searle compounds received FDA approval in 1957. However, Syntex's marketing partner backed out, fearing controversy. A new licensing agreement was negotiated with the Ortho division of Johnson & Johnson. Searle's compound came to market as Enovid in 1960 and the Syntex compound as Ortho-Novum in 1962. Syntex subsequently was purchased and a US company established in the research park adjacent to Stanford University (where Djerassi was then located).

By the late 1960s, concerns about the side effects of the Pill began to arise, the most common ones being embolisms, blood clots, and venous inflammation. These led to reduction of dosages as well as introduction of alternative compounds. For example, analogs with C-6 substituents were developed. More recently developed progestational agents include norelgestromin and etonogestrel. These are used in contraceptive transdermal patches and implants, respectively.

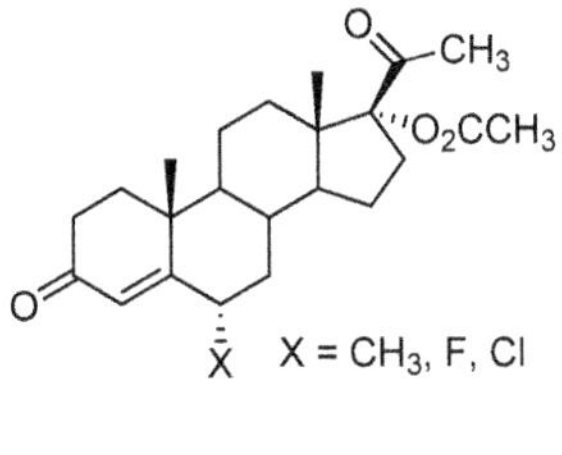

X   X = CH$_3$, F, Cl 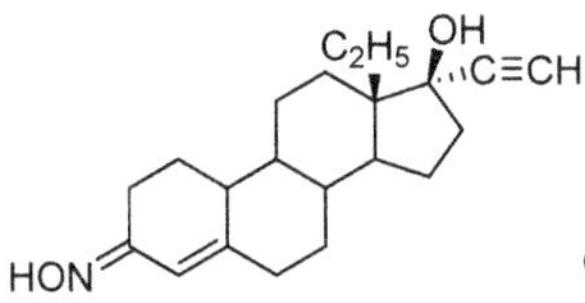   norelgestromin 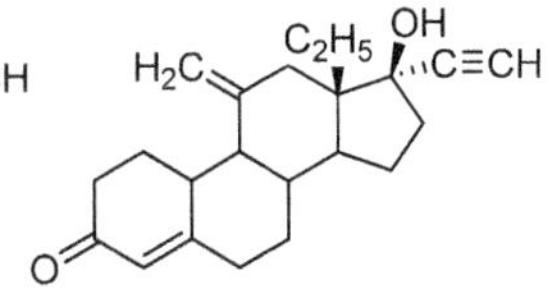   etonogestrel

In 1970, hearings were held, chaired by Sen. Gaylord Nelson (see also Topic 11.1). These hearing led to considerable media attention to the side effects associated with birth control pills. Ironically, one of the witnesses offering testimony was Dr. Hugh J. Davis, the inventor of the Dalkon Shield, an intrauterine device that was later found to cause severe infections and permanent infertility (see Topic 11.2.3). The result of the hearings and attendant publicity was the institution by the FDA of the requirement for extended tests in both beagles (7 years) and monkeys (10 years). The consequence was a greatly curtailed research effort into contraceptives, because of the daunting approval process. Liability litigation also presented an impediment to development of new contraceptives. As a result, little research is currently done in the US pharmaceutical industry. Most recent advances have come from Europe.

15.3.2 EMERGENCY CONTRACEPTIVES

Later, controversy shifted to so-called "morning-after" or emergency contraceptives. One such compound, norgestrel, is the C-18 ethyl analog of norethrindrone. Norgestrel, is marketed as "Plan B." It can prevent fertilization if taken soon after coitus and is about 90% effective. It has been available by prescription since 1999. In 2003, the FDA, under political pressure, and despite contrary advice by its expert panel, declined to permit over-the-counter access. The FDA relented in 2004 and Plan B is now available to those over 18.

norgestrel
(levonorgestrel)

The most recent controversy in fertility control in the United States has involved Mifepristone (RU486). RU486 was synthesized at Rousell-Uclaf in France. It is a progesterone analog that blocks implantation of the fertilized embryo. It is highly effective in preventing pregnancy within a few days of fertilization. It can also be used to induce abortion in the first trimester of pregnancy, in which case a prostaglandin is also prescribed. At full term, RU486 promotes cervical expansion and can be used to induce labor. It also can be used to induce labor in the case of intrauterine death of the fetus. A related drug, ulipristal, was approved by the FDA in 2010.

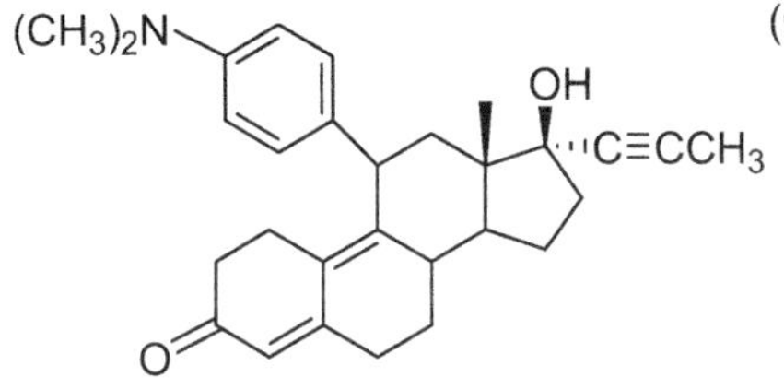

mifepristone (RU486) 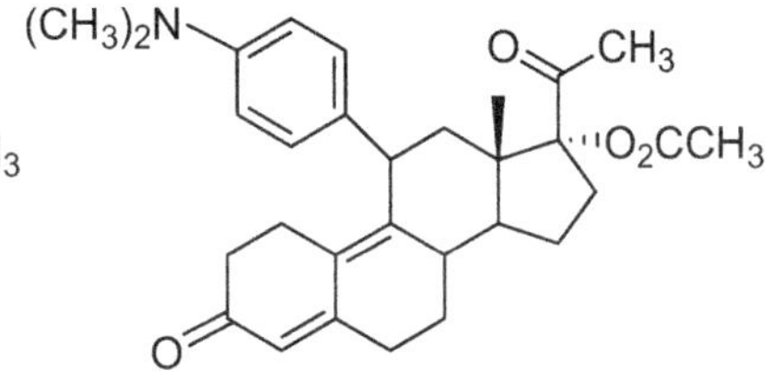   ulipristal

RU486 has been available in Europe since 1988. Its use in first trimester abortions is referred to as *medical abortion*. FDA approval in the United States occurred in 2000 after a protracted struggle influenced by antiabortion politics. However, no major pharmaceutical company was willing to undertake the manufacture and distribution. As a result a nonprofit group, the Population Council, sponsored the clinical trials and established a manufacturing company. In its final approval in 2000, the FDA imposed a number of unusual restrictions including certifications by both the patient and physician as to the term of the pregnancy. The drug is provided not by prescription, but directly from the manufacturer through the physician.

15.3.3   STEROIDS IN FERTILITY TREATMENT

Because of the crucial roles of estrogen and progesterone in reproduction, their effects in *in vitro* fertilization (IVF) have been studied extensively. The ovulatory cycle is controlled by the *gonatropins, follicle-stimulating hormone* (FSH), and *luteinizing hormone* (LH). In fertility treatments these hormones are used to stimulate ovulation. There are three stages in the fertilization process that establishes a viable embryo. These are *apposition, adhesion,* and *invasion.* In the normal cycle, there is a relatively short "implantation window" between the 6th and 10th day after the initial LH surge. One goal of IVF is to optimize the receptivity of the endometrium to the fertilized egg. The importance of LH and FSH in human reproduction has resulted in a long history of their use in fertility treatment. Several companies, including IGF, Armour Laboratories, and Searle, began producing mixtures of the hormones (and other materials) from extracts of pig and sheep in the 1930s. Preparations from pregnant horses were also commercialized. These materials tended to lose effectiveness in individual patients over time because of the development of antibodies, but achieved some success in induction of ovulation. Between 1958 and 1988, extracts of human pituitary glands were available commercially, but this was never an adequate source of material. Urine of menopausal women is also a commercial source and improvements in purification methodology eventually provided a source of essentially pure FSH. Human recombinant LH and FSH became available in the late 1980s.[3] A third related protein, chorionic gonadotrophin, is also available in recombinant form and is sometimes used in fertility treatment. Current information indicates that in most women, treatment with FSH alone is sufficient to induce ovulation.[4] There are also synthetic drugs, such as clomiphene, that can be used to stimulate FSH and LH production in fertility treatment. Clomiphene is also used in treatment of hypogonadal men, where it can increase testosterone levels. It and related compounds function as antagonists of estrogen receptors and increase testosterone levels.[5]

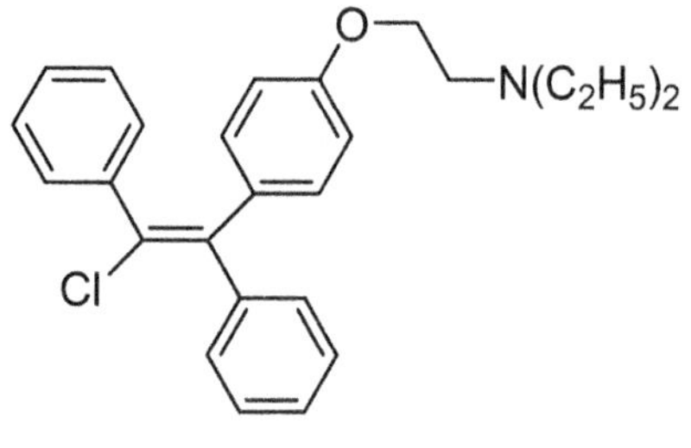

clomiphene

15.4   CARDIOVASCULAR EFFECTS OF STEROIDS

15.4.1   CHOLESTEROL AND CARDIOVASCULAR HEALTH

The association between cholesterol levels and cardiovascular disease began to be recognized around 1900, when it was observed that animals fed high cholesterol diets developed atherosclerosis. A genetic disorder, familial hyper-cholesterolemia, is characterized by both very high cholesterol levels and early onset of cardiovascular disease. The Framingham Heart Study of the National Institutes of Health provided convincing statistical evidence of a correlation between blood cholesterol levels and the incidence of heart disease.[6] It was also recognized that a high ratio of low-density lipoprotein (LDL) to high-density lipoprotein (HDL) was a risk factor for cardiovascular disease.[7] These values, in turn, are influenced by dietary intake, with saturated fat and *trans*-fat tending to increase LDL while unsaturated fats decrease LDL. Cardiovascular disease is a major cause of death in most of the developed world and accounts for about

15% of deaths in the United States, of which about 15% are <65 years of age. There are associations of cardiovascular diseases with several factors, including those that constitute the *metabolic syndrome*, namely obesity, hypertension, and insulin resistance.

The connection between high cholesterol levels and cardiovascular disease generated interest in pharmaceutical interventions. The first efforts were not very successful. The first drug to be approved was triparanol, developed by Richardson-Merrell and approved by the FDA in 1960. Backed by vigorous promotion to physicians, triparanol got off to a fast start, with some 300,000 prescriptions in 1960. The success was short-lived, however. Triparanol blocks the final step in cholesterol biosynthesis, and the accumulation of the intermediate led to development of cataracts, hair loss, and skin abnormalities. Reevaluation by the FDA found defects in the original data and a "whistle-blower" provided evidence of fabrication of data in the original submission. Eventually, a vice-president of the company and two laboratory supervisors were convicted of fraud. Lawsuits over the next several years are estimated to have resulted in payments of around $50 million. This incident contributed to the tightening of FDA safety and efficacy rules in 1962 (see Section 11.4.1).

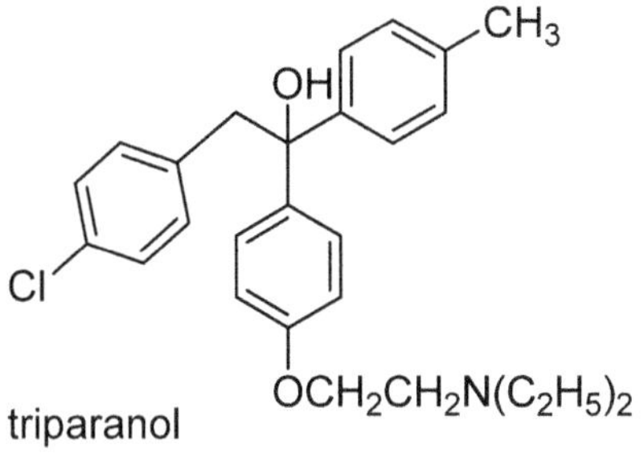

triparanol

The second drug to be tried was cholestyramine. Despite its name, it contains no steroid structure but rather is a polystyrene polymer containing quaternary nitrogen substituents. It was the subject of a study initiated in 1980. The drug had been developed at Merck, but Merck never promoted the drug. Instead it was used as a "surrogate" in a study designed to confirm the effects of lowering blood cholesterol levels. It was an extremely unpleasant drug with a foul taste and required ingestion of relatively large amounts of gritty material. It was not absorbed in the stomach, but passed into the intestines where it inhibited cholesterol uptake. The study indicated that lowering cholesterol levels did improve cardiovascular health and led to the development of the National Cholesterol Education Program. By the mid-1980s, the public was becoming well aware of the significance of cholesterol.

15.4.2   STATINS

The biosynthesis of cholesterol was worked out by Konrad Bloch, Feodor Lynen, John Conforth, and George Popjak. Bloch and Lynen received the 1964 Nobel Prize in Medicine or Physiology for their work. The biosynthesis of steroids proceeds through mevalonate as a key intermediate. *Statins* are inhibitors of 3-hydroxy-3-methylglutaryl coenzyme A reductase (HMG-reductase), which catalyzes the rate-limiting step in cholesterol biosynthesis. The statins lower LDL levels by 20–50%, with a corresponding reduction in the incidence of cardiovascular events.

Hydroxymethylglutaryl-CoA   mevalonate

The original statin, mevastatin, was isolated from the mold *Penicillium citrinium* in 1972 by A. Endo and coworkers at Sankyo Pharmaceutical Company in Japan. The compound showed promise when tested in patients with familial hypercholestemia. These results caught the attention of Merck and in 1976 Merck and Sankyo began a collaborative arrangement. A second, closely related compound, lovastatin was isolated at Merck from *Aspergillus terreus.* Sankyo suspended its development of mevastatin when it mistakenly interpreted pathology data at high doses as indicating carcinogenicity. When rumors of this spread, Merck, too, suspended its studies but pursued the matter and demonstrated that there was no carcinogenicity, even at high doses. Merck filed an NDA for lovastatin in 1986, and it was approved by the FDA in 1989. Sankyo resumed its work and launched its compound as pravastatin in 1989.

The natural statins were followed by semi-synthetic and completely synthetic analogs. Pravastatin is made from mevastatin by a microbiological transformation that introduces a hydroxyl group. Simvastatin is a semi-synthetic analog of mevastatin in which the acyl group at the C-ester group at C-1 is 2,2-dimethylbutanoyl. The synthetic statins typically contain heteroaromatic structures, as in atorvastatin, fluvastatin and rosuvastatin. Representative structures are shown in Scheme 15.2. All of the statins share the same structure and stereochemistry in a 3,5-dihydroxyheptanoic acid substituent, which is present in mevastatin and lovastatin as a lactone ring. This structural feature bears an obvious relationship to the hydroxymethylglutaric acid structure.

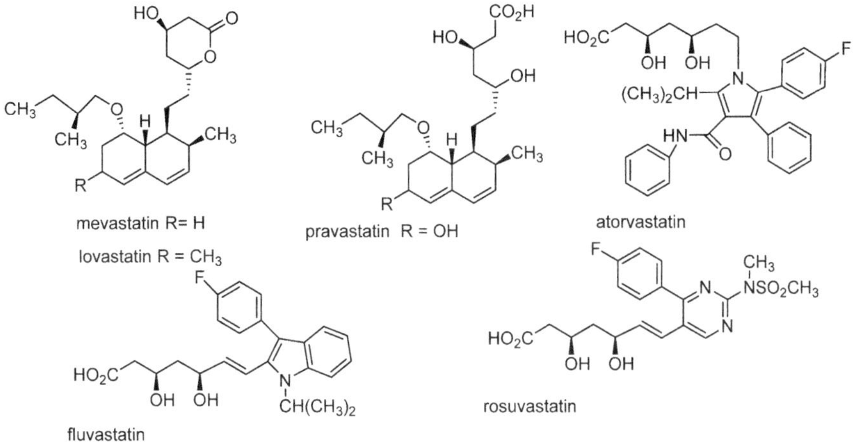

SCHEME 15.2 Structures of some statins.

There have been several studies of the statins which indicate that they achieve a 20–40% reduction is relative risk for cardiovascular incidents. A meta-analysis of 14 trials with >90,000 subjects indicates that the reduction in LDL and cardiovascular events is approximately proportional. These studies also show that use of statins is associated with a reduction of deaths from all causes.[8] These studies included patients both with and without a history of cardiovascular disease. When the data were analyzed only for patients without a prior history, the level of protection was about 10% (risk level 0.91, range 0.83–1.01), but was not statistically significant.[9]

Statins have also been investigated in the context of the connection between inflammation and cardiovascular disease.[10] The level of C-reactive protein, hs-CRP, is a predictor of cardiovascular incidents.[11] A study, JUPITER (Justification for the Use of Statins in Primary Prevention: an International Trial Evaluation of Rosuvastatin) was undertaken to determine whether statins reduced cardiovascular events in patients with normal LDL levels, but high hs-CRP levels. The study compared placebo versus rosuvastatin and found a reduction (<1.8 mmol/L) in LDL reduced cardiovascular events by 55% (from 1.11 to 0.51 events per 100 person-years). Reduction of hs-CRP below 2 mg/L resulted in a 62% reduction (0.42 events per person-year, range 0.26–0.56). LDL and hs-CRP correlated only weakly, but for patients who achieved *both* of the target levels, there was a 62% reduction in the rate of cardiovascular events (0.38; range 0.23–0.54) versus 33% for those who achieved only one of the goal levels.[12] Plaque reduction has also been observed when both LDL and hs-CRP are reduced.[13]

15.4.3   PLANT STEROLS AND STANOLS

Phytosterols and *phytostanols* are plant-derived steroids with a modified side chain having an additional ethyl or methyl group at C-24. The sterols are 5,6-unsaturated while the stanols are saturated at these positions. The main dietary phytosterols include β-sitosterol, stigmasterol, and campesterol. They are found in vegetable oils, nuts, grains, olives, and vegetables such as brussel sprouts and cauliflower. Phyosterols in the diet can reduce blood cholesterol levels by 5–10%. They evidently displace cholesterol in intestinal micelles. The optimal intake range is about 500–2500 mg/day and the effect levels off at higher intake.[14]

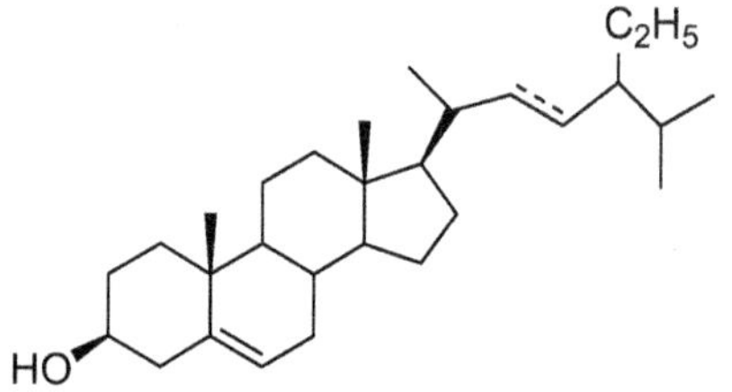

stigmasterol, C-22,23 unsaturated
β-sitosterol, C-22,23 saturated

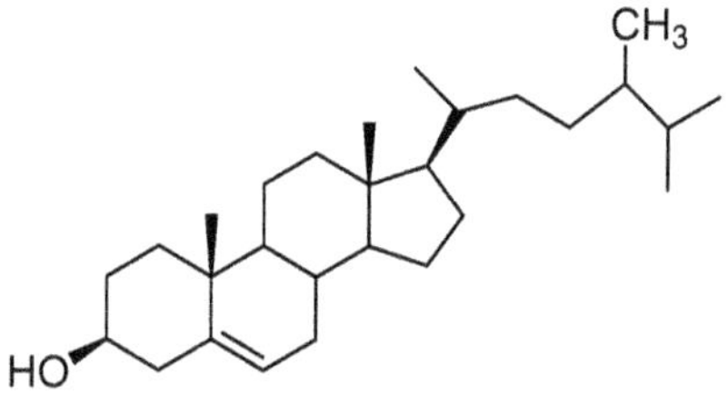

campesterol

15.4.4   CARDIOTONIC STEROIDS

There are a group of steroids, called *cardiotonic steroids*, that were initially found as the result of biological activity of natural substances. In particular the plant foxglove was noted in the 18th century to increase the strength of heart muscle contractions and came to be used in treatment of congestive heart failure. The active component in digitalis isolated from foxglove is digoxin. Related compound such as ouabain and bufalin were subsequently isolated from plants and toads, respectively.

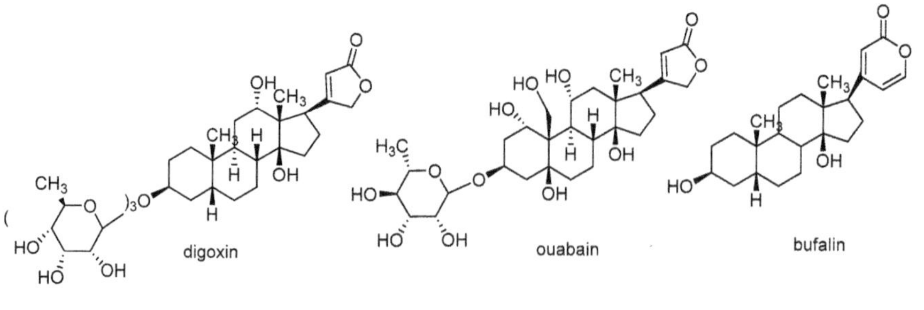

digoxin

ouabain

bufalin

The cardiotonic steroids inhibit an ATP-dependent Na^+–K^+ pump in heart muscle, which results in increased intracellular levels of both Na^+ and Ca^{2+} and strengthens the muscle contraction. Much later it was suggested that compounds similar to ouabain and digoxin may exist endogenously and be involved in regulation of heart rate and related factors. This aspect of the problem, however, has been challenged.[15] Both the therapeutic compounds and related compounds have considerable toxicity and some cause lethality when ingested by cattle or sheep.[16] They have also been used as arrow poisons.

15.5   STEROID ANALOGS IN CANCER TREATMENT

Certain forms of breast cancer are *estrogen dependent*, that is, the tumor growth is promoted by estrogen. This offers the possibility of treatment by interfering with the effect of estrogen or limiting the estrogen level. The former strategy applies to drugs such as tamoxifen, raloxifene, and fulvestrant. They block binding of estrogen to the estrogen receptor.

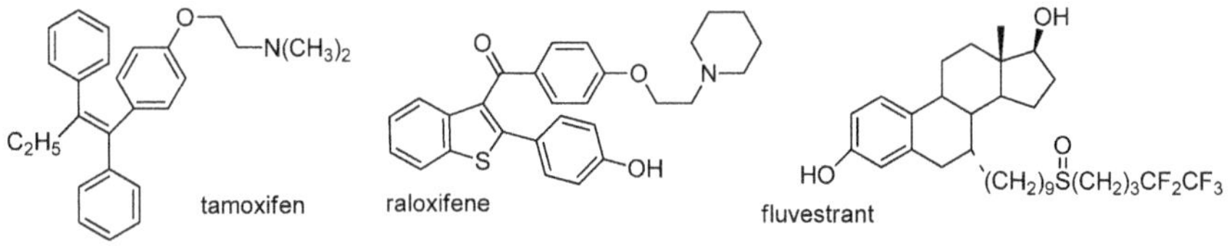

tamoxifen   raloxifene   fluvestrant

The second strategy uses *aromatase inhibitors*, drugs which inhibit the enzyme that aromatizes the A ring of estrogen. Drugs in this group include both steroid analogs such as exemestrane and also nonsteroidal compounds such as letrozole and anastrozole.

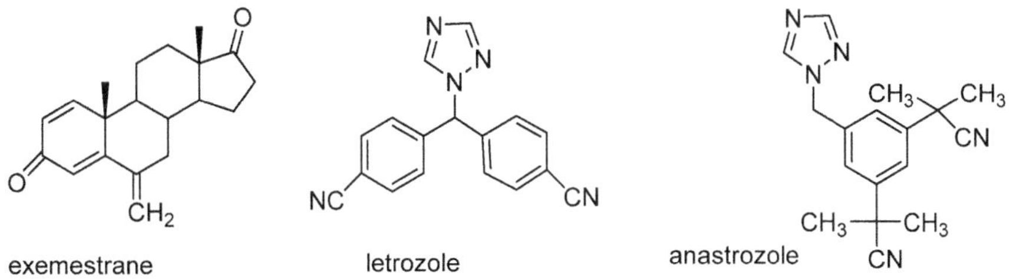

exemestrane   letrozole   anastrozole

Unfortunately, breast cancer often develops resistance to both types of drugs, allowing tumor recurrence or progression. In men, *androgen-dependent* prostate cancer is a leading cause of death. One therapeutic strategy is to limit the level of androgenic steroids. For example, abiraterone is an inhibitor of androgen biosynthesis. Like estrogen-dependent breast cancer, however, there is a very high rate of eventual resistance and recurrence of the disease.

abiraterone

15.6   STEROID ABUSE IN SPORTS

Several steroids related to testosterone promote muscle growth and weight gain. These steroids as a class are called *anabolic–androgenic steroids* (AAS). They activate various steroid receptors and are also subject to metabolic transformations to other active substances. Examples include compounds such as androstenedione, dehydroepiandrosterone (DHEA), methandrostenolone, stanazolol, and several 19-nor analogs shown in Scheme 15.3.

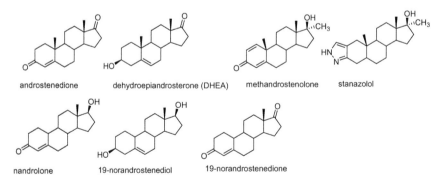

SCHEME 15.3   Examples of anabolic–androgenic steroids.

Nandrolone induces muscle growth and red blood cell production and increases bone density. Nandrolone's medical use is to enhance red blood cell production in patients with anemia resulting from renal deficiency. It is present in low concentration (<1 ng/mL) in male and nonpregnant females, but increases to >5 ng/mL in pregnant females. Because substances such as androstenedione and nandrolone occur naturally, they were available as nutritional and dietary supplements. However, in 2004, the FDA took action to ban them on the basis of the Anabolic Steroid Control Act.

The 1968 and 1972 Olympic Games were marked by both unusual records and the extraordinary physique of female athletes, particularly those from East Germany (GDR). In 1974, the IOC banned use of anabolic steroids and instituted testing procedures. Published reports in the 1960s and early 1970s indicated widespread use of AAS in the National Football League. The NCAA banned AAS in 1973, but did not institute testing until 1986. There is also evidence of extensive use of anabolic steroids by amateur athletes and men interested in body appearance. It was estimated in 1995 that anabolic steroid use might be as high as 1% in the male population.[17] Steroid use in Major League Baseball (MLB) culminated with a barrage of home runs in the late 1990s. Babe Ruth's record of 60 (154 games) stood for 37 years until Roger Maris hit 61 in 1961. By 2000, 47 players hit at least 30 home runs and a record 5693 home runs were hit. Beginning with Mark McGwire in 1998, several players hit 70, most recently Barry Bond's 73 in 2001. McGwire acknowledged use of androstenedione, which was available as a dietary supplement and not banned by MLB. Steroid use was banned by MLB in 2005. The Mitchell report in 2007 documented steroid use by many players.

The AAS enhance muscle and body mass. The role of DHEA has been studied. In healthy males it seems to have little effect on testosterone-level, probably because testosterone is produced at quite high level by the testes, so the increment is small. By a feedback mechanism, AAS tend to decrease production of testosterone by the testes. In women, in contrast, testosterone levels increase dramatically, because the normal testosterone level is low. AAS may have other effects. Surveys of male users of AAS report enhanced self-image, euphoria, increased libido, and aggression, as well as enhanced performance. It is suggested that the psychological effect of aggressiveness may result in increased intensity of training and competitiveness. Much of the knowledge of the effects of AAS in women comes from information revealed about the secret use of drugs in GDR, which began in the 1960s. The adverse effects include liver and cardiovascular damage, and in women there can be extreme disruption of the reproductive system and masculinization. Reported psychological effects include violent behavior and psychosis.[18]

Testing for doping in athletes generally involves a broad screening followed by a confirmatory analysis of the positive samples. The samples are divided into the screening and confirmatory samples at collection and sealed until analysis. Confirmatory analysis for steroids is usually done by GC–MS. The matching of both retention time and MS pattern of at least three peaks is required for a high degree of reliability. Testosterone doping is detected by the ratio to its 17-epimer, called *epi*-testosterone. This ratio increases when exogenous testosterone is present and a ratio 6/1 is taken as the reporting threshold. During the GDR program, some athletes were administered *epi*-testosterone, along with testosterone, so as to avoid detection.

15.7   ENDOCRINE DISRUPTORS; INADVERTENT FERTILITY MODIFICATION?

15.7.1   *SYNTHETIC COMPOUNDS*

There has been growing concern that there may be chemicals in the environment that have adverse effects on metabolic and reproductive systems. The suspect compounds have been called *endocrine disruptors*. The underlying hypothesis is that such chemicals may either mimic or inhibit the effect of the natural steroid hormones and thus have unintended effects on metabolic and reproductive systems. Several studies have suggested that there has been a decline in both semen quantity and sperm counts in men living in the United States or Europe and born since 1970. There has also been an increase in testicular cancer, a condition that seems to be related to hormonal regulation of development of the male fetus. A wide variety of chemicals have become suspects including the polyhalogenated compounds such as DDT and its metabolite DDE, PCBs and dioxin (for structures see Section 4.2). There have been numerous *in vitro* and animal studies demonstrating various abnormal responses to such compounds. Other compounds that have come under suspicion are phthalate esters used as plasticizers, bisphenol-A, and nonylphenol.

bis-phenol A   di-(ethylisohexy) phthalate   nonylphenol

Both the CSTEE (Scientific Committee on Toxicity, Ecotoxicity and the Environment) of the European Commission and the International Program on Chemical Safety have examined the data available for several hundred potential endocrine disruptors. Most, including nonylphenol and bisphenol-A, are roughly 10^{-6} as potent as estradiol, and on that basis, it is thought that the environmental concentration is unlikely to have significant effects.[19]

15.7.2 NATURAL PHYTOESTROGENS

There are also plant sources of estrogens, the so-called *phytoestrogens* such as genistein, daidzein, and coumestrol. These are present mainly in legumes and are found in various soy products. Genistein and daidzein are called *isoflavones*. Coumestrol is a *coumestan* and is present in soybean and alfalfa sprouts. Among the soy phytoestrogens, the estrogenic activity of daidzein is about 5×10^{-2} relative to estradiol. At the present time, there is little direct evidence as to whether these substances cause any beneficial or harmful effects.[20] There are related compounds known as lignans such as matairesinol and lariciresinol. Typical structures are shown in Scheme 15.4. These compounds have general structural similarity to estradiol and exhibit *in vitro* binding to estrogen receptors. Humans can be exposed to these compounds through food sources. In particular, it is suggested that infants fed on soy formula milk may be exposed to significant levels.[21] Reputed beneficial effects of soy-rich diets have also fueled interest in dietary supplements containing these compounds.

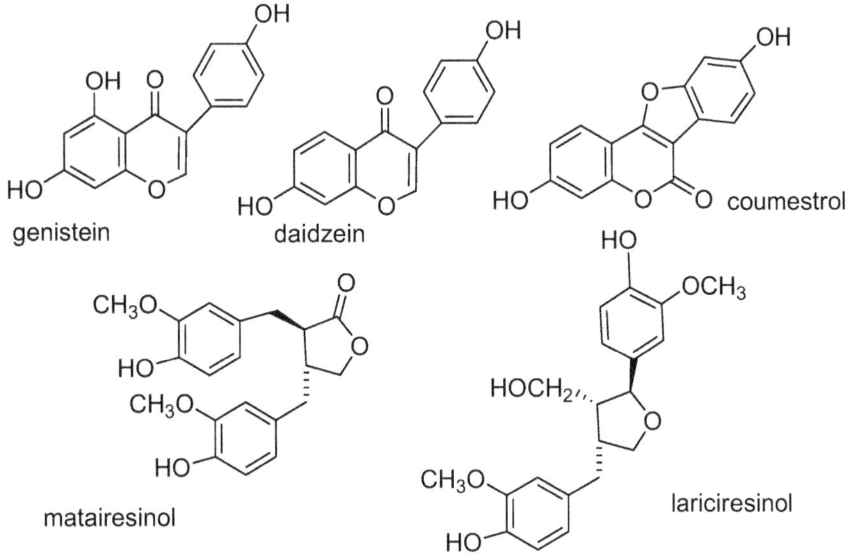

SCHEME 15.4 Examples of phytoestrogens.

15.7.3 THE CASE OF DIETHYLSTILBESTROL

A case of unintended and unfortunate circumstances indicates the great sensitivity of the endocrine-hormone system to exogenous chemical substances. In the

1950–1960s, diethylstibestrol (DES), a steroid analog, was used for the treatment of bleeding and other perceived abnormalities during pregnancy, despite the fact that a 1953 study had found no beneficial effects. In 1970, however, a rare form of cervicovaginal cancer was observed and it was found to correlate with DES use by the mother. The drug was banned for use in pregnant women in 1971. Follow-up studies have suggested a life-time risk of about 1 in 10^4–10^5 for daughters. Significant rates of other abnormalities in the reproductive system are also found in daughters. These effects are most strongly associated with DES exposure *in utero* in the first trimester. The effect on males is uncertain.

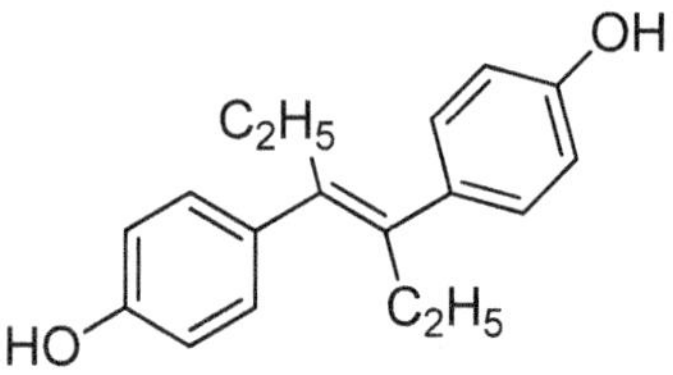

diethylstilbestrol

KEYWORDS

- steroids
- corticosteroids
- contraceptives
- cholesterol
- statins
- anabolic steroids
- phytoestrogens
- endocrine disruptors

BIBLIOGRAPHY BY SECTION

Section 15.2: Kendall, E. C. *Chem. Eng. News* **1950,** *28,* 2074–2077; Hazen, G. G. *J. Chem. Ed.* **1980,** *57,* 291–293; Djerassi, C. *Steroids* **1992,** *57,* 631–641; Rosenkranz, G. *Steroids* **1992,** 409–417; Neeck, G. *Ann. N. Y. Sci.* **2002,** *966,* 28–38.

Section 15.2.1: Fee, E.; Wallace, M. *Feminist Stud.* **1979**, *5*, 201–215; Mosher, W. D. *Fam. Plann. Perspect.* **1988**, *20*, 207–217; Edelman, D. A.; Van Os, W. A. A. *Int. J. Fertil.* **1990**, *35*, 206–210.

Section 15.3: Mosher, W. D. *Fam. Plann. Perspect.* **1988**, *20*, 207–217; Katz, E. *Trends Hist.* **1988**, *4*, 81–101.

Section 15.3.2: Baulieu, E. E. *Ann. N. Y. Acad. Sci.* **1997**, *828*, 47–58; Mahajan, D. K.; London, S. N. *Fertil. Steril.* **1997**, *68*, 967–976; Bardin, C. W.; Robbins, A.; O'Connor, B. M.; Spitz, I. M. *Curr. Ther. Endocrinol. Metabol.* **1997**, *6*, 305–311; Creinin, M. D. *Am. J. Obstet. Gynecol.* **2000**, *185*(Suppl. 2), 83–89; Joffe, C.; Weitz, T. A. *Soc. Sci. Med.* **2003**, *56*, 2353–2366.

Section 15.3.3: Lunefelde, B. *Hum. Reprod. Update* **2004**, *10*, 453–467.

Section 15.4.2: Steinberg, D. *J. Lipid Res.* **2006**, *47*, 1339–1351; Endo, A. *Proc. Jpn. Acad. Sci., Sect. B* **2010**, *86*, 484–493.

Section 15.6: Kicman, A. T.; Gower, D. B. *Ann. Clin. Biochem.* **2003**, *40*, 321–356.

Section 15.7.1: Swan, S. H.; Elkin, E. P.; Fenster, L. *Environ. Health Perspect.* **1997**, *105*, 1228–1232; Sharpe, R. M. *Pure Appl. Chem.* **1998**, *70*, 1685–1701; Irvine, D. S. *Andrologia* **2000**, *32*, 195–208; Dohle, G. R.; Smit, M.; Weber, R. F. A. *World J. Urol.* **2003**, *21*, 341–345; Sharpe, R. M.; Irvine, D. S. *Br. Med. J.* **2004**, *328*, 447–451.

Section 15.7.3: Stillman, R. J. *Am. J. Obstet. Gynecol.* **1982**, *142*, 905–921; Giusti, R. M.; Iwamoto, K.; Hatch, E. E. *Ann. Intern. Med.* **1995**, *122*, 778–788.

REFERENCES

1. Hirschmann, R. *Steroids* **1992**, *57*, 579–592.
2. Hogg, J. A. *Steroids* **1992**, *57*, 593–616.
3. Lumenfeld, B. *Hum. Reprod. Update* **2004**, *10*, 453–467.
4. Howles, C. M. *Mol. Cell. Endocrinol.* **2000**, *161*, 25–30.
5. Corona, G.; Rastrelli, G.; Vignozzi, L.; Maggi, M. *Expert Opin. Emerg. Drugs* **2012**, *17*, 239–259.
6. Wilson, P. W.; Garrison, R. J.; Castelli, W. P.; Feinleib, M.; McNamara, P. M.; Kannel, W. B. *Am. J. Cardiol.* **1980**, *46*, 649–654.
7. Gofman, J. W. *Ann. N. Y. Acad. Sci.* **1956**, *64*, 590–595.
8. Baigent, C.; Keech, A.; Kearney, P. M.; Blackwell, L.; Buck, G.; Pollicino, C.; Kirby, A.; Sourjina, T.; Peto, R.; Collins, R.; Simes, J. *Lancet* **2005**, *366*, 1267–1278.
9. Ray, K. K.; Seshasi, S. R. K.; Erquo, S.; Sever, P.; Jukema, J. W.; Ford, I. Sattar, N. *Arch. Intern. Med.* **2010**, *170*, 1024–1033.
10. Libby, P.; Ridker, P. M.; Maseri, A. *Circulation* **2002**, *105*, 1135–1143.
11. Libby, P. *Nature* **2002**, *420*, 868–874; Pearson, T. A.; et al. *Circulation* **2003**, *107*, 499–451.
12. Ridker, P. M.; Danielson, E.; Fonseca, F. A. H.; Genest, J.; Gotto, Jr., A. M.; Kastelein, J. P.; Koenig, W.; Libby, P.; Lorenzatti, A. T.; MacFayden, J. G.; Nordestgaard, B. G.; Shepherd, J.; Willerson, J. T.; Glynn, R. J. *Lancet* **2009**, *373*, 1175–1182.
13. Nissen, S. E.; Tuzcu, E. M.; Schoenhagen, P.; Crowe, T.; Sasiela, W. J.; Tsai, J.; Orazem, J.; Magioriea, P. D.; O'Shaughnessy, C.; Ganz, P. *N. Engl. J. Med.* **2005**, *352*, 29–38.
14. Marangoni, F.; Poli, A. *Pharmacol. Res.* **2010**, *61*, 193–199.

15. Nicholls, M. G.; Lewis, L. K.; Yandle, T. G.; Lord, G.; McKinnon, W.; Hilton, P. J. *J. Hypertens.* **2009**, *27*, 3–8.
16. Steyn, P. S.; van Heerden, F. R. *Nat. Prod. Rep.* **1998**, *15*, 397–413; Krenn, L.; Kopp, B. *Phytochemistry* **1998**, *48*, 1–29.
17. Bagatell, C. J.; Bremmer, W. J. *N. Engl. J. Med.* **1996**, *334*, 707–714.
18. Franke, W. W.; Berendonk, B. *Clin. Chem.* **1997**, *43*, 1262–1279.
19. Greim, H. *Angew. Chem. Int. Ed.* **2005**, *44*, 5568–5574.
20. Davis, S. R.; Dalais, F. S.; Simpson, E. R.; Murkies, A. L. *Rec. Prog. Horm. Res.* **1999**, *54*, 185–211.
21. Setchell, K. D. R.; Zimmer-Nechemias, L.; Cai, J.; Heubi, J. E. *Lancet* **1997**, *350*, 23–27.

CHAPTER 16

DIABETES AND ANTIDIABETIC THERAPY: CONTROL OF GLUCOSE

ABSTRACT

Diabetes has been known for centuries and is recognized by thirst and sugar in the urine. There was no effective treatment until insulin was isolated by Frederick Banting and Charles Best in Toronto in 1921. The first patient was treated in early 1922 and insulin was introduced commercially by Eli Lilly Company the same year. Type-1 diabetes usually occurs in children and adolescents and results from failure of the pancreas to produce sufficient insulin. It can be treated only by insulin administration, usually by injection. Type-2 diabetes results from poor sensitivity to insulin and reduced insulin generation. Type-2 diabetes responds favorably to improved diet and exercise. If necessary it can be treated with several types of drugs including sulfonylureas, biguanides, fibrates, α-glucosidase and lipase inhibitors, thiazolidinediones, and dipeptidyl peptidase-4 inhibitors.

Diabetes has been known since antiquity and was described in 10 CE by the Roman physician Celsus. Currently, it is estimated that 200–300 million people worldwide suffer from diabetes. The United States has the highest diabetes level of the large nations (about 12%), but the world rate is estimated at 6% with 50 million adult diabetics in India and more than 40 million in China. Diabetes rates are on the increase in most parts of the world. Diabetes is classified as type-1 or type-2. Type-1, also known as juvenile or early-onset diabetes, is caused by failure of the pancreas to produce sufficient insulin to properly regulate glucose metabolism. There is a major genetic component in susceptibility to type-1 diabetes, but it is uncertain as to what factors cause the pancreas to fail to produce sufficient insulin. Type-2 diabetes involves both decreased insulin production by the pancreas and impaired response to insulin, known as *insulin insensitivity* or *insulin resistance*. Type-2 develops later and in addition to genetic factors has a large life-style component. It is associated with lack of physical activity and obesity. Untreated diabetics have extreme thirst and appetite, but cannot process the food properly. Blood glucose increases to harmful levels. The key function of insulin is to regulate glucose concentration, with the basal level being about 5 mmol/l. Insulin is formed in the β cells of the pancreas. Although the weight of these cells is usually just under 1 g, they can contain about 10 day's supply of insulin. Normally, insulin is released in short pulses with a frequency of 10–15 min. After eating or administration of glucose, the insulin level rises quickly. The pancreas responds to increased insulin demand by up-regulation of transcription factors and production of enzymes. The insulin demand can range over a factor of 100 between normal and diabetic individuals. Figure 16.1 compares diurnal diabetic plasma glucose levels with a control.

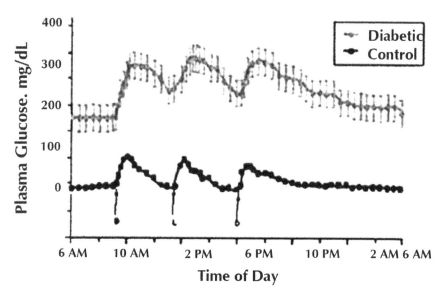

FIGURE 16.1 Comparison of normal and diabetic 24-h plasma glucose levels. (From Bhatnagar, S.; Srivastava, D.; Jayadev, M. S. K.; Dubey, A. K. *Prog. Biophys. Mol. Biol.* **2006,** *91*, 199–228 by permission of Elsevier.)

When insulin control weakens and glucose levels increase, there are deleterious effects on several organs including the pancreas, kidneys, cardiovascular system, and eyes. Advanced diabetes can lead to blindness, renal failure, and restricted circulation in the limbs. Patients eventually succumb to the resulting metabolic deterioration. Prior to the1920s, the only treatment for diabetes was extremely limited caloric intake, eliminating most fats and carbohydrates. It was literally a "starvation diet." It could extend the life of patients, but only at the cost of hunger and extreme disability. The only uncertainty was whether the patient would die from the effects of diabetes or malnutrition.

16.1 THE DISCOVERY AND CLINICAL INTRODUCTION OF INSULIN

In 1889, Joseph von Mering and Oskar Minkowski, working in Strasbourg, showed that removal of the pancreas caused diabetes in dogs. Several subsequent studies indicated that pancreatic extracts could reduce the symptoms of diabetes. As early as 1906, occasional positive responses were noted in treatment of diabetics with pancreatic extracts, but there were more failures than successes, probably because of allergic reaction to the impure extracts. E. I. Scott, a graduate student at the University of Chicago, made efforts to isolate the active material, but without success. He introduced the name *insulin* for the still hypothetical active component. Israel S. Kleiner, a biochemist working at the Rockefeller Institute, showed in 1919 that extracts of the pancreas could lower blood glucose in depancreatized (and therefore diabetic) dogs. Nicolae Paulescu, a professor of physiology in Bucharest, Romania, also had some success in treatment of diabetic dogs with pancreatic extracts prior to WWI. Paulescu resumed his experiments in Bucharest at the end of WWI and published his results in 1921.

In 1920, Dr. Frederick Banting[a] was inspired by a journal article on the relationship between the pancreas and diabetes to undertake research on pancreatic extracts. He developed a specific hypothesis. It was known that there were two general types of pancreatic secretions called "external" and "internal." The "external factor" was thought to be involved in digestion and the "internal factor" in control of carbohydrate metabolism. Banting proposed that by ligating the pancreas of experimental dogs, the external factor would be diminished faster than the internal factor, which might allow isolation of the latter. Banting discussed his idea with Prof. John J. R. Macleod[b] of the University of Toronto. He received laboratory space and financial support from Prof. Macleod, including the assignment of Charles Best[c] as a research assistant. The first successful ex-

[a]Frederick Banting graduated from the University of Toronto Medical School in 1917. He served in France during WWI and then continued his medical training in orthopedics. He opened a private practice in London, ON, but also worked as a demonstrator in anatomy at the medical school at what is now the University of Western Ontario. It was through that appointment that he knew of John J. R. Macleod, who worked on diabetes at the University of Toronto. Banting received many honors as the result of his work on insulin, but produced no other major scientific discoveries. He was killed in a plane crash in 1941, while serving as a medical consultant to the Canadian armed forces.

[b]John J. R. Macleod was born in Scotland and graduated in medicine from the University of Aberdeen in 1898. He had research training at Leipzig and Cambridge and in 1903 became a professor of physiology at Western Reserve University in Cleveland, OH. He became well known in carbohydrate metabolism and diabetes. In 1918, he moved to the University of Toronto as chair of physiology.

[c]Charles Best was born in Maine and served briefly in WWI. He then graduated from the University of Toronto as an undergraduate in physiology and biochemistry in 1921 and began graduate study under Macleod, leading to his working with Banting. Best used a recently improved glucose analysis to follow the effect of insulin on blood levels. Best succeeded Macleod as professor of physiology at the University of Toronto in 1928 and continued to make contributions to research on diabetes. On Banting's death in 1941, Best became head of the Banting–Best Medical Research Institute at the University of Toronto.

periments were done in July 1921, when a pancreatic extract lowered the blood sugar of a diabetic dog. The result was successfully repeated several times. At that point the extract was called, "isletin," the name being derived from the "*islets of Langerhans*," the part of the pancreas that produces the active component. Banting and Best followed up with a series of control experiments that showed conclusively that the activity was associated only with pancreatic extracts. On the basis of this progress, Macleod arranged for Banting to remain in Toronto as a temporary lecturer in pharmacology. The preparation of the extract was extremely time-consuming and inefficient. In order to prepare the extract more easily, Banting and Best tried other sources and found that active extract could be isolated from the pancreas of pigs or cattle. At this point Macleod engaged the services of Dr. J. B. Collip[d] to help purify and characterize the active component. Collip, a biochemist, was on academic leave from the University of Alberta. Collip was able to develop conditions for purification of insulin that involved extraction with ethanol and provided a fairly pure material from whole pancreas. It was also Collip who recognized that the convulsions and death occurring following some injections were the result of *hypoglycemia.*

The first test on a human patient was in January 1922 on a 14-year boy who was near death from juvenile diabetes. The first injection resulted in a drop in blood glucose but no obvious improvement. About 10 days later, he received a second injection and immediately improved. Six more patients were successfully treated with bovine pancreatic extract in February 1922. It had been just under 1 year between the beginning of the experiments and the first clinical success. Banting's original hypothesis was incorrect, but it had led to successful isolation of the antidiabetic factor. During this period, Banting had come to distrust and resent Macleod, feeling that Macleod was taking undue credit for the work. Banting also became angry at Collip, suspecting that he intended to take credit for and patent the improved isolation method. A truce was negotiated by the laboratory director J. G. Fitzgerald, which included an agreement than none of the group would seek to patent or commercialize the discoveries. By this time, however, Banting had developed a permanent distrust and dislike of Macleod. The efforts in Toronto to scale up production of insulin met with considerable difficulty. For a time in Spring 1922, they were unable to reliably produce insulin. The problem was eventually traced to increased temperatures involved in the scaled-up procedure. Extraction with acetone, followed by purification by precipitation from ethanol, solved the immediate problem, but the process remained

---

[d]J. B. Collip received a Ph. D. in biochemistry from the University of Toronto in 1916 and then became a professor of biochemistry at the University of Alberta. His research interests were in the physiological effects of various extracts, such as those containing adrenaline. It was this expertise that resulted in Macleod engaging him in the insulin research. Collip returned to the University of Alberta and pursued work on hormones there and later at McGill University.

exceedingly cumbersome. The first full-scale report of the clinical success was made by Macleod in a May 1922, meeting in Washington, DC.

G. H. A. Clowes[e] of Eli Lilly approached Macleod about a collaboration in March 1922. He argued that the growing publicity about insulin would lead to the extract being made by others and that without patent protection it would be difficult to assure the quality of the product. Initially, the Toronto group was resistant, intending to develop the method themselves, free of commercial involvement. But, as the magnitude and importance of the task became apparent, a decision was reached to engage Eli Lilly. It was eventually agreed that Best and Collip would file a patent and assign it to the University of Toronto. (At the time it was considered unethical for physicians, that is, Banting and Macleod, to patent potential clinical treatments). Eli Lilly was given a license to produce insulin in the United States, and the license for the United Kingdom was given to the Medical Research Council. Eli Lilly began producing insulin from porcine pancreas extracts during the Summer of 1922. Considerable batch-to-batch variation in potency caused problems and clinicians had to be alert to hypoglycemia from especially potent batches. This problem was solved in late 1922 at Eli Lilly by precipitation of insulin at its isoelectric pH. By early 1923, Eli Lilly was reliably producing insulin, although batch-to-batch potency variation was still in the range of 10%. Eli Lilly used the brand name Iletin. Insulin production began in Britain and Denmark in 1923, and, delayed by the chaos after WWI, in Germany in 1924.

Banting decided to begin a private practice using insulin, primarily supplied by Eli Lilly. Initially, he did not have accreditation at the University's Toronto General Hospital and practiced in a private clinic and at a military hospital. Later, he also received an appointment that provided access to the university hospital. Many other physicians began trials of insulin, usually supplied by Eli Lilly. Among them were prominent physicians who had used the "starvation diet" approach. The results were often spectacular. A few especially prominent patients were treated successfully. One was Elizabeth Hughes, the 14-year-old daughter of Charles Evans Hughes, the Chief Justice of the US Supreme Court, and the Republican Presidential Candidate in 1916. She had been treated by Dr. F. M. Allen, a leading proponent of the dietary approach. She was treated successfully by Banting at the newly opened Diabetes Clinic at Toronto General Hospital.

---

[e]George H. A. Clowes was born in England in 1877. He was a descendant of Sir William Clowes (1540–1604) who had been surgeon to Queen Elizabeth I. He studied chemistry at the Royal College of Science in London and then went to the University of Gottingen, where he received a Ph. D. in 1899. His research was in the area of carbohydrates. He then studied in several labs over the next few years, including those of Buchner and Ehrlich in Germany and E. Starling in London. In 1901, he came to the United States as a research chemist at the New York Institute for Malignant Diseases in Buffalo, pursuing the immunology of cancer. During WWI, he worked at the Chemical Warfare Service in Washington, DC, USA. In 1919, he moved to Eli Lilly in Indianapolis, IN, USA. Clowes had heard reports of Banting's work and attended a presentation in New Haven, CT, USA on December 26, 1921. Convinced of the importance of the discovery, Clowes initiated efforts to establish a collaboration with Eli Lilly.

Late in 1923, The Nobel Prize in Medicine or Physiology was awarded to Banting and Macleod. Banting was incensed that Macleod shared the prize. The chair of the board of the University of Toronto had to convince Banting to accept it. Banting shared his monetary award with Best and Macleod shared his with Collip. Controversy ensued and has continued. Romanians took up the cause of Nicolae Paulescu.[f] Echoes of the controversy still travel on the Internet.

The discovery and characterization of insulin, the key molecule in diabetes, was one of the most significant developments in biochemistry and medicine in the twentieth century. In addition to the 1923 Nobel Prize in Medicine or Physiology to Banting and Macleod, the 1958 Nobel Prize in Chemistry was awarded to Frederick Sanger for determining the protein sequence.[g] Insulin then became the target of synthesis, and this was accomplished in 1963 by two in-dependent groups, P. Katsoyannis in Pittsburgh and H. Zahn in Aachen. Insulin was also the subject X-ray crystallographic structure determination by Dorothy Crowfoot Hodgkins, first published in 1969, followed by several refinements. The insulin structure was also solved in China during the early 1970s.[1] The amino acid sequence of insulin is shown in Figure 16.2.

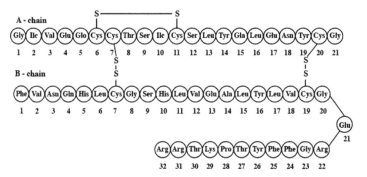

FIGURE 16.2   Amino acid sequence of insulin.

---

[f]Nicolae Paulescu was a brilliant student. After high school, he travelled to Paris and received his medical de-gree there in 1897. He returned to Bucharest in 1900 and became the head of the Physiology Department at the Medical School of the University of Bucharest. He was able to use aqueous extracts of the pancreas to lower the blood sugar of pancreatized dogs in experiments in 1920–1921. His results were published in 1921. Paulescu obtained a Romanian patent for his process in April, 1922. However, Paulescu did not proceed to any human trials. Paulescu was a controversial figure, being the author of anti-Semitic articles in the 1920s.

[g]Frederick Sanger was born in 1918, the son of a physician in Gloucestershire. He received a B. A. in natural science at Cambridge in 1939, and obtained a Ph. D. in 1943, working on amino acid metabolism. He then continued research at Cambridge, at first with his own funds but later supported by the MRC. He began work-ing on the sequencing of insulin and was the first person to sequence a natural polypeptide. He was awarded the 1958 Nobel Prize for this work. He then took up work on methods for sequencing RNA and DNA. Sanger was located at the MRC Laboratory of Molecular Biology after it opened in 1962. In 1977, his group developed the dideoxy termination method, which was much faster than previous methods. Sanger shared the 1980 Nobel Prize in chemistry with Walter Gilbert and Paul Berg for their contributions to the structure of DNA. In 1992, the Wellcome Trust and MRC founded the Sanger Center at Cambridge in his honor. It became one of the leading contributors to research on DNA, including the human genome project.

16.2 PRODUCTION AND ADMINISTRATION OF INSULIN

From 1922 to 1982, insulin was prepared by isolation and purification from pancreases of pigs and cattle obtained at meat-packing plants. There are small differences in the amino acid sequence between the human hormone and those isolated from cattle or pigs, but all are effective. Chromatography became a part of insulin purification in the 1960s and removed allergenic higher MW proteins. Production of insulin by recombinant DNA technology began in 1982 (see Section 21.3.1).

The main route of administration is by subcutaneous injection. Like most polypeptides, insulin is rapidly destroyed in the digestive tract. The dose and time of administration of insulin is critical. An insulin injection has its maximum effect in 2–4 h and a duration of 8–10 h. One objective is therefore to optimize the time course of the delivery. An alternative means of delivery, insulin pens, was introduced in the 1980s. They can involve replaceable cartridges or be entirely disposable. The pens are generally preferred by patients but are more complex than syringes and require careful glucose monitoring to detect malfunction. They are also more expensive than syringes. The most recent alternative means for administration is by automated pumps. The idea of pumps for insulin delivery began to receive significant attention in the 1960s. The first devices were about the size of a large back-pack and were not practical for patient use. The first computer-controlled pumps were introduced in the 1970s. They contained algorithms to aid the patient in computing the dose. Pumps became commercially available in the 1980s. They were only moderately reliable and weighed about one pound. New designs in the 1990s improved reliability and reduced the weight to about 100 g. Current pumps require active data and dose input by the patient. In the United States, it is estimated that about 25% of type-1 diabetes patients use pumps, whereas in Europe it is about 10%. Insulin pumps coupled with glucose-sensors for precise control of insulin levels are currently under development.[2] Other potential methods include transdermal patches and aerosols for inhalation. Aerosols were used briefly in 2006 but withdrawn after about a year.

Another approach to optimizing insulin delivery is to modify its physical form or chemical structure. Specific mutations can be accomplished by the techniques of directed mutation (see Section 21.3.1). Changes in duration of action can also be achieved by introduction of medium or long-chain fatty acid acyl groups, usually at lysine-29 in the B-chain. Both the mutations and chemical modifications affect the structure in such a way as to alter the rate of release.[3]

16.3   DIAGNOSIS AND CHARACTERISTICS OF DIABETES

16.3.1   DIAGNOSTIC TESTS FOR DIABETES

The standard test for diabetes is the *fasting glucose level*, with >7.0 mmol/l indicative of diabetes and 5.6–6.9 considered prediabetic. This is supplemented by checking glucose levels 2 h after ingestion of a standard dose of glucose, called the *oral glucose tolerance test*. The value indicating diabetes is ≥11.1 mmol/l. with 7.8–11.0 mmol/l. considered prediabetic. One problem with the oral glucose tolerance analysis is that it is very time-sensitive with respect to glucose ingestion. Various meters are available for patient self-measurement of glucose levels. An annual test for albuminuria is recommended for both type-1 and type-2 diabetes in adults. Levels >30 mg/g are indicative of increased risk for cardiovascular disease. There is no method available for routine measurement of insulin levels.

In addition to glucose measurements, in 2009, an expert committee engaged by the American Diabetes Association recommended including an assays for "glycated hemoglobin" (H_bA_{1c}) as a diagnostic criteria with a level >6.5% indicating diabetes. H_bA_{1c} is a modified form of hemoglobin that correlates with high glucose levels. There has been some controversy about this recommendation.[4] The test is not available in all parts of the world and depends on the availability of well-standardized analyses in approved laboratories. Furthermore, there appears to be considerable variation among individuals based on genetic factors, as well as among different ethnic groups. Nevertheless, the method has come into use as one of the criteria for tracking progression of diabetes and for evaluating the effectiveness of treatments.

16.3.2   TYPE-1 DIABETES

Type-1 diabetes results from impaired production of insulin by the pancreas. Type-1 diabetes usually develops in children and adolescents. The rate of progression tends to correlate with the age of onset, with younger patients progressing faster. Type-1 diabetes effects about 1.7/1000 persons <20 years old in the United States, and the rate is increasing about 5% per year. The reason for the increasing rate is not clear, although both genetic and environmental factors contribute to the disease. Several factors are believed to cause failure of pancreatic β cells, including autoimmune destruction and high rate of cell death (apoptosis).[5] Several specific genetic variations in the function of the immune system are associated with type-1 diabetes. Diabetes is most likely to occur when there are several such mutations in an individual. Genetic susceptibility allows some external factor to precipitate the destruction of β cells. The strongest evidence points to viral infections as the most frequent inducing event. There is also evidence that bacterial infections and associated inflammation can trigger diabetes. The precipitating event may occur years before the disease is diagnosed, making it difficult to connect a particular infection with the cause. Diabetes progresses gradually and is often at an advanced stage before it is detected. Typically 60–90% of pancreatic β cells have been destroyed when diabetes is diagnosed on the basis of hyperglycemia.

The only current treatment of type-1 diabetes is administration of insulin. When insulin treatment is initiated there is often a period of rebound in which the pancreas increases its production of insulin. As the disease progresses, however, continued insulin administration is essential. Well-monitored insulin-therapy can lead to a life-expectancy within 10 years of normal. So far, experimental therapeutic concepts based on attacking the autoimmune aspects of the disease have not achieved clinical use.

16.3.3 TYPE-2 DIABETES

Type-2 diabetes is characterized by poor response to insulin, which is called *insulin insensitivity or resistance.* Insulin insensitivity is the failure of insulin to adequately lower blood glucose levels. Normal control involves uptake of glucose by muscles and adipose tissue, as well as suppression of glucose production in the liver. Type-2 diabetes progresses from insulin insensitivity to decreased production of insulin by the β cells. In the diabetic state, the pancreas no longer produces sufficient insulin to control the glucose level. It is estimated that about 60–70 million American exhibit insulin resistance and at least a quarter of these will progress to diabetes. The characteristics of insulin insensitivity overlap with those associated with the metabolic syndrome and include high triglyceride levels, low HDL and increased C-reactive protein (see Section 9.8). There are other metabolic abnormalities associated with diabetes, including increased levels of free fatty acids.[6] Advanced diabetes can lead to blindness, renal failure, and restricted circulation in the limbs. Cardiovascular disease is the ultimate cause of death of 80% of type-2 diabetics.

There is a major genetic component to the etiology of type-2 diabetes. More than 40 specific gene variations have been associated with the condition. There is considerable variation in susceptibility among ethnic groups, with some at very high risk. While type-2 diabetes has a genetic component, it is also related to physical inactivity, obesity, hypertension, and aging. About 80% of type-2 diabetes patients are overweight or obese. There is also a strong genetic component in the tendency to become obese. This may operate through genes that control appetite and satiety. Psychological factors also play a role in determining food intake. There are also likely environmental influences on susceptibility to type-2 diabetes including fetal and neonatal nutrition. Maternal diabetes during pregnancy increases the risk of diabetic offspring.

Abdominal fat deposits are strongly correlated with insulin resistance and diabetes. Fat also accumulates in organs such as liver, pancreas, heart, and kidney. In obesity, the fat cells increase substantially in size. The increased deposits of visceral fat observed in obesity exhibit greater infiltration by immune cells and higher secretion of pro-inflammatory cytokines. Increased abdominal adipose tissue is associated with insulin resistance. Not all overweight individuals develop diabetes. One of the protective mechanisms is the ability to deposit excess nutrients in nonvisceral fatty deposits, sparing the liver, pancreas, and heart. Figure 16.3 summarizes some of the factors that influence the development of type-2 diabetes.

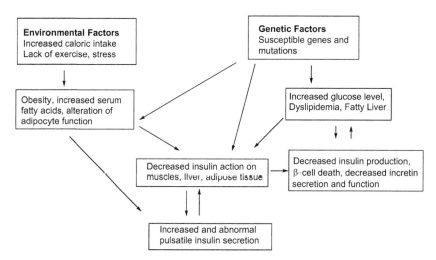

FIGURE 16.3 Summary of factors influencing development of pathogenic aspects of type-2 diabetes. MODY is "maturity onset diabetes of youth" which has a specific genetic cause. (Adapted from Freeman, J. S. *Mayo Clin. Proc.* **2010**, *85*, S5–S14).

16.3.4 BIOCHEMICAL ASPECTS OF INSULIN FUNCTION

Two types of cells, the adipocytes and endothelium, are particularly important in insulin control of carbohydrate metabolism. Adipocytes are fat-containing cells that serve both to store energy in the form of triglycerides and to release free fatty acids for oxidation when energy is required. Adipose tissue is classified as white and brown. White adipose tissue stores energy in the form of triglycerides. The brown adipose tissue regulates body temperature by metabolism of lipids in response to cold. The adipocytes generate a wide variety of cytokines that have effects on lipid metabolism, blood pressure, inflammation, and appetite. Transcription factors that are important in metabolic control are also expressed in adipocytes. Another point of metabolic control are the *peroxisome proliferation activated receptors* (PPAR), which are nuclear receptors that control aspects of both triglyceride and glucose metabolism.[7]

Insulin promotes esterification of fatty acids and retards hydrolysis, thus favoring storage of triglycerides and maintenance of low levels of free fatty acids. When adipocytes become resistance to insulin, the level of control is reduced and the circulation of free fatty acids increases. There is a close relationship between fatty acid and glucose metabolism. Increased fatty acid levels leads to decreased glucose utilization. For example, during the late stages of pregnancy, there is a shift to fatty acid metabolism, presumably to conserve adequate glucose for the fetus.

There are other polypeptide hormones besides insulin involved in glucose regulation. The α cells in the pancreas produce *glucagon*, a 29-amino acid polypeptide, which controls glucose production in the liver. In normal subjects, glucagon causes the liver to break down glucose stored as the polymer *glycogen*. The glucose level is kept in balance by regulation of endogenous glucose production and glucose use, especially by the brain. Ingestion of carbohydrates causes the glucose level to increase and results in increased insulin production and suppresses glucagon secretion. Under high glucose levels other organs such as the heart, muscle, and adipose tissue increase glucose use. There is another mechanism for glucose control that involves polypeptide hormones formed primarily in the small intestines. One is called *glucagon-like peptide-1* (GLP-1). Another is *glucose-dependent insulinotropic peptide* (GIP). These and related peptides are called *incretins*. The incretins effect the release of insulin in response to ingested glucose. They serve to connect insulin release to eating and are more sensitive to oral glucose intake than to the level of glucose in the blood. Attempts to use incretin analogs in therapy of diabetes are discussed in Section 16.4.7. These relationships are summarized in Figure 16.4.

The purpose of insulin control is integration of energy demand versus availability. A key mediator is *AMP-activated protein kinase* (AMPK). A falling energy status is reflected as a decreasing ATP level. AMPK then activates both ATP-forming catabolism and down-regulates ATP-consuming anabolic processes. AMPK also interacts with the hypothalamus to connect energy status and appetite. This is achieved through the hormonal polypeptides *ghrelin* and *leptin*. AMPK also plays a key role in the relationship between exercise and energy status. Exercise dramatically increases ATP utilization. During exercise, AMPK promotes conversion of triglycerides to fatty acids and balances fatty acid metabolism with glucose to insure adequate glucose levels in the brain. Fatty acid utilization increases with duration of exercise and exercise capacity and endurance is increased by training. AMPK upregulates mitochondrial metabolic capacity in response to regular exercise.

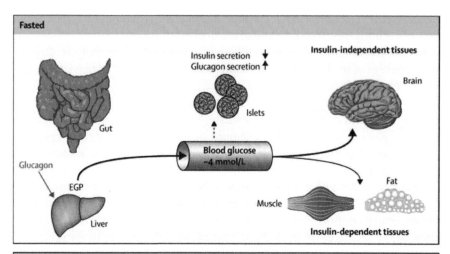

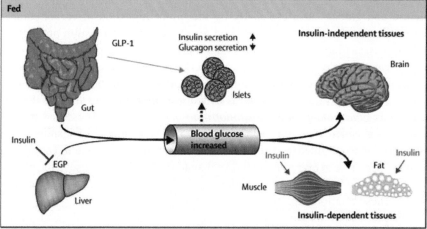

FIGURE 16.4 Overview of normal glucose homeostasis. (From Nolan, C. J.; Damm, P.; Prentki, M. *Lancet* **2011**, *378*, 169–181 by permission of Elsevier.)

The endothelium, the layer of cells lining blood vessels, also has an important role in the progression of type-2 diabetes. The healthy vascular endothelium is involved in regulating blood pressure, platelet, and leukocyte adhesion and is resistant to formation of plaque. The endothelium plays a crucial role in vascular function by releasing contracting and relaxing factors. The most important of these are NO and the prostaglandins and thromboxanes formed from arachadonic acid (see Section 13.1.1). In type-2-diabetes, NO production is reduced and this is attributed to oxidative stress and reactive oxygen species generated by glucose metabolism. There are also abnormalities in production of the prostaglandins and thromboxanes. These abnormalities make diabetics particularly susceptible to formation of plaque and the other aspects of cardio-vascular disease. The poor wound-healing associated with diabetes is related to the angiogenic function of the endothelium. Defective vascularization is also the cause of the blindness associated with diabetes. Figure 16.5 summarizes the mutual relationship between endothelial dysfunction and type-2 diabetes and emphasizes the reinforcing nature of the physiological defects.

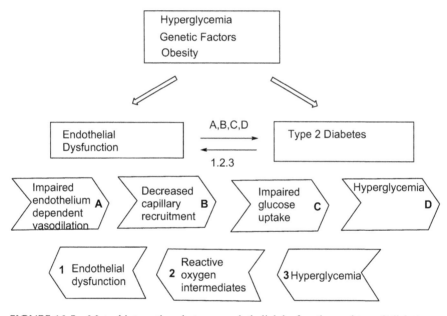

FIGURE 16.5 Mutual interactions between endothelial dysfunction and type-2 diabetes.

16.4 DRUGS USED IN TREATMENT OF TYPE-2 DIABETES

The treatment of type-2 diabetes typically begins with life-style modification, for example, diet, exercise, and weight loss. If diabetes continues to progress, drugs are used with the goal of decreasing the glycemic level. Single drugs may be replaced with combinations. When drug treatment fails, insulin is used. Because type-2 diabetes progresses with continued deterioration of β cells, one possible approach is aggressive treatment with drug combinations to forestall decrease in insulin production. To date, no drug that substitutes for the physiological effects of insulin has been found.

16.4.1   SULFONYL UREAS AND RELATED INSULINOTROPIC DRUGS

The first drugs to have some effect in treatment of diabetes were the *sulfonyl ureas*, such as tolbutamide, developed in Europe and introduced in the United States by Upjohn in 1956. A multicenter study showed that while it had no effect in type-1 diabetes, tolbutamide did lower blood sugar in type-2 patients. Other sulfonyl ureas were introduced and they became the most important drugs to treat type-2 diabetes. Structures of the sulfonylurea drugs are shown in Scheme 16.1. The sulfonyl ureas promote release of insulin from the β cells and are called *insulinotropic*. They carry some risk of inducing hypoglycemia. Most patient's responsiveness to the drugs gradually diminishes. There is also considerable genetic variation among individuals in both in the rate of metabolism and the sensitivity to the drugs.[8]

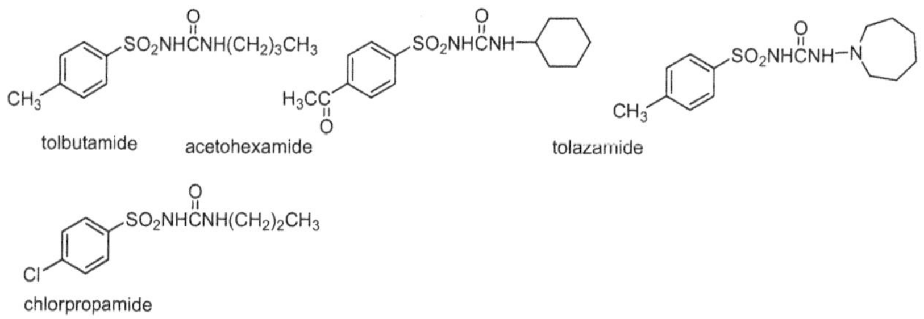

SCHEME 16.1   Sulfonyl Ureas Used in Treatment of Diabetes.

The availability of effective and convenient drugs encouraged the American Diabetes Association to promote broad screening and these efforts were supported by the American Pharmaceutical Association and the American Medical Association. The extent of prescription of tolbutamide and other oral antidiabetes medicines grew as the tests identified patients with elevated glucose levels, but that had not yet progressed to diabetes. In the mid-1960s, a large clinical study was begun that was intended to compare several treatments, including diet alone, diet and tolbutamide, diet and insulin, and a new oral antidiabetic drug, phenformin. As the results began to accumulate, they indicated an increase in cardiovascular deaths associated with tolbutamide. In 1969, the study group decided to stop the tolbutamide arm of the study. News of the decision leaked to the press prior to an official announcement, and the resulting publicity generated wide public concern and regulatory repercussions. At the time, about 800,000 people were being treated with tolbutamide. Upjohn and some parts of the medical community dissented from the conclusion of the study. Eventually, in 1975, the FDA held hearings on a restriction, under which tolbutamide would be recommended only if diet or insulin failed to control glucose level. The controversy continued, ranging from physicians' complaints of interference with their practice to consumer activists labeling tolbutamide a "bad drug." It was not until 1984, that the final version of the labeling was approved, and by then newer drugs were available.[9]

A second generation of insulinotropic drugs includes not only sulfonylureas, but also benzoic acid derivatives that act by a similar mechanism. These include glipizide, glibeclamide (also known as glyburide), glimepiride, meglitinide, and repaglinide. These drugs also improve glucose metabolism, including enhanced storage as glycogen. They lower risks of cardiac arrhythmia and heart attacks resulting from the effects of diabetes. The structures of these drugs are shown in Scheme 16.2.

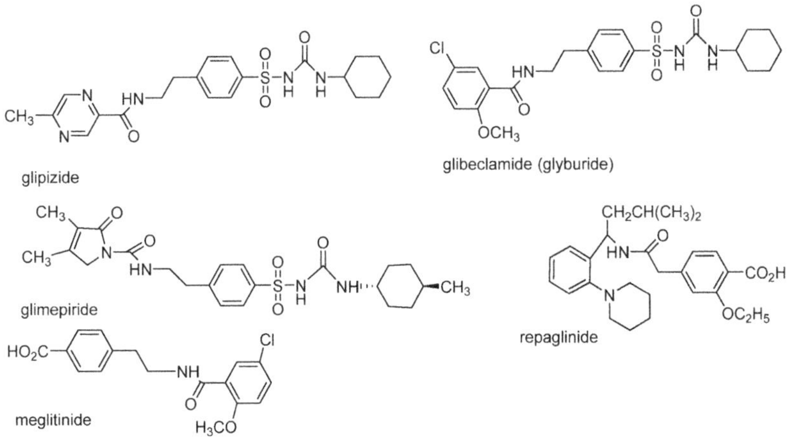

SCHEME 16.2   Second generation insulinotropic drugs.

16.4.2   BIGUANIDES

The origin of biguanides use in treatment of diabetes can be traced to use of extracts of French lilac, which contains the parent compound guanidine. Guanidine was used as a drug for treatment of diabetes in the early 1900s, prior to the discovery of insulin. The first synthetic guanidine to be used was phenformin, which was introduced in the United States in the 1950s. It was withdrawn from clinical use in the 1970s because it tended to produce lactic acidosis. It was replaced by metformin, first in Europe, and later (1995) in the United States. Metformin is currently (2010) the most widely used oral medication for type-2 diabetes. Other biguanides, such as buformin, have similar effects, but are little used clinically.

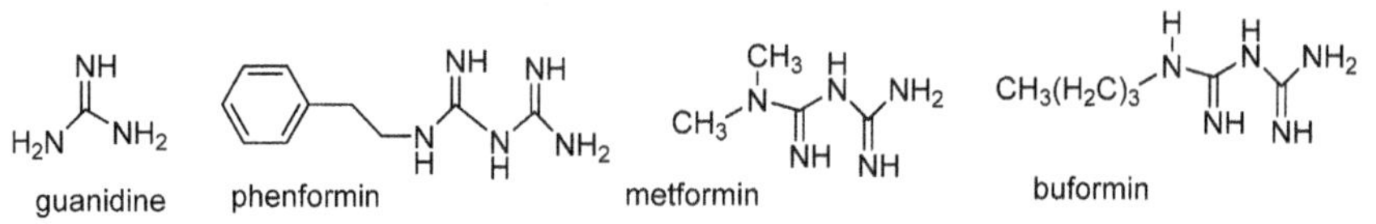

guanidine   phenformin   metformin   buformin

There is still some uncertainty about the mechanism of action of metformin. Metfomin suppresses glucose production in the liver. It also facilitates glucose uptake in muscle tissue. These effects may be the result of stimulation of AMPK, an important regulator of metabolism. Metformin also appears to have beneficial effects on endothelial function.

16.4.3   FIBRATES

The development of the fibrates followed discovery at ICI that some branched-chain fatty acids could reduce lipid and cholesterol levels. The first compound to be developed was clofibrate, which was approved in the United States in 1967, despite some concerns for liver toxicity. Other examples of this group are gemfibrozil, fenofibrate, bezafibrate, and ciprofibrate. Their structures are shown in Scheme 16.3.

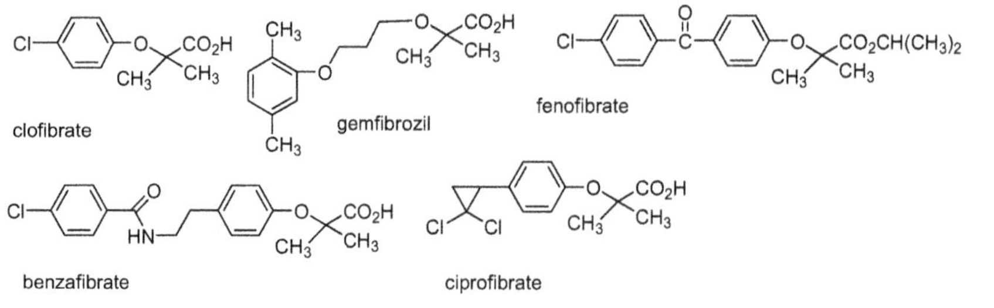

clofibrate   gemfibrozil   fenofibrate

benzafibrate   ciprofibrate

SCHEME 16.3   Fibrates.

The fibrates are believed to be activators of PPAR-α, one of a family of transcription factors that regulate lipid and carbohydrate metabolism. PPAR-α is particularly associated with the liver, muscle, and heart. The fibrates reduce triglyceride levels and modestly improve HDL/LDL ratios. They also appear to improve the resistance of the endothelium to inflammation and plaque formation.

16.4.4 α-GLUCOSIDASE INHIBITORS

The α-glucosidase inhibitors slow the breakdown of polysaccharides and thus decrease the rate of production of glucose and other monosaccharides. The first α-glucosidase to be recognized was acarbose a modified oligosaccharide isolated from *Actinoplanes* and made by fermentation. Another example is miglitol, which is available in the United States. These drugs appear to be relatively safe, but tend to have gastrointestinal side effects.

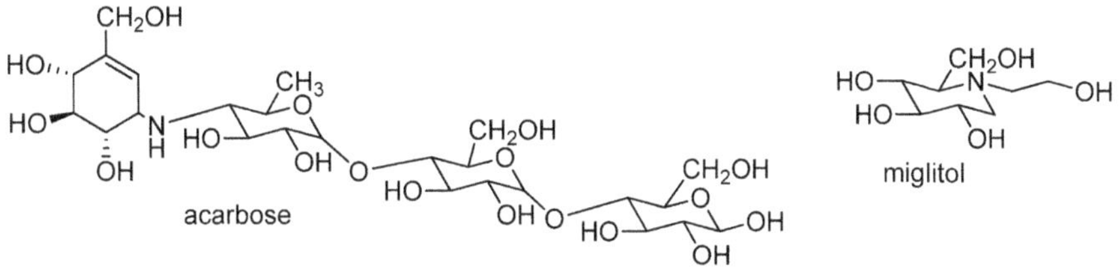

16.4.5 LIPASE INHIBITORS

Orlistat is an inhibitor of pancreatic lipase. It is produced by hydrogenation of lipstatin, which is produced by the bacterium, *Streptomyces toxytricini*.[10] It has been shown to promote weight loss, delay the progression to type-2 diabetes and reduce cardiovascular events in obese patients.

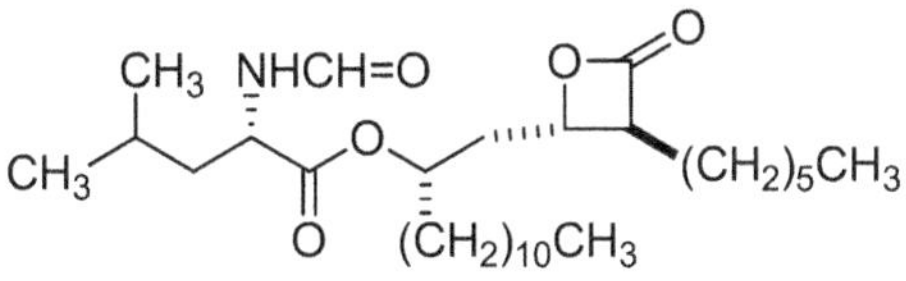

16.4.6 THIAZOLIDINEDIONES

The thiazolidindiones were first discovered in Japan in the 1980s, ciglitazone and troglitazone being the first examples. Another example, rosiglitazone was found at SmithKline in the United Kingdom in 1988. Several others, including pioglitazone, followed in the 1990s. The mechanism of action was found to involve activation of PPARs that regulate lipid and carbohydrate metabolism. The thiazolidinediones augment the effect of insulin on PPAR-γ, particularly in adipose tissue and counteract insulin resistance. Thiazolidinediones improve insulin sensitivity in several ways, including shifting the balance from muscle to adipose cells and by decreasing pro-insulin production levels. The thiazolidinediones also appear to reduce the generation of inflammatory mediators associated with atheroschlerosis and improve vascular tone. They also appear to have a protective effect on the β cells. Structures of the thiazolidinediones are given in Scheme 16.4. Several of these drugs were subsequently withdrawn, however, because of occasional but severe, liver toxicity. In the preapproval study, 1.9% of the troglitazone subjects had elevated (>3 time normal) alanine transaminase levels, an indicator of liver dysfunction, as opposed to 0.6% of the placebo group. In the group of 48 subjects showing high-transaminase levels, 25 had levels >10 times normal. Upon introduction of the drug, severe hepatotoxicity began to appear, including need for liver transplants and deaths. The drug was quickly withdrawn, as the rate of severe liver toxicity appeared to be in the range 1/8000–1/20,000 patients.[11] Rosiglitazone and pioglitazone remain in use, although rosiglitazone shows some evidence of increasing heart attacks.

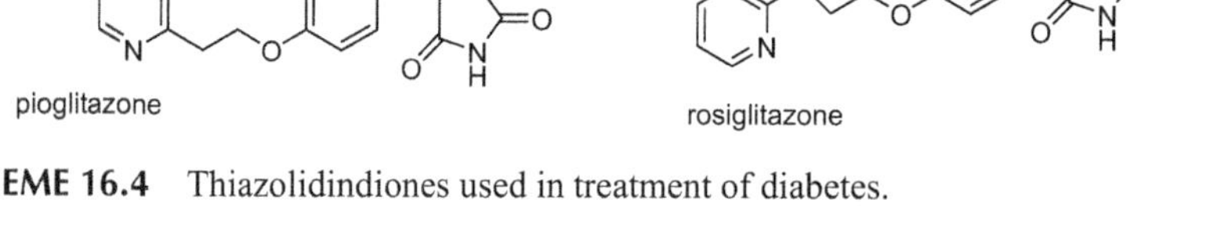

ciglitazone

troglitazone

pioglitazone

rosiglitazone

SCHEME 16.4   Thiazolidindiones used in treatment of diabetes.

Subsequent work has been aimed at compounds that might affect both PPAR-α and PPAR-γ receptors. The first examples of this class submitted for clinical studies were withdrawn because of toxic side effects.[12] Aleglitazar is an example of this group. Evaluation of this class of drugs is ongoing.[13]

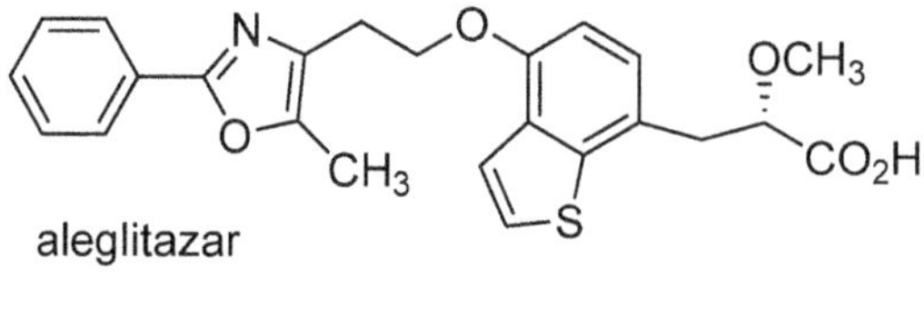

aleglitazar

16.4.7 DRUGS BASED ON THE INCRETIN EFFECT

There are two polypeptide, GLP_1 and GIP that are rapidly secreted after food is taken in and that stimulate insulin release, which is known as the *incretin effect*.[14] These hormones are produced in the digestive tract and respond more strongly to oral glucose than to glucose administered intravenously. Their action is required for normal glucose control. The hormones have short half-lives (minutes). They are degraded by an enzyme called *dipeptidyl peptidase-4* (DPP-4), which cleaves the two N-terminal amino acids. Both of the incretin polypeptides stimulate β-cell proliferation. The relationships between the incretin polypeptides and glucose control offer two mechanisms for therapeutic intervention. One is *incretin mimetics* that could exhibit the incretin effect. Two such molecules, exenatide and lirglutitide, have been approved for use in treatment of diabetes. Exenatide is a 39-AA polypeptide that was originally isolated from the saliva of a lizard, *Heloderma suspectum*. It is now produced synthetically. It must be administered by injection and has a relatively short half-life (2 h) but does bring about significant improvement in biomarkers for diabetes. Liraglutide is a 31-AA polypeptide modified by a C-16 fatty acid glutamyl group attached at an internal lysine. It has a longer half-life (13 h) than exenatide.

The other approach is to inhibit the DPP-4 peptidase, thus extending the life of the incretins. Drugs in this group include sitagliptin, vildagliptin, saxagliptin, and linagliptin. These drugs have shown promising results in extending the effectiveness the first-line drug metformin. Their structures are shown in Scheme 16.5.

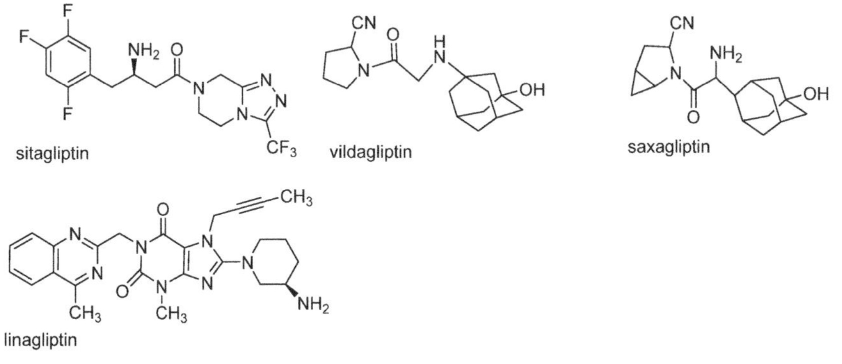

SCHEME 16.5 DPP-4 inhibitors.

Table 16.1 summarizes the characteristics of the various classes of drugs used in the treatment of diabetes. Figure 16.6 indicates the primary targets of action.

TABLE 16.1 Summary of Anti-Diabetic Drugs.

Class (date of introduction)	Examples	Mechanism of action	Advantages	Disadvantages
Insulinotropics (1946)	Glipizide, glibeclamide, repaglinide	Increase insulin secretion	Long-term safety, low cost	Can cause hypoglycemia and weight gain
Biguanides (1957)	Metformin	Decreases glucose output by liver	Long-term safety, low risk of hypoglycemia	Gastrointestinal side effects, lactic acidosis
α-Glucosidase inhibitors (1995)	Acarbose, miglitol	Inhibit carbohydrate absorption	Low risk of weight gain, low cost	Gastrointestinal side effects
Meglininides (1997)	Nateglinide, repaglinide	Increase insulin secretion	Rapid acting	Hypoglycemia and weight gain
Thiazolidinediones (1997)	Pioglitazone, rosiglitazone	PPAR agonists, increase insulin sensitivity	Low risk of hypoglycemia	Long-term safety not established
GLP-1 mimetics (2005)	Exenatide, liraglutide	Increases insulin secretion, appetite suppression	Weight loss, low risk of hypoglycemia	Long-term safety not established
DPP-4 inhibitors (2006)	Sitagliptin, vildaglipitin	Increases incretin concentration	Low risk of hypoglycemia	Long-term safety not established

Source: Adapted with permission from A. A. Tahrani, C. J. Bailey, S. Del Prato and A. H. Barrett, *The Lancet,* 378, 182 – 197 (2011).

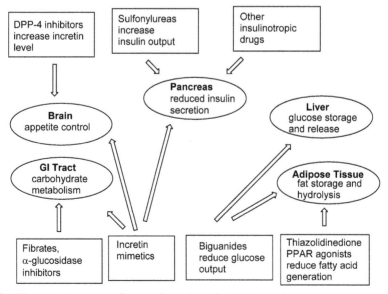

FIGURE 16.6 Summary of targets for action of antidiabetes drugs.

16.4.8   COMBINATIONS OF DRUGS

Because there are several points of intervention in the production and action of insulin, drug combinations that target different mechanisms are conceivable. Among the combinations that have shown evidence of clinical effectiveness are combinations of metformin with glyburide, sitagliptin[15], and with the thiazolidinediones.

KEYWORDS

- diabetes
- insulin
- sulfonylureas
- biguanides
- fibrates
- α-glucosidase and lipase inhibitors
- thiazolidinediones
- dipeptidyl peptidase inhibitors
- incretin

BIBLIOGRAPHY BY SECTIONS

Section 16.1: Bliss, M. *The Discovery of Insulin*; University of Chicago Press: Chicago, 1982; Rosenfeld, L. *Clin. Chem.* **2002,** *48*, 2270–2288.

Section 16.2: Pillai, O.; Panchagnula, R. *Drug Discov. Today* **2001,** *6*, 1056–1061; Brunton, S. *Am. J. Med.* **2008,** *121*, S35–S41.

Section 16.3: Chiasson, J.-L. *Expert Opin. Pharmacother.* **2007,** *8*, 3147–3158; Guilherme, A.; Vibasius, J. V.; Puri, V.; Czech, M. P. *Nat. Rev. Mol. Cell Biol.* **2008,** *9*, 367–377; Hager, G. R.; van Haeften, T. W.; Visseren, R. L. J. *Eur. Heart J.* **2008,** *29*, 2959–2971; Bakker, W.; Eringa, E. C.; Sipkema, P.; van Hinsbergh, V. W. M. *Cell Tissue Res.* **2009,** *335*, 165–189; Steinberg, G. R. *Appl. Physiol. Nutr. Metab.* **2009,** *34*, 315–322; Gustafson, B. *J. Atheroscler. Thromb.* **2010,** *17*, 332–341; Cusi, K. *Curr. Diabetes Res.* **2010,** *10*, 306–315; Nolan, C. J.; Damm, P.; Prentki, M. *Lancet* **2011,** *378*, 169–181; Hardie, D. G.; Ross, F. A.; Hawley, S. A. *Nat. Rev. Mol. Cell Biol.* **2012,** *13*, 251–262.

Section 16.3.2: Van Belle, T. L.; Coppieters, K. T.; von Herrath, M. G. *Physiol. Rev.* **2011,** *91*, 79–118.

Section 16.4: Krentz, A. J. *Expert Opin. Drug Saf.* **2006,** *5*, 827–834; Tahrani, A. A.; Bailey, C. J.; Del Prato, S.; Barnett, A. H. *Lancet* **2011,** *378*, 182–197.

Section 16.4.1: Dornhorst, A. *Lancet* **2001**, *358*, 1709–1716.

Section 16.4.2: Goodarzi, M. O.; Bryer-Ash, M. *Diabetes, Obes. Metab.* **2005**, *7*, 654–665; Stumvoll, M.; Haring, H.-U.; Matthael, S. *Endocr. Res.* **2007**, *32*, 39–57.

Section 16.4.3: Lalloyer, F.; Staels, B. *Arterioscler., Thromb., Vasc. Biol.* **2010**, *30*, 894–899.

Section 16.4.5: Guerciolini, R. *Int. J. Obes.* **1997**, *21*(Suppl. 3), S12–S23.

Section 16.4.6: Del Prato, S.; Marchetti, P. *Diabetes Technol. Ther.* **2004**, *6*, 719–731; Papaetis, G. S.; Orphanidou, D.; Panagiotou, T. N. *Curr. Drug Targets* **2011**, *12*, 1498–1512.

Section 16.4.7: Roges, O. A.; Baron, M.; Philis-Tsimikas, A. *Expert Opin. Invest. Drugs* **2005**, *14*, 705–727; Deacon, C. F. *Diabetes, Obes. Metab.* **2010**, *13*, 7–18; Gallwitz, B.; Haring, H.-U. *Diabetes, Obes. Metab.* **2011**, *12*, 1–11.

Section 16.4.8: Keskemeti, V.; Bagi, Z.; Pacher, P.; Pas, I.; Kocsis, E.; Koltai, M. Z. *Curr. Med. Chem.* **2002**, *9*, 53–71; DeFronzo, R. A. *Am. J. Med.* **2010**, *123*, S48–S58; Zinman, B. *Am. J. Med.* **2011**, *124*, 519–534.

REFERENCES

1. Vijayan, M. *Curr. Sci.* **2002**, *83*, 1598–1606.

2. Alsaleh, F. M.; Smith, F. J.; Keady, S.; Taylor, K. M. G. *J. Clin. Pharm. Ther.* **2010**, *35*, 127–138.

3. Bhatnagar, S.; Srivastava, D.; Jayadev, M. S. K.; Dubey, A. K. *Prog. Biophys. Mol. Biol.* **2006**, *91*, 199–228.

4. Kirkman, M. S.; Kendall, D. M. *Clin. Chem.* **2011**, *57*, 255–257; Malkani, S.; Mordes, J. P. *Am. J. Med.* **2011**, *124*, 395–401; Cohen, R. M.; Haggerty, S.; Herman, W. H. *J. Clin. Endocr. Metab.* **2011**, *95*, 5230–5206.

5. Sherry, N. A.; Tsai, E. B.; Herold, K. C. *Diabetes* **2005**, *54*(Suppl 2), S32–S39.

6. Wyne, K. L. *Am. J. Med.* **2003**, *115*, 29S–36S.

7. Neve, B. P.; Fruchart, J.-C.; Staels, B. *Biochem. Pharmacol.* **2000**, *60*, 1245–1250; Barbier, O.; Pineda Torra, I.; Duguay, Y.; Blanquart, C.; Fruchart, J.-C.; Glineur, C.; Staels, B. *Arterioscler., Thromb. Vasc. Biol.* **2002**, *22*, 717–726.

8. Aquilante, C. L. *Expert Rev. Cardiovasc. Ther.* **2010**, *8*, 359–372.

9. Schwartz, T. B.; Meinert, C. C. *Perpect. Biol. Med.* **2004**, *47*, 564–574.

10. Guerciolini, R. *Int. J. Obes. Related Metab. Disorders* **1997**, *21*(Suppl. 3), S12–S23.

11. Gale, E. A. M. *The Lancet* **2001**, *357*, 1870–1875.

12. Heald, M.; Cawthorne, M. A. *Handb. Exp. Pharmacol.* **2011**, *203*, 35–51.

13. Cavender, M. A.; Lincoff, A. M. *Am. J. Cardiovasc. Drugs* **2010**, *10*, 209–216.

14. Hare, K. J.; Knop, F. K. *Vitam. Horm.* **2010**, *84*, 389–413.

15. Seyoum, B. *Expert Opin. Pharmacother.* **2011**, *12*, 641–645.

CHAPTER 17

DRUGS FOR TREATMENT OF NEUROLOGICAL AND PSYCHOLOGICAL CONDITIONS

ABSTRACT

The first synthetic drugs to be used in treatment of neurological and psychological conditions were the barbiturates introduced in the early 1900s which acted as sedatives and hypnotics. Observation of the antiepileptic effects of phenobarbital led to the discovery of other antiseizure drugs, especially phenytoin. A number of other antiepileptic drugs have subsequently been introduced. The first drug with antipsychotic activity, chlorpromazine, was recognized in France in 1952. It was followed by a number of others, including chlozapine. The antianxiety benzodiazepines were discovered in 1955. Antidepressants include monamine oxidase inhibitors, tricyclic, and selective uptake inhibitors. The amphetamines, beginning with benzidrine in the 1930s, have had both medical and illicit use. Several have been used as appetite suppressants for weight-loss, but this use has been restricted because of side effects. Methyl phenidate, used in treatment attention deficit hyperactivity disorder (ADHD), is also a member of the amphetamine class. Both the discovery and evaluation of psychologically active drugs are complicated by the absence of specific bioanalytical criteria. The diagnosis and evaluation of psychological disorders are based on criteria defined by the American Psychiatric Association Diagnostics and Statistical Manual, the most recent edition of which was published in 2013.

The brain determines if we have a sense of well-being or conversely, anxiety, or depression. By adjusting the biochemical balance in the nervous system, drugs can modify these perceptions. Mental illness involving hallucinations and disconnection from reality can sometimes also be treated with drugs. In this chapter, we will consider how drugs can influence our emotional and mental status. There are other drugs that can treat conditions such as epilepsy or dementia that also have their origin in the brain and nervous system. Discovery and development of drugs for treatment of neurological and psychological conditions has been especially challenging, and in many cases advances have come from serendipitous observations. One reason for this is that there are usually not specific well-defined biological abnormalities that can be measured. Animal studies, which must be relied on to find and test compounds, also have limitations. For example, most of the tests for antidepressant drugs rely on chronic stress of rodents that can capture only some aspects of the human situation. They may not model the effect of stress or trauma during development that may correlate with later psychological and mental disorders in humans.

By the middle of the century, the role of *neurotransmitters* was beginning to be understood. The role of acetyl choline (ACh) (see Section 8.3.2) and adrenaline (see Section 14.3) were fairly well known. In the 1950s and 1960s connections were made between mental state and the neurotransmitters dopamine and serotonin. The role of amino acid neurotransmitters, including glutamate and γ-aminobutyric acid (GABA) were also recognized. These discoveries began to provide a pharmacological basis for characterizing the relationship between drugs and mental state and also permitted development of assays that might identify drug candidates. Scheme 17.1 shows the structure of the major neurotransmitters.

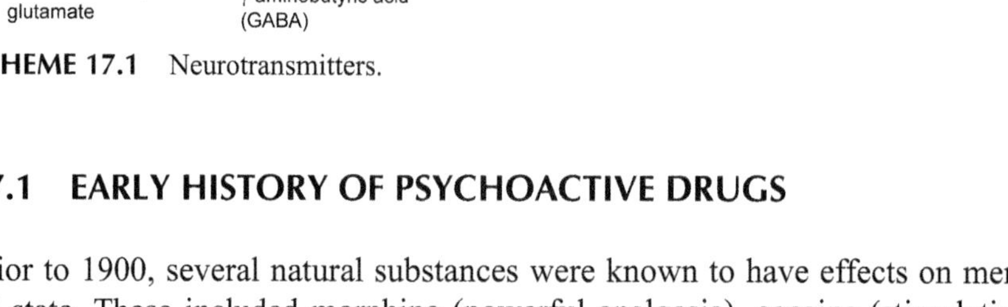

$(CH_3)_3N^+CH_2CH_2O_2CCH_3$

acetyl choline

R = CH$_3$ = epinephrine, (adrenaline)
R = H = norepinephrine, (noradrenaline)

dopamine

serotonin
5-hydroxytryptamine

$H_3N^+(CH_2)_2CHCO_2^-$
NH_3^+
glutamate

$H_3N^+(CH_2)_3CO_2^-$
γ-aminobutyric acid
(GABA)

SCHEME 17.1   Neurotransmitters.

17.1   EARLY HISTORY OF PSYCHOACTIVE DRUGS

Prior to 1900, several natural substances were known to have effects on mental state. These included morphine (powerful analgesia), cocaine (stimulation and euphoria), scopolamine (relaxant), caffeine (mild stimulant), and mescaline (hallucinogenic). A few synthetic chemicals were also in use primarily as sedatives, the most prevalent being bromide salts, paraldehyde and chloral hydrate. The first new group of synthetic drugs was the barbiturate, which were synthesized by Emil Fischer in 1903. Barbiturates are powerful hypnotics and sedatives, but with serious potential for overdose. The first synthetic drugs to exhibit effects on a psychotic state came out of studies on anti-histamines synthesized at Rhone-Poulenc in France shortly after WWII. Two lines of studies evolved around one of these compounds, chlorpromazine. Two French military physicians, Henri Laborit and Pierre Huguenard, used chlorpromazine in combination with the synthetic analgesic meperidine (see Section 13.6) in treatment of surgical and traumatic shock in French soldiers returning from Viet Nam. They noted an unusual "calming" effect. Jean Delay and Pierre Deniker at Sainte-Anne Hospital in Paris used chlorpromazine alone in treatment of manic patients. Manic and schizophrenic patient showed marked improvement. The results of Laborit/Huguenard and Delay/Deniker were both published in 1952.[a] Chlorpromazine also caused hypotension, antiadrenergic effects and temperature dysregulation. The drug received FDA approval in the United States in 1954 and was marketed by Smith, Kline French as Thorazine.

At about the same time considerable attention was focused on a natural product, reserpine. Extracts of *Rauwolfia serpentina* had been used in Indian traditional medicine for many years. Mahatma Gandhi is said to have used a tea made from *R. serpentina* as an aid to meditation. The structure of the active ingredient, reserpine, was established at Ciba in 1952. Clinical studies showed it had antiadrenergic and sedative effects, lowered body temperature, and reduced blood pressure. This pharmacological profile is very similar to that of chlorpromazine. Reserpine was used for a time in treatment of hypertension. It was not as effective as chlorpromazine in treatment of psychoses, but its effects were used in many pharmacological studies. In particular, it was shown that reserpine lowered the level of the neurotransmitters noradrenaline, dopamine, and serotonin in the brain. These results suggested that the function of chlorpromazine and reserpine might be to modify the activity of these monoamines in the brain and central nervous system.

chlorpromazine   reserpine

The crucial role of adrenaline, nor-adrenaline, dopamine, serotonin, and GABA as neurotransmitters made them appealing targets in the search for other drugs for treatment of neurological and psychological disorders. The general

[a]Henri Laborit was trained as a surgeon and was interested in creating a state of "artificial hibernation" to lower body temperature and minimize the danger of stock in surgery. His work in this area was in collaboration with Pierre Huguenard. Laborit also worked on the biochemistry of γ-hydroxy- and γ-amino-butyric acid and their role in stress and aggression. After receiving medical training, Jean Delay completed a Ph. D. in 1942 on memory disorders. He practiced psychiatry at Sainte-Anne Hospital in Paris. It was there that he and his colleague Pierre Deniker noted the effect of chlorpromazine on mental patients. Delay was also a novelist, biographer, and literary critic.

hypothesis that developed was that the levels of the neurotransmitters affected mood and behavior. In particular, deficiencies in the monoamine neurotransmitters were associated with depression. The observation that an antituberculosis drug, iproniazid had antidepressant effects, based on inhibition of the enzyme *monoamine oxidase* (MAO), led to one group of antidepressants called *monoamine oxidase inhibitors* (MAOI). Other drugs that had no MAOI activity were found to have antidepressant activity, such as imipramine. Its activity was found to be associated with the inhibition of uptake of noradrenaline at synapses. This led to discovery of other drugs that were both non-selective and selective among the neurotransmitters. They are classified as *monoamine uptake inhibitors*.

iproniazid   imipramine

During 1960s and early 1970s, investigations of the effects of resperpine and the hallucinogen lysergic acid diethylamide (LSD) focused attention on serotonin. LSD was found to block serotonin receptors. Reserpine was found to trigger depression in some patients when used as an antihypertensive agent. It also decreased the amount of serotonin at synapses, suggesting that serotonin levels might play a role in depression. Evidence for the involvement of dopamine also developed. These initial hypotheses and interpretations have undergone considerable evolution as the process of neurotransmission has become understood in more detail. In particular, it is now recognized that there are several different receptors for each of the neurotransmitters and that they can have different effects. Thus, targeting drugs to the specific receptor responsible for the desired effect can improve activity and minimize side-effects.

The antidepressant and antipsychotic drugs had a major impact on treatment of depression and other psychological conditions. At the time of their introduction, however, they were not universally welcomed by the psychoanalytical community. They were viewed by some as artificial treatments that might delay the patient's recognition of and response to the underlying stress and/or trauma that caused the condition. The effectiveness of the drugs became evident, however, and opened the area of neurochemistry to further research and it became clear that there were biochemical aspects to several of the major forms of mental illness. Drugs were also discovered for other types of psychological conditions. Lithium salts were introduced for treatment of manic/depressive disorder in 1949. The *anxiolytic drugs*, initially meprobamate, but more importantly, the benzodiazepines, were introduced in the 1960s. More recently, attention has been directed toward potential treatment of Alzheimer's disease (AD).

17.2   HYPNOTICS

The first major group of synthetic psychoactive drugs was the *barbiturates*. The parent compound, barbituric acid, was synthesized by J. F. W. Adolf von Bayer in 1863. According to legend, the name is derived from the fact that it was synthesized on the feast day of St. Barbara. The biological activity is associated with alkylated derivatives. Their calming and sleep-inducing effects were discovered by Joseph von Mering and Emil Fischer in 1903. The first compound to be commercialized was the diethyl derivative known as barbital or veronal. They were initially licensed to E. Merck but Bayer gained a share of the patent rights by litigation. Barbital quickly received acceptance both by physicians and patients. At the time, prior regulatory approval was not required. The passage of the Harrison Narcotics Act in the United States in 1914, which restricted access to opiates, led to expanded use of the barbiturates. Barbital was followed by several new and improved barbiturates, including phenobarbital, amobarbital, and thiopental. It was not long, however, before the problems of side effects, addiction, and self-medication arose. Side effects included hangovers, slurred speech, dizziness, and suicidal tendencies. These came to public attention in the 1920s, accentuated by examples of accidental over-dose and suicide among celebrities. The barbiturates were made prescription-only drugs in the 1930s. Use continued to expand however. In the immediate post-WWII period, combinations with amphetamines were used. By 1950, over 50 different barbiturates were available. There was considerable "underground" availability of the drugs, diverted from the legitimate supply channels. Further restrictions were imposed in the United Kingdom in 1968 and in the United States in 1971. Scheme 17.2 gives some of the structures.

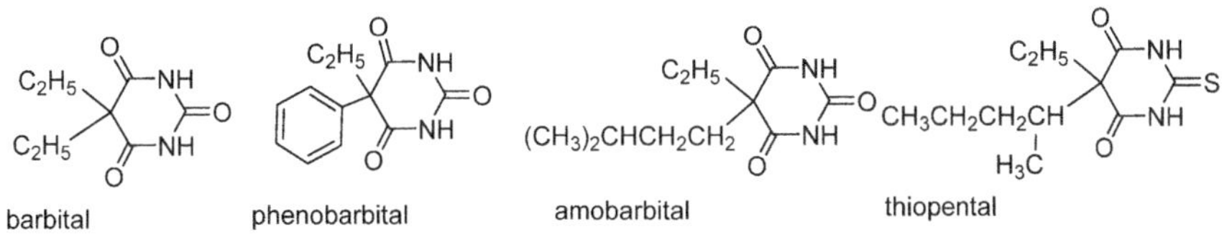

SCHEME 17.2   Barbiturate hypnotics.

The barbiturates act as modulators of GABA receptors and increase the inhibitory effects of GABA. The effects are synergistic with ethanol (alcohol) and the two substances mutually enhance their toxicity. Another group of barbiturates is used as intravenous anesthetics. The most important of these are thiopentone and methohexitone, used as their sodium salts.

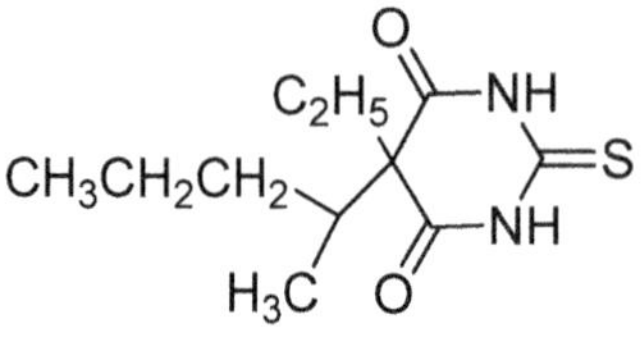

thiopentone 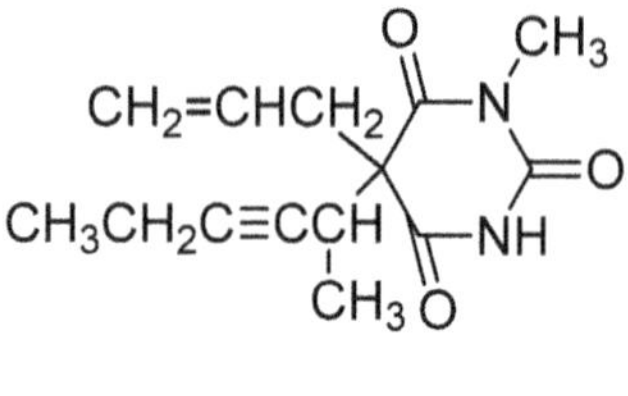   methohexitone

Another group of hypnotics is represented by zolpidem (Ambien) and zaleplon (Sonata). They are relatively short-acting sleep inducers that are prescribed for insomnia. These also target GABA receptors. Zolpidem and zaleplon are both schedule-IV controlled substances in the United States (see Section 11.4.1.8).

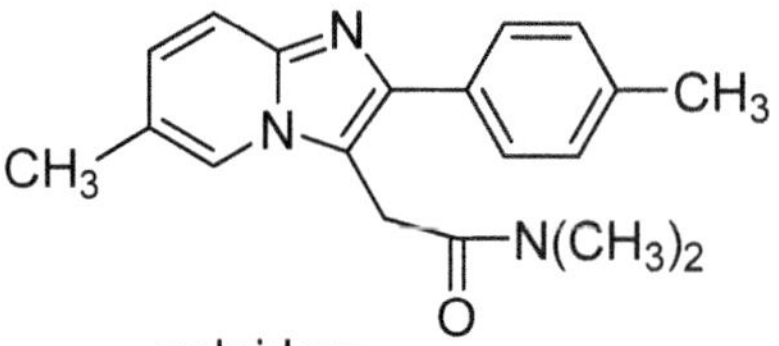

zolpidem 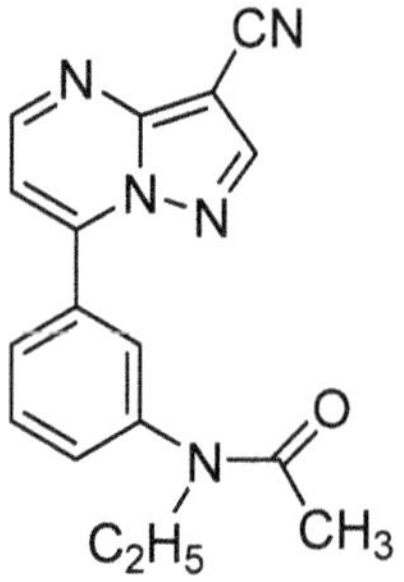   zaleplon

17.3 ANTIEPILEPSY DRUGS

Epilepsy is the result of repetitive and synchronized firing of neurons. It results in abnormal patterns in electroencephalograms. There are several types of seizures, the most common being *tonic–clonic*, which result in loss of consciousness and convulsions, and *absence seizures*, which cause brief periods of complete detachment. These were known, respectively, as "grand mal" and "petit mal" seizures in the 19th century. Some forms of epilepsy are inherited and quite a number of genes that can contribute have been identified. Epilepsy can also arise as the result of damage to the brain by trauma, tumors, or stroke.

At the beginning of the 20th century, many epileptics were institutionalized, the thought being that a controlled environment presented the best opportunity for forestalling progression of the condition. Institutional care typically included dietary and exercise regimens. The main medication used was bromide salts. Bromide salts had been introduced in the 1860s and were used extensively through the 1920s. About half the patients appeared to improve in terms of frequency of seizures, but there was no reliable information on effectiveness or on optimal dosage. Some formulation contained other materials, such as antimony arsenate. The bromides had many drawbacks. Their effect is depressive, and they induce lethargy and have side effects including digestive problems, acne, and other skin conditions. Bromide remedies were used for other ailments and were available over the counter.[1]

Phenobarbital was recognized as an antiepilepsy drug in Germany in 1912. Its use as a hypnotic in epilepsy wards was observed to reduce seizures by Alfred Hauptmann. Hauptmann followed up his observation with a careful investigation noting the frequency and severity of seizures. After WWI, phenobarbital was also adopted in the United Kingdom and United States. It was not until the 1960s that controlled studies on efficacy in comparison with other antiepileptic drugs were done. In the meantime, other drugs, particularly phenytoin, carbamazepine and sodium valoproate were introduced. In general, they all have been found to be of comparable benefit. Phenobarbital remains the drug of choice in the less developed parts of the world because it is very inexpensive.

Diphenylhydantoin (phenytoin) was introduced in 1938. It was discovered by Tracy Putnam and Houston Merritt. Putnam and Merritt were physicians who had devised a method for inducing convulsion in test animals by rapidly cycling electric currents. Phenytoin had been synthesized at Parke–Davis in 1923 during a search for hypnotics. It did not have hypnotic activity and so was not pursued further. By the 1930s, the general conclusion had been reached that the anticonvulsant and hypnotic activity were inextricably linked. Putnam undertook a search for compounds to test as anticonvulsants using the electrically induced convulsions as the screen. He obtained a sample of phenytoin from Parke–Davis and when it proved active in the assay, Merritt tested it in patients and found it effective.[2] Parke–Davis introduced it in the form of the sodium salt, Dilantin. Putnam and Merritt eventually screened several hundred compounds. The N-methyl derivative of 5-ethyl-5-phenylhydantoin (mephenytoin) and its demethylated metabolite (nirvanol) showed the best activity. In 1958, Parke–Davis introduced another drug ethosuximide. It is particularly effective for absence seizures but not tonic–clonic ones. Several of the antiepilepsy drugs have the property of inducing the P-450 type oxidative enzymes and thus can interfere with other types of drugs.

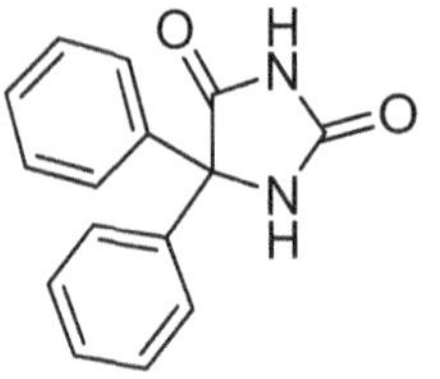

phenytoin

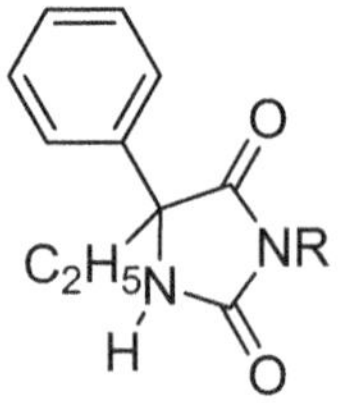

mephenytoin R = CH3
nirvanol R = H

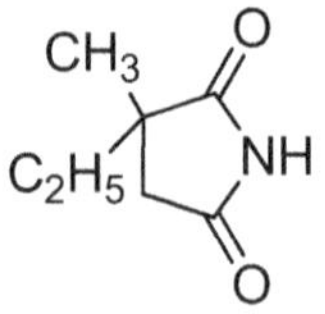

ethosuximide

In the 1960s, two other drugs that are currently recommended as first line treatments were found. These are carbamazepine and valproate. Carbamazepine was developed by Geigy and introduced in Europe in the early 1960s. It was not approved in the United States until 1974. Sodium valproate was discovered by a small Swiss pharmaceutical company when the corresponding acid was used as the solvent for a testing a series of other compounds. When all showed positive effects, the researchers suspected that valproic acid was the active constituent. This proved to be the case. The development of the drug was undertaken by Sanofi and it was introduced in Europe in the 1970s. Its structure is remarkably simple compared to most other antiepileptic drugs. Both carbamazepine and valproate have significant side effects. Carbamazepine is teratogenic and has a rate of liver failure of about 16/100,000. Valproate can lead to obesity, is teratogenic and has liver toxicity, especially in children. Nevertheless, both are among the most used antiepileptic drugs at the present time.

carbamazepine   sodium valproate

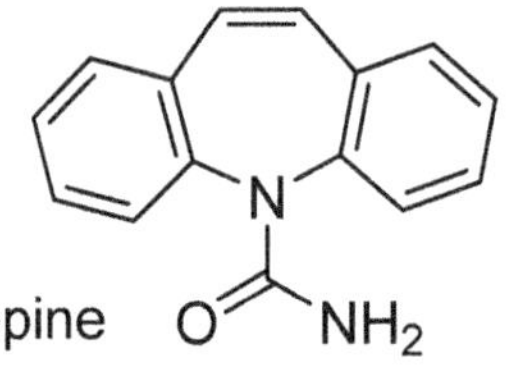

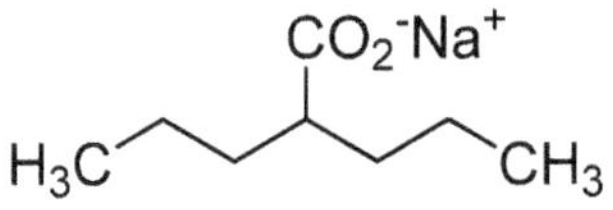

There was relative little research and development activity in antiepileptics between 1950 and 1975. In 1974, the NIH established an anticonvulsion drug screening program. Various animal models are used for preliminary screening of potential drugs for epilepsy. They include induction of convulsion by electric shock, chemical treatment, or cyclic electrostimulation. Research activity also increased in the pharmaceutical industry. As a result many new antiepileptic drugs have been introduced. Ten new drugs have been approved in the United States since 1994, including felbamate, levetiracetam, lacosamide, gabapentin, topiramate, tiagabine, zonisamide, and retigabine.[3] As with other therapeutic areas, the cost of individual drugs has increased substantially. A principal benefit of the new drugs is the prospect of improved results if the older drugs are unsuccessful. Among the most useful of the new drugs have been oxcarbamazepine, a close structural analog of carbamazepine, and lamotriginine. Despite its close structural similarity to carbamazepine, the oxo-derivative has distinct and evidently advantageous pharmacology. While carbamazepine is metabolized oxidatively to the corresponding epoxide and dihydroxy derivatives, oxcarbamazepine is metabolized by reduction to the hydroxyl derivative. These differences lead to observable differences between the two drugs, with oxcarbamazepine being generally better tolerated and with fewer side effects.[4] The structures of some of the newer antiseizure drugs are given in Scheme 17.3.

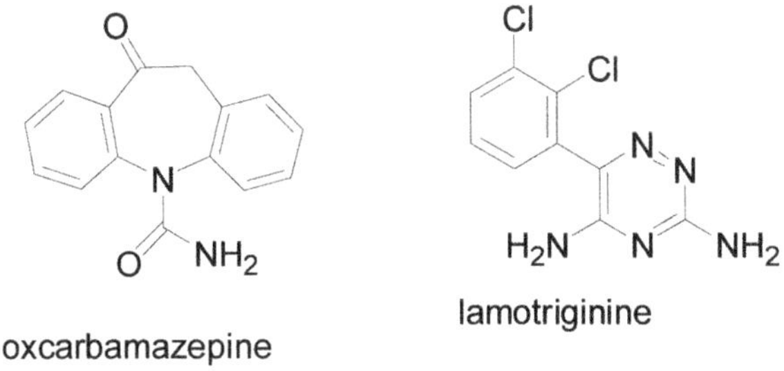

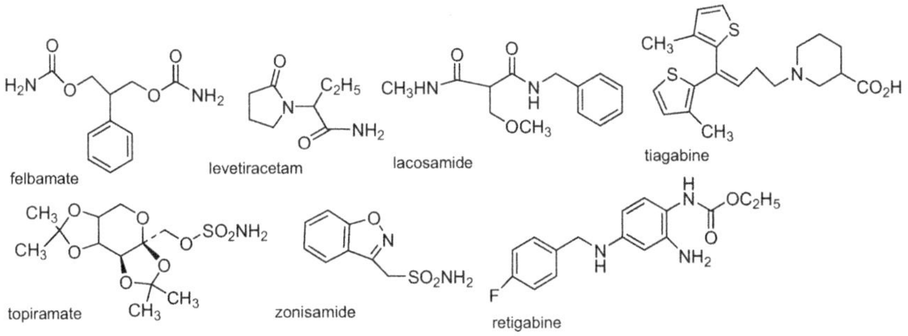

SCHEME 17.3 Examples of antiseizure drugs.

Only about half of epilepsy patients respond to initial drug treatment, and about 20% can be successfully treated with a second or a combination of drugs. The remaining 30% see little improvement with any drugs or combinations. It is thought that this is the result of epilepsy having multifaceted causes, including trauma, inflammation or specific genetic abnormalities. Most forms of epilepsy are associated with dysfunction of *voltage-gated sodium channels* (VGSC). The VGSC allow inward migration of Na^+ on depolarization and are followed by rapid closure ("fast inactivation") in the time frame of a millisecond. There is also a "slow inactivation" process that is in the range of a few seconds. Phenytoin and carbazepine bind to the inactive form of VGSC and stabilize it and prevent the rapid repetitive firing associated with seizures. There is also evidence that Ca^{2+} and K^+ ion channels can be involved in epilepsy. Lamotrigine has effects not only on Na^+ but also on Ca^{2+} channels. To some extent, the site of action of the drugs correlates with their therapeutic effects. One of the newer antiepileptic drugs, lacosamide, appears to promote "slow inactivation" of the VGSC locus.[5] The VGSC are also involved in other neurological conditions such as migraine headaches and neuropathic pain, and as a result, these drugs are used to treat some of these conditions.

Sodium valproate has a wide spectrum of activity, but its mechanism of action is uncertain. It may, at least in part, act by slowing the metabolism of the neurotransmitter GABA. Two of the more recently introduced drugs, vigabatrin and tiagabine appear to act by increasing levels of GABA. Gabapentin, pregabalin, and vigibatrin are structurally similar to one another and are thought to act to increase GABA levels by effects on enzymes and transporters. Gabapentin and pregabalin also interact with voltage-dependent Ca^{2+} channels. Pregabalin was approved for epilepsy in 2004. It has become a very successful drug and is also used in treatment of neuropathic pain and fibromyalgia (as Lyrica). Vigabatrin was withdrawn as a result of causing distortions of visual fields.

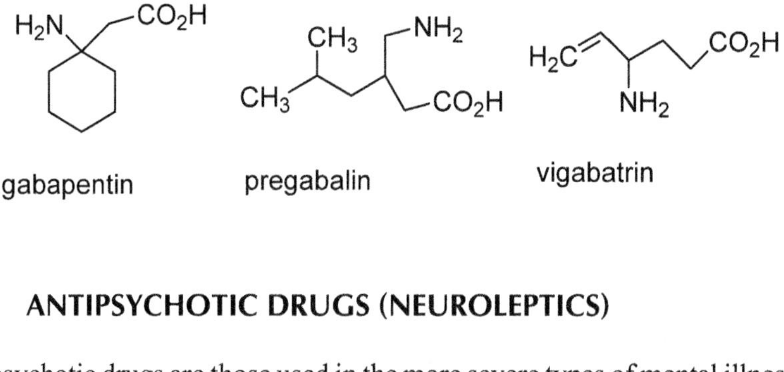

gabapentin   pregabalin   vigabatrin

17.4   ANTIPSYCHOTIC DRUGS (NEUROLEPTICS)

Antipsychotic drugs are those used in the more severe types of mental illness, including schizophrenia, manic-depressive bipolar disorder, psychotic depression, and some types of dementia. The term psychotic implies disordered thoughts and behavior and dissociation from reality, such as delusions and hallucinations.

17.4.1   FIRST GENERATION ANTIPSYCHOTIC DRUGS

The first antipsychotic drugs chlorproamazine and reserpine were found to have a calming effect on agitated mental patients, variously described as inducing "imperturbability," "equanimity," or "indifference," but without the hypnotic effects of the barbiturates (see Section 17.1). The term *tranquilizer* arose to describe these effects. The term was later subdivided with *major tranquilizer* referring to the antipsychotic drugs and *minor tranquilizers* to the antianxiety drugs. The term *neuroleptic* was also introduced for the antipsychotic drugs.

An important early antipsychotic drug was haloperidol. Haloperidol was synthesized at the Janssen Pharmaceutical Laboratories in Belgium in 1958. One of Janssen's main products at the time was an analgesic, dextromoramide, which is related to methadone. The original idea was to make derivatives of the synthetic morphine analog, meperidine (see Section 13.6). The first compounds synthesized were Mannich base derivatives, which proved to be a hundred times more active as analgesics. The chain was then extended by one carbon. This resulted in analgesia followed by a tranquilizing effect. Replacement of the ester group at C-4 of the piperidine ring by a hydroxyl group eliminated the analgesia, but retained the tranquilizing activity. Adding the halogens gave haloperidol. When it was tested in mice, it had little analgesic activity, but instead generated transient excitement, followed by a calming effect.

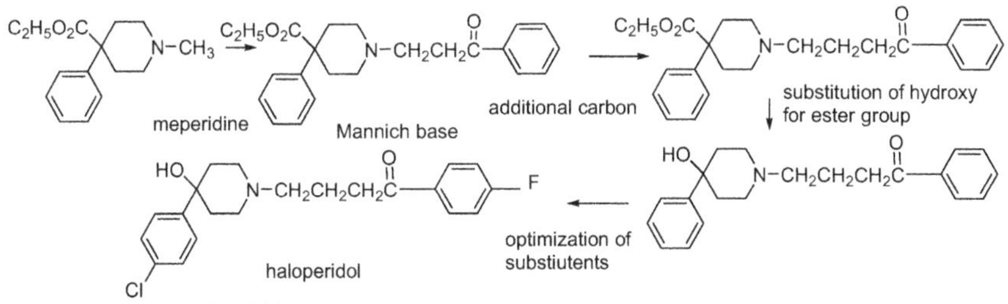

The synthetic route to haloperidol

The first trial in a human was conducted within weeks of the synthesis and animal results. A patient suffering from agitation and hypermoticity was treated and showed immediate improvement. The next results with 18 patients indicated that haloperidol was a powerful sedative without hypnotic effects, but Parkinson's-like side effects were also noted. Several groups in Europe soon reported follow-up studies including Delay and Deniker, who had first studied the effects of chlorpromazine (see Section 17.1). Most found the drug very effective in treating agitation, delusions, and hallucinations. Initial studies in the United States were less successful, but in cooperation with the McNeill Laboratories subsidiary of Johnson & Johnson, the compound was eventually successfully approved and distributed.

The Janssen Laboratory went on to develop several other antipsychotics. For example, the decanoate ester of haloperidol was introduced in 1981 as a slow release version of the drug. Another analog, pipamperone, exhibited a rather different pharmacological profile, which was later attributed to effecting both dopamine and serotonin receptors. Following this lead, risperidone was synthesized in 1984 and introduced in 1993. Its hydroxyl derivative, paliperidone, was introduced in 2007.[6]

pipamperone

risperidone X = H
paliperidone X = OH

Pharmacological interpretation of the effect of the antipsychotic drugs became available in the 1960s. The identification of serotonin as a vasoconstrictor and neurotransmitter had occurred in the mid-1950s. The effect of LSD in antagonizing serotonin was discovered. B. B. Brodie and his coworkers at the NIH showed that serotonin potentiated barbiturate-induced sleep and that this effect was blocked by the hallucinogen LSD. Chlorpromazine was also known to potentiate barbiturate-induced sleep. Reserpine was shown to have a long-lasting effect of releasing serotonin from its storage sites in the nervous system. These were the first studies that demonstrated a specific connection between a biochemical factor and mental disturbance. These observations led to explicit proposals that serotonin was implicated in mental dysfunction.

Dopamine, too, received attention. In particular, Arvid Carlsson in Sweden demonstrated that reserpine also depleted dopamine levels. The role of the amphetamines stimulants could also be interpreted in terms of their effect on dopamine. Unlike reserpine, chlorpromazine does not deplete dopamine levels. It was shown that, instead, it blocks dopamine receptors. This was first indicated in the 1960s and directly demonstrated in the 1970s. The explicit proposal that dopamine was a major factor in the etiology of schizophrenia was put forward by Jaques M. van Rossum in 1966. At about the same time, it was proposed that the catecholamines played a critical role in depression.[7] Other studies showed that serotonin and its receptors also had an important role in mental status. While many details remained to be clarified, it was becoming apparent that the antipsychotic drugs acted through effects on the monoamine neurotransmitters. In the 1970s, it became possible to identify receptors for dopamine and the other monoamines. Dopamine receptors are associated with the "reward" and "motivation" aspects of behavior and dopamine is considered to mediate the connection between experience and response. If this connection is dysfunctional, the response is inappropriate and delusional. Excessive dopamine release and receptor response is associated with psychoses. To the extent that antipsychotic drugs can correct the function of dopamine and the receptors, the patient's condition is improved.

A common side effect of the antipsychotic drugs is motor defects, somewhat similar to those observed in Parkinson's disease. These are called *extrapyramidal symptoms*, referring to their occurrence in the peripheral (extrapyramidal) parts of the nervous system. In fact, these symptoms were so common that they came to be thought of as inseparable from antipsychotic activity. The antipsychotic drugs also frequently exhibited what were called negative effects, including cognitive deficits and social withdrawal.

The discovery of the antipsychotic drugs had immediate effects on treatment of the severely mentally ill. Previously, long-term institutionalization had been the norm. The antipsychotic drugs permitted many patients to be treated on an out-patient basis and created new demands for rehabilitation and community-based care.[8] The availability of the antipsychotic drugs also provided alternatives to some of the more drastic treatments, which included electro-convulsive shock, insulin-induced coma, drug-induced convulsions, and surgery.[9]

17.4.2 ATYPICAL OR SECOND GENERATION ANTIPSYCHOTIC DRUGS

The first drug to fall into this category was clozapine. Clozapine was first synthesized in 1958 at a small pharmaceutical firm in Switerzland. Patient trials indicated it was effective in treating schizophrenia and did not have the neurological side-effects of chlorpromazine and the other existing antipsychotics. Sandoz acquired rights to the drug in 1967 and began marketing in Europe in 1972. Sandoz also began efforts to gain approval in the United States. However, in 1975, a cluster of fatal cases of agranulocytosis (loss of white blood cells) was reported in Finland. This prompted a review of previous cases and comparison with chlorpromazine, which indicated that the side-effect was observed for both drugs and in the range of 1–2%, significant but much less than found in the Finnish study. The reason for the high incidence in the Finnish study was never found. Sandoz terminated further development of the drug in the United States. However, it remained available in Europe and could be obtained in the United States under the provisions of the "compassionate need" criteria, when no other drug was suitable. As a result of this continued usage, clozapine developed a reputation of being a very effective drug, especially in the most difficult cases.

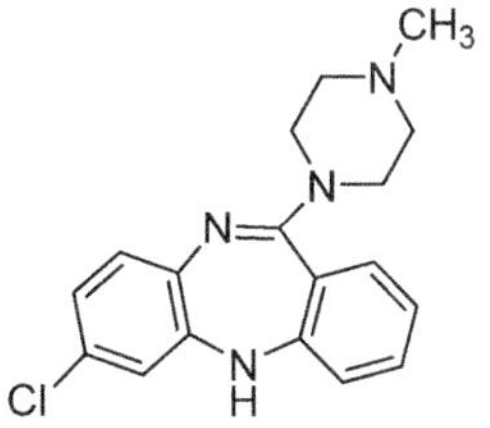

clozapine

The early 1980s brought change to FDA (see Section 11.4.1) and there was increasing concern in both the federal and state governments of the very high cost of long-term care for the mentally ill. Sandoz decided to reevaluate the drug. Clinical studies began in 1984 and were highly successful. To address the issue of agranulocytosis, Sandoz required for routine blood tests as a condition of prescriptions. FDA approval was received in September 1989. The drug was introduced in 1990. The cost, about half of which was due to the blood-testing requirements, was about $8900 per patient-year. Neither federal nor private insurance covered the blood-testing component so the drug was very expensive to patients. Several states brought antitrust suits against Sandoz. This resulted in hearings by a senate committee chaired by Howard Metzenbaum (D, OH). The outcome was to separate the drug and blood-testing but to require states to provide the drug to all eligible patients. The success of clozapine spurred the development of rival drugs. In addition to clozapine, the group now includes amisulpride, aripiprazole, olanzapine, quetiapine, risperidone, sertindole, and ziprasidone. The atypical antipsychotics minimize the extrapyramidal side effects of chlorpromazine and the other first-generation antipsychotics. Some of the more recent drugs, however, have adverse metabolic and cardiovascular side effects. The structures are shown in Scheme 17.4.

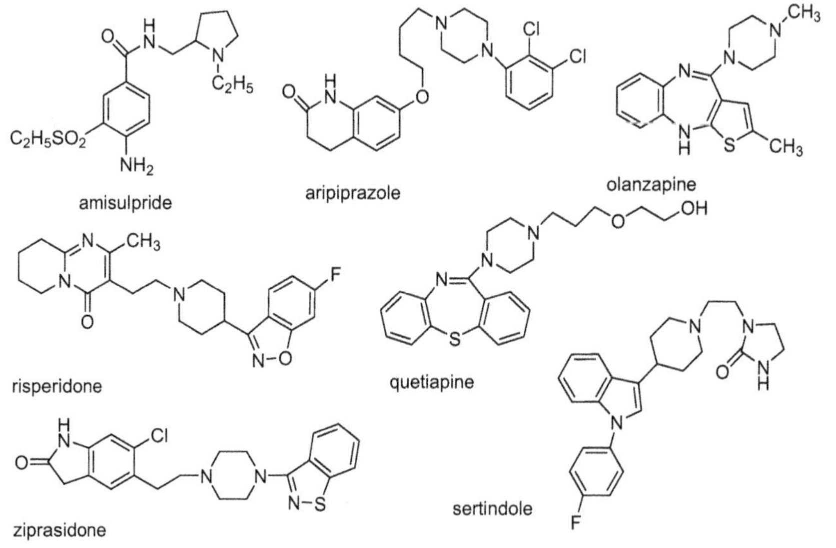

SCHEME 17.4 Second generation antipsychotic drugs.

It is not entirely clear from the point of view of mechanism as to what makes an antipsychotic drug "atypical." One interpretation suggests that relative effects at dopamine (D_2) and serotonin ($5HT_2$) receptors are important, and that the atypical antipsychotics may have relatively faster off-rates and lower affinity at some D_2 receptors. Most of the atypical antipsychotics show strong binding to the $5HT_2$ site and moderate binding to the D_2 site and this is possibly the reason for the reduced extrapyramidal side effects. Another possibility is that there are differences in the affinity for D_2 sites in various parts of the nervous system.

In the early 2000s, the NIMH sponsored a large (1460 patients) multicenter comparison of first generation and second generation antipsychotic drugs. The objective of the study was to compare the effectiveness of the second generation drugs with one another and with a first generation drug. Five drugs were evaluated. Perphenazine represented the first generation drugs and olanzapine, quetiapine, risperidone, and ziprasidone were the second generation drugs evaluated. One of the main end points of the study was the length of treatment, based on the assumption that continued treatment indicated effectiveness and tolerability, as judged by both the patient and physician. Patients who dropped their first drug had the option of trying a second drug. Causes for discontinuation included lack of efficacy, intolerability, weight gain, or other metabolic changes. Olanzapine had the lowest (64%) drop-out rate and the longest average time to discontinuation (9.2 months). The study also tracked cognitive capacity, but there was little difference among the drugs and improvements were small. The overall analysis was that there was relatively little difference among the first and second generation drugs and that if cost was included in the analysis, the first-generation drug was 20–30% less expensive.[10]

17.4.3   LITHIUM SALTS IN TREATMENT OF MANIC-DEPRESSIVE DISORDER

Lithium salts were used in the treatment of symptoms associated with gout in the latter half of the 19th century. The modern work indicating effectiveness of lithium salts in treating manic-depressive disorder was begun by John Cade in Australia in 1949. These reports were followed up by studies in Denmark that confirmed their effectiveness. These studies were not placebo-controlled, however, and therefore subject to criticism. Furthermore, earlier efforts to used lithium chloride in a salt-free diet had caused toxicity, but only under conditions of low sodium concentration. Throughout the 1960s, lithium treatment was the center of controversy. The first placebo-controlled double-blind studies were reported in 1970, and indicated long-term stabilization of manic-depressive patients. Untreated manic-depressives have a suicide rate 15 times that of the general populations. Several subsequent studies have indicated decreased suicide rates for lithium-treated patients.

17.5   ANTIANXIETY DRUGS

Anxiety conditions that are currently recognized include social anxiety, post-traumatic stress, obsessive–compulsive behavior, panic, phobias, and generalized anxiety disorders. Generalized anxiety disorder (GAD) has an overall prevalence of around 5% in the United States. The prevalence is highest in women over 40. Estimated medical costs are >$40 billion/year. GAD is characterized by chronic anxiety and related symptoms, including insomnia, muscle tension, irritability, and difficulty in concentrating. Many patients (about half) also have depressive symptoms. Patients with both disorders generally have poorer outcomes than those with only one. The severity of dysfunction and scope of remission are measured by assessment tools that judge the effect of anxiety on behavior and function.

The first compound used for treatment of anxiety was meprobamate, which has a pharmacological profile similar to the barbiturates. It was approved by the FDA in 1955 and was marketed as Miltown. It was only a marginally effective compound and was withdrawn in the mid-1960s when problems of tolerance, dependence and the potential for overdose became obvious.

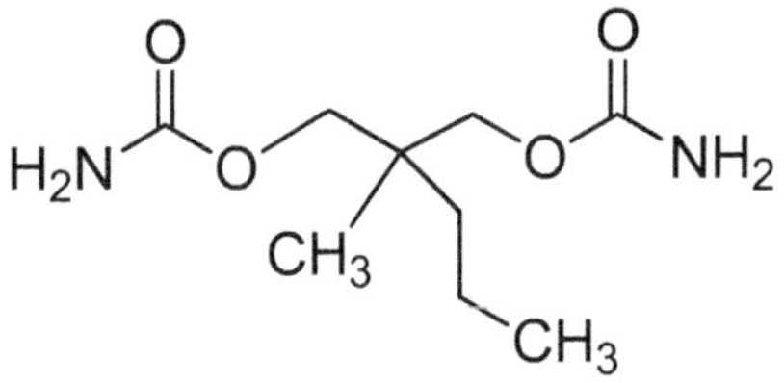

meprobamate

The first highly successful group of antianxiety drugs was the *benzodiazepines* synthesized by Leo H. Sternbach at the Hoffman-La Roche laboratory in Nutley, NJ, USA in 1955–1956. The story of the discovery and development has been recounted by Sternbach,[11] as well as others.[12] The biological activity of the compounds almost escaped detection. Sternbach was attracted to the area by the reports of the antipsychotic effects of chlorpromazine. However, most of the compounds he synthesized were inactive. It turned out that the structures that had been assigned in the literature were incorrect. Two of the last compounds made, chlordiazepoxide and diazepam, had yet another, but unknown, type of structure. When, somewhat belatedly, these compounds were evaluated they were found to have anxiolytic activity superior to meprobamate. The structures were then investigated and shown to contain the benzodiazepine ring. Clinical studies confirmed effectiveness in treatment of anxiety and related conditions such as obsessive states. Chlordiazepoxide was approved by the FDA in early 1960, and marketed as Librium. It was followed by diazepam, marketed as Valium, in 1963. Both drugs were promoted worldwide by using the rationale of providing "balance" in the face of modern sources of stress. Librium and valium were followed by a number of other related compounds both from Hoffman–LaRoche and other pharmaceutical companies. These included oxazepam (Serax, Wyeth) and alprazolam (Xanax, Upjohn). These drugs were characterized as "*minor tranquilizers*" implying their use in common forms of anxiety, as opposed to more severe conditions. A large number of such drugs have subsequently been introduced. Most prescriptions are written by general practitioners for anxiety. The popularity of the benzodiazepines led to attention by the popular culture and the media. In 1966, the Rolling Stones published *Mother's Little Helper* about a "little yellow pill" (Valium).

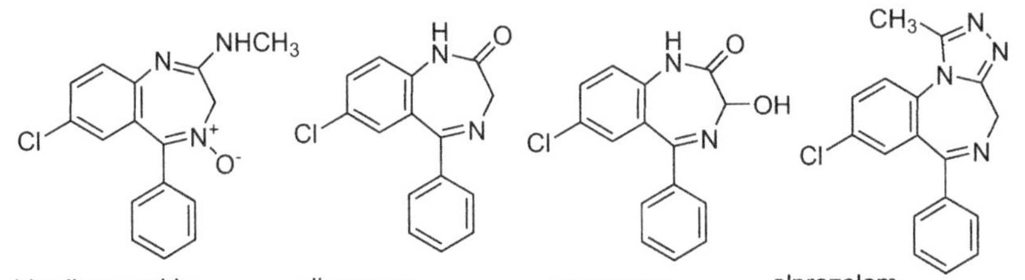

chlordiazepoxide   diazepam   oxazepam   alprazolam

Other benzodiazepines were marketed primarily as sleeping pills, including nitrazepam, triazolam (Halcion), and midazolam. One of the benzodiazepines, flunitrazepam, gained notoriety as the "date rape" drug Rohypnol, because of its powerful sedative effects, particularly in combination with alcohol. Other benzodiazepines are used as muscle relaxants and as anticonvulsants. For example, they can be used as pre-anesthetics.

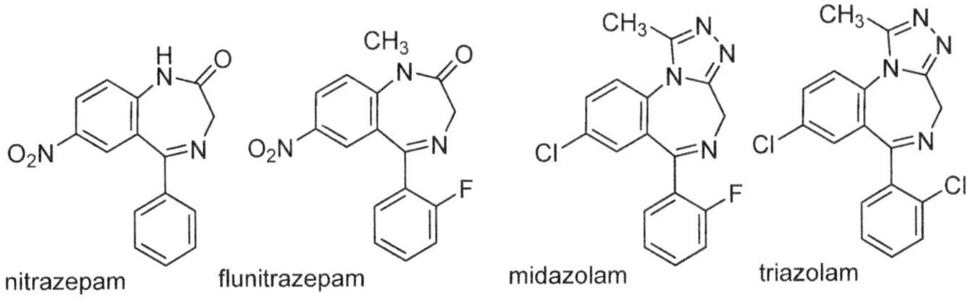

nitrazepam   flunitrazepam   midazolam   triazolam

Between 1965 and 1975, the benzodiazepines were the most widely prescribed drugs in the world. The extent of use was as high as 15–17% in the

United States and Western Europe. In the United Kingdom, the National Health Service forced Hoffman–LaRoche to repay some of the costs on the basis of exploitive pricing. Backlash to the "tranquilized society" began to develop in the public media. In the United States, the Public Citizen Health Group targeted Valium. In 1975, the FDA tightened restrictions on prescription of the benzodiazepines and classified them as Schedule IV controlled substances (see Section 11.4.1.8). The benzodiazepines tend to decrease in effectiveness with use and have withdrawal symptoms, but are not considered addictive. For that reason they are currently recommended primarily for short-term treatment of acute symptoms, such as insomnia.

The benzodiazepines are believed to function at the receptors that are the mediators of the inhibitory effects of GABA. They bind to the receptors and increase the frequency of GABA-induced chloride ion flow. There are several sub-types of the receptors and it is thought that it may be possible to separate some of the different effects of the benzodiazepines on the basis of selectivity among the subtypes. The benzodiazepines also have effects on the transporters that control the concentration of neurotransmitters at the synaptic clefts.

17.6   ANTIDEPRESSANTS

The lifetime prevalence of depression in the United States is around 15%. Depression is a major factor in both disability and health-care costs worldwide. There are a few physiological features that seem to be associated with depression, including enlargement of the pituitary and adrenal glands and increased cortisol levels, suggesting the involvement of the hypothalamic–pituitary–adrenal axis. These effects are believed to be caused by an increased level of corticotrophin-releasing factor, which is involved in the stress response (see Section 15.2). So far however, no drugs have been developed based on these relationships. The currently available antidepressant drugs are targeted to the monoamine neurotransmitters. Their effects are to increase the levels of noradrenaline, dopamine, and serotonin. The amino acid neurotransmitters, glutamate, and GABA have also been the subject of studies searching for relationships to depression, but there are no antidepressant drugs that are thought to act directly on glutamate or GABA receptors.[13] One feature of most antidepressant drugs is that the beneficial effects are not immediate, but rather require several weeks of treatment before improvement is noted. Most also have at least some undesirable side effects.

17.6.1 MONOAMINE OXIDASE INHIBITORS

The discovery of the antidepressant activity of iproniazid was the result of its use as an antituberculosis drug. It was noted that it dramatically improved the mood of some tuberculosis patients. Studies carried out at by Nathan Kline and associates at the Rockland Psychiatric Center in Orangeburg NY, showed that patients with schizophrenia and severe depression improved when treated with iproniazid. Further investigation showed that iproniazid was an inhibitor of the enzyme *MAO*, which oxidizes primary amines such as noradrenaline and dopamine. The hypothesis that developed was that these amines were deficient in depressed patients and that the inhibitors could raise their levels, thus improving the patient. Other MAOIs were soon introduced, including isocarboxazid (Hoffman–La Roche), phenelzine (Warner–Lambert) and tranylcypromine (SmithKline French). The use of MAOI declined fairly quickly for several reasons. Several exhibited heptatoxicity. Another problem was that inhibition of MAO can cause hypertensive crises triggered by eating foods such as cheese that contained relatively large amounts of tyramine. The MAOIs are currently considered second-line drugs if the various uptake inhibitors are ineffective or poorly tolerated.

iproniazid   isocarboxazid   phenelzine   tranylcypromine

17.6.2 MONOAMINE UPTAKE INHIBITORS

The first drugs in this category were found to be relatively nonselective between noradrenaline, dopamine, and serotonin. The first examples were discovered well before the concept of uptake inhibitors had developed. Later, as specific monoamine receptors were identified, they were targeted. Studies at the NIH by B. B. (Steve) Brodie and Sidney Udenfried developed spectrofluorimetric methods for detecting neurotransmitters in the brain (initially serotonin). Arvid Carlsson demonstrated that noradrenaline and dopamine could also be detected using these techniques. The method was eventually refined to permit identification and tracking of noradrenaline, dopamine, and serotonin neurons.

Julius Axelrod, also at the NIH, worked out the biosynthesis of noradrenaline and showed that one of the factors controlling its concentration was reuptake and storage by the neurons.[b] Assays were eventually developed that permitted testing of potential drugs for inhibition of the uptake process with specific monoamines.

17.6.2.1   TRICYCLIC ANTIDEPRESSANTS

In 1952, the antipsychotic effects of chlorpromazine were reported from Paris (see Section 17.1). Various phenothiazines had been synthesized at Geigy in Basel in the late 1940s and some had shown antihistamine, sedative, and/or analgesic effects. Although nothing had come of these studies, the results reported on chlorpromazine led to the compounds being reexamined for antidepressant effects. One of the compounds, imipramine, turned out to be a very effective antidepressant. Studies in Switzerland by Roland Kuhn and in Canada by Heinz E. Lehmann demonstrated dramatic improvement in severely depressed patients. The compound was introduced in Europe and North America in the late 1950s. Evidence of the therapeutic efficacy of imipramine was confirmed in a 1965 study that summarized results from about 1000 patients. Improvement was noted in about 2/3 of the patients, as compared to 1/3 on placebo. Merck introduced the second major drug in this category, amitryptyline in 1961. It was marketed worldwide and soon became the most widely prescribed antidepressant. Several related drugs soon appeared. The group is called *tricyclic antidepressants.* Current drugs that fall into the category include amoxapine, doxepin as well as imipramine and amitriptyline and their derivatives. Their structures are shown in Scheme 17.5.

Tricyclic antidepressants also have anticholinergic and antihistaminergic activity that lead to side effects. They have largely been supplanted in treatment of mild depression, because of these side effects and the relatively high risk of toxic overdoses. They have been replaced by drugs that are more selective for one or more of the monoamine neurotransmitters.

---

[b]Julius Axelrod was born in New York City in 1912. He received a B. Sc. in Chemistry from CCNY in 1933. He worked as a laboratory technician at several locations in New York, eventually joining B. B. Brodie at the Goldwater Memorial Hospital in 1946. There he and Brodie investigated the toxicity of analgesic preparations containing acetanilide and discovered its metabolism to acetaminophen. This eventually led to the introduction of acetaminophen as an analgesic (see Section 13.4). In 1950, Axelrod moved to the NIH, first at the Heart and Lung Institute, but after 1955 at the Mental Health Institute. He took leave of absence for a year to obtain a Ph. D. at George Washington University, with his research based largely on his studies at NIH. At the NIMH he showed that the catecholamines were subject to reuptake at the synapse. It was this work that was cited in the award of the 1970 Nobel Prize in Physiology or Medicine, which he shared with Bernard Katz and Ulf von Euler, who had established other facets of the process of neurotransmission at nerve synapse. Axelrod later studied the biosynthesis of serotonin and melatonin.

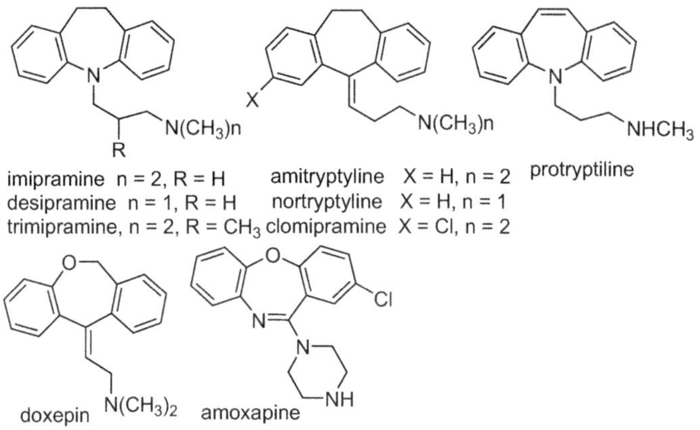

SCHEME 17.5 Tricyclic antidepressants.

17.6.2.2 REUPTAKE INHIBITORS SELECTIVE FOR NORADRENALINE

Reboxatine and atomoxetine are considered to be the most selective noradrenaline uptake inhibitors. Reboxetine is approved in much of the world, but not in the United States. A closely related compound, atomoxetine is approved in the United States. In contrast to reboxetine, which is racemic, atomoxetine is available in enantiomerically pure form. Atomoxetine is also used in treatment of attention deficit hyperactivity disorder (ADHD) (see Section 17.7.3).

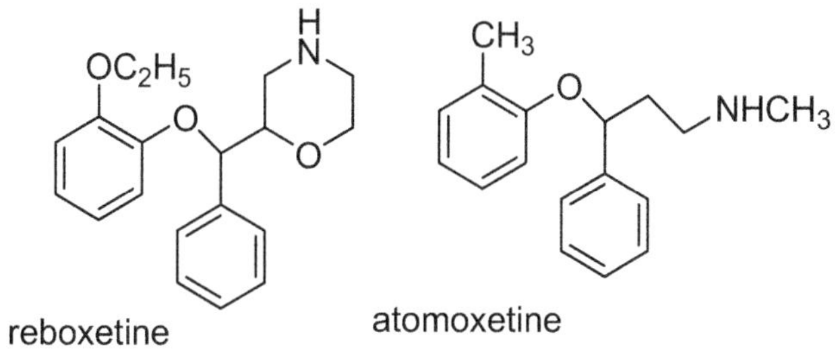

17.6.2.3 REUPTAKE INHIBITORS SELECTIVE FOR DOPAMINE

Several antidepressants are considered to be more or less selective for dopamine, including amineptine, bupropion, nomifensine, and pramipexole. Their structures are shown in Scheme 17.6.

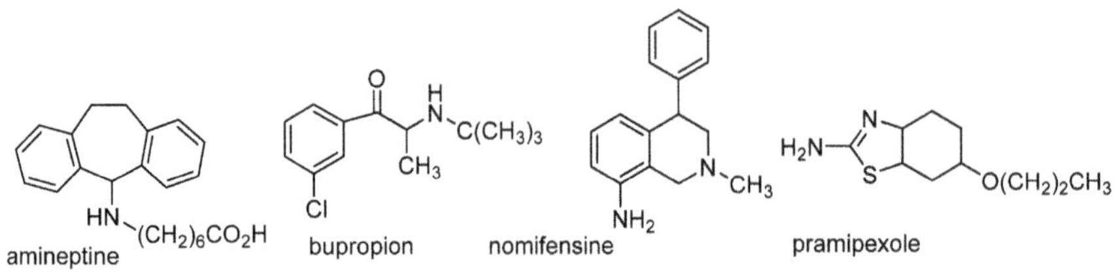

SCHEME 17.6   Selective dopamine reuptake inhibitors.

17.6.2.4   REUPTAKE INHIBITORS SELECTIVE FOR SEROTONIN

The first selective serotonin reuptake inhibitors (SSRI) to be approved in the United States was fluoxetine (Prozac) in 1987. It was developed and marketed by Eli Lilly. It became a very successful drug, including being named "Pharmaceutical Product of the Century" by Fortune Magazine in 1999. Peak annual sales in the United States were $2.8 billion in 1998. Fluoxetine was developed on the basis of the hypothesis that selective inhibition of serotonin uptake might be useful in treatment of depression. Serotonin's role in neurotransmission had become fairly widely appreciated by the late 1960s. The "serotonin hypothesis" attributed special importance to serotonin levels in depression and psychoses. The group at Lilly used "synaptosomes," a particulate fraction consisting of nerve ending that had the capacity for production and release of the monoamine neurotransmitters. The assay was used to search for selective inhibitors. Many analogs of diphenhydramine (also the original lead compound for the tricyclic antidepressants) were synthesized. Among these, fluoxetine was found to be highly selective for serotonin relative to noradrenaline. Various other assays confirmed this selectivity. The results were first reported in 1974. An IND study was approved in 1976. In clinical studies, fluoxetine did not exhibit the side effects associated with the cholinergic, adrenergic, or histamineric effects of the tricyclic antidepressants. The NDA was submitted in 1983 and final approval was obtained at the very end of 1987. In addition to treating depression, fluoxetine is used for treatment of eating disorders such as anorexia and bulimia.

Sertraline was introduced in 1992, under the brand name Zoloft as a product of Pfizer. Paroxetine (Paxil) was introduced by GlaxoSmithKline in 1993. They have similar mechanisms of action and scope of use to fluoxetine. In addition to their use as antidepressants, the SSRI have also been used for treatment of anxiety disorders, including panic, post-traumatic stress, obsessive–compulsive, and social phobia disorders. Scheme 17.7 gives the structures of these SSRIs.

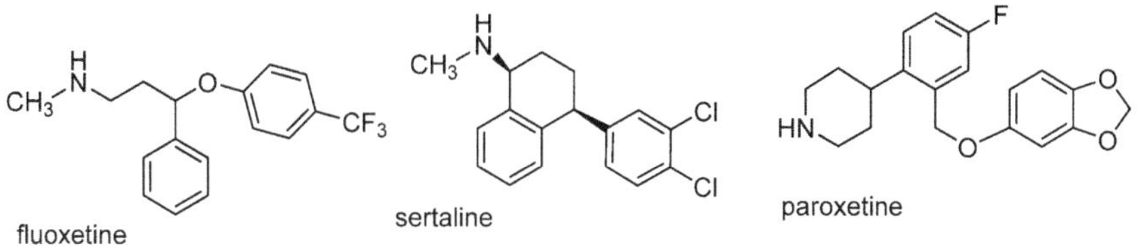

fluoxetine

sertaline

paroxetine

SCHEME 17.7 Selective serotonin reuptake inhibitors (SSRI).

The selective inhibitors of serotonin uptake (SSRI) appear to have two sequential effects. The initial action is inhibition of reuptake of serotonin at the synapse, leading to increased serotonin levels. This is followed by downregulation of serotonin auto-regulators, which also increases serotonin release. The latter mechanism may help to account for the delayed clinical response characteristic of most of the antidepressants. Figure 17.1 summarizes the idea behind the selective uptake inhibitors.

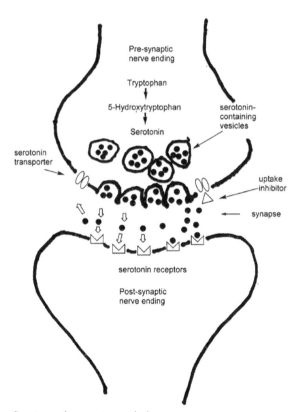

FIGURE 17.1 Serotonergic neurotransmission.

17.6.2.5   REUPTAKE INHIBITORS SELECTIVE FOR NORADRENALINE AND SEROTONIN

Duloxetine and venlafaxine inhibit both noradrenaline and serotonin uptake. This combination gives rise to some specific advantages. Duloxetine, for example, seems to be especially effective in relieving pain associated with depression. It is also approved (as Cymbalta) for diabetic neuropathic pain. Venlafaxine (Effexor) has generally similar characteristics and is considered to be an effective antidepressant. The *O*-demethyl compound, desvenlafaxine is also available.

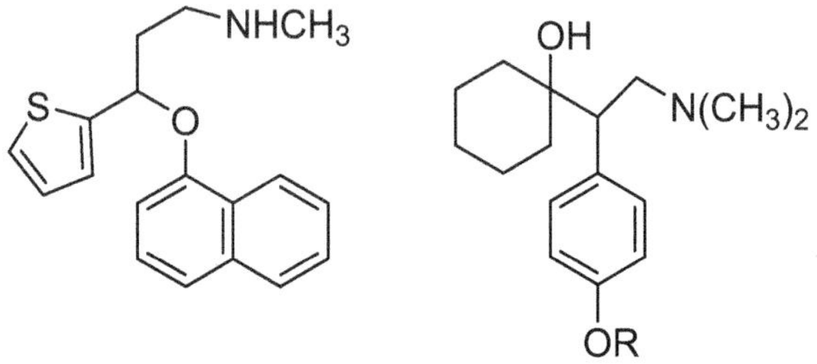

duloxetine

venlafaxine R = CH$_3$
desvenlafaxine R = H

17.7   USE AND ABUSE OF AMPHETAMINE AND ITS DERIVATIVES

17.7.1   AMPHETAMINES AS STIMULANTS

Amphetamine is closely related structurally to the catecholamine neurotransmitters noradrenaline and dopamine. It is a stimulant and also has antihistamine activity. It was introduced commercially as Benzedrine. It was first marketed as an inhalable form for asthma, but soon thereafter as a pill recommended for mild depression. It was the first of the "mood-altering" drugs to be widely distributed. In the late 1930s, Benzedrine tablets became "pep pills" among college students and in WWII they were dispensed to troops to counteract fatigue. Pilots on long-distance bombing runs were especially common users. They continued to be used in the military through the 1990s. In the early 1950s, Benzedrine came to be prescribed as an aid to weight loss. The pure dextrorotatory *S*-enantiomer was sold as Dexedrine. Benzedrine remains available as the attention-deficit drug Adderall, which contains a 75:25 mixture of the *S:R* isomers (see Section 17.7.3).

Various levels of abuse of amphetamine occurred from the beginning, but accelerated in the 1960s when it became known as "speed." Because they are easy to synthesize, the amphetamines became major drugs of abuse. Restrictions were tightened in 1965 and amphetamine was listed as a level-II controlled substance in the 1971 Controlled Substance Act. The *N*-methyl derivative, methamphetamine, is known as "crystal meth." The methylenedioxy derivative of methamphetamine is known as "ecstasy." It is estimated that around $65 billion of illegal income is generated by the amphetamines. Norephedrine is used extensively in over-the-counter medications for relief of nasal congestion. Ephedrine and norephedrine and other related compounds are also present in various natural preparations that go by the general name *ephedra*. Both the synthetic and natural materials have been restricted by the FDA because of evidence of side-effects, including occasional hemorrhagic stroke. Cathinone is the active ingredient of a plant, khat, that is chewed as a stimulant in the Middle-East and East Asia. Other amphetamine derivatives include diethylpropion, fenproporex, phendimetrazine, and mazinol. The amphetamines are structural analogs of epinephrine (adrenaline) and exhibit many effects of it and other neurotransmitters such as dopamine and serotonin. Side effects include restlessness, insomnia, cardiac arrhythmia, and hypertension. Scheme 17.8 gives the structures of some of the amphetamine derivatives.

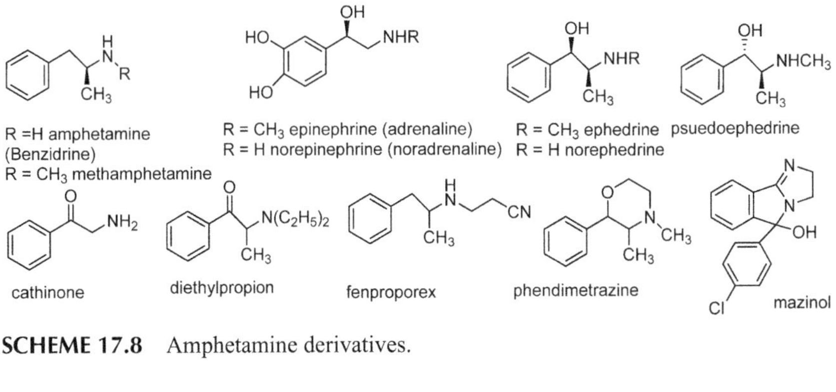

SCHEME 17.8   Amphetamine derivatives.

17.7.2 AMPHETAMINES AS APPETITE SUPPRESSANTS

The amphetamines and analogs act by releasing noradrenaline and dopamine in the hypothalamus and have a powerful effect of inhibiting appetite (*anorexigens*). The hypothalamus has receptors for neuropeptides that either enhance (*orexigenic*) or suppress (*anorexigenic*) appetite. Norephedrine was available over-the-counter in the United States and Europe as its hydrochloride salt under brand names such as Monydrin, Proin, and Propalin. It was withdrawn in 2000 as a result of studies that indicated an increased risk of hemorrhagic stroke. An amphetamine analog, aminorex, was introduced as an anorexigen in Europe in the 1960s. It was subsequently linked to an increased incidence of primary pulmonary hypertension and was withdrawn from the market in the 1970s. Phentermine was granted FDA approval in the United States in 1959. Fenfluramine was approved in 1973. They were approved only for treatment of obesity and for short-term use. These drugs act by stimulating release of serotonin and activating serotonin receptors. These effects suppress appetite through the effect of serotonin on cerebral centers. During the 1990s, the use of amphetamine weight-loss drugs grew rapidly. The pure active enantiomer of fenfluramine, called dexfenfluramine and sold as Redux, was approved for use in 1995, on a close split vote of the FDA advisory panel. The approved use was for short-term treatment of obesity. Approval was withdrawn in 1997. The combination of fenfluramine and phentermine, known as fen-phen, was introduced in 1992, although the combination was never approved by FDA. It is believed that more than 8 million people, mostly women, received prescriptions. The structures of these drugs are shown in Scheme 17.9.

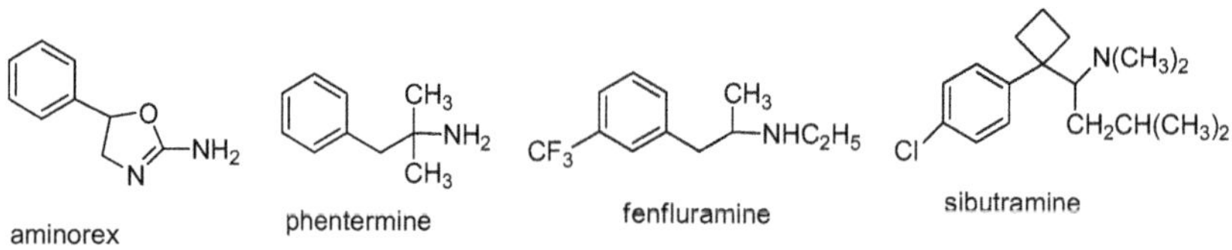

SCHEME 17.9   Amphetamines developed as anorexiants.

In 1997, reports of abnormalities in heart valve function appeared. They were first noted by Pam Ruff, an echocardiogram technician in Fargo, ND and were confirmed at the Mayo Clinic in Rochester, MN. Various diagnostic procedures confirmed abnormal function and structure in the heart valves. The incidence of abnormalities was as high as 20–30% in some studies and increased with history of extended use. Both fenfluramine and phentermine were withdrawn from the market in 1997. Eventually, 300,000 people, mainly women, joined in a class action suit against the distributor Wyeth Labs with claims amounting to $22 billion. More recently, sibutramine, which acts primarily as a serotonin reuptake inhibitor, has shown promise in inducing weight loss in type-2 diabetics.[14] It was approved for treatment of obesity by the FDA in 1997.

17.7.3   ATTENTION DEFICIT HYPERACTIVITY DISORDER

A clear description of the characteristics currently called *ADHD* was given in 1902 by Sir George F. Still, who practiced at King's College Hospital in London. He considered the "lack of moral control" to be the result of some physiological abnormality, as opposed to simply bad behavior. In the 1920s, it was noted that a number of children who had survived encephalitis associated with the 1918 influenza epidemic had similar attention and behavior problems, suggesting that it might have resulted from neurological damage. The characteristics associated with the current definition of ADHD include academic difficulty, inattention, restlessness, hyperactivity, and impulsiveness. The first official criteria for a diagnosis of Attention Deficit Disorder were adopted by the American Psychiatric Association in 1980. The H for hyperactivity was added in 1987 (see also Section 17.9.1). Both the criteria for diagnosis and the extent of medical treatment are somewhat different in Europe, with more stringent diagnostic criteria and less use of drugs being the case in Europe. It is currently estimated that ADHD effects 3–12% of children and 2–6% of adults in the United States.

The positive effect of a stimulant, Benzedrine, was first observed by Dr. Charles Bradley. He was the director of the Bradley Home, a facility that had been endowed by his great uncle and aunt, Mr. and Mrs. George W. Bradley. Their only child had suffered severe mental deterioration as the result of influenza-associated encephalitis. Dr. Bradley used Benzedrine, hoping to counteract the painful effects of spinal taps that were being used in diagnostic procedures. He observed immediate improvement in the school performance of the children receiving Benzedrine, as did the children. The positive effects disappeared on discontinuation of the drug. Bradley recognized the paradox of a beneficial effect of a simulant for a condition that involved hyperactivity. He published his results in the *American Journal of Psychiatry* in 1937. Maurice Laufer, Eric Denhoff, and Gerald Simons, all of whom also worked at the Bradley Home, described the symptoms they called "hyperkinetic impulse disorder" in 1957. They confirmed the beneficial effects of amphetamine-type stimulants. They were primarily psychiatrists, however, and regarded the drugs as aids, not substitutes, for psychotherapy.

The arrival of chlorpromazine for treatment of adult mental conditions reopened interest in use of drugs in children and adolescents. Methyl phenidate (Ritalin) was synthesized at Geigy in 1955. In 1958, the NIMH began to sponsor research into use of psychoactive drugs in children. Among the early grant recipients was Leon Eisenberg at Johns Hopkins University. He and Keith Connors undertook studies in Baltimore. Among the drugs included was Ritalin. They reported their first results of a double-blind study in 1963. The pharmaceutical industry, leery of the controversy associated with drug treatment for children, did not invest in research until the 1970s.

methyl phenidate (Ritalin)

Through the 1970s, there was little knowledge of or consensus about the origins of ADHD, although most experts considered it to be of biological origin. Nevertheless, prescription of Ritalin continued to grow. Backlash developed, in part related to the growing concern of drugs of all types. In 1970, the Washington Post reported (incorrectly) that 5–10% of the children in the Omaha, NE, USA school system were being treated with Ritalin. The 1970 revision of the Drug Abuse and Control Act put amphetamines and Ritalin in Schedule III, limiting the length of prescriptions. In 1971, they were moved to Schedule II, which placed limits of the amounts of the substances that could be produced by individual companies. Several books were published that received wide attention. They asserted that (a) Ritalin was a conspiracy to control children for the convenience of teachers and school administrators or that (b) ADHD was the result of environmental factors, or that (c) the behavior diagnosed as ADHD was in the normal range. All contributed to skepticism about the use of Ritalin. Late in the 1970s, Judith Rapoport of the NIMH found that normal children's school performance also improved on amphetamines, raising a question as to whether the effects of the stimulants were directly related to ADHD. In 1999, a 14-month study sponsored by the NIMH that compared medication, medication with behavioral intervention, behavioral intervention alone, or generalized community care indicated that medications with or without behavioral interventions were beneficial. There have been many subsequent studies of the amphetamines and methyl phenidate. Most report significantly improvement on treatment.

The effect of both methyl phenidate and amphetamines in ADHD are fast, but last for only a few hours, requiring repeated dosages. Various extended-release forms of the drugs have been developed to address this problem. The active form of methyl phenidate is the D-enantiomer and it was approved by the FDA in 2002. It seems to extend the period of effectiveness slightly. Another approach involves "pro-drugs." One such material is lisdexamfetamine dimesylate, which is the amide of D-amphetamine and lysine. It is hydrolyzed to provide D-amphetamine. It was approved in 2007. A currently approved ADHD drug that is not in the stimulant class is atomoxetine, an inhibitor of dopamine and norepinephrine transporters (see Section 17.6.2.2). Another is guanfacine, approved in 2009. It acts on adrenergic receptors.

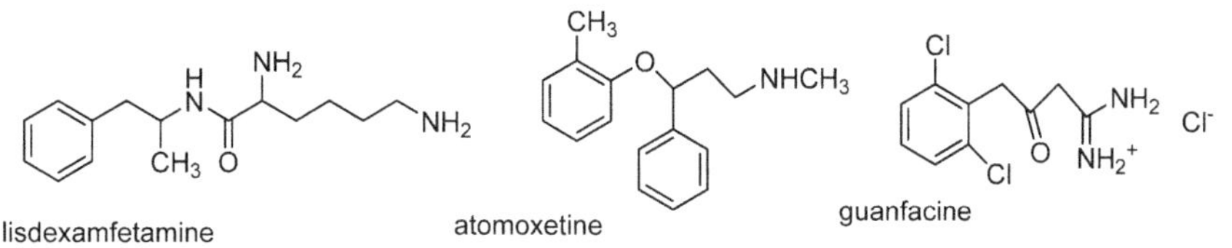

17.8 ALZHEIMER'S DISEASE AND DEMENTIA

One approach to treatment of AD is based on the neurotransmitter ACh (see Section 17.1). One of the characteristics of AD is loss of cholinergic neurons in the basal forebrain. There is a decrease in the level ACh and the enzymes involved in formation and hydrolysis of ACh (see Section 8.3.3). The decrease in cholinergic function is correlated with formation and deposition of β-amyloid peptides. This and the hyperphosphorylation of tau, a protein associated with microtubules, are the most evident biochemical manifestations of AD. The deficit of cholinergic function has resulted in investigation of cholinesterase inhibitors as potential drugs for treatment of AD patients. Several drugs appear to have at least some benefit in the mild and moderate stages of the disease. The first to be introduced was tacrine, but it has largely been abandoned because of side effects, including hepatotoxicity. The other three approved drugs are donepezil, rivastigmine, and galantamine. The structures are shown in Scheme 17.10. None of these medications slows or alters the development of the disease but they do provide some temporary improvement in cognitive function.

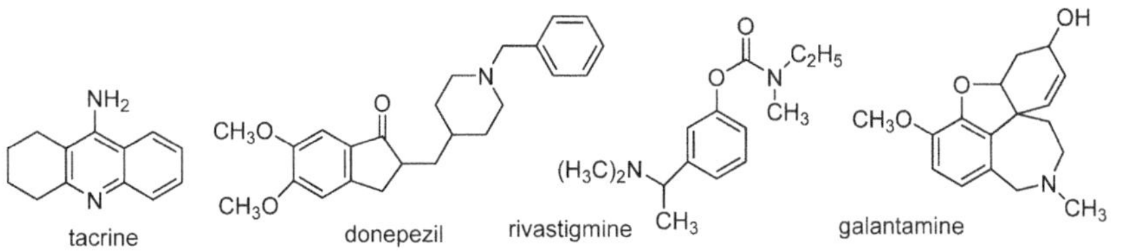

SCHEME 17.10   Drugs used in treatment of Alzheimer's dementia.

17.9 SOCIAL CONSEQUENCES AND CONTROVERSIES IN TREATMENT OF NEUROLOGICAL AND PSYCHOLOGICAL CONDITIONS

In contrast to physical disorders caused by pathogens or systemic conditions such as cardiovascular disease or diabetes, most neurological and psychological disorders are not subject to diagnosis or confirmation by quantifiable measurements such blood pressure or glucose tolerance. While the microscope and clinical chemistry can often provide definitive evidence of physical disorders, most mental and psychological conditions are not subject to such clear analysis. Furthermore, while over the course of the late nineteenth and early 20th century, the general public became well-acquainted with the cause-and-effect relationships in physical illness, understanding of the nature of mental disorders was slower to develop. Even the profession of psychiatry, dedicated to the treatment of mental conditions, remained conflicted on the origin of mental illness during the first half of the 20th century. One view was that mental illness, like other pathologies, has biochemical and genetic origins. An early proponent of this view was Emil Kraepelin.[c] Psychoanalysis, at least as practiced by Freudian psychoanalysts, did not have a strong connection to biochemistry or physiology, although the ability of drugs such as cocaine and mescaline to alter perception and mental states was well known (see Section 13.7.1). When it was found that these psychotropic drugs altered the level of various monoamine neurotransmitters, a more pharmacological approach became possible. Even today, more than half a century after the introduction of the first drugs for treatment of mental illness, many details of the workings of the mind and nervous system remain unclear. Diagnosis of mental and psychological conditions depends on evaluation against a set of criteria that, by consensus, define the conditions.

17.9.1   THE AMERICAN PSYCHIATRIC ASSOCIATION DIAGNOSTICS AND STATISTIC MANUAL

The field of psychiatry had its origins in the early 1800s, the name being introduced by Johann Christian Reil, a professor of medicine at the University of Halle in Germany. He considered it a branch of medicine equal in importance to surgery and pharmacy (internal medicine). In the mid-1800s most American psychiatrists worked in state-supported institutions for the mentally ill. The precursor organization of the American Psychiatric Association was formed in 1844. Until the early 1900s, there was little reason or effort to classify mental illness. Most psychiatrists in the United States were from the psychoanalytical tradition that associated mental illness with response ("reaction") to specific experience

---

[c]Emil Kraepelin studied medicine and neuropathology as well as psychology. He received an M. D. in 1878 at the age of 22. He continued study in both neurology and psychology. In 1883, he published a book advocating integration of psychiatry into medical science, which he updated throughout his career. He believed that mental illness had a biological basis. He first classified mental illness as "early-onset dementia" and "manic depressive psychoses." Later, he and his students added "paranoia" and "schizophrenia" as separate diagnoses. He held appointments in clinical psychiatry at Dorpat (now Tartu, Estonia) and Heidelberg. In 1903, he became professor of clinical psychology at the University of Munich. This led eventually to the founding of the Institute of Psychiatric Research in 1917.

and/or trauma and, therefore, there was no expectation that there would be general, as opposed to individual, causes or treatments of mental illness.

In the United States, the primary criterion for diagnosis of neurological, mental, and psychological disorders is the American Psychiatric Association Diagnostics and Statistics Manual. The first two editions of the Diagnostic and Statistical Manual reflected the view that the causes of mental illness were individual and diverse and gave very general descriptions of mental conditions. This changed dramatically with the introduction of the third edition (DSM-III) in 1980, which began to define conditions in terms of particular constellations of symptoms. These trends continued in DSM-IV published in 1994. Several factors drove this change. One was the evidence that mental illness was subject to medical (drug) treatment. Another was the growing understanding of the biochemistry and physiology of the nervous system. There were also economic and regulatory pressures. Government agencies and insurance companies providing funds for treatment began to demand specific diagnoses and evidence of effectiveness of treatments. These pressures have continued in the form of cost-containment efforts based on evidence-based medicine. The FDA drug approval process also required demonstration of effectiveness for specific indications, which meant that drug companies needed specific therapeutic targets. The fifth edition of the DSM was published in 2013. It lists 157 separate conditions and their characteristics. Diagnosis depends on meeting certain criteria in terms of the number and severity of symptoms. In turn, not only treatment and prescription of drugs, but also the availability of special services and insurance coverage depend on meeting these criteria.

17.9.2 METHODS FOR EVALUATION OF THE EFFECTIVENESS OF DRUGS

In Chapter 11, we discussed the general issues associated with drug development. The evaluation of drugs in the treatment of mental health patients is particularly challenging because the outcomes often are more subjectively evaluated than in other areas. In particular, the 1962 amendments to the Federal Food, Drug, and Cosmetic Act required evidence of both safety and efficacy. The earliest trials of drugs in humans usually involved only small numbers of patients, primarily in an institutional setting, and with clinical observation as the outcome. These studies were not blinded or placebo-controlled. By the end of the 1950s, placebo-controlled studies began to be introduced. The first placebo-controlled study of chlorpromazine was in 1957 and involved just 12 patients. A controlled study of imipramine was reported in 1959. With the added requirements of the 1962 FDA amendments, the scale and scope of studies expanded. Multi-site studies and specific evaluation scales were introduced. For example, in 1964, the NIMH sponsored a 9-site study involving 463 patients to compare two of the newer antipsychotic drugs with chlorpromazine. It used specific scales to evaluate patients for severity of illness and improvement. The study showed all three drugs were superior to placebo, but found no significant differences among them. By the 1980s multicenter, double-blinded, placebo-controlled studies were the norms. More data on side effects were collected. A variety of formal assessment scales were developed and much more sophisticated statistics were applied to the data. Most studies have been directed toward approval and registration of a drug. This requires demonstration of efficacy and safety, but usually does not involve direct comparison with other available drugs. As a result, there is often little data comparing the effectiveness of different drugs in a specific class. It has been argued that such studies fail to detect the subgroups within diagnostic categories who benefit from those who do not, and thus do not provide the information necessary to identify the basis for success.[15]

17.9.3   CONTROVERSY ABOUT OVERUSE AND OVER-PRESCRIPTION OF ANTIDEPRESSANT AND ANXIOLYTIC DRUGS

There has been considerable debate about the role of advertising and economic factors in the advertising and prescription of drugs for depression and anxiety. This is particularly the case in the United States, the only country (except for New Zealand) that permits direct-to-consumer advertising of prescription drugs. The issues raised include exaggeration of benefits, minimization of risks and side-effects, and oversimplification of the scientific evidence supporting use. There is also concern about the role of the industry in formulating and disseminating the therapeutic consensus through sponsorship of publications and conferences. To what extent have the incentives for developing markets for psychoactive drugs led to the expansion of what are considered treatable abnormalities?

17.9.4   CONSEQUENCES OF DRUG TREATMENT OF MENTAL ILLNESS AND PSYCHOLOGICAL DISORDERS

The consequences of drug treatment of mental and psychological conditions have been profound. At the beginning of the century institutionalization was the norm for severe mental disorders, and the treatments such as insulin coma and electro-convulsive therapy, were drastic. Mental illness carried both an aura of mystery and a stigma. By the end of the century, a much wider range of conditions was recognized and many were amenable to reasonably effective treatments. Most treatments were on an out-patient basis and institutionalization was greatly reduced. The beginnings of detailed description of conditions and their etiology were coming into place.

KEYWORDS

- **barbiturates**
- **antiseizure drugs**
- **chlorpromazine**
- **clozapine**
- **γ-aminobutyric acid (GABA)**
- **benzodiazepines**
- **monoamine oxidase inhibitors**
- **tricyclic antidepressants**
- **neurotransmitter uptake inhibitors**
- **amphetamines**
- **attention deficit hyperactivity disorder (ADHD)**
- **Diagnostic and Statistics Manual (DSM)**

BIBLIOGRAPHY BY SECTION

Section 17.1: Jacobsen, E. *Psychopharmacology* **1986**, *89*, 138–144; Ban, T. A. *Prog. Neuropharmacol. Biol. Psychiatry* **2001**, *25*, 709–727; Slattery, D. A.; Hudson, A. L.; Nutt, D. J. *Fundam. Clin. Pharmacol.* **2004**, *18*, 1–21; Lopez-Munoz, F.; Alamo, C. *Curr. Pharm. Des.* **2009**, *15*, 1663–1586.

Section 17.3: Rogawski, M. A.; Loescher, W. *Nat. Rev. Neurosci.* **2004**, *5*, 553–564; Czapinski, P.; Blaszczyk, B.; Czuczwar, S. J. *Curr. Top. Med. Chem.* **2005**, *5*, 3–14; LaRoche, S. M. *Neurologist* **2007**, *13*, 133–139; Shorvon, S. D. *Epilepsia* **2009**, *50*(Suppl. 3), 69–92, 93–130; Brodie, M. J. *Seizure* **2010**, *19*, 650–655; Mantegazza, M.; Curia, G.; Ragsdale, D. S.; Avoli, M. *Lancet Neurol. 9*, 413–404.

Section 17.4.1: Baumeister, A. A.; Francis, J. L. *J. History Neurosci.* **2002**, *11*, 265–277; Kirby, K. C. *Ann. Clin. Psychiatry* **2005**, *17*, 141–146; Lopez-Munoz, F.; Alamo, C. *Brain Res. Bull.* **2009**, *79*, 130–141.

Section 17.4.2: Kapur, S.; Mamo, D. *Prog. Neuro-Psychopharmacol. Biol. Psychiatry* **2003**, *27*, 1081–1090; Crilly, J. *History Psychiatry* **2007**, *18*, 39–60; Tanlon, R.; Nasrallah, H. A.; Keshavan, M. S. *Schizophr. Res.* **2010**, *122*, 1–23.

Section 17.4.3.: Mueller-Oerlinghausen, G. P. *Bipolar Disord.* **2009**, *11*(Suppl. 2), 10–19.

Section 17.5: Bateson, A. N. *Curr. Pharm. Des.* **2002**, *8*, 5–21; Atack, J. R. *Curr. Drug Targets—CNS Neurol. Disord.* **2003**, *2*, 213–232; Schwartz, T. L.; Nihalani, N.; Simionescu, M.; Hopkins, G. *Curr. Pharm. Des.* **2005**, *11*, 255–263; Lopez-Munoz, F.; Alamo, C.; Garcia-Garcia, P. *J. Anxiety Disord.* **2011**, *25*, 554–562; Reinhold, J. A.; Mandos, L. A.; Rickels, K.; Lohoff, F. W. *Expert Opin. Pharmacother.* **2011**, *12*, 2457–2467.

Section 17.6: Lopez-Munoz, F.; Alamo, C. *Curr. Pharm. Des.* **2009**, *15*, 1563–1586.

Section 17.6.2.2: Kasper, S.; El Giamal, N.; Hilger, E. *Expert Opin. Pharmacol.* **2000**, *1*, 771–782.

Section 17.6.2.4: Wong, D. T.; Perry, K. W.; Bymaster, F. P. *Nat. Rev. Drug Discov.* **2005**, *4*, 764–774.

Section 17.6.2.5: Bauer, M.; Moeller, H.-J.; Schneider, E. *Expert Opin. Pharmacother.* **2006**, *7*, 422–426.

Section 17.7: Iversoen, L. *Speed, Ecstasy, Ritalin. The Science of Amphetamines*; Oxford University Press: Oxford, UK, 2006.

Section 17.7.2: Gross, S. B.; Lepor, N. E. *Rev. Cardiovasc. Med.* **2000**, *1*, 80–89; Naqvi, T. Z.; Gross, S. B. *Curr. Women's Health Rep.* **2003**, *3*, 116–125; Krahenbuhl, S. In *The Pharmacotherapy of Obesity, Options and Alternatives*; Hofbauer, K. G.; Keller, U. J.; Boss, O., Eds.; CRC Press: Boca Raton, FL, 2004; pp 285–296.

Section 17.7.3: Mayes, R.; Rafalovich, A. *History Psychiatry* **2007**, *18*, 435–457; Findling, R. L. *Clin. Ther.* **2008**, *30*, 942–957.

Section 17.8: Lane, R. M.; Potkin, S. G.; Enzm, A. *Int. J. Neuropsychopharmacol.* **2006**, *9*, 101–124.

Section 17.9.4: Leon, A. C. *J. Clin. Psychiatry* **2011**, *72*, 331–340.

REFERENCES

1. Friedlander, W. J. *Arch. Neurol.* **2000**, *57*, 1782–1785.
2. Glazko, A. J. *Ther. Drug Monit.* **1986**, *8*, 490–497.
3. Brodie, M. J. *Seizure* **2010**, *19*, 650–655.
4. Ambrosio, A. F.; Soares-da-Silva, P.; Carvalho, C. M.; Carvalho, A. P. *Neurochem. Res.* **2002**, *27*, 121–130; Schmidt, D.; Elger, C. E. *Epilepsy Behav.* **2004**, *5*, 627–635.
5. Curia, G.; Biagini, G.; Perucca, E.; Avoli, M. *CNS Drugs* **2009**, *23*, 555–568.
6. Colpaert, F. C. *Nat. Rev. Drug Discovery* **2003**, *2*, 315–320; Smith, A. *Nat. Rev. Drug Discov.* **2004**, *3*, 3; Granger, B.; Albu, S. *Ann. Clin. Psychiatry* **2005**, *17*, 137–140; Awouters, F. H. L.; Lewi, P. J. *Arzneimitt. Forsch.* **2007**, *57*, 625–632.
7. Schildkraut, J. J. *Am. J. Psychiatry* **1965**, *122*, 509–522.
8. Duval, A. M.; Goldman, D. Reprinted in *Psychiatric Ser.* **2000**, *51*, 327–331; Clark, R. E.; Samnaliev, M. *Int. J. Law Psychiatry* **2005**, *28*, 532–544.
9. de Young, M. *Madness, An American History of Mental Illness and Its Treatment*; McFarland and Company: Philadelphia, PA, Chapter 6, 2002.
10. Manschreck, T. C.; Bopscles, P. A. *Harvard Rev. Psychiatry* **2007**, *15*, 245–258.
11. Sternbach, L. H. *J. Med. Chem.* **1979**, *22*, 1–7.

12. Cohen, I. M. *Discoveries in Biological Psychiatry*; In Ayd, F. J.; Blackwell, B. Eds.; Ayd Medical Communication: Baltimore, MD, 1984; pp 130–141.
13. Skolnick, P.; Popik, P.; Trullas, R. *Trends Pharmacol. Sci.* **2009,** *30*, 563–569.
14. Scheen, A. J.; Lefebvre, P. J. *Diabetes/Metab. Res. Rev.* **2000,** *16*, 114–124; Poston, W. S. C.; Foreyt, J. P. *Expert Opin. Pharmacother.* **2004,** *5*, 633–642.
15. Ban, T. A. *Prog. Neuro-Psychopharmacol. Biol. Psychiatry* **2006,** *30*, 429–441.

CHAPTER 18

ANTIMALARIAL AND OTHER ANTIPARASITIC DRUGS

ABSTRACT

Although malaria has plagued humans since the beginning of civilization it was largely eliminated from much of the world during the 20th century by use of insecticides and other means of mosquito control. Synthetic drugs joined quinine for treatment. Nevertheless, the development of drug-resistant forms followed. The most recently discovered antimalarial drugs are derivatives of the natural product artemisinin. Most of the parasitic diseases are endemic to underdeveloped tropical and subtropical parts of the world and involve transmission by vectors. Vector control provides a means of prevention. Many of these diseases, including trypanosomiasis, leishmaniasis, schistosomiasis, lymphatic filariasis, and onchocerciasis, are considered "neglected diseases" by the World Health Organization.

In this chapter, we will consider several diseases that are caused by parasites. These diseases have several things in common. They are caused by either protozoa or worm infestations and involve a vector organism as part of the infection cycle. They tend to be concentrated in tropical and subtropical parts of the world and are associated with poverty. Because of these circumstances, they have often not been high priority targets for drug development by the pharmaceutical industry. The WHO has included several in its list of "Neglected Tropical Diseases" and undertaken efforts to develop more effective treatments and preventive measures. Private and international organizations are also active in this effort.

18.1 MALARIA—ETIOLOGY, PATHOLOGY, PREVENTION, AND TREATMENT

Development of understanding and treatment of malaria parallels in many ways the development of chemistry and science. There had been speculation about the cause throughout history, and the development of the microscope led to definitive information on the cause in the late nineteenth and early 20th century. Quinine, the first effective treatment was discovered by Jesuit missionaries in the Andes Mountains of South America in the 17th century and was the mainstay until the mid-20th century. Synthetic chemistry entered the field early in the 20th century and by mid-century, a highly effective drug, chloroquine, was introduced. This, along with the discovery of the insecticide DDT in the 1940s, offered the prospect of worldwide elimination of malaria. But by the 1970s, resistance to chloroquine was widespread and DDT's adverse effects had become evident. The Viet Nam war prompted new research in both the United States and in China, and from China came a new drug discovered in an herbal medicine that had been used for millennia.

We tend to think of malaria as a tropical disease but it was endemic to much of Europe and North America until the middle of the 20th century. Malaria has probably been with humans from the beginnings of civilization. It is thought to have moved from apes, but exactly when is uncertain. Malaria incidence is believed to have increased when humans adopted permanent dwellings in villages that provided both breeding grounds for mosquitoes and increased population density. Egyptian and Chinese writings of several thousand years ago contain references to malaria and Egyptian mummies show physical evidence of the disease. Malaria spread to the Mediterranean 2000–2500 years ago. Malaria infestation was particularly high in the area around Rome. Malaria was introduced to the New World after 1500 from Europe and through slaves brought from Africa. Mortality of malaria infections in the 1600s was quite high, perhaps 50%. This was reduced somewhat by the introduction of quinine in the form of cinchona bark and then as purified quinine, but at the turn of the 20th century, it is estimated that 10% of deaths worldwide were due to malaria, and in India, it might have caused as many as half of all deaths.

The discovery of both insecticides and new antimalarial drugs during and immediately after WWII gave rise to hope for eliminating the disease worldwide. In 1957, the WHO launched a Global Eradication Strategy. The measures taken included improved drainage and other means of mosquito control, spraying of dwellings, and use of insecticide-treated sleeping nets. These measures were successful in eliminating endemic infestation in most of North America and Europe, but fell short of the goal in other parts of the world. Several problems contributed the failure to achieve the initial goal. These include resistance to insecticides (see Section 8.3), development of malaria strains resistant to drugs (see Section 18.1.7), lack of funding and medical infrastructure, and political and economic instability. Since the 1990s, the goal has been to maintain the gains that have been made. In 2002, the UN-sponsored Global Fund to Fight AIDS, Tuberculosis and Malaria (GFFATM) enlisted the support of many governmental and private agencies and focused new attention on combination chemotherapy, as well as preventive measures. The goals include the development of an effective vaccine, drug development and mosquito control. The GFFATM is also engaged in evaluating and promoting the most effective uses of the resources that are available for the effort to control malaria. At the present time there are estimated to be 300–500 million cases worldwide, with nearly one million deaths annually. The majority of the cases are in Africa, but India has 2–3 million cases annually. Most of the mortality is among children. Besides the mortality, the costs in terms of treatment and disability are very high. Travelers to endemic areas can become infected and prophylactic treatment is often recommended.

The vector for malaria is *Anopheles* mosquitoes. There are two major variations, *Plasmodium falciparum* (200–300 million cases worldwide) and *Plasmodium vivax* (100–200 million cases). *P. falciparum* causes a rapidly developing potentially fatal form of the disease and requires immediate treatment. *P. vivax* is a more chronic form of the disease. It is not usually fatal, but is prone to relapses and is very debilitating. Three other versions *Plasmodium malariae*, *Plasmodium ovale*, and *Plasmodium knowlesi* can infect humans, although in more restricted geographical areas and they resemble *P. vivax* in clinical form. Several others species of malaria infect primates, rodents, and birds. The latter two groups have been used extensively for testing of potential antimalarial drugs. All forms of malaria attack the red blood cells (*erythrocytes*) and the overt symptoms are fatigue, fevers, and chills. The fevers are periodic (about 3 or 4 days), corresponding with the life cycle of the parasites. The infection causes enlargement of the spleen, anemia, and leucopenia.

Many times throughout history malaria has been an important factor in military campaigns. Alexander the Great, after conquering most of western civilization of the time, died of malaria at age 33. Several invaders were kept from dominating Rome by the malaria-infested country-side, including Attila in 452 and a series of Holy Roman Emperors. During the American revolutionary war, it is believed that the British Army's efforts in the Carolinas and at the battle of Yorktown were severely hindered by malaria.[1] Malaria also played a role in the American Civil War. Union troops attempting to capture Vicksburg, MS and control the Mississippi River were delayed by more than a year because of disabling malaria. In WWI, British, French, and German troops suffered heavy casualties from malaria in the Balkans. During WWII, malaria was a major threat to American forces in the South Pacific. During the Viet Nam war, malaria again infected many American troops and in some cases proved to be resistant to the standard drug, chloroquine.

The construction of the Suez Canal late in the 1800s required antimalaria measures. Malaria and yellow fever were also severe problems during the construction of the Panama Canal between 1905 and 1910. In the early years, more than half the workers became infected. By then, the recognition of the role of mosquitoes permitted some reduction in the incidence by insect control. After the Spanish–American War, the United States developed an antimalaria campaign in Cuba led by William Gorgas.[a] In addition to improved sanitation, he used the natural pyrethrin insecticides in dwellings. An infestation of African mosquitoes threatened to cause an epidemic in northeastern Brazil in the late 1930s. Fred Soper[b] of the Rockefeller Foundation led efforts to combat the infection.

The highest case load in the United States occurred in 1935, when 125,000 cases and 5000 deaths were reported. In the United States, construction of major water projects, especially the Tennessee Valley Authority (TVA), incorporated malaria control measures. At the beginning of TVA in 1933, malaria infection was as high as 30% in the area. Improved drainage contributed to reduction of the level of infection. During WWII, specific antimosquito measures were taken around military bases in malaria-infested areas. At the end of WWII, these efforts were extended and by 1951 malaria had been eradicated in the continental US. At the present time, most malaria cases in the United States involve travelers returning from infected areas. Many cases involve nonimmune adults and children returning for visits to their home areas.

People living in endemic areas develop a certain level of immunity or resistance to malaria, however, there are limits to this immunity. In particular, there are numerous "strains" of the infective organism and resistance to a particular strain may not apply to another. Also, the parasite undergoes *antigenic variation*, that is, it can present a variable repertoire of surface structures that can avoid immune response. In areas where infection rates are high, as in sub-Sahara Africa, partial immunity is acquired by age 4–5, but mortality is high for young children. Even when malaria infection is not lethal, it can lead to chronic recurrence and generally poorer health status and increased mortality from other causes.

Because malaria is transmitted from human to human by mosquitoes, it requires a reservoir of infected individuals to persist. This is what offers the hope, and in many areas the reality, of eradication. Conversely, high levels of local infection can lead to epidemics. Immune individuals can harbor potentially infective forms, even though they do not exhibit any symptoms. The rates

---

[a]William Crawford Gorgas was born in Alabama in 1854, the son of a West Point graduate who eventually served as the Chief of Ordnance in the Confederate Army. He received an M. D. degree from the Bellevue Hospital Medical College in New York and joined the US Army Medical Corps in the1880s. Gorgas survived a yellow-fever infection acquired while serving in Texas. He was appointed Chief Sanitation Officer in Havana in 1898. Applying the recent findings of Ronald Ross (malaria) and Walter Reed (yellow fever) on the role of mosquitoes in transmitting these diseases, he implemented improvements in sanitary conditions that greatly reduced the incidence of infection in Havana. He served as US Army Surgeon General during WWI.

[b]Frederick L. Soper was born in Kansas in 1893 and received his undergraduate degree at the University of Kansas. He received a Ph. D. from the Johns Hopkins University School of Public Health. Soper spent most of his career with the Rockefeller Foundation. In the late 1930s, he led the organization of a campaign in Northeastern Brazil against the mosquitoes responsible for malaria and yellow fever. Soper was also among the leaders in organizing the WHO campaign for Global Eradication of Malaria.

of infection are seasonal and correlated with mosquito populations. There are inherent risks in the current situation. One is the development of wide-spread resistance to antimalarial drugs. Infections in many areas of the world are now resistance to the most valuable of the 20th century drugs, chloroquine. The development of artemisinin and its analogs (see Section 18.1.2.3) has provided a new weapon, but there are reports of artemisinin-resistance. Paradoxically, local malaria control in regions highly susceptible to infection brings its own risk. The loss of repeated exposure results in diminished immunity. Reintroduction of malaria, for example, as the result of warfare or other disruption of vector control, can then lead to epidemics.

18.1.1 RECOGNITION OF THE ETIOLOGY OF MALARIA

The association of malaria with swampy areas was known from antiquity and some ancient writing even suggested that insect bites were involved. Rice-farming areas were known to be particularly susceptible to malaria infections. The European name comes from Italian for bad (mal) air (aria). Hippocrates recognized the periodic nature of malarial fevers. Galen's theory of humors, however, attributed the disease to imbalance of humors and he recommended bleeding and purges for treatment. It was not until the 1700s that other theories began to have influence. An Italian physician, Giovanni Maria Lancisi, in 1717 summarized his observations in a book, *Noxious Emanations of Swamps and Their Cure.* He described dark pigments deposited in the spleen and other organs of malaria victims. He suggested two possible means of transmission of infection from insects, one through contamination of food or water and the other, the correct one, by insect bites.

Charles Louis Alphonse Laveran, a French Army physician working in Algeria in 1888, was the first to recognize malaria parasites in blood under the microscope. He identified them as protozoa, in contrast to the prevailing view that they must be bacteria. His work was aided by Patrick Manson, a physician in London who had long studied malaria. Camillo Golgi, an Italian physiologist, recognized the relationship between the changing forms of the parasites and the periodic fevers. Ronald Ross,[c] a British medical officer serving in India between 1881 and 1898, showed that avian malaria could be transmitted between birds by mosquitoes. Ross also recognized that specific mosquito species were involved. Ross was awarded the 1902 Nobel Prize in Physiology or Medicine. An

---

[c]Ronald Ross completed his medical training at St. Bartholomew's Hospital in 1875. He served in the Medical Service in India between 1881 and 1898. In 1899 he returned to England and continued research on malaria at the Liverpool School of Tropical Medicine. During WWI he was a consultant to the War Office on malaria. In 1911, he published *Prevention of Malaria*, which included mathematical models of malaria infections that helped lay the foundations of epidemiology.

Italian team in 1898, using volunteers, proved that human malaria could also be transmitted by mosquitoes.

During the 1930s and 1940s, the life cycle of the malaria parasite was established. The final piece in the puzzle, identification of the dormant form that causes relapse, was recognized in the 1980s. The infective form injected into the host's bloodstream by the female mosquito is called a *sporozite*. Once in the human blood stream, they develop into *schizonts*, which contain thousands of *merozites*. The merozites invade red blood cells and reproduce by asexual cell division, producing more merozites that invade additional red blood cells. Some of the merozites transform into *gametocytes* that circulate in the blood stream. When a female mosquito bites an infected person, gametocytes are transferred to the mosquito's gut where they transform into the sexual forms called *gametes*. These then produce *oocysts* that contain sporozites that can initiate a new infection. In some types of malaria (*P. vivax* and *P. ovale*), the merozites can assume a form called *hypnozoites*, that can remain dormant for up to several years. Figure 18.1 is a schematic representation of the malaria cycle.

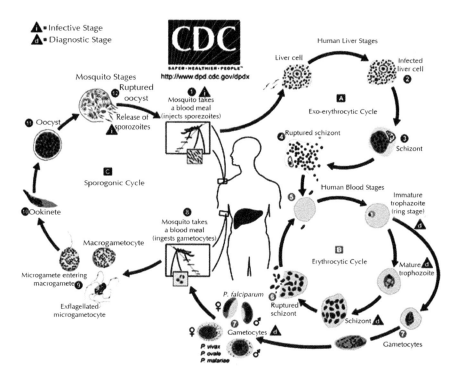

FIGURE 18.1   The life cycle of *Plasmodium*. (From http://www.dpd.cdc.gov/dpdx).

18.1.2   ANTIMALARIA DRUGS

18.1.2.1   QUININE

Quinine, initially in the form of the bark of the cinchona tree, was introduced as a treatment for malaria in Europe in the 17th century. The exact details of its discovery are uncertain. One story is that it was used to cure the wife of the governor of Spanish South America. Stronger evidence supports the view that Jesuit missionaries used cinchona to treat natives infected with malaria and then introduced it to Europe. That leaves unanswered the question of how the effect of cinchona bark was discovered. The explanation seems to be that the Indians in the high Andes, where the cinchona tree is native, used it as a muscle relaxant to stop shivering from exposure to cold. The Jesuits evidently tried it on the chills associated with malaria. In the 1640s, Juan de Lugo, a Jesuit Cardinal, investigated cinchona and successfully treated Louis XIV in 1649. His success led to Vatican support and the Jesuits in Peru organized the collection and shipment of cinchona bark to Europe, transporting it over the Isthmus of Panama. Cinchona bark, in disguise, was exploited in England by an apothecary, Robert Talbor, who prepared a secret formula containing cinchona bark and opium. He became the Royal Physician, to the consternation of the medical profession. He treated both Charles II of England and the French Dauphin with the remedy in the late 1670s. Only after his death was it revealed that cinchona bark was the active ingredient.

The active constituent of cinchona bark, quinine, was isolated first by Friedlieb F. Runge in 1819 and then again in 1820 by the French pharmacists Joseph Pelletier and Joseph Bienaime Caventou, who gave it the name quinine. Commercial production of quinine was undertaken by the Dutch in Java in 1865, using seeds collected by an English trader Charles Ledger and an expert Peruvian guide Manuel Incra Mamani. The species selected, called *Cinchona ledgeriana*, had especially high quinine content. Plantations for growing cinchona trees were established in Indonesia, Ceylon and India. Quinine is also used for its bitter taste in products such as tonic water.

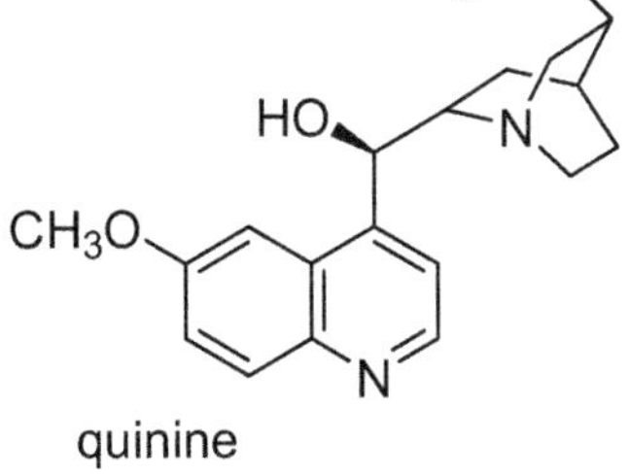

quinine

18.1.2.2   SYNTHETIC ANTIMALARIA DRUGS

The first indication of the possibility of synthetic antimalaria drugs came from work of Paul Ehrlich and Paul Guttman in 1891, who treated two malaria patients with the dye methylene blue. This was at the beginning of Ehrlich's investigation of dyes for treatment of disease (see Section 11.1). Malaria was a factor in WWI, especially in the Balkans. At the time, most of the quinine was grown in Dutch Indochina and marketed under Dutch control. No particular shortages developed, but both the Allied and German governments recognized the military importance of malaria prevention and treatment. Bayer undertook synthesis of antimalaria compounds after WWI. The first promising find was plasmoquine (later called pamaquin) in 1924. It was followed by atebrin, introduced in 1932. The program at Bayer was facilitated by an avian malaria assay developed by Wilhelm Roehl. Chloroquine was synthesized at Bayer in 1934, but it was not until the end of WWII, that its excellent antimalarial properties were recognized.

Between the two world wars, the British Medical Research Council had a Committee on Chemotherapy intended to stimulate research on antimalaria compounds, but little progress was made. The rise to power of the Nazis in Germany, and with it the prospect of war, resulted in renewed attention to the need for antimalaria drugs. At the outbreak of war in 1939, two compounds plasmoquine and atebrin (both products of Bayer) were available in addition to quinine. Production of plasmoquine and atebrin was undertaken by ICI. ICI also began a program of research that eventually led to two other drugs, proguanil and pyrimethamine, called *antifolates*. The latter was developed by Burroughs Wellcome and introduced in 1950 (see Section 18.2.4.2).

The rapid advance of the Japanese in Southeast Asia in 1942 cut off supplies of natural quinine. The United States undertook research at a number of academic and industrial labs that eventually led to two more antimalarial drugs, amodiquine and primaquine. Chloroquine was rediscovered when allied troops took control of Vichy French territory in Tunisia. French scientists there had been working on chloroquine and sontoquine. At about the same time, it was realized that information on these drugs was available at Sterling–Winthrop as a result of its relationship with Bayer. Cloroquine went into production near the end of the war and was the preferred drug between 1950 and the 1980s. The most recently developed of the quinoline group is tafenoquine, which is similar to primaquine in the scope of its activity, but requires a somewhat shorter course of treatment.[2]

Several of the new antimalaria drugs were synthesized in China. Lumefantrine was first synthesized at the Academy of Military Medical Sciences in Beijing in the 1970s. Another quinoline derivative, piperaquine, was used as a primary malaria treatment and preventive drug in China during the 1980s but resistance emerged by the end of the decade.[3] Piperaquine was also developed by Rhone-Poulenc in France. Aablaquine was developed at the Central Drug Research Institute in India and is available in that country. Scheme 18.1 gives the structures of some of the most important antimalaria compounds.

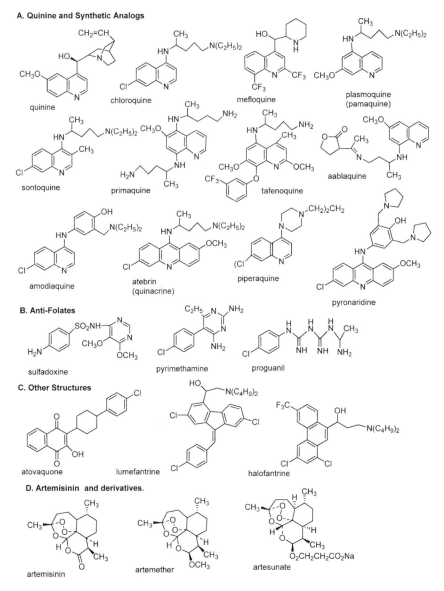

SCHEME 18.1   Structures of antimalaria drugs.

TOPIC 18.1 THE WWII AND VIET NAM ANTIMALARIA PROGRAMS IN THE UNITED STATES

The initial efforts to seek better antimalarial drugs in the United States began in the Division of Chemistry and Chemical Technology of the National Resource Council and focused on the inadequacies of the existing drugs. With war on the horizon, the matter took on some urgency. In 1941, The Committee on Medical Research (CMR), chaired by Alfred N. Richards (see Section 12.1.1) was formed under the umbrella of the Office of Scientific Research and Development. The CMR undertook studies in several areas including: parasite life cycle, synthesis of new compounds, immunology, toxicology, and cytology of the existing drugs. The program eventually engaged a number of academic and industrial laboratories. The testing was done primarily in ducks and chicks, following the avian model used at Bayer. Eventually, nearly 14,000 compounds were screened.

One of the major challenges to the US program was the uncertain relationship between tests conducted in birds and the results to be expected in humans. Prison volunteers were initially used for human tests. However, the most extensive studies were carried out under the direction of Neil H. Fairley[d] of the Australian Medical Corp in sub-tropical Queensland. Chloroquine and sontoquine were tested there in 1944 with very promising results against *P. falciparum*, although the drugs did not prevent relapses of *P. vivax*. Other tests were done using plasmoquine, but it had a very low margin of safety. Eventually primaquine was found to have the best activity against relapses of *P. vivax* and it was introduced in 1950. Primaquine was important during the Korean War in the treatment of recurrent *P. vivax*. Primaquine was also found to be effective in treating chloroquine-resistant *P. vivax* in New Guinea and Java in the early 1990s.

The antimalaria program, along with the penicillin and synthetic rubber programs, were the largest of the chemical efforts directly connected with WWII. The malaria program in many ways became a model for subsequent develop of large scale government-funded programs in biomedical research, particularly at the National Institutes of Health. The US Army again supported a major effort to synthesize antimalaria compounds during the Viet Nam war, when chloroquine resistance became a problem. Two compounds, mefloquine and halofantine, developed from that effort.

---

[d]Neil Hamilton Fairley joined the Australian Army Medical Corps in 1915. He served in Cairo during WWI and became familiar with schistosomiasis. Between the wars he became an expert in tropical diseases, working in England and India, as well as Australia. When Japan entered WWII, he returned to Australia and was responsible for prevention and treatment of malaria. He was a strong spokesman for taking steps to prevent malaria in the South Pacific during WWII. In June 1943, he began research at Cairns in Queensland on the effectiveness of the new drugs. After the war, Fairley continued research on malaria at the London School of Hygiene and Tropical Medicine.

Sources for Topic 18.1: Slater, L. B. *Ambix* **2004**, *51*, 107–134; Ockenhouse, C. F.; Magill, A.; Smith, D.; Milhous, W. *Military Med.* **2005**, 12–16; Slater, L. B. *Bull. History Chem.* **2006**, *31*, 75–80.

18.1.2.3 ARTEMISININ AND DERIVATIVES

Artemisinin was discovered as the result of an antimalarial program undertaken in China in support of the North Vietnamese during the Vietnam War. Reportedly, Mao Zedong himself decreed that the program should include Chinese traditional medicines. Artemisinin was isolated from a plant *Artemisia annua* called "*Qinghao*" used as an antimalaria treatment in parts of China. The active compound has a complicated and very unusual structure. The content in the crude material is 0.5–1.2% and the extraction yield is 50–80%. There is some prospect for increased yield by classical plant selection techniques and there is also the possibility of producing artemisinin microbiologically.[4] Three semi-synthetic derivatives, artemether, arteether, and sodium artesunate, have been used as antimalarials. They are all made from dihydroartemisinin. This compound, too, has antimalarial activity, but there are questions about it stability in pharmacological preparations.[5] The ethers are hydrophobic and adsorption is variable, but artesunate is water soluble and can be given by injection. So far, no totally synthetic drugs of the artemisinin group have been developed.

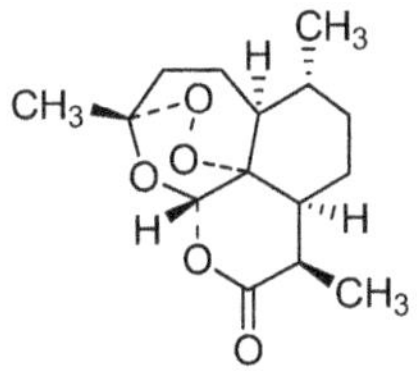

Artemisinin

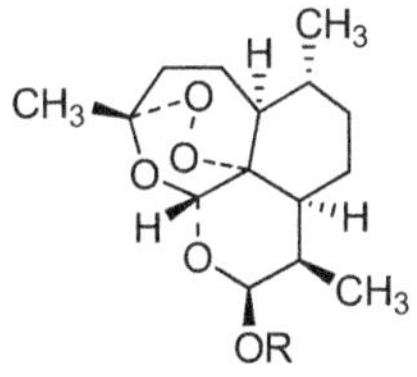

Artemether R = CH_3

Arteether R = C_2H_5

Sodium Artesunate R = $\overset{\displaystyle O}{\overset{\|}{C}}CH_2CH_2CO_2Na$

The Chinese results were not published in the open literature until 1979. The work of isolation and structure confirmation was reproduced in the West, but there was no immediate development by Western pharmaceutical companies. The artemisinin derivatives are fast-acting, although when used alone, they have a substantial recurrence rate of 10–25%. A three-day course of artemisinin can reduce the number of parasites by a factor of 10^8 and if a combination drug or the immune response can completely clear the parasitemia, the likelihood of resistance is greatly reduced. There have been some reports of resistance, particularly in Cambodia, where the drug has been available from China for some 30 years. In 2006, the WHO accepted artemisinin derivatives as first-line antimalarials, but recommended use only in combinations to slow the development of resistance. The artemesinins are very fast acting but have short half-lives so that combinations with longer acting drugs are preferred. The combination artemether–lumefantrine was developed by Novartis and is currently available as Coartem. It received FDA approval in the United States in 2009. Two other combinations, artesunate–mefloquine and dihydroarteminsin–piperaquine are also under study.

18.1.3   TARGETS AND MECHANISM OF ACTION OF ANTIMALARIA DRUGS

Based on the point of attack, antimalarial drugs are classified as *schizontocides*, *gametocides*, or *sporontocides*. The schizontocides are further classified as tissue or blood schizontocides. The blood schizontocides attack the erythrocytic stage. Quinine, its quinoline analogs and the artemesinins are in this group. Drugs that attack the tissue forms include primaquine and some of the antifolates. Only primaquine has activity against the dormant hypnozoites, and thus has antirelapse activity against *P. vivax*. Drugs that can eliminate these forms and thus prevent relapses are called "radical curative." Drugs that can attack the liver stage are called "causal prophylactics."

18.1.3.1   QUININE, QUINOLINES, AND PHENANTHRIDINES

Quinine, and the 4-aminoquinolines are active against the blood schizonts, but not the tissue forms. This is also the site of action of artemisinin and its derivatives and the antifolates. Primaquine, and to some extent proguanil, have activity against the tissue forms. An important aspect of primaquine is its ability to kill the gametocyte form, including those of drug-resistant strains. Primaquine does not affect the endoerythrocytic forms and has a rather short half-life and requires administration over an extended period (e.g., 14 days). The mode of action is not entirely clear, but seems to be directed at the mitochondrion. Two analogs of primaquine, tafenoquine and aablaquine, are currently under investigation. Tafenoquine, which is being developed by GlaxoSmithKline, was identified at the Walter Reed Army Institute of Research and can be used for treatment of both *P. falciparum* and *P. vivax*. It also has potential use as a chemoprophylactic.[6] It has a longer half-life than primaquine and requires a shorter treatment cycle. The most troublesome side-effect of the 8-aminoquinolines is to cause hemolysis, especially in persons with low 6-glucose phosphate levels, a fairly common occurrence.

18.1.3.2   ANTIFOLATES

The drugs in this group include sulfadoxine, pyrimethamine, biguanide, and proguanil (see Scheme 18.1 for structures). Proguanil is actually a prodrug being converted to the active form cycloguanil.

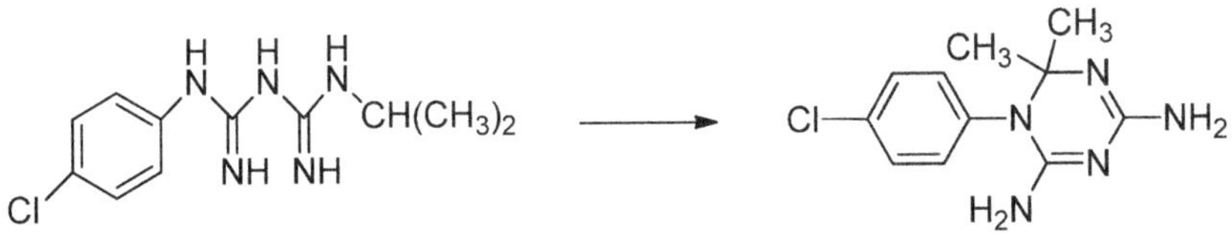

proguanil   cycloguanil

Malaria parasites must synthesize folic acid. Drugs that interfere with this metabolism have antimalarial activity. (Recall that antifolate activity is also the basis of the sulfonamide antibiotics, see Section 11.1.) Pyrimethamine, biguanide, and proguanil inhibit dihydrofolate reductase (see Section 10.1.2.2). These two types of antifolates are frequently used in combination with one another. An importance characteristic of the antifolates is their ability to attack immature gametocytes, as well as the erythrocytic forms.

18.1.3.3   QUINONES

Atovaquone used alone is not very effective and resistance develops rapidly. However, it is very effective in combination with proguanil.[7] Atovaquone appears to interfere with mitochondrial electron transport and one specific site of action may be dihydroorotate dehydrogenase, which is involved in both pyrimidine biosynthesis and mitochondrial electron transport. Atovaquone is also active against a number of other protozoal parasites, including those that affect immune-compromised patients.[8]

18.1.3.4   ARTEMISININ AND DERIVATIVES

Like the quinolines, artemisinin and its derivatives are primarily active against the erythrocytic forms. There have been several proposals for specific mechanisms of action, most suggesting generation of reactive free radicals from the peroxide group in artemisinin. However, many details are unsettled. Like the quinolines, artemisinin derivatives lead to the formation of heme pigments. The artemisinin derivatives are believed to act through a common metabolite, dihydroartemisinin. One of the important differences between the artemisinins and quinolines is the relatively short half-life of the artemisinins. For this reason, combination therapy with one of the longer acting drugs is now recommended.

18.1.3.5   COMBINATION THERAPIES

As evidence of malaria resistance accumulated, one approach was to use combinations of drugs such as sulfadoxine–pyrimethamine and atovaquone–proguanil.[9] Although these treatments can be effective, they require adequate medical infrastructure. Several combinations of artemisinin derivatives are also available. The most widely used is artemether–lumefantrine. It is made by Novartis as Coartem and is part of WHO malaria control programs. Several other drug combinations are available.

18.1.3.6   PROPHYLACTIC DRUGS

Prophylactic drugs are used for travelers to regions where malaria is endemic. There are an estimated 30,000 cases annually among travelers and about 150 deaths. There is considerable variation in the likelihood of infection, depending on the season of the year and length of stay. Natives of endemic areas can lose their immunity after spending several years in non-infested areas. It is estimated that 65–90% of cases in the United States involve people who have returned to their native country for a visit. Nonimmune children of returning families are especially at risk. The malaria mosquitoes feed from dusk to dawn, so night protection is important, including screens in housing, bed-netting, and insect repellants.

The combination atovaquone–proguanil, marketed as Malarone, is effective as a prophylactic drug and is licensed for this purpose in North America and Europe. It is also recommended by the US Center for Disease Control.[10] The tetracycline antibiotic doxycycline is also recommended as a prophylactic. Mefloquine was recommended for a time, but incidents of psychiatric side-effects have limited its use. Primaquine is also used on an off-label basis as a prophylactic. A newer 8-aminoquinoline, tafenoquine, is also effective. Both of these drugs can cause life-threatening hemolysis in patients who are deficient in glucose-6-phosphate dehydrogenase and the patient's status with respect to this enzyme must be checked prior to use.

18.1.4 DRUG-RESISTANT MALARIA

Resistance of *P. falciparum* to chloroquine was first noted in the 1950s and was wide-spread by the 1970s. A combination of two drugs, sulfadoxine and pyrimethamine, then became the first line therapy. Mefloquine was introduced in the 1970s, but resistance developed fairly quickly. Current treatments include a combination of atovaquone and proguanil or combinations including artemisinin or derivatives. Chloroquine-resistant to *P. vivax* is somewhat less wide-spread, although it is high in some areas, particularly Indonesia and New Guinea. There have been reports of resistance to artemisinin in Cambodia. This is viewed with grave concern, because artemisinin and its derivatives are currently the most reliable antimalarial drugs available. The WHO has undertaken efforts to contain the resistance, in particular recommending against continued use of the artemisinins as monotherapies.[11]

18.1.5 PROSPECTS FOR FUTURE DRUG DEVELOPMENT

Development of antimalaria drugs has generally not received high-priority attention from the pharmaceutical industry because development costs are high and prospects for cost recovery poor. Military-oriented programs have been the source of most of the current drugs. More recently, drug discovery has been supported by research funded by private–public organizations, with industrial cooperation in the development of the most promising drugs. A particularly important area is drugs that would kill mature gametocytes. Currently, only primaquine and pyronaridine are effective on this stage. Mature gametocytes are required for transmission so effective elimination at this stage could break the transmission cycle.

18.2 TRYPANOSOMIASIS

There are three major forms of trypanosomiasis, two centered in Africa, caused by *Trypanosoma brucei rhodesiense* and *T.b. gambiense*, and the other in South America, caused by *T. cruzi* and called Chagas's disease.

18.2.1 AFRICAN SLEEPING SICKNESS—T.B. RHODIESENSE AND T.B. GAMBIENSE

African trypanosomiasis, or sleeping sickness, is transmitted by the tsetse fly (*Glossina*). It is limited to the range of the tsetse fly in Africa, between about

14°N and 20°S latitude. The *T.b. rhodiesnse* variety is an acute form of the disease and is fatal in a matter of a few months if untreated, while *T.b. gambiense* is more chronic and victims can survive for several years. The infections become fatal when the central nervous system is involved. The disease can be transferred from animals (cattle in the case of *T.b. rhodesiense*) or humans. Civil unrest which interferes with medical treatment can permit resurgence of the diseases. Figure 18.2 shows the life cycle of *T. brucei*.

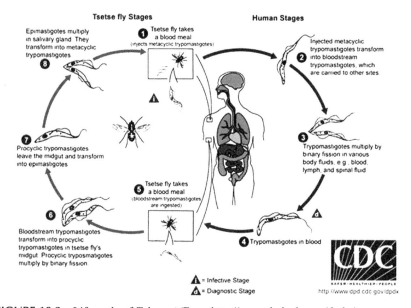

FIGURE 18.2   Life cycle of *T. brucei* (From http://www.dpd.cdc.gov/dpdx.)

A major epidemic that occurred in the late 1800s and early 1900s prompted the European colonial powers to investigate means of preventing and curing the disease. Among those who took up the problem was Paul Ehrlich, who tested many dyes for activity. This eventually led to the synthesis of suramin, the first acceptable drug at Bayer by Wilhelm Roehl and his colleagues in 1916. It was introduced into use in 1923 and is still in use today.

Ehrlich and coworkers pursued another lead, an arsenic compound called "atoxyl" that had been synthesized many years earlier. This line of investigation was also pursued by others and led to tryparsamide, melarsen oxide, and eventually in 1949 to a mixture of melarsen oxide with the sulfur compound dimercaprol, which greatly reduced the toxic effects of the arsenic compounds.

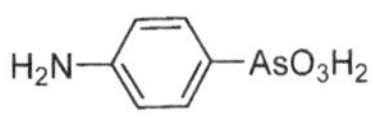

atoxyl

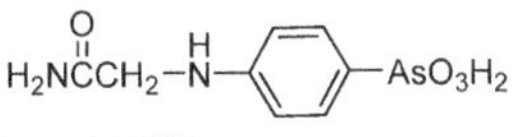

tryparsamide

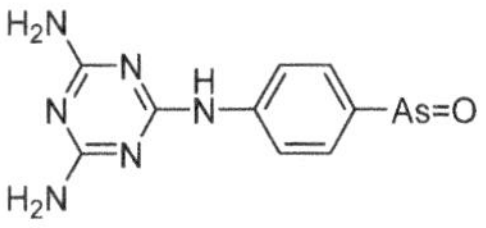

melarsen oxide

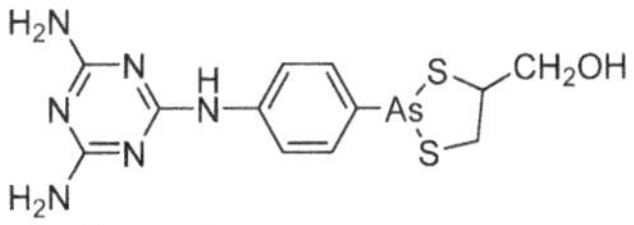

melarsoprol

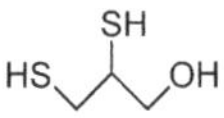

dimercaprol

Additional drugs were developed from studies in England in the 1930s. These included the cationic drugs ethidium bromide and isometamidinium.

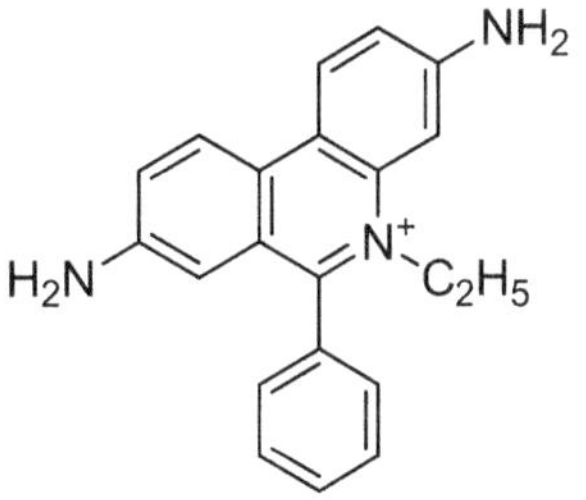

ethidium bromide

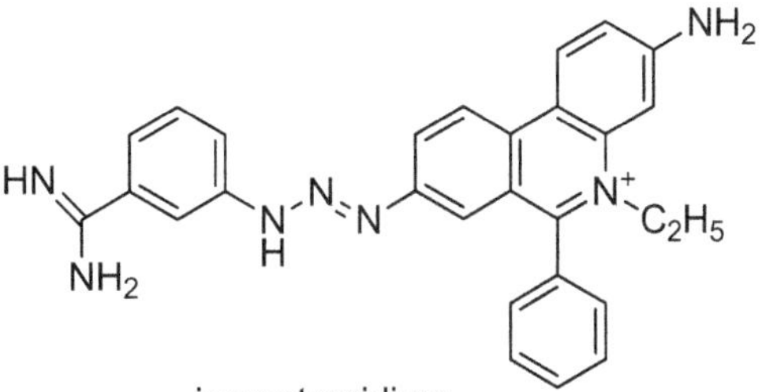

isometamidium

The *bis*-amidines berenil and pentamidine also have trypanocidal activity and pentamidine is currently used. Many other related *bis*-amidines have activity, but none has met the criteria for introduction as drugs.

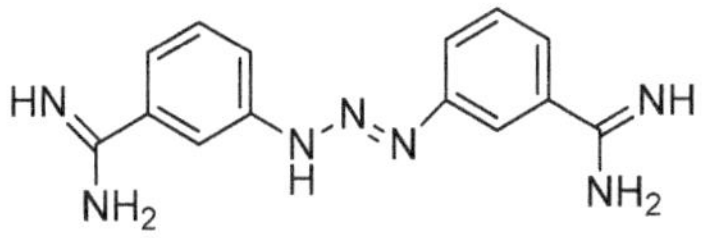

berenil

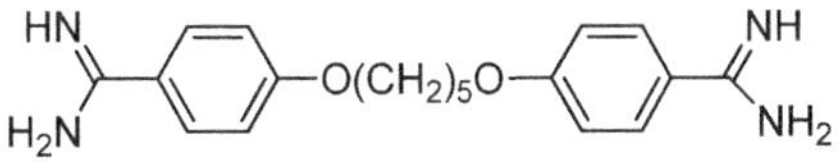

pentamidine

The most recently introduced drug is an ornithine decarboxylase inhibitor, eflornithine, which was initially investigated as an anticancer drug. It has good activity against *T.b. gambiense* but not *T.b. rhodesiense*.

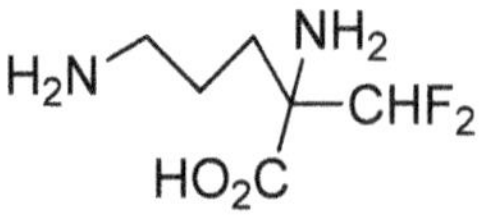

$$H_2N\diagdown\diagup\diagdown\diagup\overset{\displaystyle NH_2}{\underset{\displaystyle HO_2C}{C}}-CHF_2$$

eflorinithine

None of these drugs has ideal characteristics. Suranim has significant side effects and melarsoprol requires administration in a hospital because of the severity of potential side effects. Pentamidine is administered by intramuscular injection. Eflornithine is administered by multiple infusions, which also requires substantial medical support. The drugs are currently donated by their makers through medical aid agencies.

18.2.2 CHAGAS DISEASE—T. CRUZI

Chagas disease is caused by the protozoa *T. cruzi*. It is endemic in much of Central and South America and is also known as American Trypanosomiasis. The WHO estimated that in 2000 there were about 18 million infected people and about 20,000 deaths annually. It is considered third in significance, behind malaria and schistosomiasis, among the parasitic diseases of the world. The disease is seen occasionally in the United States, mainly in immigrants from endemic areas. It is associated with poverty and poor housing. Efforts to eliminate the disease have made progress, and it has largely been eliminated in Uruguay, Chile, and Brazil. The highest level of infection is in Bolivia (6–7%), while in Central American countries, it is 2–3%.

The disease was carefully described in 1909 by a Brazilian physician Carlos Chagas, who recognized the symptoms, pathology, and the vector. The vectors are insects of the *Triatomine* and *Rhodnius* genera. The transmission can be from human–insect–human or animal–insect–human. It is also possible to transfer the infection through blood (e.g., transfusion) or even contaminated food. The parasite has three life stages called *epimastigote*, *trypomastigote*, and *amastigote*. The epimastigotes undergo cell division in the insect gut and can transform to trypomastigotes. The trypomastigote is the infectious form transferred by the insect. The amastigote can undergo differentiation in the host, generating more trypomastigotes. The life cycle is summarized in Figure 18.3.

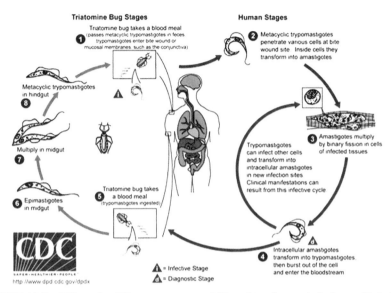

FIGURE 18.3   Life cycle of *Trypanosoma cruzi* (From http://www.dpd.cdc.gov/dpdx.)

A bite by an infected insect results in a quite high level of parasitemia but only mild symptoms, such as swelling at the bite site and fever. The symptoms are often unrecognized and untreated. The acute phase is followed by a dormant phase that is asymptomatic. About 20–30% of the patients develop symptoms 10–20 years after the original infection involving cardiac and digestive tract abnormalities. The cardiac pathology can lead to congestive heart failure.

There are only two recognized drugs for Chagas disease, nifurtimox and benznidazole, both of which were introduced prior to 1960. Neither of the drugs is approved by the US FDA, and they can only be obtained in the United States from the CDC under "investigational drug" protocols. Only benznidazole is currently available commercially. These drugs have limited effectiveness and significant side effects. They can be effective if administered during the acute phase, but have little effect on the dormant or chronic phases. The modes of action of both drugs seem to involve *in vivo* reduction of the nitro group that provides the active species. There are several other biochemical targets that have been identified but no satisfactory drugs have been found.

nifurtimox   benznidazole

18.3   LEISHMANIASIS

Like trypanosomiasis, leishmaniasis is caused by a parasitic protozoa transmitted by insects. There are several forms. The most severe, called visceral leishmaniasis, also known as kala-azar, attacks internal organs. It is caused by *L. donovani* and is transmitted by a sandfly of the *Phlebotomine* family. It is found primarily in India, Bangladesh, Nepal, Ethiopia, Sudan, and Kenya. There are estimated to be 500,000 cases of visceral leismaniasis annually with 50,000 deaths. The other forms are called cutaneous and mucocutaneous leishmanisis and are caused by other *Leishmania* species. Cutaneous lesiamaniasis caused by *L. major* is found mainly in the Middle-East, Afghanistan, and Pakistan. It is estimated that there are 1–1.5 million cases each year. Mucocutaneous leishmania in the Western Hemisphere is caused by several species, including *L. braziliense* and *L. mexicana.* This form of the disease attacks the oral and nasal cavities. Leishmania is usually transmitted from human to human by the vectors, but there are animal reservoirs in some areas, including dogs and rodents. There are regional variation in the form and severity of the infections. The risk of infection is particularly high in immunocompromised individuals, such as AIDS patients. Travelers and tourists, as well as military personnel, can acquire the infection in endemic areas.[12] The life cycle for leishmaniasis is shown in Figure 18.4.

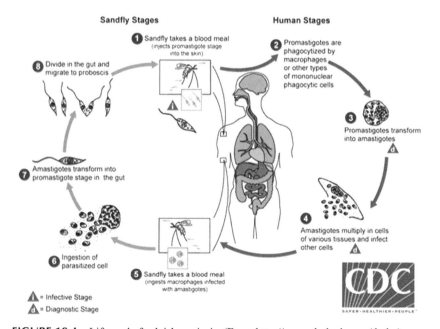

FIGURE 18.4   Life cycle for leishmaniasis. (From http://www.dpd.cdc.gov/dpdx.)

The most common drugs used for treatment of leishmaniasis are antimony compounds that were introduced in the 1920s. Their activity was recognized in an evolutionary way from antimony drugs used for other conditions. They can only be administered by injection over a long treatment period that usually requires hospitalization. While the drugs themselves are inexpensive, the treatment regimen is not. The next drug to be used, amphotericin B was introduced in the 1960s. It, too, is quite toxic and is administered by injection in a hospital setting. More recently paromomycin has shown some activity. The most recently developed drug is miltefosine. It is orally active. The structures are shown in Scheme 18.2. Drug combinations that can shorten treatment and perhaps delay resistance are under study.

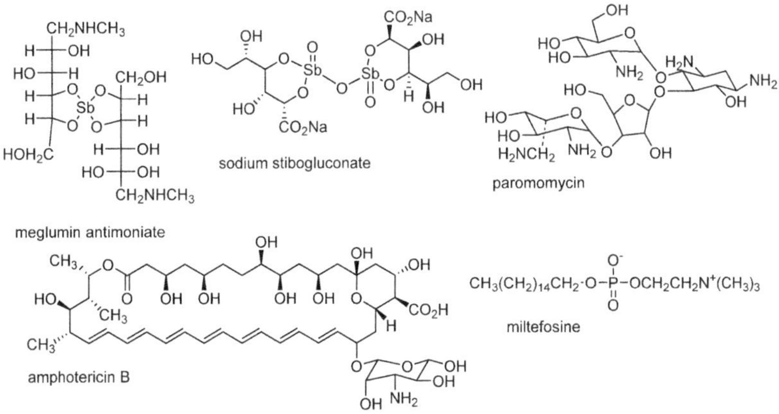

SCHEME 18.2   Drugs used in treatment of leismaniasis.

18.4   SCHISTOSOMIASIS

Schistosomoiasis, sometimes called snail fever, is cause by a trematode worm that is transmitted to humans through contact with water containing an infective larval form called *cecariae*. The worm has an intermediate stage in snails. There are several variations including *Schistosoma japonica*, *S. mansoni* (primarily intestinal), and *S. haematobium* (primarily urogenital). It is estimated that there are 200 million cases worldwide, mainly in subtropical or tropical parts of Africa, Asia, and South America. About 80% of the cases are in Africa. In some cases, the infestation causes *Katayama fever*, a systemic hypersensitivity several weeks after the primary infection. This is most common among people who have not had previous exposure. The chronic infection is characterized by swollen organs and anemia and can result in growth retardation and restricted cognitive development in children. Urogenital schistosomiasis results in lesions that increase the likelihood of HIV infection, and rates of schistosomiasis and AIDS are correlated. Squamous cell bladder cancer rates are also correlated with schistosomiasis and have declined significantly, where the incidence of schistosomiasis has been reduced.

Several species of snails are involved. The snails release larvae into the water that can penetrate human skin. The larvae differentiate to male and female forms in the human host and reproduce in the digestive or urinary tract. The eggs produced within the human host cause most of the symptoms of the disease through an inflammatory response. The adult worms have life spans of 3–10 years but can live much longer. The adult worms can develop an antigenic surface that is not recognized by the host immune system. The eggs are released back into the water through feces or urine. The eggs released in water hatch to give *miricidia*, which reinfect snails. The miricidia reproduce asexually to give first sporocysts and then the larvae that infect humans. Domestic animals, can also serve as mammalian hosts and thus as reservoirs for the disease. Cattle and water buffalo are particularly significant reservoirs. Because of the crucial role contamination of water by human waste plays in the transmission cycle, sanitation procedures, and facilities can prevent the disease. The life cycle is shown in Figure 18.5.

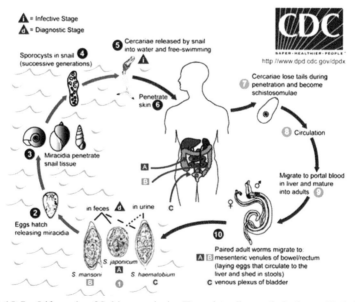

FIGURE 18.5   Life cycle of Schistosomiasis. (From http://www.dpd.cdc.gov/dpdx.)

The antimonial drugs were introduced in 1907, but were difficult to administer and quite toxic. A single drug praziquantel has been the main treatment since the 1970s. The drug was discovered at E. Merck in Germany and then developed by Bayer for use in treatment of animals. It is currently manufactured by several companies worldwide, including Shin Poong in Korea and Xixing Xinyu in China. The drug is inexpensive and relatively free of side effects. Its mechanism of action is not known in detail, but seems to involve disruption of calcium ion channels. Praziquantel is not active against all stages of the parasites and must be administered periodically. The artemisinin antimalarials have shown some success in treatment of schistosomiasis. They tend to be most effective against juvenile forms of schistosomes that are the least susceptible to praziquantel, and a combination of the two drugs has been considered. The artemisinin derivatives also have prophylactic activity and have been used, for example, with flood workers in endemic areas.[13]

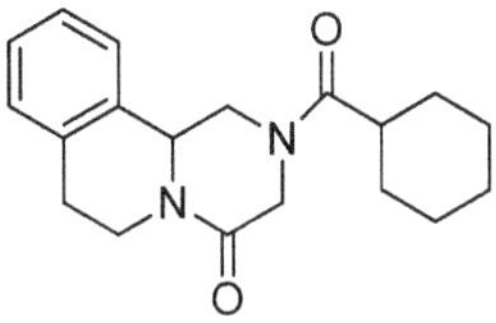

praziquantel

In China, the prevalence of schistosomiasis has been reduced by about 90% since the 1950s, when 10–12 million people were infected. Beginning in the 1950s, efforts in education, sanitation and hygiene, and snail control were put in place. Many waterways and irrigation channels were modified and/or treated with molluscicides to control the snail population. These efforts reduced infections to about 1.5 million cases by 1990. The availability of praziquantel then permitted the emphasis to shift to a chemotherapeutic approach and reduced the level of infection by about another 50% by 2002. Since then, there has been some indication of slight increases in infection.[14] As a result, new emphasis has been given to minimizing the prevalence of infection in endemic areas. These include improved water supply, sanitation practices, and livestock management to minimize transfer of infection.[15] Considerable progress has also been made in Egypt, where rural infection rates have been reduced from around 60% to <5%. Currently, WHO recommends chemotherapy with praziquantel, rather than snail eradication. This approach has been quite successful in countries such as Egypt, Brazil, and the Philippines, but less so in tropical Africa. Currently, efforts are being made to provide annual preventive treatment to school children. The current dependence on a single drug raises the issue of potential development of resistance. Another area of concern in China is the effect of massive water control projects, such as the Three Gorges Dam and the South-to-North water transfer project.[16] The dam, in particular, has created a new habitat for snails in an area that is currently not infested, but which lies between two areas that are. Also, the change in downstream flow may improve the habitat for the infective snails. There is also concern that global warming may expand the snail habitat. It remains to be seen how these environmental and ecological factors play out.

18.5   LYMPHATIC FILARIASIS

Lymphatic filariasis results from infection by helminthic worms. There are several species with different primary sites of infection around the world. *Wucheria bancrofti* is dominant in Southeast Asian and *Brugia malayi* is the main form in Indonesia. The infection is transmitted by several species of mosquitoes, including *Anopheles*, *Aedes*, and *Culex*. Infected individuals have both immature (microfilaria) and adult worms (macrofilaria). Humans are the only significant mammalian reservoir. The worms infect the lymph system and spleen and can also involve lung, liver, and kidneys. The clinical manifestations are the result of immune response to dead worms and secondary bacterial infection of the damaged tissue. In some cases, obstruction occurs in the lymph system resulting in enlargement of limbs or reproductive organs, which is known as *elephantiasis*. As many as 120 million people are thought to be infected in areas of Africa, Bangladesh, India, Southeast Asia, Latin America, and the Pacific Islands.

The first drug to be recognized as having activity was diethylcarbamazine (DEC), which was introduced in the 1940s. It kills microfilaria, but not the adult worms. In heavily infested patients, it can trigger a very strong immune response to the dead worms. It was introduced prior to development of modern drug standards throughout the world, and originally was administered over 2 weeks. Only later was it recognized that a single dose was just as effective. The mechanism of action of DEC is unknown, but it probably works in concert with the immune system. Albendazole is another important drug in control of filariasis. Unlike DEC, it is able to kill the adult macrofilaria. Albendazole was originally developed for veterinary use by SmithKline & French and like similar drugs, for example, mebendazole can control a variety of intestinal helminthes.

diethylcarbamazine   albendazole   mebendazole

Ivermectin was originally introduced as a veterinary drug (heartworm) by Merck, but subsequent investigation showed it had activity against lymphatic filariasis. It was approved for human use in 1987. It has a long-lasting effect of suppressing production of microfilaria, although it does not eliminate macrofilaria. Its action may target chloride channels, but definitive information is unavailable. The structure of ivermectin is shown in Section 18.7.

In some countries, particularly China, table salt treated with DEC has been used as a control measure. The approach is effective if the availability of nontreated salt can be limited. Using screening and treatment, as well as treated salt, from the 1970s until 1994, the disease has been essentially eliminated in China. The vectors remain in place, however, so that imported cases from neighboring countries remains a possibility. In the 1990s, the WHO sponsored several studies that indicated that combinations of albendazole, with DEC or ivermectin were safe and effective in eliminating filarial infections.[17] These combinations are now the main tool of the Global Programme to Eliminate Lymphatic Filariasis.[18] There is a symbiotic relationship between the filarial and a bacterial species, *Wolbachia*. This has opened another chemotherapeutic route involving treatment with the antibiotic doxycycline (see Section 12.4), which interferes with reproduction of the filaria.

At the present time, concerted efforts to eliminate lymphatic filariasis are underway, funded in part by the Bill and Melinda Gates Foundation, and managed through the WHO. The efforts are also supported by governmental funding from developed countries. The effort is based on annual treatment of at-risk communities with albendazole and either ivermectin or DEC for a period of 4–6 years.[19] The drugs are donated by Glaxo (albendazole) and Merck (ivermectin). The albendazole–ivermectin combination also kills other helminthic worm infections, such as hookworm. Treatment is expected to bring ancilliary benefits to children, because intestinal worm infections are known it retard growth and cognitive development. The current goal is to eliminate transmission of lymphatic filariasis by 2020.

18.6 OTHER HELMINTHIC INFECTIONS

There are a variety of parasitic worms of the digestive tract with common names such as round worms, hook worm, pin worm, whip worm, and tape worm. As many as 2 billion people worldwide are affected. These infestations can lead to anemia, malnutrition, and retarded growth and cognitive development of children. Infections are acquired from soil contaminated with fecal material, rather than through a vector species. Many of the drugs available for treatment were originally developed as veterinary drugs. The benzimidazoles mebenzadole, albenzadole (see Section 18.5) thiabenzdazole, and triclabendazole are an important group. These drugs are tubulin binders that show selectivity for the helminthic tubulin.

thiabendazole

triclabendazole

Other drugs such as pyrantel, oxantel, amidantel, and metrifonate (also called chlorofos and trichlorofon) appear to be directed at the cholinergic system of the worms.

pyrantel   oxantel   amidantel   metrifonate

18.7   ONCHOCERCIASIS

Onchocerciasis is transmitted by a black fly, *Simulum damnosum*, that resides in fast-flowing tropical streams. It leads to damage of the eye and eventually to blindness, called river-blindness. There is a related infection, loasis, caused by *Loa loa*, which is less severe. Figure 18.6 shows the life cycle for onchocerciasis.

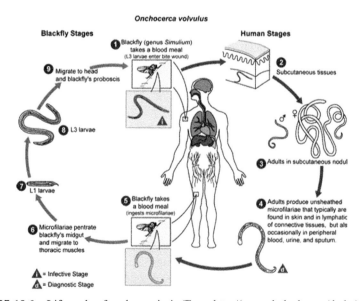

FIGURE 18.6   Life cycle of onchocerciasis (From http://www.dpd.cdc.gov/dpdx.)

The implementation of a control program in 1974, and the discovery of a very effective drug, ivermectin, has greatly reduced its incidence. Ivermectin, however, does not kill adult worms, so periodic retreatment is necessary. There is one important limit on ivermectin use. In areas where both onchocerciasis and loasis are present, ivermectin can cause a severe adverse reaction in patients infected with loasis. Treatment with doxycline is recommended in these circumstances.

Ivermectin arose as the result of a collaboration between Merck in the United States and the Kitasato Institute in Japan.[20] A microorganism, *Streptomyces avermectinins*, isolated in Japan produced a compound found at Merck to have anthelminthic activity. The active compound was isolated and identified as a 16-membered lactone, avermectin. The 22,23-dihydro derivative, called ivermectin, was developed as an agent against parasitic helminths, arachnids, and insects. Marketing began in 1981, and it quickly became successful in uses such as treatment of heartworms in dogs and in other veterinary applications. Ivermectin also kills fleas, ticks, lice and mites. One of the mechanisms for selectivity of ivermectin is the multidrug efflux system that is present in mammals but not worms. This system selects for exogenous chemicals and prevents certain drugs from entering cells. Certain dog breeds, collies especially, tend to lack this system and therefore are susceptible to ivermectin toxicity.[e] Related macrocyclic lactones, such as derivatives of milbemycin have biological activity similar to ivermectin and have been introduced into veterinary medicine.

avermectin   milbemycin

Merck scientists also examined the activity of ivermectin against *Onchocerca volvulus*, the parasitic worm that is the causal agent of river-blindness in Africa. In cooperation with the WHO, ivermectin was tested and approved for human

---

[e]Dowling, P. *Can. Vet. J.* **2006,** *47*, 1165–1168. Two lines of evidence resulted in the elucidation of this problem. One was the observation that MDR-knockout mice were sensitive to ivermectin (discovered accidentally during anti-mite treatment of mouse facilities) and the second was that collies were over-represented in reports of drug toxicity records. Genetic analysis of collies and related breeds revealed a low level of the MDR gene.

use in 1987 under the name Mectizan. Because infected persons are normally impoverished, Merck donated the drug. Since 2002, the program has been administered by the African Programme for Onchoceriasis Control of the WHO. Ivermectin has very low toxicity and can be administered with little medical supervision. Ivermectin kills only immature filarial and must be administered periodically.[21] It is used in combination with two other antiparasitic drugs albendazole and DEC.

KEYWORDS

- **Malaria**
- **quinine**
- **quinoline derivatives**
- **artemisinin**
- **drug-resistance**
- **trypanosomiasis**
- **leishmaniasis**
- **schistosomiasis**
- **lymphatic filariasis**
- **onchocerciasis**
- **helminthic infections**

BIBLIOGRAPHY BY SECTIONS

Section 18.1: Rocco, F. *The Miracle Fever Tree: Malaria and the Quest for a Cure that Changed the World*; Harper Collins: New York, 2003.

Section 18.1.1: Carter, R.; Mendis, K. N. *Clin. Microbiol. Rev.* **2002**, *15*, 564–594; Cox, F. E. G. *Parasites & Vectors* **2010**, *3*, 5.

Section 18.1.2.2: Greeenwood, D. *J. Antimicrob. Chemother.* **1995**, *36*, 857–872; Wiesner, J.; Ortmann, R.; Jomaa, H.; Schlitzer, M. *Angew. Chem. Int. Ed. Engl.* **2003**, *42*, 5274–5293; Bray, P. G.; Wood, S. A.; O'Neill, P. M. *Curr. Top. Microbiol. Immunol.* **2005**, *295*, 3–35; Muregi, F. W. *Curr. Drug Discovery Technol.* **2010**, *7*, 280–316.

Section 18.1.2.3: Kindermans, J.-M.; Pilloy, J.; Olliaro, P.; Gomes, M. *Malar. J.* **2007**, *6*, 125–131; Schlitzer, M. *Chem. Med. Chem.* **2007**, *3*, 944–986; Whitty, C. J. M.; Chandler, C.; Ansah, E.; Leslie, T.; Staedke, S. G. *Malar. J.* **2008**, *7*(Suppl. 1), S1–S7; White, N. J. *Science* **2008**, *320*, 330–334; Cui, L.; Su, X. Z. *Expert Rev. Anti-Infect. Ther.* **2009**, *7*,

999–1013; Butler, A. R.; Khan, S.; Ferguson, E. *J. R. Coll. Surg. Edinburgh* **2010**, *40*, 172–177; Maude, R. J.; Woodrow, C. J.; White, L. J. *Drug Dev. Res.* **2010**, *71*, 12–19.

Section 18.1.3: Klayman, D. L. *Science* **1985**, *228*, 1049–1055; Haynes, R. K.; Vorwiller, S. C. *Trans. R. Soc. Trop. Med. Hyg.* **1994**, *88*, S23–S26; Peters, W. *J. R. Soc. Med.* **1999**, *92*, 345–352.

Section 18.1.4: Olliaro, P. *Pharmacol. Ther.* **2001**, *89*, 207–219; Rieckmann, K. H. *Ann. Trop. Med. Parasitol.* **2006**, *100*, 647–662; D'Alessandro, U. *Expert Opin. Pharmacother.* **2009**, *10*, 1291–1306; Eastman, R. T.; Fidock, D. A. *Nat. Rev. Microbiol.* **2009**, *7*, 864–874; Martinelli, A.; Moreira, R.; Cravo, P. V. L. *Mini-Rev. Med. Chem.* **2008**, *8*, 201–212.

Section 18.1.5: Petersen, E. *Expert Rev. Anti-Infective Ther.* **2004**, *2*, 119–132; Shanks, G. D.; Edstein, M. D. *Drugs* **2005**, *65*, 2091–2110; Castelli, F.; Oddini, S.; Autino, B.; Foca, E.; Russo, R. *Pharmaceuticals* **2010**, *3*, 3212–3239.

Section 18.1.7: Sweeney, A. W. *Parasitologia* **2000**, *42*, 33–45; Rieckman, K. H. *Ann. Trop. Med. Parasitol.* **2006**, *100*, 647–662.

Section 18.2.1: Seed, J. R. *Int. J. Parasitol.* **2001**, *31*, 434–442; Wainwright, M. *Biotechnol. Histochem.* **2010**, *85*, 341–354; Steverding, D. *Parasites Vectors* **2010**, *3*, 15–24; Burri, C. *Parasitol.* **2010**, *137*, 1987–1994.

Section 18.2.2: Sanchez-Sancho, F.; Campillo, N. E.; Paez, J. A. *Curr. Med. Chem.* **2010**, *17*, 423–432; Urbina, J. A. *Acta Trop.* **2010**, *115*, 55–68.

Section 18.3: Murray, H. W.; Berman, J. D.; Davies, C. R.; Saravia, N. G. *Lancet* **2005**, *366*, 1562–1577; den Boer, M.; Davidson, R. N. *Expert Rev. Anti-Infect. Ther.* **2006**, *4*, 187–197; Mishra, J.; Saxena, A.; Singh, S. *Curr. Med. Chem.* **2007**, *14*, 1153–1169; Sundar, S.; Rai, M. *Expert Opin. Pharmacother.* **2008**, *6*, 2821–2829.

Section 18.4: Gryseels, B.; Polman, K.; Clerinx, J.; Kesters, L. *Lancet* **2006**, *368*, 1106–1118; Doenhoff, M. J.; Pica-Mattoccia, L. *Expert Rev. Anti-Infect. Ther.* **2006**, *4*, 211–222; Fenwick, A.; Rollinson, D.; Southgate, V. *Adv. Parasitol.* **2006**, *61*, 567–623; Doemling, A.; Khoury, K. *ChemMedChem* **2010**, *5*, 1420–1434.

Section 18.5: Gyapong, J. O.; Kumaraswami, V.; Biswas, G.; Ottenson, E. A. *Expert Opin. Phramcother.* **2005**, *6*, 179–200; Tisch, D. J.; Michael, E.; Kazura, J. W. *Lancet Infect. Dis.* **2005**, *5*, 514–523.; Lammie, P. J.; Fenwick, A.; Utzinger, J. *Trends Parasitol.* **2006**, *22*, 313–321; Ottesen, E. A. *Adv. Parasitol.* **2006**, *61*, 394–441.

Section 18.6: van den Emde, E. *Expert Opinion Pharmacother.* **2009**, *10*, 435–451; Geary, T. G.; Woo, K.; McCarthy, J. S.; Mackenzie, C.; Horton, J.; Prichard, R. K.; de Silva, N. R.; Olliaro, P. L.; Lazdins-Helds, J. K.; Engels, D. A.; Bundy, D. A. *Int. J. Parasitol.* **2010**, *40*, 1–13.

Section 18.7: Molyneux, D. H.; Bradley, M.; Haoerauf, A.; Kyelem, D.; Taylor, M. J. *Trends Parasitol.* **2003**, *19*, 516–522; Peters, D. H.; Phillips, T. *Trop. Med. Int. Health* **2004**, *9*(suppl.), A4–A35; Omura, S.; Crump, A. *Nat. Rev. Microbiol.* **2004**, *2*, 984–989; Crump, A.; Otoguro, K. *Trends Parasitol.* **2005**, *21*, 126–132; Taylor, M. J.; Hoerauf, A.; Bockari, M. *Lancet* **2010**, *376*, 1175–1185.

REFERENCES

1. McNeill, J. R. *Washington Post*, Oct. 18, 2010.
2. McIntyre, J. A.; Castener, J. A.; Bayes, M. *Drugs Future* **2003**, *28*, 859–869; Crockett, M.; Kain, K. C. *Expert Opin. Invest. Drugs* **2007**, *16*, 705–715.

3. Myint, H. Y.; Ashley, E. A.; Day, N. P.; Nosten, F.; White, N. J. *Trans. R. Soc. Trop. Med. Hyg.* **2007,** *101*, 858–866.
4. Liu, C.; Zhao, Y.; Wang, Y. *Appl. Microbiol. Biotechnol.* **2006,** *72*, 11–20.
5. Jansen, F. H. *Malaria J.* **2010,** *9*, 212–216.
6. Peters, W. *J. Roy. Soc. Med.* **1999,** *92*, 345–352; Crockett, M.; Kain, K. C. *Expert Opin. Invest. Drugs* **2007,** *16*, 705–715.
7. Looareesuwan, S.; Chulay, J. D.; Canfiled, C. J.; Hutchinson, D. B. A. *Am. J. Trop. Med. Hyg.* **1999,** *60*, 533–541.
8. Baggish, A. L.; Hill, D. R. *Antimicrob. Agents Chemother.* **2002,** *46*, 1163–1173.
9. Boggild, A. K.; Parise, M. E.; Lewis, L. S.; Kain, K. C. *Am. J. Trop. Med. Hyg.* **2007,** *76*, 208–223.
10. CDC. *CDC Health Information for International Travel 2010*. CDC: Atlanta, GA, 2009.
11. Maude, R. J.; Woodrow, C. J.; White, L. J. *Drug Dev. Res.* **2010,** *71*, 12–19.
12. Androula, P.; Maltezou, H. C. *Int. J. Infect. Dis.* **2010,** *14*, 1032–1039; Weina, P. J.; Neafie, R. C.; Wortmann, G.; Polhemus, M.; Aronson, N. E. *Clin. Infect. Dis.* **2004,** *39*, 1674–1680.
13. Utzinger, J.; Shuhua, X.; Keiser, J.; Minggan, C.; Jiang, Z.; Tanner, M. *Curr. Med. Chem.* **2001,** *8*, 1841–1859.
14. Utzinger, J.; Xiao-Nong, Z.; Ming-Gang, C.; Bergquist, R. *Acta Trop.* **2005,** *96*, 69–96.
15. Wang, L.-D.; Guo, J.-G.; Wu, X.-H.; Chen, H.-G.; Wang, T.-P.; Zhu, S.-P.; Zhang, Z.-H.; Steinmann, P.; Yang, G.-J.; Wang, S.-P.; Wu, Z.-D.; Wang, L.-Y.; Hao, Y.; Bergquist, R.; Utzinger, J.; Zhou, X.-N. *Trop. Med. Int. Health* **2009,** *14*, 1473–1483.
16. McManus, D. P.; Gray, D. J.; Li, Y.; Feng, Z.; Williams, G. M.; Stewart, D.; Rey-Ladino, J.; Ross, A. G. *Clin. Microbiol. Rev.* **2010,** *23*, 442–466.
17. Horton, J. *Ann. Trop. Med. Parasitol.* **2009,** *103* (Suppl. 1), S33–S40.
18. Tisch, D. J.; Michael, E.; Kazura, J. W. *Lancet Infect. Dis.* **2005,** *5*, 514–523.
19. Lammie, P. J.; Fenwick, A.; Utzinger, J. *Trends in Parasitol.* **2006,** *22*, 313–321; Gyapong, J. O.; Kumaraswami, V.; Biswas, G.; Ottenson, E. A. *Expert Opin. Pharmacother.* **2005,** *6*, 179–200.
20. Omura, S.; Crump, A. *Nat. Rev. Microbiol.* **2004,** *2*, 984–989.
21. Hopkins, A. D. *Eye* **2005,** *19*, 1057–1066.

CHAPTER 19

RESPONSE TO THE AIDS PANDEMIC

ABSTRACT

The recognition of AIDS as a fatal communicable disease came as a shock in 1981. It challenged the scientific and political community to respond effectively. Various classes of drugs were developed based on the recognition of the biochemical aspects of the retrovirus mechanism of infection and reproduction. By 2000, effective drug combinations called highly active antiretroviral therapy had decreased the rate of infection and mortality substantially in the United States. The cost of the various anti-AIDS drugs has slowed the progress in less developed parts of the world.

The first cases of AIDS, *acquired immunodeficiency syndrome*, in the United States appeared in 1981, concentrated among homosexual men. Soon thereafter, cases began to appear among intravenous drug users and hemophiliacs, the latter infected by blood transfusions. In 1984, a retrovirus, HIV-1 (human immunodeficiency virus) was identified as the cause. Epidemiological studies eventually traced the disease to a transfer of a monkey (simian) virus to humans in central Africa. The earliest case documented by DNA evidence dates to 1959 in the Democratic Republic of the Congo. The disease probably had smoldered for years, presumably transmitted primarily heterosexually. The untreated disease is devastating, exposing victims to a variety of pathologies, including specific forms of pneumonia, cancer, and general wasting due to an impaired immune response. It is estimated that in 2005 there were 40 million active cases, with 4.9 million new infections and 3.1 million deaths, annually. The incidence of HIV infection in sub-Saharan Africa is especially high, as high as 25% in South Africa.

The AIDS epidemic arose at the beginning of the Reagan administration. For both ideological and political reasons, the response by the administration was slow. An exception was Dr. C. Everett Koop, the US Surgeon General at the time, who began distribution of a pamphlet outlining protective measures.[a]

---

[a]C. Everett Koop, M. D., was surgeon general of the United States at the time of the AIDS outbreak. In contrast to the indifference of the political wing of the administration, Koop recognized the critical nature of the disease and mounted a vigorous campaign of public education. In 1986, the Public Health Service issued a report on the disease and 20 million copies were distributed to local governments, schools, and physicians. In 1988, an eight-page booklet describing the disease and prevention was distributed to about 107 million households, representing the largest mass mailing ever undertaken.

Agencies in the federal government responsible for public health were also prompt to respond. The CDC, NIH, and FDA were instrumental in establishing the cause of the epidemic and instituting such safeguards as screening of blood for transfusion. However, the administration resisted initiation of a major research effort and recommended only transfer of funds from other sources. In contrast to most previous public health crises, such as polio, the AIDS epidemic was accompanied by strident controversy. Two years passed before a concerted course of action began. Public and congressional pressure eventually led to allocation of research funds, but as late as 1986, the administration was recommending reduced expenditures.

It is fortunate that the AIDS epidemic appeared at a point in time when considerable understanding of the mechanism of retroviruses existed, because this permitted a quick and rational approach to the chemotherapy of the disease. The target cells in humans are $CD4^+$ T cells, which are a critical component of the immune system. Figure 19.1 shows the basic mechanism of HIV infection. The retrovirus is incorporated into the target cells at the CD4 receptor, which involves co-receptor proteins. Inside the cell, the retrovirus sheds its coat and reverse transcriptases transcribe the viral DNA and integrate it into the DNA of the host cell. The proteins produced are processed by proteolytic enzymes and reassemble into the virus, which is released and capable of reinfection of other cells.

19.1   DRUGS FOR TREATMENT OF AIDS

Each of the steps in the HIV life cycle provides an opportunity for chemotherapy. The first drug discovered to have anti-HIV activity was 3-azidothymidine, now known as zidovudine. It is a substrate for HIV reverse transcriptase. It and other drugs of its type function by incorporation into the viral DNA. They lack the 3'-hydroxy group and thus terminate DNA extension. (The principle is analogous to the Sanger dideoxynucleotide sequencing method, see Section 20.2.) The drugs are called *reverse transcriptase inhibitors* (RTI). A number of additional drugs that function on this basis have been developed. They are called *nucleoside analogs*, and as part of their mechanism of action they are phosphorylated at the 5'-position. There are also *nucleotide analogs*, which contain the phosphate or analogous groups in place. HIV reverse transcriptase is also susceptible to *nonnucleotide inhibitors* that bind at a site that becomes exposed when a nucleotide is bound. Among drugs and potential drugs of this class are nevirapine, delaviridine, and efavirenz. The RTI drugs are very susceptible to resistance caused by mutation of the binding site.

A third group of drugs are *protease inhibitors (PIs)*. In the life cycle of HIV, the proteins synthesized by translation of the viral DNA are cleaved into smaller

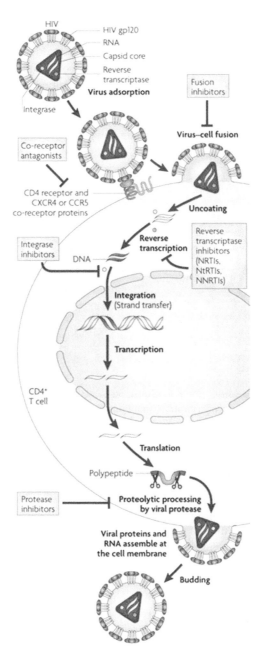

FIGURE 19.1 Simplified life cycle diagram for HIV-I virus showing points of drug action. (From De Clerq, E. *Nat. Rev. Drug Discov.* **2007,** *6*, 1001–1018. Reprinted with permission of Nature Publishing Group.)

proteins by a *protease*. The HIV protease is called an *aspartyl protease* because its function depends upon an aspartic acid in the catalytic site. The protein structure was determined in 1988. It is a homodimer. It was also demonstrated in 1988 that inactivation of the protein led to immature, non-infectious virus particles. The three dimensional structure of this protease permitted the design and synthesis of inhibitors called *peptidomimetics* that are bound to the protease as if they were substrates, but which cannot be processed and inhibit the function of the enzyme (see Fig. 19.2). Promising structures were then synthesized and modified to address such issues as bioavailability, metabolic stability, and toxicity.

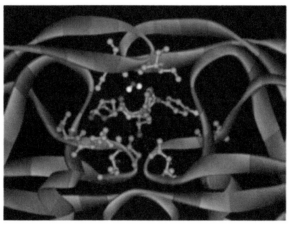

FIGURE 19.2   Structure of a protease inhibitor bound in dimeric HIV protease. (From Pauwels, R. *Antiviral Res.* **2006**, 71, 77–89 by permission of Elsevier.)

Other points in the HIV life cycle are potential targets for drug inhibition. One is the entry of the virus into the target cell, which depends on the CD-4 receptor and co-receptor proteins. Currently, enfuvirtide, a 36 AA peptide that interferes with the fusion is used. It is extremely expensive and must be administered by injection.[1] Another target is the integrase that integrates the viral DNA into the host DNA (see Fig. 19.1). One such drug, raltegravir, has been approved as of 2006. Two other integrase inhibitors, elvitegravir and dolutegravir are under clinical investigation.[2]

raltegravir

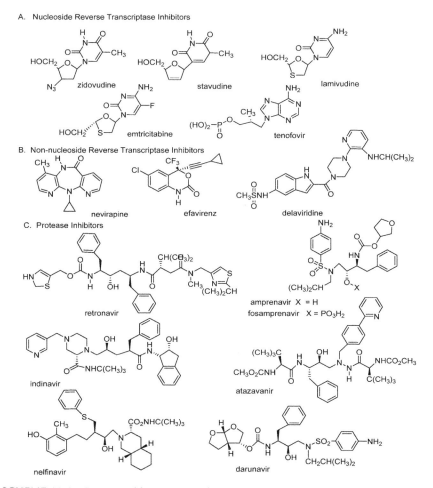

elvitegravir

dolutegravir

Examples of the major classes of the drugs are shown in Scheme 19.1. Each of the classes of HIV drugs has pronounced side effects. The RTI class can result in reduced mitochondria function with resulting changes in pyruvate metabolism that can lead to lactic acidosis. The PIs cause accumulation of fat in the abdomen and are also associated with precipitation of diabetes.

A. Nucleoside Reverse Transcriptase Inhibitors

zidovudine stavudine lamivudine

emtricitabine tenofovir

B. Non-nucleoside Reverse Transcriptase Inhibitors

nevirapine efavirenz delaviridine

C. Protease Inhibitors

retronavir

amprenavir X = H
fosamprenavir X = PO₃H₂

indinavir atazavanir

nelfinavir darunavir

SCHEME 19.1 Drugs used in treatment of AIDS.

Current HIV chemotherapy uses a combination of these types of drugs. The combined chemotherapy is called HAART, *highly active antiretroviral therapy*. The HAART combination often includes two RTI and a PI. There are a number of fixed combinations of drugs that have been approved in the United States. Some of the recently approved combinations include an integrase inhibitor. Treatment includes monitoring the $CD4^+$ cell count and viral load. After introduction of the HAART regimen in 1996, the death rate from AIDS in the United States dropped by 50% within 3 years. Because it often decreases viral load to nondetectable levels, the HAART regimen also reduces transmission and new cases. As mentioned above, HIV is very prone to mutation that leads to development of resistance to drugs. There are subtypes of HIV that may differ in susceptibility to particular drugs. The combination of potential resistance and side-effects requires careful implementation of the HAART combinations. Monitoring of $CD4^+$ lymphocytes and HIV RNA levels can be used to indicate the need for initiation of the chemotherapy. Patient adherence is also an important variable in the level of success achieved. There is a strong correlation between both viral load and $CD4^+$ lymphocyte count and disease progression.[2]

19.2   ECONOMIC ASPECTS OF AIDS TREATMENT

The high cost of treatment of AIDS has raised issues both in the developed and developing parts of the world. In the United States, annual cost is estimated as $22,000 per patient, including $12,000 per year for drugs.[3] Despite their expense, however, the drugs considerably reduce the overall cost of treatment and hospitalization for conditions such as infections and cancer that are associated with AIDS. In consideration of long-term costs, the cost of lifetime treatment must be balanced with the economic benefits of better health and potential productivity. Although currently manageable in most cases in the developed world, the cost and sophistication of the most advanced treatments are a deterrent to use in the less developed world. The economic impact has been highest in Africa where the infection rate is the highest and the economic situation is fragile. Several international programs are attempting to address this situation. The World Bank and UN operate the Multi-country AIDS Programme. A Global Fund to Fight AIDS, Tuberculosis and Malaria is financed primarily by the EU and private donors. The United States has funded the "President's Emergency Plan for AIDS Relief." There are also numerous private organization involved in AIDS treatment and prevention.

In 2003, the Indian generic drug firm CIPLA introduced a generic form of a drug combination of stavudine, lamivudine, and nevartrapine as "Triomune."

Both the reduced cost and simplicity (single pill) greatly expanded the availability of combination therapy in the developing world. Another combination was introduced as "Viraday." It contains efavirens, emtricitabine, and tenofovir. Neither of these combinations contains a PI. Nevertheless, there are problems with use of the drugs in very resource-poor regions. These include (1) lack of physicians adequately trained to administer the regimens; (2) need for programs to encourage patient adherence; (3) expense of tests to monitor the progress of the treatment, and (4) potential interference with drugs used for other conditions, such as tuberculosis. Patent law also impacts the availability of generic drugs. Most countries adhere to international patent law, which restricts access to drugs while under patent. Some countries have instituted mandatory licensing of patented drugs under a provision of the WTO Doha Declaration of 2001, which provides for compulsory licensing of drugs under crisis conditions.

KEYWORDS

- **AIDS**
- **HIV-1**
- **reverse transcriptase inhibitors**
- **protease inhibitors**
- **integrase inhibitors**
- **highly active antiretroviral therapy (HAART)**
- **drug costs**

BIBLIOGRAPHY

Werth, B. *The Billion Dollar Molecule: One Company's Quest for the Perfect Drug*; Simon and Schuster, New York, 1994; Demars, S. In *Truth Lies and Public Health*; Finkel, M. L., Ed.; Praeger: Westport, CT, 2007; pp 32–54; Volberding, P. A.; Sande, M. A.; Greene, W. C.; Lange, J. M. A. *Global HIV/AIDS Medicine*; Elsevier-Saunders: Philadelphia, PA, 2008; DeClerq, E. *Adv. Pharmacol.* **2013,** *67*, 317–358.

BIBLIOGRAPHY BY SECTION

Section 19.1: Ogden, R. In *Viral Infections and Treatment*; Rubsamen-Waigmann, H.; Deres, K.; Hewlett, G.; Welker, R., Eds.; Marcel Dekker: New York, 2003; pp 523–553; Locatelli,

G. A.; Cancio, R.; Spadari, S.; Maga, G. *Curr. Drug Metab.* **2004**, *5*, 283–290; Johns, S.; Furtex, K.; Looney, D. J. *Curr. Med. Chem. Immunol. Endocrinol. Metab. Agents* **2004**, *4*, 27–47; Pommier, Y.; Johnson, A. A.; Marchand, C. *Nat. Rev. Drug Discov.* **2005**, *4*, 236–248; De Clerq, E. *Nat. Rev. Drug Discov.* **2007**, *6*, 1001–1018.

Section 19.2: Moore, R. D. *Pharmacoeconomics* **2000**, *17*, 324–330; Kumarasamy, N. *Lancet* **2004**, *364*, 3–4; Gogtay, J. A. *IDrugs* **2007**, *10*, 881–884; van Oranje, M. *Global HIV/AIDS Medicine*; In Volberding, P.; Sande, M. A.; Greene, W. C.; Lange, J. M. A. Eds.; Elsevier-Saunders: Philadelphia, PA, 2008; pp 793–800.

REFERENCES

1. Matthews, T.; Salgo, M.; Greenberg, M.; Chung, J.; De Masi, R.; Polognesi, D. *Nat. Rev. Drug Discov.* **2004**, *3*, 215–225.

2. Metitiot, M.; Marchand, C.; Pommier, Y. *Adv. Pharmacol.* **2013**, *67*, 75–105.

3. Mellors, J. W.; Munoz, A.; Giorgi, J. V.; Margolick, J. B.; Tassoni, C. J.; Gupta, P.; Kingsley, L. A.; Todd, J. A.; Saah, A. J.; Detels, R.; Phair, J. P.; Rinaldo, Jr., C. R. *Ann. Intern. Med.* **1997**, *126*, 946–954.

4. Bozzette, S. A.; Berry, S. H.; Duan, H.; Frankel, M. R.; Leibowitz, A. A.; Lefkowitz, D.; et al. *N. Engl. J. Med.* **1998**, *339*, 1897–1904; Moore, R. D. *Pharmacoeconomics* **2000**, *17*, 324–330.

PART V
Molecular Biology and Its Applications

DNA STRUCTURE, SEQUENCING, SYNTHESIS, AND MODIFICATION: MAKING BIOLOGY MOLECULAR

ABSTRACT

The investigation of deoxyribonucleic acid (DNA) culminated in the recognition of its base-paired double helical structure by James D. Watson and Francis Crick in 1953. This led to the understanding of DNA control of protein synthesis through the genetic code. The methods of molecular biology developed rapidly and included sequencing of DNA, biological and chemical methods for DNA synthesis, splicing of DNA to form recombinant DNA, cloning, and mutagenesis. These methods had many practical applications including identification of individuals and detection of mutations. The sequence of the human genome was announced in 2003 as the result of the Human Genome Project and the Whole Genome Shot-Gun Approach.

The most spectacular advances in molecular modification in the last quarter of the twentieth century had their origin in understanding the nature of DNA. DNA structure and sequence information underpins the fields of biotechnology and genetic engineering. The crucial concept is the specificity of base-pairing that depends on hydrogen bonding. This fundamental concept was established in 1953 and gave rise to a range of techniques and methods now known collectively as *molecular biology*. Understanding of DNA structure and function led to new ways for production and manipulation of DNA. Usually, the changes are effected with biological catalysts. The ability to utilize DNA for protein synthesis enabled the production of specific proteins for a variety of applications. We want to understand the chemical principles that are involved in these techniques. We also want to recognize how the processes have been adapted from nature. These methods have been applied to many aspects of biochemistry, pharmacology, and drug development. Information about DNA is also the foundation for efforts to tailor drug therapy to individuals based on genetic information, which is called *molecular medicine*.

20.1 STRUCTURE OF DNA AND THE GENETIC CODE

The first characterization of DNA is credited to Friedrich Miescher, a Swiss physician working in Tubingen, Germany. In 1869, he isolated the material from the nuclei of white blood cells and called it *nuclein*. He noted its significant phosphorus content. The composition was recognized as containing heterocyclic bases, deoxyribose and phosphate by Phoebus Levene in 1899. Levene also correctly identified the bases as adenine, cytosine, guanine, and thymine. Levene incorrectly believed that the DNA existed as a polymer of the four nucleotides as tetramers. At that time there was no hint that the material contained the fundamental genetic information.

adenine   cytosine   guanine   thymine

An important aspect of heredity was recognized by Frederick Griffith, a British geneticist. Griffith was working with *Streptococcus pneumoniae* in the aftermath of the great flu epidemic of 1918. He observed two different strains that appeared rough (R) or smooth (S) under a microscope. The S strain was much more virulent in the test mice he was using. He found that the S strain could be killed and its virulence ended by heating. However, when he mixed the dead S strain with the living nonvirulent R strain, the bacteria became virulent. He concluded that there must be some "transforming factor" that survived the heating and had the capacity to induce virulence in the R strain. In 1944, Oswald Avery,[a] Colin MacLeod, and Maclyn McCarty investigated the "transforming factor" and concluded that it must be DNA. They ruled out lipids, carbohydrates, proteins, and ribonucleic acids as possibilities by showing that destruction of each did not eliminate the "transforming factor." The conclusion was confirmed by Alfred Hershey and Martha Chase, who showed that bacteria incorporated radioactive P from ^{32}P-labeled viral particles. In 1950, Erwin Chagraff noted that while there was variation in the amount of the individual DNA heterocyclic

---

[a]Oswald Avery was born in Canada in 1877 but moved with his family to New York in 1887. He received an M. D. from Columbia in 1904. He began his research career in 1907 at the Hoagland Laboratory, a private bacteriological lab. Perhaps because of the deaths of close associates, he turned his research interests to tuberculosis. He moved to the Rockefeller Institute in 1913 and remained there until he retired in 1943 at the mandatory age of 65. He continued to work at the Institute. His paper on the transforming factor was published in 1944, after his retirement.

bases among species, there was always a 1:1 relationship between the pyrimidines (cytosine, thymine) and the purines (adenine and guanine). Furthermore the amount of A corresponded to the amount of T and G corresponded with C, indicating some kind of combination of A–T and C–G pairs.

This set the stage for the base-paired double helix structure conceived by James Watson and Francis Crick in 1953.[b] Watson and Crick were both working on crystallography at Cambridge University, but were not engaged in the study of DNA *per se*. They had partial access to crystallographic data being collected by Maurice Wilkins and Rosalind Franklin who were working on DNA at King's College in London. This crystallographic work implied a helical structure and imposed some dimensional limits. Watson and Crick began building models that could explain the existing structural information. One key aspect to solving the problem was recognizing the hydrogen-bonding relationship between the A–T and C–G pairs. At the time, the thymine, cytosine and guanine structures were written in a way that depicted the oxygens as being in hydroxyl groups. The actual structures have the hydrogen on nitrogen so that the oxygens are carbonyl, not hydroxyl groups. When Watson and Crick used the correct structures, the hydrogen bonding pattern became apparent.

---

[b]James D. Watson was born in Chicago in 1928. He entered the University of Chicago at the age of 15 and received a B. Sc. in Zoology in 1947. He then studied at Indiana University, where he received a Ph. D. in Zoology in 1950. He then held a postdoctoral position in Copenhagen. When he became intensely interested in DNA structure, he moved to Cambridge University, where he worked with John Kendrew. It was there that he met Francis Crick and they began their efforts to model the structure of DNA, and in 1953 proposed the double helical structure. Watson later studied the organization of viruses by X-ray crystallography at both Caltech (1952–1955) and Cambridge (1955–1956) before becoming a member of the Harvard faculty in 1956. In 1968, he became director of the Cold Spring Harbor Laboratory and remained there until he retired in 2007, except for 1990–1992, when he headed the Human Genome Project. He authored *The Double Helix* (1968) as well as several text books on Molecular Biology.

Francis H. C. Crick was born in 1916 at Northampton England. He studied Physics at University College, London and obtained a B. Sc. in 1937. During WWII, he worked on mines at the British Admiralty. In 1949, he began working with M. F. Perutz and John Kendrew at Cambridge University. He began to learn the chemical aspects of proteins and DNA, as well as their crystallography. It was during this period that Crick and Watson collaborated on their model of DNA structure. Crick's research focused on the mechanism of transfer of DNA sequence information in protein synthesis. Crick formulated these ideas as the "central dogma" describing the information flow from DNA to protein structure. From 1976 until his death in 2004, he was a research professor at the Salk Institute in California.

Maurice Wilkins was born in New Zealand but returned with his family to England as a young boy. He received a Ph. D. in Physics from the University of Birmingham in 1945. He moved with his Ph. D. mentor, John Randall, first to St. Andrews and then King's College in London where Randall established one of the first laboratories for application of physics to biological problems. Wilkins undertook X-ray crystallography on samples of calf thymus DNA that had been drawn into long threads. Wilkins was a friend of Francis Crick and shared information on his project with Crick. Also working at King's College was Rosalind Franklin. She had received a bachelor's degree in chemistry from Cambridge in 1941 and a Ph. D. in physical chemistry. She had studied X-ray crystallography in Paris before joining the King's College group in 1951. She and Wilkins worked more or less independently on the crystallography of DNA and were not always on the best of terms. Both Wilkins and Franklin's work on DNA pointed toward a helical structure and it was on the basis of their data that Watson and Crick constructed their DNA model. Crick, Watson, and Wilkins shared the 1962 Nobel Prize in Physiology or Medicine. Franklin died of cancer in 1958 at the age of 37, or she, too, probably would have been one of the Nobel awardees.

Adenine–thymine and cytosine–guanine base-pairing

Watson and Crick's model corresponds to what is now called B-DNA. There are other forms including A-DNA and Z-DNA, as shown in Figure 20.1. All show the characteristic base pairing. Many other structures are present in functioning cells, including regions of triplet DNA and loops that are not base-paired.

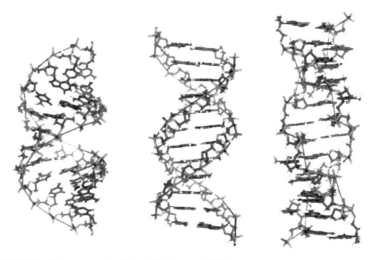

FIGURE 20.1   Structure of A, B, and Z forms of DNA.

The concept of the double helix structure of DNA led to rapid development of understanding of how genes could be *replicated*, *transcribed*, and *translated* to determine the structure of proteins. The replication, which occurs during cell division, preserves genetic information from generation to generation by the synthesis of complementary DNA by *DNA polymerase*. Among the critical features of DNA synthesis is that it requires a short double-stranded region formed by a complementary *primer* and that it is *unidirectional*, building the new strands only in the 5'-to-3' direction. Thus a single strand of DNA can be converted to its complementary strand by a DNA polymerase using the four deoxynucleotide building blocks.

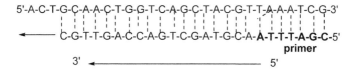

5'-A-C-T-G-C-A-A-C-T-G-G-T-C-A-G-C-T-A-C-G-T-T-A-A-A-T-C-G-3'

◄──── C-G-T-T-G-A-C-C-A-G-T-C-G-A-T-G-C-A-**A-T-T-T-A-G-C**-5'
                                                         primer

3' ◄──────────────────────────── 5'

The next major advance was the determination of the 3-letter code corresponding to each of the 20 amino acids in proteins. The first two codes were deciphered by Marshall Nirenberg[c] working at the NIH. He developed cell-free systems that could synthesize polypeptides from synthetic strands of polyribonucleotides (RNA). The first successful experiment was with poly-U, which incorporated only phenylalanine into protein, thus providing the first identified code. Similarly, poly-C gave polyproline, indicating that its codon must be CCC. H. G. Khorana,[d] working at the University of Wisconsin, determined most of the rest of the code by using synthetic RNA of defined sequence. For example, poly-UC contains alternating UCU and CUC codons. The polypeptide found was polyserylleucine, giving the codons for serine and leucine, respectively. Similar experiments unraveled the code for all of the 3-base sequences. The corresponding DNA code in the gene is given in Figure 20.2.

		T		**C**		**A**		**G**	
							Second base		
F	**T**	TTT	Phe (F)	TCT	Ser (S)	TAT	Tyr (Y)	TGT	Cys (C)
i		TTC	Phe (F)	TCC	Ser (S)	TAC		TGC	
r		TTA	Leu (L)	TCA	Ser (S)	TAA	STOP	TGA	STOP
s		TTG	Leu (L)	TCG	Ser (S)	TAG	STOP	TGG	Trp (W)
t	**C**	CTT	Leu (L)	CCT	Pro (P)	CAT	His (H)	CGT	Arg (R)
		CTC	Leu (L)	CCC	Pro (P)	CAC	His (H)	CGC	Arg (R)
b		CTA	Leu (L)	CCA	Pro (P)	CAA	Gln (Q)	CGA	Arg (R)
a		CTG	Leu (L)	CCG	Pro (P)	CAG	Gln (Q)	CGG	Arg (R)
s	**A**	ATT	Ile (I)	ACT	Thr (T)	AAT	Asn (N)	AGT	Ser (S)
e		ATC	Ile (I)	ACC	Thr (T)	AAC	Asn (N)	AGC	Ser (S)
		ATA	Ile (I)	ACA	Thr (T)	AAA	Lys (K)	AGA	Arg (R)
		ATG	Met (M) START	ACG	Thr (T)	AAG	Lys (K)	AGG	Arg (R)
	G	GTT	Val (V)	GCT	Ala (A)	GAT	Asp (D)	GGT	Gly (G)
		GTC	Val (V)	GCC	Ala (A)	GAC	Asp (D)	GGC	Gly (G)
		GTA	Val (V)	GCA	Ala (A)	GAA	Glu (E)	GGA	Gly (G)
		GTG	Val (V)	GCG	Ala (A)	GAG	Glu (E)	GGG	Gly (G)

FIGURE 20.2   The DNA genetic code in DNA.

---

[c]Marshall Nirenberg was born in New York in 1927, but grew up in Florida. He received B. S. and M. S. degrees in Zoology at the University of Florida in 1948 and 1952, followed by a Ph. D. from the University of Michigan in 1957. He then went to the NIH Institute of Arthritis and Metabolism, first as post-doctoral fellow and then as a permanent staff member. He used defined RNA polymers to determine which amino acids were incorporated into polypeptides under conditions of cell-free protein synthesis. Nirenberg remained at the NIH his entire career.

[d]H. Gobind Khorana was born in the Punjab region of India in 1922, while it was under British control. He received a scholarship to study at the Punjab University in Lahore, and received B. S. (1943) and M. S. (1945) degrees there. He then went to England where he received a Ph. D. in Chemistry in 1948 from the University of Liverpool. He then did postdoctoral work in Switzerland and at Cambridge. In 1952, he came to the University of British Columbia. He moved to the University of Wisconsin in 1960 and to MIT in 1970. Khorana's contribution to deciphering the genetic code involved the synthesis of defined polyribonucleotides which were then used to synthesize polypeptides.

Another piece of the puzzle was solved by Robert W. Holley[e] working at Cornell University, who determined the structure of the first molecule of transfer RNA (*t*-RNA). These are the molecules that guide individual amino acids to the messenger RNA template during protein synthesis. These results came together to form what was called by Francis Crick, the *central dogma*, the description of how genetic information is transferred from generation to generation and converted into functional proteins. Holley, Khorana and Nirenberg shared the 1968 Nobel Prize in Physiology or Medicine.

Other aspects of the picture became clear later when gene sequence information became available (see Section 20.2). In particular, it became apparent that genes occur as separated segments that are "spliced" and that genes can encode for more than one protein. It was also found that DNA contains large amounts of nontranslated sequences called *introns*. Other later developments included the recognition of the ability of retro-viruses to carry genetic information in RNA, which then uses host cell DNA polymerase to use viral genetic information to direct synthesis of proteins. The broad picture is summarized in Figure 20.3.

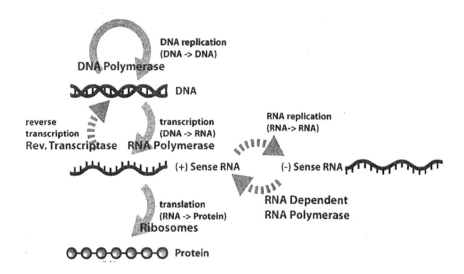

FIGURE 20.3   Information flow in DNA replication, transcription and translation. (Source: https://en.wikipedia.org/wiki/Central_dogma_of_molecular_biology)

---

[e]Robert W. Holley was born in Urbana, IL, USA in 1922 and graduated from the University of Illinois in 1942. He received a Ph. D. in chemistry from Cornell University in 1947. He joined the USDA laboratory at Cornell University in 1948, later transferring to the Department of Biochemistry and Molecular Biology. He isolated and determined the structure of the first transfer RNA molecule in 1964. In 1968, he joined the Salk Institute in La Jolla, CA, USA.

20.2 DNA SEQUENCING

From a chemical point of view, one of the first goals in the characterization of DNA was to establish the order (sequence) of the individual bases. Historically, the first DNA sequencing method was entirely chemical, but was developed by Walter Gilbert,[f] a physicist turned molecular biologist, and his student Allan Maxam. They found that the various DNA bases showed somewhat differential sensitivity to chain cleavage by different reagent combinations. Dimethyl sulfate with piperidine reacts at purines and with hydrazine at pyrimidines. Selectivity between the two pairs was achieved with a third reagent. Adding formic acid to the dimethyl sulfate–piperidine combination distinguished the two purines, as this combination cleaves mainly at guanine. Similarly, distinction between cytosine and thymine was achieved by including 1.5 M NaCl, which selectively cleaved at cytosine. The DNA to be sequenced was end-labeled with radioactive ^{32}P, and samples subjected to each of the four cleavage conditions. The reaction was stopped after only a small amount of cleavage so that upon electrophoresis a ladder array of the fragments can be observed. Detection was done by exposing film to the electrophoresis gel, which results in detection of each radioactive spot. By this method about 200–300 bases could be determined with a weeks effort if all went well. Figure 20.4 shows the concept of the Maxam–Gilbert method.

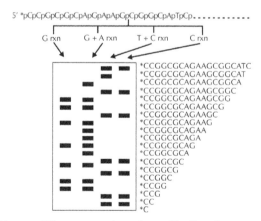

5′ *pCpCpGpCpGpGpCpApGpApApGpCpGpGpCpApTpCp- - - - - - - - - - - -

| G rxn | G + A rxn | T + C rxn | C rxn |

*CCGGCGCAGAAGCGGCATC
*CCGGCGCAGAAGCGGCAT
*CCGGCGCAGAAGCGGCA
*CCGGCGCAGAAGCGGC
*CCGGCGCAGAAGCGG
*CCGGCGCAGAAGCG
*CCGGCGCAGAAGC
*CCGGCGCAGAAG
*CCGGCGCAGAA
*CCGGCGCAGA
*CCGGCGCAG
*CCGGCGCA
*CCGGCGC
*CCGGCG
*CCGGC
*CCGG
*CCG
*CC
*C

FIGURE 20.4 Maxam–Gilbert sequencing scheme. The four cleavage samples are analyzed by electrophoresis and detected by autoradiography. (From Integrated DNA Technologies Tutorial: DNA Sequencing.)

---

[f]Walter Gilbert graduated from Harvard with majors in Chemistry and Physics. After receiving his Ph. D. in Physics at Cambridge University, he returned to Harvard. In 1960, he began experimental work in Molecular Biology that led to discoveries in the area of mechanism of protein synthesis and the first isolation of a gene repressor. He was also one of the first to achieve synthesis of insulin by the methods of genetic engineering (see Section 21.3.1). He was one of the founders of the biotech company Biogen.

A second method was developed at about the same time by Frederick Sanger,[g] working at Cambridge in the United Kingdom. The concept is similar but the method relies on *synthesis of a complementary strand* of DNA, rather than cleavage of the strand. Each of four tubes is set up with the single-stranded DNA to be sequenced, a primer, DNA polymerase and the four deoxynucleotides. In each tube is placed a small amount of *one* of the four *dideoxy* analogs of the nucleotides. Each time a dideoxynucleotide is introduced into the growing strand it terminates, because there is no 3-hydroxy group for continuation. A radioactive label is incorporated into the primer that is used to start the synthesis. The four lanes are subjected to electrophoresis and the sequence can be read. It is the *complement* of the strand that is being sequenced. This method is several times faster than the Maxam–Gilbert method. Figure 20.5 shows the concept of the Sanger dideoxy method. Gilbert and Sanger shared the 1980 Nobel Prize in Chemistry with Paul Berg, who pioneered the formation of recombinant DNA (see Section 20.4).

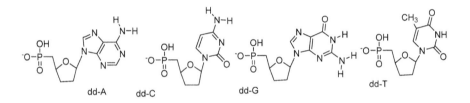

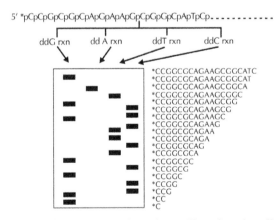

FIGURE 20.5 Sanger Dideoxy sequencing scheme. The primer is radio-labeled and each of the four samples is analyzed by electrophoresis and detected by autoradiography. The sequence is the *complement* of the original sequence. (From Integrated DNA Technologies Tutorial: DNA Sequencing.)

---

[g]For biographical data on Frederick Sanger, see Section 16.1.

Several improvements of the Sanger methodology have been commercialized. The first was the use of fluorescent dyes in place of ^{32}P radioactivity to detect the DNA fragments. This innovation was introduced by Leroy Hood, a biochemist at the California Institute of Technology. Each of the deoxy bases is labeled with a different fluorescent dye. Each terminated fragment can then be recognized on the basis of the fluorescence and the identification of the base made on the basis of the color. The necessary instrumentation was developed at Applied Biosystems (ABS). These changes had several advantages. The fluorescence detection increased sensitivity. Because a different dye is used for each base, only one lane is necessary, instead of the four in the original Sanger method. The need for radioactivity is eliminated. The analysis can be done and recorded by computers that detect the fluorescence characteristic of each base, so the sequence reading is automated. The gels were also made thinner, which allows for a faster separation. The original versions of these instruments could sequence 500–1000 bases in a period of 4–10 h. Simultaneous sequencing of up to 100 lanes (different samples) can be done. Another major improvement in sequencing technology involved use of capillary electrophoresis. Each sample is run in an individual capillary and the bands are detected and identified by their fluorescence. Deconvoluting software identifies the bases. Instruments that operate with 96 or 384 capillaries were developed. Figure 20.6 shows the output from this method.

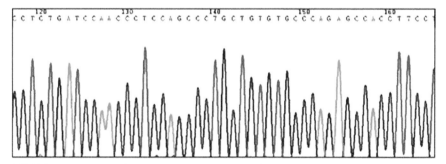

FIGURE 20.6 Output from DNA sequencing using fluorescent dye labeling.

The most recently developed sequencing methods are based on the "shotgun" approach. The DNA to be sequenced is fragmented into random pieces averaging about 2500 bp. The fragments are dispersed into an excess of microbeads, such that any bead captures a single fragment. These fragments are than amplified within water-in-oil microreactors, so that each bead contains millions of copies of the original fragment. The four DNA nucleotides are then sequentially passed over the beads, and the addition of a nucleotide is recognized by light emission triggered by the formation of pyrophosphate in the coupling step. Computer analysis can reassemble the sequence of the random fragments into a single sequence. This method can sequence 25 million bases in a four-hour run.[1] This method is summarized in Figure 20.7.

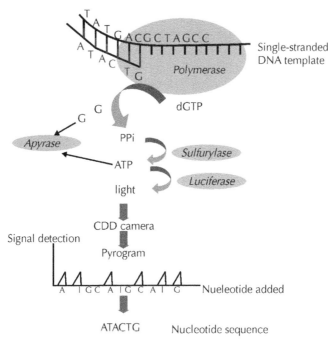

FIGURE 20.7 Operation of the pyrophosphate–luminescence coupled method for DNA sequencing. (From © Armougom, F.; Raoult, D. Journal of Computer Science & Systems Biology, 2009, 2, 74–92. Used with permission via the Creative Commons License.)

Methods for DNA sequencing have continued to improve beyond the pyrosequencing technique. As of 2010, rates of sequence acquisition have increased by 10^{12} and costs per genome has been reduced by a factor of at least 10^{3}.[2]

20.3   SYNTHESIS OF DNA

20.3.1   CHEMICAL SYNTHESIS

The ability to synthesize DNA by chemical methods permitted study of fundamental aspects of DNA. The first example was in the determination of the DNA code for the amino acids (see Section 20.1). Synthetic DNA sequences are also used as primers in replication and for generation of specific mutations.

Subsequently, chemical synthesis was used to provide small genes for the first successful production of insulin by biotechnology (see Section 21.3.1).

The first method applied to synthesis of DNA was the *phosphodiester method*. It was applied, primarily by H. G. Khorana, first to synthesize defined sequences for breaking the DNA code and then in the early 1970s for the first synthesis of small genes. DNA synthesis is approached by the sequential coupling of the individual units by repetitive coupling/deprotection steps. It is an iterative method that adds each nucleotide one by one. In the phosphodiester method, coupling is accomplished by converting the 3'-phosphate group to a reactive sulfonate ester, usually a mesitylenesulfonate. This is then coupled with the 5'-hydroxy of the next nucleotide. The protecting group is the 4,4'-dimethoxytrityl group, which provides optimum conditions for deprotection. This method required prodigious effort and extensive purification of intermediates but demonstrated that functional genes could be synthesized.

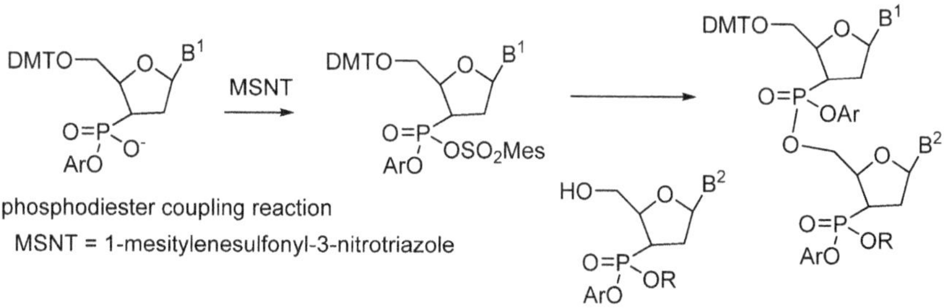

phosphodiester coupling reaction

MSNT = 1-mesitylenesulfonyl-3-nitrotriazole

A second method is the *phosphotriester* method. It uses the individual nucleotide components as the phosphate esters protected by *o*-chlorophenyl and 2-cyanoethyl groups. The coupling is done with *tris*-isopropylbenzenesulfonyl tetrazolide. This method results in fewer impurities than the phosphodiester method. It was applied by making short oligonucleotide sequences and then coupling them together to make longer sequences. This was the method used in the first successful synthesis of the insulin gene.[3]

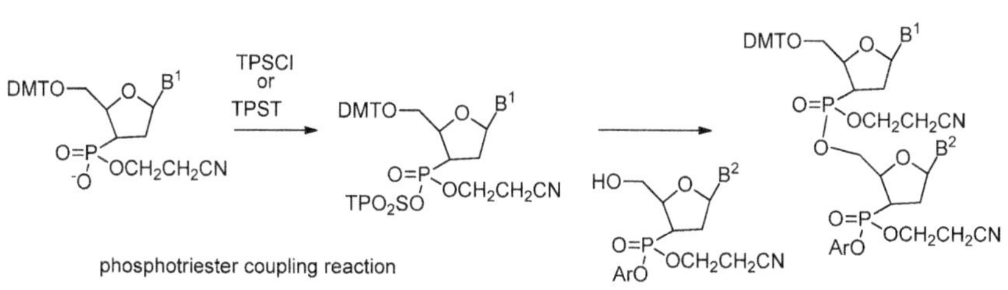

phosphotriester coupling reaction

TPSCl = tris-isopropylbenzenesulfonyl chloride
TPST = tris-isopropylbenzenesulfonyl tetrazolide

Further improvements were made with the introduction of the *phosphora-midite method.*[4] The phosphoramidite method is usually used as a solid-phase synthesis, with the growing DNA chain attached to a silica support. This method takes advantage of two fundamental aspects of phosphorus chemistry. It uses nitrogen-substituted phosphorus(III) units for the individual nucleotides. This takes advantage of the greater reactivity of the P(III) centers and uses the preference for O- over N-substitution at phosphorus to drive the reaction. After each coupling step, the P(III) center is oxidized to P(V), which is inert to any further substitution under the reaction conditions. A capping step is incorporated in the cycle. This prevents any incorporation of a gap (missing nucleotide) if coupling is not 100% efficient. The "capped" material is easily removed during later purification. The final stages involve deprotection of the cyanoethyl group and removal of amide protecting groups on the nucleotide bases. This is done with ammonia. The sequence of transformation is summarized in Figure 20.8.

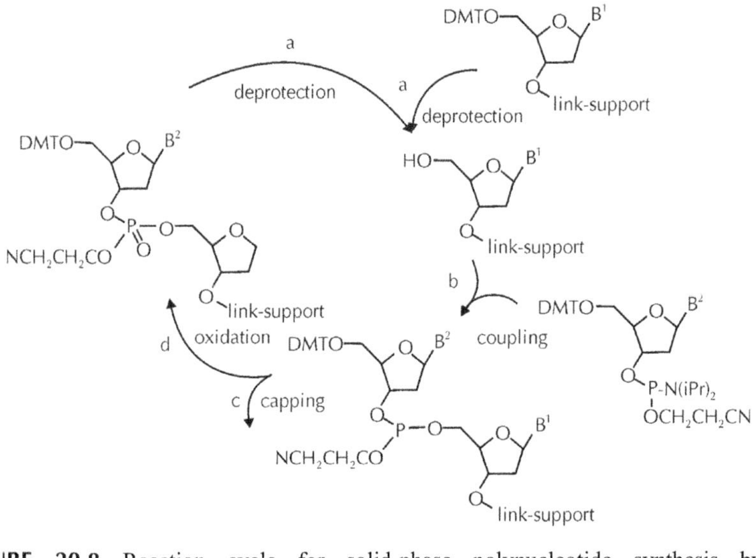

FIGURE 20.8 Reaction cycle for solid-phase polynucleotide synthesis by the phosphoramidite method. Reagents: (a) 3% Cl_3CCO_2H in CH_2Cl_2; (b) 3% tetrazole in CH_3CN; (c1) 10% $(CH_3CO)_2O$ and 10% lutidine in THF; (c2) 7% imidazole in THF; (d) 3% I_2, 2% H_2O 2% pyridine in THF.

20.3.2   THE POLYMERASE CHAIN REACTION

The replication of DNA can be conducted in the laboratory and an important means for doing this is the *polymerase chain reaction.* The polymerase chain reaction was invented by Kary Mullis[h] in 1983, while working at Cetus Corporation. Reputedly, the idea came to Mullis as he drove along the California coast. The polymerase chain reaction provides a means of amplifying a specific segment of DNA. The process uses primers to delineate the start points for both strands of a double-stranded DNA sample. A few years later, in 1986, Mullis made a major improvement by introducing the use of a thermally stable DNA polymerase. The two primers and the four deoxynucleotides are mixed together with the DNA-polymerase and subjected to thermal cycling. At the lower temperature, the DNA area between the primers is replicated. At the higher temperature, the strands separate. When the temperature is lowered again the primers bind and another cycle of the polymerization ensues. After a few rounds almost all of the DNA being synthesized is the region between the two primers. Figure 20.9 outlines the concept of the polymerase chain reaction. The polymerase chain reaction is used in many ways including forensic investigation, identification of pathogens, genetic screening, and investigation of ancient DNA samples.

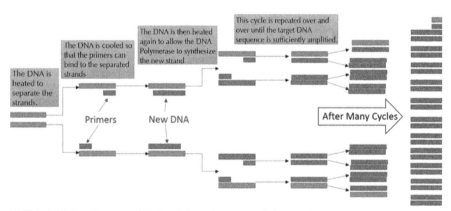

FIGURE 20.9 Conceptual basis of the polymerase chain reaction.

---

[h]Kary Mullis was born in North Carolina in 1944 and received his undergraduate degree in Chemistry at Georgia Tech. He then received a Ph. D. in Biochemistry at UC, Berkeley. After short alternative careers as a writer and a baker, Mullis joined Cetus Corporation, one of the new biotech companies in the San Francisco area. Mullis conceived the idea of using a pair of primers to bracket an area of DNA for replication in 1983. He introduced repetitive thermal cycling based on Taq polymerase in 1986. The patents on the polymerase chain reaction were purchased by Roche Molecular Systems. Mullis left Cetus in 1996. Mullis received the 1993 Nobel Prize in Chemistry, along with Michael Smith, who had developed the techniques of site-directed mutagenesis. Mullis has become something of a scientific iconoclast and expressed opinion contrary to the prevailing views in areas such as climate change and the cause of AIDS.

20.4   MANIPULATION, MODIFICATION, AND ANALYSIS OF DNA

20.4.1   RESTRICTION ENDONUCLEASES

Restriction endonucleases are a key tool for manipulation of DNA. They are enzymes that can cleave DNA at particular sequences of 4–8 base pairs. They were first discovered and their importance recognized by Hamilton Smith, Daniel Nathan, and Werner Arber, who received the 1978 Nobel Prize in Physiology or Medicine.[i] Dozens of such enzymes have been isolated and characterized. Because the cleavage occurs at specific sites, the sequence at the cleavage site is known. Furthermore, other DNA cleaved by the same endonuclease will terminate in the same matching sequence. The cleavage often occurs with a short overhang on each strand. These termination sites are called "sticky," that is they can base-pair to another matching segment. If the enzyme is used to cleave both the DNA to be inserted and the host DNA, the sequences will attach by hydrogen bonding. In the presence of a DNA ligase, the ends will become attached and incorporated into the original cleaved strand.

20.4.2   RECOMBINANT DNA

Recombinant DNA (rDNA) refers to DNA that contains segments from two or more sources. The resulting DNA is sometimes called a *chimera*, referring to mythological creatures that were combinations of two different animals. Efficient and selective recombination was reported in 1973, and the key to success was a restriction endonuclease, EcoRI, that had the property of asymmetric cleavage of DNA, thus leaving cohesive ("sticky") termini that could be reconnected. DNA recombination was known to occur in nature and small amounts of incorporation had been noted by other methods, but the recombination based on EcoRI was much more efficient.

---

[i]Hamilton O. Smith was born in 1931 and grew up in Champaign-Urbana, IL, USA. He received a B. A. in mathematics from UC Berkeley in 1952 and an M. D. from Johns Hopkins in 1956. He served as a Navy physician from 1957 to 1959 and then completed his medical residency. In 1962, he began a postdoctoral fellowship at the University of Michigan. He began his independent career at Johns Hopkins in 1967. He studied the enzymology of genetic recombination and transformation and discovered the first selective endonuclease, HindII.

Daniel Nathans was born in Wilmington, DE, USA, the youngest of nine children. He received a B. S. in Chemistry at the University of Delaware in 1950 and an M. D. from Washington University in St. Louis in 1954. He received further training and experience in medical research at Columbia University, the NIH, and Rockefeller Institute before joining the faculty of Johns Hopkins in 1962. He recognized and demonstrated several of the important applications of the selective endonucleases.

Werner Arber was Swiss and received his undergraduate degree at the ETH in Zurich. He then worked on electron microcopy at the University of Geneva, earning a Ph. D. 1958. After postdoctoral study, he returned to Switzerland and was on the faculty first at the University of Geneva and then Basel. He discovered the first nonselective endonuclease and explained their basic function.

Construction of chimeric, that is, recombinant DNA involves the introduction of genetic material into some *vector* that can then insert the gene into the target organism. The vectors are usually *plasmids*. Plasmids are moderately sized cyclic forms of double-stranded DNA. The first examples of recombinant DNA were reported in 1973, based on the EcoRI enzyme. The enzyme had been discovered in the laboratory of Herbert Boyer[j] at the University of California, San Francisco (UCSF). Stanley N. Cohen[k] of Stanford University was studying the mechanism by which bacteria acquire resistance, a process that occurs in nature (see Section 12.4). Cohen and Boyer undertook a collaboration and demonstrated that an antibiotic resistance gene could be transferred into both *Escherichia coli* and *Staphylococcus aureus*. The latter observation was significant in that it showed that a gene could be transferred between different species. At the same time, experiments were being conducted in the laboratory of Paul Berg,[l] who was also at Stanford. Berg's experiments showed that a bacterial virus could be incorporated into a monkey virus, known as SV40, which can also live in mouse or human cells. This research was suspended, until safety guidelines were established.[m] Berg and coworkers went on to demonstrate that DNA from the frog *Xenopus laevis* could be incorporated into *E. coli*. Stanford and UCSF were awarded a broad patent for the methodology of DNA recombination and the technology was adopted by the newly formed companies that constituted the biotechnology industry. The process of incorporating DNA into a host organism was called *cloning*. The gene or group of genes of interest can be incorporated into a host organism and direct the synthesis of the desired protein product. The methodology was soon applied in the preparation of polypeptides,

---

[j]Herbert Boyer was born in Derry, PA, USA in 1936 and received a B. S. in Biology and Chemistry from St. Vincent College. He then received a Ph. D. from the University of Pittsburg in 1963 and spent 3 years at Yale University as a postdoctoral. He joined the faculty of UCSF in 1966. His research focused on the endonuclease restriction enzymes. Boyer was a major factor in the development of the biotechnology industry. He was a co-founder of Genentech and contributed to its development of the commercial production of insulin in collaboration with the Eli Lilly Company (see Section 21.3.1).

[k]Stanley N. Cohen was born in Perth Amboy, NJ, USA in 1935 and received his undergraduate degree at Rutgers and an M. D. from the University of Pennsylvania in 1960. He then pursued research as a postdoctoral fellow at NIH and the Albert Einstein College of Medicine. He joined the faculty of Stanford in 1968, where his research interest was the mechanism of transfer of antibiotic resistance.

[l]Paul Berg was born in Brooklyn in 1926. His undergraduate education was interrupted by service in WWII, but he received his B. S. at Penn State in 1948, where he majored in Biochemistry. He received a Ph. D. from Case Western Reserve in 1952. His original work was on metabolism but his interest shifted to DNA during postdoctoral work in Copenhagen. He was on the faculty at Washington University in St. Louis before moving to Stanford in 1959. He shared the 1980 Nobel Prize in Chemistry with Walter Gilbert and Frederick Sanger for work characterizing DNA.

[m]The original safety regulations were developed under the auspices of a committee of the National Academy of Sciences and were quite stringent, including the requirement that any human recombinant DNA be handled only in so-called P4 facilities that were self-contained and required extensive decontamination procedures for use. The regulations in Europe were somewhat less severe and some of the early work on human rDNA was done there. The regulations were gradually relaxed as experience in handling the material was gained.

including insulin, granulocyte stimulating factors and human growth hormone, as will be described in the next chapter. The process is outlined in Figure 20.10.

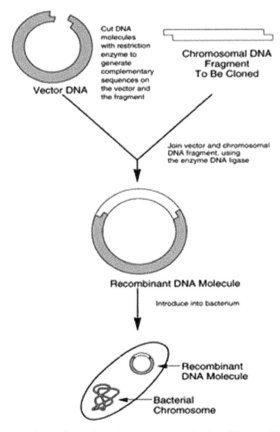

FIGURE 20.10   Outline of gene insertion process in cloning. (Source: U.S. Department of Energy Genomic Science program, http://genomicscience.energy.gov)

20.4.2.1   CLONING FROM DNA

In DNA cloning, the fragment of DNA to be reproduced is inserted into the plasmid by cleaving the circular plasmid with a restriction endonuclease. The fragment is then incorporated into the plasmid with a *DNA ligase*. The plasmids used for cloning are usually modified from natural ones and are considerably smaller, on the order of a few thousand base pairs. Also incorporated into the plasmid is a gene for drug resistance. The modified plasmid containing the gene being transferred is then inserted into an organism, typically *E. coli*. The nontransformed bacteria do not have the drug-resistant gene and are killed by the antibiotic. This strategy is necessary to identify the small fraction of the cells that have been successfully transformed.

For cloning that involves larger DNA a bacterial virus, bacteriophage λ, is used. This virus can reproduce in large numbers in the host bacterial cells. The viral DNA is modified so that it incorporates only the minimal DNA necessary to reproduce. The foreign DNA is then incorporated using the same strategy of cleavage and recombination as for plasmids. The foreign DNA can be introduced in fragments that contain an entire DNA sequence. The fragments typically can contain on the order of 20,000 base pairs. The virus is then allowed to reproduce in *E. coli*, providing a large number of colonies, each containing one of the fragments. This is called a *library* and is capable of containing an entire species genome in less than 50 separate cultures.

Clones coding for a particular protein are usually identified by *hybridization*, also sometimes called *annealing*. Hybridization can occur with either a matching sequence of RNA or DNA. The hybridization occurs by base-pairing sequences that are complementary between the clone and the probe. The matching clone is thereby identified and recovered from the plate. The probes can be obtained in several ways. If part of the protein's AA sequence is known, the corresponding DNA sequence can be synthesized. A sequence of at least 20 bp is needed to insure a unique hybridization. This corresponds to about seven AA. One complication, however, is that since many of the AAs have degenerate codes, any of the sequences might be present. To solve this problem, the probe can be synthesized as a mixture containing all of the potential matching sequences.

Even larger DNA sequences can be cloned using *yeast artificial chromosomes* (YACs) or *bacterial artificial chromosomes* (BACs). YACs can contain 100–1000 kb sequences. They are, however, prone to inserts and rearrangements. BACs can contain inserts of 80–300 kb. They are less prone to rearrangement and insertion problems. Libraries of BACs can be created by inserting random sequences.

20.4.2.2   CLONING FROM RNA

DNA can also be cloned from RNA. This approach is particularly useful in obtaining the genes connected to specific proteins by targeting the corresponding *m*-RNA. In this case a *reverse transcriptase* is used to obtain the complementary DNA sequence, called *cDNA*. The process begins with isolation of all of the RNA from a particular cell line. The *m*-RNA is then separated, taking advantage that it contains a "polyadenylate tail" and therefore can be bound by a poly-T matrix. The purified collection of *m*-RNA molecules are then converted into the corresponding DNA by a reverse transcriptase. The c-DNA corresponds only to the protein sequence, because the non-coding portions (introns) have been removed during *m*-RNA formation. The c-DNA, in turn, can be converted to double stranded DNA by a DNA polymerase. Finally, the double stranded DNA is modified to add restriction sites that will permit its incorporation into the bacteriophage vector. Cloning then proceeds as described for DNA.

Expressed sequence tags are cDNA copies of *m*-RNA. They typically contain 600–800 bp sequences. Because they are derived from *m*-RNA, they correspond to processed genes, that is, they do not include introns and are spliced corresponding to the final protein sequence. One method of relating cDNA sequences to particular proteins is to use the expressed sequence tag database, which includes an enormous number of random 200–400 bp cDNA sequences. If part of the protein's AA sequence is known, the database can be searched for a match. If a match is found, as is almost certain, a corresponding DNA sequence can be synthesized as a probe.

20.4.2.3   POSTTRANSLATIONAL MODIFICATION

Many functional proteins are subject to additional modification after the polypeptide is formed. This is called *posttranslational modification* and can include such changes as glycosylation, acylation, phosphorylation, and sulfation. The capacity for posttranslational modification depends on the organism in which the protein is produced. Bacteria do not effect posttranslational modification. Fungi and yeast can affect glycosylation and phosphorylation. Mammalian cell lines, such as Chinese hamster ovary, can effect the entire range of posttranslational modifications.

20.4.3   DNA PROFILING FOR IDENTIFICATION

Humans have 46 chromosomes. There are 22 pairs with one coming from each parent and the XX or XY sex chromosomes. The genes that code for the various proteins are found within these chromosomes. The regions of the genes that code for protein are called *exons*. Although the sequence of DNA is 99.9% the same for all individuals, there is variation in the remaining 0.1%. One type of variation is called an *allele*, which is a small variation in a gene that can lead to a characteristic change in structure and function of protein, called a *phenotype*. Other variations are *polymorphisms* that have no recognizable effect, either because there is no change in peptide sequence, or because there is no change in protein function. Most of the variations are in *introns* that do not contain genes. This part of DNA contains *short tandem repeats* (STR), which are 3, 4, or 5-base sequences that are repeated a number of times. Related individual have very similar sequences in these regions, but that are different from those of unrelated individuals. DNA profiling depends on comparing cleavage patterns from samples of DNA.

The first demonstration of the use of DNA for identification was by Sir Alec Jeffrey of the University of Leicester in the United Kingdom. The method was commercialized by ICI. This method depended on use of *restriction endonucleases*, which can cleave at specific sequences of base pairs. The endonucleases are targeted at a region called the *variable number tandem repeats*. Identical DNA strands give identical cleavage patterns and closely related individuals have very similar patterns. This method requires a relatively large sample and can be confounded by contamination. Current profiling methods depend on STRs. In the United States, a standard set of 13 regions has been selected and incorporated in the CODIS (Combined DNA Index System) identification system. The STR sites are targeted by primers and can be amplified by PCR. The amplified DNA is then subjected to cleavage by endonucleases. A match of all 13 sites corresponds to a random chance of only 1 in 10^{13}. There are several important applications of DNA profiling. One is forensic analysis, such as in the law enforcement investigations or identification of disaster victims. Another is determination of parentage and familial relationships. Profiling is also used to detect pathogenic organisms and in matching organ donors.

20.4.4 DETECTION OF MUTATIONS

Mutations are changes in DNA sequence that lead to modified protein sequence, structure, and function. These can lead to disease or predisposition to diseases. Some of these mutations are well characterized; for example, cystic fibrosis, sickle cell anemia, and hemophilia. In 1988, Bert Vogelstein of Johns Hopkins University identified several mutations associated with colon cancer, including the *p*53 gene, which was later associated with many other types of cancer. Currently, several thousand disease-associated mutations have been identified. In many cases, clusters of a mutation in family groups or isolated populations have been instrumental in detecting mutations. DNA sequencing is used to detect and identify mutations. Figure 20.11 shows the comparison of a reference gene sequence with a gene carrying a single point mutation. The difference presentation along the bottom line readily detects the mutation.

There are many tests now available for detection of mutations. In all states, newborns are screened for phenylketonuria and congenital hypothyroidism.

Prenatal screening for inherited diseases such as hemophilia and cystic fibrosis, as well as a number of others, are available. The results of screens can be used for genetic counseling. Screening for the breast cancer genes *BRAC1* and *BRAC2* is available. Genetic screening is available not only through the health care system, but also commercially. In general, tests for mutations are not regulated at the federal level in the United States, although some states have regulations.

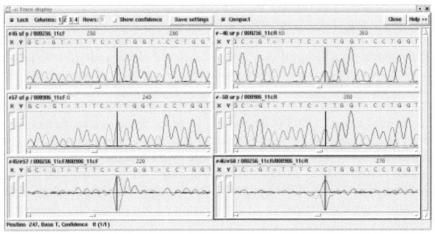

FIGURE 20.11   Detection of a point mutation by fluorescence sequencing. (Source: Medical Research Council, Laboratory of Molecular Biology; http://staden.sourceforge.net/ mutations/)

TOPIC 20.1 SICKLE CELL ANEMIA—THE FIRST MOLECULAR DISEASE

Sickle cell anemia is an inherited condition that results from a mutation in one of the protein constituents of hemoglobin. The term "sickle cell" comes from the abnormal shape of red blood cells during a crisis of the disease, which occurs when the degree of oxygenation is low. The molecular basis of the condition was first recognized by Linus Pauling who showed that the mutated protein had different pH-dependent properties from normal hemoglobin. When it became possible to sequence the DNA, it was found that it results from a mutation of GAG to GTG for the codon of amino acid 6 in the hemoglobin β-chain, which results in substitution of a valine for glutamic acid. The change in charge type is the basis for the physical and physiological differences with normal hemoglobin.

When a person inherits the mutated gene from both parents (*homozygous*) they are susceptible to the condition. The mutation is particularly high in regions of Africa that are heavily infested with malaria. Heterozygous individuals are to some extent protected from the effects of malaria. The same mutation is found in areas of South India with a high incidence of malaria. The protection provided by the mutation reinforces the protection provided by developed immunity to malaria and increases the likelihood of survival to adulthood. There is also a developmental aspect to the relationship between hemoglobin status and survival of malaria. The fetal form of hemoglobin is not subject to the mutation and new born infants are not susceptible to the sickle cell condition until about age 2, when the hemoglobin content shifts to the adult version. Subsequently, a number of other hemoglobin mutations have been identified and related to particular conditions and syndromes. For examples, there are high levels of mutations in β-hemoglobin in parts of the world (Mediterranean, Southeast Asia) that lead to β-thalassemia, which, like sickle cell anemia, affects the oxygen-carrying capacity of hemoglobin.

Sources for Topic 21.1: Allison, A. C. *Genetics* **2004**, *166*, 1591–1599; Ingram, V. M. *Genetics* **2004**, *167*, 1–7.

20.4.5   MUTAGENESIS

20.4.5.1   SITE-SPECIFIC MUTAGENSIS

Site-specific mutagenesis permits the incorporation of a specific modified (mutated) DNA sequence into a vector for subsequent production of the modified protein. Site-specific mutagenesis was introduced by Michael Smith[n] working in collaboration with C. A. Hutchinson, III. He shared the 1993 Nobel Prize with Kary Mullis for this work. One approach is to cleave DNA in the area targeted for mutation. A synthetic polynucleotide having the desired sequence can then be inserted and the modified DNA cloned. The process is summarized in Figure 20.12. This method has been used extensively to study the effect of substitution at specific sites in proteins.

---

[n]Michael Smith was born in England in 1932 and received his elementary and secondary education near his home in Blackpool. He received a scholarship to the University of Manchester and earned both his B. S. and Ph. D. degrees there, the latter in organic chemistry. He then went to work with H. G. Khorana at the University of British Columbian (UBC) and moved with Khorana to the University of Wisconsin in 1960. His work with Khorana involved the phosphotriester method of oligonucleotide synthesis. He returned to British Columbia, where he worked first at the Fisheries Research Board of Canada, with partial support from an NIH grant to study nucleotide chemistry. In 1966, he transferred to the Dept. of Biochemistry at the UBC, and it was there that he developed the techniques for site-specific mutation. In the 1980s, he co-founded a biotechnology company, Zymos and participated in research to produce insulin by biotechnology supported by Novo-Nordirsk. In 1986, he became the director of a new institute for biotechnology at UBC.

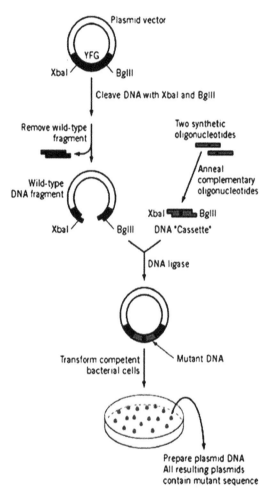

FIGURE 20.12   Site-directed mutagenesis. (Source: http://what-when-how.com/molecular-biology/site-directed-mutagenesis-part-1-molecular-biology/)

20.4.5.2   RANDOM AND SATURATION MUTAGENESIS

Another approach is to generate a variety of random mutations and then use high-throughput assays to identify those of interest. One way to do this is by *error-prone polymerase chain reactions.* Another is to determine potential sites of interest for mutation and then manipulate the codons in that region to permit formation of various amino acids at each site. This can involve only one or several sites. If all 20 of the amino acids are included, the mutations increase from 20 to 400 to 8000 for one, two, and three sites respectively.

20.4.5.3   GENE DELETION OR REPLACEMENT

More drastic mutations can be constructed by complete removal, inactivation or replacement of a gene in organisms ranging from yeast to mice. Species in which a gene has been removed or inactivated are called "knock-out." Such lines allow observation of the effect of a particular gene, which can range from no observed effect to absolute lethality. In higher organisms, such as mice, mutations can be introduced into the embryonic stem cells in such a way that they are incorporated into the germ line. Subsequent mating of mice carrying the mutation will produce off-spring that carry the modified gene. Replacement of genes can be accomplished by the same type of techniques, and species with inserted genes are called "knock-in."

The application of gene transfer to treat human disease is called *gene therapy*. The first experiments in humans were conducted in the 1990s with mixed results. A well-publicized case in 1998 resulted in the death of an 18-year-old patient when gene therapy triggered a massive immune response. In 2000, results from France and Italy reported successful treatment of very young children suffering from immune deficiency caused by lack of the enzyme adenosine deaminase. Although several of the patients showed improved immune response capacity, some developed leukemia attributed to mutagenesis associated with the gene insertion. The first gene therapy to receive EU approval for clinical use occurred in 2012 and was for a treatment of a severe inherited condition, lipoprotein lipase deficiency. Several gene therapies have also been approved in China. There remain high expectations that the solution for the many technical and ethical issues associated with gene therapy will eventually be found.

20.5   SEQUENCING THE HUMAN GENOME

20.5.1   THE HUMAN GENOME PROJECT

Discussion of the feasibility of sequencing the human genome began in the mid-1980s. In 1988, a budget of $200 million per year for 15 years was proposed, although it was recognized that technological advances would be necessary to achieve success. It was proposed that the project be jointly managed by the NIH and the Department of Energy (DOE). The latter had some expertise in genetics from its work on the effects of radiation and also had facilities, funds and personnel that could be diverted from waning efforts in nuclear weapons technology. The *Human Genome Project* (HGP) got underway in 1989. James D. Watson was named director of the project and $53 million was appropriated. It was intended that the project would be international in scope, with about 2/3 of the work to be done in the United States and the rest internationally. Work got underway in 1990. A disagreement between the NIH director Bernadine Healy and Watson developed when Healy approved patent applications for sequenced genes. Watson resigned in 1992. Eventually, the applications were denied by the patent office and Healy's successor, Harold Varmus, decided not to pursue further efforts to obtain patents.

Francis M. Collins° was appointed as Director in 1993 and a 2005 target for completion was set. At that time there were believed to be about 100,000 human genes. The goal was to sequence all 3.1 billion base pairs, store the data in an accessible form for scientific study and investigate the social, ethical, and legal consequences of the genetic information. There was extensive cooperation with other countries, especially the United Kingdom, where the Wellcome Trust and the Medical Research Council (MRC) were involved. There were collaborating groups in France, Germany, and Japan. The project also incorporated efforts then underway to sequence the genes of other organisms, including *Saccharomyces cerevisiae* (yeast), *Caenorhabitis elegans* (a nematode worm), and *Drosophila melanogaster* (fruit fly). In terms of relative size, yeast is about 12 million bases, nematode about 100 million, the fruit fly about 160 million and human about 3 billion. The strategy to be used was to isolate the individual chromosomes of the organisms and begin by mapping the location of various genes. The genes were first incorporated into YACs as relatively large fragments, which permitted the mapping process. They could then be fragmented and incorporated into bacterial plasmids (BACs) for cloning and sequencing.

In 1996, the sequence of *S. cerevisiae* was completed. It contains about 12 million bases and about 6000 genes. *C. elegans*, a tiny round worm that has less than 1000 cells and six chromosomes was successfully sequenced by a group led by John Sulston at the Sanger Center in Cambridge, with financial support from the Wellcome Trust. It contained 97 million bases and 19,099 genes. Although progress was being made, it was not fast enough to meet the 2005 target set for the HGP. Each sequencing machine could sequence about 50,000 bases/ day, but by early 1998, only about 2–4% of the sequence had been completed.

°Francis Collins received his B. S. degree in chemistry from the University of Virginia and then a Ph. D. in physical chemistry from Yale. He then entered medical school, receiving an M.D. from the University of North Carolina. Working at the University of Michigan, Collins devised techniques that speeded up the location of genes on chromosomes and applied them to location of the gene for cystic fibrosis, in collaboration with Lap-Chee Tsui of the Centre for Applied Genomics at the University of Toronto. Collins also found the gene for neurofibromatosis and was one of the team that located the gene for Huntington's disease.

Controversy was beginning to arise as to whether the HGP was being managed efficiently.

20.5.2   THE WHOLE GENOME SHOT-GUN APPROACH

In the 1990s Craig Venter[p] and Mark Adams, working at the NIH Institute for Neurological Diseases and Stroke, developed a new sequencing approach that promised faster results. The method was called the *whole genome shotgun* (WGS) approach. One of the crucial features of Venter's approach was use of *expressed sequence tags*, or EST for short. These are derived from the *m*-RNA corresponding to active protein production. The complementary DNA strand, called *c-DNA* can be cloned into bacterial DNA, and, when sequenced provides information on functional DNA, that is, genes. The concept was originally introduced by Paul Schimmel of MIT. It was further developed by Sydney Brenner at the MRC laboratory in Cambridge and has the advantage of focusing attention on expressed gene sequences, while avoiding the introns that do not contain coding information. The EST are only partial fragments of genes, however, and do not provide any direct information about the function of the gene. However, they can be identified if other information on the gene sequence is available. Using this approach, by 1992, Venter had sequenced parts of about 2500 human genes.

The issue of patenting genes from the HGP had raised a broad question as to how sequence information should be shared and disseminated. This issue arose again in the context of Venter's efforts after he left NIH. Although much of the community was skeptical about the WGS approach, Venter along with his wife Claire Fraser founded The Institute for Genomic Research (TIGR) with funding from Wallace Steinberg, a biotech entrepreneur. The understanding was that a private company, Human Genome Sciences (HGS), would hold the intellectual

---

[p] J. Craig Venter grew up in California and served as a medical corpsman during the Vietnam War, where he dealt with hundreds of wounded and dying men. After his military service, he received a B. S. in biochemistry and a Ph. D. in Physiology and Pharmacology at UC, San Diego. He began his professional career at the Roswell Park Cancer Institute in Buffalo, NY in 1976, and then in 1984 he moved to the NIH Institute for Neurological Diseases and Stroke. His research there involved a search for the adrenaline receptor in heart muscle cells. Venter began an association with Applied Biosystems which manufactured automated DNA sequencers, then able to sequence some 12,000 bases per day. He also used robotics for sample handling. Nevertheless, Venter was interested in finding faster methods. He began using cDNA, generated from RNA actively engaged in protein synthesis as a means of focusing on genes, as opposed to the less informative "junk" DNA. This approach was capable of much faster sequencing than the existing methods but generated only partial sequence information and did not necessarily relate the sequence to any particular biological function. Nevertheless, Venter results were a bombshell in the sequencing field, as his first paper in 1991 increased the number of identified genes by 10% and a second paper in 1992 reported parts of around 2500 human genes. The approach, however, generated considerable skepticism and criticism. Frustrated by his inability to receive adequate funding at NIH, Venter founded TIGR, with the financial support of a biotech entrepreneur, Wallace Steinberg. A related company, HGS, held intellectually property rights to the data developed at TIGR.

property that was developed. The EST data began to be used to identify significant mutations included some associated with cancer and Alzheimer's disease. Recognizing the importance of EST data, Merck funded an effort at Washington University in St. Louis to assemble additional data independent of the TIGR/ HGS effort. Another major pharmaceutical firm, SmithKline–Beecham, purchased 7% of HGS and rights to the sequence data for $125 million in 1993. In 1995, *Nature* published a compendium of 175,000 EST sequence data collected at TIGR. Another 118,000 were available from other sources. Together, they accounted for about 30,000 whole or partial genes and at the time, Venter estimated the final count of genes would be in the range of 60,000–70,000.

The concept of the WGS method was further advanced by Hamilton O. Smith, a Nobel Laureate at Johns Hopkins (see Section 20.4.1), in conjunction with Venter and TIGR. The idea was to shear the c-DNA into relatively small segments, sequence them, and then rely on computers to reassemble the numerous fragments by matching overlapping sequences. The essential difference from the methods adopted by the HGP is that the DNA is not mapped to individual chromosomes before the sequencing is undertaken. This makes the deciphering of the code a larger problem in that there are many more fragments to be reassembled. The data must be reassembled by computer algorithms that find sequence matches. The excess of data is large, so the probability of correct sequencing is very high. The approach was first applied to *Haemophilus influenzae*, an organism Smith had used during his work on restriction endonucleases. The effort was successful and the sequence was completed in about a year, whereas the existing methods for similar size genomes required 10 years or more. The sequence was completed in 1996 giving 1830,137 bases in 1743 genes. The total cost was around $1000,000, just over $0.50 per base.

The TIGR group continued to sequence other microbial genes, including pathogenic organisms such as the cholera-causing *Vibrio cholerae.* Among the organism sequenced was *Methanococcus jannaschii*, a methane-producing organism that lives at thermal vents in the deep Pacific. It was the first of the *Archea* to be sequenced and showed that it contained many genes unrelated to any other type of organism. The sequence confirmed that *Archaea* represented a third branch of the evolutionary tree, along with prokaryotes and eukaryotes. By 2000, TIGR had sequenced 20 different microbes. These efforts revealed that there were many unsuspected genes and that there were many examples of "whole gene tranposons," that is, incorporation of entire genes into an organism.

Venter and associates proposed to do the sequencing of the human genome by the WGS method. Celera Genomics was then founded to sequence the human genome, in collaboration with Applied Biosystems, which provided $300 million. This effort required development of the computer science aspect, as well as enhanced sequencing capacity. Eventually, the facility involved hundreds of sequencing machines operating in highly automated environment, running 24 hours 7 days a week. The computer network stored, updated and analyzed the massive data produced. Celera's efforts got underway in early 1999 and focused first on *D. melanogaster*, the fruit fly, a classical system for study of genetics in biology. At the time about 20% of the sequence was known. Its genome is about 1/20th the size of the human genome. Celera completed the sequence by the summer of 1999. The results were published in early 2000. It contains 13,601 genes, somewhat fewer than had been expected. The Celera sequence depended on a total of 3 million bases in fragments of average length of 500 that had been reassembled by the computer programs. There were a few remaining gaps. There were 16 conflicts with previously determined sequences and all of these turned out to be errors in the reported sequences. On the basis of these successes, Celera proceeded to the sequencing of the human genome.

20.5.3 COMPETITION AND THEN COOPERATION

With the success of the WGS approach, the genetics community began to consider its applicability to the much larger human genome. This interest coincided with development of a new sequencing machine by Applied Biosciences. The ABI Prizm 7000 used capillaries for electrophoresis and could sequence 96 samples simultaneously. It had the ability to sequence about a million bases per day. Venter decided to attack the human genome by the WGS approach at Celera. There was considerably rivalry and mutual recrimination between HGP and Celera. A major issue was the availability of the sequence data that Celera generated. The HGP data was updated daily, but Celera planned to release data only every 3 months and to restrict some data of potential pharmaceutical application. The Celera plan was a challenge to the HGP, especially by raising the prospect that the publicly funded project might be outstripped, risking congressional ire. However, funding for the HGP was continued and to some extent increased. The Wellcome Foundation doubled funding for the Sanger Center at Cambridge. The HGP resources were focused on six US centers, the Sanger Center and groups in France, Germany, and Japan. The chromosomes were assigned to the groups. The HGP leadership decided to publish a "rough draft" of the genome by the Spring of 2000. Orders for 500 of the new model sequencers were placed by HGP labs. By December, 1999, the HGP was able to announce the completion of its first human chromosome, 22. It contained 33.4 million bases, but with some gaps. There were nearly 5500 identifiable genes, but 42% of the sequence was so-called "junk" DNA. In April, 2000, the DOE component of the HGP announced "rough drafts" of chromosomes 5, 16 and 19, totaling 300 million bases, about 10% to the total genome. In May, the sequence of chromosome 21, the smallest, was announced by the collaborating German and Japanese laboratories. It contained about 33.5 million bases and was the most complete and accurate sequence to date. It accounted for all but 0.3% of the sequence and was judged to be 99.995% accurate. It contained about 225 genes, only about a quarter of the number that had been expected. A little over half were already known.

Early in 2000, there was some discussion of cooperation between the HGP and Celera, but progress stalled on the issue of Celera's rights to the commercial application of the data. Controversy continued, and in the meantime Celera stock, as well as that of other biotech companies, fluctuated widely as investors speculated on the ultimate value of its database. The controversy ended in June 2000 when Venter and Collins in the presence of President Clinton jointly announced their rough drafts, with British Prime Minister Tony Blair and the British research team connected by remote hookup. Clinton had ordered that the controversy to end, and the agreement was negotiated by Aristedes Patrinos, the head of the DOE sequencing project. At that point, HGP had some 22 billion bases, 95% of the genome at an estimated 99.99% accuracy. Celera had sequenced about 14.5 billion bases and had about 99% of the genome.

20.5.4 RESULTS AND COMPARISONS BETWEEN THE HGP AND WGS METHODS

By April, 2003, the 50th anniversary of the Watson–Crick double helix, the HGP had sequenced about 99.9% of the human genome with 400 gaps. There were some discrepancies between the HGP and WGS versions. According to analysis by Celera, about 2.7 million bp disagree. The HGP data is said to contain about 140 million base pairs that are redundant, while the Celera contains 50 million bp of such data. There are discrepancies between the two sets of data for about 200 bp. Continued refinement is being done. The final results provided some surprises. It turned out there were only around 25,000 genes, not 100,000. By way of comparison, *C. elegans* has about 20,000 and *D. melanogaster* about 14,000. The mouse genome is about 89% identical with that of humans. As the sequencing work proceeded, it became apparent that individual genes can make several proteins, overturning the one gene–one protein relationship. It also became apparent that all the sequenced genomes, human included, incorporated genes from ancient bacteria and single-cell organisms. The genetic material also contains many sequences that are not transcribed, which includes the extensive areas called tandem repeats.

20.5.5   THE GENOMES OF J. CRAIG VENTER AND JAMES D. WATSON

A post-script to the human genome sequencing is the fact that both J. Craig Venter and James D. Watson's genomes have been sequenced and published. A modified version of the Celera approach was used to sequence the gene of Venter.[5] The sequence was compared with both the HGP and WSG consensus sequences. The sequence contained 3213,401 single nucleotide polymorphisms, 53,823 block substitution of 2–206 bp, 292,201 heterozygous insertions/deletions and 559,473 homozygous insertion/deletions. The genome of James D. Watson was determined using the pyrophosphate-linked luminescence method (see Section 20.2). Watson's genome was sequenced in a period of 2 months and published in the April 2008 edition of *Nature*.[6] The cost was estimated to be about 1/100th of that by the capillary electrophoresis method used for Venter's genome. Watson's genome contained 3.3 million single nucleotide polymorphisms, of which 10,654 cause an amino acid substitution. Comparison of the Venter and Watson genomes differed from the reference genome by 3766 and 3882 nonsynonymous SNPs, respectively.

KEYWORDS

- **DNA bases**
- **double helix genetic code**
- **DNA sequencing**
- **DNA synthesis**
- **polymerase chain reaction (PCR)**
- **restriction endonucleases**
- **recombinant DNA**
- **cloning**
- **profiling**
- **mutation**
- **mutagenesis**
- **human genome sequencing**

BIBLIOGRAPHY

Davies, K. *Cracking the Genome*. Free Press: New York, 2001.

BIBLIOGRAPHY BY SECTIONS

Section 20.2: Marziali, A.; Akeson, M. *Ann. Rev. Biomed. Eng.* **2001**, *3*, 195–223; Karger, B. L.; Guttman, A. *Genomic/Proteomic Technol.* **2003**, *3*, 14–16; Ziebolz, B.; Droege, M. *Biotechnol. Ann. Rev.* **2007**, *13*, 1–26.

Section 20.3.1: Urbina, G. A.; Gruebler, G.; Weiler, A.; Echner, H.; Stoeva, S.; Schernthaner, J.; Gross, W. Voelter, W. *Zeitschr. Naturforsch.* **1998**, *B53*, 1051–1068.

Section 20.4.1: Mundy, C. *Pharmacogenetics* **2001**, *2*, 37–49; Green, E. In *The Genomic Revolution*; Yudell, M.; DeSalle, R., Eds.; John Henry Press: Washington, DC, 2002; pp 35–47.

Section 20.4.2: Berg, P.; Mertz, J. E. *Genetics* **2010**, *184*, 9–17; Cohen, S. N. *Proc. Natl. Acad. Sci. U.S.A.* **2013**, *110*, 15521–15529.

Section 20.4.3: Lynch, M. *Endeavour* **2003**, *27*, 93–97; www.ornl.gov.sci/techresources/ Human_Genome/elsi/forensic (accessed 4-14-2010); Giardina, E.; Spinella, A.; Noveli, G. *Nanomedicine* **2011**, *6*, 257–270.

Section 20.4.4: Botstein, D.; Risch, N. *Nat. Genetics Suppl.* **2003**, *33*, 228–237.

Section 20.4.5.1: Brannigan, J. A.; Wilkinson, A. J. *Nat. Mol. Cell Biol.* **2002**, *3*, 964–970.

Secton 20.4.5.3: Wirth, T.; Parker, N.; Yla-Herttuala, S. *Cell* **2013**, *525*, 162–169; Kastelein, J. J. P.; Ross, C. J. D.; Hayden, M. R. *Hum. Gene Ther.* **2013**, *24*, 472–478.

Section 20.5.2: Venter, J. C. In *The Genomic Revolution*; Yudell, M.; DeSalle, R., Eds.; John Henry Press: Washington, DC, 2002; pp 48–63.

REFERENCES

1. Ronaghi, M.; Uhlen, M.; Nyren, P. *Science* **1998**, *281*, 363–365; Margulies, M.; Egholm, M.; Altman, W. E.; Attiya, S.; Bader, J. S.; Bemba, L. A.; Berka, J.; Braverman, M. S.; Chen, K.-J.; Chen, Z.; Dewell, S. B.; Lei, D.; Fierro, J. M.; Gomes, X. V.; Godwin, B. C.; He, W.; Helgesen, S.; Ho, C. H.; Irzyk, G. P.; Jando, S. C.; Alenquer, M. L. I.; Jarvie, T. P.; Jirage, K. B.; Kim, J.-B.; Knight, J. R.; Lanza, J. R.; Leamon, J. H.; Lefkowitz, S. M.; Lei, M.; Li, J.; Lohman, K. L.; Lu, H.; Makhijani, V. B.; McDade, K. I.; McKenna, M. P.; Myers, E. W.; Nickerson, E.; Nobile, J. R.; Plant, R.; Puc, B. P.; Ronan, M. T.; Roth, G. T.; Sarkis, G. J.; Simons, J. F.; Simpson, J. W.; Srinivasan, M.; Tartaro, K. R.; Tomasz, A.; Vogt, K. A.; Volkmer, G. A.; Wang, S. H.; Wang, Y.; Weiner, M. P.; Yu, P.; Begley, R. F.; Rothberg, J. M. *Nature* **2005**, *437*, 376–380.
2. Metzker, M. L. *Nat. Rev. Genetics* **2010**, *11*, 31–46; Mardis, E. R. *Nature* **2011**, *470*, 198–203.
3. Itakura, K.; Rossi, J. J.; Wallace, R. B. *Ann. Rev. Biochem.* **1989**, *66*, 577–580.
4. Caruthers, M. H.; Barone, A. D.; Beaucage, S. L.; Dodds, D. R.; Fisher, E. F.; McBride, L. J.; Matteucci, M. D.; Stabinski, Z.; Tang, J.-Y. *Methods Enzymol.* **1987**, *154*, 287–313.

5. Levy, S.; Sutton, G.; Ng, P. C.; Feuk, L.; Halpern, A. L.; Walenz, B. P.; Axelrod, N.; Huang, J.; Kirkness, E. F.; Denisov, G.; Lin, Y.; MacDonald, J. R.; Pang, A. W. C.; Shago, M.; Stockwell, T. B.; Tsiamouri, A.; Bafna, V.; Bansal, V.; Kravitz, S. A.; Busam, D. A.; Beeson, K. Y.; McIntosh, T. C.; Remington, K. A.; Abril, J. F.; Gill, J.; Blrman, J.; Rogers, Y.-H.; Frazier, M. E.; Scherer, S. W.; Strausber, R. L.; Venter, J. C. *PLoS Biol.* **2007,** *5*, 2113–2144.

6. Wheeler, D. A.; Srinivasan, M.; Egholm, M.; Shen, Y.; Chen, L.; McGuire, A.; He, W.; Chen, Y.-J.; Makhijani, V.; Roth, G. T.; Gomes, X.; Tartaro, K.; Niazi, F.; Turcotted, C. L.; Irzyk, G. P.; Lupski, J. R.; Chinaljult, C.; Song, X.; Liu, Y.; Yuan, Y.; Nazareth, L.; Qin, X.; Muzny, D. M.; Margulies, M.; Weinstock, G. M.; Gibbs, R. A.; Rothberg, J. M. *Nature* **2008,** *452*, 872–876.

CHAPTER 21

APPLICATIONS OF BIOTECHNOLOGY: BIOLOGY DOING CHEMISTRY

ABSTRACT

Biotechnology can be broadly defined as the use of biological catalysts or systems to effect biological or chemical transformations. Classical examples include fermentation to produce alcohol and the use of enzymes in starch hydrolysis and cheese production. Molecular biology provided new methods based on genetic modification. Various organisms including yeasts, bacteria, plants, and animals can be modified to produce defined substances. The products include small molecules such as amino acids, vitamins, and antibiotics. The application of molecular biology to agriculture led to plant varieties that are resistant to herbicides, insects, and diseases. Herbicide-resistant varieties of corn, cotton, and soy beans were rapidly adopted between 1995 and 2000. Insect-resistant crops were produced by inserting genes for *Bacillus thuringensis* toxins. Bovine growth hormone was used to increase milk production. The introduction of genetically modified food-producing plants and animal has generated both a regulatory system and public controversy. Molecular biology has been applied in medicine to produce biologic drugs. Insulin was the first example in 1983, followed by many others including human growth hormone, granulocyte and granulocyte-macrophage factors, follicle-stimulating hormone, cytokine inhibitors, and botulinum toxin.

In previous sections, we have considered how chemical transformations can provide many vital and useful materials such as fuels, food, polymers, and medicines. Developments in molecular biology during the last quarter of twentieth century provided an entirely new set of tools for making and modifying molecules. In this section we will consider some examples of application of *biotechnology*. We can begin by asking "What is Biotechnology?" We will use a broad definition that includes any process in which a structural transformation is carried out under the agency of a biological catalyst or system. There are two broad distinctions that can be made between individual processes. In one group, the transformation is effected by a specific biological molecule, usually an enzyme, acting as a catalyst. The catalyst effects a chemical transformation but is not otherwise engaged in the process. The other broad category requires a series of biological functions such as transcription of a gene and protein synthesis. A lipase that is used to hydrolyze esters would be an example of the first category, while *Escherichia coli* producing human insulin is in the second category. In *either case*, the biological system might be a genetically modified version, the result of *genetic engineering*.

Biotechnology's origins go back before recorded history, with production of wine, beer, and other alcoholic beverages by fermentation being the obvious example. Production of cheese and soy sauce also involves fermentation with microorganisms. Dextrin, made by enzymatic hydrolysis of starch, has been produced commercially since the 1830s. A purified enzyme for cheese-making, rennet, was introduced in 1874, although at that time there was no understanding of the nature of enzymes. In the late 1800s, Eduard Buchner proved that a soluble enzyme was responsible for fermentation of glucose to ethanol. The fact that enzymes were proteins became clear in the early part of the twentieth century.

We have already encountered some examples of biotechnology. In Section 1.2.2, we mentioned the production of acetone used during WWI by fermentation of the organism *Clostridum acetobutylium*. The conversion of starch or sugar to ethanol for fuel discussed in Section 2.2.3 relies on fermentation by *Saccharomyces cerevisiae*. Further advances in conversion of cellulosic biomass will require organisms that can effect hydrolysis of cellulose to small carbohydrates. In Section 6.2.1, we noted that many laundry detergents now incorporate enzymes to assist in removing soil and stains. They are produced by biotechnology. The production of high fructose corn syrup by enzymatic catalysis was described in Section 9.1. In the antibiotics section, we noted that many of the important antibiotics are produced by microorganisms. These are all examples of production of useful materials using biological catalysts or organisms that have been selected for their favorable characteristics. Some have been improved by genetic modification.

At the present time, biotechnology is used to produce many substances; among them are small molecules such as amino acids, vitamins, sweeteners and flavors, as well as antibiotics and other drugs. There is a fairly wide range in the nature of the changes that can be effected. The transformation can involve modifying existing molecules or synthesizing new ones. Among the methods are separation of enantiomers (kinetic resolution and related methods), specific chemical changes such as hydrolysis, oxidation, reduction or introduction of new groups, for example, by acylation. Other processes assemble new molecules. Amino acids and vitamins can be made by modifying microorganisms to over-produce the desired substances. Many of the recent developments have been focused on therapeutic substances. Recombinant technology is used to produce complex molecules, especially peptides and proteins such as insulin, erythropoietin, human growth hormone and botulinum toxin (Botox). Various enzymes including amylases, glucose isomerase, penicillinase, and subtilisin are produced by recombinant technology. Genetic modification has also been used to develop herbicide and insect-resistant crops. In this chapter, we will look at several kinds of materials that are produced by a biologically mediated transformation. The possibilities and successful examples expanded rapidly in the last two decades of the twentieth as the techniques of molecular biology allowed genes to be transferred and/or modified.

21.1   EXAMPLES OF SMALL MOLECULES PRODUCED BY BIOTECHNOLOGY

21.1.1   AMINO ACIDS, NUCLEOTIDES, AND VITAMINS

Several of the amino acids, nucleosides and vitamins are produced by biotechnology. The production of amino acids by biotechnology began in Japan in the 1950s when glutamic acid for making mono-sodium glutamate (MSG) was produced by fermentation. The organism used, *Corynebacterium glutamican*, has remained a mainstay for amino acid production through the genetic engineering era. About 1.2 billion pounds of glutamic acid is produced for making MSG (value of about $900 million). The same organism is used to make L-lysine for use as a supplement in cereal grains, especially for animal and poultry food. The annual value is around $450 million. Several other amino acids are produced both for animal feeds and for dietary supplements for humans (see Section 10.3.1). The starting materials include various carbohydrate sources such as grains, isolated starch, or molasses from sucrose production. The organism chosen overproduce the desired amino acid and can be obtained either by mutation and selection or by genetic manipulation. *E. coli* has been used in amino acid production, including phenylalanine, threonine, and cysteine. A few of the amino acids are made by enzymatic processes, for example, L-aspartic acid is made by enzyme-catalyzed addition of ammonia to fumaric acid. The nucleotides guanosine monophosphate and inosine monophosphate are produced as flavor enhancers.

Vitamins for both animal and human use are obtained in several ways. Both vitamin A (retinol) and its precursor carotene are produced in multi-ton quantities. Vitamin D is produced from steroids by UV radiation (see Section 10.1.1.8). Vitamin E is mainly semisynthetic, as described in Section 10.1.1.9, but about 10% is extracted from natural sources. Vitamin B_1 (thiamine) is produced mainly by chemical methods. Vitamin B_2 (riboflavin) is made in substantial quantities for use in both human and animal food. It is produced by several of the major agricultural and pharmaceutical companies, including Aventis,

BASF, and Roche. Genetically engineered version of both *Bacillus subtilis* and *Corynebacterium ammoniagenes* are used. Both vitamin B_3 (niacin) and B_6 (thiamine) are produced mainly by chemical methods, but biological alternatives are under study. Vitamin B_{12} is produced only by prokaryotic microorganisms. Humans require the vitamin in the amount of about 1 ng/day, and it is adsorbed from the digestive tract where it is produced by microorganisms. Production of vitamin B_{12} by fermentation technology began at Merck in 1952, using *Pseudomonas denitrificans*. There have been many subsequent improvements. Currently, vitamin B_{12} is produced commercially using microorganisms by Aventis. The producing organisms have been improved both by random mutation and genetic engineering.[1] The annual value of vitamin B_{12} production is around $70 million. Vitamin C (ascorbic acid) is produced from D-glucose by a sequence consisting of five chemical steps and one fermentation step, the latter being conversion of D-sorbitol to L-sorbose.

21.1.2 ANTIBIOTICS

The production of antibiotics has involved biotechnology from its beginnings with penicillin in the 1940s (see Section 12.1). Originally, increases in yield and other process improvements were achieved by screening for favorable strains. The intentional introduction of random mutations also began in the 1940s. Beginning in the 1960s, the use of acylases to convert penicillin G into penicillanic acid began, permitting the generation of semisynthetic penicillins (see Section 12.1.2). These approaches have been supplemented by recombinant DNA technology since the 1980s. Currently, most of the penicillanic acid is produced using bioengineered *E. coli*.

The cephalosporin antibiotics (see Section 12.1) share a common biosynthetic pathway with the penicillins, being derived from them by a ring expansion. The key intermediate in the ring expansion is 7-(D-5-aminoadipoyl)penicillanic acid, also known as penicillin N. A chemical transformation was introduced in the early 1970s, providing material that could be synthetically modified to various cephalosporin antibiotics.

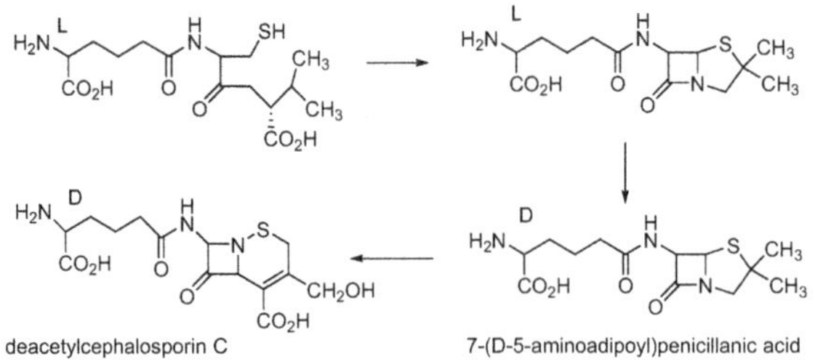

deacetylcephalosporin C   7-(D-5-aminoadipoyl)penicillanic acid

Cephalosporin C is subjected to oxidative deamination, catalyzed by a D-amino oxidase. The product, 7-(glutarylamido)cephalosporanic acid, is then deacylated enzymatically. More recently, this process has been accomplished by use of a recombinant organism that contains genes for both of the requisite enzymes.

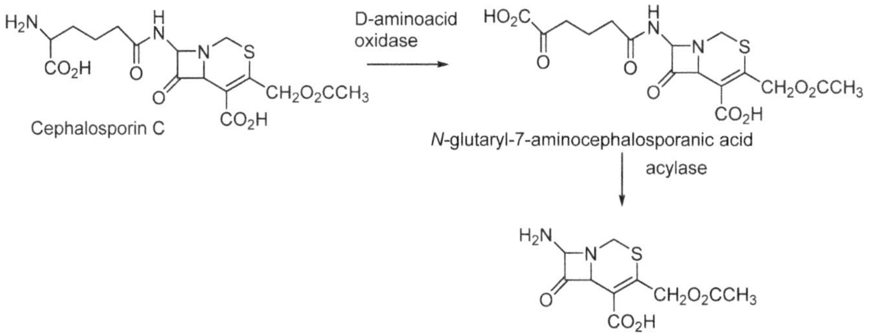

Cephalosporin C

N-glutaryl-7-aminocephalosporanic acid

D-aminoacid oxidase

acylase

21.1.3   ASPARTAME

Aspartame is an artificial sweetener approved by the FDA in 1981. Worldwide consumption in 2000 was about 20,000 t (see Section 9.2.2.3). Aspartame is produced enzymatically from methyl phenylalanate and N-Cbz L-aspartic acid by the Tosoh process. Aspartame is also produced by a chemical route developed by the Nutrasweet Company. A main advantage of the enzymatic process is that it can use racemic methyl phenylalanate, whereas the chemical process requires the pure L-enantiomer. The enzyme that is used is a metalloprotease, thermolysin. This protease is more selective for amide bond formation over methyl ester hydrolysis than other industrial proteases. The catalyzed reaction is the *reverse* of the reaction usually catalyzed by proteases. The basis of this transformation is the selective precipitation of the coupled N-protected product as a complex with the unreacted methyl D-phenylalanate. The methyl D-phenylalanate is recovered, racemized by base, and recycled. Figure 9.2 gives a flow sheet for the process.

21.2   BIOTECHNOLOGY IN AGRICULTURE

As discussed in Chapter 8, crop yields and properties have been extensively improved by use of cross-breeding (hybridization) methods. Traditional cross-breeding is limited to closely related species. Genetic engineering introduces an entirely new dimension in that genes can be transferred from a different organism (see Section 20.4). Furthermore, genetic engineering permits high specificity. A gene with a specific property is introduced or altered, whereas in cross-breeding, several genes are likely to be modified and the combinations with the most desirable traits must be selected. Another difference is that GM crops can transfer the desired characteristic to subsequent generations, while some hybrids must be produced by controlled breeding in each generation.

Several GM plants have been introduced with brand names. In the mid-1980s, the Flavr-Savr tomato was introduced, but its success was very limited and it was soon discontinued (see Topic 21.1). Several types of squash are modified for resistance to mosaic viruses. Hawaiian papayas, called *Rainbow* and *SunUp*, contain resistance to papaya ringspot virus. Several varieties of tomatoes and potatoes with improved storage properties have been approved. The most important examples in terms of acreage planted are HR and IR crops. The main examples are canola, corn, cotton, and soybeans, with cotton and corn being the main examples of a combined HR–IR crops. Both HR and IR rice have also been introduced recently. There are six major producers of GM crops seeds in the United States, all of which have their origins in the chemical industry. These are Monsanto (United States), DuPont (United States), Dow Agrochemical (United States), Bayer Cropsciences (Germany), BASF Crop Products (Germany) and Syngenta (Switzerland). Several crops currently grown in the North and South America have high percentages of GM acreages. These include soybeans, corn, canola oil and cotton. In contrast, in Europe only limited acreage is currently planted to GM crops.

21.2.1   HERBICIDE-RESISTANT CROPS

From 1946, when the first herbicide, 2,4-D (see Section 8.3) was introduced until 1995, when the first HR plants were commercialized, herbicide discovery and development was almost entirely a function of chemistry. Large numbers of compounds were synthesized and screened, and promising lead compounds were subjected to optimization. Gene transfer techniques provided an opportunity to create crops that were HR. There are several steps in the creation of an HR plant. Candidate genes must be identified and introduced into the crop plant. Correct expression and subcellular targeting of the desired tolerance gene are necessary. Typically, the transformations are carried out on tissue cultures, from which the whole plant is regenerated. The transformed plant must meet or exceed the yield and vigor of the untransformed plant. It is also important to consider the potential for dissemination of the gene into wild plants, with the result being HR-resistant weeds.

21.2.1.1   ROUND-UP READY SOY BEAN, CORN, AND COTTON

The herbicide for use with Roundup Ready crops is glyphosate (see Section 8.2.3.3). It is a powerful non-selective herbicide that was first introduced in the 1970s. It is a post-emergent herbicide and was originally used primarily in areas where it could be physically directed at the weeds, as between trees in orchards. It was also used in non-crop areas, such as for weed control along road-sides and irrigation channels. It could be used in crops only if it could be selectively delivered on the weeds. The basis for creation of Round-Up Ready plants is control of the enzyme 5-enylpyruvylshikimate-3-phosphate synthase (EPSPS) (see Section 8.2.3.3). The enzyme catalyzes formation of chorismic acid, which is required for biosynthesis of the aromatic amino acids. There is no comparable enzyme in animals so glyphosate has very low toxicity, but is highly toxic to all plants, because all have a similar enzyme. The role of EPSPS in amino acid synthesis is summarized in Figure 21.1.

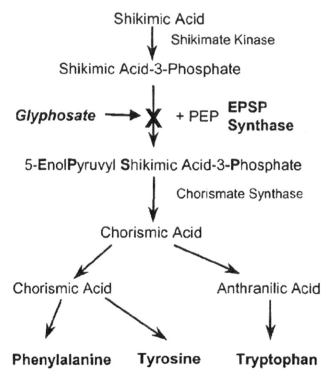

FIGURE 21.1   Mechanism of herbicidal action of glyphosate. (From Dill, G. M. *Pest Manage. Sci.* **2005**, *61*, 219–224 by permission of Wiley. © 2005, Society of Chemical Industry.)

Several possible strategies were explored to make plants glyphosate resistant, including overexpression of EPSPS or introduction of an enzyme that would detoxify glyphosate in the protected plant. The successful strategy was introduction of an alternative form of the enzyme called CP4 from an *Agrobacterium* strain. The particular strain was found in an environment exposed to high concentrations of glyphosate from a production waste stream. The bacterial enzyme is about 50% similar in amino acid sequence to that in higher plants but is not inhibited by glyphosate. It is however functional in the shikimate to chorismate conversion and thus bypasses the glyphosate blockade, as summarized in Figure 21.2. The DNA insert also contains promoter and regulatory sequences.

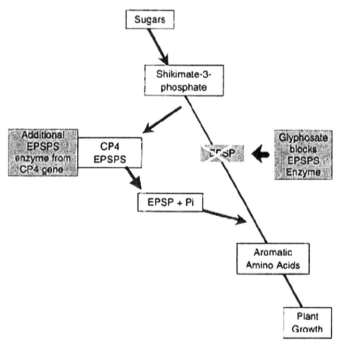

FIGURE 21.2 Mechanism for CP4 bypass in glyphosate-resistant crops. (From Dill, G. M. *Pest Manage. Sci.* **2005**, *61*, 219–224 by permission of Wiley. © 2005, Society of Chemical Industry.)

The adoption Round-Up Ready soybeans was very rapid in the United States and by 2005 almost 90% of the acreage was planted with HR-resistant varieties. Round-Up Ready cotton use rapidly increased after its introduction in 1997 and by 2008, 95% of the crop was glyphosate-resistant as shown in Figure 21.3. Adoption of Round-Up Ready corn has been slower, although GM-IR corn has been widely adopted.

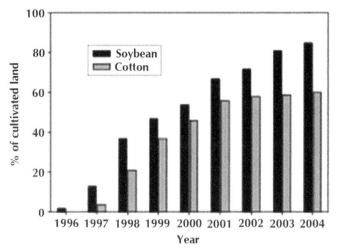

FIGURE 21.3 Adoption of round-up ready soybeans and cotton in the United States. (From Duke, S. O. *Pest Manage. Sci.* **2005**, *61*, 211–218 by permission of Wiley. © 2005, Society of Chemical Industry.)

One of the initial advantages of the use of glyphosate-resistant crops was a reduction in use of other herbicides, although, as shown in Figure 21.4, total application remained in the 1.0–1.25 kg/ha range.

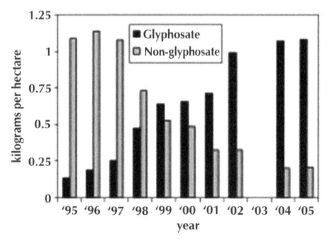

FIGURE 21.4 Comparison of glyphosate versus other herbicide used on soy beans in the United States, 1995–2005. Kleter, G. A.; Bhula, R.; Bodnaruk, K.; Carazo, E.; Felsot, A. S.; Harris, C. A.; Katayama, A.; Kuiper, H. A.; Racke, K. D.; Rubin, B.; Shevah, Y.; Stephenson, G. R.; Tanaka, K.; Unsworth, J.; Wauchope, R. D.; Wong, S.-S. *Pest Manage. Sci.* **2007**, *63*, 1107–1115 by permission of Wiley. © 2007, Society of Chemical Industry.)

In some areas of the United States, glyphosate-resistant soybeans and corn are rotated as crops and glyphosate is applied annually. Weed resistance in areas of heavy use of HR crops was first noted in 2001, and has increased since then. In Canada, only HR-canola is common and it tends to be rotated with non-HR crops, especially wheat and barley. Emergence of glyphosate-resistance is likely to be slower under these circumstances.[2] A particularly troublesome resistant weed species in cotton in the US South is pigweed (*Amaranthus palmer*). Whereas previously a variety of herbicides were used, the advantages of glyphosate resulted in its pervasive use, which sped the development of resistance. The resistance trait can be spread from field to field by pollen from resistant species. The incursion of resistant pigweed is expensive for the grower and has been estimated to be in the range of $100/ha. The strategies used to combat resistant weeds are alternative herbicides, increased tillage and cover crops, all of which are intended to minimize the extent of germination of resistant seeds.[3]

21.2.1.2   IMIDAZOLINONE-RESISTANT CROPS

Gene transfer is not the only way of developing HR plants. Another approach is natural or induced mutation. The latter approach was used to develop resistance to the imidazolinone group of herbicides (see Section 8.2.3.1). The resistant strains were developed by intentional mutagenesis and selection for resistance to the imidazolinone herbicides. This allows for a somewhat simpler regulatory approval process. Corn, canola, rice, wheat, and sunflower have been made HR in this way. The products were commercialized by BASF under the name "Clearfield."[4] Although the "Clearfield" products are not produced by genetic modification, steps must still be taken to prevent the transfer of the resistance to weeds. These include growing the seed in fields that are free of weeds that are likely to be cross-pollinated. Also, using seeds for a second year crop is prohibited to prevent inadvertent inclusion of resistant weeds. The HR crop should be rotated regularly to minimize repeated use of the same herbicide.

21.1.1.3   HERBICIDE-RESISTANT RICE

Rice is the main food crop for 3 billion people in most of Asia, including China and India. As such, it plays a major role in feeding parts of the world with increasing populations. The importance of weed control in rice has increased as labor-saving seeding has replaced transplanting in many rice-growing areas. Rice presents a particular problem in that one of the main weeds is wild or "red" rice, a close relative that is easily susceptible to gene transfer. This weed has spread to most rice-growing areas in Asia over the past few decades. There are also other grasses that are significant problems. Imidazolinone-resistant (Clearfield) rice was introduced in the United States in 2002 and has also been introduced in Central and South America. Resistant wild rice has become a problem rather rapidly. There are conceivable biotech solutions to the problem. One approach is to incorporate two unrelated genes that both produce resistance. The concept is that two simultaneous mutations conveying resistance to both is unlikely. Another strategy is the incorporation of both resistance genes and genes that make the rice *particularly susceptible* to a different herbicide. A series of such modifications, rotated sequentially through several years, has the potential to eliminate lines that have become resistant to other herbicides. However, there would need to be several such genes used over period of years.[5]

21.1.1.4 HERBICIDE-RESISTANT CANOLA OIL

Canola oil is the name given to the oil from several varieties of *Brassica*. It is the fourth most important plant in the world in terms of acreage grown. The oil is also known as rapeseed oil. The "canola" designation is an invented word for a hybrid developed in Canada. There is another version referred to as "high oleic" that has a relatively high concentration of oleic acid and a low concentration of linoleic acid. Both of these varieties were developed by classical plant breeding. Canola oil presents an especially interesting case in that three separate HR-versions are available. Two are GM, one resistant to glyphosate (Round-Up Ready) and the other to glufosinate (Liberty Link). The third is resistant to imidazolinones (Clearfield) herbicides and was created by classical breeding. Figure 21.5 shows the rapid shift between 1995 and 2000 from conventional to HR-resistant varieties. In 2008, 99%

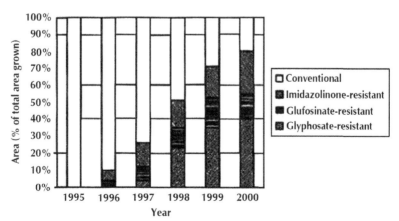

FIGURE 21.5 Shift from conventional to herbicide-resistant canola in Canada from 1995–2000. (From Brimmer, T. A.; Galivan, G. I.; Stephenson, G. R. *Pest Manage. Sci.* **2005,** *61,* 47–52 by permission of Wiley. © 2005, Society of Chemical Industry.)

of the *Brassica* grown in Canada was HR. The glyphosate and glufosinate-resistant variety each accounted for about 45% of the market, while the imidazolinone-resistant (Clearfield) variety had about 10%.

21.2.2   INSECT-RESISTANT CROPS

The idea of protecting crops from insects by biological means goes back to the nineteenth century. In 1835, Agostino Bassi, an Italian entomologist, showed that the microorganism *Beuveria bassiana* could be transferred from infected to healthy silkworm larvae. In the 1870s, J. L. Le Conte suggested this organism might be used in insect pest control, but no practical application developed. The first field tests were carried out in Russia, where Elie Metchnikoff, a renowned zoologist, and Isaak Krassilistschik used a fungus, *Metarhizium anisophae*, in attempts to control grain beetles. It is not clear how successful the trials were. In the late 1880s, under the leadership of Francis H. Snow of the University of Kansas, *B. bassiana* spores were produced and distributed to wheat farmers in Kansas and surrounding states. Again, it is unclear how successful the program was. During the 1940s, various fungi were tried for insect control in citrus groves in Florida. Similar efforts were made in Russia between 1958 and 1973 to control insect pests in citrus in the Black Sea region. A particular target in the United States was the Japanese beetle, which was introduced inadvertently in 1916. The USDA undertook a program to control and limit the spread of the beetle. This led to the identification of *Bacillus popillaie* and *Bacillus lentimorbus* as potential pathogens. Commercial preparation of spore dusts from these became available in the 1940s. The most widely used organism for insect control is *Bacillus thuringiensis* (Bt), which was initially identified in diseased flour moths in Europe. Sprays containing Bt spores have been used for a long time. The organism was tested against the European corn borer, *Ostinia nubilalis* in Eastern Europe in the 1920s and 30s. A commercial product was produced in the United States by Pacific Yeast Products and marketed as "Thuricide" in the 1950s. Various other companies introduced related products, including improved strains.

21.2.2.1   BACCILLUS THURINGIENSIS-MODIFIED CORN AND COTTON

Bt is a widely distributed soil microorganism that produces a series of protein toxins. The toxins affect insect larvae and there is considerable species specificity. This is a potential advantage, because, in principle, it allows targeting of specific pests. Further selectivity comes from the need for the toxin to be ingested and digested by the insect. The toxins are cleaved to proteins that then bind to receptors in the larvae, again exhibiting selectivity. The multilevel specificity provides a high level of safety to the toxins. There have been no authenticated cases of poisoning of fish, birds, domestic animals, or humans from Bt toxin administered as either sprays or in GM plants.[6]

Plants genetically modified to produce Bt toxins were introduced in the 1990s. Bt-resistant cotton was developed independently in China and introduced as early as 1994.[7] Several other plants have been modified in this way, but only corn and cotton are produced commercially. The gene in corn is targeted at corn borers and root worms, while in cotton the targets are armyworms, bollworms, and budworms. One of the genetically modified corn varieties, known as "Star-Link" contained a somewhat different form of the Bt toxin. In the tests for regulatory approval this particular protein was somewhat more slowly digested than other forms and there was some concern about allergenicity. As a result, the EPA did not approve Starlink corn for human consumption. However, because of the difficulties in segregating the corn during commercial handling, it soon appeared in human food products, resulting in an FDA ban. Reports of allergenic reactions were investigated by the CDC, but no conclusive confirmation could be found. Nevertheless, Star-Link corn was removed from the market in 2000.

Resistance to Bt can develop in insects. One strategy used to slow the rate of development of resistance involves planting areas of unmodified crop, called refuge areas, in conjunction with the GM crop. These "refuge areas" permit the survival of nonresistant insects, which by interbreeding with resistant ones, can slow the development of resistance. Compliance is believed to be about 90%.

21.2.2.2   INSECT-RESISTANT RICE

Rice is the major crop in most of Asia and much effort has been devoted to exploring the potential for IR-rice. In China effort is being devoted to both insect-resistance and bacterial blight resistance.[8] The first two lines were approved for production in 2009. Both lines incorporate Bt genes. They are directed primarily at boring insects such as stem borers. These have become a major pest in recent years as a result of insecticide resistance, changed cropping practices and climatic warming. The potential advantages of insect-resistant rice are similar to those with other crops, including reduced insecticide use and cost, reduced environmental contamination and reduced pesticide residues in food. The potential pitfalls are also the same, including development of resistant insects. This is a particular concern because the level of toxin in the modified rice is not as high as in corn and typically doesn't exceed 90% in lethality. This has the potential for accelerating the development of resistance.

21.2.3   BOVINE AND PORCINE GROWTH HORMONES

Bovine growth hormone (bGH), also called *bovine somatotropin*, is a 191-AA polypeptide that is analogous to human growth hormone in its mechanism of action and effects. Early studies in the 1920s and 1930s showed that extracts containing bGH could increase milk production. Extraction methods, however, could not produce sufficient material for commercial application. This changed with the advent of genetic engineering in the mid-1980s, which permitted the production of recombinant hormone (rbGH) in *E. coli*. Use of rbGH was approved by the FDA in the United States in 1994, and it is currently (2007) estimated that 15–20% of dairy cows are treated with rbGH. Injection in dairy cows results in increased milk production by 10–15%. There is also an extension of the peak production of milk. There has been some controversy about the safety of milk from rbGH-treated cows. There is no evidence of change in composition of milk with respect to carbohydrate, fat or protein. The rbGH polypeptide is present in fresh milk but is destroyed by pasteurization and digestion. Furthermore, the bovine GH differs significantly in AA-sequence from the human hormone and is not active in humans. Besides the United States, rbGH is approved in most of Latin America, but not in the EU, Canada, Australia, New Zealand, or Japan. The basis of the Canadian and EU disapproval is animal, not human, health and is based on evidence that treated cows have a higher incidence of mastitis. As a result of consumer concerns, there have been some efforts at labeling and advertising milk and derived dairy products as being free of rbGH. Most regulatory and legal rulings, however, have supported the view that there is no difference in the products.

Recombinant porcine growth hormone (rpGH) is also available by genetic engineering. Its use results in a significant shift from fat to protein content in the meat produced, which is potentially advantageous in the diet of consumers. It also improves the efficiency of feed use in pork production. At the present time, however, there is no approved commercial use of rpGH in pork production.

21.2.4   POTENTIAL BENEFITS OF GM CROPS AND FOODS

GM products may have indirect benefits. For example, Bt-modified corn tends to have lower insect damage and this correlates with lower concentration of mycotoxins. HR plants permit reduction in use of herbicides. HR crops also permit use of no-till methods that save energy and reduce soil erosion. The no-till methods in turn permit closer spacing of row crops, which can lead to increased yields. Several potential properties might be engineered in the future to improve crop productivity and efficiency. For example, increased efficiency of photosynthesis, higher efficiency of nitrogen utilization, drought and cold or heat tolerance, and increased vitamin content are all potentially valuable traits. Considerable effort is being devoted to these goals. However, the efforts are complicated, because traits such as drought-resistance and yield typically involves several genes.

It is generally recognized that poverty, not lack of food, is the main cause of hunger and poor nutrition around the world. It is projected that world food production will continue to outpace population growth.[9] Nevertheless, malnutrition is widespread. It is estimated that 25% of children in Southeast Asia and 33% in Sub-Sahara Africa are malnourished, with vitamin A deficiency being widespread. Among the factors causing this deficiency is that natural food sources for vitamin A are seasonal, relatively expensive, and not always easily digested and/or palatable, especially for children. Diets that depend mainly on white rice, are associated with vitamin A deficiency. Golden Rice, a rice variety that incorporates a gene for β-carotene production has generated considerable interest.[10] Several questions remain however. Foremost is probably acceptance, but the problems of monitoring ecological effects, if any, are also challenging. Perhaps most significant is that several countries have introduced vitamin supplement distribution patterned on successful vaccination programs and these may be more cost effective.

A wide-spread human mineral deficiency is iron. Most plants have limited capacity to absorb and store iron from the soil, legumes such as soybeans being the most important exception. Thus modification of plants to increase iron content is a potential goal. Another crop that might be subject to GM is cassava (*Manihot esculenta*), which is an important energy source for some 600 million people, many in Sub-Sahara Africa. The root contains mainly carbohydrate, but it is low in protein. The plant also contains cyanogenic glycosides that must be removed in processing. These traits could conceivably be changed by genetic modification.

Corn is a major feed crop for poultry and swine, but is low in lysine content. As a result, diets are frequently supplemented by lysine that is produced by fermentation using microorganisms (see Section 21.1.1). Corn has been modified to increase lysine content.[11]

21.2.5 REGULATORY AND ECONOMIC ASPECTS OF GENETICALLY MODIFIED CROPS AND FOODS

The FDA regulates GM food and animal feed in the United States. It also regulates material produced for therapeutic purposes. It can prohibit materials containing toxic or allergenic substances. Allergenicity is a particular concern, especially for genes related to commonly allergenic foods such as milk, fish, shellfish and nuts. Food from GM sources must be shown to be "substantially equivalent" to the unmodified food. This means it must be similar in content of the major food components such as protein, fats, carbohydrates, amino acids, and other nutrients. This is usually the case because the modified DNA *per se* is a negligible component of the total DNA and is rapidly degraded by digestive processes. The derived proteins can have a higher content but are also subject to digestion. There have been several studies that demonstrate the nutritional equivalence of GM-modified crops such as Roundup Ready soy beans and Bt-corn as animal feed.[12] Of potentially more concern are minor constituents that conceivably might be increased or decreased. For example, there has been an issue as to whether levels of isoflavones such as genistein and daidzein (see Section 15.6) might be modified in GM soybeans. There is no current require-ment that GM food be labeled as such, unless there is a significant change in composition. For example, GM canola oil with intentionally modified fat com-position must be so labeled.

Sponsors of a GM plant can apply for *non-regulated status* on the basis of evidence that it will have no adverse environmental impact. This is administered by the USDA. When non-regulated status is granted, restrictions on produc-tion are removed. This approval is separate from that of the FDA which apply specifically to use as food. Both commercially used and development-only GM plants have non-regulated status. There are currently (2014) 106 GM plants for which approval has been granted, beginning in 1992 with the Flavr-Savr tomato (see Topic 21.1).[13] Most are large acreage crops such as corn, cotton, soybeans and canola, but there are also several varieties of tomatoes and potatoes, as well as a few fruits. The USDA is responsible for monitoring field tests of geneti-cally modified plants. The EPA is responsible for approval of materials having pesticide-resistant properties.

21.2.5.1   ISSUES OF CONCERN WITH GENETICALLY MODIFIED CROPS

Genetic modification is widely accepted in the United States in applications such as production of drugs and detergents. The public reaction is more ambivalent with respect to food. Various factors can limit the prospects for development of new GM crops. One is the sociopsychological issues that limit acceptance. The general public's unfamiliarity with the concepts and operation of GM crops and the active opposition of advocacy groups often leaves the public confused as to the benefits, risks and wisdom of accepting GM crops. Another factor has been the ambivalence of food processors toward using GM crops directly as human food, for fear of consumer backlash. Some food companies have begun to advertise products as "GMO-free." Europeans in general have been more resistant to GM-modified foods and this has limited adoption in crops that are grown in or exported to Europe. The Europeans have generally insisted on segregation and labeling of GM food. The European Union established a 0.9%-limit on the GM content for food. There is no food shortage in Europe so there is no immediate consumer benefit from expanded production. The occurrence of mad cow disease (bovine spongiform encephalopathy) raised fears that pathogens might travel through the food supply. The political "Green" movement has been effective in public campaigns against biotechnology.[14] There is also pressure to protect local agricultural producers.

The most thoroughly studied aspect of GM crops is their compositional and nutritional equivalence with the corresponding unmodified crops. There have been a considerable number of studies with a variety of laboratory and commercial (e.g., chickens, pigs) animals that indicate that there are no significant nutritional differences. There have been few studies on long term, for example, multigenerational effects.[15] There are two potential areas of concern. One is allergenicity which would result if a component of the GM crop had enhanced ability to trigger allergies. There are no documented cases, but one of the general review criteria is similarity to known allergens. The other general concern is that portions of the modified DNA, whether the modified gene itself or a marker-gene, might be incorporated into bacteria, thereby generating new, possibly pathogenic species. Such incorporation is known to occur, so the issue is whether there is any potential for adverse consequences.

Several other general concerns have been expressed about GM crops. One is that their use might generate resistant insects and, similarly, that herbicide-resistant super-weeds might emerge. Another is that insect-resistant varieties might wipe out beneficial insects. The development of super-weeds or super-bugs has the potential to upset the local ecological balance. The development of resistant pests, being a biological adaptation, is self-propagating, or, as it has been called, "irrevocable." Experience has shown that resistant species do arise and present problems for effective control. The same problems exist with conventional herbicides and insecticides but experience indicates that very intense use of a particular product can speed the development of resistance.

The International Food Biotechnology Committee of the International Life Sciences Institute (ILSI) has made six general recommendations with respect to evaluating the safety of genetically modified foods and animal feeds.[a] They can be summarized as follows: (1) The safety assessment of a nutritionally improved food or feed requires that a comparative assessment of the new food

[a]The International Life Sciences Institute is a private organization funded largely by agricultural, food, beverage, chemical, and pharmaceutical companies.

or feed crop with an appropriate comparator crop that has a history of safe use. (2) To evaluate the safety and nutritional impact of modified food and feed crops, it is necessary to develop data on a case-by-case basis in the context of the proposed use of the product in the diet and consequent dietary exposure. (3) The safety of any protein(s) newly introduced into a crop needs to be assessed. (4) Compositional analysis of crops with known toxicants or anti-nutrient compounds should include analysis of those specific analytes. (5) The phenotypic properties of the nutritionally modified crop need to be assessed when grown in representative production locations as part of the overall comparative safety assessment process. (6) Animal nutritional feeding studies need to be performed with a suitable species and should follow the guidance by the ILSI Task Force on conduct of animal studies to evaluate genetically modified crops.[16]

21.2.5.2   LEGAL AND ECONOMIC IMPLICATIONS OF GENETICALLY MODIFIED CROPS

GM plants are generally covered by patents and other intellectual property restrictions that give the developer control over usage of the seeds. Typically the user must agree not to use seed produced from the GM variety. This is not a particularly restrictive situation as compared with hybrid seeds, because the latter are prevented from use since the hybrid traits are not carried through to the next generation. More controversial has been the potential use of "terminator technology" by which the GM crop is modified so that seeds will be sterile or otherwise defective. This approach has faced public and legislative opposition and is not currently in practice. An economic requirement for success of GM crops is synergy between the seed producer and the pesticide manufacturer. In the United States, this synergy has been accomplished by mergers or acquisitions between pesticide manufacturers and seed producers. Monsanto acquired DeKalb in 1998 and the vegetable seed producer Seminis in 2005. DuPont acquired Pioneer in 1999. Several factors have limited development of HR crops beyond the corn, canola, cotton and soybeans that have been widely adopted. One factor is that to be economically viable, the crop must have a relatively high volume of production, as is the case for the four above. Another element of success is the ability to target the GM to a single, proprietary pesticide. This has been possible for glyphosate and glufosinate because there are no analogs that have been developed.[17,18] The costs of development and regulatory approval must be weighed against possible benefit. Figure 21.6 gives some of the aspects and time requirements for full development of a HR crop plant.

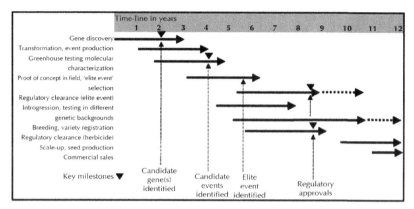

FIGURE 21.6 Components and approximate timeline for development of GM crops. (From. Devine, M. D. *Pest Manage. Sci.* **2005**, *61*, 312–317 by permission of Wiley. © 2005, Society of Chemical Industry.)

Specific political and economic considerations also enter into consideration. In certain countries, especially Western Europe, the importation of GM products requires meeting relatively stringent regulations, which may deter development. The success of HR crops has probably slowed the development of new herbicides by diminishing the market for conventional herbicides. However, the potential for the eventual development of HR-resistant weeds implies that new herbicides will be needed.

TOPIC 21.1 THE FLAVRSAVR TOMATO. THE FIRST GENETICALLY MODIFIED FOOD

The development of the FlavrSavr tomato was undertaken by Calgene, a biotech start-up company located in Davis, CA, USA in the 1980s. At the same time Calgene was working on herbicide-resistant forms of cotton and canola. The goal for the FlavrSavr project was to produce a tomato that could be ripened naturally but still survive shipping (see Section 9.5). The approach taken was to inactivate one of the genes responsible for the ripening and softening of tomatoes, *polygalacturonase.* The enzyme breaks down the pectin in the plant as part of the ripening process. When the gene is inactivated the level of the enzyme drops to less than 90% of that in the normal tomato. Calgene was granted a patent in January 1989 and later received a patent for the broad concept of use of antisense genes in plants (1992). The award of the 1989 patent set the stage for efforts to obtain approval from the FDA for use as a food.

Among the issues that needed to be addressed in the approval process was the fact that the modified gene included an anti-biotic resistance gene (see Section 20.4.2), in this case a gene that confers resistance to several aminoglycoside-type antibiotics (see Section 12.2). Calgene approached the approval process by submitting a request for approval of a "process aid," that is a production procedure. The safety issues that needed to be addressed were whether either the modified gene, its enzyme product, or the gene for antibiotic resistance constituted a hazard. The first effort at Calgene was to analyze the potential for these risks. Calgene scientist were able to establish that the genes in question did not survive the digestive process and that the maximum amount of modified enzyme would be <0.1% of total protein produced. Another argument made was that any antibiotic-resistance generated would be negligible in comparison to the existing level of such organisms. The application for approval was filed in November 1990. The application required documentation of both intended and unintended consequences of the genetic modification. This included demonstration that the vitamin content, particularly of A and C, was the same as unmodified tomatoes. Potential toxins, in particular glycoalkaloids, were also shown to be in the same range as for unmodified tomatoes. The absence of acute toxicity was demonstrated by rodent feeding experiments. Eventually eight prospective version of the tomato were characterized to establish that there was a single location for the modified gene. These studies were completed and submitted to the FDA in August 1991. In late 1993, the FDA required that the approval be resubmitted as a "food additive" rather than a "process aid."

The FDA scheduled a food safety advisory panel review for April 1994 at which the food additive application was considered. All the previously submitted safety data were reviewed. Approval was granted on May 18, 1994, and the first sale of a GM-modified food took place on May 21, 1994. The tomatoes were given the brand name MacGregor (after the gardener in the story of Peter Rabbit). They sold at about $2.00/lb around $0.70 above competing standard tomatoes, but were enthusiastically purchased. The supply, however, was limited and even at the premium price, the tomatoes were being sold at a substantial loss. Calgene lost nearly $43 million in fiscal 1994. It was, however, the first company to bring a GM food to market and was a year or more ahead of the competition, which included Monsanto. The plans were for increased marketing for the 1994–1995 winter tomato season. Unfortunately, the supply remained very limited. Furthermore, while the tomatoes did have the expected longer shelf life after delivery, they were not any better than other vine-ripened tomatoes in terms of sturdiness to shipment and handling. By January 1995 these problems were becoming evident to the financial community and Calgene stock started to drop. Patent litigation also went to trial in March 1995. The suit was brought based on antisense gene insertion into *E. coli* that had been done at the State University of New York (SUNY). The suit claimed that the SUNY patent applied to *all organisms*, including plants. The suit also claimed that a competing patent claim from the Fred Hutchinson Institute had been fraudulent. The suit was eventually settled in Calgene's favor in 1996.

By May 1995, the MacGregor brand of FlavrSavr tomatoes were on sale in 1700 stores in several parts of the country. Supply remained a problem however. For fiscal 1995, Calgene reported a loss of $30 million and negative stories in the financial press kept the stock price under pressure. In June 1995, Calgene agreed to be partially purchased by Monsanto. Monsanto increased its share in 1996 and completed the purchase of Calgene in April 1997. Monsanto decreased the emphasis on fresh tomatoes and the FlavrSavr disappeared from the market place.

Source for Topic 21.1: Martineau, B. *The Creation of the FlavrSavr Tomato and the Birth of Genetically Engineered Food*; McGraw Hill, 2002.

21.3   GENETICALLY ENGINEERED PRODUCTS FOR MEDICINE

Biotechnology has had many important applications in medicine. It is especially well-suited to synthesis and modification of polypeptides. One or more genes can be inserted into an organism and provide the enzymes necessary to carry out the synthesis of the desired polypeptide. Substitution of particular amino acids can be achieved by the techniques of selective mutation (see Section 20.4.4). In the sections that follow, we describe a few examples of these applications.

21.3.1   INSULIN

Insulin is currently needed by about 150 million diabetes patients worldwide and the number is growing (see Section 16.3). From its introduction in 1922 until 1982, insulin was produced by isolation from bovine or porcine pancreases. Various longer acting forms were introduced in the 1940s and 1950s. "*Hagedorn Insulin*" is a mixture of insulin with a basic protein, portamine. "*Lente insulin*" is a hexameric complex of insulin with zinc. Used in combination with unmodified insulin, these materials can more closely match the normal physiological profile of insulin release. This is an extremely important consideration, because of the dangers of hypoglycemia associated with excessive insulin levels. Most insulin regimens combine use of a fast-acting version prior to meals with a slow-acting form one or two times per day.

Insulin was the first drug to be produced commercially by genetic engineering and as such blazed the trail for both production methodology and regulatory approval. The first product was developed by a collaboration that began in the late 1970s between Eli Lilly, the major US insulin producer, and the biotech firm Genetech. Recombinant human insulin was introduced in 1982. Insulin is a favorable substance for production by genetic engineering. It is relatively small and contains no non-peptide posttranslational modifications such as glycosylation. Another crucial feature of insulin is that it can spontaneously assume its native (active) shape and form the disulfide bonds between its two chains without biological catalysis. Human insulin is synthesized using genetically altered *E. coli*. In addition to incorporating the insulin gene, the organism must be modified to inactivate enzymes that would otherwise hydrolyze the insulin as it is produced. The two polypeptide chains are produced separately and combined by a chemical process. Insulin and modified insulins are also produced in yeast systems. Eli Lilly, Novo Nordisk, and Aventis are the main worldwide producers of recombinant insulin and its derivatives.

A number of structural modifications of insulin have been made to improve the pharmacological profile. Several fast-acting forms that can be injected immediately before a meal have been introduced and are called *Lyspro*, *Aspart*, and *Glulisine*. In Lyspro, the two amino acids at β-28 and β-29 are exchanged. In Aspart, an Asp is substituted for Pro at position β-28. In Glulisine β-3 Asn is replaced by Gly and β-29 Lys by Glu. The substitutions are accomplished using synthetic nucleotide sequences coding for the mutations. These forms have reduced tendencies toward aggregation and are therefore more rapidly adsorbed. Two longer lasting forms have also been developed. *Glarcine* is a modification with a Gly for Arg substitution at A-21 and two C-terminal Arg residues added in the B-chain. This modifies the pI of the polypeptide in such a way that it is less rapidly adsorbed from the injection site. *Detemir* is modified by excision of the N-terminal β-30 Thr and the β-29 Lys is acylated by a C-14 fatty acid. This form is bound by serum albumin and also is long-acting. These modifications are summarized in Figure 21.7.

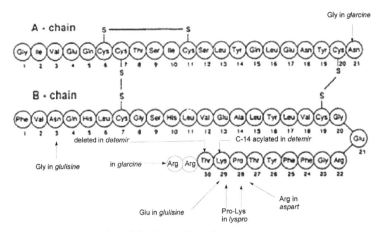

FIGURE 21.7 Structural modifications of insulin.

TOPIC 21.2 CLONING AND COMMERCIALIZATION OF HUMAN INSULIN. A NEW FRONTIER FOR BIOTECHNOLOGY

Plans to produce insulin by recombinant DNA technology began to be seriously considered in the mid-1970s. At that point, insulin was produced by isolation from the pancreases of cattle and pigs. Eli Lilly had 80–85% of the US market, but the Danish firm Novo-Nordisk was a major competitor in Europe. Furthermore, there was beginning to be concern that in the long run, the demand for insulin from the increasing incidence of diabetes would outstrip the supplies available from the animal sources. Also, there are small differences in the insulin amino acid sequences (one in porcine, three in bovine), although both are effective. By 1976, three academic groups had begun to explore cloning human insulin. One was the group of Herbert Boyer at UCSF. Another group, also located at UCSF, was led by William J. Rutter and Howard M. Goodman. The third group was that of Walter Gilbert at Harvard. The Rutter–Goodman and Gilbert groups were pursuing the isolation of the *c*-DNA for the insulin gene (see Section 20.4.2). The Boyer group chose an alternative approach, the synthesis of the insulin gene from the individual nucleotides. At the time there was considerable public controversy about the safety of recombinant DNA. In particular, any studies that involved human DNA had to be carried out in the highest level of isolation facility, called P4. There were no academic P4 facilities; the only ones in the United States were used by the military for germ warfare studies. Efforts at Harvard to establish the next highest level of security, P3, met with considerable public resistance and delay.

The Gilbert group searched for human insulin *c*-DNA in a particular type of tumor, called an insulinoma, that produced large amounts of insulin, and presumably large amounts of insulin m-RNA from which the *c*-DNA could be prepared. The Rutter–Goodman group focused on isolating rat insulin *c*-DNA by isolating and processing the pancreases of a large number of rats. They eventually succeeded in isolating the *c*-DNA, but early in 1977, it was inadvertently cloned into a plasmid (called pBR322), which was not then authorized for use. The first clones had to be destroyed. The *c*-DNA was successfully cloned into an alternative plasmid in April 1977.

Boyer and Robert Swanson, a biotech entrepreneur, formed the new firm Genentech with the ultimate goal of commercializing recombinant human insulin. They decided to use a synthetic insulin gene, rather than isolating a *c*-DNA. Toward this end, they recruited Keiichi Itakura, an expert in DNA synthesis as part of the team. He was located at the City of Hope Medical Center in southern California. Itakura used the newly developed phosphotriester method for oligonucleotide synthesis, which was much faster than the previous method (see Section 20.3.1). Genentech established a research laboratory in converted warehouse space in the San Francisco area, but also supported the research be-ing done at both UCSF and City of Hope. To test the feasibility of the synthetic approach they first synthesized the gene for somatostatin. The 14-AA peptide requires only the two complementary 42 bp DNA strands. The synthesis was ap-proached by making the various trimeric nucleotides that corresponded to each of the amino acids in the somatostatin sequence. The somatostatin gene was then attached to a gene for β-galactosidase in such a way that the somatostatin could be cleaved from the larger enzyme by the reagent cyanogen bromide. The group succeeded in producing somatostatin and publicly announced the success in late 1977.

Two of Gilbert's co-workers, Argiris Efstratiadis and Lydia Villa-Komaroff, successfully incorporated the rat insulin gene into the pBR322 plasmid in early 1978 and filed a patent on the method in June 1978. Gilbert, several other promi-nent American and European molecular biologists, and entrepreneurs Daniel Adams and Raymond Shaefer, formed another biotech company, Biogen, in 1978. An effort to clone the human insulin c-DNA was made in the early fall, 1978. This required a P4 facility, and none was available in the United States. Gilbert made arrangement to have access to a P4 facility operated by the British military for 4 weeks. The cDNA and necessary reagents and apparatus were transported to the United Kingdom and work began. It was exceedingly slow because of the arduous restrictions applied in a P4 facility, which essentially required sterilization of all materials going into or out of the facility. It turned out that the c-DNA had somehow become contaminated with rat c-DNA and the experiments failed to produce any human insulin.

At almost the same time, Genentech succeeded in producing recombinant human insulin. In their approach they had made the α- and β-insulin chains by separately using the synthetic genes and needed to recombine the chains. There was a report in the literature that this could be done, but it was not en-tirely clear how efficient it would be. The problem was establishing the cor-rect disulfide bonds linking the two chains. After carefully purifying the chains, they were successfully coupled in September 1978. Genentech's approach had a regulatory advantage. Because it used the synthetic, rather than an isolated human gene, it was not subject to the requirement for operating in a P4 facility. Immediately after succeeding in the isolation of the insulin, Genentech conclud-ed an agreement for further research and development with Eli Lilly. The results were announced at a press conference on September 6, 1978. The agreement with Lilly stipulated several further progress deadlines toward commercializa-tion. Genentech met the various Lilly deadlines. The yield was dramatically increased by using an enzyme involved in tryptophan synthesis in place of the much larger β-galactosidase. In 1980, Genentech turned the project over to Lilly for commercialization and marketing. Genentech went public in October 1980 with the share price more than doubling on the first day of trading. On literally the same day, Walter Gilbert learned he had been awarded the Nobel Prize in chemistry for his method of DNA sequencing, sharing it with Frederick Sanger and Paul Berg (see Section 20.4.2). Lilly had also been supporting the research of the Rutter–Goodman group on use of the human insulin c-DNA. In the summer of 1978, the work was moved to a Lilly-owned facility in France to overcome the P4 restrictions in the United States. The work eventually succeeded in cloning the proinsulin gene containing both the α- and β-chains.

Source for Topic 21.2: Hall, S. S. *Invisible Frontiers, The Race to Synthesize a Human Gene*; Atlantic Monthly Press: New York, 1987; Johnson, I. S. *Nat. Rev.* **2003**, *2*, 747–751.

21.3.2 HUMAN GROWTH HORMONE

Human growth hormone (hGH) has a MW of about 22 kDa. It is a 191 AA single strand peptide having two disulfide bonds, as shown in Figure 21.8. There are several closely related materials (isotypes), one of which is expressed in the placenta of pregnant women. Regulation of hGH involves both a stimulatory protein (growth hormone releasing hormone, GHRH) and an inhibitor (somatostatin). The production of hGH is closely related to development and is highest during gestation and puberty. hGH production in individuals occurs in bursts and decreases with age after puberty. The hGH receptors are membrane-bound proteins and the highest concentration is found in the liver. The hormone regulates several aspects of metabolism and organ function. Deficiencies in hGH lead to diminished growth and delayed puberty, while excess levels can lead to excessive growth.

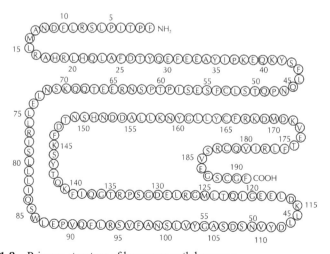

FIGURE 21.8 Primary structure of human growth hormone.

Until 1985, the only hGH available was from cadavers. It was produced by governmental agencies, such as the National Pituitary Agency in the United States. Commercial production by genetic engineering began in the early 1980s and there are now several materials available, produced in genetically engineered microorganisms or mice. The availability of recombinant hGH coincided with concern that hGH from cadavers might be contaminated with prions of Creutzfeld–Jakob Disease, a fatal neurodegenerative condition. Recombinant hGH was first developed at Genentech and approved by the FDA in 1985.[19] Current sources of commercial product include Genentech, Lilly, Novo Nordisk, Sandoz, Teva, Pfizer and Merck Serono. The major clinical indication for rhGH is for growth deficiency in children. There are also a number of other specific developmental disorders, usually associated with gene deletions, that can be treated by rhGH. Adults who are hGH deficient because of surgical removal of the pituitary gland (hypophysectomy) can also be treated. The material is administered by subcutaneous injection. Other currently approved uses include treatment of such conditions as chronic kidney disease and AIDS-associated wasting. There are also several rare genetic abnormalities associated with growth for which rhGH is authorized. There has also been interest in whether rhGH could be used to treat effects associated with aging, which tend to parallel specific hGH deficiency conditions. The evidence is not strong in this case. In fact, there is a curious anomaly. Mice that are GH-deficient or GH-resistant have significantly increased (25–60%) lifespans. This suggests that GH must have "aging" and well as "antiaging" activity.[20] In hGH-deficient patients, hGH increases lean body mass and reduces HDL levels. It also has anabolic effects. These properties have led to the suspicion that it may be abused by athletes. Detection, however, is difficult because of the endogenous presence of hGH. Testing was introduced in the 2006 Olympic Games based on differing ratios of isotypes present in endogenous and recombinant material.[21]

21.3.3   GRANULOCYTE COLONY AND GRANULOCYTE-MACROPHAGE COLONY STIMULATING FACTORS

Granulocyte colony stimulating factor (G-CSF) stimulates production of neutrophils and is used to counteract neutropenia (reduced white blood cells) associated with cancer chemotherapy. Human G-CSF is a 174 AA polypeptide that is partially glycosylated in its natural state, although this is not required for activity. Three variants are currently in use, Filgrastrim, Lenograstim, and Pegfilgrastim. Filgastim was first approved in 1991. It is produced in *E. coli* and contains a terminal methionine. Lenograstim is produced in a hamster ovary cell-line and is glycosylated. Pegfilgrastim is a modified peptide having a PEG chain that slows clearance through the kidneys and increases its half-life. The PEG chain is introduced on the N-terminus by chemical methodology.

The related *granulyte macrophage colony-stimulating factor* (GM-CSF) stimulates formation of neutrophils, macrophages, and eosinophils by bone marrow. It is a 127 AA polypeptide. Three forms of recombinant GM-CSF have been developed including Molgramostim (produced in *E. coli*), Regramostim (produced in Chinese hamster ovary cells) and Sargramostim (produced in yeast). The latter also has a leucine substitution at position 23. In addition to indications similar to G-CSF, GM-CSF was originally approved to accelerate myeloid recovery in patients after bone marrow transplants. There is some evidence that GM-CSF stimulates the immune response toward tumor cells. The benefits of treatment with G-CSF and GM-CSF include reduced risk of infection, increased likelihood of completion of the entire chemotherapy course, and, in some cases, more frequent chemotherapeutic treatments. There is only very limited data that the use of G-CSF has favorable long-term outcome, and it is currently recommended only for patients at significant (>20%) risk for infection.

21.3.4   FOLLICLE-STIMULATING HORMONE AND LUTEINIZING HORMONE FOR FERTILITY TREATMENT

Follicle-stimulating hormone (FSH) and luteinizing hormone (LH) play crucial roles in the establishment of pregnancy. As a result they are used both to induce ovulation in anovulatory women and to promote egg formation for *in vitro* fertilization (see Section 15.3.3). FSH is also used in treatment of men with low sperm production due to hypogonadism. Both FSH and LH are now available from recombinant technology. Both FSH and LH are also collected from menopausal urine. The recombinant material offers advantages in terms of highly defined composition and reduced likelihood of an allergenic effect. The precise role of LH in fertilization procedures remains under investigation and the availability of both hormones may permit optimization of the use of LH.

21.3.5   TUMOR NECROSIS FACTOR AND CYTOKINE INHIBITORS IN TREATMENT FOR ARTHRITIS

Tumor necrosis factor (TNF) is one of the primary cytokines that initiates and controls the inflammatory process (see Section 13.1.1). As such, it is of interest in treatment of arthritis and other autoimmune conditions. Particular attention has been given to treatment of arthritis, including juvenile arthritis which affects children and adolescents. Several types of agents have been developed by biotechnology and applied to the treatment of arthritis. The main side effect of the TNF inhibitors is to suppress immunity to infectious diseases, including relative common infections such as sinusitis, but also tuberculosis. An important current issue with these drugs is whether they must be continued or if remission can be maintained after reduction or elimination of the drugs. The three best-selling biologic drugs are in this group are Enbrel, Humira, and Remicade, all of which had sales exceeding $6 billion in 2010. Etanercept (brand name Enbrel) is a fusion protein of AA 1–235 of TNF and AA 236–467 of human immunoglobu-lin. It shows significant improvement in treatment of juvenile arthritis and can be used in combination with other drugs used to treat that condition, including methotrexate. It functions by binding to TNF-α, thus blocking its pro-inflamma-tory effects. While original recommendations were for use if standard therapies failed (e.g., methotrexate or anti-inflammatory steroids), there are now recom-mendation that aggressive therapy be started early. Adalimunmab (Humira) is a monoclonal antibody to human TNF. There are several other monoclonal anti-bodies that are used in treatment arthritis and other auto-immune conditions. In addition to the original approval for arthritis (2002) and juvenile arthritis (2008), adalimumab has also been approved for several other auto-immune diseases. Other examples include infliximab (Remicade), certolizumab (Cimzia), and golimumab (Simponi). These materials act by inhibiting the pro-inflammatory effects of TNF. Other drugs are targeted at other pro-inflammatory cytokines, for example, tocilizumab (Actemra),[22] which is targeted at interleukin-6 (IL-6). Tocilizumab binds to the IL-6 receptor, blocking the effect of the inflammatory effect of the cytokine.

21.3.6   BOTULINUM TOXIN

Botulinum toxin (BoTN) is produced by an anaerobic bacterium *Clostridium botulinum*. The spores of the organism are widely distributed in the environment and contamination of food results in poisoning known as *botulism*. The disease was first associated with sausage in the seventeenth century and the name comes from the Latin word for sausage, "botulus." The association with food was con-firmed by Justinius Kerner in the 1820s, who described the symptoms. The early symptoms include disappearance of tear fluid, paralysis of eye muscles, and decreased salivation. Fatal poisoning proceeds to paralysis of skeletal, gut and diaphragm muscles. Late in the 1800s, Emile P. M. van Ermengen, a student of Robert Koch, isolated and characterized the causative organism *C. botulinum*. He found the organism in raw salted ham that had been served at a wedding and caused 23 poisonings, of which 3 were fatal. As commercial food preparation and shipping increased in the early 1900s, botulism became a problem in the

United States. K. F. Meyer[b] was able to define the conditions necessary to kill the organism in food canning.

The toxin is a polypeptide consisting of three 50 kDa domains. One is a zinc-dependent endopeptidase that specifically cleaves protein involved in exocytosis of acetyl choline. The toxin also suppresses release of other neurotransmitters, including epinephrine, norepinephrine, and histamine. Thus the toxin interferes with neurotransmission and can lead to paralysis and death. The effects are slowly reversible so that patients who can be maintained by artificial ventilation may ultimately survive.

The toxin is thought to be the most toxic substance known on a weight basis, with a lethal intravenous dose of about 100–150 ng and 10–20 μm orally. There are seven known forms of BoTN, A-G. The types differ in the complexes they form with other high MW proteins. Two (A and B) are produced commercially. Botox (Allergen, USA), Dysport (Ipsen, UK) and Xeomin (Merz Pharma, Germany) are type A, while Myobloc and Neurobloc (Elan Phramaceutical, Ireland) are type B. The individual products differ somewhat in the methods of purification and the amount of other proteins present. BoTN was also produced as a potential bio-weapon in the United States, United Kingdom, Russia, and perhaps elsewhere.

Medical uses of BoTN have evolved as physicians learned to use it to control muscle functions. The use of BoTN was introduced by an opthamologist, Alan B. Scott. It was first used in treatment of blepharospasm (involuntary blinking). It can be used to treat strabismus (the inability to maintain binocular focus) due to an imbalance in eye muscle strength. BoTN can also be used in treating muscle dystonia, the loss of proper muscle tonicity. Use of the recombinant toxin for treatment of strabismus and blepharospasm was approved by the FDA in 1989. J. S. Elston pioneered introduction of BoTN in the United Kingdom, working with material supplied by national Public Health Service. This was the source of material eventually commercialized as Dysport. In 2000, BoTN was approved in the United States by the FDA for uses in cervical, pharyngolaryngeal, and oromandibular dystonia. These conditions are caused by malfunction of particular muscles. By targeting the specific muscle, the contractions can be relieved. The beneficial effects usually last several months. BoTN has several favorable

---

[b]Karl F. Meyer grew up in Switzerland and completed his undergraduate education at the University of Zurich. He studied infectious disease at the University of Munich and then completed a Ph. D. at the Veterinary School of the University of Bern. He also studied in South Africa where he was stricken with malaria. He came to the United States in 1910 to teach at the Veterinary School of Pennsylvania, and in 1914 he moved to California. He was affiliated with the Hooper Foundation Institute for Medical Research which had connections with both the Berkeley and San Francisco campuses of the University of California. He was director of the Institute from 1921 until his retirement in 1954. He investigated many aspects of communicable diseases involving animals, such as brucellosis and equine encephalitis. He investigated several aspects of food contamination and helped to establish criteria for proper food processing after several incidents of botulism from both home and commercial canning during WWI.

characteristic for these and related therapies. (1) It is extremely potent, meaning that small (ng) quantities are effective. (2) It can be injected into localized areas targeting specific muscles and generally does not relocate. (3) Its duration of action is quite long, usually in the range of months. (4) Because it is used in very small doses and a high state of purity, development antigenic response is rare. Currently the most frequent medical use of BoTN-A is for pharmacocosmetic purposes. Following its introduction for use in blepharospasm and related facial spasms, it was noticed that treatment led to the disappearance of wrinkles. This observation led Alistair and Jean Carruthers, a dermatologist and opthamologist, respectively, to investigate the potential for cosmetic use.[23] Injections are used to tighten facial muscle, reducing the wrinkled appearance of facial skin. In 2006, there were more than 3 million treatments of patients with BoTN-A for cosmetic purposes.[24]

KEYWORDS

- biotechnology
- enzymes
- herbicide resistance
- insect resistance
- growth hormones
- insulin
- granulocyte and granulocyte-macrophage factors
- follicle-stimulating hormone
- cytokine inhibitors
- botulinum toxin

BIBLIOGRAPHY

Schefer, T.; Borchert, T. W.; Nielsen, V. S.; Skagerlind, P.; Gibson, K.; Wenger, K.; Hatzack, F.; Nilsson, L. D.; Salmon, S.; Pedersen, S.; Heildt-Hansen, H. P.; Poulsen, P. B.; Lund, H.; Oxenboll, K. M.; Wu, G. F.; Pedersen, H. H.; Xy, H. *Adv. Biochem. Eng. Biotechnol.* **2007,** *105,* 99–131.

BIBLIOGRAPHY BY SECTIONS

Section 21.1.1: Demain, A. L. *TibTech* **2000**, *18*, 26–31; Hermann, T. *J. Biotechnol.* **2003**, *104*, 155–172; Ikeda, M. *Adv. Biochem. Eng. Biotechnol.* **2003**, 791–735; Survase, S. A.; Bajaj, I. B.; Singhal, R. S. *Food Technol. Biotechnol.* **2006**, *44*, 381–396; Webdisch, V. F.; Bott, M.; Eikmanns, B. J. *Curr. Opin. Microbiol.* **2006**, *9*, 268–274; Wustenberg, B.; Stemmler, R. T.; Letinois, U.; Bonrath, W.; Hugentobler, M.; Netscher, T. *Chimia* **2011**, *65*, 420–428; Ledesma-Amaro, R.; Santos, M. A.; Jiminez, A.; Revuelta, J. L. *Food Sci. Technol. Nutr.* **2013**, *246*, 571–594.

Section 21.1.2: Diez, B.; Mellado, E.; Rodriguez, M.; Fouces, R.; Barredo, J.-L. *Biotechnol. Bioeng.* **1997**, *55*, 216–226; Bruggink, A.; Roos, E. C.; de Vroom, E. *Org. Process. Res. Dev.* **1998**, *2*, 128–133; Bornschueuer, U. T.; Buchholz, K. *Eng. Life Sci.* **2005**, *5*, 309–323; Srirangan, K.; Orr, V.; Akawi, L.; Westbrook, A.; Moo-Young, M.; Chou, C. P. *Biotechnol. Adv.* **2013**, *31*, 1319–1332.

Section 21.1.3: Hanzawa, S. In *Encylopedia of Bioprocess Technology: Fermentation, Biocatalysis and Biofermentation*; Flickinger, M. C.; Drew, S. W., Eds.; Wiley: New York, 1999; pp 201–210.

Section 21.2.1.1: Franz, J. E.; Mao, M. K.; Sikorski, J. A. *Glyphosate: A Unique Global Herbicide, American Chemical Society Monograph Series*; 1997, vol. 189, 1–638.

Section 21.2.1.2: Dill, G. M. *Pest Manage. Sci.* **2005**, *61*, 219–224.

Section 21.2: Lord, J. C. *J. Invertebr. Pathol.* **2005**, *89*, 19–29.

Sources 21.2.3: Etherton, T. E.; Bauman, D. E. *Physiol. Rev.* **1998**, *78*, 745–761.

Section 21.2.4: Jauhur, P. P.; Khush, G S. *Food Security and Environmental Quality in the Developing World*; CRC Press: Boca Raton, FL, 2002; pp 107–128.

Section 21.2.5: Lemaux, P. G. *Ann. Rev. Plant Biol.* **2008**, *59*, 771–812; Lemaux, P. G. *Ann. Rev. Plant Biol.* **2009**, *60*, 511–559.

Section 21.3.1: Chance, R. E.; Frank, B. *Diabetes Care* **1993**, *16*, 133–142; Kjeldsen, T. *Appl. Microbiol. Biotechnol.* **2000**, *54*, 277–386; Walsh, G. *Appl. Microbiol. Biotechnol.* **2005**, *67*, 151–159.

Section 21.3.2: Marian, M.; Oeswein, J. Q. *Pharmaceutical Biotechnology: Fundamentals and Applications*, 3rd ed., 2008; pp 281–291; Kemp, S. F.; Frindik, J. P. *Drug Des. Dev. Ther.* **2011**, *5*, 411–419.

Section 21.3.3: Sylvester, R. K. *Am. J. Health-System Pharm.* **2002**, *59*(Suppl 2), S6–S12; Molineux, G. *Anti-Cancer Drugs* **2003**, *14*, 259–264; Renwick, W.; Pettengell, R.; Green, M. *Biodrugs* **2009**, *23*, 175–186.

Section 21.3.4: Shoham, Z. *Expert Opin. Pharmacother.* **2003**, *4*, 1985–1994; Pang, S. C. *Women's Health* **2005**, *1*, 87–95; Gibreel, A.; Bhattacharya, S. *Biol. Targets Ther.* **2010**, *4*, 5–17.

Section 21.3.5: Gatto, B. *Reumatismo* **2006**, *58*, 94–103; Ma, X.; Xu, S. *Biomed. Rep.* **2013**, *1*, 177–184.

Section 21.3.5.1: Hunt, L.; Emery, P. *Expert Opin. Biol. Ther.* **2013**, *13*, 1441–1450; Berard, R. A.; Laxer, R. M. *Expert Opin. Biol. Ther.* **2013**, *13*, 1623–1630; Marotte, H.; Cimax, R. *Expert Opin. Biol. Ther.* **2014**, *14*, 569–571.

Section 21.3.5.2: Sfikakis, P. P. *Curr. Dir. Autoimmunity* **2010**, *11*, 180–210; Marzan, K. A. B. *Adolesc. Health Med. Ther.* **2012**, *3*, 85–93; Voulgari, P. V.; Drosos, A. A. *Expert Opin. Biol. Ther.* **2014**, *14*, 549–561.

Section 21.3.6: Scott, A. B. *Dermatol. Clin.* **2004,** *22,* 131–133; Markey, A. C. *Dermatol. Clin.* **2004,** *22,* 213–219; Aoki, K. R. *Curr. Med. Chem.* **2004,** *11,* 3085–3092; Montecco, C.; Molgo, J. *Curr. Opin. Pharmacol.* **2005,** *5,* 274–279; Bigalke, H. *Botulinum Toxin; Therapeutic Clinical Practice and Science;* Jankovic, J. Ed.; 2009; pp 389–397.

REFERENCES

1. Martens, J.-H.; Barg, H.; Warren, M. J.; Jahn, D. *Appl. Microbiol. Biotechnol.* **2002,** *58,* 275–285.
2. Powles, S. B. *Pest Manage. Sci.* **2008,** *64,* 360–365.
3. Webster, T. M.; Sosnoskie, L. M. *Weed Sci.* **2010,** *58,* 73–79.
4. Tan, S.; Evans, R. R.; Dahmer, M. L.; Singh, B. K.; Shaner, D. L. *Pest Manage. Sci.* **2005,** *61,* 246–257.
5. Gressel, J.; Valverde, B. E. *Pest Manage. Sci.* **2009,** *65,* 723–731.
6. Federici, B. A. *Genetically Modified Crops: Assessing Safety*; Taylor and Francis: London, 2002; pp 183–185.
7. Zhang, B. H.; Liu, F.; Yao, C. B.; Wang, K. B. *Curr. Sci.* **2000,** *78,* 37–44.
8. Wang, Y.; Xue, Y.; Li, J. *Trends Plant Sci.* **2005,** *10,* 610–614; Chen, M.; Shelton, A.; Ye, G.-y. *Ann. Rev. Entomol.* **2011,** *56,* 81–101.
9. van Wijk, J. *Phytochem. Rev.* **2002,** *1,* 141–151.
10. Ye, X.; Al-Babili, S.; Kloti, A.; Zhang, J.; Lucca, P.; Beyer, P.; Porrykus, I. *Science* **2000,** *287,* 303–305.
11. Glenn, K. C. *J. AOAC Int.* **2007,** *90,* 1470–1479.
12. Hammond, B. G.; Vicini, J. L.; Hartnell, G. F.; Naylor, M. W.; Knight, C. D.; Robinson, E. H.; Fuchs, R. L.; Padgtette, D. R. *J. Nutr.* **1996,** *126,* 717–727; Clark, J. H.; Ipharraguerre, I. R. *J. Dairy Sci.* **2001,** *84,* E9–E18.
13. www.aphis.usda.gov/biotechnology/petitions.
14. Arntzen, C. J.; Coghlan, A.; Johnson, B.; Peacock, J.; Rodemeyer, M. *Nat. Rev. Genet.* **2003,** *4,* 839–843.
15. Domingo, J. L. *Crit. Rev. Food Sci. Nutr.* **2007,** *47,* 721–733.
16. International Life Sciences Institute. *Compr. Rev. Food Sci. Food Safe.* **2004,** *3,* 35–104.
17. Franz, J. E.; Mao, H. K.; Sikorski, J. A. *Am. Chem. Soc. Monogr.* **1997,** *189,* 1–638.
18. Lydon, J.; Duke, S. O. *Plant Amino Acids, Biochemistry and Biotechnology*; Singh, B. K., Ed.; Marcel Dekker: New York, 1999; pp 445–464.
19. Cronin, M. J. *J. Pediatr.* **1997,** *131,* S5–S7.
20. Bartke, A. *Clin. Intervent. Aging* **2008,** *3,* 659–665.
21. McHugh, C. M.; Park, R. T.; Sonksen, P. H.; Holt, R. I. G. *Clin. Chem.* **2005,** *51,* 1587–1593; Sergura, J.; Gutierrez-Gallego, R.; Ventura, R.; Pascual, J. A.; Bosch, J.; Such-Sanmartin, G.; Nikolovksi, Z.; Pinyot, A.; Pichini, S. *Ther. Drug Monit.* **2009,** *31,* 3–13.
22. Kishimoto, T. *Int. Immunol.* **2010,** *22,* 347–352.
23. Carruthers, A.; Carruthers, J. *Dermatol. Surg.* **1998,** *24,* 1168–1170; Carruthers, J.; Carruthers, A. *Dermatol. Clin.* **2009,** *27,* 417–425.
24. De Boulle, K. L. V. *Expert Opin. Pharmacol.* **2007,** *8,* 1059–1072.

INDEX

Printed and bound by CPI Group (UK) Ltd, Croydon, CR0 4YY

23/10/2024

01777696-0018